CU00664128

BIRDS OF PARADISE AND BOWERBIRDS

HELM IDENTIFICATION GUIDES

BIRDS OF PARADISE AND BOWERBIRDS

Phil Gregory

Illustrated by Richard Allen

HELM

LONDON · OXFORD · NEW YORK · NEW DELHI · SYDNEY

Dedicated to the memory of Ian Burrows, instigator of the project and sadly missed.

HELM
Bloomsbury Publishing Plc
50 Bedford Square, London, WC1B 3DP, UK

BLOOMSBURY, HELM and the Helm logo are trademarks of Bloomsbury Publishing Plc

First published in the United Kingdom 2019

A catalogue record for this book is available from the British Library

Library of Congress Cataloguing-in-Publication data has been applied for.

ISBN: HB: 978-0-7136-6027-2; ePDF: 978-1-4729-7584-3; ePub: 978-1-4729-7585-0

2 4 6 8 10 9 7 5 3 1

Design by Susan McIntyre
Maps by Gregory, P. (2017) *Birds of New Guinea. Including Bismarck Archipelago and Bougainville.* Lynx Edicions.
Barcelona, with the exception of: p. 218–226, 241, 231, 325–327 and 332–338 (by Julian Baker)

Printed and bound in India by Replika Press Pvt. Ltd.

To find out more about our authors and books visit www.bloomsbury.com and sign up for our newsletters

CONTENTS

The species and subspecies of birds of paradise (total 71 taxa)

Birds of paradise (45)	Family Paradisaeidae	
	Genus *Lycocorax*	Bonaparte, 1853
Halmahera Paradise-crow	*Lycocorax pyrrhopterus pyrrhopterus*	(Bonaparte, 1850)
	L. p. morotensis	Schlegel, 1863
Obi Paradise-crow	*L. (p.) obiensis*	Bernstein, 1865
	Genus *Manucodia*	Boddaert, 1783
Glossy-mantled Manucode	*Manucodia ater*	(Lesson, 1830)
Tagula Manucode	*Manucodia alter*	(Lesson, 1830)
Jobi Manucode	*Manucodia jobiensis*	Salvadori, 1876
Crinkle-collared Manucode	*Manucodia chalybatus*	(J. R. Forster, 1781)
Curl-crested Manucode	*Manucodia comrii*	P. L. Sclater, 1876
	Genus *Phonygammus*	R. Lesson, & Garnot, 1826
Trumpet Manucode	*Phonygammus keraudrenii keraudrenii*	(Lesson, & Garnot, 1826)
	P. k. neumanni	Reichenow, 1918
	P. k. purpureoviolaceus	A. B. Meyer, 1885
	P. k. hunsteini	Sharpe, 1882
	P. k. jamesii	Sharpe, 1877
	P. k. gouldii	(G. R. Gray, 1859)
	Genus *Paradigalla*	Lesson, 1835
Long-tailed Paradigalla	*Paradigalla carunculata*	Lesson, 1835
Short-tailed Paradigalla	*Paradigalla brevicauda*	Rothschild & Hartert, 1911
	Genus *Astrapia*	Vieillot, 1816
Arfak Astrapia	*Astrapia nigra*	(J. F. Gmelin, 1788)
Splendid Astrapia	*Astrapia splendidissima splendidissima*	Rothschild, 1895
	A. s. helios	Mayr, 1936
Ribbon-tailed Astrapia	*Astrapia mayeri*	Stonor, 1939
Princess Stephanie's Astrapia	*Astrapia stephaniae stephaniae*	(Finsch & A. B. Meyer, 1885)
	A. s. feminina	Neumann, 1922
Huon Astrapia	*Astrapia rothschildi*	Förster, 1906
	Genus *Parotia*	Vieillot, 1816
Western Parotia	*Parotia sefilata*	(J. R. Forster, 1781)
Carola's Parotia	*Parotia carolae carolae*	A. B. Meyer, 1894
	P. c. chalcothorax	Stresemann, 1934
	P. c. meeki	Rothschild, 1910
	P. c. chrysenia	Stresemann, 1934
Bronze Parotia	*Parotia berlepschi*	Kleinschmidt, 1897
Lawes's Parotia	*Parotia lawesii*	Ramsay, 1885
Eastern Parotia	*Parotia helenae*	De Vis, 1897
Wahnes's Parotia	*Parotia wahnesi*	Rothschild, 1906
	Genus *Pteridophora*	A. B. Meyer, 1894
King of Saxony Bird of Paradise	*Pteridophora alberti*	A. B. Meyer, 1894

	Genus *Lophorina*	Vieillot, 1816
Superb Lophorina	*L. s. latipennis*	Rothschild, 1907
	L. s. feminina	Ogilvie-Grant, 1915
Western Lophorina	*Lophorina superba*	(J. R. Forster, 1781)
	L. (s.) niedda	Mayr, 1930
Eastern Lophorina	*L. (s.) minor*	Ramsay, 1885
	Genus *Ptiloris*	Swainson, 1825
Paradise Riflebird	*Ptiloris paradiseus*	Swainson, 1825
Victoria's Riflebird	*Ptiloris victoriae*	Gould, 1850
Magnificent Riflebird	*Ptiloris magnificus magnificus*	(Vieillot, 1819)
	P. m. alberti	Elliot, 1871
Growling Riflebird	*Ptiloris intercedens*	Sharpe, 1882
	Genus *Epimachus*	Cuvier, 1816
Black Sicklebill	*Epimachus fastosus fastosus*	(Hermann, 1783)
	E. f. atratus	(Rothschild & Hartert, 1911)
Brown Sicklebill	*Epimachus meyeri*	Finsch & A. B. Meyer, 1885
	Genus *Drepanornis*	P. L. Sclater, 1873
Black-billed Sicklebill	*Drepanornis albertisi albertisi*	(P. L. Sclater, 1873)
	D. a. cervinicauda	P. L. Sclater, 1884
	D. a. geisleri	A. B. Meyer, 1893
Pale-billed Sicklebill	*Drepanornis bruijnii*	Oustalet, 1879
	Genus *Cicinnurus*	Vieillot, 1816
Magnificent Bird of Paradise	*Cicinnurus magnificus magnificus*	(J. R. Forster, 1781)
	D. m. chrysopterus	Elliot, 1873
	D. m. hunsteini	A. B. Meyer, 1885
Wilson's Bird of Paradise	*Cicinnurus respublica*	(Bonaparte, 1850)
King Bird of Paradise	*Cicinnurus regius regius*	(Linnaeus, 1758)
	C. r. coccineifrons	Rothschild, 1896
	Genus *Semioptera*	
Wallace's Standardwing	*Semioptera wallacii wallacii*	G. R. Gray, 1859
	S. w. halmaherae	Salvadori, 1881
	Genus *Seleucidis*	
Twelve-wired Bird of Paradise	*Seleucidis melanoleucus*	(Daudin, 1800)
	Genus *Paradisaea*	Linnaeus, 1758
Greater Bird of Paradise	*Paradisaea apoda*	Linnaeus, 1758
Raggiana Bird of Paradise	*Paradisaea raggiana raggiana*	P. L. Sclater, 1873
	P. r. intermedia	De Vis, 1894
	P. r. salvadorii	(Mayr & Rand) 1935
	P. r. augustaevictoriae	Cabanis, 1888
Lesser Bird of Paradise	*Paradisaea minor minor*	Shaw, 1809
	P. m. jobiensis	Rothschild, 1897
Goldie's Bird of Paradise	*Paradisaea decora*	Salvin & Godman, 1883

Red Bird of Paradise	*Paradisaea rubra*	Daudin, 1800
Emperor Bird of Paradise	*Paradisaea guilielmi*	Cabanis, 1888
	Genus *Paradisornis*	Finsch & A. B. Meyer, 1885
Blue Bird of Paradise	*Paradisornis rudolphi rudolphi*	(Finsch & A. B. Meyer, 1885)
	P. r. margaritae	Mayr & Gilliard, 1951

Formerly classified as birds of paradise

Satinbirds (3)	**Family Cnemophilidae**	
Loria's Satinbird	*Cnemophilus loriae*	(Salvadori, 1894)
Crested Satinbird	*Cnemophilus macgregorii macgregorii*	De Vis, 1890
	C. m. sanguineus	Iredale, 1948
Yellow-breasted Satinbird	*Loboparadisaea sericea sericea*	Rothschild, 1896
	L. s. aurora	Mayr, 1930

Honeyeaters (187)	**Family Meliphagidae**	
MacGregor's Honeyeater	*Macgregoria pulchra*	De Vis, 1897

The species and subspecies of bowerbirds (total 37 taxa)

Bowerbirds (28)	**Family Ptilonorhynchidae**	
	Genus *Ailuroedus*	Cabanis, 1851
White-eared Catbird	*Ailuroedus buccoides*	(Temminck, 1836)
Tan-capped Catbird	*A. (b.) geislerorum*	Meyer, A. B., 1891
Ochre-breasted Catbird	*A. (b.) stonii*	Sharpe, 1876
	A. s. cinnamomeus	Mees, 1964
Green Catbird	*Ailuroedus crassirostris*	(Paykull, 1815)
Black-eared Catbird	*Ailuroedus melanotis melanotis*	(G. R. Gray, 1858)
	A. m. facialis	Mayr, 1936
	A. m. joanae	Mathews, 1941
Arfak Catbird	*A. (m.) arfakianus*	A. B. Meyer, 1874
Northern Catbird	*A. (m.) jobiensis*	Rothschild, 1895
Huon Catbird	*A. (m.) astigmaticus*	Mayr, 1931
Black-capped Catbird	*A. (m.) melanocephalus*	Ramsay, 1883
Spotted Catbird	*A. (m.) maculosus*	Ramsay, 1875
	Genus *Scenopoeetes*	Coues, 1891
Tooth-billed Bowerbird	*Scenopoeetes dentirostris*	(Ramsay, 1876)
	Genus *Archboldia*	Rand, 1940
Archbold's Bowerbird	*Archboldia papuensis papuensis*	Rand, 1940
	A. p. sanfordi	Mayr & Gilliard, 1950
	Genus *Amblyornis*	D. G. Elliot, 1872
MacGregor's Bowerbird	*Amblyornis macgregoriae macgregoriae*	De Vis, 1890
	A. m. mayri	Hartert, 1930
Huon Bowerbird	*A. (m.) germana*	Rothschild, 1910

Streaked Bowerbird	*Amblyornis subalaris*	Sharpe, 1884
Vogelkop Bowerbird	*Amblyornis inornata*	(Schlegel, 1871)
Yellow-fronted Bowerbird	*Amblyornis flavifrons*	Rothschild, 1895
	Genus *Prionodura*	De Vis, 1883
Golden Bowerbird	*Prionodura newtoniana*	De Vis, 1883
	Genus *Sericulus*	Swainson, 1825
Masked Bowerbird	*Sericulus aureus*	(Linnaeus, 1758)
Flame Bowerbird	*Sericulus ardens*	(D'Albertis & Salvadori, 1879)
Fire-maned Bowerbird	*Sericulus bakeri*	(Chapin, 1929)
Regent Bowerbird	*Sericulus chrysocephalus*	(Lewin, 1808)
	Genus *Ptilonorhynchus*	Kuhl, 1820
Satin Bowerbird	*Ptilonorhynchus violaceus violaceus*	(Vieillot, 1816)
	P. v. minor	Campbell, 1912
	Genus *Chlamydera*	Gould, 1837
Western Bowerbird	*Chlamydera guttata guttata*	Gould, 1862
	C. g. carteri	Mathews, 1920
Great Bowerbird	*Chlamydera nuchalis nuchalis*	(Jardine & Selby, 1830)
	C. n. orientalis	Gould, 1879
Spotted Bowerbird	*Chlamydera maculata*	(Gould, 1837)
Yellow-breasted Bowerbird	*Chlamydera lauterbachi lauterbachi*	Reichenow, 1897
	C. l. uniformis	Rothschild, 1931
Fawn-breasted Bowerbird	*Chlamydera cerviniventris*	Gould, 1850

ACKNOWLEDGEMENTS

Particular thanks are owed to John Cantelo, who read multiple drafts of the text, patiently fielded queries and suggested many improvements. Nigel Redman instituted the *Birds of Paradise and Bowerbirds* project and was very understanding over various unforeseen interruptions, later followed by Jim Martin and Alice Ward, who took the project on to its conclusion. My wife Sue and son Rowan Gregory patiently supported and encouraged my work and research on the project over many years, and without them I could not have completed it. Clifford and Dawn Frith are also thanked for their exemplary scholarship with birds of paradise and bowerbirds, as also are Bruce Beehler and Thane Pratt, whose New Guinea field guides have been an essential companion for many years and whose classic *Birds of New Guinea: Distribution, Taxonomy and Systematics* (2016) will be the default reference for years to come. The inspirational *Birds of Papua New Guinea* by Brian Coates was also an essential reference, with much insight into the two families.

It is unusual to have two quite distinct families in a single volume, but there is a long historical precedent for it, with many famous titles in both the 19th and the 20th centuries. In 1977, Forshaw and Cooper produced a wonderful volume entitled *The Birds of Paradise and Bowerbirds*, with the large-format illustrations conceivably the best since the days of lithography in the previous century. The two families are singularly attractive and unique groups from the same zoogeographical region, and together they exhibit some of the most elaborate and bizarre plumages and behaviours in the avian kingdom. For the present work Richard Allen was available as the artist, using his remarkable flair for depicting iridescent plumages and both accurately and attractively depicting the birds themselves. Together with the beautiful photographs sourced from many different people, I believe we have a winning combination that celebrates and details these two wonderful bird families.

Many other people helped in various ways, from field companionship and discussion on taxonomy, identification, sound recordings and general advice to travel logistics, and my thanks go to all: Stephen Ambrose, Joseph Ando, Kim & Paul Arut, Kenneth & Tanya Arut, Bob Bates, Untu Baware, Tony Baylis, Andrew Bowes, the late Ian Burrows, the late Barbara Burrows, Dominic Chaplin, Brian Coates, David Donsker, Jack Dumbacher, Brian Finch, Chris Eastwood, Andy Elliott, Mat Gilfedder, Judith Giles, Frank Gill, Will Goulding, Benson Hale, Roger Hicks, Betty & Peter Higgins, Dion Hobcroft, the late Jon Hornbuckle, Josep del Hoyo and Arnau Bonan Barfull for the marvellous resource that is *HBWAlive* and the *Internet Bird Collection* (IBC), Phil Hurrell, the late Steven Ipai, David James, Pak Jamil at Nimbokrang, Markus Lagerqvist, Tim Laman, Mary LeCroy, Murray Lord, Sharon Mackie, Max Male, Jun Matsui, David Mitchell, Lloyd Nielsen, Moyang Okira, Max Pakao, Jan Pierson, Shita Prativi, Pamela Rasmussen, Rose-Ann Rowlett, Ed Scholes, Dave Stejskal, Joseph Tano, Mike Tarburton, Kisea Tiube, Karen Turner, Philip Veerman, Leonard Vaieke, Daniel Wakra, Peggy Watson, Richard Webster, Bret Whitney, Zeth Wonggor, and Iain Woxvold; and I have appreciated working over many years with Samuel Kepuknai, Jimmy Woram, Edmund Woram and Kwiwan Sibu of what is now Kiunga Nature Tours. Sue and Rowan Gregory of Sicklebill Safaris/Cassowary Tours and the staff at Field Guides Inc. are also thanked for their great tour logistics, which enabled me to gain field experience of most of the species in this book. The late Steve Mead at the International Education Agency (IEA) in Port Moresby was a very understanding and supportive employer regarding my various New Guinea birding diversions from educational administration during 1991–97. The Victoria's Riflebirds and Spotted Catbirds at Cassowary House enlivened my authorial duties by their frequent visits while I was working on the text, and provided some useful behavioural and vocalisation insights. David Christie was enlisted as the copy editor and I have greatly benefited from his vast experience, patience and knowledge. Errors and omissions are inevitable and are the author's responsibility; please do advise of any which you come across, so that they may be corrected in any future edition.

This project began back in 1998 when my friend and colleague the late Dr Ian Burrows gave a talk to the British Ornithologists' Club in London about 'Birds of Paradise in Papua New Guinea'. Nigel Redman, from what was then Pica Press, was in the audience, and he promptly commissioned a book about the family. Ian knew that he could not do this alone, so I was invited to be a co-author and duly began putting together an overall framework for each species. Ian and our artist, Richard Allen, began work on examining specimens from the Natural History Museum at Tring (BMNH, now NHMUK), and some four years later we had a basic outline of all species prepared but awaiting plumage details. Life, families, careers and professional responsibilities then intruded, and the project was essentially placed on hold for some years. Tragedy then intervened, with Ian's untimely death in October 2009 an immense setback, which was very distressing for all concerned. Following some changes at the publishing house, Jim Martin became the editor and proved keen to get the project restarted. My own touring and publishing commitments precluded an immediate start, but things gradually resolved and I was happy to be able to conclude the long-gestating work. I dedicate it to Ian and his brilliance and just wish that he had been able to take part in it. I sincerely hope that readers find it worthy of his memory.

The information in these introductory sections is derived from a wide variety of sources, both electronic and printed, but I gratefully acknowledge the huge debt owed to the many researchers and scientists who are gradually unearthing the details of the biology of these fascinating and still quite poorly known families. Foremost among them have to be Drs Clifford and Dawn Frith, who have devoted their working lives to research on both the birds of paradise and the bowerbirds, and Dr Bruce Beehler, whose contributions to the biology of these families from New Guinea are immense. Their monographs on *The Birds of Paradise* (Frith & Beehler 1998, OUP) and *The Bowerbirds* (Frith & Frith 2004, OUP) are the fountainheads of knowledge, and we refer readers to these works for the vast amounts of detailed information and background data painstakingly acquired over many years and which are not appropriate for a more general overview such as this current title. We also acknowledge the contributions from the monumental *Handbook of the Birds of Australia and New Zealand and Antarctica* (HANZAB), the late E. T. Gilliard's classic work *Birds of Paradise and Bowerbirds* (1969), Peter Rowland's invaluable short work on *Bowerbirds*, and of course the phenomenal resource *Handbook of the Birds of the World*, now in electronic format as *HBWAlive*. Here, the work of the Friths and Arnau Bonan has been most instructive and we again acknowledge our debt to them in making this marvellous archive accessible.

No work on the birds of paradise is complete without acknowledgement of the brilliant work of Tim Laman and Ed Scholes in the *Bird of Paradise Project*. They have for many years been photographing and making videos of the paradisaeids, and their recordings of the displays of most species are a fantastic resource and archive. Their book *Birds of Paradise* (National Geographic 2012) is quite simply the most astonishing collection of photographs of these extraordinary birds ever. Thanks to funding from the Cornell Lab of Ornithology, National Geographic Expeditions Council and Conservation International, they have forayed to the most remote parts of New Guinea in quest of these species. By 2011, they had archived every single paradisaeid, the project by then having taken some eight years, with 18 expeditions to 51 different field camps. By their own account they had climbed hundreds of trees, built dozens of hides, and made thousands of video and audio recordings, with more than a year and a half spent in the field, and taken more than 39,000 photographs.

One of the great discoveries that has emerged from their studies is that female birds of paradise often watch males from very specific viewpoints. A proper understanding of the displays ideally requires the capacity to see what the female sees, and an innovative camera set-up has now allowed them to film this viewpoint for the Wahnes's Parotia. The plan now is to document this perspective for all the other species, and the project continues.

Their latest contribution (2017) was a landmark paper concerning the courtship and displays of the poorly known *Astrapia* species, with some remarkable discoveries arising from the careful study of thousands of hours of audiovisual material vouchered at the Cornell Lab of Ornithology.

Phil Gregory

INTRODUCTION

What is a bird of paradise? The family Paradisaeidae

The very name 'bird of paradise' has an exotic and rather romantic ring to it, and this is surely one of the most iconic, striking and simply extraordinary bird families in existence. The name has gone through various iterations, such as bird of paradise, Bird-of-paradise, Bird-of-Paradise and Bird of Paradise, the last a form that was in use early in the last century and which is perhaps the most pleasing to the eye, shorn of extraneous hyphens and featuring capitals for both nouns. Nevertheless, the modern style dictates that we should restrict the capitalisation to the level of species names alone, and this is, with some reluctance, followed here.

The birds of paradise are a very well-known and truly remarkable radiation of species, justly celebrated for their extraordinary diversity of courtship behaviours and bizarre exotic plumages derived from sexual selection. The 43–45 species evolved over roughly 20 million years, with most ornamental phenotypic evolution occurring within the core birds of paradise (all but the basal *Manucodia/Lycocorax* clade) over the last 15 million years, and sometimes considerably more recently, as with the *Paradisaea* group (Laman & Scholes 2012; Scholes *et al.* 2017).

The question of what makes a bird of paradise a bird of paradise is, however, frequently asked, and the answer is perhaps surprisingly not particularly easy to formulate. The family Paradisaeidae is very diverse in form and coloration and includes some of the most spectacular and striking species of bird, but what defines them? Several features are shared by all the species: they all have a chunky and quite stout body shape, but with huge variation in bill and tail size and shape. None of them can be considered a great songster, either. Most have harsh, resonant or strident calls, a few more melodious than others, but there is none that can be considered an outstanding musical vocalist. Most of them are primarily frugivorous, from environments where this resource is often abundant and allows plenty of time for other activities such as courtship and displays. Many have unusually striking displays, be it from dancing grounds, songposts or canopy display sites. Apart from the earliest evolved species, the manucodes and paradise-crows, which are monogamous, the birds of paradise are polygynous, the promiscuous males performing displays to attract females, which choose which male to mate with. Most (but not all) of the polygynous species show marked sexual dimorphism.

Birds of paradise were formerly considered closely related to bowerbirds, and sometimes even combined with that family in either Paradisaeidae (Stresemann 1934a; Gilliard 1969) or Ptilonorhynchidae (Sharpe 1891–98; Schodde 1975; Cracraft 1981), owing to certain similarities in zoogeography, chunky body shape, sexual dimorphism and polygynous breeding strategy. As the sum of knowledge grew, however, it was realised that major differences exist, though the birds-of-paradise are, anatomically, broadly typical of the oscine passerines, and are generally somewhat crow-like in many ways.

- Adult male plumage of many of the paradisaeids includes striking iridescence, with an amazing range of extremely bright plumages not otherwise found among the corvoid assemblage.

- Egg coloration is different in each family, the birds of paradise laying attractively coloured eggs, typically decorated with broad dark brush-like streaks.

- Birds of paradise have a distinctive flap-and-glide undulating flight, and the wings of adult males in many genera (especially *Astrapia* and *Ptiloris*) produce a characteristic rustling sound in flight.

- Eggs are laid on consecutive days by many birds of paradise, but on alternate days by many bowerbird species.

- Nestling plumage differs: young birds of paradise lack down and have dark skins, whereas bowerbird nestlings are very downy and pale-skinned.

- Birds of paradise feed their young by regurgitation, a method not used by bowerbirds when feeding their young.

- There are large differences in skull osteology between the two families, birds of paradise skulls technically having a small or non-existent lachrymal and, consequently, a large ectethmoid plate solidly fused with the frontal bone, and a short orbital process of the quadrate with an expanded distal tip (Frith *et al.* 2017).

- Birds of paradise are not known to cache food, whereas this behaviour is habitual in some bowerbird species.

- Birds of paradise do not use vocal mimicry, whereas many bowerbirds are accomplished mimics.

- The loud bugling or far-carrying calls of many birds of paradise have no equivalent in bowerbirds.

- Nest structure differs between the two families, with the birds of paradise constructing nests made of the long supple stems of epiphytic orchids and vines; in contrast, bowerbirds use a woody stick base when nesting.

- Many birds of paradise use arenas or courts, and none of them uses sticks or vegetation for construction purposes; the use by the males of stick bowers for display is restricted to bowerbirds.

- Elements of the displays differ greatly, and the hide-and-peek display of many bowerbirds is unknown among the paradisaeids.

- Tool use is a very rare habit among birds and unknown among paradisaeids, while several species of bowerbird are known regularly to use a form of tool to paint their bowers.

- Bill size and shape have a very diverse range in the paradisaeids, but are relatively uniform among the bowerbirds.

- Both birds of paradise and bowerbirds have 12 rectrices and ten primaries, but bowerbirds have an atypically large number of secondaries (11, or 14 including the tertials). Paradisaeids have eight secondaries (11 if one includes the tertials).

- Hybridisation, rare to uncommon in birds generally, is quite widespread among birds of paradise, with a variety of hybrids known (including some intergeneric pairings), but it is very rare among bowerbirds.

- Birds of paradise habitually use their large and strong feet to hold and manipulate food items, something not done by bowerbirds.

Stonor (1937) found no evidence that the two families were closely related, and classified the birds of paradise as a distinct family with no close relatives. Sibley & Ahlquist (1985, 1990) and their pioneering studies of DNA–DNA hybridisation also found no close relatives, but showed that the bowerbirds were in fact part of the great corvoid songbird lineage. Christidis & Schodde (1991) showed a relationship with Artamidae (butcherbirds, woodswallows etc.), and later studies also reveal them as a monophyletic family (Kusmierski *et al.* 1993).

Ornithologists formerly treated the bird of paradise family, Paradisaeidae, as being closest to the bowerbirds (Ptilonorhynchidae), even after avian anatomists published contrary views. Indeed, some combined the bowerbirds and the birds of paradise into a single family, the Paradisaeidae. Research reveals that the majority of Australasian passerines, about 85% of the total, derives from a southern, or Gondwanan, origin. This major group, referred to as the 'Corvida', radiated in Australia and New Guinea, and today comprises three major lineages or superfamilies:

- Menuroidea, which includes the lyrebirds and Australasian treecreepers (Climacteridae), as well as bowerbirds.

- The Meliphagoidea, which includes honeyeaters and their allies.

- The Corvoidea, which includes the birds of paradise (Frith *et al.* 2017).

Genetic research revealed the major distinctions between the paradisaeids and the bowerbirds, in tandem with increasing biological evidence that also showed a major dichotomy between the two families. Birds of paradise were found to be not only distinct from the bowerbirds but also relatively distant from them in evolutionary terms. Current genetic research indicates that the separation of the birds of paradise from the bowerbirds occurred some 28 million years ago, this finding being based on molecular-clock calibrations, which are still being refined and resolved.

The ground-breaking genetic research of Sibley & Ahlquist (1990), based on what was then a new technique of avian DNA–DNA hybridisation, showed the Paradisaeidae as the sister-group to the radiation now comprising the woodswallows and butcherbirds (Artamidae), the Old World orioles (Oriolidae) and the cuckooshrikes (Campephagidae). Going back one level, the crows and their allies (Corvidae) form the sister-group to this lineage. Later passerine phylogeny, based on two single-copy nuclear-gene sequences, places Paradisaeidae close to the apex of what is termed the 'core Corvoidea', along with Australian mudnesters (Struthideidae), monarchine flycatchers (Monarchidae) and true corvids.

Besides bowerbirds, a number of species or groups, including the silktails (*Lamprolia*) of Fiji, the Kokako (*Callaeas*) and saddlebacks (*Philesturnus*) of New Zealand (wattlebirds) and the Lesser Melampitta (*Melampitta lugubris*) of highland New Guinea, have been suggested as members of the Paradisaeidae family. More recent work has placed the New Zealand wattlebirds as their own ancient family, Callaeidae, while the melampittas, too, are now allocated an ancient family of their own, Melampittidae, and the silktails look likely to be placed in their own family of Lamproliidae, along with the Pygmy Drongo-fantail (*Chaetorhynchus*) of New Guinea.

There have been major adjustments within the family in the past two decades, and four species that were formerly classified as birds of paradise have been reallocated, following genetic studies revealing their true placement.

For decades the Crested (*Cnemophilus macgregorii*), Loria's (*C. loriae*) and Yellow-breasted Birds of Paradise (*Loboparadisea sericea*) comprised the 'wide-gaped' paradisaeid species of the subfamily Cnemophilinae. These three, which we now call satinbirds, were always an uneasy and atypical fit among the true Paradisaeinae, as they possess a very different skull morphology and a rather small, delicate bill and feet. Their behaviours are also completely different, the satinbirds not using their feet to hold prey items, and having the wide gape indicative of an almost entirely frugivorous diet. Their vocalisations also are very different; they build domed nests with a side entrance; and they do not hold court or have obvious displays. In addition, they have a distinct juvenile plumage, which is lacking among the birds of paradise.

For many years the classification of the satinbirds was uncertain, some authorities interpreting the cnemophiline cranial and mandibular osteology as indicative of a basal sister-group to the main paradisaeine lineage. The uncertainty was resolved through genetic studies, which showed that the cnemophilines form a quite distinctive group, part of a clade which also contains the New Guinea berrypeckers and longbills (Melanocharitidae) and the New Zealand wattlebirds (Callaeidae). They are thus quite distant from the true birds of paradise, being rather a basal part of the oscine assemblage and indeed more closely related to bowerbirds, with which they share some skull characters. The three cnemophilines now comprise the satinbird family, endemic to New Guinea and named for the rather glossy textured plumage of the males.

The other species reclassified from the Paradisaeidae is a large and striking one formerly called 'MacGregor's Bird of Paradise' (*Macgregoria pulchra*), a highly prized species difficult for birders to find, and previously another anomaly in the family. This spectacular high-altitude and restricted-range montane endemic was shown by genetic work (Cracraft & Feinstein 2000) to belong to the family of honeyeaters (Meliphagidae), and not to the birds of paradise. As with many honeyeaters, it is largely frugivorous. During fieldwork in 1991, its active behaviour had been noted as resembling that of honeyeaters and the yellow facial wattle was seen to blush orangey (Gregory & Johnstone 1993), reminiscent of Smoky Honeyeaters (*Melipotes fumigatus*). It was formerly treated as one of the basal members of the Paradisaeinae, and was very little known. The common name has proven vexatious in recent times, and both 'Ochre-winged Honeyeater' and 'Giant Wattled Honeyeater' have been used, neither being particularly appropriate and the latter inviting confusion with the giant honeyeaters of Fiji. Here we maintain the historic vernacular name, which commemorates Sir William MacGregor, the first governor of the Territory of Papua and New Guinea, and treat the species in an appendix along with the satinbirds. All four of these former birds of paradise are covered briefly in the addendum, and also illustrated for the sake of completeness and as a recognition of the historic links between the species.

The family Paradisaeidae as treated here comprises 45 species (following the split of the former 'Superb Bird of Paradise'), in 16 genera, with a current total, including subspecies, of 71 taxa. Most occur in New Guinea and its satellite islands, which have 40 species, two of these, the Trumpet Manucode (*Phonygammus keraudrenii*) and Magnificent Riflebird (*Ptiloris magnificus*), extending also to Cape York, in extreme north-east Queensland, Australia. Two of the four *Ptiloris* riflebirds are found only in east Australia, where they are endemic in eastern Queensland and New South Wales. Away from New Guinea and Australia, there are three other members of the family in the North Moluccas of Indonesia, one in the monotypic genus *Semioptera* (Wallace's Standardwing *S. wallacii*) and the two species of paradise-crow (*Lycocorax*).

The six species of manucode and the two paradise-crow species have a monogamous breeding strategy. This is in contrast to the other 33 polygynous species in the family, the promiscuous males of which attract females to traditional solitary courts, display perches or communal leks, where they court and mate with them.

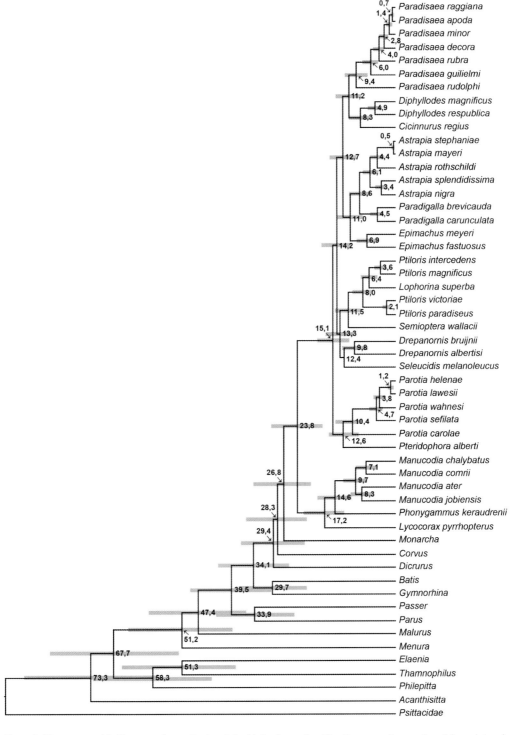

Figure 1. Chronogram with divergence-time estimates of the birds of paradise. The divergence times and confidence intervals (grey bars) were estimated under a relaxed clock model implemented in Beast 1.4.7. For the calibration of the chronogram the postulated separation of *Acanthisitta* from all other passerines in the phylogeny was used. (Reproduced by kind permission from Irestedt *et al.* 2009)

From the mid-20th century there was a long-standing stability in respect of what constituted the bird of paradise family, and the number of constituent species stayed much the same. Over the past 25 years or so, however, there have been gradual changes and some drastic re-evaluations of what species are involved, and how they are defined. The well-established and generally accepted biological species concept (BSC), defined by whether or not different species can interbreed and raise fertile young, held sway for much of the latter part of the last century. Some difficulties arise, though, when attempting to apply this concept to allopatric groups where similar-looking taxa do not actually come into contact, so the interbreeding hypothesis remains hypothetical. Modifications have been proposed and broader, more flexible definitions are now often utilised. When the phylogenetic species concept (PSC) was applied to the group, based on often minor physical features weighing very heavily as taxonomic signifiers, the result was a dramatic increase in the number of constituent paradisaeid species recognised, to about 90 (Cracraft 1992). This taxonomic treatment has not received wide recognition, but it does shed light on relationships within the family, and further refinement and taxonomic research continue apace.

Origins and species diversification

The family Paradisaeidae comprises two well-defined lineages, one being the manucodes and the paradise-crows (*Lycocorax*) and the other being the 25 species of plumed paradisaeid. *Paradigalla* seems also to be a basal form, while *Astrapia* may be a sister-group to the plumed birds. Manucodes and paradise-crows are distinctive, with dark plumage, special syrinx morphology and monogamous nesting habits, and can be regarded as constituting at least a valid subfamily. A study by Irestedt *et al.* (2009) revealed five main clades in their phylogeny.

· **The first clade (A)** consists of manucodes (*Manucodia* and *Phonygammus*) and the paradise-crows (*Lycocorax pyrrhopterus*). This clade is supported as the sister-group to the core birds of paradise (remaining genera).

A very significant finding here was that the split between the *Manucodia–Lycocorax* clade and the core birds of paradise clade was estimated to have occurred about 24 million years ago, while the basal divergence of the polygynous core birds of paradise was suggested to be around 15 million years old. 'The age estimates further suggest that generic separations occurred between six and 14 million years ago, while speciation within genera mostly occurred between 0.5 and ten million years ago.' The origin seems, therefore, to date back to late Oligocene and early Miocene periods, with much speciation later in the latter epoch.

Other studies have indicated that the manucodes and the plumed paradisaeine lineages may have diverged from each other from about 20–18 million years ago to as recently as about seven million years ago. Calibrating the genetic clock remains a work in progress, but much has been learned of the evolutionary history thanks to these important studies.

The plumed paradisaeids

Genetic studies (Irestedt *et al.* 2009) indicate that all the plumed paradisaeid species seem to derive from a single radiation. These plumed taxa can be divided into four morphologically well-defined clades, with manucodes and *Lycocorax* as the first clade, although genetic work gives a slightly different arrangement (see below).

· **The second clade (B)** consists of the King of Saxony Bird of Paradise (*Pteridophora alberti*) and the parotias (*Parotia*). Within this clade Carola's Parotia (*Parotia carolae*) is highly divergent and warrants separate generic treatment. These are sometimes referred to as the 'flagbirds', the males with spectacular erectile occipital plumes, velvety-black dorsal plumage, an iridescent throat or breast patch, and highly distinctive courtship displays.

· **The third clade (C)** consists of the Twelve-wired Bird of Paradise (*Seleucidis melanoleucus*), the *Drepanornis* sicklebills, Wallace's Standardwing, the riflebirds (*Ptiloris*) and the lophorinas (*Lophorina*). Support for more basal nodes within this clade, however, is low, and the affinity of the Twelve-wired Bird of Paradise, *Drepanornis* sicklebills and the Wallace's Standardwing should be regarded as provisional. Although bootstrap values are low, New Guinean *Ptiloris* and *Lophorina* together form a clade separate from the Australian *Ptiloris*. This apparent relationship suggests a significant phylogenetic break between Australia and New Guinea, although a study by Irestedt *et al.* (2017) found no support for this seeming divergence. Riflebirds and lophorinas all have a striking large iridescent blue breast shield, displays based around

King of Saxony Bird of Paradise, adult male, PNG (*Shutterstock*).

song-posts atop vertical snags or logs, and harsh but relatively simple display songs. They appear to have diverged about four million years ago and are sister to the parotias and King of Saxony group, with more distant links to sicklebills and the plumed paradisaeids. Note that the recent HBW–BirdLife taxonomy (del Hoyo & Collar *et al.* 2016) merges *Ptiloris* into *Lophorina*.

The sicklebills (*Epimachus* and *Drepanornis*) constitute two closely related but distinct lineages, the two very long-tailed *Epimachus* ('sabretails') having marked sexual dimorphism, iridescent crown and mantle feathering, red or pale blue iris, and the chin and throat lacking barring in both sexes. The Irestedt *et al.* (2009) phylogeny placed *Epimachus* as sister to a clade that includes *Paradigalla* and *Astrapia*, implying that the long, curved bill has been independently acquired in relation to *Epimachus*.

Characteristic of the *Drepanornis* Black-billed (*Drepanornis albertisi*) and Pale-billed (*D. bruijnii*) Sicklebill species pair are a much shorter tail, much less marked sexual dimorphism and bare facial skin, and both sexes have a brown iris and barred chin and throat. Irestedt *et al.* (2009) positioned *Drepanornis* as sister to a clade that includes *Ptiloris*, *Lophorina* and *Semioptera*.

· **The fourth clade (D)** includes the long-tailed *Epimachus* sicklebills, paradigallas (*Paradigalla*) and astrapias (*Astrapia*).

The two rather poorly known *Paradigalla* species constitute a lineage of obscure affinities. The black plumage and egg morphology link them to basal taxa such as *Manucodia*, but there are also resemblances to *Astrapia* regarding habit, plumage, and body and bill morphology. *Paradigalla* was shown to be sister to *Astrapia* by Irestedt *et al.* (2009), and subsumed within *Astrapia* by Dickinson & Christidis (2014), but, given the substantial morphological, behavioural and vocal distinctions, we feel it best to retain *Paradigalla*.

Astrapia seems to be allied also to the plumed lineage in terms of the male's ornate erectile head and mantle feathering, the barred underparts of females, and the egg morphology. Conversely, the mainly black plumage of the two sexes may link *Astrapia* to the basal paradisaeid taxa, although the black plumage, iridescent head and breast feathering and simple harsh vocalisations may equally suggest a link to *Ptiloris* riflebirds.

· **The fifth clade (E)** consists of the *Paradisaea* birds of paradise and the three sickletails – Wilson's Bird of Paradise (*Cicinnurus respublica*) and Magnificent Bird of Paradise (*C. magnificus*), both formerly placed in

Diphyllodes, and the King Bird of Paradise (*C. regius*). Within this clade the genera *Diphyllodes* and *Cicinnurus* were sister lineages (with *Diphyllodes* now subsumed into *Cicinnurus* in subsequent treatments), while the Blue Bird of Paradise (*Paradisornis rudolphi*) is sister to other *Paradisaea* species (and was formerly placed in *Paradisaea*).

These plumed species form a complex clade with several sublineages. It comprises the four genera *Cicinnurus, Semioptera, Seleucidis* and *Paradisaea*, linked by several shared characters. Adult males have iridescent green and brilliant red and/or yellow areas in the plumage, emit loud harsh, scolding *Paradisaea*-type advertisement vocalisations, and have display postures. The three diminutive *Cicinnurus* species (sickletails) have the central pair of rectrices modified into recurved 'wires', with iridescent green plumage on the underparts, red dorsal plumes, and bright blue legs and feet. We follow recent authorities in merging *Diphyllodes* into *Cicinnurus*; see Dickinson & Christidis (2014), Gill & Donsker (2017) and Beehler & Pratt (2016).

The *Paradisaea* genus is defined by the presence in adult males of both filamentous elongated flank plumes and grossly elongated wire-like or tape-like central rectrices. This sublineage contains the seven 'true' plumed bird of paradise species, namely the Lesser Bird of Paradise (*Paradisaea minor*), Greater Bird of Paradise (*P. apoda*), Raggiana Bird of Paradise (*P. raggiana*), Goldie's Bird of Paradise (*P. decora*), plus three more aberrant plumed species, namely the Red Bird of Paradise (*P. rubra*), Emperor Bird of Paradise (*P. guilielmi*) and the rather divergent Blue Bird of Paradise which is now placed in a monospecific genus.

The monotypic genera *Semioptera* and *Seleucidis* are the final members of the clade. Wallace's Standardwing is an aberrant member of the family restricted to the Moluccas, and appears to be a sister form of the *Cicinnurus–Paradisaea* group, supporting evidence being the green adult male breast plumage and the zoogeographical proximity to those genera. The Twelve-wired Bird of Paradise is also quite aberrant, showing characters associated with two other sublineages: the black adult male plumage, heavily barred female, long decurved bill and tree-stump display sites link it to the riflebirds, while the filamentous yellow flank plumes, *Paradisaea*-type advertisement vocalisations and displays, plus pale leg and iris colours suggest links to the *Paradisaea* lineage, but quite how to rank the evolutionary significance of these characters remains problematic.

Irestedt *et al.* (2009) propose that the main reason why the core birds of paradise, unlike other corvoid bird families, have diversified only within New Guinea and islands in the immediate vicinity is linked with their promiscuous breeding system in which just a few males acquire most of the mating opportunities. A similar sedentary life strategy, with limited dispersal outside mainland areas (and islands which have been connected with these areas) is evident in several other families with elaborate male plumages, such as the pheasants (Phasianidae), and several New World families such as manakins (Pipridae) and the polygynous species within the cotingas (Cotingidae).

Unlike the situation with bowerbirds, where hybridisation is very rare, this phenomenon is very well documented among the birds of paradise, most of the resultant forms being given vernacular names and some having historically been recognised as species. These hybrids are listed under the various species accounts. At least seven different intrageneric hybrid combinations are known, and, unusually, 13 different variations of intergeneric crosses have been recorded. This reflects the close genetic relationships within the family despite the often extreme morphological distinctions among the species. The only other family with such variety would be among the wildfowl, again where many species are closely related. It is only among the polygynous paradisaeid species that hybridisation in the wild is known, but this is not too surprising given the proclivities of the promiscuous males to mount any female-plumaged birds that come close to their display areas. Young or inexperienced birds may presumably mate with the wrong species in error, or where at the extremity of habitat or range there may be no other mate available.

Patterns of speciation and diversification

It is suggested that the reproductive strategies used by birds of paradise may have constrained their capacity to disperse, as they are so dependent upon habitats that have a great diversity of fruiting tree species. In addition, it is likely that the complex ornamental plumes and intensive displays are a limiting factor, as is the strong attachment to display sites. Furthermore, males and females generally live separately, except during the mating period, which makes it less likely that successful long-distance dispersal and establishment of new breeding populations can occur.

The genera *Manucodia*, *Phonygammus*, *Paradigalla* and *Lycocorax* are monomorphic, with the sexes very similar to each other in appearance, and they are monogamous, forming pair bonds, quite unlike the promiscuous behaviour of the rest of the family. This simpler and less complex life strategy perhaps enabled these birds to have a higher dispersal capacity, which could explain the present-day occurrence of *Manucodia* and *Phonygammus* on several islands off the New Guinea coast and on Australia's Cape York Peninsula, and of *Lycocorax* on the North Moluccas. Conversely, *Paradigalla* is a genus restricted to the main New Guinean mountain ranges and does not occur in the coastal lowlands, which renders it less likely to disperse. The present-day distribution of Wallace's Standardwing in the North Moluccas is more difficult to explain. A late-Miocene origin seems probable for this species, at a time when the Moluccas were located much closer to what is now the Vogelkop Peninsula. These monomorphic groups are believed to be among the most ancient in the family, so the timespans involved may have permitted more dispersal.

The paradisaeid genera *Astrapia*, *Ptiloris* and *Paradigalla* have distinct allopatric clades in both the east and the west of New Guinea, which separated about 3–6 million years ago. This **allopatric speciation** model appears also to be the major source of diversification in *Paradisaea*, the mainland lowland species complex of which originated quite recently in the Pleistocene, while the island endemic Red and Goldie's Birds of Paradise appear slightly older. In this genus all lekking species are morphologically very homogeneous, although the age of the genus is more than six million years (the split between Emperor Bird of Paradise and other lekking *Paradisaea* species). The Emperor Bird of Paradise and the highly divergent montane Blue Bird of Paradise are much older divergences. Similarly, the morphological variation is modest within the genus *Parotia*, for which the exploded lekking system is estimated to be around ten million years old. The divergences within the New Guinea and Australian riflebirds (*Ptiloris*) correspond to broad land connections in the upper Miocene and marine transgression in the Pliocene when the land bridges were flooded.

Heads (2001b, 2002) postulated that the present-day distributions of birds of paradise in New Guinea are difficult to explain simply by Pleistocene refugia processes, and relate rather to historic terrane movements over a longer timespan. He used three bird of paradise genera (*Astrapia*, *Parotia* and *Paradisaea*) in support of his thesis, assuming sister relationships with a strong biogeographical connection between the species occurring in the Vogelkop, West Papua, and those in the Huon Peninsula, Papua New Guinea. The phylogeny of Irestedt *et al.* (2009), however, does not provide support for this biogeographical scenario. A sister relationship between the Western Parotia (*Parotia sefilata*) in the Vogelkop and Wahnes's Parotia (*P. wahnesi*) from the Huon Peninsula is only weakly supported, while within the genus *Astrapia* their phylogeny conversely strongly indicates sister relationships between species that occur in closely connected geographical areas (between the Huon Peninsula

Western Parotia, adult male in display, West Papua, Indonesia (*Marek Stefunko*).

and the Central Highlands, and between the Vogelkop and the Star Mts). Oddly, though, some elements of the display of the Huon Astrapia (*Astrapia rothschildi*) do bear some resemblance to what is known of the Arfak Astrapia (*A. nigra*). Within the genus *Paradisaea*, the Emperor Bird of Paradise from the Huon Peninsula is sister to all other *Paradisaea* species that occur in most of lowland New Guinea (excluding the morphologically rather divergent montane Blue Bird of Paradise, which is now allocated to a different genus).

The sicklebills present what may well be an instance of quite ancient altitudinal speciation, with the two *Drepanornis* species separating ten million years ago and the two *Epimachus* species about seven million years ago. Black-billed and Pale-billed Sicklebills (*Drepanornis*) occupy different elevations in lowland and mid-montane forests, while the Black (*Epimachus fastosus*) and Brown Sicklebills (*E. meyeri*) replace each other altitudinally in mountain forests. Likewise, the montane and patchily distributed Lawes's Parotia (*Parotia lawesii*) and Eastern Parotia (*P. helenae*) may represent a more recent altitudinal speciation event, similar instances of which are exhibited by various other bird families, e.g. the honeyeaters (Meliphagidae) and the scrubwrens and gerygones (Acanthizidae).

Irestedt *et al.* (2009) conclude that divergence-time estimates for birds of paradise indicate an older clade than previously suspected. Diversification within several genera of birds of paradise seems to have been a continuous process through the Tertiary period, the younger divergences (as in *Paradisaea*) being geographically quite closely linked. Besides allopatric speciation, there appear to be examples also of altitudinal speciation.

The observation that sexually dimorphic polygynous genera are morphologically homogeneous is particularly interesting, despite divergences between species being suggested as being several million years old. These calculations of diversification rates indicate that the speciation rate is in fact not excessively high. It can be argued that sexual selection in birds of paradise appears not to have generated a particularly rapid change in sexual ornaments or a markedly high speciation rate. This tentative explanation remains uncertain, although the long generation time of polygynous male birds of paradise coupled with extensive hybridisation may have constrained morphological diversification and speciation. Ecological factors such as abundant food sources may also explain why male birds of paradise have been able to develop and maintain their promiscuous breeding systems, magnificent plumages and elaborate courtship displays.

Birds of paradise have undergone some recent taxonomic changes, with the Growling Riflebird (*Ptiloris intercedens*) a long overdue split and the Obi Paradise-crow (*Lycocorax obiensis*) another, while the three-way split of the Superb Lophorina (*Lophorina latipennis*) has now been proposed and is tentatively accepted in the present work. In the manucode subfamily, the Tagula Manucode (*Manucodia alter*) is recognised here as

Brown Sicklebill, female feeding, Kumul Lodge, PNG (*Shane P. White*).

distinct, and it is clear that the Trumpet Manucode consists of at least four cryptic species, which await further clarification regarding their vocalisations. We indicate where further splits are likely. Generic realignments also have occurred, with *Diphyllodes* merged into *Cicinnurus* and the Blue Bird of Paradise moved into *Paradisornis*, although we have resisted the trend to merge *Ptiloris* with *Lophorina* and we maintain *Paradigalla* as distinct from *Astrapia* and the Trumpet Manucode *Phonygammus* as distinct from *Manucodia*.

Plumage aberrations

Albinism as such is unknown among the birds of paradise, while leucism – formerly often termed partial albinism, which is a genetic impossibility (van Grouw 2012) – is extremely rare and known from just ten paradisaeid species. C. B. Frith (Frith & Frith 1998) examined more than 6,000 museum specimens in preparation for the family monograph (F&B 1998), finding plumage abnormalities in less than 1% of the total sample (which at that time included cnemophiline satinbirds and MacGregor's Honeyeater). The most common type of aberration was white feathering, ranging from usually a few to (rarely) more extensive white areas. The sample size may be slightly skewed by collectors obtaining the more aberrant forms, but it is clear that plumage aberrations are very rare in the family.

The following species showed minor abnormalities with varying small amounts of white feathering: Paradise Crow, Curl-crested Manucode (*Manucodia comrii*), Princess Stephanie's Astrapia (*Astrapia stephaniae*), Pale-billed Sicklebill, Blue Bird of Paradise.

More extreme examples were:

- an extensively white Splendid Astrapia (*Astrapia splendidissima*), detailed under the species account;
- a leucistic Magnificent Riflebird *sensu lato* (of unspecified taxon) with coloration entirely a pale smoky brownish-grey, with just a trace of iridescent purple-blue on the lores;
- a Magnificent Bird of Paradise that was largely white except on the breast and underparts, which were of the normal iridescent green with a few brownish back and wing feathers, with another individual that was almost entirely white;
- a Wilson's Bird of Paradise with all the underparts and most of the upperparts a deep rich fawn, with silvery-buff head and throat feathers and brownish head skin, although the yellow cape and red centre of back are present. This fits the character of non-eumelanic schizochroism, where the black pigment eumelanin is reduced, resulting in a washed-out appearance (per Harrison 1995).

Wilson's Bird of Paradise, adult male displaying, Waigeo, West Papua, Indonesia (*Huang Kuo-wei*).

Habitat

Birds of paradise primarily inhabit tropical rainforests from Wallace's Line eastwards through New Guinea and extending to the east-coastal wet tropics and temperate rainforest zone of Australia. The four Australian species occur in rainforest habitats, the Paradise Riflebird (*Ptiloris paradiseus*) extending also into adjacent wet sclerophyll forest and in addition utilising nearby dry sclerophyll forest, and the Magnificent Riflebird of Cape York occupying relict rainforest and vine-forest patches. The Glossy-mantled Manucode (*Manucodia ater*) inhabits open savanna woodland, as well as rainforest, over large areas of New Guinea, while the Curl-crested Manucode lives in similar formations on the south-east islands of Papua New Guinea.

Manucodes in New Guinea have a broader habitat tolerance than other birds of paradise, some lowland forests supporting three manucode species. Each one presumably occupies a specialist niche or has specific dietary preferences, but it is not known how they co-exist ecologically. 'Habitat dictates distribution' is the rule, with most paradisaeid species in the 1,000–2,000-m zone. The Trumpet and Curl-crested Manucodes occupy wider altitudinal ranges than do any other family members, but the Trumpet Manucode appears to be an assemblage of cryptic species with at least one a montane representative. The distribution patterns of the Crinkle-collared (*Manucodia chalybatus*) and Jobi Manucodes (*M. jobiensis*) show the importance of altitude, the former being adapted to upland forest and the latter to lowland forest, though with a wide and poorly understood geographical overlap, perhaps with little or no sharing of specific habitat. Indeed, most species are habitat specialists, this being true especially of the montane ones. In Western Province of Papua New Guinea, the Crinkle-collared is the species of lowland and hill forest, with Glossy-mantled Manucode in the lowlands and Trumpet Manucode both there and in the lower hill forest to 500m.

Habitat choice seems to be quite limited among the paradisaeids, and only a few species utilise more than three habitats, the Raggiana Bird of Paradise being fairly catholic in its choice and occurring in both savanna and rainforest up to about 1,600m. The species with the most restricted ranges are those found in the mid-montane forests at 1,200–2,500m, such as the Long-tailed Paradigalla (*Paradigalla carunculata*), Huon Astrapia and various parotias. It is in this zone that the largest numbers of species occur sympatrically. Good examples of this are found in Papua New Guinea, where at the Tari Gap area at 2,100–2,500m some five paradisaeid species overlap, and at around 1,480m on Mount Missim, north of Wau, a remarkable nine species co-exist.

Congeneric species pairs are another phenomenon among various groups, including paradisaeids in New Guinea, whereby one species largely replaces another at different altitudes with only marginal overlap. Magnificent and King Birds of Paradise are a good example, the former living in upper hill and lower montane zones and the King Bird of Paradise in the lowlands and hills. Black and Brown Sicklebills demonstrate a similar example of limited overlap in mid-montane forest, the former at lower elevations and the latter at higher ones. Two Papua New Guinean endemic astrapias have partially discrete altitudinal ranges, the Ribbon-tailed (*Astrapia mayeri*) at higher altitude and Princess Stephanie's at lower levels, but they come into contact and hybridise fairly frequently where the ranges abut. Raggiana and Blue Birds of Paradise are a similar instance among closely related plumed birds of paradise, the Raggiana occurring mainly below the Blue but with overlap in some areas. The Raggiana experiences a similar situation in the upper hill forests of the Huon Peninsula of north-east New Guinea, where the Emperor Bird of Paradise inhabits the uplands and the Raggiana the lowlands.

Relatively few of the Paradisaeidae have demonstrated an ability to colonise islands across deep-water barriers, with their absence from the Bismarck Archipelago a notable example. Only ancestral forms of *Manucodia* and *Lycocorax*, plus *Cicinnurus*, *Semioptera*, *Seleucidis* and *Paradisaea* from the plumed birds of paradise, occur on islands in such areas. There are quite a few examples of seemingly anomalous distributions not readily explicable by the usual theories of biogeographical dispersion. Much remains to be learned about the influence of past tectonic events and geological processes on the present distributions of New Guinea taxa.

General habits

The behaviour of the Paradisaeidae as a whole is not well known, and many of the more restricted-range species are almost unknown, although some other paradisaeid species have been quite well studied by scientists. Where species occur in sympatry, the mechanisms that segregate them and enable co-existence are poorly understood. A handful of species, one such being the Raggiana Bird of Paradise, have been quite well studied when breeding, but vast gaps in our knowledge remain and many research and doctoral-

Raggiana Bird of Paradise, male displaying to female, Varirata National Park, PNG (*Shutterstock*).

thesis opportunities beckon for those able to gain funding. Most species tend to be seen singly, although monogamous species are sometimes in pairs and there are at times aggregations at fruit sources. Up to five Victoria's Riflebirds (*Ptiloris victoriae*) have been seen loosely associating at feeders at Kuranda, in Queensland, and similar numbers have been reported in orchards. A fruiting tree at Ambua Lodge, in Papua New Guinea, is noteworthy as, when it is in peak fruit, up to five species have been seen there at the same time; Blue Bird of Paradise, Lawes's Parotia, Black Sicklebill, Brown Sicklebill and Superb Lophorina occurred simultaneously in July 2016, with up to five individual parotias at once.

From what little is known, it would appear that most birds of paradise are sedentary within small, permanent home ranges. They tend to be rather shy, flying readily and often difficult to observe. Their vocalisations are a good means of locating them, the males when breeding often calling from regular sites, particularly but not exclusively in the early morning and late afternoon. All members of the family are markedly arboreal, keeping mostly to the middle and upper levels of the forest. They will also descend to lower levels, particularly when foraging in mixed flocks. New Guinea species are very seldom seen on the ground when foraging, although the Ribbon-tailed Astrapia will descend close to it to get at the red fruits of some ginger plants. Two Australian species, the Paradise and Victoria's Riflebirds, are reported as foraging occasionally on the ground, one of the latter coming to within a few centimetres of it when piercing the nectary of some flowers at Kuranda.

A well-known phenomenon in New Guinea is the occurrence of mixed feeding or foraging flocks, often known as 'brown-and-black bird flocks' owing to the predominance of species of those colours (Diamond 1987). There is presumably some advantage to these colour combinations, perhaps involving social cohesion, lessening the chances of predation and probably disturbing more insects. The core species are often the Papuan Spangled Drongo (*Dicrurus carbonarius*), Papuan Babbler (*Garritornis isidorei*), Little Shrike-thrush (*Colluricincla megarhyncha*), Rusty Shrike-thrush (*Pseudorectes ferrugineus*), and both the Hooded Pitohui (*Pitohui dichrous*) and the Northern Variable (*P. kirhocephalus*) and Southern Variable Pitohuis (*P. uropygialis*). Pitohuis are now known to be slightly toxic and distasteful to the human palate, and there may thus be some advantage to other birds in associating with them. In the lowlands and hills such flocks will often have also Raggiana, Magnificent or King Birds of Paradise or Growling or Magnificent Riflebirds present. Other species recorded include manucodes, Princess Stephanie's Astrapia, Western, Carola's (*P. carolae*) and Lawes's Parotias, Superb

Lophorina, all the sicklebills, Twelve-wired Bird of Paradise, and other *Paradisaea* species. Flocking behaviour occurs much more among lowland-forest species than with those in upland forest, the much scarcer food resources in the latter habitat reducing flock frequency and species composition.

One interesting theory that arose from the phenomenon of brown-and-black flocks postulates that the toxic nature of some pitohuis has led to Batesian mimicry of their plumage pattern by *Paradisaea* species. The female and young of these are not typical of most paradisaeids as they lack the prominent barring shown by other birds of paradise in such plumage, but they do have rufous-brownish upperparts and a dark head, analogous with the pattern of pitohuis. Quite how this could ever be proven is uncertain, but it makes for fascinating speculation, and the apparent mimicry by various species of others here is a potentially very rewarding field of study: e.g. friarbirds (*Philemon*) with orioles, Streak-headed Honeyeater (*Pycnopygius stictocephalus*) with New Guinea Friarbird (*Philemon novaeguineae*), and immature Long-billed Cuckoo (*Rhamphomantis megarhynchus*) with Tawny-breasted Honeyeater (*Xanthotis flaviventer*).

Paradisaeids are somewhat variable in their degree of intraspecific aggression, but overall do not seem unduly pugnacious. The fruiting tree at Ambua can, as mentioned above, host five species of bird of paradise at the same time, all co-existing quite peaceably on the whole. Victoria's Riflebirds at Kuranda practise a time-share at the various feeders, almost taking it in turns to forage, albeit with males tending to dominate and fights and chases erupting occasionally, though never for very long. They are also noted as co-existing quite happily together outside the breeding season, with up to five males seen in the same area without strife. Interestingly, Spotted Catbirds (*Ailuroedus maculosus*) do dominate them at feeders.

Predation on birds of paradise by raptors and other non-human predators is remarkably little documented, a surprising fact given the bright plumage and loud noise of adult male paradisaeids when displaying. In some 25 years of visiting rainforest habitats in Australia and New Guinea, the author has never witnessed any predation on a bird of paradise, nor found any remains of such, and other observers have reported similar experiences. Perhaps they are mostly of sufficient size and mass to deter predators, and also equipped with strong and stout feet and bills which may be a deterrent. No doubt predation by raptors does occur opportunistically, as the following indicates: 'Earlier this year I experienced first hand a Black-mantled Goshawk (*Accipiter melanochlamys*) hunting for singing King-of-Saxony BoPs at the summit clearing above Rondon Ridge – fortunately we scared the bird off before it had the opportunity to make contact, but it was very clear what it was trying to do' (T. Palliser *in litt.*). Certainly when perched and calling or displaying, paradisaeids will react strongly to any passing raptor, even innocuous species such as a Whistling Kite (*Haliastur sphenurus*) or Black Kite (*Milvus migrans*), and it is likely that all the larger accipiters would opportunistically prey on them. Rand & Gilliard (1967) reported a Lesser Bird of Paradise being eaten by a Doria's Hawk (*Megatriorchis doriae*), and C. Frith (1998) has seen a Grey Goshawk (*Accipiter novaehollandiae*) take a Victoria's Riflebird. A male Black-billed Sicklebill was preyed on by what was probably an accipiter in 1974 (Pratt in F&B), but was replaced within days by a female-plumaged singing male.

Snakes, particularly pythons, must also pose a threat, and again at Kuranda it is noteworthy that Victoria's Riflebird has a particular alarm call heard only when one of the local pythons is located and is being encouraged to move off by vocal mobbing behaviour, with the riflebird always a leading protagonist. Presumably, active nests would constitute the main target at risk of predation, but again there is a lack of information. Dr Stephen Ambrose (*in litt.*) commented as follows: 'I wonder if predation is that common? In forests and other dense vegetation where brightly coloured birds are found, there would be dimmed ambient light, particularly close to the ground. Bright colours are harder to see in low light conditions, especially from a long distance away. Conversely, there are other bright colours near the top of a forest canopy, e.g. fruits and flowers, so bright coloration may be an advantage under those conditions. But my overall feeling is that the usual cryptic behaviour of these birds (except when displaying) and their stealth of movement probably counter the risk of predation from being brightly coloured. There are lots of brightly coloured bird species in global tropical and subtropical forests, particularly in South America, so either bright colour doesn't disadvantage them greatly with respect to predation risk, or enough survive to breed and keep populations going.

'Reptiles do have colour vision, but I suspect that many snakes would rely more on body heat, odour and vibration to detect prey, and goannas rely more on odour, especially in dark forest situations.'

Human predation on birds of paradise has been the subject of various books over the years, such as that by Swadling *et al.* (1996), and is largely outside the scope of the current work. Suffice to say that very large quantities of plumes were taken in the 19th and early 20th centuries, for display in glass cases, in cabinets of curiosities, and as part of the millinery trade when plumes were considered desirable fashion accoutrements

Victoria's Riflebird, male displaying to female, Malanda, Queensland, Australia (*Martin Willis*).

by wealthy women. 'The aim of every ordinary collector… seems to be… to see how many of these beautiful creatures he can procure for the decoration of the hats of the women of Europe and America' (Sharpe 1898). Employees of the German New Guinea company had shot all the fully plumed males of '*Paradisaea (minor) finschi*' near the coast of German New Guinea. Sharpe writes that this taxon is 'so rare in museums that we may yet be compelled to study its characters by permission of our wives and daughters, whose hats are decorated with its mutilated bodies'. Most of the very rare named hybrid varieties were also derived from the plume trade: to give an idea of the scale involved, from 1905 to 1920 as many as 30,000–80,000 bird of paradise skins were exported *annually* to the feather auctions of London, Paris and Amsterdam. There is currently a somewhat worrying resurgence in the trading of plumes and skins to certain overseas destinations, and a disturbing indication that some of the Asian markets may be acquiring a taste for such beautiful, unusual and valuable status objects, with potentially very harmful outcomes. It is to be hoped that the relevant customs authorities can crack down on it. Happily, foreigners in Papua New Guinea at least are supposed not to purchase the skins or plumes, and most visitors seem to abide by the rules.

Within New Guinea, certain species are still highly prized as cultural adornment (or *bilas* in Tok Pisin language) for ceremonies, shows and sing-sings. A visit to the Mount Hagen, Wabag or Goroka shows will reveal hundreds of skins and plumes being worn for these major ceremonies. Commonly used items include the plumes of Lesser and Raggiana Birds of Paradise, the iridescent triangular blue breast shields of the male Superb Lophorina, which are in great demand as a centrepiece for head or chest ornaments, and the long tail plumes of Princess Stephanie's Astrapia, used in the spectacular head-dresses. The tail plumes of the Ribbon-tailed Astrapia and the male Black Sicklebill are also highly prized and expensive to purchase, as are the head plumes of King of Saxony males and, less commonly, the plumes of Greater and Blue Birds of Paradise or the head wires from Lawes's Parotia.

When such skins were harvested by means of traditional hunting methods the take would have been relatively small, but the advent of shotguns and perhaps even mist nets has seen the quantities rise. Without local regulation the adult males can be eliminated quite quickly, but certainly in some areas the catch is restricted, and the species concerned seem to survive quite well so long as the habitat is intact, a kind of sustainable harvest.

Daily routine maintenance activities have been little documented. All three species of riflebird in Australia are known to drink and bathe at pools in the forest, the Paradise and Victoria's Riflebirds using knotholes holding rainwater. A Paradise Riflebird was seen to utilise a garden pool during a period of drought, ceasing to visit when the rain returned, and Victoria's Riflebirds at Kuranda regularly use the birdbath for bathing and drinking during the drier periods. The Black-billed Sicklebill has been noted to rain-bathe at Mount Missim, sitting 10m up in a sapling and preening for some eight minutes in a heavy downpour (Beehler in F&B), while solitary males have been seen bathing in small pools (Pratt in F&B).

Preening is likely to follow bathing, the birds readjusting their plumage after getting wet. Hunters in New Guinea will wait by forest pools to snare or spear paradisaeids, both for food and for their plumes, and more recently photographers have erected hides by such sites in order to obtain remarkable photographs and videos. Some paradisaeids occupy prominent perches early in the morning, both for display purposes and, probably, to catch the morning sunlight.

Roosting behaviour of the family is also very little known and primarily anecdotal or based on incidental observation, although finding roosting birds in the dense forest habitat is a very difficult task. It is reported that male Wallace's Standardwings will roost above their leks, and Twelve-wired Birds of Paradise appear to roost relatively low down.

Food and foraging

Paradisaeids have an unusually diverse range of body sizes and bill shapes, far more so than the bowerbirds do, and this reflects a quite wide range of dietary strategies. Most species are frugivorous and will also take arthropods, although a few, such as the Black-billed Sicklebill and the Superb Lophorina, are primarily insectivorous. The frugivores tend to range more widely in search of fruiting trees, shrubs, vines and plants, undertaking much foraging in the canopy of such trees. Many species are non-territorial, or defend only their display site, whereas many of the more insectivorous species maintain all-purpose territories as befits their more sedentary lifestyle.

Frugivores tend to occur more in small groups or in foraging flocks, while insectivores are more solitary, probing and gleaning for prey in bark and foliage in the forest canopy or middle stratum. The long decurved bills of riflebirds allow for probing in rotten wood, and sicklebills use theirs to probe among pandan leaves. There is also some opportunist supplementing of the diet with the taking of small vertebrates such as frogs and skinks, and also flowers and nectar. Several species are known to eat leaves, too, in captivity, and this may well be an occasional habit in the wild.

Much of the foraging for fruit is carried out solitarily, but many species will join in mixed-species bird flocks, although, when feeding from fruiting trees, they show little interaction with other frugivorous species such as bowerbirds, fruit-doves (*Ptilinopus*) or satinbirds. Unusually, a few frugivorous paradisaeids will defend a fruit source, resting and feeding by it for much of the day. Dietary preferences are still not very well known and, despite the availability of hundreds of different types of fruit, it appears from some studies that each paradisaeid species eats the fruit of only a few of the potentially available plant species. This varies according to species, some, such as the Trumpet and Crinkle-collared Manucodes, being obligate eaters of figs (*Ficus*) over much of their range, whereas other species prefer drupes or berries (Lawes's Parotia) or capsular fruits (Raggiana and Magnificent Birds of Paradise and the Superb Lophorina).

Unlike bowerbirds, paradisaeids use their feet for holding and manipulating foods, extracting the nutritional content from fruit if it is not swallowed whole. They generally pluck fruit when perched, rarely while in flight, and are among the most significant dispersers of fruits/seeds in both Australia and New Guinea. They do not digest seeds, which pass unharmed through the digestive tract, maybe even deriving some benefit from being excreted among ready-made fertiliser. The only other avian family in Australia and New Guinea with anything like this number of seed-dispersers is that of the pigeons and doves (Columbidae).

None of the paradisaeids is exclusively insectivorous, as even the primarily insectivorous ones take some fruit. A great range of arthropods is preyed on and no doubt the full spectrum will prove to be even wider, but it includes caterpillars, beetles (Coleoptera) and their larvae, katydids (Tettigoniidae) and other Orthoptera, cockroaches (Blattidae) and ants (Formicidae). Three foraging strategies are used by paradisaeids when taking such prey, namely bark-gleaning, the probing and tearing of dead wood and foliage, and the more generalised gleaning of twigs and foliage; the last of those, more generalised gleaning, is somewhat oddly more characteristic of the rather large manucodes. Foraging is done by hopping while searching foliage, boughs and tree trunks, checking dead leaves, bark and epiphytes; mossy clumps are torn into and pulled apart, bits of moss flying about as the birds probe vigorously. Most birds of paradise utilise their feet to hold prey in order to tear the item apart, but manucodes employ a different and more generalised strategy of upright gleaning among twigs and foliage, quite unusual for species as large as these.

Riflebirds are noteworthy for occupying a niche analogous to that of certain woodpeckers (Picidae), a family which is absent from this region. They cling on to trunks and branches with their powerful feet and claws, using the long, strong and decurved bill to probe and tear bark and dead wood, often delicately. They will also use the partially opened bill to probe into soft wood or debris, a technique known as *Zirkeln* (a German word meaning 'to measure with a pair of compasses'). Black-billed Sicklebills have been seen to employ their narrow sickle-shaped bill to probe into wormholes and knotholes, at times opening the bill wide and using just the lower mandible in order to spear prey, a neat adaptation with the curious slender, decurved bill.

Most paradisaeid genera will regurgitate food into the bill before swallowing it again or feeding it to their offspring. They will also regurgitate seeds and simply drop them, another facet of their role as seed-dispersers. An opportunistic feeding association occurs when pigeons and doves forage for seeds dropped or excreted near the display courts of parotias or Magnificent and Wilson's Birds of Paradise, also cleaning up the area at the same time.

As with some bowerbirds, it appears that paradisaeids will vary the dietary content fed to their young, initially provisioning the chicks with the much more protein-rich and presumably calcium-rich arthropods rather than fruit, which becomes more significant later. Arthropods form a rich dietary supplement helpful for growth and successful development, which can be met with a more fruit-based diet later once the formative stages are past. An unexpected finding from a study of the breeding biology of Trumpet Manucodes was that they seem to be obligate fig specialists at this stage, feeding the young very large quantities of relatively low-nutritional-value figs.

It is widely accepted that a frugivorous diet based on abundant and readily available fruit resources permits the adoption of polygyny as a breeding strategy, freeing the males from nest duties and permitting promiscuity and the evolution of court- or lek-based lifestyles, where the females are the primary drivers of mate choice. Recent fieldwork has focused on dietary ecology and specialised frugivory in the evolution of such complex and elaborate breeding systems, where it would appear that females require a richer diet with many more arthropods when nesting, while the males can minimise foraging time, and therefore maximise display time, by being primarily frugivorous.

Breeding

With the exception of a handful of better-studied species, the breeding cycles of birds of paradise are still relatively poorly known, but across their range it is evident that nesting ideally coincides with a peak in the availability of fruit and arthropods. Nesting has been reported in every month of the year, with a peak from August to January and least activity in March–June, often peaking at the end of the wetter months but with great local variation depending on the differing microclimates and habitats available.

All the manucodes and paradise-crows are believed to be **monogamous**, and the sexes are alike. The pair-members share nest duties and do not maintain defended territories, as they are frugivorous and need to forage widely.

The remaining 35 species of bird of paradise are **polygynous**, all of them being sexually dimorphic, usually (except for *Paradigalla*) to a quite marked extent. The females alone perform the nest duties, while the males are focused wholly on display and courtship, spending the maximum possible time at their courts in order to increase the chances of successful mating. These promiscuous males use traditional sites year after year. Display sites range from branches in the forest canopy or understorey to vertical tree trunks or stumps, vine tangles, fallen logs, or the more elaborate terrestrial courts or dancing grounds of *Parotia* and the *Cicinnurus* species, which are cleared by the birds themselves. Males of many of the polygynous species modify their display sites by removing leaves growing around the immediate area, this gardening thus improving visibility and sight lines for spotting not only potential mates but also rivals and predators, making the site more obvious and enhancing lighting conditions, which are important for optimum views of the plumage.

The individual species' displays are described in some detail under each species account, as are nesting details in so far as these are documented. Numerous gaps remain, with the nests of several species still undescribed at the time of writing; for example, the nest of the Arfak Astrapia was first found in 2017, while the nests of the Black Sicklebill and Wilson's Bird of Paradise remain unknown. For nearly half of the other species, including such widespread ones as the King Bird of Paradise, the Twelve-wired Bird of Paradise and most of the parotias, only the barest details of nests and eggs are known.

Some recent discoveries have been made concerning iridescence and the role of ultraviolet in paradisaeid displays, the colour blue in particular being potentially very significant, but this is as yet little known. It may be that females have a particular preference and adult males with the greatest UV reflectance may be at an advantage compared with immatures, which lack or have much-reduced UV coloration. The iridescent blue/ magenta/violet tips of the breast-shield feathers of sicklebills and astrapias, some of which display in fairly dim light, may have unexpected dimensions in how they are perceived. It is now realised that the bright white narial shields of the parotias and the extraordinary head pattern of Wilson's Bird of Paradise are very striking when viewed from directly above, as happens when the females are watching the males display on their courts. The various head wires and long, curved iridescent rectrices are also fantastic signalling devices, and it would be good to know more precisely how these are perceived by the females.

Another recent study (McCoy 2018) has revealed that the structure of paradisaeid feathers is very distinctive, with myriad tiny protrusions that stick up with no exposed flat surfaces, unlike the usual black feather structure, which has flat surfaces that reflect varying amounts of light. These protrusions combine to absorb 99.6% of direct incident light, which means that the black feathers of paradisaeids such as riflebirds (and presumably *Lophorina* and *Parotia*) are the blackest of the black (the blackest material so far known has an absorption rate of 99.6%). This has implications for the displays, as the contrast with the coloured iridescent feathers is thus greatly accentuated, making the males even more striking to the females. Quite what this means with the entirely black manucodes, and whether their feathers are structured in the same way, is not yet known. As always, one line of research opens up yet more lines to be explored.

There is great variety in the display strategies of the polygynous species, these ranging from solitary and non-territorial displays through a range of intermediate strategies to true communal lekking. Solitary and non-territorial display is the most frequently used, with considerable variation even within that: Magnificent and Wilson's Birds of Paradise and the parotias use cleared terrestrial display courts, while the *Ptiloris* riflebirds display from a simple tree snag or branch. Most solitary species display in the understorey, but a few, such as the Twelve-wired Bird of Paradise, use prominent perches high above ground or, as in the case of the King of Saxony and Blue Birds of Paradise and the lophorinas, sing from exposed perches and display much lower down.

Those species which display solitarily have greater space between their territories, thus minimising or avoiding rivalry but often within auditory contact of competitors, as with the Twelve-wired Bird of Paradise. Their display periods tend to be quite limited, with less time spent displaying compared with the court-based species, often with a short early-morning display and frequently a repeat later in the afternoon. The Twelve-wired Bird of Paradise at Kiunga typically spends around 20 minutes in the early morning on the display perch, and has finished at around 07:00 hours, but much depends on female activity. This type of courtship by solitarily displaying promiscuous males, such as those of riflebirds and the Twelve-wired Bird of Paradise, includes sudden advances at females, sometimes buffeting them with the wings, pecking at them, changing size and shape by erecting various plumes, showing off colourful bare parts such as mouth-linings, legs or even the bare thigh patches in this latter species, and using different vocalisations.

In an interesting variation, males are loosely clustered over a wider area and in auditory contact with one another in what is termed an 'exploded lek'. The strategy used by the King Bird of Paradise is a good example of this, with two to four males situated 50–70m apart. Both Lawes's and Carola's Parotias seem to behave similarly at some sites.

The species that form what can be termed true leks include Princess Stephanie's Astrapia, Wallace's Standardwing and most of the *Paradisaea* species, while up to five male Ribbon-tailed Astrapias have occasionally been seen together in some form of display association. These leks vary in the numbers of individuals involved, which can range from just a couple to more than 40 in the case of Wallace's Standardwing and the Raggiana Bird of Paradise, all in close proximity in a single or sometimes several tall trees. Leks tend to be scattered and can be a few hundred metres or a kilometre or more apart, but much depends on the habitat, population density, levels of disturbance and hunting pressures, with many formerly large leks now sadly much depleted. Some of them have been in existence for many years, with one lek of Greater and Raggiana Birds of Paradise at Kiunga active for more than 20 years from March/April through to October/ November, but again this depends on many variables.

There is fierce competition among the males at the leks, the older birds generally having dominance and getting what may be termed the prime or alpha positions for attracting females for successful mating. Much squabbling goes on, with chasing and pecking, and with great vocal interplay among the males of the various age classes, all coupled with attempts to lure females at the same time. Studies suggest that older leks have more males, less territorial aggression and more female visits, the converse holding for newer, smaller leks, which have more fights between males and fewer visits by females. A distinctive and unique feature within these leks is the existence of a **convergence display**, whereby males will suddenly and noisily arrive or converge as females show up.

Polygynous species advertise their presence by means of their colourful plumage and loud calls, but the type of display which they have evolved will depend on the sort of perch which they utilise. The terrestrial-court species have elaborate and complex ritualised dances involving astonishing shape transformations, as does the Superb Lophorina using a log or bare ground, whereas the King of Saxony simply bounces up and down on understorey vines. Those species using vertical snags have much more limited and static repertoires. Such species will drastically alter their shape by erecting flank plumes or breast shields, and zigzag up and down (Twelve-wired) or sway rhythmically back and forth (Black Sicklebill), or form a striking fan shape with swishing noises as Victoria's Riflebird does. The congeners of the last-mentioned species have a much more linear hopping format, still forming a wing-fan with associated sounds but on horizontal limbs or thick vine stems, rather than a display post.

The more complex and ornate the plumage, the more complex and elaborate the display may be a good generalisation in this family, which also has probably the most extreme examples of sexual dimorphism. For each specially modified feather, be it plumes of varying sorts, erectile fans, wing feathers or tail feathers, there is a corresponding ritualised display pose. For example, the King of Saxony waves his flashing head plumes, erects his mantle cape and vibrates the wings to show the pale buff patches. Black Sicklebills erect the axe-shaped epaulettes on the shoulders and flash the iridescent feather tips while swaying rhythmically and showing off the large size and long tail, whereas astrapias have specific poses which show their long tail to great advantage. It is hardly surprising that many of the historic illustrators got the poses of the birds completely wrong!

There is a strong auditory but non-vocal component to some displays. The riflebirds, for instance, show off their specially modified wings, which make a swishing sound when hoisted into the fan shape, and the Superb Lophorina produces curious clicking sounds by rapidly opening and closing its wings in display. The wings are quite prominent in many of the *Paradisaea*, which hold them in specific striking poses, flutter them and also make sounds by striking the carpal joints together while displaying. Wallace's Standardwings also strike their carpal joints together to make quite loud cracking or snapping noises, a louder sound than the *Paradisaea* produce. Bill-rattling is known from Pale-billed Sicklebills during courtship, and riflebirds are well known for the rustling plumage of the males which, when displaying, produce loud swishing noises with the wings as well as wing-beating the female. Male astrapias are similarly very noisy in flight, making loud rustling sounds, and some incorporate elements of wing noise into their displays.

Scholes *et al.* (2017) produced a fascinating and important paper on the displays and courtship of the genus *Astrapia*, derived from detailed analysis of the audiovisual material lodged with the Cornell Lab of Ornithology. They state that the behavioural components of courtship for species in the genus *Astrapia* are very poorly known (Healey 1978; F&B 1998; Laman & Scholes 2012). There is no inventory of systematically named and described courtship behaviours or vocalisations for species in the genus. Summaries from species accounts in the secondary literature indicate a few simple behaviours, including a form of hopping back and forth between branches and an inverted display posture by the Huon Astrapia (Healey 1978; F&B 1998; Frith & Frith 2010).

One remarkable display, the **inverted tail-fan display** of the Huon Astrapia and Arfak Astrapia, is one of the most distinctive and specialised of all the *Astrapia* courtship behaviours and, indeed, of all paradisaeid displays. In this family, true inverted displays – i.e. those that involve hanging upside-down from a horizontal branch – are known for only three species from the *Cicinnurus–Paradisaea* clade: the King Bird of Paradise, the Blue Bird of Paradise and the Emperor Bird of Paradise (F&B 1998; Laman & Scholes 2012). Several other species adopt head-down inverted postures from vertical or sloping branches (e.g. Twelve-wired Bird of Paradise and Greater Bird of Paradise), but they do not hang fully inverted in the manner of those mentioned above and the two species of *Astrapia*.

The overall similarity of the inverted displays of Huon and Arfak Astrapias is quite striking, and such strong convergence in display behaviour is uncommon even among paradisaeids in the same genus. The parotias are an example: even though the **ballerina dance display** is shared by all the species in the genus *Parotia*, the details of the displays are nevertheless different, with diagnosably distinct components, for each *Parotia* species (Scholes 2008c). Because data documenting the inverted tail-fan display of the Arfak Astrapia are so few and likely incomplete, we feel confident that there are additional components still to be discovered which will distinguish it from the inverted display of the Huon Astrapia. These differences are likely to be most noticeable in displays given to females, which is something that has not yet been documented for the Arfak Astrapia.

Both the Huon and the Arfak Astrapias have highly iridescent abdominal plumage, which appears dark (almost black) when the birds are in normal upright perched posture. When the abdominal feathers are oriented skywards during inverted display, however, they become highly visible, another notable feature of the displays of these two species. Interestingly, the Splendid Astrapia has highly iridescent green abdominal plumage, which raises the question of whether it, too, has an as yet undocumented inverted display. According to the best-supported recent phylogeny available, *A. nigra* and *A. splendidissima* are sister-species, the two together being sister to a clade that includes *A. rothschildi*, *A. mayeri* and *A. stephaniae*, with *rothschildi* as the basal member (Irestedt *et al.* 2009). This means that the possession of green abdominal plumage is likely to be ancestral and that the darker abdomens of *A. mayeri* and *A. stephaniae* are derived. It further implies that the inverted-display behaviour either evolved twice, independently, in *A. nigra* and *A. rothschildi* or that it is also, like green abdominal plumage, the plesiomorphic state and was present in the common ancestor of the extant *Astrapia* clade. If the second scenario is true, we can predict that *A. splendidissima* is likely have an inverted courtship display similar to those of *A. rothschildi* and *A. nigra*. Documenting the presence of or confirming the absence of an inverted display in *A. splendidissima* should be a high priority for future field observers (Scholes *et al.* 2017).

Mouth-lining colour is also quite vivid and significant in many species of paradisaeid, ranging from bright orange or yellow to lime-green and shown prominently when the bird is calling, though seemingly not with *Paradisaea* or Wallace's Standardwing. Leg colour, too, may be utilised in visual displays, *Cicinnurus* having bright blue legs, Twelve-wired possessing colourful reddish-pink legs and thighs, and Wallace's Standardwing exhibiting orange-yellow legs.

Inverted display poses are used by several species, including *Paradisaea* and *Astrapia*. Especially noteworthy is the Blue Bird of Paradise, which hangs upside-down and flashes the iridescent breast markings while making an amazing pulsing 'electronic' sound, *kisim pawa* ('go electric') in the local Tok Pisin tongue. The Emperor

Wallace's Standardwing, male at lek, Halmahera, North Maluku, Indonesia (*Chien Lee*).

Bird of Paradise differs not only vocally, but also in holding the wings open when inverted and flashing the deep green breast colour and white flank plumes. The King Bird of Paradise is one of the most remarkable of all, swaying backwards and forwards like a pendulum, resembling a glistening red tomato in a gale, when inverted. Other strange adaptations are the bare bright blue head skin of both sexes of Wilson's Bird of Paradise, elongated tail wires that are very obvious in the *Paradisaea*, and the flank wires of the Twelve-wired Bird of Paradise. The male of this last-mentioned species uses these wires to brush the face of the female, an unusual tactile element of the display, but the function of the elongated central rectrices of the *Paradisaea* is not known beyond the visual effect of the long, twisted shiny black ribbons of the Red Bird of Paradise. The Blue Bird of Paradise, in the monotypic genus *Paradisornis*, also has long central rectrices, the tips of these with iridescent mirror-like blobs which bob and sway.

None of the paradisaeids is known to be a colonial nester, and none has helpers at the nest, while almost all known nests have been built in branches of varying tree species. Some, such as Victoria's Riflebird, the Trumpet Manucode and the Twelve-wired Bird of Paradise, nest in the heavily foliaged crowns of small saplings, palms or pandan-type species, where there is some protection from arboreal predators such as snakes. The Trumpet Manucode may have a nesting association with the Black Butcherbird (*Melloria quoyi*) on Cape York, at least, the manucode building its nest near that of the predatory butcherbird, although how it avoids having its own nest preyed on remains unknown.

Both of the smaller Australian riflebirds are known to build sometimes atop the previous season's nest or at least nearby, and use a traditional site, and the same holds true for the Short-tailed Paradigalla (*Paradigalla brevicauda*) and Ribbon-tailed Astrapia, at least. Whether the same male is the parent is not known, but it must be a strong likelihood given that the same display sites are used year after year.

There are two basic types of **nest structure** within the paradisaeids, none of which is known to use sticks in its nests, a very different situation from that of the bowerbirds, where sticks are commonly used. The more ancient basal group of manucodes and paradise-crows builds shallow open cups of mainly vine tendrils, sometimes with a lining of dead leaves or dead wood, these being constructed in a branch fork. All the other known paradisaeid nests are deep and bulky bowls of orchid stems, leaves, mosses and ferns, also built in branch forks. An unusual and well-documented habit in Australia is that the females of both Victoria's and Paradise Riflebirds typically decorate the nest rim with sloughed snakeskin, which may perhaps deter some potential predators.

Data on the **clutch size** of birds of paradise are limited, but clutches typically consist of one or two eggs (rarely three), laid on consecutive days, although the monogamous manucodes typically have two-egg clutches (rarely three) and this may reflect the greater time available for foraging and provisioning nestlings when both sexes perform nest duties. Interestingly, what little is known of the highland paradisaeids seems to indicate that single-egg clutches are the norm, perhaps reflecting the poorer availability of food in the cooler, damper climates – although this, of course, conflicts with ideas about the diversity of species here being due to a greater variety of foods. So much remains to be learned. Data on the eggs of paradisaeids and brief egg descriptions are given under each species where known.

Incubation periods are known for only some of the polygynous species of this family, and range from 14 to 27 days, the longer periods occurring at higher altitudes. Passerines typically spend 60–80% of each day incubating, and this holds true for both polygynous species and those in which both sexes incubate so far as is known. Unlike the downy pale-skinned young of bowerbirds, paradisaeids hatch naked or with only very sparse down and they have, or develop, dark skin.

Nestling or fledging periods are known for only a few species and vary from 14 days to 30+ days; as with incubation, they seem longer for higher-altitude species than for lowland ones. The eyes open at around six days of age, and at about 8–10 days the body feathers of *Paradisaea* nestlings emerge from the pin tips, the primary coverts and primaries emerging at about two weeks of age. Frith *et al.* (2017) state that 'Growth curves for nestling birds-of-paradise in the wild show that the larger species grow at a fast rate, and also that all such curves are virtually linear over much of their length, except for the downturn just prior to the chicks' departure from the nest. Nestlings regularly lose some weight at this time because of the energetic cost of rapid growth of the tail feathers and flight-feathers.'

Data on **nest-provisioning** and nestling-brooding are scant. Females can spend anything between 14% and 48% of their time in brooding young, but much may depend on weather conditions, food availability, the physical condition of the female, and so on. Similarly, and again subject to many variables, feeding of single-chick broods varies from one to three meals per hour, while Trumpet Manucodes feeding two young over four days averaged 3.4 feeding visits per hour.

Paradisaeids, unlike bowerbirds, feed their young by regurgitation, the exception to this being that sometimes a last item is carried in the bill and fed directly to the chick. The faeces of the nestling seem to be swallowed, although this behaviour is as yet poorly known, and they may be carried away from the nest late in the cycle as fledging becomes imminent.

Distraction displays and aggressive nest defence occur but are very little known. Females will chase away conspecifics that venture into the general nest area, and Victoria's Riflebird and the Raggiana Bird of Paradise simply 'freeze' on a perch at the sight or sound of a potential predator.

Information on **parental care** once the young have left the nest is very limited. Juvenile Victoria's Riflebirds are certainly fed by the female for some weeks after fledging, and one timed instance was up to 74 days. A female Short-tailed Paradigalla was seen to feed an immature for as many as 108 days, but she apparently ceased doing so a week later.

Nest parasitism is very rare, the only documented example being of an Eastern Koel (*Eudynamys orientalis*) that laid an egg in an active nest of a Victoria's Riflebird, but the hosts for many of the New Guinea cuckoos are remarkably poorly known.

Voice

Birds of paradise are among the most spectacular avian species in the world in terms of plumage and displays, but their vocalisations are not quite in the same league. Many make loud, harsh sounds vaguely reminiscent of the calls of corvids, but none has anything that can be described as a conventionally beautiful song. Unusually, both sexes of the monogamous species, the manucodes and paradise-crows, call regularly, the Trumpet Manucode having sex-specific notes. Conversely, among the polygynous species, it is typically only the males that vocalise, giving loud and often far-carrying advertisement calls. The females appear to be largely silent except when alarmed, although the picture is complicated by the fact that young males have female-like plumage for some years and are quite vocal, rehearsing their repertoire for later.

Some truly extraordinary sounds are made by paradisaeids. One of the most unusual is the bizarre and electronic-sounding pulsing intense-display call of the male Blue Bird of Paradise, given while the bird is hanging upside-down in a dense thicket, and quite distinct from the sonorous bugling of the advertisement call, which carries up to 2km across the mountain valleys. The King of Saxony Bird of Paradise, as well as having one of the longest vernacular names, has a very strange fizzing rattling series, given by the males on their songposts, while the Brown Sicklebill is famous for the loud machine-gun-like rattle produced by the males. Other odd calls are emitted by manucodes, the Trumpet Manucode somewhat imaginatively named for trumpeting its advertising call, the Glossy-mantled having a rising call resembling the sound made by a tuning fork, and some manucode species having quiet deep sonorous calls which may have an infrasonic element that helps the sounds to carry through dense forest habitat, a field that merits research.

When properly courting, as opposed to advertisement calling, males of the plumed *Paradisaea* species are relatively quiet, some species making sharp clicking sounds with their mandibles or giving just a few brief soft notes. Magnificent Birds of Paradise also give soft, quiet contact notes when courting, and make a similar clicking sound, while Wallace's Standardwing emits quiet twittering noises in display and the King of Saxony utters much quieter vocalisations when bouncing up and down while displaying on a vine.

There are no convincing records of **mimicry** by birds of paradise in the wild, in complete contrast to the bowerbirds, which are famous for it. Some quiet *chip*-type notes may resemble those of other species but are not convincing demonstrations of mimicry. Some examples published on websites have proven to be misidentifications of different species, and it is essential that, for a convincing record, the bird must be seen to be producing the sound.

Movements

The birds of paradise are a little unusual in that there are no migratory species in the family. All are basically resident and sedentary, with some very limited and local dispersal in relation to food abundance or scarcity, drought or fire. In Australia, both Victoria's and Paradise Riflebirds undertake some dispersal into adjacent drier woodlands after breeding and in winter, the latter species being known also to wander down to the coastal lowlands in winter, but otherwise occurring at elevations above about 200m. Victoria's Riflebird similarly has some limited dispersal to drier woodlands west of the Great Divide, but there are no records of vagrancy as such for any of the family.

Conservation

New Guinea still has a great abundance of forest, much of it in extremely challenging terrain, and this has thus far enabled many of the island ecosystems to remain relatively intact. Hunting has been continuous for thousands of years, formerly at low-impact and largely sustainable levels until the advent of modern technology and the vastly increased human-population pressure adversely tipped the balance.

Timber extraction remains a major threat, with the consequent degradation/devastation of the forest, while wholesale replacement of lowland-forest ecosystems with oil-palm (*Elaeis guineensis*) monocultures is a developing problem in certain parts of the mainland lowlands. Copper- and gold-mining has also adversely affected substantial local areas, with gas extraction the latest extractive industry to have an impact, opening up previously remote and inaccessible areas to exploitation. The rapidly expanding human population is also a mid-term problem, with around 2% rates of annual increase very fast and exerting increasing pressure on fragile resources.

National parks, which might offer some degree of protection, are problematic in New Guinea (and Melanesia in general), since traditional modes of land ownership render government actions in establishing national parks either impossible or largely irrelevant. Local communal decision-making in Papua New Guinea, at least, makes centralised direction difficult. Despite this, there are three quite large national parks designated in West Papua. The largest, the Lorentz National Park, dates from colonial times and extends from the Snow Mts to the Arafura Sea. Teluk Cendrawasih, in the lowland coastal Vogelkop, is largely irrelevant here as it is essentially a marine reserve. The third, Wasur, in the Trans-Fly, protects the lower Trans-Fly drainage. There are 30 other designated reserves scattered about the western half of the region, but once again many of them for marine areas or islands. Where the local community is involved in the running and maintaining of them, Wildlife Management Areas in Papua New Guinea seem moderately effective. Changing demands, expectations or local conditions, however, mean that they are not necessarily the answer in the long term.

Ecotourism is in its infancy and income generated from it is so far low and at subsistence levels, although some initiatives are quite meaningful at a local level. Much can be done to expedite wildlife-based tourism, be it centred on birds, tree-kangaroos, fish, plants or butterflies. As a result of these initiatives, increasing numbers of local people are working as skilled and knowledgeable guides. This not only directly helps the individuals involved, but also gives an important local voice to environmental concerns. The training of a cadre of guides, by aid organisations and the international conservation groups, has great potential for the future.

The great unknown with conservation these days is, of course, the impact of climate change, which can be expected to be very significant in the medium to long term. This is of particular concern for the higher-altitude species that have a restricted habitat, as it is likely that changing wind, temperature and rainfall patterns will gradually modify the flora and fauna of the forest communities, with huge impacts on all species. The potentially increased frequency and intensity of cyclones will also have serious consequences, especially for species on islands, where habitats are limited and thus very prone to catastrophic-event damage. Species at higher altitudes are likely to find their habitats shrinking, while changes to salinity, sea-level rises and vegetation modification will have adverse impacts on coastal and island species. Immense and profound changes to ecosystems can be expected over the short term, with worrying conservation implications for specialised or restricted-range species.

In Australia, all the higher-altitude rainforest taxa must be at risk as they have nowhere else to which they can relocate, and species such as the Golden (*Prionodura newtoniana*) and Tooth-billed Bowerbirds (*Scenopoeetes dentirostris*) could therefore be in trouble. In New Guinea, the Fire-maned Bowerbird (*Sericulus bakeri*), Archbold's Bowerbird (*Archboldia papuensis*) and both Long-tailed Paradigalla and Arfak Astrapia are species with restricted high-elevation ranges that are potentially vulnerable. Much depends on the next few years, but the overall prognosis has to be bleak, with political will lacking and vast socio-economic pressures driving development irrespective of environmental costs, which are never included in the economic-cost packages.

All members of the paradisaeid family in Australia are classified by BirdLife as being of Least Concern (**LC**), with none currently rare or endangered. In New Guinea, on the other hand, the picture is not quite so rosy, this due primarily to habitat loss or damage coupled with some small-scale hunting for certain highly prized species:

VU Vulnerable: Black Sicklebill, Blue Bird of Paradise and Goldie's Bird of Paradise

NT Near Threatened: Wahnes's Parotia, Pale-billed Sicklebill, Long-tailed Paradigalla, Ribbon-tailed

Astrapia, Wilson's Bird of Paradise, Emperor Bird of Paradise, Red Bird of Paradise. All of these are of relatively restricted range and several are island endemics.

Membership of conservation organisations is becoming increasingly important as the human population inexorably rises and the 'Great War' on nature intensifies. There is a continuing loss and degradation of forest habitats even in such relatively prosperous countries as Australia, and far worse in both Papua New Guinea and Indonesia, the other states having custodianship of these two spectacular families. Important organisations that have achieved significant conservation outcomes are BirdLife International, based in the UK and with regional branches in Australia and Indonesia, and Conservation International from the USA, which has done a great deal of conservation work with ecologically sound credentials and centred on local people in Papua New Guinea.

The BirdLife affiliate *Burung Indonesia* has some worthwhile goals, which neatly encapsulate conservation aims in the context of that important country for birds of paradise and bowerbirds. Its mission statement is as follows.

Aim
· To be the guardian of Indonesia's wild birds and their habitats through working with people for sustainable development.
· To achieve this, Burung Indonesia has been dedicated to:
· Promoting conservation of sites, species and habitats.
· Working with communities to promote collaborative conservation and natural-resource management for sustainable development.
· Developing the organisational capacity for improved management of habitats, sites and species.

Key activities
· Comprehensive conservation action for species, sites and habitats through working in protected areas, sustainable productive landscape management and ecosystem restoration in production forests.
· Policy advocacy at local and national levels, utilising multi-stakeholder approaches.
· Research and monitoring work on priority bird species (globally threatened, endemic, or species of parrot) in the Wallacea region.
· Management of data, information and knowledge to set priorities, support conservation actions, and function as a provider of information services to external constituencies.
· Public involvement through local conservation groups, membership of as well as working with the media, NGOs, individuals, and private-sector and government agencies.
· Promotion of bird conservation to the public.

Contact details
Burung Indonesia
Jl. Dadali 32, Bogor, ID, 16161
Tel: +62 251 835 7222 ext. 101
Fax: +62 251 835 7961
Birdlife@burung.org
www.Burung.org

BirdLife International
The David Attenborough Building, 1st Floor, Pembroke Street,
Cambridge, CB2 3QZ, UK
Tel: +44 (0) 1223 277 318
Fax: +44 (0) 1223 281 441
birdlife@birdlife.org

BirdLife Australia
Suite 2-05, 60 Leicester Street,
Carlton, VIC 3053
info@birdlife.org.au

What is a bowerbird? The family Ptilonorhynchidae

Bowerbirds and the *Ailuroedus* catbirds are all members of the family Ptilonorhynchidae, and include species that build some of the most striking and complex structures in the avian kingdom. The scientific name refers to the partial covering of the base of the bill by feathers, from the Greek *ptilon* (feather) and *rhunkos* (bill). They constitute a family that currently, with the latest taxonomic adjustments, comprises some 28 species in eight genera, restricted to Australia and New Guinea. In New Guinea they live primarily in the rainforests, with a couple of forest-edge and savanna species. In Australia they occupy rainforest and drier woodland habitats. Bowerbirds and catbirds do not occur west of Wallace's Line in Sulawesi, and are also absent from Halmahera, the Bismarck and Solomon Archipelagos and parts of the arid interior of Australia.

The family is divided into two subfamilies, Ailuroedinae (catbirds) and Ptilonorhynchinae (bowerbirds), an ancient separation dating back around 24 million years (Irestedt *et al.* 2105). The molecular studies of Barker *et al.* (2004) indicated that this family is a sister-group to the Australasian Treecreepers (Climacteridae), a placement corroborated by Jønsson *et al.* (2011).

Although fairly compact, the family supports eight generic lineages in four major groups:

- **The catbirds (*Ailuroedus*) and Tooth-billed Bowerbird (*Scenopoeetes*):** All the catbird species are most obviously more closely related to one another than they are to any other bowerbirds; recent biomolecular research indicates that they diverged from other bowerbird genera about 24 million years ago, and that they are quite distinct from the builders of avenue bowers. The Tooth-billed Bowerbird is, however, as distinct from the catbirds as are other bowerbird subgroups, and appears to be genetically closer to the New Guinea *Amblyornis* gardener bowerbirds than it is to catbirds (Kusmierski *et al.* 1997).

- **The *Amblyornis* grouping, the gardener bowerbirds:** including Golden Bowerbird (*Prionodura*) and Archbold's Bowerbird (*Archboldia*) and perhaps Tooth-billed Bowerbird (*Scenopoeetes*).

- ***Sericulus* and *Ptilonorhynchus*:** the avenue-bower-building silky bowerbirds.

- ***Chlamydera*:** the avenue-bower-building grey bowerbirds.

The *Ailuroedus* catbirds (the Spotted and White-eared Catbird assemblages) are now classified as some ten species, after recent research revealed more of the genetic history of this cryptic group (Irestedt *et al.* 2015). This group and the Tooth-billed Bowerbird are the basal members of the family, but somewhat divergent in breeding behaviour.

These ten species of the green *Ailuroedus* catbird group all lack marked sexual dimorphism, form pair bonds in which the male helps to build the nest but only the female incubates, and have rather simple arboreal chasing displays without the use of bowers or stages. In contrast, the sexually monomorphic but polygynous Tooth-billed Bowerbird has a terrestrial display and builds a simple stage of large leaves on the forest floor as a display arena, the males not forming pair bonds and not helping to build nests or to raise the young.

Gilliard (1969) pointed out that the bowerbird species with the most beautifully plumaged males exhibit the smallest and least-developed bowers, all of the sexually dimorphic species being polygynous and bower-builders. The *Sericulus* species build fairly small avenue-type bowers with limited decoration, while the Satin Bowerbird (*Ptilonorhynchus violaceus*) builds a much more elaborate type of avenue bower, lavishly adorned with found objects (often of a bright blue colour). Members of the *Amblyornis*/*Prionodura*/*Archboldia* assemblage build some of the most remarkable structures in the avian world, the species of the genus *Amblyornis* constructing maypole-type bowers. These include the Vogelkop (*Amblyornis inornata*) and Streaked Bowerbirds (*A. subalaris*), which build a highly distinctive hut-style structure, as well as the Golden Bowerbird, which constructs a double maypole bower. *Archboldia* makes a strange mat-type formation and uses unique decorative items, including beetle wings and the plumes of the King of Saxony Bird of Paradise.

An anomaly from New Zealand

There is one anomalous genus, *Turnagra* from New Zealand, of what are now considered to be two species, the New Zealand Thrush or North Island Piopio (*T. tanagra*) and the South Island Piopio (*T. capensis*). The piopio has been the subject of an extraordinary amount of taxonomic speculation over many years, having been variously allied with thrushes (Turdidae) and then accorded a new family Turnagridae (Buller 1887), before being placed in Corvidae, and later allied to Ptilonorhynchidae on the basis of external morphology

(Finsch 1874). Mayr & Amadon (1951) allied it with Pachycephalidae and later workers placed it with both bowerbirds and the cnemophilines, which were then considered a subfamily of birds of paradise. *Turnagra* languished as basically *incertae sedis* for many more years, but the most recent detailed genetic research now places it with Oriolidae (Zuccon & Ericson 2012) and it may be that the long uncertainty has finally been resolved.

The piopio was formerly fairly tame and common in some areas but declined rapidly in the 1880s, the last North Island specimen dating from 1902, although it was rumoured to exist right up to the 1930s or even later. The population of the South Island Piopio on Stephens Island – infamous for the eponymous flightless Stephens Island or Lyall's Wren (*Xenicus lyalli*), claimed to have been extirpated by the dreaded 'Tibbles', the lighthouse-keeper's cat – was common until 1895, but had gone by 1898, presumably a victim of the same or another noxious feline. The decline has been attributed to land clearance, the introduction of numerous feral predators and disease, although the bird was also found by settlers to be tasty eating: 'it makes a savoury broil for those who bring the right sauce – hunger' (Hutton & Drummond 1904).

Bowerbird biology

Most of the New Guinea bowerbird species are found mainly in the hills and mountains, although five – the Flame (*Sericulus ardens*) and Fawn-breasted (*Chlamydera cerviniventris*) Bowerbirds and three of the catbirds – are primarily lowland-dwelling. All of the *Chlamydera* species (Great, Fawn-breasted, Yellow-breasted, Spotted and Western Bowerbirds) prefer non-forest drier scrub and edge habitats, but the remaining members of the family inhabit the forest interior, ranging upwards in elevation to the edge of the treeline. All are primarily frugivorous and some are fig specialists. The males of the polygamous species are noted vocalists and are adept at mimicking sounds, including the songs of other bird species and sounds of human origin.

The name 'bower-bird' was first coined in 1840 by the renowned naturalist John Gould, who had seen bowers of Spotted (*Chlamydera maculata*) and Satin Bowerbirds in New South Wales in 1839. He was talking about the very first bower ever seen in England, that of a Satin Bowerbird, which was exhibited at the Zoological Society of London. Gould saw a structural resemblance to a wooded bower, which in formal landscape gardening at that time was a romantic secluded trysting place with a seat. The name has been used ever since to cover the great variety of bowers constructed by most species of the family.

The majority of bowerbirds are stout-bodied and lack a long tail, are primarily frugivorous, the males build varyingly complex bower structures, and the female alone constructs the nest and carries out the incubation and nestling-care duties. The exceptions are the *Ailuroedus* catbirds, which are monogamous, share some of the nest duties between the sexes (though the female incubates the eggs) and do not build bowers or stages, and the rather aberrant Tooth-billed Bowerbird, which makes a simple stage of upturned leaves on the forest floor but is not monogamous and does not share nest duties.

Bowerbirds were once widely considered closely related to birds of paradise, and were sometimes even combined with that family in either Paradisaeidae (Stresemann 1934; Gilliard 1969) or Ptilonorhynchidae (Sharpe 1891–98; Schodde 1975; Cracraft 1981) on the grounds of certain similarities in zoogeography, chunky body shape, sexual dimorphism and a polygynous breeding strategy. As the sum of knowledge grew, however, it was realised that major differences exist:

· Egg coloration differs between the two families.

· Eggs of a clutch are laid on alternate days by many bowerbirds, but on consecutive days by paradisaeids.

· Nestling plumage differs, that of bowerbirds being very downy and pale-skinned, not bare-skinned and dark like birds of paradise.

· Parent bowerbirds do not feed the young by regurgitation.

· Differences in skull osteology exist, an enlarged lacrymal (part of the cranium bone structure) near the eye socket being a feature unique to the Ptilonorhynchidae and the lyrebirds (Menuridae).

· Some bowerbirds are known to cache food, a trait not shown by birds of paradise.

· Many bowerbirds are accomplished mimics, another trait not found in birds of paradise.

· The loud bugling or far-carrying calls of many birds of paradise have no equivalent in bowerbirds.

· Nest structure differs between the two families, the bowerbirds (unlike paradisaeids) using a woody stick base.

· The use of bowers for display by the males is restricted to bowerbirds, whereas birds of paradise are mostly landscape gardeners rather than architects (using arenas or courts, instead of using sticks or vegetation for construction purposes).

· Elements of the displays differ greatly, and the hide-and-peek display of some bowerbirds is unknown among the paradisaeids.

· Bowerbirds include several species known to use a tool to paint their bowers, tool use being a very rare habit among birds.

· Bill size and shape are relatively uniform among the bowerbirds, lacking the great range of shape and size found in the paradisaeids.

· Birds of paradise and bowerbirds both have 12 rectrices and ten primaries, but bowerbirds have an atypically large number of secondaries (11–14 including the tertials).

· Hybridisation is very rare in bowerbirds, but much more widespread in birds of paradise.

· The bowerbirds do not use their feet to hold or manipulate food items, whereas paradisaeids are quite adept at this.

Stonor (1937) found no evidence that the two families were closely related, and classified bowerbirds as a distinct family with no close relatives. Sibley & Ahlquist (1985, 1990), in their pioneering studies of DNA–DNA hybridisation, also found that bowerbirds had no close relatives and were in fact part of the great corvoid songbird lineage. Christidis & Schodde (1991) showed a relationship with Artamidae (butcherbirds, woodswallows etc.), and later studies also reveal them as a monophyletic family (Kusmierski *et al.* 1993). Bowerbirds are a part of a major radiation in the Australasian region over the past 60 million years, diverging from lyrebirds (Menuridae) and scrub-birds (Atrichornithidae) about 45 million years ago (Sibley & Ahlquist 1985). The divergence from the other corvine lineages seems to have occurred around 28 million years ago, the main lineages of the bowerbird family evolving about 24 million years ago (Helm-Bychowski & Cracraft 1993), although these findings were based on molecular-clock calibrations, which are still being resolved.

The exact placing of the family Ptilonorhynchidae within the oscine passerines was somewhat unsettled until recent decades. The results of several biomolecular studies show it to be a highly distinctive family relatively distant from, and basal to, the Paradisaeidae within the oscines. Recent studies have shown that the majority of Australasian passerines, about 85% of the total, are derived from a southern, or Gondwanan, origin (Ericson *et al.* 2002). This major group, referred to as the 'Corvida', radiated in Australia and New Guinea, and today comprises three major lineages:

· Menuroidea, which includes the lyrebirds (Menuridae) and Australasian treecreepers (Climacteridae), as well as the bowerbirds; lyrebirds appear to be bowerbird relatives and are among the most ancient of the oscines (Frith *et al.* 2017).

· Meliphagoidea, which includes honeyeaters (Meliphagidae) and their allies.

· Corvoidea, which includes the birds of paradise.

The 28 species of bowerbird represent a basal group of Australasian oscine passerines, this categorisation supported by certain osteological characters such as the presence of a single corvoid pneumatic fossa (a depression or hollow) in the humerus bones of bowerbirds. Bowerbird anatomy is otherwise broadly typical of that of the higher passerines, although bowerbirds have characters of the legs, feet, palate, syrinx and spermatozoa which, in combination, can define them as a distinct family (Frith *et al.* 2017).

Catbirds are not related to the American Grey Catbird (*Dumetella carolinensis*), which is a member of the mockingbird family (Mimidae). They are, however, as with that species, named for their cat-like calls, although the loud squalling Siamese-cat-like calls of the *Ailuroedus* catbirds bear scant resemblance to the quieter raspy calls of their North American namesake. They are strongly territorial and reside in the same patch of forest year after year, using these loud distinctive calls to establish their territory (Schodde & Tidemann 1988). All the catbirds are coloured a bright leaf-green above, and have a spotted or scaled breast. The thick, short bill is pale in colour, adapted for their primary diet of fruit in their rainforest habitat. They are among the most ancient basal members of the bowerbird family and do not build bowers, which suggests that this trait evolved later. Further, the catbirds have a quite different lifestyle from that of bowerbirds, and form monogamous pairs. There are three species in Australia (one shared with New Guinea) and a further seven in New Guinea (three from the former White-eared Catbird group and four from the Spotted/Black-eared Catbird group).

These ten (formerly three) *Ailuroedus* catbird species are most obviously more closely related to one another than they are to any other bowerbirds; recent biomolecular research indicates that they diverged about 24 million years ago, and that they are quite distinct from the builders of avenue or maypole bowers. Another basal taxon is the Tooth-billed Bowerbird, which is as distinct from the catbirds as are other bowerbird subgroups and appears to be genetically closer to New Guinea gardener bowerbirds than it is to catbirds.

All bowerbirds have ten primaries and 12 rectrices, but the number of secondaries varies from 11 or 12 in *Ailuroedus* catbirds and *Amblyornis* to 12 or 13 in *Sericulus* and 14 in most *Chlamydera*. Birds occupying the more arid areas tend to be less colourful than those living in the denser forest habitats, as with the *Chlamydera* grey bowerbirds. This may reflect the greater concealment opportunities within forests and consequently the lesser need for a drab plumage. The family can be divided into three broad groupings:

· The monogamous and **non-bower-building** Spotted/Black-eared, Green (*Ailuroedus crassirostris*) and White-eared (*A. buccoides*) Catbird groups (*Ailuroedus*). The Tooth-billed Bowerbird, known also as the Tooth-billed Catbird, does not construct as bower as such, but just a simple leaf stage on the forest floor. One of the old and very appropriate names for this species was 'Stagemaker', a reference to this unique habit in the family.

· The **avenue-bower**-builders, these being the spectacularly bright-plumaged male silky bowerbirds (*Sericulus*) and the grey bowerbird (*Chlamydera*) group plus the Satin Bowerbird (*Ptilonorhynchus*).

· The third group is the largely much plainer-plumaged **maypole-bower**-building *Amblyornis* (plus *Archboldia* and *Prionodura*).

Bowerbird males display near their bower structure in order to attract females, which then choose which male they want. Males also collect display objects, the composition and quantity varying with the species, and some favouring particular colours, such as blue in the case of the Satin Bowerbird.

Bowerbird morphology

Most bowerbirds are quite plump and stocky, with broad, rather rounded wings, and, being closely related to one another, they are morphologically less diverse than many passerine families of comparable size. They vary in body length from 22cm to 37cm, with Great Bowerbird (*Chlamydera nuchalis*) and Archbold's Bowerbird the largest, the former weighing *c.*200g. The Golden Bowerbird is the smallest, at around 80g, and unusual in being quite slender and relatively long-tailed, which makes it slightly longer than the plumper, heavier (95–140g) and shorter-tailed *Amblyornis* gardener bowerbirds. Plumage abnormalities in bowerbirds are extremely rare. The sole documented case refers to a female Satin Bowerbird exhibiting leucism, whereby the eumelanin in the plumage is incompletely oxidised and remains brown, causing a brown aberration. The aberrant bird in this case was a cream-coloured individual with bleached flight-feathers, as this mutation is prone to rapid fading (van Grouw 2012).

Typically, bowerbirds have a juvenile plumage, then a very similar immature plumage, and sometimes a distinct subadult dress before the adult plumage. Among all the catbirds, the Tooth-billed Bowerbird, the Vogelkop Bowerbird and all the *Chlamydera* grey bowerbirds the sexes are the same or very similar, and there is no distinctive immature plumage. Polygynous bowerbirds, however, generally exhibit a distinct sexual dimorphism, males of most *Amblyornis* gardener bowerbirds (except the Vogelkop Bowerbird) and of Archbold's Bowerbird having a short erectile crest of orange or yellow, this lacking in the dull brownish females and immatures. In the monotypic Golden Bowerbird the males are primarily bright yellow and greenish-yellow, whereas the females and immatures are grey-brown and whitish. The Satin Bowerbird also is monotypic, and likewise dimorphic, with a striking glossy blue-black male, while females and immatures are green above and have varying amounts of greenish scaling on the underparts. The *Sericulus* silky bowerbirds all have very striking male plumages of orange, yellow and black, the females and immatures being sombre in plumage with ventral barring.

Montane species in New Guinea are often good examples of Bergmann's Rule, which states that species resident in colder environments will average larger than those in warmer zones. Bowerbirds in both New Guinea and Australia reflect this dictum. Looking at the various genera, the upland Black-eared, MacGregor's (*Amblyornis macgregoriae*), Fire-maned and Masked Bowerbirds (*Sericulus aureus*) are all larger and heavier than their lowland counterparts in the White-eared Catbird group, the Streaked Bowerbird and the Flame Bowerbird. The bowerbird species resident at the highest altitude of all, Archbold's Bowerbird, is considerably longer than even other montane bowerbirds. The Regent Bowerbird (*Sericulus chrysocephalus*) in Australia is the largest *Sericulus* and it, too, is a hill-forest species.

Most adult bowerbirds have very similar body size and shape, but adult male Archbold's and Golden Bowerbirds develop a slightly longer tail than that of their females. The converse is true for adult female *Sericulus* bowerbirds, the Satin Bowerbird and most of the *Chlamydera* grey bowerbirds, which have a slightly longer tail than the adult males. Adult female catbirds average nearly 10% lighter in weight than the adult males, although the body proportions remain the same. Juvenile and immature bowerbirds are generally slightly smaller than the adults. Exceptions are the younger males of the *Sericulus* bowerbirds, the Satin Bowerbird and the Spotted and Western (*Chlamydera guttata*) Bowerbirds, all of which have a slightly longer tail than that of the adults. Most polygynous bowerbirds have adult females slightly smaller than adult males, except for three of the four *Sericulus* species, the females of which are the same size as or slightly larger than the adult males.

Passerines with proportionately longer legs are generally more terrestrial in habits, whereas those with shorter legs are more arboreal. Within the Ptilonorhynchidae, the grey bowerbirds in the genus *Chlamydera* have long legs, as befits their often terrestrial lifestyle, while the more arboreal Tooth-billed Bowerbird as may be expected has shorter legs.

Most bowerbirds have a relatively short, deep, robust bill, but the four *Sericulus* silky bowerbirds have a longer and finer bill. The Tooth-billed Bowerbird was for many years a puzzle, with various theories advanced to explain the unique notch or 'tooth' on both mandibles. It was eventually discovered that during winter, especially, this bird is a folivore, eating considerable quantities of leaves, and the tooth cusp and notch structure are an adaptation to help it to tear or snip off leaf pieces and wad them for consumption.

Polygynous bowerbirds generally have a dark bill, but *Amblyornis* gardener bowerbirds and *Sericulus* silky bowerbirds exhibit a pale bluish base to the lower mandible. The Satin Bowerbird has a striking bluish-white bill (which appears short owing to some feathering covering the loral area), while all the catbirds have a whitish bill with a black mouth-lining. Unlike other species in the family, adult male Regent Bowerbirds have a yellow or pinkish-yellow bill. Adult male Tooth-billed Bowerbirds have a black bill which has a contrasting white front portion on the inside of the upper mandible, whereas females and immature males of this species have the mouth pale yellowish-flesh, unlike the orange-yellow or yellow of most other bowerbirds.

Soft-part coloration is quite variable within the bowerbird family, most adults having a dark to pale brown iris, that of the juveniles being paler and greyer. Adult catbirds, however, have a red iris, which is blue-grey in juveniles, while adults of both sexes of the Golden Bowerbird have a pale iris, as do all the male *Sericulus* silky bowerbirds, though the females of the *Sericulus* species show a brownish iris. Satin Bowerbirds are very different, the males having an astonishing violet-blue iris while the females have striking blue eyes. The coloration of the legs and feet also is quite variable, ranging from black to shades of grey or olive-brown, but the Vogelkop Bowerbird stands out in having distinctly blue legs and feet.

Most male passerines develop adult plumage and breed within their first year or two of life. Males of the brightly plumaged polygynous bowerbird species, however, do not acquire full adult dress until five to six or even seven years of age. There may be a subadult plumage, with a gradual transition of some male plumage characters into the female-type dress over a year or two. Like other promiscuous male passerines that use display sites, promiscuous males of the bowerbirds have delayed acquisition of their distinctive colourful adult male features, and so can live among their conspecifics for more than five years before these become apparent.

Importantly in this family, a subadult plumage is a clear visual signal that the bird is not yet fully part of the highly competitive adult male community, although it may, of course, try opportunistically to mate with females. It is thought that a prolonged period in immature plumage may enhance a young male's opportunity to move within otherwise aggressive adult male society while gaining valuable experience. A dull plumage may also perhaps reduce the risk of attack by a predator, although predation remains remarkably little documented for either paradisaeids or bowerbirds.

Moult

Adult bowerbirds have an annual moult, mainly during December to March, peaking during the wetter months after the young have fledged. They are in prime condition in terms of body mass and fat levels at the end of this period, just before the austral winter when it is drier and cooler and when food resources become much less abundant. The grey *Chlamydera* bowerbirds also moult towards the end of the year, after courtship and nesting, but the timing seems less clear-cut as some also shed the wing and tail feathers at other times of the year.

Detailed knowledge of bowerbird moult is available only for some Australian species, the Satin Bowerbird being quite well known, as are, to a lesser extent, the Spotted Catbird and Tooth-billed and Golden Bowerbirds. Bowerbirds follow the pattern common to most small passerines. The head-and-body moult is the most protracted, starting before that of the wings and tail but often not finishing until after the flight-feathers have been moulted. Critical flight-feathers are moulted symmetrically (presumably helping flight). Hence, the primaries in each wing are renewed sequentially from the inner wing outwards towards the tip, and this pattern is mirrored by the primary-covert moult, too. During this process the tail feathers also are renewed, starting with the central tail feathers and again working outwards (although this process is completed before the primary moult has finished). The moult of the secondaries proceeds only once four or five primaries have been fully replaced (or nearly so); in contrast, replacement of the tertials is irregular and completed before that of the secondaries has finished – see Frith *et al.* (2017).

Bowerbird behaviour

Climate conditions and food resources, with sufficient fruit and arthropods available to sustain the heavy energy requirements of the breeding regime, are the fundamental determinants of the life cycle of the bowerbirds. Within Australia, the peak in courtship and nesting activity is primarily from late September to December (the austral spring and early summer), when temperatures and rainfall increase and food resources are at their most abundant. The start of the rains determines when the bower and nest activities finish, nestlings fledging just before or during the early wet season. This is a time of abundant resources and it is when the females, or both parents in the case of catbirds, are still feeding the nestlings. There is considerable seasonal variation in rainfall totals, even in the rainforest habitats, and in the drier country the avenue-building grey *Chlamydera* bowerbirds rely heavily on the seasonal abundance of grasshoppers (Orthoptera). Their breeding cycles are dependent upon the somewhat unpredictable rainfall. The picture in New Guinea is far less well known and more complicated, with a far greater altitudinal range and many microhabitats and microclimates making for greater variability in timing.

Behaviours known for some members of the family include the usual **preening** to keep the feathers clean and in good order, but also the less usual anting and sunning behaviour. **Anting**, using live ants (Formicidae) with which to brush the feathers, is a curious and relatively rarely observed activity of wild birds, but Tooth-billed, Golden and Satin Bowerbirds are known to indulge occasionally, and it is suggested that this may lessen parasite loads. **Sunning** is a well-known, widespread, but again still relatively mysterious and little-understood aspect of avian behaviour, indulged in by only a few of the bowerbird species. These include Spotted Catbirds and male Golden Bowerbirds, which are known to **sunbathe** by adopting a distinctive and unusual posture which resembles a sick or injured bird: they lie on the ground with the feathers of the breast, rump, head and nape erect, the tail pressed down and the wings drooped. One unusual observation was of some eight adult male and at least six 'female-plumaged' Regent Bowerbirds which sunned themselves in a semicircular area of 5m × 1.5m of open, directly sunlit leaf litter and on leafy perches above this (Frith *et al.* 2017). At any one time, up to eight of these individuals were squatting or lying on the litter in typical passerine sunning postures, this activity lasting for up to 15 minutes.

Another unusual bowerbird behaviour is the **caching** of fruits. Golden Bowerbirds conceal fruits among forest-floor litter and beneath fallen timber; these fruits, which are usually (but not always) ripe, sometimes germinate if not retrieved. Fruit-caching for later consumption is well known also for Green and Spotted Catbirds. It seems that cached fruits do not serve as bower decorations, but some male Golden and MacGregor's Bowerbirds do 'store' bower decorations around their bower sites.

Bowerbirds utilise various water sources for **drinking** and **bathing**, usually creeks and water-filled holes in trees for the rainforest species, or waterholes and cattle troughs for the grey bowerbirds of the drier habitats. Males also drink water droplets from forest-floor leaf litter and wet foliage around their bower sites. An adult male Golden Bowerbird has been observed foliage-bathing by flying into wet foliage and briefly fluttering and hovering among the leaves, following a brief shower of rain (Frith *et al.* 2017).

Bowerbirds have quite broad wings and a strong and purposeful **flight**, which is undulating over longer distances. The wings can produce an audible rustling sound in normal flight, and loud wing noises are known for some species when flying about their bowers. Foraging flights are typically rapid and direct, going straight to the food source. A few species perform courtship flights, the most remarkable being the striking butterfly-like hovering display of the Golden Bowerbird. In New Guinea, *Amblyornis* species make some simplistic 'flight' displays between adjacent vertical saplings at their bowers.

Roosting behaviour is very little documented, the forest species in particular being very hard to locate in the dense habitat, but J. Leonard (*in litt.*) reports Satin Bowerbird 'all-violet males roosting singly low down (within a metre of the ground) on a thin quite exposed branch within a few metres of a bower. I have observed this four or five times in one spot in Canberra, an arboretum type plantation in Weston Park which has had an active bower for many years.' D. James (*in litt.*) reports 'a female-plumaged Satin Bowerbird roosting about eight feet up in crown of a ten foot large-leaved privet, about an hour after dark. It was in a narrow strip of remnant riparian forest (Sydney Sandstone Gully Forest) adjacent to Epping Railway Station in suburban Sydney. Very weedy, with privet making a dense understory. Satins seem to be occasional visitors to the site, not resident.' P. Veerman (*in litt.*) reports an interesting sighting that may well pertain to flocks going to roost near Canberra: 'Just as the sun was setting, many Pied Currawongs (*Strepera graculina*) were flying over the river… to perch in the Bullen Range. That happens almost each time I am there. Very easy to see, as they fly high. This time I happened to look down at one point… and in about two minutes I would have seen 50 + Satin Bowerbirds flying the same route as the Currawongs, except all low within the canopy. It is odd that I have never noticed that before.'

Habitat

Rainforest is the major habitat for most bowerbirds, including all the catbirds, the Tooth-billed, Golden and Archbold's Bowerbirds and the *Amblyornis* gardener bowerbirds. *Sericulus* silky bowerbirds and the Satin Bowerbird also occupy rainforest, but are more closely associated with rainforest edges, with the latter species also in adjacent wet sclerophyll woodlands. The *Chlamydera* grey bowerbirds are different, being adapted to drier, more open environments such as riparian forests, open woodlands, savanna, the forest–grassland ecotone and, in Australia, semi-desert conditions, where the four species of grey bowerbird inhabit drier habitats, predominantly to the west of the Great Divide.

Some species in Australia exploit man-made environments such as urban parks, gardens, caravan parks, orchards and farms, where water and food are plentiful. Great, Spotted and Western Bowerbirds and Satin Bowerbirds frequently have bowers in such sites, while in New Guinea the Fawn-breasted Bowerbird occurs in gardens and open spaces in some major cities, such as Port Moresby.

New Guinea rainforests sometimes have two species co-existing, as evidenced, for example, by MacGregor's and Archbold's Bowerbirds overlapping in a part of their range. It is in the Australian rainforests, however, where most members of this family are sympatric, with both monogamous and polygynous species. In North Queensland, the tropical upland rainforests support Spotted Catbird, Tooth-billed and Golden Bowerbirds, plus the small northern subspecies of the Satin Bowerbird. Farther south, the Green Catbird shares subtropical rainforest with Regent and Satin Bowerbirds, where they occur also in both subtropical uplands and subtropical lowlands.

The *Chlamydera* grey bowerbirds provide some interesting examples of sympatry in Australia, where Fawn-breasted and Great Bowerbirds are sympatric in the northern Cape York Peninsula. Where the two occur together, the Fawn-breasted Bowerbird is found mostly in mangroves or adjacent *Melaleuca*-dominated habitats, with Great Bowerbirds in the open woodland. The Great Bowerbird is widely distributed through much of the Australian tropics, often building its bowers in riverine vegetation adjacent to seasonally dry riverbeds. Sympatry with the Spotted Bowerbird is very local in some parts of north-central Queensland, and some very rare instances of hybridisation are known.

The species pair of Spotted and Western Bowerbirds, which were formerly lumped as one species, occurs in the more semi-arid to arid habitats of northern Queensland and the subtropics of central and western tropical Australia. Spotted Bowerbirds occupy open woodland with low dense shrubbery, which provides good cover for their bowers. Central and Western Australia have the Western Bowerbird replacing the Spotted Bowerbird, and favouring areas close to water, such as river gorges with shady copses and Rock Figs

(*Ficus platypoda*). The availability of the fruit of this fig species may influence the distribution of Western Bowerbirds in more arid areas, with figs a significant food resource.

Within New Guinea, one or two catbird species may occur together with one or more species of polygynous bowerbird. Around Kiunga, Flame Bowerbirds can be found in the same area as both Black-eared (*Ailuroedus melanotis*) and Ochre-breasted (White-eared) Catbirds, which are locally sympatric in this area despite generally being altitudinally separated. Four pairs of closely related polygynous bowerbird species meet in parts of New Guinea where their otherwise typically separate altitudinal ranges abut or overlap, and they share the habitat.

Another such species pair is represented by the Masked and Flame Bowerbirds, the former replacing the latter at higher altitudes in their narrow zone of abutment, with one or two examples of hybrids known (which for years led to the two being lumped as one species). A similar situation occurs with the Streaked Bowerbird of lower montane forest, which is replaced by MacGregor's Bowerbird at higher elevations, again with some very limited hybridisation.

Mid-montane forest habitats have the Vogelkop and Yellow-fronted (*Amblyornis flavifrons*) Bowerbirds alone in their restricted ranges. Archbold's Bowerbird occurs at altitudes above 1,750m, very locally sharing habitat with MacGregor's Bowerbird around the Tari Gap, in east-central New Guinea. The former species is the bowerbird that occurs at the highest elevation, being known from the montane moss forest at up to 3,660m above sea level in the Ilaga valley of the Snow Mts, in West Papua.

Food and foraging

The **diet** of bowerbirds contains a heavy preponderance of fruit, but these birds eat a great diversity of plant and animal foods, too. The seasonal availability of food is, as with all species, the major determinant of the life and breeding cycles. Bowerbirds will eat flowers, including buds, petals, stamens and nectar, and also consume leaves, including buds, shoots, stems and petioles, as well as taking some sap and (rarely) seeds. During the nesting period, protein-rich arthropods can form a significant dietary component. Spotted Catbirds at Kuranda are known to take sugar water regularly during the breeding season, presumably as a valuable energy supplement, but they never touch the various seeds on offer at the feeding trays, where they happily consume bananas and paw-paws. In the tropical rainforests of North Queensland, fruits represent 80–90% of the diet of Spotted Catbirds and Tooth-billed and Golden Bowerbirds. In subtropical rainforests, for comparison, fruits comprise around 65–80% of the diets of Green Catbirds and Regent and Satin Bowerbirds, while these species eat more flowers, leaves and/or small animals than the tropical-rainforest bowerbirds do. This presumably reflects the differences in the two habitats and the relative availability of fruits. Little is yet known about the diets of the New Guinea species beyond a similar heavy dependence on seasonal fruit. It is clear, however, that the diet permits long periods of attendance at courts or bowers, food being sufficiently abundant to minimise foraging times.

Different sympatric bowerbirds in the rainforests utilise different niches or microhabitats, and forage at different heights and at varying sites. Fruit seems to be chosen for its nutritional value, size, texture or colour, or perhaps for the ease of obtaining it. Interestingly, it is mostly the comparatively easily accessible drupes or berries that are chosen as opposed to the much-harder-to-obtain capsular fruits. Since bowerbirds do not use their feet to hold and manipulate foods, they are unable easily to extract the desired edible contents from the protective husks of capsular fruits, waiting until these are very ripe before tackling them. It is also possible that the presence of insect larvae may add to the protein content of fruits, as bowerbirds will opportunistically take such prey.

Frith *et al.* (2017) postulate that the catbird breeding system of a defended and exclusive all-purpose territory could not work for the primarily fruit-eating polygynous bowerbird species. This is attributed to dietary limitations, whereby a feeding area may contain one or more fruiting trees in one month but have none in the next. Fig trees, on the other hand, are quite well distributed through the habitat of catbirds in Australia and are a reliable year-round source of food, permitting catbirds to have a socially monogamous breeding system. There would be many variables at work here, such as species composition, fruit productivity, habitat, altitude, dietary preferences etc., and it may be that a complex of factors, perhaps involving the more basal position of catbirds in the bowerbird family, is also involved.

Foraging is often undertaken solitarily or by two individuals together, and sometimes by pairs and loose flocks depending on the species, although Regent and Satin Bowerbirds occasionally form very large flocks. Foraging takes place in the canopy and subcanopy foliage, on branches, epiphytes and trunks. The Australian species, especially the catbirds and the Satin Bowerbird, also forage on the ground among leaf litter. The feeding habits in New Guinea are far less well known, but the *Amblyornis* gardener-bowerbird species seem to be largely arboreal, whereas the Fawn-breasted and Yellow-breasted (*Chlamydera lauterbachi*) Bowerbirds certainly spend some time in terrestrial foraging.

Bowerbirds obtain fruit while perching upright, or while leaning forwards or sideways, plucking and swallowing the fruit *in situ*. How much is eaten depends on the size of the fruit. Some species, notably the Golden Bowerbird, sally for fruits, snatching them from plants while in flight; this method is used particularly for vine or capsular fruits, where the ripe husks split to expose the contents. Large ripe fruits, such as those of some fig species and climbing pandans (*Freycinetia*), may be torn apart and eaten on the spot. Some larger fruits, or bunches of smaller ones, may be plucked and carried in the bill to be eaten at a perch, or are taken away and cached for later consumption.

In Australia, figs are a very important food in many areas, while hatching of the eggs is timed so as to coincide with the peak seasonal availability of larger insects, another very valuable and nutritious food resource. Fruit-caching for later consumption is well known for Green and Spotted Catbirds, as well as males of MacGregor's, Golden and Great Bowerbirds. These species use tree and root crevices, branch forks, the tops of tree stumps, the gaps between vines and tree trunks, fallen trees or branches, tree-fern crowns and other suitable forest sites for concealing the fruit.

Australian bowerbirds swallow complete flowers, petals, buds, stamens, pollen or nectar immediately where they take them. The Regent Bowerbird, with its longer, finer bill, is better adapted to nectar-eating than are other bowerbirds, and nectar forms a much greater part of its diet. Satin Bowerbirds, with their quite stout bill, also feed on nectar from *Banksia* and *Grevillea* flowers, as well as from various herbaceous plants. Spotted and Western Bowerbirds, in the more arid habitats, eat the flowers of *Acacia* trees. All of the Australian species, except possibly the rather delicate-billed Golden Bowerbird, are recorded as eating some leaf material, as well as succulent leaf buds, stems and/or vine tendrils. The specialised notched bill of the Tooth-billed Bowerbird enables it to tear, manipulate and chew up pieces of leaves and leaf stems. This species is unusual in being primarily folivorous (leaf-eating) over the winter months, staying within the dense forest canopy, but reverting to a mainly fruit diet when the season changes. Similarly, Satin Bowerbirds are known to eat much more foliage during the winter, and will 'chew' leaf matter before ingesting it. Spotted Catbirds and Satin Bowerbirds are known to take tree sap occasionally.

Adult and young bowerbirds take a wide variety of arthropods, mostly insects, especially during courtship and nesting periods. Insect groups taken by bowerbirds include grasshoppers and their relatives (Orthoptera), caterpillars, cicadas (Cicadidae), beetles (Coleoptera), ants (Formicidae), cockroaches (Blattodea), mantids (Mantodea) and also stick-insects (Phasmida). Larger prey, such as cicadas and phasmids, is dismembered by tearing it into pieces, often on the ground. Other, less usual bowerbird dietary items include termite alates (Isoptera), worms (Annelida), frogs (Anura), skinks (Scincidae), and also sometimes the eggs and nestlings of birds, with Spotted Catbirds known to take those of several species. The Australian dry-country *Chlamydera* grey bowerbirds have a greater proportion of arthropods, especially grasshoppers and locusts, in their diet than is found in the food of the more fruit-based rainforest species, for which cicadas and beetles are more abundant than grasshoppers and feature much more in the diet. Food items are not regurgitated to the young as with the birds of paradise, the bowerbirds instead simply carrying them in the bill.

Foraging methods used by bowerbirds include snatching or sallying for insects, or grabbing them in the air by hawking. The typical strategy involves searching and gleaning for invertebrates in living and dead foliage, on tree branches and trunks, epiphytic plants, flowers, and forest-floor leaf litter. The primarily frugivorous rainforest species feed more insect prey to the young just after these hatch and when protein needs may be highest, but revert later to a more fruit-based diet. Very significant during the breeding season are fruiting trees, especially figs, and multiple individuals of several species may congregate at such a food source along with other frugivores such as Wompoo Fruit-doves (*Megaloprepia magnifica*) and Australian Figbirds (*Sphecotheres vieilloti*), usually without rancour. Regent Bowerbirds, however, are notably possessive and will often try to drive away competitors, such as the Green Catbird, Satin Bowerbird and Topknot Pigeon (*Lopholaimus antarcticus*), at a food source. In New Guinea, MacGregor's and Archbold's Bowerbirds seem to be able to feed in one and the same tree without problems, but because of their shy nature little is known of the interactions of the other bowerbirds here.

Breeding

Bowerbird breeding strategies

All three Australian catbird species are socially **monogamous**, but, following the recent taxonomic splits, data for most of the New Guinea species are lacking, although there is no reason to expect anything different. They maintain an all-purpose territory, the females performing all nest duties but with the mate helping to feed her and the nestlings. Catbirds maintain no court or bower, and their courtship is much less complex compared with that of bowerbirds; males hop rapidly back and forth between perches, usually carrying food in the bill, and have no ritualised posturing or use of their dark mouth pigmentation in such simple displays.

Stealing of decorations from the bowers of rival males is commonplace among all the *Chlamydera* grey bowerbirds, and is particularly well known with the Satin Bowerbird, as well as with Tooth-billed, Regent, Golden and Vogelkop Bowerbirds and occasionally with MacGregor's Bowerbird. The practice is sometimes known as **marauding**, and bower damage as well as the stealing of objects and decorations is also quite frequent. It is thought that the better quality and variety of decorations will influence the female's mate choice, so that the more impressive the spread on offer, the better the chances of mating. Stealing from a dominant male may damage the latter's prospects and enhance those of the rival, and the converse also applies where dominant males improve their own status and limit the prospects of rivals. Male Satins typically repair the damage before going to replace the missing items, and marauders are therefore able to maximise the time during which they can retain stolen material. The scale of destruction varies from complete levelling of the bower to the removal of a few sticks, and at one study site 51% of structures were damaged in some way (Borgia 1985). Destruction or damage is usually effected while the owner is absent, and the study showed that each of the damaged bowers was destroyed an average of 4.0±1.9 times per year, the figure rising to an astonishing 8.2±2.4 times in the next year. There is also a correlation between the number of times a male destroys the bower of rivals and how often he is himself a victim of the same. Stealing by Satin Bowerbirds was much more frequent than actual destruction. The Golden Bowerbird does occasionally steal the odd stick from rival bowers, but the fungus that fuses the sticks together in these wet habitats makes them less susceptible to damage, although theft of decorations such as the beard lichen is much more frequent.

Bower destruction is known for all the *Chlamydera* grey bowerbirds, and for MacGregor's, Vogelkop, Regent and Satin Bowerbirds. Such activities reflect the social dominance of particular males and will detrimentally affect the success of rivals.

Display and nesting

One of the most notable characteristics of bowerbirds is their extraordinarily complex courtship and mating behaviour, one of the most complex in the natural world, based on the construction by the male of a bower to attract mates. These bowers can be divided into two broad groups, one clade of bowerbirds building **maypole bowers**, which the bird constructs by placing sticks around a sapling and which vary from simple maypole types to double maypoles and even extraordinary hut-like structures which are some of the most incredible designs in the kingdom of birds. The other major bower-building clade is the **avenue-builders**, which construct bowers consisting of variations on two or more walls of vertically placed sticks. The males decorate the areas in and around their bower with a variety of brightly coloured objects, the composition and layout varying with each species. Collected items include fruits, berries, flowers, leaves, shells, feathers, stones and even at times, for some species, discarded colourful plastic or paper items (bottle tops, plastic strips, toothbrushes etc.), coins, nails and screws, rifle shells, or pieces of glass. The males will spend many hours in arranging this collection and are constantly adjusting it and keeping it clear of extraneous debris. Each species has its own bower morphology, but there is often some individual or regional variation. The decorative objects reflect the preferences of the males and their ability to procure items from the habitat, often actually stealing them from neighbouring bowers. It has now been shown in several studies of different species that the colours of the decorations which the males use at their bowers are chosen to match the preferences of females. This itself raises questions as to how these biases were acquired: were they female-driven or did the male choices originally drive the female choice? Research is continuing and there remains a lot to be learned about these complex behaviours.

Uy *et al.* (2000) noted that females in search of a mate commonly visit numerous bowers, often returning to the male several times, watching his elaborate courtship display, inspecting the quality of the bower and pecking at the paint which the male has placed on the bower walls. Following this review process, the same male may be chosen by several different females, and those males deemed 'under-performing' can be left without copulations. Females which have mated with high-performing males also tend to return to the same male in the following year, and expend less time and energy in the search for a prime mate.

The famous ornithologist E. T. Gilliard (1956b) proposed a phenomenon which he called the **'transfer effect'**, suggesting an inverse relationship between bower complexity and the brightness of plumage. He suggested that there is an evolutionary 'transfer' of ornamentation in some species, from their plumage to their bowers, which reduces the visibility of the male and thereby its vulnerability to predation. This hypothesis has not, however, gained general acceptance and is not well supported, since species with vastly different bower types have similar plumages.

It has also been suggested that the bower functioned initially as a device that benefits females by protecting them from forced copulations, so giving them a better opportunity to choose males, and which benefits males by enhancing female willingness to visit the bower without undue threat. Evidence supporting this hypothesis could be that, in the modified bowers of Archbold's Bowerbird, the courtship itself is greatly modified, with the male limited in his ability to mount the female without her cooperation. Tooth-billed Bowerbirds have no bowers, and males may capture females out of the air and forcibly copulate with them. Once this initial function was established, the bowers were then co-opted by females for other functions, such as assessing males on the basis of the quality of the bower construction, painting and decoration. Patricelli *et al.* (2006) used robot female bowerbirds(!) and showed that males react to female signals of discomfort during courtship by reducing the intensity of their potentially threatening courting behaviour. Coleman *et al.* (2007) found that young females tend to be more easily threatened by intense male courtship, and these females tend to choose males on the basis of traits not dependent on male courtship intensity.

The expenditure of so much effort by females regarding mate choice, and the heavy preponderance of mating success with males showing the required quality in displays, indicate that females gain important benefits from mate choice. Males play no role in parental care, their primary overriding function being to fertilise the eggs and thus pass on their genes, and it is suggested therefore that females gain genetic benefits from their mate choice. This would perhaps seem self-evident, but it is in fact the subject of a long-term study that is very hard to prove and has not been established as such, in part because of the difficulty of following offspring performance since males may take seven years to reach sexual maturity.

The evolution of the bowerbird displays remains a major field for scientific investigation. One such avenue of research is the 'bright bird' hypothesis of Hamilton & Zuk (1982), who state that sexual ornaments are indicators of general health and heritable disease resistance. Doucet & Montgomerie (2003b) concluded that the state and quality of the plumage of the male bowerbird indicate degree of internal parasitic infection, whereas the bower quality is a measure of external parasitic infection. This would suggest that the bowerbird mating display evolved as a consequence of parasite-mediated sexual selection, although this is a controversial conclusion and has so many variables that identifying precise causes is fraught with difficulties.

Bowerbirds are often regarded as being one of the most behaviourally complex species of bird owing to their intricate mating behaviour, with elaborate structures and designs and the use of a varied assortment of objects and colours. **Mimicry**, too, is well known for some species of bowerbird, but not for catbirds. MacGregor's Bowerbird is known to mimic local bird species, the sound of water, human voices and domestic animals, while Satin Bowerbirds commonly mimic other local bird species during courtship display.

Phenotype denotes such characteristics as physical structure, development, biochemical or physiological properties, behaviour and the products thereof, such as a bird's nest or bower. The phenotype derives from an organism's genetic code or genotype, plus environmental factors and the interactions between these two. Dawkins (1982) postulated a biological concept that the phenotype be not limited just to biochemical processes but should include behavioural effects as well. An example is bower-building among the Ptilonorhynchidae, which modifies the environment by means of what may be termed architectural constructions. This **extended-phenotype** concept provides some of the most compelling evidence that this can play a role in sexual selection, and also act as a powerful evolutionary mechanism, with many parallels with human behaviours using the same mechanism.

Bowerbirds have also been observed to create a kind of optical illusion by arranging objects in their bower's court area from smallest to largest, creating a perspective which apparently holds the attention of the female for longer (Maxmen 2010; Endler *et al.* 2010). It would seem that males with objects arranged in a way that creates this striking optical illusion are likely to have higher mating success.

All the bower-building bowerbirds are **polygynous**, each promiscuous adult male using advertisement calls and/or colourful plumage to attract females to his court or bower site. There the male will court and mate with as many females as possible, although mate choice is female-driven and choosing the desired male involves her in a lengthy process of checking out other bowers. The female builds the nest and raises the young alone (unlike the situation with catbirds). Males do not assist, as they are too involved with the complex and sophisticated processes of building, decorating, collecting, arranging and maintaining, raiding other bowers, and singing, displaying and dancing at their court or bower so as to attract and impress females and facilitate mating. These polygynous species are strikingly dimorphic, the female being much drabber in coloration, with the exception of a few species in the genus *Amblyornis* and all the *Chlamydera*. Female bowerbirds build a nest by laying soft materials, such as leaves, ferns and vine tendrils, on top of a loose foundation of sticks. The males court the females by displaying at the bowers with a variety of complex movements and ritualised decoration. All the New Guinea bowerbird species customarily lay one egg, while Australian species lay from one to three (often two) with laying intervals of two days.

The term '**bower**' is used to describe the complex structures that the birds build. The Tooth-billed Bowerbird is the odd one out, his court (not bower, despite the bird's name) being simply a cleared area of ground with a leaf mat, without any construction as such. The other 17 polygynous species build bowers of varying complexity and longevity, and these can be divided into two basic structural types: maypole bowers are constructed by *Amblyornis*, *Archboldia* and *Prionodura*, and avenue bowers by *Ptilonorhynchus*, *Sericulus* and *Chlamydera* species. Males have nothing to do with nesting or raising the young, but focus solely on their bowers or courts. These may be in the same place as they were in the previous season, or they may be at a different spot but close by the same traditional court or bower site. Courts and bowers are carefully sited at appropriate spots, with suitable terrain, building materials, foodplants, foliage cover, light conditions, saplings and perches. Available data indicate that traditional bower sites are evenly dispersed through suitable habitat, the fairly regular distances between them being maintained by the social interactions of competitive males. Court and bower sites, and even specific structures, of these long-lived species are often attended by generations of males over many years, even decades.

Males' **bower sites** are the focal point for sexual and courtship activities of all bowerbirds of both sexes, including immatures. Bowers or courts may be simple and rudimentary ones, generally used for just a season and often built by immatures or subadults. Tooth-billed Bowerbirds once again are unlike the rest of the family, their leaf courts being less evenly distributed and with a tendency to be closer together and within auditory contact of rival males, rather like the exploded lek of certain birds of paradise. It seems likely that those birds holding prime sites within such aggregations are the older and more experienced, dominant males with more successful matings than their peripheral rivals. Female Tooth-billed Bowerbirds, as well as females of other species in the family, may return to the site of previous successful courting, which is another big advantage to the owners of such locations.

The rudimentary bowers or courts built by immatures or subadults are generally visited only occasionally, being what we might term 'practice bowers'. The owners of practice bowers or courts may be evicted by the resident male of an adjacent territory, who indeed may take over such a site. Studies of Tooth-billed, MacGregor's, Archbold's, Golden, Regent, Satin, Spotted and Great Bowerbirds indicate that female-plumaged, immature males undergo an 'apprenticeship' of 5–6 years, during which they will build and visit rudimentary or practice sites while also visiting the bowers or courts of older adult males. They gain experience of decoration and site maintenance, as well as finding and/or stealing display items and practising displays and vocalising, all good preparation for taking over a site or establishing their own. Males show a very high degree of site-fidelity, which is not surprising given how hard it is to achieve ownership of such a premium-status object, one that is instrumental in mating success.

Female **mate choice** is extremely important, as the only contribution from the male is the fertilisation of the eggs. Acquiring an alpha male in good condition with good genetic inheritance will enhance the prospect of such genes being passed on, and it is no wonder therefore that the females are decidedly choosy. It may be that experienced females choose to nest near the bower site of older and more successful males, although the age at which the females start to breed is still uncertain (probably in their second or third year). It is also likely that subadult males not yet in full plumage may mate opportunistically.

Bowerbirds and catbirds often use the same **location**, sometimes the very same site, in which to nest each year, old nests often remaining in the vicinity. The nesting season is dependent on habitat, weather conditions and altitude, but most Australian bowerbirds nest from late September to December, often just before the onset of the main wet season. Mating has been observed both before and after nest-building and up to a month prior to egg-laying.

Bowerbird **nest sites** are usually in the forks of trees, saplings, bushes or vine tangles, although the Golden Bowerbird builds its nest in a tree crevice, a habit unique to this species. Nests are well hidden and sited, often avoiding overhanging vegetation, to minimise the potential threat of arboreal predators. Females alone build the nest, and also do the brooding alone, irrespective of the breeding strategy utilised. Their nests are usually located within hearing distance of a male calling at his court or bower.

Catbirds build the most substantial **nests** within the family. These are large, deep, open bowl-shaped structures with some four basic layers: a stick foundation (unlike paradisaeids), and then a deep nest cup with, uniquely, a layer of decaying wood and/or bits of mud atop the central leafy bowl, with an inner egg-cup lining. Bowerbirds proper lack the wood-and-mud component, the nests of the *Amblyornis* gardener bowerbirds and of Archbold's and Golden Bowerbirds being substantial deep bowl-shaped structures having a stick foundation, a nest cup and a lining. Tooth-billed, Regent and Satin Bowerbirds and the *Chlamydera* grey bowerbirds, however, build relatively small, slightly concave, loosely constructed shallow nests, with just an egg-cup lining.

All the Ptilonorhynchidae lay clutches of from one to three eggs, notwithstanding their different breeding strategies of monogamy as opposed to polygyny. The New Guinea species seem to have generally a single-egg clutch, the Australian ones averaging almost two. Egg details are given under each species, but as a generalisation the rainforest species of catbird, along with Tooth-billed, Archbold's and the *Amblyornis* gardener bowerbirds, have pale and unmarked eggs. Uniquely within the family, the Golden Bowerbird lays white eggs, and this may be an adaptation to its habit of nesting in tree crevices, as opposed to the more open sites of other bowerbird species. The other species of more open or drier habitats – Satin Bowerbird, the *Sericulus* silky bowerbirds and the *Chlamydera* grey bowerbirds – lay eggs that are coloured and are marked with vermiculations, streaks or blotches.

All of this family of bowerbirds, irrespective of breeding strategy, typically lay the eggs on alternate days, incubation usually starting once the clutch is complete. Data on incubation in the wild are somewhat limited and presently known for just six species, primarily from Australia, and were until very recently unknown for any *Amblyornis*. The monogamous catbirds incubate for 22–24 days, polygynous species averaging one day fewer at 21–23 days, whereas Archbold's Bowerbird, the highest-altitude species, has an incubation period of 26–27 days. It is noteworthy that, with both the monogamous and the polygynous species, it is only the female that incubates and she devotes about 70% of the diurnal time to the task. When there is more than a single egg, hatching is nearly synchronous, the siblings emerging within a day of one another. Nesting female Green and Spotted Catbirds are supplied by the male during both nest-building and incubation periods, with more protein-rich arthropod food (as opposed to primarily figs) provided in the egg-laying period as shown in studies of the former species.

Data on the **nestling periods** of the bowerbird family in the wild are available for just eight species, and they are similar for the monogamous (20–21 days) and the polygynous groups (17–21 days). On the other hand, Archbold's Bowerbird of the high montane zone, where food is scarcer and conditions harsher, has a nestling period of some 30 days.

Much remains to be learned about **brooding periods** of bowerbirds, and quite when they cease varies from species to species. Australian data suggest little difference between brooding times of the primarily frugivorous rainforest-inhabiting Green and Spotted Catbirds and Golden Bowerbird and those of the Satin, Spotted and Great Bowerbirds of the drier areas, which have a more varied fruit and arthropod diet.

Provisioning rates appear to be variable. Green and Spotted Catbirds average five feeds per hour, while some of the polygynous species feed their brood from three to seven times per hour, but the rates are obviously dependent on food availability. Parent bowerbirds swallow the nestlings' faeces throughout most of the nestling period, as with the paradisaeids, not carrying them away until late in the nesting process, along with any seeds voided into the nests.

The time spent by the female in brooding is influenced by the size of the brood. Studies have shown that Spotted Catbirds and Golden Bowerbirds brood single nestlings for longer periods than they do with two or more chicks. Presumably, multiple chicks keep themselves warmer than single ones do, and smaller broods may be less demanding overall. Quite when bowerbirds cease brooding their nestlings varies slightly from species to species. There appears to be little difference between nestling periods of the frugivorous catbirds and Golden Bowerbird and those of the Satin, Spotted and Great Bowerbirds, which feed more insect and arthropod items to the nestlings.

The mean length in minutes of incubation sessions for six polygynous bowerbirds was 30.9 for Huon Bowerbird (*Amblyornis germana* Donaghey *in litt.*), 27.7 for Satin Bowerbird (Donaghey 1981), 40.8 for Tooth-billed Bowerbird, 18.7 for Golden Bowerbird, 16.5 for Archbold's Bowerbird and 24.5 for Great Bowerbird (Frith & Frith 2004). Nest failure rate was high (>70%) and productivity low in the Satin Bowerbird (Donaghey 1981) and the Golden Bowerbird, but nest success of Archbold's Bowerbird was high at 88% (Frith & Frith 2004). For nesting songbirds in which only the female incubates, there is a trade-off between sessions spent in warming the eggs and absences from the nest to gather food. The daily number of bouts off the nest is determined by ambient temperature, the female's energy needs, food availability, and the risk of nest predation (Conway & Martin 2000).

Many northern-temperate songbirds at a high risk of nest predation may favour increased nest attentiveness, with long off-nest bouts and fewer trips to the nest (Conway & Martin 2000). In the case of the Satin Bowerbird, the high risk of nest predation appears to have favoured a high incubation constancy with relatively long on-bouts. Other factors influencing incubation constancy and the duration of on-bouts and off-bouts may include ambient temperature, the energy requirements of the female, food availability and diet. This may account for variation among different species at different elevations and habitats, but the high risk of nest predation may be an ultimate factor. For example, the Great Bowerbird's high incubation constancy of 78% in an arid environment may be influenced more by diet (proportions of plant and animal food) and foraging behaviour. Little is known about bowerbird nest-predation rates and types of predator in New Guinea. For Archbold's Bowerbird in montane habitats in New Guinea, high nest success may be related to fewer or different predators compared with lowland areas and the fewer, shorter incubation bouts related to cool weather and less insect food (Frith & Frith 2004.)

The hatching of the eggs is nearly synchronous, the chicks emerging within 24 hours of each other. From what is known of a few species, the eyes start to open at around 7–9 days of age and are fully open by 9–12 days. Unlike the dark-skinned and largely naked hatchlings of the birds of paradise, **bowerbird hatchlings** have varying shades of pale orangey skin, a yellow to orange gape, a white egg tooth, and conspicuous long fluffy down. The rather long, dense down is present in patches on the crown, wings and body, and is reddish grey-brown on catbirds, greyish-brown on the Tooth-billed Bowerbird and *Amblyornis* gardener bowerbirds, dark greyish-brown on Archbold's and Golden Bowerbirds, and a much paler grey on Satin and Regent Bowerbirds and the *Chlamydera* grey bowerbirds. The emergence of the flight-feathers varies, those of young Spotted Catbirds and Golden Bowerbirds appearing at 5–6 days, and of Archbold's Bowerbird at eight days; the primaries emerge from their pin sheaths on days 13–14 in the Spotted Catbird and 11–12 days in the Golden Bowerbird, but not until 14–15 days in Archbold's Bowerbird. Once fledged, the juveniles have a much shorter tail than the parents, sometimes only 10% of that of the adult, but otherwise their plumage is very similar to that of the adult female. How long they remain dependent on the parents is not well known, but at least 40–60 days seems the usual period.

Bowerbirds are typically **single-brooded**, nesting once every season, but should the nest be destroyed or abandoned early in the season they can build and lay again. As nest-building usually takes 1–3 weeks, incubation and nestling periods each last about three weeks, and at least 2–3 months are required in order to raise a fledgling to independence, there is unlikely to be sufficient time to raise a second brood in the same season. The monogamous Australian catbird species, in which both parents perform nest duties, have far greater nesting success than the polygynous Golden, Regent and Satin Bowerbirds, which produce far fewer young, the females alone carrying out the nest duties. What are the long-term implications of such an apparently more successful and productive breeding strategy? There are just so many unknown variables operating that it is impossible to say without very long-term and wide-ranging studies. A clue may be supplied by the high nesting-success rate found during the Frith pair's study of Archbold's Bowerbird in Papua New Guinea, which may, as mentioned above, be related to longer incubation and nestling periods and fewer predators in the cool montane habitat.

Predation on bowerbirds and their nests is a remarkably poorly known field with very few or no data, despite the plethora of bird, mammal or reptile species that might avail themselves of such a food resource. It is known that many species have distraction displays. They may attack or chase the intruder, and they may freeze when sighting a potential threat, or use mimicry to distract the predator and also summon help from other species to mob it. Spotted and Green Catbirds are themselves known to prey on the nests of other passerines. Bowerbirds can fall victim to diurnal raptors, with the Grey Goshawk (*Accipiter novaehollandiae*) known to take them opportunistically, and the Lesser Sooty Owl (*Tyto multipunctata*) is possibly another predator. New Guinea data regarding such predation are notably scant or lacking. Large lizards such as the Lace Monitor (*Vanellus varius*) and pythons are also potentially serious predators. Bowerbird anti-predator posture involves simply freezing on the spot, and remaining perfectly motionless until the danger has passed. Some of the Australian species also mimic predator alarm calls, which may encourage other species to come and mob the intruder and potentially help to drive it away.

Humans are, of course, significant predators for bowerbirds in some areas, with children known to wreck bowers of Fawn-breasted and MacGregor's Bowerbirds in New Guinea, while Huli and Engan people occasionally acquire the plumes of the male King of Saxony Bird of Paradise from the bowers of Archbold's Bowerbird, with multiple plumes sometimes taken. The skins of male Flame Bowerbirds have been seen as ornaments dangling from drivers' mirrors, and such a skin adorned the altar in Kiunga Catholic Church for some years. The orange crest of the male MacGregor's Bowerbird is also sometimes used as *bilas* (personal ornamentation for cultural events) in Papua New Guinea. None of this trade is overly harmful as it is on such a small scale and is localised. In Australia in the past, Satin Bowerbirds were shot in some numbers by fruit-growers when they raided orchards.

Life expectancy of bowerbirds is almost unknown for the New Guinea species. It is better documented in Australia, where individual Spotted Catbirds are known to have survived for at least 19 years, and some of the polygynous male bowerbirds can live for 20–30 years, with ringed Satin Bowerbirds surviving 10–18 years and a Golden Bowerbird known to have lived for 23 years. Related to this is the use of bower sites over long periods, sometimes up to 20 years.

Nest parasitism is another barely known field, with no data from New Guinea, where even the host of some of the potential cuckoo parasites remains unknown. Within Australia, there are no records of parasitism of the Ptilonorhynchidae by cuckoos, which exploit primarily the smaller passerines such as honeyeaters and the acanthizids (scrubwrens and gerygones) as hosts.

Voice

The calls of the bowerbirds and catbirds are quite distinct but generally not musical, with strange, harsh and often mechanical-sounding clicking, buzzing, scolding and chirping, and with some species known to mimic other birds or ambient sounds. The Satin Bowerbird does have a short whistled song, but all other species are far less musical. All of the socially monogamous, pair-bonding catbird species give loud, squalling territorial songs throughout the year, although far more so when breeding, with peaks in the early morning and the late afternoon. Contact calls of all the catbirds are sharp, high-pitched *zik*-type notes and none of the catbirds is known to mimic. Parent catbirds disturbed at the nest produce harsh scolding notes. Tooth-billed Bowerbirds call from habitually used perches, typically immediately above or near the court, and this species is a remarkably skilled mimic of bird calls as well as of ambient sounds.

Bowerbirds also call from habitually used perches above or near their bowers, advertising the location of their bower primarily in the early morning and late afternoon, but also at other times of day during the peak courting months. Regent Bowerbirds are unusual in that the males advertise their bower locations from the forest canopy, and it may be that other species in the still poorly known *Sericulus* silky bowerbird group do likewise.

Bowerbirds are well known for their mimicry of avian sounds, as well as ambient sounds, including human voices and sounds made by machinery. They use mimicry at the bowers and during courtship, and also in the presence of potential predators or congeners. It is likely that all the polygynous species can mimic; although this remains little known for the *Sericulus* species, it is well documented for Satin, Tooth-billed and Regent Bowerbirds, and also for the Golden Bowerbird and all the *Chlamydera* grey bowerbird species. In New Guinea, MacGregor's and Vogelkop Bowerbirds are good mimics, and mimicry is known for both Streaked and Archbold's Bowerbirds, too.

Vocal mimicry is a learned behaviour and can, therefore, be passed on to descendants, and the repertoire and quality of vocal avian mimicry improve with age. Older males of the Satin Bowerbird produce longer bouts of higher-quality avian vocal mimicry than do younger males, and consequently gain higher mating success. Astonishingly, during courtship mimicry, males of both Tooth-billed and Satin Bowerbirds repeat their repertoire of calls of other bird species in a specific order, recalling both the calls and the sequence. Females may preferentially select those males with good mimicry produced in the preferred order, so sexual selection by females may act to improve the males' song repertoire and quality.

Vocal mimicry by female bowerbirds at the nest is reported as typical only of the avenue-builders, particularly of the Satin Bowerbird and the *Chlamydera* bowerbirds. This mimicry is often of predatory birds, presumably as a form of deterrence, but calls of non-predatory bird species have also been noted, as has the imitation of non-avian sounds. These latter imitations include a cat-like meowing by females of the Spotted, Western and Great Bowerbirds near their nests, cats being another potential predator.

It is possible that the mimicking by female bowerbirds near their nests of the calls of predatory birds, and the same behaviour by males near their bowers, may be useful for distracting a potential predator. It may also attract other birds, which may then mob the visiting predator. In addition, nestlings may learn the calls of potential predators and increase their chances of survival in the future.

Movements

All members of this family are essentially resident, although there is some winter dispersal to lower altitudes by some of the hill-forest species and those of lower-montane forest during times of food scarcity. Tooth-billed, Golden, Regent and Satin Bowerbirds and Spotted Catbirds are known to make some localised movement down to lower altitudes, and limited wandering is recorded also for MacGregor's Bowerbird in New Guinea. There is some evidence of a large decline over the past few years in the number of Satin Bowerbirds in winter on the Sunshine Coast of south Queensland, where the species is now being seen far less frequently than was earlier the case, for unknown reasons.

Home ranges of Green and Spotted Catbirds average 1–2ha, but their all-purpose territories during breeding diminish as the parents forage closer to the nest site. This may apply also to the New Guinea species, but more fieldwork is required.

All of the polygynous bowerbird species are territorial only in the immediate area of the bower or nest site. Their females and immature males maintain relatively large and overlapping home ranges throughout the year, and will forage in the same areas as the adult males. How far they need to travel to find food will depend on the time of year and how the fruiting season turns out, although catbirds, with their dependence on reliably fruiting *Ficus* species in their home territory, may not need to move very far.

Young males in female-like immature plumage appear to range widely for several years, but subsequently the range covered becomes increasingly limited and they check out adjacent bower sites before concentrating on one. They may eventually challenge the dominant adult male for ownership of the prized bower. While all this is occurring, female bowerbirds move freely about the habitat and check out many bowers in order to assess their suitability prior to choosing a mate.

In the winter period in Australia, the woodland-dwelling Satin Bowerbird and also, to a lesser extent, the Regent move to more open habitats and form flocks containing as many as 100 birds, with up to 200 individuals known for the former species. These flocks eat shoots and the leaves of grasses, herbs, shrubs and trees, including those of eucalypts, but also (and more problematically) cultivated green vegetables and soft fruits. Larger flocks can consist of both adult males and female-plumaged individuals, up to 40 adult males having been observed in one flock. *Chlamydera* bowerbirds in Australia also flock in winter, with some very local dispersal to adjacent areas, but in much smaller groups.

Bowerbird conservation

All members of the bowerbird family in Australia are classified by BirdLife as being of Least Concern (**LC**), with none currently rare or endangered. Habitat destruction and degradation have resulted in quite significant local range loss for some species, such as the Green Catbird and the Regent, Golden and Spotted Bowerbirds. The last of those is now extinct in South Australia and probably also in Victoria, although having always been of limited range and numbers in both states. The subspecies *carteri* of the Western Bowerbird is listed as Near Threatened (**NT**) owing to its highly restricted range and estimated total population of 2,000 breeding individuals.

New Guinea has a number of range-restricted bowerbird taxa, such as the Yellow-fronted Bowerbird, restricted to the Foya (or Gauttier) Mts, in West Papua, and the Fire-maned Bowerbird, confined to the Adelbert Mts of Papua New Guinea. The Yellow-fronted is now classified as Least Concern, being restricted to a remote, inaccessible and lightly populated region. The Fire-maned Bowerbird, on the other hand, is Vulnerable (**VU**) and generally rather uncommon within its small range, but quite numerous locally. It occupies a relatively narrow elevational zone, mainly at 1,200–1,450m, which is also that used by the sparse local human population for agriculture. Fortunately, the local people seem not to hunt this species and much of its range is comparatively inaccessible, making logging less likely. In addition, this bowerbird seems as if it may be adaptable to the large areas of secondary forest. Throughout New Guinea the human population is growing fast, and information on forest status and hunting pressure in the Adelbert Mts is limited, so the species does need to be monitored.

The only other bowerbird of immediate conservation concern is the range-restricted and rather rare Archbold's Bowerbird, which occurs at low density and is classified as Near Threatened (**NT**). Habitat loss and climate change resulting in changes to the flora are two main threats, and, since this species is a high-altitude moss-forest endemic, it is possible that climate breakdown could have very serious effects indeed. A drier climate could mean a heightened risk of burning, and a change in tree species composition may mean loss of food resources, with no alternatives available. The eastern race *sanfordi* appears to have a total range of no more than about 800km^2, and on two mountains within this tiny range it is threatened by logging activities. Populations of the western, nominate race appear to be larger in number and are considered probably secure, at least in the short term, although logging is an ever-increasing local problem.

No bowerbird species is presently threatened with extinction or under threat of major population decline, and most bowerbird populations appear currently to be stable. Global warming is an unquantified threat that could in the longer term be seriously detrimental to upland species. As ever, it is loss and degradation of habitat and predation by exotic animals, particularly feral cats, that potentially threaten some bowerbirds. Hunting has not been a significant problem, although fruit-growers did in the past shoot considerable numbers of Spotted Bowerbirds.

An as yet unknown but, it is hoped, only small-scale potential threat is the export of bowerbirds and birds of paradise for aviculturists in China, America, Europe and the Middle East. Several live captive bowerbirds, including catbirds and Fawn-breasted and Flame Bowerbirds, have appeared in such collections, the majority likely as illegal exports from West Papua. Presumably these are trophy species, the presence of rare and unusual taxa in collections accruing status for their owners, much as with the 19th-century specimen collections. As the final draft of this book was being written, disturbing news came of the reported disappearance of some five male Golden Bowerbirds from their bowers, with suspicions of poaching for some nefarious purpose, and investigations are under way. Conserve the habitat, preserve the bird, is the gold standard, though.

Flame Bowerbird, male at bower, Gusiore, Western Province, PNG (*Markus Lilje*).

LAYOUT AND SCOPE OF THE BOOK

The plates

Richard Allen is a well-known artist with various ornithological books to his credit, notably the *Sunbirds and Flowerpeckers* title in the Helm series, and his work for the monumental *Handbook of the Birds of the World*. He has been the artist since the start of the present work back in 1998, and has great expertise with illustrating iridescent species such as sunbirds. This was invaluable for this project, with so many wonderfully iridescent species, and he made numerous visits to museums to check skins and specimens. Both sexes are illustrated where appropriate, along with immature or subadult if required, and any significantly different subspecies are also illustrated.

The photographs

These form an important supplement to the splendid plates by Richard Allen, illustrating most of the species in the book, usually as adult males but with females and juveniles, too, where possible. All the photos are of wild birds, and we are greatly indebted to the many photographers who submitted material for possible inclusion here. A list of those whose photographs appear in the book is given in the Acknowledgements section.

Taxonomy and systematics

Detailed information is given in the preamble to the book, reflecting the complex and changing understanding of what actually constitutes a bird of paradise and a bowerbird. Species-level taxonomy and sequence follow the International Ornithological Congress (IOC) treatment, but with changes where this was felt appropriate, such as the splitting of the Paradise-crow into two species and the recognition of the Tagula Manucode, as well as the splitting of the genus *Lophorina* into three species. Future potential changes, such as the splitting of Trumpet Manucode, are also suggested. The treatment of subspecies generally follows Beehler & Pratt (2016), who, thankfully, greatly reduced the number of minor subspecies of New Guinea birds, and this treatment is now adopted also by the IOC, with variations in a couple of species where new information has come to light. We recognise here some 45 species of bird of paradise, with 72 taxa including subspecies; and some 28 species of bowerbird, with 37 taxa including subspecies.

English names

The perennially vexatious matter of English vernacular names naturally arose, but with these two families there was a broad degree of consensus among the various modern authorities. Generally we use the IOC names from its frequently updated checklist of the birds of the world (Gill & Donsker 2017, IOC World Bird List v 7.3), while taking account also of the usage in the various regional field guides. Fortunately, inconsistencies are relatively few and come down to a matter of personal preference – here we retain the vernacular name 'Wallace's Standardwing' despite moves to remove the names of people from specific epithets, and we have opted to avoid hyphens where feasible. We have also chosen to follow the common practice from the early part of the last century until quite recently, and use the term 'bird of paradise' rather than 'bird-of-paradise', thereby removing what many now consider rather clumsy hyphenation.

Distribution

The geographical range of the species is summarised in each account. All species in both families are basically resident, with some very local seasonal or food-resource-driven wandering in a few cases. (see example on page 53).

The maps

These are derived from the updated maps produced by Lynx Edicions for the *Birds of New Guinea including the Bismarck Archipelago and Bougainville* and *HBWAlive*, with thanks in particular to Arnau Bonan, who worked with me to make extensive revisions of the originals from the *Handbook of the Birds of the World*, and to Josep del Hoyo and Lynx Edicions for permission to utilise them.

Genera

We follow primarily the IOC based on version 7.3 (2017) but with some minor amendments, such as the use of *Paradisornis* for the Blue Bird of Paradise.

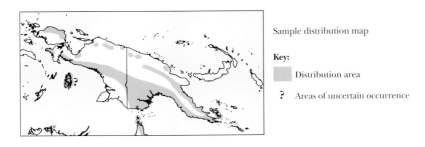

Sample distribution map

Key:

Distribution area

? Areas of uncertain occurrence

Species accounts

The main heading for each species consists of the English name followed by the scientific name. This is followed by citation of the original description with the original scientific name, and then by the name of the describer, date of first description, the original publication and the type locality from where the bird was described (usually but not always where it was first collected). The *Howard and Moore Complete Checklist of the Birds of the World* (4th Edition), Volumes 1 (Dickinson & Remsen 2013) and 2 (Dickinson & Christidis 2014), gives a comprehensive and recent review of such citations, and we follow them, with subsequent amendments up to early 2017. Alternative or obsolete English vernacular names are also given, with no attempt made to be totally comprehensive.

FIELD IDENTIFICATION This introductory section gives the size and weight of the species and the most obvious features of plumage, shape or behaviour that aid identification in the field. Geographical range is often of value, as many species do not overlap with congeners in distribution and/or may be separated altitudinally from others. Fortunately, most species are reasonably distinctive in good views, although these are not always easy to obtain in the dense shady forest habitats, but female or immature plumages can be more problematic. **SIMILAR SPECIES** are listed if appropriate, with a brief summary of how to distinguish them from the taxon being described; if there are no similar species, this, too, is mentioned.

RANGE covers the geographical distribution of the species in New Guinea, Indonesia (West Papua, Halmahera and nearby islands such as Obi) and Australia. A section on **MOVEMENTS** is sometimes included. Since none of the species covered undertakes migrations as such, this covers local movements in response to seasons, food type, rainfall or altitude. All species within the two families are largely, if not entirely, resident or sedentary, and for most even local movements are unknown. Accordingly, there is no evidence of long-distance vagrancy, but merely some local or altitudinal wandering, mostly from Australia, among a few of the bowerbirds and the Paradise Riflebird, Superb Lophorina and Carola's Parotia.

DESCRIPTION This includes measurements and weights, taken from standard references in the literature and given in centimetres (cm), millimetres (mm) and grams (g). Detailed summaries based on long series' of museum specimens are available in Frith & Beehler (1998) for the paradisaeids and in Frith & Frith (2004) for the bowerbirds. For polytypic species, the nominate subspecies is always the one described in detail, with any other taxa covered under the following section (dealing with taxonomy and geographical variation). Both sexes are described in some detail but not exhaustively, and colours of soft parts are given. Following this are brief descriptions of juvenile and immature plumages where these are known.

TAXONOMY AND GEOGRAPHICAL VARIATION This section gives the subspecies and their characteristics where appropriate, or, if not, mentions the species' monotypic status. It may cover questions regarding taxonomy and relationships if required.

VOCALISATIONS AND OTHER SOUNDS The major calls known for each species, and also display calls, are covered here, with references given to sound recordings published on xeno-canto, the Internet Bird Collection or the Macaulay Library of the Cornell Lab of Ornithology, along with the name of the recordist. Other, non-vocal sounds are also detailed, such as the swishing wing sounds of riflebirds and the wing noises made by some *Paradisaea* and *Astrapia* species and Wallace's Standardwing.

HABITAT This section gives an outline of the habitat typical of the species, as well as the altitudinal range in metres, with some brief definitions of standard habitat type given in the glossary, e.g. rainforest, moss forest and savanna.

HABITS The species' general behaviour, excluding breeding, is outlined here. For many species relatively little information is available, and this fact is mentioned where relevant.

FOOD AND FORAGING This section covers the primary diet and foraging methods when these are known, but many gaps in our knowledge remain, and we have chosen not to give exhaustive lists of plant species readily available in other references such as HANZAB or HBWAlive.

BREEDING BEHAVIOUR This is a very significant part of the account for each species, as the breeding behaviours in both families are so extraordinary and complex. We have necessarily relied heavily on published sources, both old and new, and are particularly indebted to the work of Drs Clifford and Dawn Frith and their impressively comprehensive, detailed and painstaking studies of many of the species in both families. Courtship/Display Rather than trying to describe in minute detail every single aspect of displays, which can quickly become very turgid and tedious, we make reference to the major aspects and try to summarise them. We also include references to the marvellous video archive now available on the Internet at the Macaulay Library of the Cornell Lab of Ornithology and the Internet Bird Collection, where many incredible display sequences are now available to be viewed. Certain TV programmes have also been invaluable, such as the work by Laman & Scholes on birds of paradise and David Attenborough's programmes on both that family and the bowerbirds. The various bowers of the bowerbirds are described in some detail and illustrated on the plates, with photographs of most, and their broad type classification is indicated, e.g. avenue, maypole or stage. Nest, Eggs & Nestlings Nestlings and eggs, when known, are briefly described. Much remains to be learned here as the eggs and nests of some species, such as the newly split Tagula Manucode and many of the Spotted/Black-eared Catbird complex, are still undescribed or known from just one or two examples. In general, the nesting behaviour of the species that occur in Australia are far better known, while the New Guinea taxa which are rare and/or remote are very poorly known, examples of the latter including, among others, the Bronze Parotia (*Parotia berlepschi*) and Eastern Parotia, Fire-maned Bowerbird, Yellow-fronted Bowerbird and Huon Bowerbird. It is astonishing that the nesting cycle of so widespread and common a species as the King Bird of Paradise is virtually undocumented, and manucodes as a whole remain largely unknown despite such species as Glossy-mantled and Crinkle-collared Manucodes being common and widespread.

HYBRIDS Mention is made here of any hybrids between the species being treated and others.

MOULT Data on moult have been kept brief, in many cases because very little is known, or because detailed summaries for some of the better-known species are available in standard references and there seems little point in repeating them.

STATUS AND CONSERVATION This section commences with a short summary of the current status of the species and gives an indication of whether it is common (and whether or not it is likely to be seen), uncommon or rare. The current IUCN conservation status is given as defined below, and also indications of current and perhaps future trends where possible. These are derived from published sources, various correspondents and fieldworkers, personal observations and judicious use of Internet data where appropriate.

IUCN conservation-status categories
CR Critically Endangered
EN Endangered
VU Vulnerable
NT Near Threatened
LC Least Concern
DD Data Deficient
NA Not Assessed

References

Significant references are mentioned within the individual species texts when considered helpful. All pertinent references are combined in a lengthy bibliography at the end of the book. This lists the primary overall references for the two families and the generally most significant books, papers, notes and articles relevant to the species included within them. It is not intended to be exhaustive, but the major sources are listed. Three frequently cited core references are shortened to their initials and dates, as they are significant for many of the species covered:

Beehler, B. M., & Pratt, T. K. (2016) *Birds of New Guinea: Distribution, Taxonomy and Systematics.* Princeton University Press, Princeton NJ. (B& P 2016)

Coates, B. J (1990) *The Birds of Papua New Guinea: Volume 2.* Dove Publications, Alderley. (Coates 1990)

Frith, C. B., & Beehler, B. M. (1998) *The Birds of Paradise.* Oxford University Press, Oxford. (F&B 1998, usually shortened to simply F&B)

Frith, C. B., & Frith, D. W. (2004) *The Bowerbirds – Ptilonorhynchidae.* Oxford University Press, Oxford. (F&F 2004)

Frith, C. B., & Frith, D. W. (2010) *Birds of Paradise: Nature, Art, History.* Frith & Frith, Malanda. (F&F 2010)

Frith, C., Frith, D., & Bonan, A. (2017) Bowerbirds (Ptilonorhynchidae). In: del Hoyo, J., Elliott, A., Sargatal, J., Christie, D. A., & de Juana, E. (eds). *Handbook of the Birds of the World Alive.* Lynx Edicions, Barcelona. (Frith *et al.* 2017)

Frith, C., Frith, D. W., & Bonan, A. (2017) Birds of Paradise (Paradisaeidae). In: del Hoyo, J., Elliott, A., Sargatal, J., Christie, D. A., & de Juana, E. (eds) *Handbook of the Birds of the World Alive.* Lynx Edicions, Barcelona. (Frith *el al.* 2017)

Rowland, P. (2008) *Bowerbirds.* Australian Natural History Series. CSIRO Publishing, Collingwood, Victoria. (Rowland 2008)

Schodde, R., & Mason, I. J. (1999) *The Directory of Australian Birds: Passerines.* CSIRO, Collingwood.

Glossary

adaptive radiation evolution from an ancestral form to various new taxa, usually in a fairly short geological timespan, as with, e.g., *Zosterops* species

advertisement song song that advertises the singer's presence, often given from a territory

agonistic describes behaviour involving some sort of conflict or aggression, and can include submission, appeasement and retreat behaviour

albinism complete loss of colour pigmentation in both feather and body parts, resulting in white plumage with pink eyes and pale pink legs and bill, and basically unknown in both families. Partial albino is impossible, and refers to leucism (see below).

alkaloidal of primarily plant-derived chemicals rich in nitrogenous basic compounds (typically physiologically active and insoluble in water)

allopatry (n.) describing geographical distribution whereby different populations inhabit distinct ranges without any overlap (adj. **allopatric**) (antonym sympatry/sympatric)

allopreening behaviour in which one bird preens the feathers of another

allospecies one of a species comprising a superspecies, used in the present work to describe taxa that are treated as full species but are closely related to others without overlapping ranges

allozyme electrophoresis a method of estimating paternity using polymorphic proteins, frequently from the blood, as genetic markers

altricial a developmental term referring to bird species the nestlings of which hatch helpless and dependent on their parent(s) (antonym precocial)

analogous similar in function but not in origin and structure (antonym homologous)

anastomosing the branching and fusing of structures to form a reticulate pattern where the branch angles are acute

angiosperm a flowering and seed-bearing plant (of Angiospermae) in which ovules are enclosed in an ovary and develop into fruit once fertilised (*cf.* gymnosperm)

anisodactyl describing a foot structure whereby all toes are free and mobile, with three forward and one (the hallux) backward, as in the feet of passerines

anthropogenic human-made, man-made

anting stereotyped bird behaviour whereby the bird anoints its plumage with the defence (formic acid) and other fluids of worker ants – presumably to reduce ectoparasites and help in feather maintenance

antiphonal song sung in two responsive, alternating parts, e.g. duetting

apterium an unfeathered patch of skin

arachnids spiders and their relatives

Araneida the true spiders

archaeology the study of past human cultures

aril (n.) the edible covering of all or part of some seeds, developing from the ovule stalk; **arillate** (adj.) possessing an aril

arthropods insects and their allies (spiders, centipedes etc.)

articular cavity a cavity as a structural component in a bone joint

atavistic showing a resemblance to an earlier form or relative, a throwback

attenuate to reduce in size, strength, density or value, or to weaken or become weak

Australasia the zoogeographical faunal region which includes Australia, New Guinea, New Zealand, New Caledonia, the Solomon Islands, Vanuatu, and the Moluccas

Australo-Papuan the Australia–New Guinea region

autochthonous faunal elements fauna endemic to the region concerned

avenue bower a terrestrial structure consisting basically of sticks and/or grass stems placed upright to form two parallel walls, constructed by male *Sericulus* silky bowerbirds, Satin Bowerbird and *Chlamydera* grey bowerbirds as the centrepiece for courtship and mating

Bassian (biotic region) of the biota inhabiting temperate moist sclerophyllous eucalypt woodland and forest of south-western and south-eastern Australia, including the mallee zone

Batesian mimicry the phenomenon of a palatable and vulnerable species, usually an insect, resembling a relatively unpalatable species and thus gaining protection (frequent among butterflies, perhaps with some friarbirds and orioles)

Bergmann's Rule an ecogeographical trend whereby populations of a species in warmer environments at lower latitudes and altitudes have a smaller body mass than those at colder, higher ones

biodiversity the diversity of organisms (synonym biological diversity)

biogeography the study of the geographical distribution of life on earth

biological species fundamental taxonomic unit based on the reproductive isolation of the constituent populations from other species

biological species concept (BSC) introduced by Ernst Mayr and based on the reproductive isolation of populations, which do not interbreed if they are in or come into contact with other species (*cf* PSC); becomes impossible to test with allopatric taxa and now somewhat modified, with hybridisation no longer a total barrier to species status

biometrics measurements of morphological features of an animal

biota the plant and animal life of a region

BirdLife International a global conservation alliance, with branches in many countries, working for the conservation of birds and their habitats

Blattidae the insect family of cockroaches

BoP/bop vernacular term for bird of paradise (used also for bird of prey)

bower a structure made from sticks and other vegetable matter, constructed by male bowerbirds for the purpose of mate attraction and courtship

bower basal column or cone the basal part of the central sapling support of gardener bowerbirds' bowers that is modified into a mossy column, or cone (Vogelkop Bowerbird only), and which may be decorated by males

bower mat the ground-level area of gardener bowerbirds' and Archbold's Bowerbird bowers altered by males using mosses, rootlets, fern fronds and/or other vegetation to form a flat mat, upon which the birds place decorations

bower perch(es) a pre-existing horizontal or near-horizontal perch protruding from or connecting with a bower structure (Golden Bowerbird) or directly above a bower mat (Archbold's Bowerbird) which males may variously decorate

bower platform an area of varying complexity of sticks and/or grass stems at the ends of one or both entrances of avenue bowers, or at the end(s) of a bower perch (Golden Bowerbird only), upon which males may place decorations

cache (n.) a hidden store of provisions, or the place where such a store is hidden, quite frequent among some bowerbirds but not among birds of paradise; **cache** (v.) to conceal food or other items for future use

capsule, capsular fruit a fruit that opens along a line or lines of weakness to release the seeds

Cerambycidae the coleopteran (beetle) insect family of longicorn or longhorn beetles

Chilopoda the arthropod order of centipedes

clade a group of species descended from a common ancestor and therefore comprising a monophyletic group

cladistic, cladistics a methodology for delineating evolutionary relationships among lineages based on establishment of monophyletic groups (clades), practically defined by the presence of uniquely shared derived characters

cline a gradation in a characteristic (such as plumage) within a species, across the breadth of its range

cloud forest a montane forest type at 1,000–3,500m and upwards, frequently fog-shrouded and with an abundance of mosses, ferns and epiphytic growths

the cnemophilines (n.) the wide-gaped satinbirds, formerly classed as birds of paradise of the subfamily Cnemophilinae but now in their own family (adj. **cnemophiline**)

Coleoptera the insect order of beetles

condyle the rounded projection on the articulating end of a bone

congener (n.) a species of the same genus as one being compared with it; **congeneric** (adj.)

conspecific (adj.) referring to subspecies belonging to the same species, or to a former species now merged with a species under discussion

convergence (n.), **convergent** (adj.) in evolutionary biology, the independent evolution of like forms or like features in two unrelated lineages (e.g. court-clearing by the Tooth-billed Bowerbird, some birds of paradise and the Neotropical fruitcrows is a convergent behaviour)

cordillera (n.) a linear chain of mountains, usually referring to the principal mountain range of a continent but used also for the main New Guinea mountains (adj. **cordilleran**)

corvine, corvid (adj.) of the avian family that includes the crows and their allies

Corvine Assemblage a large and diverse assemblage of the bird groups constituting the Parvorder Corvida

Corvida an avian Parvorder consisting of the three primary lineages: lyrebirds and allies (Menuroidea), honeyeaters and allies (Meliphagoidea), and crows, monarch flycatchers and allies (Corvoidea)

court a terrestrial area cleared and sometimes decorated by a male bird or birds as a focal point for courtship and mating, as with some bowerbirds and parotias

culmen the dorsal ridge running lengthwise down the midline of the upper mandible from where it joins the forehead to the tip of the bill

cultivar a horticultural variety of plant or crop

Curculionidae the coleopteran insect family of weevils

cytochrome *b* a mitochondrial DNA gene, used in determining phylogenetic relationships

Dacrycarpus a montane tree belonging to the Podocarpaceae family, and a significant food resource for the rare MacGregor's Honeyeater, formerly classified as a bird of paradise

dehiscent with a tendency to split (as in capsular fruits that burst or split open when ripe)

demography, demographics the study of the dynamics (birth/death rates, age/sex ratios and numbers) of populations

dichromatism (n.) possession of two colours (adj. **dichromatic**)

dimorphic appearing in two distinct forms

dipteran of the fly insect order Diptera

displacement chase the behaviour of a bowerbird when evicting another from its court or bower site

display tree a tree from which a bird of paradise may display; used also in some bowerbird courts

distraction display active parental anti-predator strategy that aims to deflect or divert a predator from the parent birds' eggs or young, and known for some bowerbirds but not paradisaeids

distally towards the tip of an appendage

DNA fingerprinting a method of establishing identity and estimating paternity by comparing patterns of fragments of genetic material (deoxyribonucleic acid); an individual's unique DNA fingerprint includes elements from each parent

DNA–DNA hybridisation a molecular technique employed to estimate evolutionary relatedness by comparing base-pair similarity in the nuclear DNA of two species

drupe a succulent fruit that contains a stony seed within it

Ebbinghaus Effect or illusion is an optical illusion of relative size perception (named for its discoverer, the German psychologist Hermann Ebbinghaus), and seemingly used in the structure of some bowerbird bowers

ecotone the edge, boundary or transition habitat created by the juxtaposition of distinctly different habitats such as grassland and forest

ectethmoid a bone in the forepart of the orbital space of the skull

ectoparasite a parasite that lives externally on its host

ectoskeleton (exoskeleton) the external skeleton of invertebrates

electrophoresis a technique for separating taxon-specific organ molecules (commonly proteins) by their different mobilities within an electric field

elfin forest high-elevation stunted windswept forest, the trees often gnarled and twisted and commonly festooned with epiphytes

ellipse an oval; a form shaped like a flattened circle

El Niño (now formally termed **ENSO**: El Niño Southern Oscillation) a periodic global climactic phenomenon typically appearing every 4–6 years, produced by the establishment of anomalously high surface-water temperatures in the central Pacific; an El Niño produces severe drought conditions in Australasia westward into the Greater Sundas

elytra the horny front wings of beetles and some other insects

endemic native to and confined to a specific area, which can be large or small in extent: e.g. birds of paradise are endemic to Australasia, but Victoria's Riflebird is endemic to North Queensland.

epiphyte (n.), a non-parasitic plant that lives entirely upon another plant, but receives nourishment independent of the host plant, as with many orchids, ferns, mosses and lichens (adj. **epiphytic**)

exuvia the cast skin of an insect nymph; cicadas often leave them on trees after the adult has emerged (pl. **exuviae**)

Eyrean the (biotic region) of the Australian arid interior zone

facultative able to exist under more than one set of environmental conditions

femur the bone of the upper leg, which articulates with the pelvis

filamental, filamentous (adj.) slender in structure or part (e.g. hair-like)

folivore (n.) an animal that eats foliage, **folivory** being the eating of foliage (adj. **folivorous**)

foramen (pl. **foramina**) a small opening, e.g. a hole in a bone, through which a nerve passes

Formicidae the hymenopteran insect family of ants

fossa a depression or hollow in a bone

founder effects the genetic divergence shown by small isolate populations, bringing with them only a limited amount of the genetic diversity present in the parent population. Such a small reduced gene pool is likely in island populations and may be part of the process of evolution of species, along with natural selection, genetic drift and sexual selection

frass insect or larva excrement or other refuse, sometimes used in *Amblyornis* gardener bowerbird maypole-bower decorations

frontal the forehead bone of the skull

frugivore (n.) a fruit-eater; **frugivorous** (adj.) fruit-eating

funicles the stalk of an ovule or seed

gene flow exchange of genetic traits between populations by movement of individuals, gametes or spores

genus the taxonomic category immediately above the species level, where species of common phylogenetic origin are grouped

geometrid belonging to the moth family Geometridae

geographical isolation isolation of populations from each other by barriers such as deserts, oceans, rivers or other unsuitable habitats – a crucial stage in species formation

gleaning the capture of arthropods by birds, sometimes by hovering, probing or snatching

gonad an animal organ in which gametes are produced, e.g. testis or ovary

Gondwana (n.), **Gondwanan** (adj.) a massive supercontinent of the geological past that began to break up around 160 mya. It included South America, Africa, Antarctica, Australia, New Zealand, India, and parts of China

granivorous seed- or grain-eating

Gryllidae the orthopteran insect family of field crickets

gymnosperm 'naked seed' plants, a woody plant lacking flowers, and a major plant lineage including pines, araucarias, podocarps, and other primitive tree families (*cf.* angiosperm)

hallux the hind toe of a perching bird

hawk (v.) to pursue, attack, or take on the wing

herbaceous (of plants) having no woody stem above ground

herbivore (n.) an organism that consumes living plants or their parts (adj. **herbivorous**)

herp (plural herps) reptiles and amphibians in general

herpetofauna the amphibian and reptilian fauna

heterochrony the differential rate of development in the two sexes

hide a small structure from which concealed observer(s) can watch wildlife (in the USA known commonly as a 'blind')

home range the area occupied by an individual, pair or group of animals

homogeneous similar in kind or nature

homologous of the same evolutionary and structural derivation (ant. analogous)

imitative learning the process of learning by imitating another, as with bowerbird and riflebird displays

inflorescence a flower or flower cluster

insectivore (n.) insect-eating (also commonly used to include arthropod-eating) (adj. **insectivorous**)

intergeneric hybridisation hybridisation between two species of different genera; frequent in wildfowl, rare in birds of paradise, very rare in bowerbirds

Internet Bird Collection (IBC) free-access web database archive from Lynx Edicions, publishers of the *Handbook of the Birds of the World*, with a vast collection of video, photographic and sound recordings

intersexual selection sexual selection between the sexes of a species

intrageneric hybridisation within a genus, as in some bird of paradise hybrids such as *Paradisaea apoda* × *P. raggiana*

intrasexual selection sexual selection within one sex of a species

introgression the incorporation of the genes of one lineage into the lineage of another

Irian a low-altitude biotic element that invaded New Guinea from the north, primarily of palaeotropic origin

iridescence glossy or metallic appearance of structural colours that change with the angle of light, produced by refraction and not pigmentation, frequent in birds of paradise (adj. **iridescent**)

isolating mechanism a feature differing between species (or populations) that tends to prevent cross-mating and so bring about or maintain species isolation/integrity

karst the sharp, jagged topography of some limestone regions

katydids bush-crickets (referred to also as long-horned grasshoppers) of the orthopteran insect family Tettigoniidae

Kina the main currency denomination in Papua New Guinea, named after a type of golden crescentic seashell

kinetic full of movement

lachrymal a part of the cranium near the orbit, two fused small bone plates near the tear glands in most birds

land-bridge island an island that has in the past been linked to the mainland (or another island), usually in a period of lowered sea level during ice ages or glacial periods, the most recent of which ended about 10,000 years ago

lectotype the single type specimen chosen subsequently from a series of syntypes to establish the name of a taxon

lek (n.) a place where males traditionally cluster in order to attract and mate with visiting females; also the term for the actual cluster of males in display and the act or habit of forming in clusters to display (v. **to lek**)

Lepidoptera the insect order of butterflies and moths

leucism partial loss of pigmentation reducing colour intensity, or complete pigment loss leaving the feathers white but the body parts normally coloured; very rare in both families (adv. **leucistic**)

lineage a group of species allied by common descent

lipid biochemical term for any of a large group of organic compounds that are esters of fatty acids

lump vernacular term denoting the act of regarding one species as belonging to another, e.g. Western and Spotted Bowerbirds were formerly lumped but are now *split* (the antonym of lump)

Macaulay Library huge archive of bird, herp and mammal vocalisations and videos at Cornell University's Laboratory of Ornithology; see macaulaylibrary.org

macropod a member of the kangaroo family (Macropodidae) as in wallabies, tree-kangaroos and kangaroos

male dominance polygyny behaviour whereby males gather together during the breeding season and females select mates from these aggregations

mandible the lower jaw

marauder (n.) term applied to a male bowerbird when it inflicts damage upon the bower of a rival male and/or steals decorations from it (v. **maraud**)

maxilla the upper jaw, or upper mandible of the bill

maxillopalatine bony structure on each side of the vomer that extends out and attaches to the maxilla, forming part of the palate

maypole bower structure built by a male gardener bowerbird (*Amblyornis, Archboldia, Prionodura*) consisting of sticks, orchid stems or ferns, based upon and supported by one or more central vertical saplings or (Archbold's Bowerbird) on low horizontal branches above a bower mat

Melanesia, Melanesian the Pacific island region centred on the island of New Guinea, and including the Admiralty, Bismarck, Solomon and Vanuatu archipelagos and New Caledonia; inhabited by mainly dark-skinned island people

melanin the most common pigment in the feathers of birds, typically appearing as black, but also as brown, red-brown or yellow

Meliphagidae the family of honeyeaters, now including MacGregor's Honeyeater, formerly thought to be a bird of paradise

Menuroidea the avian superfamily that includes the bowerbirds and allies

mesophyll a wet-forest habitat consisting on average of medium-sized (12.5–25cm long) leaves

microphyll a wet-forest habitat consisting on average of small-sized (2.5–7.5cm long) leaves

microsympatry the occurrence of two or more populations or species in the same habitat (*cf.* sympatry)

midden a pile of refuse

millinery (trade) relating to hats and hat trimmings, sold by a milliner and formerly significant for large-scale usage of paradisaeid plumes in the late 19th and early 20th centuries

mimicry see vocal mimicry

Miocene epoch in the Tertiary period dating from 23.5–5 mya

mitochondrial DNA non-nuclear DNA (passed on by females) that resides in the mitochondrion, used in some DNA-testing but more recently superseded by more detailed techniques using nuclear DNA

monobrachygamy implies that females mate with but a single male in each breeding season

monochromatic possessing a single colour pattern

monogamy (n.), **monogamous** (adj.) mating with only one mate

monomorphism (n.) the state in which the sexes are indistinguishable by plumage; a **monomorphic** species is one in which the sexes appear phenotypically identical as adults (more correctly called **monochromatic**, as sexual size dimorphism is frequent)

monophyly (n.), **monophyletic** (adj.) of a single evolutionary lineage, comprising all descendants of the most recent common ancestor

monotypic of a single type (with no variation): hence a monotypic species includes no subspecific variation, e.g. Golden Bowerbird (*Prionodura newtoniana*)

morph plumage-colour variant within a species (known also as a phase), and not to be confused with leucism or albinism; Loria's Satinbird has a grey morph

morphology (n.) the science or study of form and shape – in ornithological terms usually refers to both external (including coloration) and internal characters; (adj. **morphological**)

morphometrics measurement features of morphology, such as wing length and body weight

moult (molt in North America) in birds, the sequential loss and replacement of individual feathers of entire plumages

Müllerian mimicry the convergent evolution of similar colour patterns by two or more noxious or unpalatable species that mutually reinforces avoidance effects upon potential predators

mutualism a mutually beneficial relationship between two species, maintained by special behaviours or habits

nares the external nostrils (adj. **narial**, of the nares)

natural selection elimination of an inferior genetic trait from a population through differential survival and reproduction of individuals bearing that less effective trait

nectary a glandular plant structure that produces nectar

nectarivore (n.) an animal that takes flower nectar as food, **nectarivory** (n.) the eating of nectar (adj. **nectarivorous**)

Neotropical pertaining to the New World tropics

Newcastle disease an avian disease caused by a global multi-strain paramyxovirus; frequently affects domestic fowl and also impacts most avian orders

niche all the components of the environment with which the organism or population interacts

nidification nest-building, manner of nesting

nomenclature the terminology or system of names used in a particular science, art or activity

nominate the first named form of a species and therefore having the subspecies name the same as the species name, e.g. *Paradisaea raggiana raggiana* is the nominate subspecies

notophyll a wet-forest habitat consisting on average of medium-sized (7.5–12.5cm long) leaves

nuchal of the back or nape of the neck, as in nuchal crest of some bowerbirds

nuptial pertaining to the breeding season, especially to plumage and display

occipital pertaining to the large bone at the base of the skull to which the vertebral column is attached; the back part of the head (**occiput**)

omnivore (n.) a species which eats both animal and plant matter, also fungi and the like; **omnivorous** (adj.) eating a wide range of foods

oocyte(s) immature female germ cell(s) giving rise to eggs in birds

ootheca a capsule containing eggs produced by some insects and molluscs

orbit the cavity on each side of the skull that holds the eye

orbital process (also postorbital process) a protuberance on the cranium above the temporal fossa on the margin of and often protruding into the orbital cavity

Orthoptera the insect order of crickets and their relatives

oscines the higher songbirds, one of two suborders of the Passeriformes and constituting most passerines, typified by having more than three pairs of syringeal muscles, important in songs and other vocalisations

ossification the process by which tissue is converted to bone

osteology (n.) the study of skeletal anatomy and bones (adj. **osteological**)

ovaries the two female reproductive organs which produce ova

oviduct the tube through which ova are conveyed from an ovary

Pachycephalinae the avian subfamily (of the family Corvidae) that includes the whistlers and allies

Palaeotropic/paleotropic the Old World tropics

palate the roof of the mouth, or ventral face of the upper jaw (maxilla), of differentiated osteological structure in birds

palatine either of two bones forming the hard palate

panmictic describing a situation in which the entire population or species is freely interbreeding

Papua general term pertaining to New Guinea, originally the south-eastern British and then Australian territory of what is now PNG; now also the easternmost Indonesian province of western New Guinea, bordering Papua New Guinea

Papuan pertaining to the island of New Guinea

paradisaeid (n.), **paradisaeinine** (adj.) pertaining to the family of birds of paradise, Paradisaeidae; pertaining to the subfamily of true birds of paradise, Paradisaeinae

parapatry a distributional phenomenon in which the ranges of two species meet but do not overlap (*cf.* allopatry, sympatry)

paraphyletic descriptive of a group of organisms with a single evolutionary origin but not including all descendants of the last common ancestor

parvorder a subordinate taxonomic classification between infraorder and superfamily

passerine pertaining to the perching birds, the largest order of birds, the Passeriformes, divided into the monophyletic oscines (Passeres) and the likely paraphyletic suboscines (Tyranni)

Passerida a Parvorder of passerines at the apex of the avian evolutionary tree

perennial a plant that dies back each year after flowering but the roots of which survive for subsequent years

pericarp the ripened ovary wall; a collective term to describe the outer layers of fleshy fruits

peticole a leaf stalk; a footstalk of a leaf, connecting the blade with the stem

phase a colour variation within a species, more usually termed morph (which does not carry the possible time constraints of the term phase, which in avian biology do not really apply)

phasmids of the phasmid insect family Phasmatidae, the 'walking sticks' or stick-insects

phenology the seasonal rhythms of plant growth and of seed or fruit production, or the relation of migration and breeding to the seasons

phyllode flattened leaf stalk that resembles and functions as a leaf

phyletic of or pertaining to evolutionary lineages (*cf.* phylogenetic)

phylogenetic, phylogenetic species concept (or PSC) a species concept in which each morphologically diagnosable population is treated as a species (*cf.* biological species concept)

phylogeny an evolutionary tree, with branching to indicate evolutionary divergence of lineages and showing the history of the taxa

phylum (pl. **phyla**) taxonomic rank in the hierarchical classification of biology: kingdom, phylum, class, order, family, genus, species, subspecies. A phylum is the category immediately below kingdom and above class, containing a group of related classes

physiognomy the physical features and structure of habitats in the biological sense

physiology the metabolic functions and physical processes of living organisms, or the science of studying them

plantar the hind section of the horny tarsal sheath

plesiomorphic a primitive character state within a group inherited from a common ancestor and retained by some of the group, but carrying no information about group evolutionary relations

Ploceidae the widespread avian passerine family of weaver finches, of Old World origin

pneumatic refers to air cavities in bones

polyandry (n.) a mating system in which females mate concurrently or sequentially with two or more males during a single reproductive cycle, either by monopolising critical resources (resource-defence polyandry) or by limiting sexual access by other females to males (female-access polyandry) (adj. **polyandrous**)

polybrachygyny indicates that individual males may mate with more than one female each season without prolonged social bonding

polychotomous having multiple branches originating from the same point (n. **polychotomy**)

polytomy is a term for an internal node of a cladogram that has three or more branches (i.e. sister taxa); in contrast, any node that has only two branches (usually called a dichotomy) is said to be incompletely resolved

polygamy (n.) a mating system in which either individual males gain sexual access to more than one female (polygyny), or individual females gain sexual access to more than one male (polyandry) (adj. **polygamous**)

polygyny (n.) a mating system in which males mate with two or more females during a single reproductive cycle, either establishing pair bonds with them simultaneously or pairing with several different females in rapid sequential (sometimes overlapping) succession; frequent among birds of paradise and bowerbirds (adj. **polygynous**)

polyphyletic a group of organisms derived from more than one common ancestor, thus having different evolutionary origins (antonym monophyletic)

polytypic of more than one type, as in a genus that possesses two or more species, or a species that has two or more subspecies

postorbital posterior to the orbit or eye socket

postorbital process see orbital process

primaries the major flight-feathers of the wing, which are attached to the carpometatarsus and digital phalanges (or hand); they vary in number between nine and 11 (usually 10), and now numbered from the innermost outwards (descendantly) and termed P1 to P9/10 or 11, the 'p 'sometimes not capitalised

promiscuity (n.) a polygamous mating system in which breeding adults form no pair bonds but associate with the opposite sex only ultimately for mating; similar to polyandry and polygyny, but without implying temporary pair bonding (adj. **promiscuous**)

protein electrophoresis a laboratory technique that measures the movement of proteins suspended in a solution under the influence of an electrical field; used in ornithology to map variation in protein electro-types in egg albumin as a means of developing biochemical characters for systematic analysis or the study of population genetics

proximal, proximally close to the base of an appendage or process (antonym distally)

pseudo-copulation the performance of actions and associated behaviour typical of copulation by a male, but lacking a female

pterylosis the arrangement of contour feathers on the skin of a bird

Ptilonorhynchidae the bowerbird family

pyriform pear-shaped

quadrate the bone at the base of the cranium to which the mandible is attached

quadruped an animal that has all four limbs specialised for walking

Quaternary a geological age of the Cenozoic era, from about 1.8 mya to the present, and divided into the Pleistocene and Holocene (from 10,000 years ago) epochs

race synonymous with subspecies

radiation the simultaneous evolutionary divergence of several related lineages

rain shadow the relatively dry area leeward of high ground in the path of rain-bearing winds

rectrices the tail feathers

relict a population or species of an animal or plant existing as a remnant of a formerly more widely distributed lineage

remiges the primary and secondary flight-feathers of the wing

resource-based polygyny a mating system whereby males control access to females indirectly, by monopolising resources of potential use to females (**non-resource-based polygyny** is a mating system in which females obtain no resources other than sperm from males)

rictal bristles stiff hair-like feathers originating from each side of the gape, at the bill base

riparian along the bank of a waterway, as in, e.g., riparian vegetation

rupiah the basic unit of Indonesian currency

sagittal the median longitudinal plane dividing a body into right and left halves – following the backbone and including the midline of the crown from forehead to occiput

sally to pounce upon or fly to a substrate to snatch prey

scansorial adapted for or related to climbing or creeping

scapulars the feathers above a bird's shoulder

Schefflera a widespread plant in the tropics and, when flowering and fruiting, an important food source for birds of paradise and bowerbirds; known also as **umbrella tree**

sclerophyll a plant having hard, stiff leaves

scutellate bearing scutes (large scales) on the tarsi

secondaries the inner flight-feathers of the wing, attached to the bone of the 'forearm' of the wing (ulna); they include the tertiaries or tertials (sometimes referred to as the innermost secondaries)

secondary sexual characters any sex-related trait (usually limited to adults of only one sex) associated with obtaining a mate or facilitating reproduction, but usually not directly required for such reproduction or for parental care (these latter structures being primary sexual characters)

semi-species one of a member of a superspecies group; a well-marked geographical population obviously related to one or more allopatric forms; a distinct population at or below the species level

sexual dimorphism the presence of morphological differences such as colour and size between the sexes

sexual selection the evolutionary product of competition among members of one sex for reproductive access to members of the other sex, usually resulting in distinctive behavioural and morphological differences between the sexes

sickletail(s) the three species of *Cicinnurus* (Wilson's, Magnificent and King Birds of Paradise) have elongated often curved tail feathers, and are sometimes known as sickletails

social monogamy the mating system in which pair-members rear chicks together but may partake in mating with one or more individuals other than the original mate; more than 90% of bird species are monogamous, but it is surprisingly common for seemingly monogamous females to have some chicks sired by a male other than their partner (this is known as social monogamy)

sociobiology the study of social behaviour in animals

songbird any member of the oscine birds. See oscines

Southern Oscillation see El Niño

spatulate resembling a utensil with a broad flat blade for spreading, used for describing some tail feathers

speciation the process by which two or more populations differentiate to the level of becoming distinct species

spermatogenesis the seasonal formation of sperm within the avian testes

split vernacular term used when one species is regarded as no longer belonging to the parent species, e.g. Spotted (Black-eared) Catbird has now been split into six species (antonym lump)

squab a young unfledged bird, especially a pigeon (Columbidae)

subcutaneous beneath the skin

subspecies morphologically distinct populations but considered to belong to one and the same species and probably capable of interbreeding. Some are far more distinct than others and certain species require revision as to their subspecific limits, some currently delimited taxa being part of a cline instead, while other subspecies are in fact unrecognised or cryptic species. Subspecies are known also as races or, in the past, varieties. (The threshold of distinction of subspecies is still a matter of opinion, but many older named taxa are really too slight to meet a more rigorous definition.) Beehler & Pratt (2016) synonymised a number of weakly defined paradisaeid subspecies, and we have to a large extent followed their findings.

suboscines the lower perching birds, including all families of Passeriformes not included in the oscines, and primarily but not entirely Neotropical

subsong avian song produced softly or faintly and carrying a far shorter distance (a few metres) than the normal song

sunning intentional 'sunbathing' in sunlight by birds, typically involving specific postures, and perhaps helping to reduce parasite loads

superspecies a monophyletic group of two or more distinct but closely related largely or entirely allopatric species

sympatry (n.) the situation in which two or more species inhabit the same geographical area (adj. **sympatric**) (antonym allopatry, allopatric)

synergy, synergies the combined action of two distinct forces

synonym in biological taxonomy, a scientific name that refers to a taxon that already has a valid name

syntype specimens upon which a new and undescribed taxon was based, without any being designated as the type (= holotype)

syringeal of the syrinx

syrinx the avian organ producing voice or song; an evolutionary distinct structure from the larynx of mammals

systematics the science of objectively ordering/arranging the diversity of life

tarsus (sing.), **tarsi** (pl.) in birds, the common name for the lower leg bone, technically the tarsometatarsus, formed by the fusion of the tarsal and the metatarsal elements; includes the foot, and is sometimes called the lower leg or shank

taxon (sing.), **taxa** (pl.) general term for any category used in biological classification (species, genus, etc.); used also in the specific sense to refer to a population or cluster of related populations (a clade)

taxonomy the science of naming and classifying organisms (performed by a taxonomist)

tectonic plate a distinct section of the earth's crust, one or more of which form each of the earth's continents and major island arcs (some are very large, others much smaller)

temporal fossa(e) a transverse depression at the 'temple' in the skull of birds

Tenebrionidae the coleopteran insect family of darkling beetles

tertials the innermost secondaries, usually moulted as a distinct group

Tertiary in geology, the first period of the Cenozoic (Cainozoic) Era, following the Cretaceous period; it began about 65 million years ago and ended at the start of the Quaternary, about two million years ago

testis (sing.), **testes** (pl.) alternate words for testicle and testicles

Tettigoniidae the orthopteran insect family of katydids

topology the study, description or mapping of the surface features of a region

Torresian the tropical and subtropical eucalypt biota (biotic region) of northern Australia, extending from the Kimberley across Arnhem Land and Gulf of Carpentaria region to all of eastern Queensland south to north-east New South Wales

trachea the windpipe, leading from the glottis to two bronchi that lead to the lungs

transferral effect a hypothesis stating that species- and sex-specific characteristics of male birds may sometimes be transferred to species-specific male-controlled environmental objects or structures (such as bowers), thereby reducing the need for these males to exhibit more elaborate and possibly costly secondary sexual characteristics. Often cited with bowerbirds.

type specimen the single designated specimen upon which a new and undescribed taxon is based

vermiculated having wavy or worm-like (sinuous) markings

vicariance the process of geographical speciation in which a once-continuous population is broken into two or more (vicariant) subpopulations, which subsequently differentiate in geographical isolation

vicariant a geographical population that has been isolated from a parent population by some environmental process

visual display a display performed as a visual stimulus to the recipient, often part of the breeding cycle

vocal appropriation the act of incorporating the call of one bird by another, in vocal mimicry (*cf.* Tooth-billed Bowerbird); such mimicry is not done by birds of paradise.

vocal mimicry the imitation by birds of various sounds other than their specific vocalisations, frequent in some bowerbirds but not known for birds of paradise

Vogelkop alternative name for the Bird's Head Peninsula of West Papua, named as the shape resembles a bird's head (Dutch *vogelkop* means bird head)

vomer the fused medial bone of the palate

wallaby small species of kangaroo (Macropodidae)

Wallace's Line a boundary between the Asian and Australasian vertebrate fauna originally delineated by A. R. Wallace. It runs north–south between the islands of Bali and Lombok (in the south) and Borneo and Sulawesi (in the north)

wattles bare, unfeathered, often brightly coloured fleshy facial flaps or appendages; sometimes nuptial and/or species- and sex-specific recognition marks

West Papua since a 2003 administrative reorganisation the westernmost province of western New Guinea, centred on the Vogelkop or Bird's Head Peninsula. It is also the term used by many for the entire Indonesia-occupied west of the island of New Guinea. Formerly the province of Irian Jaya, the western half of the island of New Guinea was a Dutch colony.

xeno-canto free-access citizen-science sound archive of most of the world's bird species; see xenocanto.org

Zingiberaceae the family of wild gingers, many of which have colourful edible fruits, an important food source for birds of paradise

zoogeography (n.), **zoogeographical** (adj.) pertaining to the geographical distribution of animals

zygomatic process in cranial anatomy, a part of the temporal bone which forms part of the zygomatic arch

List of abbreviations

AMNH	American Museum of Natural History, New York
BMNH	British Museum (Natural History), now the Natural History Museum at Tring (NHMUK)
c.	(Latin *circa*) about or approximately
cf.	(Latin *confer*) compare with
e.g.	(Latin *exempli gratia*) for example
et al.	(Latin *et alii*) and others, additional authors
F&F	C. B. & D. W. Frith (2004) *The Bowerbirds – Ptilonorhynchidae.*
	C. B & D. W. Frith (2010) *Birds of Paradise: Nature, Art, History.*
F&B	Frith & Beehler *The Birds of Paradise* (1998), Oxford
g	gram(s)
ha	hectare(s)
HANZAB	*Handbook of Australian, New Zealand and Antarctic Birds*
HBW	*Handbook of the Birds of the World*
HBWAlive	Internet version of HBW (see https://www.hbw.com/)
I, Is	Island, Islands
IBA	Important Bird Area (BirdLife designation)
IBC	Internet Bird Collection database of videos, photos and sound recordings
Ibid.	(Latin *ibidem*, in the same place) in the same source (referring to previously cited references)
i.e.	(Latin *id est*) that is
in litt.	(Latin *in litteris*) in correspondence, usually referring to an unpublished observation received in writing
IOC	International Ornithological Congress (world bird checklist)
IUCN	International Union for the Conservation of Nature
km	kilometre(s)
Lab	Laboratory, as in Cornell Lab of Ornithology
m	metre(s)
mm	millimetre(s)
mtDNA	mitochondrial DNA
Mt./Mts, mts	mountain(s)
mya	million years ago
NG	New Guinea
NHMUK	Natural History Museum at Tring
NP	national park
NSW	New South Wales (state)
NT	Northern Territory (state)
pers. comm.	personal communication
PNG	Papua New Guinea
Pt	Point (as in Corny Pt)
R	river

SA	South Australia (state)
sp./spp.	species (singular)/species (plural)
ssp.	subspecies
UK	United Kingdom
USA	United States of America
WA	Western Australia (state)
WP	West Papua, formerly Dutch New Guinea then Irian Jaya, now the westernmost Indonesian province of the island of New Guinea
XC	xeno-canto sound archive
YUS	conservation area on Huon Peninsula

BIRDS OF PARADISE

PLATE 1: GENUS *LYCOCORAX* – PARADISE-CROWS

Halmahera Paradise-crow *Lycocorax pyrrhopterus*   Map and text page 152

42cm. Sexes alike. Two subspecies. Endemic to forest, woodlands and gardens of the North Moluccan (Maluku) islands of Halmahera, Bacan (Batjan), Kasiruta, taxon *morotensis* on Morotai and Rau, from sea level to 750m. Monogamous, but breeding behaviour almost unknown; possible that female alone attends the single egg in the large cup-shaped nest.

a   **Nominate race** at rest and in flight. Ochre underwing.

b   **Adult** *morotensis* at rest and in flight. Larger, paler, with most white in the wing.

Obi Paradise-crow *Lycocorax obiensis*   Map and text page 154

42cm. Sexes alike. Monotypic. Occurs in forest, woodland and farmland with scattered trees from sea level to around 1,220m on the North Moluccan (Maluku) islands of Obi and Bisa. Presumed to be monogamous, but not known if both sexes incubate the single egg in the large cup-shaped nest.

c   **Adult** at rest and in flight.

Halmahera Paradise-crow

Obi Paradise-crow

a

b

c

Glossy-mantled Manucode *Manucodia ater*

Map and text page 161

33–42cm. Sexes alike, female slightly smaller. Monotypic. Locally sympatric with Crinkle-collared, Jobi and Trumpet Manucodes. Entirely blue-black with variable iridescence on smooth (not crinkled) neck feathers, with a smooth head profile. Endemic to NG, Aru Is, WP islands of Misool, Batanta, Salawati, Waigeo and Gebe, and islands off the coast of south-east NG (Samarai, Sariba, Mailu and Yule). Widespread in lowlands and hills, from sea level, locally to about 900m (rarely 1,100m). Vocally distinctive: loud, mournful, musical rising *zheeeee*, reminiscent of a tuning fork being struck. Breeding behaviour poorly known but, unlike most others in the paradisaeid group (except *Lycocorax, Phonygammus* and probably other *Manucodia* species), it is monogamous, both sexes nest-building and performing parental duties.

a **Adult** (top middle).

b **Immature**.

Tagula Manucode *Manucodia (ater) alter*

Map and text page 164

43cm. Large and heavy-billed, with longer tail than allopatric Glossy-mantled and with very distinctive voice. This taxon from Tagula (Sudest) I, in Louisiade Archipelago, is usually classified as *Manucodia ater alter*, but vocal and morphological evidence suggests treatment as an insular allospecies appropriate, thus making both this species and Glossy-mantled Manucode monotypic. Breeding behaviour unknown.

c **Adult** (top left).

Jobi Manucode *Manucodia jobiensis*

Map and text page 165

31–34cm. Sexes alike, female slightly smaller. Monotypic. Quite a small, compact species, neater-looking and shorter-tailed than either Glossy-mantled or Crinkle-collared Manucodes but readily misidentified. Head bumps (supraorbital tufts) are not so obvious as those of Crinkle-collared. Occurs in forest from sea level up to about 750m, more usually to 500m; replaced at higher elevations by Crinkle-collared Manucode. Occurs in north NG lowlands eastwards to Ramu R, PNG, and south-west WP, also Yapen I in Geelvink Bay, where it is the only manucode. Apparently absent from Vogelkop and Bird's Neck Peninsula. Breeding almost unknown, but expected to be monogamous.

d **Adult**.

e **Immature**.

Crinkle-collared Manucode *Manucodia chalybata*

Map and text page 167

33–36cm. Sexes alike, female slightly smaller. Monotypic. Sympatric with Glossy-mantled, Jobi and Trumpet Manucodes. Distinctive head bumps above eye are a good field character, giving a different profile from the much smoother head shape of Glossy-mantled Manucode. The bumps (supraorbital tufts) can be erected to become very obvious. The ruffled crinkly texture of breast and neck feathers is visible at close range. Hooting call also distinctive. Occurs over much of NG, also on Misool, mainly in hill forest to 1,500m, locally in lowlands. Breeding little known, presumed to be monogamous.

f **Immature**.

g **Adult**.

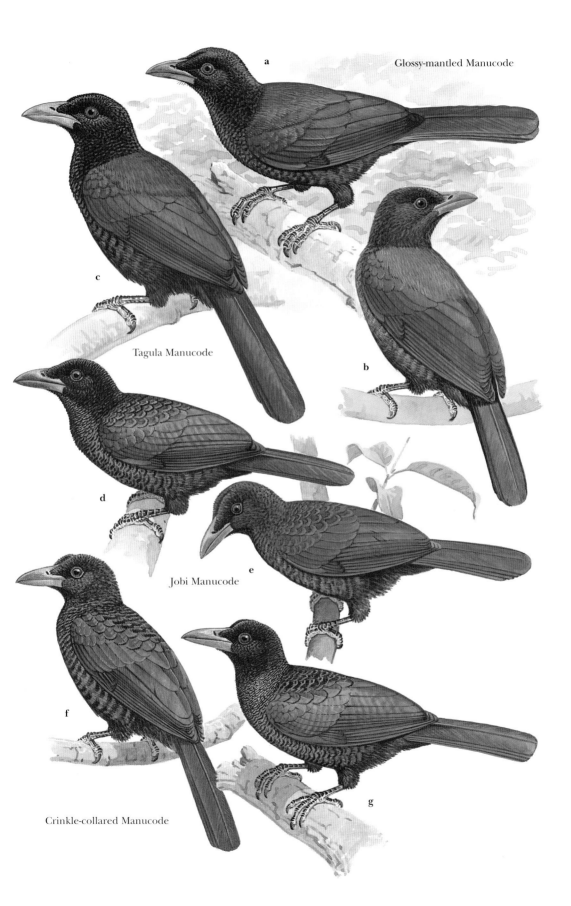

Glossy-mantled Manucode

Tagula Manucode

Jobi Manucode

Crinkle-collared Manucode

Trumpet Manucode *Phonygammus keraudrenii*   Map and text page 156

28–31cm. Sexes alike, female slightly smaller. Six subspecies. Sympatric with Crinkle-collared, Jobi and Glossy-mantled Manucodes, and also with Curl-crested Manucode in some areas, but has noticeable head plumes, long neck hackles and a very different voice. Colour of iridescent plumage varies with subspecies and light conditions. Inhabits lowland and hill forest, from lowlands to about 1,600m; likely that several species involved, the lowland, highland, island and Queensland taxa differing in vocalizations. Breeding behaviour of NG taxa almost unknown. Queensland birds are monogamous, both sexes nest-building and performing parental duties.

a   **Juvenile** *keraudrenii* (Vogelkop, Onin Peninsula and Weyland Mts).

b   **Adult** *keraudrenii*.

c   **Adult** *purpureoviolaceus* (montane form from Astrolabe and Owen Stanley Mts of south-east NG and montane areas of Wau, Morobe Province, north-east NG).

d   **Adult** *gouldii* (north Cape York Peninsula, Queensland).

e   **Adult** *hunsteini* (D'Entrecasteaux Archipelago).

Curl-crested Manucode *Manucodia comrii*   Map and text page 169

41–43cm. Sexes alike, female slightly smaller. Monotypic. Trobriand I birds smaller but now synonymized. A distinctive very large blue-black manucode with large mop of short curled feathers on the crown and thickly crinkled and curled feathers on the neck; tail fairly long, terminal third of central pair of feathers twisted and recurved to expose underside. Endemic to forest, woodland and gardens of D'Entrecasteaux Archipelago (Goodenough, Fergusson and Normanby, also Wagifa and Dobu I) and Trobriand Is (Kiriwina and Kaileuna), from sea level to c.2,000m. Breeding behaviour little known.

f   **Adult**.

g   **Immature**.

Trumpet Manucode

Curl-crested Manucode

PLATE 4: GENUS *PARADIGALLA*

Long-tailed Paradigalla *Paradigalla carunculata* Map and text page 171

35–37cm. Sexes almost alike; female slightly smaller, slightly paler, duller plumage. Monotypic. Both sexes have distinctive fairly long graduated tail. Inhabits montane forests in the Vogelkop (and perhaps Fakfak Mts?) at 1,250–2,100m. Breeding behaviour very little known; likely to be polygynous, promiscuous males, with female alone performing nest duties. Nest a large mossy cup, incubation details unknown. Classified as Near Threatened by BirdLife International in the IUCN Red List of Threatened Species.

a   **Adult male**.

b   **Adult female**.

Short-tailed Paradigalla *Paradigalla brevicauda* Map and text page 173

22–23cm. Sexes almost alike, female slightly smaller and duller-plumaged. Monotypic. Highly distinctive shape and plumage. Inhabits montane forest at 1,400–2,580m in west and central sections of the cordillera, from Weyland Mts east to the Karius and Bismarck Ranges of PNG; not found in the Vogelkop, and easternmost limits are not well known. Juveniles have noticeably longer tail than adults and duller, paler wattles. Appears to be polygynous, with promiscuous males, and females alone performing nest duties. Nest a bulky cup of leaves held together by creeper tendrils, and camouflaged with moss, orchids and ferns. Incubation about 19 days and fledging period about 25 days, but few data.

c   **Adult female**.

d   **Adult male**.

e   **Immature**.

Long-tailed Paradigalla

Short-tailed Paradigalla

Arfak Astrapia *Astrapia nigra*

Map and text page 176

50–76cm. Sexually dimorphic, male much larger and longer-tailed than female. Monotypic. No other astrapia in range. Inhabits moss forest at 1,700–2,250m in Arfak and Tamrau Mts of Vogelkop Peninsula, WP. Breeding behaviour almost unknown, but presumably polygynous, with promiscuous males and maybe some lekking behaviour, but display still remarkably little known. Nest, first discovered in 2017, cup-shaped and made from leaves, moss and twigs; no data on incubation or eggs.

a   **Female plumage.**

b   **Male.**

Splendid Astrapia *Astrapia splendidissima*

Map and text page 178

37–39cm. Sexually dimorphic, female slightly smaller. Two weakly defined subspecies. No other astrapia in range. Endemic to west cordillera of NG, from Weyland, Snow and Nassau Mts of WP eastwards as far as the Star, Hindenburg and Victor Emanuel Ranges of westernmost PNG. Inhabits montane and subalpine forest and forest edge, up to the tree line on the higher summits, from 1,750–3,800m, mainly 2,100–3,300m. Occurs also in secondary growth. Expected to be another polygynous species with promiscuous males, and some lekking behaviour observed. Nesting details almost unknown; clutch a single egg, female performs nest duties.

c   **Male nominate race.**

d   **Female plumage.**

e   **Immature with chestnut crown and hindneck.**

f   **Subadult male.**

Arfak Astrapia

Splendid Astrapia

Ribbon-tailed Astrapia *Astrapia mayeri*

Map and text page 181

Male 32cm, plus tail of up to 1m+; female 35cm, with tail 53cm. Monotypic. Sexually dimorphic. Occurs at higher altitudes than sympatric Princess Stephanie's Astrapia; female-plumaged birds similar to latter but with narrow tail feathers. Endemic to moss forest of central cordillera of west PNG, ranging from Mt Hagen and Mt Giluwe west to both sides of the Strickland R, including the south Karius Range. Polygynous with promiscuous males and, as is typical for the genus, females alone perform nest duties; males may have favourite display areas and can display solitarily or in groups of up to five adults. Cup-shaped nest, with a single egg; fledging period believed to be 22–26 days. Classified as Near Threatened by BirdLife International in the IUCN Red List of Threatened Species.

a   **Male in flight**.

b   **Adult male**.

c   **Subadult male**.

d   **Female plumage**.

e   **Subadult male**.

Ribbon-tailed Astrapia

Princess Stephanie's Astrapia *Astrapia stephaniae*   Map and text page 185

Male 37cm, with tail included 84cm; female 53cm (including tail). Sexually dimorphic, with two weakly defined subspecies. Male distinctive; females told from Ribbon-tailed by much broader tail feathers. Restricted to central cordillera of west PNG, where it is the easternmost representative of the genus, ranging from the Tari Gap and the Bismarck, Schrader, Kubor, Tondon and Doma Peaks Ranges to Mt Hagen and Mt Giluwe and east to the Owen Stanley Mts. Inhabits montane forest up to subalpine moss forest, from 1,280m to 3,500m, but where sympatric with Ribbon-tailed Astrapia is usually not found above 2,450m, higher elevations being occupied by that species. It will inhabit disturbed and selectively logged areas, also regenerating burnt areas, and readily visits gardens and forest patches. Polygynous, with promiscuous males, also known to lek and, as usual with this genus, the female alone performs the nest duties. Nest cup-shaped; single egg. Data on incubation and fledging in the wild still unknown.

a   **Subadult male**.

b   **Adult male nominate**.

c   **Female plumage**.

d   **Immature**.

Huon Astrapia *Astrapia rothschildi*   Map and text page 190

47–69cm. Sexually dimorphic, female much smaller. Monotypic. Distinctive, the only astrapia in its range. Inhabits montane forests at 1,460–3,500m on the Huon Peninsula in the Saruwaged, Rawlinson, Cromwell and Finisterre Ranges. A restricted-range PNG endemic. Believed to be polygynous, with promiscuous males, and females alone performing nest duties. Unknown if this species leks, and some display aspects (such as hanging inverted and tumbling off a perch) appear very unusual, but still very little known. Nest a shallow cup of roots, vines and creepers built on broad flat leaves, moss and leaf skeletons; nestling period 25–27 days, single clutch only.

e   **Female plumage**.

f   **Adult male**.

Princess Stephanie's Astrapia

Huon Astrapia

PLATE 8: GENUS *PAROTIA* 1

Western Parotia *Parotia sefilata*

Map and text page 194

30–33cm, female slightly smaller. Sexually dimorphic. Monotypic. Endemic to the Tamrau, Arfak and Wondiwoi Mts of the Vogelkop and Wandammen Peninsulas, in west WP. The only parotia in its range. Inhabits mid-montane forests around 1,100–1,900m, being often found in old secondary forest with a fairly open canopy and many small saplings, much as with Lawes's Parotia. Males may prefer the denser parts of the forest, with females and young at lower elevations. Males perform elaborate terrestrial courtship dances at courts, but the breeding cycle remains unknown, with nest and eggs as yet undiscovered.

a   **Subadult male.**

b   **Female plumage.**

c   **Adult male in display.**

d   **Adult male.**

Lawes's Parotia *Parotia lawesii*

Map and text page 199

25–27cm, female slightly smaller. Sexually dimorphic. Monotypic. Lawes's Parotia male differs from Eastern Parotia male in having a very short erect supranarial tuft of silvery-white (not bronzy-golden brown), also a slightly different shape. Females of the two are very similar, but Lawes's Parotia has a blunter culmen with convex (not concave) ridge, though the two species are almost, if not entirely, allopatric. Endemic to the mountains of the central ranges of PNG, from Hela west as far as Oksapmin in Western Province, the Schrader and Bismarck Ranges, Baiyer and Jimi valleys, Mt Hagen, Mt Giluwe and Ubaigubi in Eastern Highlands; eastern limits uncertain and may come into contact with Eastern Parotia. Inhabits lower montane forest and well-wooded gardens from 500m to 2,300m, primarily 1,200–1,900m. Polygynous, promiscuous males holding territorial terrestrial courts in an exploded-lek type of system. Probably only females nest-build and rear the young; clutch size appears to be a single egg.

e   **Subadult male.**

f   **Female plumage.**

g   **Adult male.**

Eastern Parotia *Parotia helenae*

Map and text page 204

25–27cm. Sexually dimorphic, female slightly smaller. Monotypic. Male differs from Lawes's Parotia in having very short erect supranarial tuft of bronzy-golden brown (not silvery-white). Females of the two are very similar; the best diagnostic characters may be a sharply keeled culmen, and the unfeathered distal portion of culmen in *helenae*, the maxilla of which is slightly concave dorsally (convex for *lawesii*). Restricted to the eastern sector of the Southeast Peninsula, west to the Waria R and only to the low, dry Keveri Hills, just west of Mt Suckling, which separate it from Lawes's Parotia there. Occurs primarily in the east Owen Stanley Range in mixed montane rainforest, east into Milne Bay Province around 1,100–1,500m (recorded down to 500m). Nesting unknown, presumably much as for Lawes's Parotia.

h   **Adult male.**

Western Parotia

c

a

b

d

e

f

Lawe's Parotia

g

Eastern Parotia

h

PLATE 9: GENUS *PAROTIA* 2

Wahnes's Parotia *Parotia wahnesi*   Map and text page 197

36–43cm. Sexually dimorphic, male notably larger and longer-tailed than female. Monotypic. The only parotia in its range. Endemic to the mid-montane forests of the north coastal ranges of PNG – Cromwell, Rawlinson, Saruwaged and Finisterre Mts of the Huon Peninsula, and the Adelbert Mts of Madang, at altitudes of 1,100–1,700m. Believed to be polygynous, as other parotias; uses terrestrial display courts for dancing, but nest and eggs still unknown. Classified as Near Threatened by BirdLife International in the IUCN Red List of Threatened Species.

a   **Subadult male.**

b   **Adult male.**

c   **Female plumage.**

d   **Adult male display posture.**

Carola's Parotia *Parotia carolae*   Map and text page 205

25–26cm. Sexually dimorphic, female slightly smaller. Four subspecies. Females are surprisingly similar to the much longer-billed and smaller Superb Lophorina, which may be a case of mimicry by the smaller species of the larger and more aggressive female Carola's. Female-plumaged Lawes's Parotia is dark-headed without the facial pattern, and has a striking yellow (female) or violet-blue (male) eye. Inhabits western two-thirds of the central cordillera (absent from the Vogelkop), from the Weyland Range eastwards to Mt Giluwe and Hagen Range, the Sepik–Wahgi divide, Schrader and Bismarck Ranges, and Crater Mt in the Eastern Highlands; known also from Mt Bosavi. Inhabits primary and secondary mid-montane forests; will visit adjacent gardens and regrowth areas. Males display at terrestrial courts, as do other parotia species, implying a promiscuous polygynous species, again as with congeners. Adult males maintain terrestrial exploded leks, females alone attend to nesting duties; nesting details unknown.

e   **Head of female-plumaged** *meeki*.

f   **Subadult male** *meeki*.

g   **Female plumage.**

h   **Adult male nominate.**

i   **Adult male in display.**

Bronze Parotia *Parotia berlepschi*   Map and text page 209

25–26cm. Female slightly smaller, sexually dimorphic. Monotypic. A rather aberrant *Parotia* resembling Carola's; females very similar to latter, but males much more distinctive. No other parotia occurs within the remote limited range in WP. Inhabits the interior of montane forest in the Foja Mts, east of upper Mamberamo R, at 1,200–1,600m. Polygynous, males forming exploded leks, maintaining a terrestrial display and courtship site with a floor cleared of vegetation and a suitable horizontal perch therein. Only females build and attend nests; nesting details unknown.

j   **Adult male.**

k   **Adult female-plumaged head.**

Wahnes's Parotia

Carola's Parotia

Bronze Parotia

PLATE 10: GENUS *PTERIDOPHORA*

King of Saxony Bird of Paradise *Pteridophora alberti*   Map and text page 213

20–22cm. Sexually dimorphic, female slightly smaller. Monotypic. Males distinctive; female-plumaged birds can be puzzling, but no other species is greyish above with dark chevrons below and cinnamon undertail-coverts. Endemic to montane forests of the west and central main cordillera of NG. Extends from the Weyland Mts of WP eastwards perhaps as far as the Kratke Range of central PNG, including the Snow Mts, Star Mts, Hindenburg, Victor Emanuel, Schraderberg (Sepik Mts), Bismarck, Karius and Kubor Ranges, Mt Hagen and Giluwe and the mountains of Ambua and the Tari Gap; eastern limits as yet uncertain. Ranges at 1,400–2,850m, mainly 1,800–2,500m. Inhabits middle and upper montane forest, also forest edge and clearings; seems to avoid small forest islands in montane grassland, but can be found around logging areas and trails so long as good forest exists nearby. Polygynous, promiscuous males singing solitarily from traditional arboreal songposts. Nest duties presumed to be undertaken by female alone, as is usual with the family. Nesting cycle very little known; clutch two eggs, in shallow cup nest.

a   **Adult male**.

b   **Adult female-type plumage**.

c   **Immature male**.

Superb Lophorina *Lophorina latipennis*   Map and text page 220

25–26cm. Sexually dimorphic. Two subspecies. Male is distinctive; male of Loria's Satinbird is entirely black and lacks the breast shield and cape, while male Lawes's Parotia has head wires, a white narial tuft, no cape and a vivid violet eye. Female's black-barred underparts preclude all except other paradisaeids, with females of local *Parotia* species similar in some areas: female Lawes's Parotia somewhat larger, with more elongated head profile, violet eye, no superciliary stripes; female Wahnes's Parotia has limited range, longer tail, and much less well-marked head pattern. Magnificent Bird of Paradise females much smaller and shorter-tailed, with bluish post-ocular stripe and bill, and blue legs; all the astrapias have much longer tail. Endemic to NG montane forests, through the entire central cordillera from Weyland Range eastwards including Snow and Star Mts, also the Hindenburg and Victor Emanuel Mts, and Central Highlands as far east as Herzog, Kuper and Ekuti Ranges of Eastern Highlands; also Hunstein Mts, Adelbert Mts, the Huon Peninsula, and Mt Bosavi. Ranges from 750–2,300m, mostly 1,650–1,900m. Mid-montane to upper montane forest, forest edge, and remnant patches, particularly those with *Casuarina* trees, in gardens in montane valleys. Polygynous, promiscuous males singing solitarily from traditional arboreal songposts. Nest duties presumably by female alone, as is usual with the family; nesting cycle little known, clutch one egg, sometimes two, in open cup nest.

a   **Adult male displaying to female.**

b   **Female plumage** *feminina.*

Western Lophorina *Lophorina superba*   Map and text page 217

25–26cm. Sexually dimorphic. Two subspecies. Similar female of Western Parotia is chunkier, slightly larger, has flat head shape and violet eye, and is richer buff below than the black-barred quite pale whitish of black-headed western races of Curl-caped Lophorina. Magnificent Bird of Paradise females are considerably smaller and shorter-tailed, with bluish post-ocular stripe and bill, blue legs. All the astrapias have much longer tail. Mid-montane to upper montane forest, forest edge, and remnant patches in gardens in montane valleys of Arfak, Tamrau and Wondiwoi Mts. Altitudinal range similar to that of Superb Lophorina, from around 750m up to about 2,300m, most common at middle ranges. Polygynous, promiscuous males holding solitary display areas within auditory or visual contact of others. Nesting biology unknown.

c   **Adult male** *superba.*

d   **Female plumage** *niedda.*

Eastern Lophorina *Lophorina minor*   Map and text page 224

25–26cm. Sexually dimorphic. Monotypic. Similar species as for Superb Lophorina. Ranges from south-east PNG through mountains of Papuan Peninsula (probably from south-east of Wau) and north-west to Owen Stanley Range. Habitat as for Superb Lophorina. Breeding biology almost unknown, presumably much as for *L. superba.*

e   **Subadult male.**

f   **Head of female-type plumage.**

Superb Lophorina

Western Lophorina

Eastern Lophorina

PLATE 12: GENUS *PTILORIS* – RIFLEBIRDS 1

Paradise Riflebird *Ptiloris paradiseus*

Map and text page 226

27–30cm. Female slightly smaller, sexually dimorphic. Monotypic. Male distinctive; female sometimes thought to be a large honeyeater (Meliphagidae). This species is the sole member of the family to occur outside the tropics; ranges along the east coast of Australia from the Calliope Range south of Rockhampton, Queensland, to just north of Newcastle, NSW, in the Great Dividing Range, and in some lowlands to the east. Southern limit is the north Hunter valley near Barrington Tops. Generally, this species is most numerous above 500m, extending up to about 1,100m. Inhabits the subtropical and temperate rainforests of south Queensland and north NSW, but may be found also in nearby wet sclerophyll forest and during the winter months in dry sclerophyll forest. Males are sedentary and promiscuous, displaying from songposts and holding territories around them. Usual clutch two eggs. Nest is a shallow bowl, larger and bulkier than that of Victoria's Riflebird, made from vine stems and lined with finer vines, fibres and rootlets. Fresh fern fronds are used to decorate the rim, as often are snakeskins.

a   **Adult male in display.**

b   **Subadult male.**

c   **Adult male.**

d   **Female-type plumage.**

Victoria's Riflebird *Ptiloris victoriae*

Map and text page 231

27–30cm. Female slightly smaller, sexually dimorphic. Monotypic. Male distinctive; female sometimes misidentified as a large honeyeater (Meliphagidae). Endemic to the wet tropics of North Queensland, where it is the only bird of paradise. Occurs in the wet tropics from just south of Cooktown to Mt Elliot, just south of Townsville. Extends from sea level to about 1,200m, most frequent at 250–1,100m. Inhabits tropical rainforests in lowlands and hills, and found also on some small offshore islands; also in adjacent wet sclerophyll eucalypt or melaleuca woodland, and landward edge of mangroves, particularly in winter when some local dispersal to such habitats occurs (probably from higher altitudes, as with a number of other species). Males sedentary and promiscuous, displaying from songposts and holding territories around them. Usual clutch two eggs. Nest cup-shaped, made from fern stems and moss and lined with dead leaves, often decorated with snakeskins.

e   **Adult male in display.**

f   **Subadult male.**

g   **Adult male.**

h   **Female-type plumage.**

Paradise Riflebird

Victoria's Riflebird

Magnificent Riflebird *Ptiloris magnificus*

Map and text page 236

28–33cm. Sexually dimorphic, female slightly smaller. Two subspecies. Distinctive, as quite a large species. Endemic to west and central parts of mainland NG, from Vogelkop eastwards into PNG, where it ranges to the Wewak area of Sepik drainage in north, and in southern watershed as far as Purari R, Gulf Province. Appears to be absent from forests and savannas of the lower and middle Trans-Fly, and from any of the satellite islands. East PNG is occupied by the allospecies Growling Riflebird, while an outlying population on Cape York, in extreme North Queensland, bears more vocal resemblance to the western Magnificent Riflebird than to the far closer geographically occurring Growling Riflebird. Inhabits rainforests in lowlands, hills and lower montane zones, also monsoon, swamp and gallery forests. Occurs from sea level to as high as 1,450m (exceptionally to 1,740m), but mainly to 700m. It is presumed that female alone carries out the nest duties, as is usual for the sexually dimorphic species in this family. Nesting habits are remarkably little known.

a   **Adult male in display.**

b   **Adult male nominate.**

c   **Subadult male nominate.**

d   **Adult female plumage nominate.**

e   **Adult male *alberti*** (Cape York Riflebird).

f   **Adult female plumage *alberti*** (Cape York Riflebird)**.**

g   **Female-type head, nominate.**

h   **Female-type head *alberti*** (Cape York Riflebird).

Growling Riflebird *Ptiloris intercedens*

Map and text page 244

28–34cm. Sexually dimorphic, female slightly smaller than male. Monotypic. Male is unlikely to be confused with anything else in a reasonable view, and is allopatric with its congeners except for an ill-defined zone of abutment or sympatry with the Magnificent Riflebird on the Huon Peninsula and perhaps Eastern Highlands of PNG. Telling the two species apart would be difficult as field characters are subtle and of limited use, but fortunately the voice is diagnostic. The extensive feathering at base of culmen of this species, however, may be of some use in a good view (both sexes). The flank plumes on males of this species project only level with or short of the tail, whereas those of Magnificent Riflebird project beyond the tail. Endemic to east PNG, from Adelbert Mts of the northern slope and Purari–Kikori R of southern watershed eastwards to Milne Bay; it is also absent from all of the satellite islands. Inhabits rainforests in lowlands, hills and lower montane zones, also monsoon, swamp and gallery forests. Polygynous, promiscuous males displaying solitarily at songposts and attracting female-plumaged birds. It is likely that females alone are responsible for nest duties, as is typical of the family. Nesting habits are remarkably little known.

i   **Female-type plumage.**

j   **Adult male.**

k   **Female-type head.**

g   h   k

Magnificent Riflebird

Growling Riflebird

PLATE 14: GENUS *EPIMACHUS* – SICKLEBILLS

Black Sicklebill *Epimachus fastosus*

Map and text page 248

Male 63cm (110cm including central tail), the largest of the family; female 55cm. Sexually dimorphic. Two similar subspecies. Similar in shape to Brown Sicklebill, but the two species have only a narrow range of altitudinal overlap. Male Black has blacker underparts, and a red (not pale blue) eye. Females of the two species are similar, but Black has a duller chestnut cap and a red eye. The calls of the males are diagnostic. Astrapias are smaller, lack decurved bill, have dark eyes, and males have specifically distinct and distinctive tail shape. Endemic to lower mountains of west and central NG, from Weyland Mts of central ranges east to Bismarck and Kratke Ranges, Mt Bosavi, and Bewani and Torricelli Ranges of north coast; also in Vogelkop and Wandammen Peninsulas. Inhabits lower montane and mid-montane forest in a narrow altitudinal band at 1,280–2,550m (mainly 1,800–2,150m), sometimes at forest edge and in second growth or forest patches. Polygynous; promiscuous but solitary and sedentary adult males occupy a home range, with one or more prominent display perches. Breeding behaviour remains little known, with nest and eggs unknown. Classified as Vulnerable by BirdLife International in the IUCN Red List of Threatened Species.

a   **Adult male in a display pose**.

b   **Adult male nominate**.

c   **Subadult male**.

d   **Female-type plumage**.

Brown Sicklebill *Epimachus meyeri*

Map and text page 252

Male 49cm (96cm including central rectrices), female 52cm. Sexually dimorphic. Monotypic. Central mountain ranges of NG from Weyland Mts of WP to south-east Owen Stanley Mts, extending farther east than Black Sicklebill but not occurring in Vogelkop. Occurs at 1,500–3,200m, mainly 1,900–2,900m, being found at the lower altitudes in eastern part of range where Black Sicklebill absent. Upper to mid-montane forest, mostly above the range of Black Sicklebill, sympatric only in narrow elevational bands in a few areas. Black Sicklebill has a red or brown eye but is otherwise quite similar and can be hard to separate if not seen well in the limited zone of overlap, but the songs of the two species are very distinct. Polygynous, promiscuous male maintaining large territory within auditory contact of other males; breeding behaviour still poorly known. Nest a cup of stringy living moss and small vines, lined with slender rootlets and dried leaves; single egg.

e   **Adult male**.

f   **Subadult male**.

g   **Female-type plumage**.

h   **Subadult male feeding on a pandan**.

Black Sicklebill

a

b

c

d

Brown Sicklebill

e

f

g

h

PLATE 15: GENUS *DREPANORNIS* – SICKLEBILLS

Black-billed Sicklebill *Drepanornis albertisi*

Map and text page 256

33–35cm. Slight sexual dimorphism. Three little-known subspecies. Much smaller than *Epimachus* sicklebills, with much shorter and sometimes paler buff tail. *Melidectes* honeyeaters also have a decurved bill, but are smaller, with pale or bluish bill, bare greenish-yellow or blue skin around the eye, lack barring below, and are very vocal with quite different vocalisations. Endemic to the lower montane zone of NG, and with curiously disjunct range, seemingly absent or undiscovered in a huge stretch of the central cordillera from west of Hindenburg Range, PNG, to Wissel Lakes, Vogelkop, and Fakfak, Kumawa and Foya Mts (Gauttier Mts) of WP (not yet recorded from Adelbert Mts or northern coastal ranges of PNG); range extends to the mountains of south-east PNG, including Herzog Range and west as far as Huon Gulf in north and Nipa/Tari areas in south, reaching to Mt Giluwe. Inhabits lower montane and mid-montane forest, and occasionally seen in selectively logged areas, but rarely at forest edge. Breeding behaviour little known; builds cup-shaped nest and lays one or two eggs; presumably female alone performs nest duties.

a **Female-type plumage.**

b **Adult male nominate.**

c **Immature male.**

d **Male foraging on tree trunk.**

Pale-billed Sicklebill *Drepanornis bruijnii*

Map and text page 260

34–35cm. Sexually dimorphic. Monotypic. Not known to overlap with its congener, a distinctive species. Endemic to the lowland rainforest of north-west NG, from east side of Geelvink Bay eastwards through basin of the Meervlakte and northern coastal lowlands to vicinity of Vanimo, in PNG, there extending to north-west reaches of the Sepik drainage at Utai on southern side of the Bewani Mts. Occurs to 180m in the limestone hills near Vanimo. Polygynous, males with display sites and occupying a large display and foraging territory, similar to Black-billed Sicklebill. No details of nesting cycle are known.

e **Female-type plumage.**

f **Subadult male.**

g **Adult male.**

Black-billed Sicklebill

Pale-billed Sicklebill

Magnificent Bird of Paradise *Cicinnurus magnificus*   Map and text page 267

19cm, male with central tail feathers adding a further 7cm (making 26cm). Sexually dimorphic. Three weakly differentiated subspecies. Confusable only with female-plumaged King Bird of Paradise where the two happen to overlap. Present species has a feathered bill base and rather flat crown, giving a strange head shape; it lacks the rusty wings of the King Bird, and has more heavily barred underparts. Shows the typical short tail, quite broad rounded wings and tubby body of the family. The legs and feet are purplish-blue, less intensely coloured than those of the King Bird. Endemic to the hill forests of NG, from the WP island of Salawati (doubtfully Misool), and from Yapen, in Geelvink Bay, eastwards across entire mainland in all mountain ranges. Absent from the unsuitable lower-altitude savanna habitats of the Trans-Fly, and replaced by the sister-species Wilson's Bird of Paradise on Waigeo and Batanta (in WP Is). Extends from about 400–1,200m, locally as high as 1,600m and exceptionally to 1,780m; occurs also at some lower altitudes where hill forest extends to the foothills, and adjacent lowlands. Polygynous, promiscuous males attending terrestrial display courts, and females alone responsible for nest duties. Nesting details unknown; only four nests known, clutch one or two eggs.

a   Adult male *hunsteini* in display pose.

b   Adult male nominate.

c   Subadult male.

d   Female-type plumage.

e   Juvenile plumage.

Wilson's Bird of Paradise *Cicinnurus respublica*   Map and text page 273

16cm, male including central rectrices 21cm. Sexually dimorphic. Monotypic. Distinctive, the smallest paradisaeid, endemic to Waigeo and Batanta (in WP Is), where it occurs in hill forest from about 300m (occasionally as low as 60m) up to summits of the hills at 1,000–1,200m. Polygynous, the promiscuous males occupying terrestrial courts. No breeding information; nothing is known about the nest, clutch, incubation or fledging periods. Classified as Near Threatened and potentially Vulnerable by BirdLife International in the IUCN Red List of Threatened Species.

f   Adult male.

g   Male in display pose.

h   Immature male.

i   Female-type plumage.

Magnificent Bird of Paradise

Wilson's Bird of Paradise

King Bird of Paradise *Cicinnurus regius*   Map and text page 263

Male 16cm, 31cm including tail wires; female 19cm. Two subspecies. The only similar species is the congeneric Magnificent Bird of Paradise in female plumage, but that has darker plumage without rusty wings, a bluish bill and a pale streak behind the eye, and is also not found in lowland forest, although the King's range does extend into its hill-forest habitat in some areas. The King's bright violet-blue legs and feet are also a striking character, being duller purplish-blue in the Magnificent. Endemic to mainland NG and the Aru Is, the WP Is of Misool and Salawati (doubtfully Batanta), and Yapen I, in Geelvink Bay. A widespread species but absent from much of the lower Trans-Fly, where the eucalypt and *Melaleuca* savanna habitat is unsuitable. Occurs from sea level up to a maximum of 1,150m, but mostly to 300m, regularly but sparsely to 420m in Western Province, inhabiting rainforest, vine thicket, monsoon forest, gallery forest, tall secondary forest and selectively logged forest. Promiscuous males attend courts in thick vines in the middle to upper strata, usually in shade but occasionally in quite well-lit places. Nesting behaviour almost unknown; only one nest found, in a hole some 18 inches (45cm) deep filled with palm fibres. Female alone performs nest duties; clutch appears to be two eggs.

a   **Subadult male nominate**.

b   **Adult male nominate**.

c   **Immature**.

d   **Female-type plumage**.

e   **Adult male** *coccineifrons*.

PLATE 18: GENUS *SEMIOPTERA*

Wallace's Standardwing *Semioptera wallacii*   Map and text page 276

23–26cm. Sexually dimorphic, female slightly smaller. Two subspecies. A unique isolated species, one of the three most westerly of the family and the only plumed paradisaeid in its range. Both sexes have distinctive rather pale, washed-out plumage, with broad wings and bright orange legs. Confined to Indonesian islands of Bacan, Halmahera and Kasiruta, in North Moluccas (Maluku), where it is a sedentary species of the hill forests from 200–1,500m on Bacan and from 350m to at least 900m on Halmahera. Wallace's Standardwing is polygynous, promiscuous males displaying at leks and nest duties likely to be undertaken by the females alone, as with other paradisaeids that lek. Nesting almost unknown; the only recorded nest was an open cup with a single egg. Classified as Near Threatened by BirdLife International in the IUCN Red List of Threatened Species.

a   **Adult male in display posture.**

b   **Adult male in display posture.**

c   **Subadult male.**

d   **Adult male.**

e   **Female-type plumage.**

Wallace's Standardwing

PLATE 19: GENUS *SELEUCIDIS*

Twelve-wired Bird of Paradise *Seleucidis melanoleucus* Map and text page 280

33–35cm. Sexually dimorphic. Monotypic. The female resembles a female riflebird but has a blackish head and a red eye; moreover, the shape is very odd, with very broad rounded 'butterfly' wings, a short tail and a long decurved bill. This is a species endemic to the coastal lowland plains, lowland forests and great river systems of NG. It ranges from Salawati I eastwards along the southern Vogelkop and throughout the lowland forests of NG, in the south to as far as the Port Moresby area, in the north to the Ramu R valley. It is absent east of these areas, where the coastal plain is lacking or narrow. Occurrence on the Onin Peninsula and southern Trans-Fly is as yet uncertain, but it is known from Lake Murray. Widespread at or near sea level in swamp, monsoon, riparian and lowland forests, extending up to about 180m in uplifted karst hill forest in the West Sepik, PNG. This species is fond of seasonally flooded forest like that along Fly and Sepik Rs, and around Vanapa and Brown R near Port Moresby. Much of its habitat may be very difficult of access, being swampy and low-lying with sago swamp and *Pandanus* thickets. Female alone undertakes the nest duties, as is usual for the family. Nests and eggs little known; clutch may be a single egg.

a   **Female-type plumage**.

b   **Adult male in flight**.

c   **Adult male in flight**.

d   **Adult male on display perch**.

e   **Subadult male**.

Twelve-wired Bird of Paradise

PLATE 20: GENUS *PARADISAEA* – PLUMED BIRDS OF PARADISE 1

Lesser Bird of Paradise *Paradisaea minor* Map and text page 293

31–32cm. Sexually dimorphic. Two similar subspecies. Similar to Greater Bird of Paradise in the south-west, both having yellow plumes, but lacks the well-defined breast shield of Greater and has a yellowish back, and minimal/ no known range overlap. Female is very distinctive, with white underparts contrasting with dark head. Occurs from sea level to about 1,800m in lowland and hill forests of north and far west NG, one of the most widespread of the genus. Occurs on Misool and in the Vogelkop, then across the northern watershed to the Gogol and Ramu R and the Finisterre Range of the north-west Huon Peninsula. Hybrids with Raggiana known where the two overlap in Ramu and Baiyer valleys. Polygynous, promiscuous males display in leks; females alone perform nest duties. Nest cup-shaped, with usually a single egg.

a **Subadult male *minor*.**

b **Adult male *minor*.**

c **Adult male *minor* in display.**

d **Female-type plumage.**

Greater Bird of Paradise *Paradisaea apoda* Map and text page 284

35–43cm. Sexually dimorphic. Monotypic. A yellow or apricot-orange plumed species of south-west NG, barely extending into PNG around Kiunga and Tabubil, in Western Province. Overlaps narrowly only with Raggiana, hybrids with which (with pinkish-tinged yellow or orangey flank plumes and often the yellow throat strap of Raggiana) are known from inland around Kiunga. Female very distinctive, being entirely dark chestnut-brown and lacking yellow on nape, though subadult males can be confusing. Inhabits lowland and hill forest from sea level to around 850m in the Ok Tedi region and at to least 950m in WP. Polygynous, promiscuous males display at lek sites, while females carry out nest duties alone. Nest an open shallow cup, usually with a single egg; incubation and nestling care unknown.

e **Adult male.**

f **Female-type plumage.**

g **Adult male in display.**

h **Subadult male.**

Lesser Bird of Paradise

Greater Bird of Paradise

PLATE 21: GENUS *PARADISAEA* – PLUMED BIRDS OF PARADISE 2

Goldie's Bird of Paradise *Paradisaea decora*   Map and text page 297

29–33cm, female smaller. Sexually dimorphic. Monotypic. Endemic to the hill forests of Fergusson and Normanby, in the D'Entrecasteaux Archipelago, south-east PNG, where it is the only member of the family. Occurs from the lowlands to about 600m, primarily above 350m. Presumed to be polygynous, males displaying in loosely formed arboreal leks; nest and eggs unknown. Classified as Vulnerable by BirdLife International in the IUCN Red List of Threatened Species.

a   **Adult male in flight**.

b   **Female-type plumage**.

c   **Subadult male**.

d   **Adult male**.

Red Bird of Paradise *Paradisaea rubra*   Map and text page 300

Male 33cm, with twisted tail wires 56cm; female 30cm. Sexually dimorphic. Monotypic. Endemic to WP Is, where found only on Batanta, Waigeo and Gam and the two small islets of Gemien and Saonek (off Waigeo). Distinctive in being the only *Paradisaea* on the islands. Inhabits lowland and hill forest to about 600m. Polygynous, with promiscuous communally displaying males. Females presumably build nest and rear the young unaided, as with other species in the genus. No data on nesting in the wild; clutch size one or two eggs. Classified as Near Threatened by BirdLife International in the IUCN Red List of Threatened Species.

e   **Adult male**.

f   **Subadult male**.

g   **Female-type plumage**.

Goldie's Bird of Paradise

a

b

c

d

e

f

Red Bird of Paradise

g

Raggiana Bird of Paradise *Paradisaea raggiana*   Map and text page 288

33–34cm, tail wires of male extending up to 36cm beyond tail tip. Sexually dimorphic. Four subspecies. Greater, Lesser and Emperor Birds of Paradise all have marginal overlap and essentially are the geographical replacements for the Raggiana. Lesser Bird of Paradise male has yellow plumes, with yellow of nape extending more onto mantle, and lack the dark breast cushion and yellow throat strap of Raggiana, while male Greater similarly lacks breast strap and has dark chestnut mantle and apricot or yellow flank plumes. Females and immatures of Raggiana lack the whitish underparts of corresponding plumages of Lesser Bird of Paradise, and are not dark chestnut-brown below like female and immature Greater Bird of Paradise. Inhabits lowland and hill forest of south, central and south-east PNG, barely extending into WP, replaced by closely related sibling species the Greater Bird of Paradise in far west PNG, and by Lesser Bird of Paradise in north-west Sepik–Ramu area. Extends from sea level to about 1,800m in some highland valleys. Occurs in lowland, hill and lower montane forest, and found also in second growth, gardens and *Casuarina* groves, even occasionally visiting mangroves. Polygynous, males are very promiscuous. Nesting behaviour not well documented; clutch one or two eggs.

a   **Adult female-type plumage.**

b   **Adult male *salvadorii.***

c   **Adult female *augustaevictoriae.***

d   **Adult male *augustaevictoriae.***

e   **Subadult male nominate.**

f   **Adult male nominate in display.**

Raggiana Bird of Paradise

PLATE 23: GENERA *PARADISAEA* & *PARADISORNIS* – PLUMED BIRDS OF PARADISE 4

Emperor Bird of Paradise *Paradisaea guilielmi*   Map and text page 303

31–33cm, average length of male central tail wires 56cm. Sexually dimorphic. Monotypic. Males distinctive. Females resemble Raggiana (which can be sympatric in some lower montane areas), but have a dark crown and more extensively dark face and chest, dark eyes and brighter richer yellow on hindneck and mantle than local *augustaevictoriae* race of Raggiana. Endemic to lower mountains and hills of Huon Peninsula, PNG. Polygynous, promiscuous males assembling at leks. Females alone perform nest duties, nest a deep cup of creeper tendrils and vines built on large, broad leaves; clutch size one or two eggs, incubation details unknown. Classified as Near Threatened by BirdLife International in the IUCN Red List of Threatened Species.

a   **Subadult male.**

b   **Adult male.**

c   **Female-type plumage.**

Blue Bird of Paradise *Paradisornis rudolphi*   Map and text page 306

30cm. Sexually dimorphic. Two similar subspecies. Distinctive species, endemic to PNG highlands. Occurs at 1,100–2,000m in lower montane forest and wooded gardens from the Owen Stanley Range westwards through Mt Hagen area to Hela Province. Promiscuous males establish a solitary calling station, and do not lek. Females alone perform nest duties, building bowl-shaped nest and laying clutch of one egg (sometimes two); incubation details unknown. Classified as Vulnerable by BirdLife International in the IUCN Red List of Threatened Species.

d   **Adult male in display.**

e   **Adult male nominate.**

f   **Adult male nominate.**

g   **Female-type plumage nominate dorsal view** (east PNG).

h   **Female-type plumage *margaritae* front view** (cental and west PNG).

i   **Subadult male.**

Emperor Bird of Paradise

a

b

b

c

d

Blue Bird of Paradise

e

f

g

h

i

BOWERBIRDS

PLATE 24: GENUS *AILUROEDUS* 1 – WHITE-EARED CATBIRD COMPLEX

White-eared Catbird *Ailuroedus buccoides* Map and text page 325

25cm. Sexes alike, no marked sexual dimorphism; female slightly smaller, crown and perhaps leg colour may be paler. Monotypic. Endemic to WP on islands of Waigeo, Batanta and Salawati, also the Vogelkop and Bird's Neck up to about 800m, and lowland north-west NG east to Siriwo R in north and Mimika R in south. Inhabits dense tropical rainforest up to about 800m. Breeding almost unknown, but unlikely to differ significantly from that of the other members of this complex. This species does not make any kind of stage or bower.

a **Immature.**

b **Adult.**

Ochre-breasted Catbird *Ailuroedus stonii* Map and text page 327

25cm. Sexes alike, no marked sexual dimorphism; female slightly smaller, crown and legs perhaps paler. Two similar subspecies. Heavily built stout catbird, crown dark blackish-brown, tinged olive-green; the pure white of ear-coverts extends forward onto lower lores. Black-eared Catbird has limited range overlap, lacks white ear-coverts and has crown heavily spotted with white. Southern watershed of Eastern Range and Southeast Peninsula east to Amazon Bay; not recorded from Trans-Fly. Inhabits rainforest, hill forest and monsoon forest in lowlands, occurring mostly from sea level up to about 800m, but locally to higher levels (as at Baiyer R, 1,200m). Sympatric with Black-eared Catbird in some areas, as at Kiunga and Tabubil, and with Black-capped Catbird at Sogeri Plateau. Nest cup-shaped; single egg, incubation details unknown. This species does not make any kind of stage or bower.

c **Immature.**

d **Adult.**

Tan-capped Catbird *Ailuroedus geislerorum* Map and text page 326

25cm. Sexes alike, no marked sexual dimorphism; female slightly smaller, crown and leg colour may be paler. Two similar subspecies. Inhabits rainforest, hill forest and monsoon forest in lowlands, occurring mostly from sea level to about 800m, but locally to higher levels. Yapen I and north-west lowlands, Sepik–Ramu, Huon Peninsula, and northern watershed of Southeast Peninsula. Breeding almost unknown, but unlikely to differ significantly from that of the other members of this complex. Maintains a territory during nesting period, but does not make any kind of stage or bower.

e **Adult.**

White-eared Catbird

b

a

Ochre-breasted
Catbird

c

d

Tan-capped Catbird

e

PLATE 25: GENUS *AILUROEDUS* 2 – BLACK-EARED CATBIRD COMPLEX

Green Catbird *Ailuroedus crassirostris*

Map and text page 329

31cm. Sexes alike, female slightly smaller than male. Monotypic. Distinctive, with no other catbird in range. Endemic to south-east Australia, inhabiting rainforest of temperate and subtropical lowlands in south Queensland and NSW, also adjacent eucalypt sclerophyll forest and woodland. Occasionally visits gardens and orchards. Forms monogamous pairs; female alone builds nest and incubates, both sexes feed young. Nest a bulky cup of sticks; clutch one or two eggs. This species does not make any kind of stage or bower.

a **Adult *crassirostris*.**

b **Head study *crassirostris*.**

Black-eared Catbird *Ailuroedus melanotis*

Map and text page 332

29cm. Sexes alike. Three taxa in south NG and Cape York Peninsula of North Queensland: nominate in Aru Is and Trans-Fly lowlands east to Oriomo R; *facialis* in the west-central mountains in hill forest at 600–1,700m; and *joanae* in lowland rainforest/vine forest of Cape York Peninsula. Breeding very poorly known or unknown for all NG taxa of the former Spotted Catbird complex, which do not make any kind of stage or bower. Nest a bulky cup of sticks; clutch one or two eggs.

c **Adult *melanotis*.**

Spotted Catbird *Ailuroedus maculosus*

Map and text page 337

29cm. Sexes alike, female smaller. Monotypic. Tropical North Queensland, primarily in hill forest. Tablelands rainforest and tall secondary forest from just south of Cooktown south as far as the Seaview and Paluma range north of Townsville. Occurs to sea level in a few areas, but mainly a hill-forest species at 350–1,700m. Monogamous; only female builds nest and incubates, both sexes feed young. Nest a bulky cup of sticks; clutch one or two eggs. Does not make any bower.

d **Adult.**

Green Catbird

Black-eared Catbird

Spotted Catbird

PLATE 26: GENUS *AILUROEDUS* 3 – BLACK-EARED CATBIRD COMPLEX

Black-capped Catbird *Ailuroedus (melanotis) melanocephalus*   Map and text page 336

29cm. Sexes alike. Monotypic. May overlap marginally with Ochre-breasted Catbird (Plate 24), from which the lack of white ear-coverts and the blackish cap should easily separate it, and call very different. Mountains of Southeast Peninsula of PNG west to Mt Karimui, and in northern watershed to Herzog Mts and head of Huon Gulf. A montane taxon found in middle- and upper-level forests of the south-east mountains, from about 600–1,700m, occasionally higher. Nesting details unknown.

a   Adult.

Northern Catbird *Ailuroedus (melanotis) jobiensis*   Map and text page 335

29cm. Sexes alike. Monotypic. Like Black-eared Catbird, but spotting on blackish crown more pale buff. Chin, throat and upper breast blackish with fine buff spotting; underparts darker than on Black-eared. Easily told from Ochre-breasted or Tan-capped Catbirds (Plate 24) by lack of white ear-coverts and blackish cap, also call very different. A bird of northern slope hill forest and lower montane forest, inhabits northern slopes of Western, Border and Eastern Ranges; also mountains of the Sepik and Jimi R. Breeding very poorly known or unknown; presumably much as for Spotted Catbird.

b   Adult.

Arfak Catbird *Ailuroedus (melanotis) arfakianus*   Map and text page 334

29cm. Sexes alike. Two similar subspecies. Black ear spot enclosed by a large white post-ocular patch, dark crown with variable pale (often whitish) spotting, paler than in allopatric Black-eared Catbird (nominate *melanotis*: Plate 25). White-eared Catbird (Plate 24) may be locally sympatric, but has no black on ear-coverts and has very different call. Inhabits hill forest on Misool I and Arfak Mts in the Vogelkop up to at least 1,700m. Kumawa Mts population apparently referable to this form, while birds of Fakfak and Wandammen Peninsula mountains are currently of unknown affinity but may likewise belong to this form. Breeding very poorly known or unknown for all the NG taxa of former Spotted Catbird complex, and no details known for present taxon. Does not make any kind of stage or bower.

c   Head of *arfakianus*.

Huon Catbird *Ailuroedus (melanotis) astigmaticus*   Map and text page 336

29cm. Sexes alike. Monotypic. Crown blackish with a few small and narrow thin white lines rather than spots, and dark blackish collar with abundant pale buff spotting. Partially sympatric Tan-capped Catbird (Plate 24) has a tan cap, extensive white on ear-coverts, very different call. Inhabits montane forest in Huon Peninsula up to at least 1,800m. Breeding very poorly known or unknown for all NG taxa of former Spotted Catbird complex. Males with enlarged testes in Oct, but nothing else as yet recorded; presumably much as Spotted Catbird in its nesting behaviour. Does not make any kind of stage or bower.

d   Head of *astigmaticus*.

Black-eared Catbird *Ailuroedus melanotis*   Map and text page 332

e   Head of *melanotis* for comparison.

Black-capped Catbird

Arfak Catbird

a

c

Huon Catbird

Northern Catbird

d

b

e

PLATE 27: GENUS *SCENOPOEETES* – TOOTH-BILLED BOWERBIRD

Tooth-billed Bowerbird *Scenopoeetes dentirostris*   Map and text page 340

27cm. Sexes alike, but female has yellowish to pinkish-flesh (not blackish) mouth coloration. Monotypic. Spotted Catbird (Plate 25) is of similar size and shape but is green, with white spots beneath, and has a pale bill. Endemic to lower montane tropical rainforests of far North Queensland, from near Townsville to Mt Lewis massif north-west of Cairns. Occurs from about 600–900m (rarely to 1,200m), but has been known to wander to the coast during times of drought, with some winter dispersal to lower altitudes around 350m. It is likely that female alone incubates and performs nest duties, but behaviour still not well known. Nest a frail dish of thin twigs, clutch size one or two eggs; nestling period not known. No bower as such, but male makes unique stage of upturned leaves on forest floor.

a   Juvenile.

b   Adult.

c   Male and female at stage-type bower.

Tooth-billed Bowerbird

Archbold's Bowerbird *Archboldia papuensis*

Map and text page 343

35–37cm, female slightly smaller. Two subspecies. Mountains of west-central NG and (race *sanfordi*) east-central NG; status in Star Mts and Border Ranges not known. The relatively short bill and smaller size readily distinguish this species from sicklebills or astrapias, while black male Loria's Satinbird (Plate 40) is quite differently shaped, smaller and shorter-tailed. Much more elongated and less rotund than MacGregor's Bowerbird (Plate 29), with longer slightly notched tail and a curiously flattened head shape. Archbold's Bowerbird is found mainly at 2,300–2,900m, but ranges from 1,750–3,660m, the highest altitude for any member of the family. Occurs in frost-pockets in high montane mossy Antarctic beech (*Nothofagus*) forest with coniferous (*Podocarpus*) tree species in canopy, and understorey including pandanus (*Pandanus*), umbrella trees (*Schefflera*) and scrambling bamboo. Nominate race of WP occurs in subalpine habitat. Large and highly distinctive mat bower decorated with beetle elytra, snail shells and fruits, often embellished with King of Saxony head plumes. Classified as Near Threatened by BirdLife International in the IUCN Red List of Threatened Species.

a   **Adult male *sanfordi* at bower**, with King of Saxony plumes.

b   **Immature**.

c   **Female-type plumage**.

d   **Adult male nominate**.

a

Archbold's Bowerbird

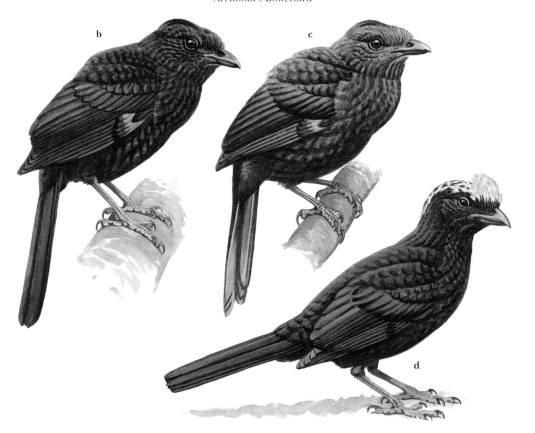

b   c

d

MacGregor's Bowerbird *Amblyornis macgregoriae*

Map and text page 346

26cm, male with orange crest usually not erect. Two similar subspecies. Narrowly sympatric Streaked Bowerbird differs from MacGregor's in that both sexes are streaked buffy ochraceous on throat and breast, but would be hard to separate from it in bad light or high in a tall tree. The bower morphology, however, is totally distinct, and if a bird is seen at or near the bower this is a good clue to its identity. A more problematic confusion species is female/immature Crested Satinbird (Plate 41), which has a similar stocky structure and quite short bill, and is olive-brown above; the underparts of this species have a much more pronounced yellowish cast, the crown is more olive, and it is slightly smaller and has smaller thinner bill which is brownish or bluish-grey, not black. Black Pitohui (*Melanorectes nigrescens*) has a larger and heavier, slightly hooked bill, much more rufous coloration below, and often a much greyer head. Inhabits mountain forests throughout most of NG (except the Vogelkop and Wandammen Mts) from Western Ranges and Border Ranges east to Hindenburg Mts of extreme west PNG, and presumably farther east to Strickland R gorge and Eastern Range and the mountains of Southeast Peninsula east to Milne Bay, also Adelbert Mts and Mt Bosavi. Occurs mostly at 1,600–3,300m, with lowest recorded altitude around 1,050m. Resident, although immatures and females may descend to lower elevations in winter months. Polygynous, promiscuous male seasonally decorates complex maypole bower; female builds and attends nest alone. Nesting poorly known, but single-egg clutch typical of this genus. Distinctive stick-tower bower with mossy circular rim.

a **Male at bower**.

b **Adult male**.

c **Adult female-like plumage**.

Huon Bowerbird *Amblyornis germana*

Map and text page 350

26cm. Formerly classed as part of MacGregor's Bowerbird complex, but split on basis of distinctive bower location and morphology, smaller size with shorter crest and wings, and genetic divergence. Monotypic. The only montane bowerbird on Huon Peninsula, where no satinbirds present to cause possible confusion. Black Pitohui (*Melanorectes nigrescens*) has larger and heavier, slightly hooked bill, and much more rufous coloration below, often with much greyer head. Occurs in montane forests of Huon Peninsula mountains, including Rawlinson Range and Saruwaged Mts, at 1,660–2,940m. Nesting little known; clutch size one. Bower siting distinctive, being not on or near ridgetop but lower down, and with perimeter of sticks (not moss), and tower more like that of Streaked Bowerbird (Plate 31).

d **Adult male**.

e **Adult female-type plumage**.

f **Male at bower**.

a

MacGregor's Bowerbird

b

c

Huon Bowerbird

d

e

f

Vogelkop Bowerbird *Amblyornis inornata* Map and text page 355

25cm. Sexes alike. Monotypic, but plain-plumaged and sexually monochromatic *Amblyornis* populations from the Kumawa and Fakfak Mts (south Bird's Neck region) are of uncertain taxonomic status. Apart from female Black Pitohui (Shrike-thrush) (*Melanorectes nigrescens*), no other major confusion species occurs here; female of the local race of Black Pitohui has a heavy hook-tipped bill and greyish cap. Occurs in the Arfak and Tamrau Mts of the Vogelkop and on Onin Peninsula (Fakfak), Bomberai Peninsula (Kumawa) and Wandammen Mts, in north-west NG. Inhabits rainforest with canopy 25–30m tall and with emergents of *Agathis labillarderi* and *Araucaria cunninghamii* at 1,000–2,075m, mainly 1,200–2,000m. The maypole bowers of the Fakfak and Kumawa populations are more like the bower of *A. macgregoriae* than those of *A. inornata*, but lack the mossy rim or parapet. The bower type from the Tamrau, Arfak and Wandammen Mts is one of the most complex structures made by a bird, an extraordinary piece of avian architecture resembling the hut-like bower of the Streaked Bowerbird, with a roof of epiphytic-orchid stems or sometimes sticks or fern fronds. Polygynous, promiscuous male seasonally decorates complex maypole bower; female builds and attends nest alone. Nesting biology little known; single-egg clutch as usual for the genus. Traditional bower sites are located on ridge spines and slopes.

a   **Adult at hut-style maypole bower** (from Arfak Mts).

b   **Adult**.

c   **Adult at different-style hut/maypole bower** (from Fakfak Mts).

Vogelkop Bowerbird

Streaked Bowerbird *Amblyornis subalaris*

Map and text page 353

24cm. Sexually dimorphic. Monotypic. Very similar to MacGregor's Bowerbird (Plate 29), with which it has very limited local sympatry in south-east PNG. Streaked Bowerbird is slightly smaller, the crest of male is shorter, and both sexes show indistinct streaking on throat and breast. Female or immature Crested Satinbird (Plate 41) has similar stocky structure and quite short bill, and is olive-brown above, but its underparts have a much more pronounced yellowish cast, the crown is more olive, and it is slightly smaller with a smaller, thinner brownish or bluish-grey (not black) bill. Black Pitohui (Shrike-thrush) (*Melanorectes nigrescens*) has a larger and heavier slightly hooked bill, much more rufous coloration below, and often a much greyer head. Streaked Bowerbird is a restricted-range species of the lower montane forest of the eastern sector of the Southeast Peninsula of PNG, occurring in the southern watershed from the Angabanga R (north-west of Port Moresby) east to Milne Bay and in the northern watershed from Mt Suckling east to the mountains of Milne Bay Province. Exact range still undetermined, and the most westerly limits uncertain. Inhabits primary forest and taller secondary forest with *Lithocarpus* and *Castanopsis* oaks in lower montane zone at about 650–1,300m, occasionally to 1,500m. Within its limited range it is very locally sympatric with MacGregor's Bowerbird, which occupies forest at higher elevation. Polygynous, promiscuous male seasonally decorates complex maypole bower; female builds and attends nest alone. Nest and eggs poorly known, but single-egg clutch as usual with this genus. Distinctive hut-style bower resembles that of Vogelkop Bowerbird (Plate 30), one of the most complex of the family.

a   **Adult male at bower.**

b   **Immature.**

c   **Adult male.**

Yellow-fronted Bowerbird *Amblyornis flavifrons*

Map and text page 359

24cm. Sexually dimorphic. Monotypic. No other *Amblyornis* bowerbird occurs in the range of this species, male of which should be unmistakable. Female may be confused with Little Shrike-thrush (*Colluricincla megarhyncha*), which is paler and more slender in build and has a more slender bill; Crested Satinbird (*Cnemophilus macgregorii*) is not found in same region. This bowerbird occurs in Foya (Gauttier) Mts of north-west WP between 940m and 2,000m, adult males rarely below 1,600m and bowers at 1,650–1,800m; females, subadult males and immatures tend to inhabit the lower altitudes. It lives in montane moss forest dominated by *Araucaria*, southern beech (*Nothofagus*), *Podocarpus*, and *Lithocarpus* oaks. Polygynous, promiscuous male seasonally decorates maypole bower. Nesting biology unknown. Bower a simple maypole type, with basal circular moss mat of about 1m in diameter which lacks elevated perimeter rim of the bower mat of MacGregor's Bowerbird (Plate 29).

d   **Adult female** *flavifrons.*

e   **Adult male.**

f   **Adult male at bower.**

Streaked Bowerbird

Yellow-fronted Bowerbird

Golden Bowerbird *Prionodura newtoniana*

Map and text page 361

23–25cm. Sexually dimorphic. Monotypic. The smallest bowerbird. Sympatric Golden Whistler (*Pachycephala pectoralis*) male has yellow underparts, but easily separated by its distinctive white chin and throat and black pectoral band and head; female similar in size but grey-brown above and greyish-buff below, the nominate race (found in same habitat as Golden Bowerbird) showing pale yellow undertail-coverts. Bower's (*Colluricincla boweri*) and Little Shrike-thrushes (*C. megarhyncha*) are far more rufous or greyish above than female and immature Golden Bowerbird, and rusty below. Restricted-range endemic, occurs only in uplands of the wet tropics of north-east Queensland from Thornton Range and Mt Windsor Tableland southwards through tablelands area to the Seaview–Paluma Range north of Townsville. Inhabits upper montane tropical rainforest with a marked wet season, from 600–1,500m. Polygynous, promiscuous male attends bower display site. Nesting behaviour well studied, females alone construct nests; clutch one to three eggs, incubation 21–23 days, nestling period 17–20 days. Male constructs distinctive complex twin-maypole bower.

a   **Male and female at twin-maypole bower**.

b   **Immature**.

c   **Female**.

d   **Adult male**.

Golden Bowerbird

PLATE 33: GENUS *PTILONORHYNCHUS* – SATIN BOWERBIRD

Satin Bowerbird *Ptilonorhynchus violaceus*   Map and text page 365

24–35cm. Sexually dimorphic. Two subspecies. Distinctive. Occurs in a broad band of coastal eastern Australia from the Bunya Mts of South East Queensland south to the Otway Ranges in southern Victoria, occurring from sea level to about 1,000m. There is also an isolated population (race *minor*) in the uplands of the wet tropics in North Queensland from Cooktown south through the tablelands to the Paluma Range near Townsville, occurring above about 500m in tall wetter forests and woodlands and nearby open areas. This species shows a strong preference for forest edges, as well as adjacent tall sclerophyll woodlands with a sapling understorey. Winter flocks occur also in more open habitats such as parks, fruit orchards and gardens. Some may move to lower altitudes in winter, vacating the higher levels. Nesting biology well known, clutch one to three (usually two) eggs. Avenue-bower structure also well studied; this is the most familiar of the family, with several well-known public sites.

a   **Adult male nominate race at bower.**

b   **Adult female**.

c   **Subadult male**.

Masked Bowerbird *Sericulus aureus* Map and text page 371

24–25cm. Sexually dimorphic. Monotypic. Distinctive and unlikely to be confused. Golden Myna (*Mino anais*) differs in having a black mantle and a black patch on nape, plus yellow eye and bill, and a white wing patch in flight. Flame Bowerbird overlaps marginally in range in south-west WP, but male of that species lacks the black face of present species, and female is less mottled on chin, throat and upper breast (see below). Inhabits canopy of hill forest at 850–1,400m in west and north NG: mountains of the Vogelkop (Bird's Head), Bird's Neck (mountains of Wandammen Peninsula and likely to occur in Fakfak Mts), north scarp of Western, Border and Eastern Ranges (Weyland Mts east to Jimi R), Foja Mts, North Coastal Range, Torricelli Mts and Prince Alexander Mts. Nest and eggs unknown. Simple bower of avenue type, much as with congeners, but display very little known.

a   **Subadult male inspecting bower.**

b   **Adult female.**

c   **Adult male.**

Flame Bowerbird *Sericulus ardens* Map and text page 373

25–26cm. Sexually dimorphic. Monotypic. No confusion species in range except for a marginal overlap with Masked Bowerbird in south-west WP, the latter readily told by the black face (male) or more mottling on throat (female). Inhabits south NG from Wataikwa R to Mimika R and upper Noord R–Endrich R east to upper Fly R, Strickland R and Nomad R, and to near Ludesa Mission (Mt Bosavi); also in Tarara–Morehead area and inland from Merauke (between Kumbe and Merauke R). Inhabits lowland and foothill rainforest and also tall secondary forest, including mosaic-type habitat where some trees have been cleared, also savanna in Trans-Fly, mainly beneath paperbarks (*Melaleuca*). Nest and eggs unknown. Simple bower of avenue type, distinctive display now on film.

d   **Subadult male inspecting bower.**

e   **Adult male.**

f   **Adult female.**

Masked Bowerbird

Flame Bowerbird

Fire-maned Bowerbird *Sericulus bakeri*

Map and text page 375

27cm. Sexually dimorphic. Monotypic. The male is a very striking bird, perhaps momentarily reminiscent of Golden Myna (*Mino anais*), and appearing quite myna-like in its structure as well. It shows a broad yellow band across the flight-feathers when seen from below. The female is distinct, the only similarly sized sympatric species having barred underparts being Magnificent Bird of Paradise, which has much darker upperparts, a blue-grey bill and a flattened head shape. Inhabits interior of lower montane forest in the Adelbert Mts of north-east NG north-north-west of Madang, in a narrow elevational band between 900m and 1,450m. Breeding biology little known; only one nest described, eggs and incubation/fledging details unknown. Builds a small avenue bower. Classified as Vulnerable by BirdLife International in the IUCN Red List of Threatened Species.

a   **Female.**

b   **Male.**

c   **Adult male at bower.**

Regent Bowerbird *Sericulus chrysocephalus*

Map and text page 377

24–25cm. Sexually dimorphic, female slightly larger. Monotypic. The male is distinctive; the female is less obvious but still distinctive, and unlikely to cause confusion. Endemic to the coastal ranges and foothills of CE Australia, one population in the Connors and Clarke Ranges of the Eungella Plateau inland from Mackay, then a gap across the Fitzroy R basin inland of Rockhampton before the main distribution, which extends through South East Queensland and NSW to just north of Sydney. Inhabits subtropical rainforest, associated sclerophyll woodland, and more open habitats including cultivated country and urban gardens. Occurs from sea level to 900m, altitudinal limits varying across range. The polygynous, promiscuous male builds and decorates a small avenue bower; female builds and attends the nest alone. Nesting biology fairly well known; clutch one to three (usually two) eggs, but details of incubation and fledging periods lacking.

d   **Subadult male.**

e   **Female.**

f   **Adult male.**

g   **Male at avenue bower.**

h   **Immature male.**

Fire-maned Bowerbird

Regent Bowerbird

Great Bowerbird *Chlamydera nuchalis*

Map and text page 380

32–37cm. Slight sexual dimorphism. Two subspecies. The largest of all bowerbirds, and quite distinctive. Fawn-breasted Bowerbird (Plate 39) is considerably smaller and much less grey-brown, with rufous on underparts, and overlaps only on Cape York; Spotted Bowerbird (Plate 37) is smaller still, and much more richly coloured, with large buffy spots above. Present species is endemic to tropical Australia in North Queensland, the north of NT and northernmost WA, ranging from sea level to around 600m, in areas with annual rainfall between 500mm and 1,500mm. Inhabits the edges of rainforest, vine thickets, riparian woodlands, open savanna, eucalypt and *Melaleuca* woodlands, also suburban gardens and mangroves. Polygynous, promiscuous male builds and decorates elaborate avenue bower; female builds and attends nest alone. Non-territorial except for defence of bower sites by male, as is usual in the family. Nesting biology well known, and large avenue bowers often locally available for viewing.

a   **Adult male at bower**.

b   **Female-type plumage** *nuchalis*.

c   **Subadult male**.

d   **Adult male**.

Great Bowerbird

PLATE 37: GENUS *CHLAMYDERA* 2

Western Bowerbird *Chlamydera guttata*   Map and text page 385

24–28cm. Slight sexual dimorphism. Two subspecies. A distinctive species. Occurs in a broad band in two seemingly disjunct regions of mid-western and central lowland Australia, from Everard Ranges of north SA to about 300km north-east of Alice Springs, and extending west to the Pilbara in the north and Leonora in the south of WA, with a gap apparently in centre of this range (absent from much of Simpson Desert). This species and Spotted Bowerbird are allopatric, both reportedly absent from a band of about 100km between 137° and 138° east, at inland extremities of their respective ranges. It was believed that present species' range was coincident with that of the Rock Fig (*Ficus platypoda*), but it has subsequently been found to occur beyond that into the south-east interior of WA. Occurs near gum-fringed creeks, in wooded savanna and dense vegetation such as *Casuarina* and *Acacia* in rocky areas and near water sources such as bores and dams, as well as gardens, parks, camping sites and picnic areas. Breeding behaviour not well known, appears similar to that of Spotted Bowerbird. Nest-building and care of young entirely by female, as is usual with the bowerbirds; clutch usually two eggs, incubation and fledging periods unknown. Builds an avenue bower decorated with green or white objects. Restricted-range smaller subspecies *carteri* of WA classified by federal authorities in Australia as Near Threatened.

a   **Adult male.**

b   **Adult male at bower.**

c   **Adult female.**

d   **Immature.**

Spotted Bowerbird *Chlamydera maculata*   Map and text page 387

28–31cm. Sexes alike. Monotypic. A distinctive species. Widely distributed in the arid and semi-arid zones of east Australia from about Rockhampton, Queensland, west to the Georgina R and Eyre Creek catchment, extending south through central NSW. Now extinct in SA, where formerly occurred along Murray R; decline due to extensive land modification and clearance. This species shows a strong preference for riverine vegetation within open eucalypt savanna, also areas of the much-reduced Brigalow (*Acacia harpophylla*) habitat in north-east Queensland. Bower sites seem often to be within fruit-bearing or thorny habitats such as Wild Lemon (*Canthium oleifolium*) and Currant Bush (*Apophyllum anomalum*) in NSW. Polygynous, promiscuous male builds and decorates avenue bower; female builds and attends nest alone. Non-territorial except for defence of bower sites by male, as is usual in the family. Nesting biology fairly well known; clutch one to three eggs, usually two.

e   **Adult male at bower.**

f   **Adult male.**

g   **Adult female.**

h   **Immature.**

Western Bowerbird

Spotted Bowerbird

Yellow-breasted Bowerbird *Chlamydera lauterbachi*   Map and text page 390

26.5–29cm. Sexes similar, female slightly duller. Two subspecies. Fawn-breasted Bowerbird (Plate 39), sympatric with present secies in a few areas at upper altitudinal limit of its range, is a slightly larger and less gracile species with rufous on underparts, entirely lacking the rich yellow coloration. Nominate subspecies occurs in Ramu valley east to the north scarp of the Huon Peninsula, also perhaps the eastern Sepik and Baiyer R; subspecies *uniformis* locally distributed in the west and central ranges of NG. Localities include Digul R, Puwani R, the north scarp of Bewani Mts, the middle Sepik–Wahgi valley, Sepik–Wahgi Divide, Minj, Awande, Asaro valley and Mt Rondon. The two subspecies appear to meet in the Baiyer valley. Polygynous, promiscuous male builds and decorates distinctive elaborate avenue bower; female builds and attends nest alone. Non-territorial except for defence of bower sites by male, as is usual in the family. Clutch one egg; no information on incubation and nestling periods.

a   **Adult male at H-shape avenue bower.**

b   **Adult male race *uniformis*.**

c   **Adult female *uniformis*.**

d   **Adult male nominate race with rusty cap.**

Yellow-breasted Bowerbird

Fawn-breasted Bowerbird *Chlamydera cerviniventris* Map and text page 394

28–30cm. Sexes alike, female slightly smaller. Monotypic. A distinctive species; has very local range overlap with Yellow-breasted Bowerbird (Plate 38), which is bright yellow (not rufous) below. Largely confined to coastal areas of east NG, west to Jayapura in north and, in south, to the Trans-Fly (Dolak I east to Wasur NP, extending north to the Bian Lakes); in PNG ranges from Bensbach east to the Aramia Range and north to Lake Daviumbu, and is locally common around Port Moresby, and also at Alotau, in Milne Bay Province. Some isolated and poorly known populations live in the Vogelkop (the Ransiki and Kebar valley), and a few interior populations exist in mountains of the Huon and Southeast Peninsula. Some of the upland populations may involve colonisation of areas now modified by agriculture. In Queensland, it is local around the coastal lowlands of northern Cape York south to about Silver Plains (east of Coen). The Action Plan for Australian Birds (2010) upgraded the conservation status of the Fawn-breasted Bowerbird from Least Concern to Near Threatened in Queensland. The polygynous, promiscuous male builds and decorates his avenue bower, while the female alone builds and attends the nest, as is usual with the family. Breeding biology quite well known; clutch one egg, sometimes two.

a   Adult male.

b   Immature.

c   Male and female at avenue bower.

146

Fawn-breasted Bowerbird

Formerly considered a distinct subfamily Cnemophilinae within the birds of paradise, the satinbirds were shown by Cracraft & Feinstein (2000) to be only distantly related to the latter. Despite the superficial similarity of the species, they were always considered somewhat anomalous members of the family Paradisaeidae.

Loria's Satinbird *Cnemophilus loriae* Map and text page 312

22cm. Sexually dimorphic. Now considered monotypic. This species and Crested Satinbird (Plate 41) are very similar in shape, and separating the females and immatures is sometimes difficult, although female-type Crested lacks the olive-green plumage of Loria's Satinbird and is rather more yellowish on underparts. MacGregor's Bowerbird (Plate 29) is another pitfall, but this is a much darker brown bird with no hint of green, and has a more robust black bill. The male Black Pitohui (Shrike-thrush) (*Melanorectes nigrescens*) is smaller and not so dumpy, and unlikely to be seen perched high up in a tree, being a shy dweller of the lower strata of the forest. Loria's Satinbird is found in montane forests, forest edge and second growth of the central ranges of NG, from Weyland Mts of WP to the Owen Stanley Range, at 1,500–3,000m but mostly 2,000–2,400m; absent from the Vogelkop and the Huon Peninsula. Breeding biology very poorly known.

a **Juvenile**.

b **Female**.

c **Male**.

d **Subadult male**.

Yellow-breasted Satinbird *Loboparadisea sericea* Map and text page 318

17cm. Sexually dimorphic, female slightly heavier. Two subspecies. A unique and distinctive species; Crested (Plate 41) and Loria's Satinbirds lack both the enlarged wattles of the male and the yellow underparts, rump and lower back, and are quite differently coloured above. Endemic to the mountains of NG, from the Weyland Mts through the Snow Mts, Victor Emanuel and Kubor Range eastwards to the Kuper Range north of Wau and Bulolo, but very patchy in occurrence. Appears to be absent from the Vogelkop, Huon Peninsula and south-east PNG. Confined to lower montane and mid-montane forest at 625–2,000m, mainly above 1,200m and seemingly mostly in the interior of the forest, not at the edges (which are favoured by Crested and Loria's Satinbirds). Breeding biology almost totally unknown. Classified as Near Threatened by BirdLife International in the IUCN Red List of Threatened Species.

e **Adult male nominate**.

f **Adult male *aurora***.

g **Female**.

h **Immature**.

Loria's Satinbird

Yellow-breasted Satinbird

Crested Satinbird *Cnemophilus macgregorii*
Map and text page 315

24cm. Sexually dimorphic. Two subspecies, treated by BirdLife International/HBW as species. Males are unmistakable in a good view, but care is needed to separate females and immatures from similar plumages of Loria's Satinbird, or even more so from female-plumaged MacGregor's Bowerbird (Plate 29). Loria's Satinbird (Plate 40) is a much greener bird than the Crested, but often has a browny-yellow patch on the secondaries and can be quite yellowish-brown below, leading to a tricky identification problem unless seen well. The bowerbird is even more problematic, but tends to feed higher in trees and has a slightly different head shape with a stouter blackish bill, and duller and darker brown plumage, especially on the underparts, which lack the yellowish tinge often found in Crested Satinbirds. Bill coloration is not a useful field character as both Loria's and Crested Satinbirds are dark-billed, but bill shape of the bowerbird is different. A grey-phase juvenile would be very difficult to identify, but this is seldom seen in either of the *Cnemophilus* species. Endemic to the highest mountains of the eastern half of the central ranges, extending into WP in the Star Mts and with recent sightings from north of Lake Habbema in upper Ibele valley, an offshoot of the Baliem valley. Absent from Huon Peninsula, with race *sanguineus* (Red Satinbird) in Southeast Peninsula. Builds a domed nest, incubation by female alone, hints of territorial behaviour by males, but breeding biology remains remarkably little known. This species was for long thought to be either a bowerbird or a member of an anomalous subfamily of birds of paradise.

a   **Immature male**.

b   **Adult male race** *sanguineus* (Red Satinbird).

c   **Female**.

d   **Adult male nominate race** (Yellow Satinbird).

e   **Immature** *sanguineus.*

MacGregor's Honeyeater *Macgregoria pulchra*
Map and text page 320

35–40cm. Sexes alike, female slightly smaller. Two similar subspecies. A relict species of the highest altitudes of NG; absent from huge tracts of seemingly suitable country. Occurs in small remnant populations in the west Snow and Oranje Mts (Mt Carstenz, Mt Wilhelmina, Carstenz Meadow, Kemabu Plateau, Lake Habbema) and the Star Mts at Mt Capella and Dokfuma, then disjunctly again in south-east PNG in the Wharton (Mt Albert Edward, Mt Strong/Chapman and Batchelor) and Owen Stanley Ranges (Mt Scratchley–English Peaks, Mt Victoria). Formerly known as MacGregor's Bird of Paradise, but now shown to be a honeyeater. Classified as Vulnerable by BirdLife International in the IUCN Red List of Threatened Species.

f   **Adults** at rest and in flight.

Crested Satinbird

MacGregor's Honeyeater

BIRDS OF PARADISE

FAMILY PARADISAEIDAE

Genus *Lycocorax*

The genus *Lycocorax*, here split into two species, is often seen as being most closely allied to the manucodes, with which it shares some characteristics. Frith & Beehler (1998) regard it as sister form to the manucodes and hypothesise that, during a period of lower sea levels, an ancestral form to both lineages ranged throughout New Guinea and the northern Moluccas, subsequently diverging into the distinct modern taxa. Further genetic analysis is desirable to help to elucidate its true position within the paradisaeid lineage, as a case could be made for according family status to the two species of this genus and the manucodes. It was originally thought to be a crow (Corvidae), and was then reassigned to the birds of paradise, where it is the earliest known offshoot from the paradisaeid family tree, dating back approximately 17 million years in the Miocene period. Wallace thought that it was an odd starling (Sturnidae), but recent genetic work indicates that it is a basal member of the birds of paradise family.

Etymology *Lycocorax* is derived from the Greek *lycos*, a wolf, and *korax*, a raven.

HALMAHERA PARADISE-CROW
Lycocorax pyrrhopterus   Plate 1

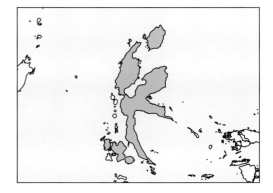

C[orvus] pyrrhopterus Bonaparte, 1850, *Conspectus Genera Avium* 1, 384. Gilolo and Halmahera.
Other English names Brown-winged Bird of Paradise, Brown-winged Paradise-Crow, Paradise-Crow, Mortay Island Paradise-Crow, Silky Crow
Etymology *Pyrrhopterus* means red-winged, from the Greek *pyrrhos*, a flame or the colour red, and *pteros*, wing.

FIELD IDENTIFICATION The Halmahera Paradise-crow is a restricted-range, rather crow-like forest dweller from Halmahera and Morotai, in the northern Moluccas of Indonesia, which behaves somewhat like a manucode. The stout dark bill, black plumage lacking iridescence and cinnamon-brown wings, red eye and pale wing-flash make it readily identifiable. Along with the Obi Paradise-crow and Wallace's Standardwing, this is the most westerly member of the bird of paradise family. **SIMILAR SPECIES** The pale wing-flash should make the Halmahera Paradise-crow readily identifiable. The similar Obi Paradise-crow, previously treated as a race of this species, is allopatric and restricted to Obi and Bisa. The Torresian Crow (*Corvus orru*) is readily told by the adults having pale eyes and lacking the wing-flash and brown wings of the present species, as well as by its quite different voice. The Long-billed Crow (*C. validus*) has a distinctive long bill and short tail and also lacks the brownish appearance and pale wing-flash of the paradise-crows.

RANGE Endemic to the North Moluccan (Maluku) islands of Halmahera, Bacan (Batjan), Kasiruta, Morotai and Rau. **MOVEMENTS** This species is resident.

DESCRIPTION This species, like the manucodes, lacks the sexual dimorphism of many of the paradisaeids. *Adult* Length 42cm. The head is a slightly glossy dark dusky brown, the upperparts somewhat paler and glossed dull blue-grey, and the brown wings show variably white concealed bases of primaries (obvious in flight) extending along the inner half of the primaries and narrowing on the bases of the secondaries. The underparts are a paler brown, darker on the belly, vent and undertail-coverts, and glossed blue-green. *Iris* deep red in male, the female dull crimson (Hartert on birds from Batjan, in Iredale 1950); *bill* blackish; *legs and feet* black. *Immature* appears duller and paler below, perhaps with paler brown wings, but this plumage is not well known. *Juvenile* is much browner, virtually lacking gloss, and has brownish downy underparts, but again very little known.

TAXONOMY AND GEOGRAPHICAL VARIATION Polytypic, with two races.
1. *L. p. pyrrhopterus* (Bonaparte, 1850) of Halmahera, Bacan and Kasiruta.
2. *L. p. morotensis* Schlegel, 1863, of Morotai and Rau Islands. Larger than the nominate subspecies and *L. obiensis*; resembles nominate but paler, slightly more brownish above and darker below, with more white on inner half of the primaries narrowing to bases of the secondaries.

The form **obiensis**, previously considered a subspecies of *L. pyrrhopterus*, is here treated as a full species, the Obi Paradise-crow: see below.

VOCALISATIONS AND OTHER SOUNDS The nominate race has calls distinct from those of the species found on Obi (Lambert 1994; Bishop in F&B). Heinrich (1956) recorded a short hoarse dog-like bark, and Coates & Bishop (1997) noted a loud *wu-wnk*, the first note short and non-musical, the second nasal and hollow and repeated about every ten seconds, perhaps as an advertising call. In flight, the wings produce a rustling noise. Ripley reported a harsh *tschak.........tschak*, which is probably the double calls noted above, but also heard a male give a deep low *om* seemingly as a display call. Coates & Bishop record a contact note as a single quite loud, dry and slightly upslurred guttural bark or croak, *ekk*, repeated at short intervals. A short frog-like croak or bark, *ech*, is also mentioned, but this may be a different transcription of the previous note. On the recordings at xeno-canto XC216743 (by P. Åberg) and XC151734 (by M. Nelson) the call sounds like a raspy, dry *wek*. A recording by J. Mittermeier at the Macaulay Library at the Cornell Lab of Ornithology ML182282 presents what are presumed to be alarm calls, in which the loud raspy *wek* is repeated and then run into a short series. Birds of the taxon *morotensis* give a raspy disyllabic *wek wek* call, as at xeno-canto XC195362 (by M. Thibault); there is also an emphatic ringing hollow *wook* as at XC196589 (by F. Lambert), which may be the song.

HABITAT The nominate form is a forest dweller on Halmahera, Bacan and Kasiruta, being found also in gardens, forest edge and regrowth, from sea level to at least 750m. It prefers the taller trees of the forest interior and is not typically seen in the more open agricultural areas dominated by scrub (Lambert 1994). Seldom found in swamp forest or mangroves, it sometimes occurs in coconut plantations and orchards, and frequents the mid-level to the canopy of the vegetation. Bernstein (in Iredale 1950) described the race *morotensis* as inhabiting thick woods and living in trees at a moderate height.

HABITS Some accounts suggest that this is a shy bird keeping in dense foliage and difficult to see well (Heinrich 1956; Lambert 1994), but others have found it to be much more obliging on Halmahera. Coates & Bishop (1997) report it as being inconspicuous and secretive, usually found singly or in twos, but sometimes in groups of up to five, and rarely with mixed-species foraging flocks or pigeons. The flight is direct and moderately fast, with frequent glides and an audible rustling noise, through the forest mid-storey and just above the canopy (Coates & Bishop 1997). The tail is usually held slightly fanned both when the species is perched and in flight, and the birds move from tree to tree in a follow-my-leader fashion, with a pause between each bird's departure (Frith & Poulsen 1999). Bernstein (in Iredale 1950) described the race *morotensis* as being a very difficult bird to see, and best observed by waiting near a fruiting tree. The birds come in by gliding from the top of one tree to the crown of another, pausing briefly on the outer branches before diving into the thickest of the foliage.

FOOD AND FORAGING This species appears to be primarily frugivorous, but it will also take arthropods, foraging in the dense foliage of the mid-storey to canopy (Coates & Bishop 1997). Heinrich (1956) noted the fruit of Pinang palm *Pinanga* to be favoured and found a seed of this palm in a nest containing a nestling.

BREEDING BEHAVIOUR Breeding is known between December and June, with eggs laid over the same period (F&B). Two nests on Halmahera each contained a single young in the second half of April. As the sexes are similar, it is likely that the species is monogamous, both parents attending the nest, but this is not known for certain. **COURTSHIP** Nothing definitely noted, although some of the described calls may be a feature of courtship display. **NEST** On Halmahera, at the end of April, Heinrich (1956) found a deep cup-shaped nest suspended like an oriole's and containing a single chick. Poulsen (in F&B) found one, also with a single chick, on 19th April at 750m. Halmahera nests are described as a large basin-shaped structure made of roots and moss and lined with soft chips of wood (Shaw-Mayer in Parker 1963). They are sited 4–15m above ground, usually in a fork. Old nests found on Halmahera were sited well below the forest canopy, hanging in the fork of a branch (Poulsen in F&B). **EGGS & NESTLINGS** The clutch appears to consist of just a single egg, and this may become stained by the wood chips of the nest lining (Parker 1963). Egg colour is greyish-white or pinkish-stone with a fine erratic tracery of brown, violet and pale lilac wiggly lines, some thicker than others. Average size (sample of six) is 37.8 × 27.2mm (Frith in F&B 1998). Nothing is known of the incubation period and nestling development other than that the nestling is naked and pale in colour (see photo in F&B). It is not certain that both sexes attend the nest as is usual with paradisaeids lacking or with little sexual dimorphism, but it is worth noting that with both *Paradigalla* species (sexes very similar) only the female attends.

HYBRIDS As may be expected, there are no records of hybrids with other species.

MOULT Museum specimens show evidence of moult throughout the year.

STATUS AND CONSERVATION Classified as Least Concern by BirdLife International. This paradise-crow is described by Coates & Bishop (1997) as locally common on Halmahera. Surveys on Halmahera in 1994 found an average of 0.2 birds per hectare in both sedimentary and limestone habitats, but 0.5/ha in montane habitats, while Frith and Poulsen found an average of 0.4/ha in logged lowland forest (F&B). The population is believed to be stable, but, as ever, logging on these small islands can be expected to have negative impacts. There are no records of the species being used in cultural adornment.

Halmahera Paradise Crow, sexes similar, Sidangoli, Halmahera (*Gareth Knass*).

Halmahera Paradise Crow, sexes similar, Weda Resort area, Halmahera (*Huang Kuo-wei*).

OBI PARADISE-CROW
Lycocorax obiensis   Plate 1

Lycocorax obiensis Bernstein, 1865, *Journal für Ornithologie* **12**, 410. Obi Island.
Etymology The name *obiensis* is derived from the island name, Obi.

FIELD IDENTIFICATION The Obi Paradise-crow is a restricted-range, rather crow-like forest dweller from Obi and Bisa, in the northern Moluccas of Indonesia, which behaves somewhat like a manucode. The stout dark bill, black plumage lacking iridescence and cinnamon-brown wings, red eye and pale wing-flash make it readily identifiable. Compared with the Halmahera Paradise-crow, it is a larger and rather distinctive darker taxon with more blue-black on the crown and upper tail; the bases of the primaries show only a trace of white. Lambert (1994) reported Obi birds as having a whitish streak behind the eye, and several observers remark that in appearance it resembles a manucode more than it does the neighbouring taxon, and is one of the more visible and vocal species of forest and degraded forest areas. Along with the Halmahera Paradise-crow and Wallace's Standardwing, this is the most westerly of the bird of paradise family. **SIMILAR SPECIES** The pale wing-flash, which is much smaller than that of the Halmahera species, should make the Obi Paradise-crow readily identifiable. The Torresian Crow (*Corvus orru*) is readily told by the adults' pale eyes and the lack of the wing-flash and brown wings of the present species, as well as by its quite different voice. The Long-billed Crow (*C. validus*) has a distinctive long bill and short tail and also lacks the brownish appearance and pale wing-flash of the paradise-crows.

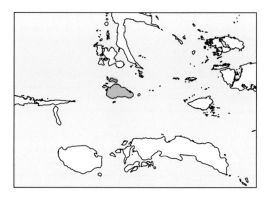

RANGE Confined to the North Moluccan (Maluku) islands of Obi and Bisa. **MOVEMENTS** Resident.

DESCRIPTION This species, like the manucodes, lacks the sexual dimorphism of most of the paradisaeids. *Adult* Length 42cm. The head is a slightly glossy dark dusky brown, the upperparts paler with dull blue-grey gloss, and the wings brown with concealed white bases to the primaries (barely visible in flight). The underparts are a paler brown, browner on the belly, vent and undertail-coverts, and glossed blue-green. *Iris* deep red in male; *bill* blackish; *legs and feet* black. *Immature* appears duller and paler below, perhaps with paler brown wings, but this plumage is not well known. *Juvenile* is much browner, virtually lacking gloss, and has brownish downy underparts, but again very little known.

TAXONOMY AND GEOGRAPHICAL VARIATION Monotypic. A large and rather distinctive taxon, darker and with more blue-black on the crown and upper tail than the Halmahera species. Taxon *obiensis* is quite distinct from *L. pyrrhopterus* in both morphology and calls, and is here given species status as first suggested by Cracraft (1992), and later by del Hoyo & Collar (2016), summarised as follows: Previously treated as conspecific with *L. pyrrhopterus*, although its distinctiveness and potential for separation have for long been noted. Differs in its darker, glossier, greener body plumage; dark brown *vs* pale brown outer webs of flight-feathers, creating no contrast with body; longer, thicker feathering on the lores and superciliary area, resembling heavy 'eyebrows'; slightly larger size, with longer wing (mean male 204mm *vs* 190mm); strikingly different behaviour ('demonstrative' *vs* 'secretive'); more catholic habitat selection; seemingly unique calls.

VOCALISATIONS AND OTHER SOUNDS Bernstein (1864) reported a *wuhk* or *wunk* note, which may be the one described by Lambert (1994) as a loud *who-up*, often answered with a loud *hwhoo*. Linsley (1995) noted a similar but rising *woo-up*, often followed by a rasping *rek* or *krek*. Coates & Bishop report *obiensis* from Bisa as having a very distinctive advertisement call, a loud, deep, resonant and disyllabic *OO-lip* or a deeper *OO-lee*, with a sound quality like that of a woodwind instrument; the first syllable is deep and descending and the second high-pitched and rising. A recording on xeno-canto at XC68604 (by F. Lambert) documents a rapid incisive disyllabic clicking *tiktik*, while another, at XC68602, has a hollow resonant *whoo* sometimes followed by a deeper-toned *whuuo*, which may be the song. The wings produce a rustling noise in flight.

HABITAT This species differs somewhat from the paradise-crow of Halmahera and Morotai in that it frequents all wooded habitats, including agricultural land with scattered scrub and trees, from sea level to around 1,220m on Obi.

HABITS The Obi Paradise-crow is reportedly much more demonstrative than the Halmahera species. It gives loud calls, and is one of the most vocal and visible species on Obi (Bishop in F&B; Lambert 1994; Mittermeier *et al.* 2015). The flight is direct and moderately fast, with frequent glides and an audible rustling wing noise, through the forest mid-storey and just above the canopy (Coates & Bishop 1997). This species usually holds its tail slightly fanned both when perched and in flight, and the birds move from tree to tree in a follow-my-leader fashion, with a pause before each individual's departure (Frith & Poulsen 1999).

FOOD AND FORAGING Although it appears to be primarily frugivorous, this species will also take arthropods, foraging in the dense mid-storey to canopy foliage (Coates & Bishop 1997). Linsley (1995) reported these birds as frequenting fruiting trees along with imperial-pigeons (*Ducula*) and fruit-doves (*Ptilinopus*) on Obi.

BREEDING BEHAVIOUR Breeding is known between December and June, with eggs laid over the same period (F&B). The sexes are similarly plumaged, so it is likely that the species is monogamous, both parents attending the nest, but this requires confirmation. **COURTSHIP** Nothing definitely noted, although some of the described calls may be a feature of courtship display. **NEST** Lambert (1994) found a nest on Obi in primary forest at 500m. It was at the edge of the lower canopy and seemingly attached by pieces of vine, so that it hung below the branches; the sitting adult was obvious, but the nest itself was well camouflaged by mosses on the trunk and by branches. Nests from Obi are large cup-shaped structures of dead leaves and moss, or of lightweight woody vines, with dry wood chips between the lining and the outer vines, resembling manucode nests (Frith in F&B). They are built 4–15m above ground, usually in a tree fork. **EGGS & NESTLINGS** The clutch appears to consist of a single egg, which may become stained by the wood chips of the nest lining (Parker 1963). Eggs from Obi are larger than those of the Halmahera Paradise-crow, but otherwise similar in colour and pattern. Nothing is known of the incubation period and nestling development of this species. Whether or not both sexes attend the nest, as is usual among paradisaeids exhibiting little or no sexual dimorphism, is not known.

HYBRIDS As may be expected, there are no records of this species hybridising with others.

MOULT Museum specimens show evidence of moult throughout the year.

STATUS AND CONSERVATION Classified as Least Concern by BirdLife International. The Obi Paradise-crow is described by Coates & Bishop (1997) as moderately common to common on Obi and Bisa. Linsley (1995) saw up to ten in a day in north-west Obi in December. The population is believed to be stable, but logging on these small islands can be expected to have negative impacts. There are no records of the species being used for cultural adornment.

Obi Paradise Crow, adult male, Obi Island (*James Eaton*).

Genus *Phonygammus*

This is a rather distinctive monospecific genus endemic to Cape York, in North Queensland, and the New Guinea region. It is distinguished from the related species of *Manucodia* by the occipital tufts, which form two horns at the back of the head, the throat and neck hackles and the unique coiled (not looped) trachea situated atop the breast. For the past few decades, this genus has been subsumed in *Manucodia* by numerous authors (e.g. Diamond 1972; Frith & Beehler 1998), but the trend recently, influenced by molecular evidence, has been to regard *Phonygammus* as generically distinct from its sister lineage, *Manucodia* (Cracraft & Feinstein 2000; Irestedt *et al.* 2009). There are major differences in male vocalisations among the different subspecies, which also have both lowland and highland forms, very unusual in a single species and likely to be indicative of cryptic species as yet unrecognised. This requires further genetic testing and fieldwork; the poorly known vocalisations alone would be worthy of detailed study. Tentative ideas for the long overdue break-up of the Trumpet Manucode complex are detailed in the account below.

Etymology *Phonygammus* is derived from the Greek *phone*, a cry or sound, and *gamos*, wedlock or marriage, i.e. a bride of sound (Jobling 1991).

TRUMPET MANUCODE
Phonygammus keraudrenii   Plate 3

Barita Keraudrenii Lesson and Garnot, 1826, *Bulletin des Sciences Naturelles* (Férussae) **8**, 110. Dorey (Manokwari), north-western New Guinea.
Other English names Trumpetbird, Trumpet Bird, Gould's Manucode
Etymology Named for M. Pierre F. Keraudren, Inspector General of Medical Services.

The species as currently recognised contains some nine subspecies, occurring from sea level to high in the mountains and with varying degrees of morphological and vocal differentiation. It is likely that there are several distinct species involved, but museum material is limited for some taxa, and further field studies are required to help in the unravelling of this fascinating taxonomic problem. DNA and vocal studies would prove particularly rewarding in what is clearly a complex assemblage, likely to have lowland, insular and highland species. The description of up to nine races of this species (Cracraft 1992) was based on size, head-plume length, and the degree of iridescent coloration of the plumage, but many of the characters seem minor or variable.

FIELD IDENTIFICATION This is a typical manucode with broad wings and red eyes, but is shorter-tailed than others. In addition, it has distinctive occipital tufts that appear as spikes on the head, and long lanceolate neck feathers. **SIMILAR SPECIES** The Trumpet Manucode is readily told from other manucodes by the spiky head tufts, the long neck feathering and the shorter tail. The voice is also distinctive. For distinctions from other medium-sized black forest birds see the Glossy-mantled Manucode account.

RANGE This species is endemic to the Australasian region, where it is the most widespread of all the birds of paradise. The Trumpet Manucode is widely but patchily distributed in the lowlands and some hill forest regions of New Guinea, though very local in the northern watershed, where it appears to be absent from the lowlands. It is not recorded from the Cyclops Mts or from the northern slopes of the Snow Mts; it is, however, known from the outlying mountains of the Van Rees, Foya (Gauttier), Bewani, Torricelli and Adelbert Ranges, although its occurrence on the Huon Peninsula is not as yet confirmed. It is found also on some of the northern Torres Strait islands belonging to PNG (Boigu and Saibai), in the Aru Islands, and in the D'Entrecasteaux Archipelago on the islands of Goodenough, Fergusson and Normanby (Coates 1990). It is present also (race *gouldii*) in the relict rainforest on far northern Cape York Peninsula, in Queensland, extending south as far as north of Coen in the west and to Silver Plains in the east (Frith 1994a). In addition, it is known from Albany and Mai Islands, in the southern Torres Strait adjacent to the tip of Cape York Peninsula (Schodde & Mason 1999). There are no confirmed records of this manucode's occurrence or wandering farther south than this, any such claims being likely the result of misidentifications of other red-eyed black species such as koels or drongos (see below), and no extralimital records were made during the current Birds Australia Atlas project (A. Silcocks *in litt.*). **MOVEMENTS** This species is sedentary. The Australian subspecies *gouldii* was reported as being a migrant in some of the early Australian literature. Banfield (in Iredale 1950), from Dunk Island (off Townsville, well south of the species' range), describes the birds as arriving in September with a disyllabic loud rich call and giving rise to an onomatopoeic Aboriginal name, *Calloo-calloo*. It is clear, however, that he was in fact describing the Australian Koel (*Eudynamys orientalis cyanocephalus*), which does indeed arrive then and has a call very like its local name.

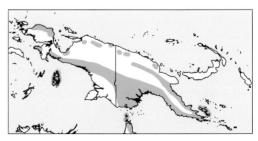

156

DESCRIPTION Sexes are similar but distinguishable. *Adult Male* averages 31cm in length and weighs 130–240g, but size varies with taxon. The plumage is very similar to that of the Glossy-mantled Manucode, being a glossy blue-black, with distinctive elongated occipital tufts and elongated pointed feathers or hackles on nape and neck, and with variable colour iridescence. The convoluted trachea in males is made up of five or six coils atop the breastbone. *Iris* red; *bill* blackish, *mouth-lining* blackish; *legs and feet* blackish. *Adult Female* is about 28cm long, with weight 130–182g, size varying with taxon. Compared with the males it appears slightly duller in coloration below, with shorter head and neck hackles. Soft parts as those of males. Trachea not looped in females or juveniles. *Immature* is dull blackish, especially on the head, with only slight variable iridescence on upperparts and breast, but this may vary with taxon and remains poorly known. Time taken to attain adult plumage uncertain. *Juvenile* is blackish with variable bluish gloss but no purple on upperparts, but again this may vary with taxon. *Iris* greyish-brown, becoming redder with age.

TAXONOMY AND GEOGRAPHICAL VARIATION Polytypic, with a wide yet highly fragmented range. I predict that this complex of taxa, because of the great variation in vocalisations, habitat and morphology, will prove to consist of several biological species once more information becomes available. The matter is greatly complicated by poorly understood sex- and age-related differences, and large areas where taxa may or may not be allopatric are still poorly known. Much work remains to be done. Cracraft (1992) recognised nine seemingly largely allopatric species under the phylogenetic species concept. The present arrangement of six subspecies in four groups represents basically a work in progress. Gilliard (1969) provided a useful initial overview. Here, following Beehler & Pratt (2016), we synonymise the more recently named taxa *aruensis* and *diamondi* with *jamesii*, and also synonymise *mayri* with *purpureoviolaceus* as per Cracraft (1992), while Beehler & Pratt (2016) further synonymise *adelberti* with *neumanni*. No hybrids are known. The nominate taxon and *jamesii* seem to occur in both lowland and highland areas, and the little-known *neumanni* and *purpureoviolaceus* may be similar. In support of the distinctness of the lowland and upland species, Beehler (in Beehler & Pratt 2016) noted that the upland form exists in the upper Biaru River at 1,000m, a day's walk from the alluvial lowlands of the Lakekamu basin, where the lowland form is found (in the same watershed, with no distributional barriers separating the two).

Lowland and hill-forest taxa: Lowland Trumpet Manucode

1. *Manucodia keraudrenii keraudrenii* (Lesson and Garnot, 1826). Far western West Papua on the Vogelkop and Onin Peninsula and the Weyland Mts; altitudinal limits uncertain but seems to be hill and lowland forest, as well as up to 1,600m. A small, steely greenish-blue form with short occipital tufts and purplish wings and tail.
2. *M. k. jamesii* Sharpe, 1877, *Catalogue of Birds in the British Museum*, **3**, 181. Type locality Aleya, Hall Sound, south-eastern New Guinea. Ranges from the Mimika River, West Papua, along the southern-watershed lowland forest east to at least Hall Sound, Central Province (and conceivably eastwards as far as Milne Bay?); also on islands of Boigu and Saibai, in northern Torres Strait. In Western Province this taxon is found below about 400m, but may occur higher elsewhere. Larger than the nominate form and with longer occipital tufts; throat and breast dark metallic blue, washed greenish, and lacking the purple of the nominate (F&B).

   Synonyms Aru Islands taxon *M. k. aruensis* (Cracraft 1992). *Cladistics* **8**, 10. Type locality Wanoem Bay, Kobror Island, Aru Islands. Reported as differing from adjacent *jamesii* in being dark dull metallic purplish-blue, strongly suffused purple, with less green than either that or nominate form; upperparts suffused purple, with lanceolate head feathering a deeper and purpler cobalt-blue. Occipital tufts much shorter than on *jamesii*, and similar to nominate, but synonymised with *jamesii* by Beehler & Pratt (2016) as reported differences are quite minor. Similarly, *M. k. diamondi* (Cracraft, 1992), *Cladistics* **8**, 12. Type locality Awande, near Okapa, Eastern Highlands, PNG. This form, too, is synonymised by Beehler & Pratt. Occurs in hill forest at 1,000–2,000m on the southern watershed of the eastern highlands, perhaps to the Kratke Range. It was described as being similar to *purpureoviolaceus* (see below) but having back, wings and tail with strong violet-purple sheen, breast and belly dark metallic blue with little or no violet-purple, and lanceolate head feathers bluish-green. This taxon also has the longest occipital tufts.

Montane taxa: Montane Trumpet Manucode

3. *M. k. purpureoviolaceus* Meyer, 1885, *Zeitschrift für die gesammte Ornithologie* **2**, 375, pl. 15. Astrolabe Mts. Synonym *M. k. mayri* (Greenway, 1942). A montane form known from the Astrolabe and Owen Stanley Mts of south-eastern New Guinea and the montane areas of Wau, in Morobe Province, north-east New Guinea. Found predominantly in hill forest from 950m to 2,000m, but occurs also at lower altitudes; presumably this taxon at Varirata NP at around 600m, but this requires confirmation. Similar to the nominate form but larger, with back, breast and belly an intense purplish-violet rather than steel-green, and has feathers of hindcrown, nape and sides of neck very elongate and glossed greenish blue-purple, and occipital tufts far longer.
4. *M. k. neumanni* Reichenow, 1918, *Journal für Ornithologie* **66**, 438. Type locality Lordberg. Known from the Lordberg, Sepik Mts, and the Jimi and Baiyer River montane regions up to about 1,500m. Reported from the Bewani Mts (via Pratt in F&B). Sightings of Trumpet Manucodes in the lowlands at Karawari, on the Sepik (Price & Nielsen 1991), and from Aiome at 100m, near the Schrader Mts (M. LeCroy in F&B), may be referable to this form, but photos or specimens are desirable since these are non-montane areas. Similar to the nominate but with the lower back, wings, rump and tail dark

bluish-violet (not greenish), sharply demarcated from metallic dark blue of crown, nape and upper back; breast and belly are dark metallic blue, and the head has short occipital tufts.

M. k. adelberti (Gilliard and LeCroy, 1967), *Bulletin of the American Museum of Natural History* **138**, 72. Type locality Nawawu, Adelbert Mts. This form, found on the Adelbert Range up to about 1,200m, is reported as having short occipital tufts and wings and tail green (not blue to purple), with dark metallic blue feathers on hindcrown, nape, sides of neck and throat. Differences from *neumanni*, however, appear minor and it is now synonymised by Beehler & Pratt (2016), a treatment which seems appropriate.

Insular taxon: Island Trumpet Manucode

5. *M. k. hunsteini* (Sharpe, 1882), *Journal of the Linnean Society*, London **16**, 442. Type locality 'East Cape, New Guinea' (Milne Bay Province), in error for Normanby Island, D'Entrecasteaux Archipelago. An island taxon, found on Normanby, Fergusson and Goodenough in the D'Entrecasteaux Archipelago, where it is sparsely distributed in the upland forests. This is the largest taxon, resembling the nominate but with longer, more greenish and less bluish occipital tufts; back, rump and tail dark bluish-purple (less steel-green), and throat dark metallic greenish. The central tail feathers are somewhat twisted, reminiscent of *Manucodia comrii* (Beehler & Pratt 2016).

Australian (Cape York) taxon: Australian Trumpet Manucode

6. *M. k. gouldii* Gray, 1859, *Proceedings of the Zoological Society of London*, note, p.158. Type locality Cape York. Known from the northern part of Cape York Peninsula, Queensland, discontinuously in relict lowland rainforest from Pajinka/Bamaga, at the tip of the cape, south to just north of Coen in the west and at Iron Range as far south as Silver Plains in the east. Like the nominate race, but with iridescence much more green and less purple, especially on the wings and tail. The occipital tufts are narrower and more pointed and much longer than those of the nominate, and the relative tail length also is much longer. These morphological differences, plus the distinct vocalisations, make this taxon a good candidate for species status. It was formerly known as Gould's Manucode.

VOCALISATIONS AND OTHER SOUNDS The vocals of this manucode are extremely diverse, with some calls peculiar to specific taxa, and some sex-specific, and these will have a bearing on the taxonomy in due course. The **nominate race** in the Arfaks has a loud, quite deep-toned, harsh *skowup* call resembling that of the Yellow-faced Myna (*Mino dumontii*); see xeno-canto XC62467 (by A. Spencer) and XC23208 (by G. Wagner). Around Kiunga, in Western Province, birds of **race jamesii** have a loud gruff, coughing, almost retching *hwaaugh* call, which cannot under any circumstances be likened to the sound of a trumpet! This seems to be the advertising song, as individuals have been seen with drooped or raised wings and erect neck hackles when

making it; see XC186079 (by P. Gregory) and XC268361 (by F. Lambert). Glossy-mantled Manucodes here also have a quite loud *awwwoooo* call, but it is not so harsh and retching as the note of the Trumpet Manucode. Rand (1938) noted a loud, harsh squawk reminiscent of a call of the Yellow-faced Myna but given by *jamesii* at the Wassi Kussa River. Calls of similar type are known from males and females of both lowland and upland taxa. There is also a beautiful melodious, fluty whistled *ee-loo* call, quite unlike the normal retching call and not given very often (it is in fact heard only in the hills near Tabubil at about 400m, and never at Kiunga in the lowlands [pers. obs.]). This call may be the one transcribed by Coates as a high and clear *whuu-oh*, which Beehler (in F&B) further describes as being a quite harsh *kee-yowk* when very close, but the *ee-loo* call at close quarters is still melodious and fluty, and it may well be that we are describing different vocalisations here. The montane taxa have not hitherto been reported as giving any call analogous to this. Presumed *purpureoviolaceus* individuals from Varirata have a loud ringing and very melodious *wow-oo*; see XC187310 (by P. Gregory). Male nuptial song is described as a hollow, low, reverberating and tremulous single note, *wodldldldldldldl*, at even pitch, sometimes preceded by a harsh *kaup* note. Coates (1990) gives a comprehensive list of vocalisations, presumably from the Port Moresby area and thus presumably of this taxon, including a short loud, belching bark, *chaw, chak* or *chow*, a longer *cheow* or *krouw*, given with tail spread and wings drooped, and also a short, nasal, rather Yellow-faced Myna-like *wha*, which is often given with the barking *chow*-type series; another call is a high-pitched hollow hoot, *whOu* or *whOOUw*, reminiscent of a Raggiana Bird of Paradise call and given by the bird when leaning forward with body feathers raised; also a vibrating mammal-like *crrraaa*. The alarm call is a short, deep, harsh nasal note given 2–8 times in a variable-speed series, *chu-chu-chu* or *ch-ch-ch*, and repeated many times. The display call is a tremulous low, resonant, throaty growl, gradually increasing in volume and then dying away, *rrrRRROw*, and resembling a snore if heard from some distance away. Trumpet Manucodes of the **race gouldii** from Cape York have a harsh *wak-chaw* note (see XC158164, by H. Krajenbrink) and a loud, resonant, deep-toned *skaw*, as well as a loud ringing, almost woodwind-like elongated *choooo* note and a mewing *skeeow*, which may be given as a reply to the deeper note (F. van Gessel 2013, *Birds of Australia* mp3 sound collection; and H. Plowright 2006, *A Field Guide to Australian Birdsong* CD9). Presumed male *gouldii* gives repeated *skowlp* notes for periods of c.30 seconds or more during the early breeding season, possibly as nest-site advertisement. A nesting pair made a strong, brief, 'creaking-door' alarm call, rather like that of a cicada (Cicadidae) but deeper (Frith *et al.* 2017a). These calls are quite varied, seemingly more so than those of other taxa and rather distinct from the New Guinea ones, but the complex remains woefully understudied vocally. Pratt (in F&B) notes form *adelberti* as making a forceful, rasping *ha...*, a clear and ringing *who!* and, as display song, a whistled far-carrying note like that of a peacock (*Pavo*), and a displaying male of this taxon gave a *hrarrr* call followed by a pigeon-like, bell-like musical, downslurred, slightly nasal *oo* or *kyeu* (Opit in F&B); but Beehler & Pratt (2016) synonymise this taxon with *neumanni*. There is no mention of the species making a rustling sound in flight as some of the *Manucodia* species do. This would be worth verifying in field observations.

HABITAT This manucode occupies various forest types, including monsoon and gallery forest and rainforest, and forest edge, as well as wooded areas in gardens, ranging from sea level to 2,000m. See also under TAXONOMY AND GEOGRAPHICAL VARIATION (above). It occurs in the mountains mainly from above 900m to 1,800m in the north. In the Ok Tedi area of Western Province it is very uncommon and found only up to about 400m, not occurring higher in the hills, where the Crinkle-collared is the dominant species (Gregory 1995).

HABITS Throughout its range the species is sympatric with up to two other manucodes and various other paradisaeids, such as the King Bird of Paradise, Magnificent Bird of Paradise and both Growling and Magnificent Riflebirds, Greater and Raggiana Birds of Paradise and the Twelve-wired Bird of Paradise in Western Province, PNG. It will join brown-and-black bird flocks in the lowlands, along with Papuan Babblers (*Garritornis isidorei*) and Rusty Shrike-thrush (Pitohui) (*Pseudorectes ferrugineus*), and with paradisaeids such as King, Greater and Raggiana Birds of Paradise in the Kiunga area. Trumpet Manucodes are unobtrusive and easily overlooked, but have the habit (race *jamesii*, at least) of perching out on tall dead snags for minutes at a time, when they may be readily seen. Near Kiunga, in Western Province, PNG, they have been observed with Glossy-mantled Manucodes in the same tree without any interactions. They frequent the lower canopy and middle levels and are seldom seen low down. Bell (1982a) estimated densities in the Port Moresby area to be five birds per 10ha (presumably the taxon *purpureoviolaceus*), while on Mount Missim the home range was estimated at about 200ha (Beehler 1985). The now synonymised taxon *aruensis* was reported as being a nest parasite (W. Frost in Rothschild 1930), laying its eggs in nests of Greater Birds of Paradise in the Aru Islands, but this was considered incorrect by Rand (1938) and Gilliard (1969), and it does seem highly implausible. Maybe a nest was taken over by the manucodes or, perhaps more likely, a misidentification of Eastern Koel (*Eudynamys orientalis*) was involved. Other such reports are based on doubtful identifications of single eggs found in Raggiana nests and in the nest of a Blue-grey Robin (*Peneothello cyanus*) (Coates 1990). The voice of these manucodes is one of the best means of finding them, but this varies greatly across the range of the species. Fruiting trees are another likely place to see them, and in such places they may gather in groups of up to six or seven individuals, although they are more usually seen singly or in twos. Trumpet Manucodes at Iron Range (*gouldii*) are shy and difficult to find unless they are calling or at fruiting trees. There is no evidence of any of the taxa being other than sedentary, moving only within the local area to fruit sources. Thorpe (in Iredale 1950) records the

manucode as being 'plentiful in dense bushes close to Somerset (on Cape York), usually met with in pairs high up in fruit of berry bearing trees' with other (frugivorous) species. 'The males make a very loud and deep guttural note, and it astonished me that such a comparatively small bird could make so much noise.' They 'evince more curiosity than timidity and I have frequently shot them by trying to imitate their notes, or by making a strange noise, they would hop down from branch to branch in an inquisitive kind of way.... The bird is particularly fond of the fruit of a certain species of fig; but the stomachs I examined contained insects, as well as fruits and berries of various kinds.'

FOOD AND FORAGING Figs are very important for this species, accounting for up to 80% of the diet at Mount Missim (Beehler 1985), with just 1% insect prey, although snails, too, have been recorded in the diet of far western birds of the nominate race (Hoogerwerf 1971). Trumpet Manucodes glean insects from branches, epiphytes and leaf debris (Bell 1983), and will come to forest pools to drink during dry weather (Coates 1990). This species forages mostly in the middle to upper levels of forest, often as a pair, but aggregations of up to ten will gather in fruiting trees. They glean fruits in much the same manner as do much smaller insectivorous species, clambering about in the upper-level foliage, and they also feed the fledglings on great quantities of figs. At one nest, fig pieces accounted for at least 90% of the nestling diet. Unlike most birds of paradise, they do not use their feet to manipulate fruit or prey, instead simply plucking or gleaning it.

BREEDING BEHAVIOUR The breeding season is at least May–January over the entire range. Males with enlarged gonads are recorded in February–April and July–December and females in June and October–November; displays have been noted in May and August, and calls thought indicative of display are produced in all months. Copulation has been observed in mid-November. The species is monogamous, some partners remaining together for more than one season. It is reported as being non-territorial, but the existence of regularly utilised songposts may contradict this. **COURTSHIP/DISPLAY** Individuals display solitarily on mid-sized tree branches, often bare ones, and in auditory contact with neighbouring conspecifics, responding to the display calls of the latter. The display may involve a chase, but seems usually to consist of the calling bird, presumably male, bowing forwards on his calling perch with wings drooped and half-opened, fanned tail raised, neck hackles erected and very prominent, the occipital tufts raised and very obvious, and emitting a variety of loud harsh calls. In a higher-intensity version, the wings are fanned behind the back. Intervals between displays may be quite short, about 15 seconds, but more often they appear to last for several minutes and such displays may continue for up to half an hour. Display calls seem most frequent in middle to late afternoon in Western Province, and are heard in all months of the year. **NEST** The nest is an open basin-shaped structure of curly vine tendrils, lined with fine creeping-fern tendrils, sparser than in those of congeners and lacking larger leaves or pieces of dry wood. It is suspended 6–27m above ground in a horizontally forking tree branch, around which the nest-rim vine tendrils are entwined. The taxon *gouldii* in Cape York is known to nest near breeding Black Butcherbirds (*Melloria quoyi*), presumably deriving some form of protection from the presence of this aggressive and predatory species. **EGGS & NESTLINGS** The clutch consists of 1–2 eggs, pale reddish-buff and with rich red-brown, purplish and grey streaks and blotches; their size averages 36 × 25mm, but race unspecified (presumably *gouldii*). Incubation is performed by both parents, and both also brood and feed the nestlings; no information is available on duration of incubation and nestling periods.

HYBRIDS No hybrids are known.

MOULT Specimens show evidence of active moult throughout the year, but mostly in January–July.

STATUS AND CONSERVATION Classified as Least Concern by BirdLife International. With the expected break-up into several species, however, some populations will need to be re-evaluated on the grounds of small range and habitat degradation. This manucode is sparse in occurrence in the somewhat fragmented and limited rainforest habitat in Cape York (*gouldii*), while the D'Entrecasteaux taxon (*hunsteini*) also has a rather small and fragmented range. The species as currently recognised has a very large, albeit unusually disjunct range, with a population trend that appears to be stable, and hence it does not approach the thresholds for Vulnerable under the range-size and population-trend criteria (>30% decline over ten years or three generations). The population size has

not been quantified, but it is believed not to approach the thresholds for Vulnerable under the population-size criterion (<10,000 mature individuals and with a continuing decline estimated to be >10% in ten years or three generations, or with a specified population structure). For these reasons, the species is evaluated as Least Concern.

Trumpet Manucode, sexes similar, calling, Iron Range NP, Queensland, Australia (*Jun Matsui*).

Trumpet Manucode, sexes similar, Iron Range NP, Queensland, Australia (*Jun Matsui*).

Genus *Manucodia*

The genus *Manucodia* is endemic to the New Guinea region and is characterised by the lack of sexual dimorphism, black plumage glossed purple or blue, a red iris, strong legs and powerful feet, an elongated and looped (not coiled) trachea for making loud vocalisations, and a specialised frugivorous diet primarily of figs. Unlike most birds of paradise, the members of this genus do not use their feet to manipulate fruit or prey, instead simply plucking or gleaning items. All species are inhabitants of forest, and shy but quite vocal. The morphology of the looped trachea needs further analysis in respect of some taxa, and may aid in species identification (e.g. *jobiensis vs. chalybatus*). The taxonomy is still unsettled within the Glossy-mantled Manucode group, and here we recognise the Tagula Manucode as a new species, *Manucodia atra*, on the basis of the very distinctive (and only recently discovered) vocalisations and some morphological differences.

Etymology Manucode is derived from the Malay words *manuk dewata*, meaning 'bird of the gods', assigned to this group by Brisson, 1760 (Jobling 1991). Quite why these birds should be associated with gods is uncertain.

GLOSSY-MANTLED MANUCODE
Manucodia ater   Plate 2

Phonygamma ater Lesson, 1830, *Voyage of the Coquille, Zoology*, **1**, 638. Dorey (Manokwari), north-western New Guinea.
Synonym: *Manucodia ater subalter* Rothschild and Hartert, 1929, *Bull. Brit. Orn. Club* **49**: 110. Dobo, Aru Islands.
Other English names Glossy Manucode, Black Manucode
Etymology The Latin word *ater* means black or matt black, the feminine form being *atra*.

FIELD IDENTIFICATION This is a large, glossy black forest-dwelling bird with a fairly long tail and a red eye. The flight is loose and floppy, dipping and uncoordinated-looking on rather broad wings.
SIMILAR SPECIES Readily confused with the Crinkle-collared Manucode, which is sympatric over much of the range. The Crinkle-collared has distinctive ruffled velvet-looking feathering on the neck and breast, and prominent eyebrow ridges giving a more bumpy head shape than the smoother profile of the Glossy-mantled Manucode. Viewed head-on, the Crinkle-collared shows a V-shaped indentation in the crown outline, with a bump at each side (almost as if somebody has hit it and caused two swellings), quite lacking in its congener. The tail length of the Glossy-mantled may be longer in some areas, but there is much overlap in Western Province, PNG, and this can be no more than a guideline. The blue-green iridescent gloss on the neck and breast may be useful in distinguishing it from the Crinkle-collared in good light as the latter species has a bronzy yellow-green gloss, but this is probably of little use in the field. Vocalisations of the two are, however, quite different, the Glossy-mantled having a rising, high-pitched, almost tuning-fork-like *zheeeee* advertising note and harsh coughing notes, whereas the Crinkle-collared is much more silent but has a distinctive, rather quiet hooting call, as well as a rich incisive *tuk* or *chuk* contact note similar to that of the present species. The little-known Jobi Manucode of northern and western New Guinea is also sympatric and can be a major identification problem with both the Glossy-mantled and the Crinkle-collared Manucodes.

It is a smaller and more compact species with reduced head bumps, giving a more bumpy head profile than Glossy-mantled, and with reduced crinkling on the breast. It may also appear shorter-tailed. The voice is not well known, and there has been much confusion with other manucodes, but a reported hooting song would at least distinguish it from Glossy-mantled. The sympatric Trumpet Manucode is a smaller and shorter-tailed species altogether, with distinctive spiky head plumes (occipital tufts), longer neck feathering and very different vocalisations; the colour of the plumage gloss is very variable and would be of little value as a distinguishing field character. Other black species that may cause confusion include: Black Butcherbird (*Cracticus quoyi*), which has a big blue-grey bill with a dark hooked tip and a dark eye; Papuan Spangled Drongo (*Dicrurus carbonarius*), which is far smaller and has a distinctively forked tail, although it, too, has a red eye; and perhaps even the Metallic Starling (*Aplonis metallica*), which has a crimson eye, metallic plumage and a long diamond-shaped tail, besides being much smaller and having a rapid and direct flight. Males of the Australian Koel (*Eudynamys orientalis cyanocephalus*) have a quite different shape, with longer tail, pointed wings and a whitish bill. Greater Black (*Centropus menbeki*) and Lesser Black Coucals (*C. bernsteini*) also have a long tail and are either very much or considerably larger, the former with a whitish bill and red eye, the latter with a dark bill and eye.

RANGE Endemic to New Guinea, the Aru Islands, the West Papuan islands of Misool, Batanta, Salawati, Waigeo and Gebe, islands off the coast of south-eastern New Guinea (Samarai, Sariba, Mailu and Yule).
MOVEMENTS This species is sedentary.

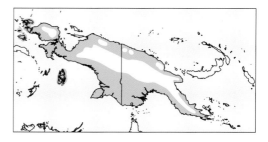

DESCRIPTION A large, black, sexually monomorphic paradisaeid of the lowlands with a long, graduated tail; base of culmen ridge broadened and flattened to a degree intermediate between those of the Jobi and Crinkle-collared Manucodes. **Adult Male** 38–42cm, weight 170–315g. The male's tracheal loop is short, extending less than halfway down the sternum (Beehler & Pratt 2016). The head and neck are blue-black, with iridescent blue-green tipping on smooth (not crinkled) feathering, and may show a plum-purple sheen in some lights; the smooth-plumaged blue-black mantle and the back, rump, upperwing and uppertail have variable iridescent glosses of blue and purple to magenta. The tail may show subtle blackish barring (age-related), and the rectrices have tiny hair-like central points. The feathers of the chin and throat have an iridescent tipping forming an inverted V-shape; the breast is similar in coloration to the upperparts, but may show more blue-green iridescence, more marked on the belly, vent and undertail-coverts. *Iris* is blood-red with a brown inner ring; *bill* dark grey to black; *legs* black. **Adult Female** 33–37cm, weight 155–252g. Averages smaller than the male, with greener and less purple gloss, particularly on the underparts. *Immature* is duller than the adult, with the tail on average slightly shorter. *Iris* orange-yellow to yellowish-orange (changing with age). Rather difficult to differentiate in this plumage from immature *M. chalybatus*. The length of time taken to assume adult plumage is uncertain, but likely to be within two years. *Juvenile* is dull brownish-black, gradually developing the glossy feathers of immature plumage, which initially show more on the upperparts.

TAXONOMY AND GEOGRAPHICAL VARIATION
Monotypic, with proposed subspecies *M. a. subalter* now synonymised owing to its being a curious disjunct assemblage which is biologically inexplicable on current information. Known from the Western Papuan Islands, Aru Islands and the South-east Peninsula of PNG westwards to the Gulf of Papua in the south and, in the north, to the watershed of the Kumusi River, the described form *subalter* was supposed to exhibit major plumage differences at each extremity of the range and to differ in average size (Gilliard 1969; F&B), but Cracraft (1992) could find no diagnosable characters when using the phylogenetic species concept, although this method does not work well with clinal variation. This form was said to be on average more purple and violet in coloration, with oil-green gloss rare on the adults (Gilliard 1969). It averages larger than the 'nominate' form, but the contact zone between the two is little known. The former nominate subspecies *M. a. ater* inhabits the lowlands of the western two-thirds of New Guinea, the eastern extent being not well known, but, according to Gilliard (1969), extending in the north to the Huon Gulf and in the south to the Purari River. The population on Tagula Island is usually classified as *Manucodia ater alter*, but vocal and morphological evidence suggests that treatment as an insular allospecies is appropriate. This treatment is followed here, thereby rendering the Glossy-mantled Manucode monotypic.

VOCALISATIONS AND OTHER SOUNDS The advertising call of what are presumably the males is a distinctive loud, mournful, musical rising *zheeeee*, reminiscent of a tuning fork being struck, and not unlike some notes of Blue Jewel-babblers (*Ptilorrhoa caerulescens*) (Coates 1990; pers. obs.). It is also rather like a longer and stronger, more vigorous version of the advertising note of Loria's Satinbird. It is given throughout the day, but with greatest frequency in the early morning, with a resurgence in the late afternoon. The call is documented on xeno-canto at XC148670 (by P. Gregory) and XC215190 (by P. Åberg). Glossy-mantled Manucodes also utter a very loud, almost yodelling husky *aa-aaawh-oo*, at least in the Kiunga area, and this call may be given by loose assemblages of birds, almost suggestive of some kind of display behaviour. This note is easily dismissed as being an odd call from the Trumpet Manucode, which is probably why it is hitherto unrecorded. The contact note is a simple, quite incisive and deep-toned *tuck* or *tuck tuck*, similar to but somehow simpler and less rich than that of the Crinkle-collared Manucode. A nesting individual gave a loud *ack, ack, ack* when alarmed (Bergman 1961), while Rand (1938) noted a deep *chug* or *chook* from alarmed nesting birds, and an undisturbed nesting individual uttered a low chattering call. Hoogerwerf (1971) noted a subdued *korrrrr-korrrrr*, a soft *kwek*, a hoarse *kek-kek-kek* and a not very loud screeching from Vogelkop birds, and these appear unlike other transcriptions of calls of this species and perhaps suggest a regional or racial dialect. This species seems not very interested in playback as a rule, but Coates (1990) has recorded them as responding to imitations of the advertising call and coming close to a partly concealed observer, which may imply some degree of territoriality. Like its congener the Crinkle-collared Manucode, the Glossy-mantled produces a rustling sound in flight, but this is nothing like so loud as that made by a riflebird or an astrapia.

HABITAT This species inhabits lowland and hill forest, including swamp and monsoon forests, second growth, riparian scrub and even mangroves. It seems to avoid the interior of closed primary forest, where the Crinkle-collared Manucode is predominant. On the mainland of New Guinea it is widespread in the lowlands and hills, ranging from sea level locally in the hills to about 900m. Frith (in F&B) records a specimen from 1,100m in the Adelbert Mts. In the Ok Tedi area of Western Province it is unrecorded above c.400m (pers. obs.), but is very common near sea level at Kiunga in the nearby lowlands of the same province.

HABITS This is often a conspicuous and sometimes fairly confiding species, given to perching out in the open on tall bare limbs. The distinctive tuning-fork-like rising whistled call is a typical sound of the lowland and hill forests. Glossy-mantled Manucodes are generally seen singly or in pairs, or in small parties of up to five individuals. They readily join mixed-species foraging flocks along with pitohuis, babblers, drongos and *Paradisaea* species. The King Bird of Paradise may also join such assemblages, as at Kiunga. At the latter locality the present species is seen in close proximity to both Trumpet Manucodes and Crinkle-collared Manucodes, and sometimes in the same tree as the former. These are active birds with a strange loose, floppy flight. They are often seen dipping across roads and tracks into forest cover, and in flight, as mentioned above, the plumage makes a swishing or rustling sound like a quiet version of a riflebird. They are noisy and quite aggressive, chasing other species from fruit trees. Small parties seem to congregate at times in the forest, giving a loud yodelling, coughing *aa-aaawh-oo* call and flying up into the air like a displaying pigeon, gliding back down from the crest of the rise, which is presumably some form of display behaviour (see under BREEDING BEHAVIOUR, below).

FOOD AND FORAGING Glossy-mantled Manucodes forage for fruit and insects, gleaning from branches and debris accumulations and also from epiphytes. They favour the middle stratum for the most part, but will ascend into the canopy and come down into the shrubbery layer at times. Wallace (1869) recorded it on the Aru Islands as 'a very powerful and active bird; its legs are particularly strong, and it clings suspended to the smaller branches while devouring the fruits on which alone it appears to feed'. Bishop (in F&B) noted a flock of eight individuals at a fruiting tree there in April 1988. Bell (1984) recorded this species as hovering occasionally, and also coming to the ground to forage. Beehler (in F&B) saw one eating the bright orange-red fruits of a *Piper* vine some 10m above ground. These manucodes will also visit flowers, but it is not yet certain what they are taking (F&B).

BREEDING BEHAVIOUR The breeding behaviour of this manucode remains poorly known, but it is evident that, unlike most others in the paradisaeid group except *Lycocorax*, *Phonygammus* and probably other *Manucodia* species, it is monogamous, both sexes nest-building and performing parental duties. It is reported in F&B as being apparently non-territorial, but this is open to question. **DISPLAY** When giving the advertising song, the presumed male raises itself on its legs, stretches its neck and, as it calls, expands the breast plumage, raises the mantle and back feathers, and half-opens and droops the wings, ceasing these actions when the calling finishes (Coates 1990). It calls from high up in tall trees, using bare branches or sometimes hidden among the leaves. Calling may persist for lengthy periods, with one call given every couple of minutes, and the bird is often within auditory and sometimes visual contact of other vocalising birds. The species seems to form clusters around Kiunga, in Western Province, PNG, perhaps in areas of optimal habitat. One strange call is the almost coughing, yodelling *aa-aaawh-oo* mentioned above, which is given from within dense cover at low levels, often with several birds in the area. In a kind of parachute display observed here in July and August, the birds flap slowly up over a clearing or river, wings fully expanded with spread, fingered primaries and deep heavy wingbeats, and the tail also spread and held slightly depressed; they flap heavily and very slowly for a few beats in a shallow downward trajectory before dropping suddenly on opened wings back into

cover. This is quite different from the normal floppy flight and seems to be performed where clusters of individuals occur, and is presumably some kind of display. It is curiously reminiscent of a Marsh Whydah (*Euplectes hartlaubi*) display in Africa. A presumed male was seen to shake its slightly spread wings and tail and momentarily erect its body feathers as a presumed female landed nearby (Rand 1938). Calling birds near Kiunga (presumably males) have been seen to open out the wings like a huge black butterfly, and then hold them partially opened and level with the head while fluffing out the body feathers; this may be an agonistic display directed at other males, as two individuals are often involved or another is calling nearby. Wahlberg (1992) noted similar behaviour but he interpreted it as being between pair-members, despite the second bird giving a higher-pitched version of the first bird's *zheeeee* note, which rather suggests a young male. Display has been noted in January, July, August and September, and it is likely that, somewhere over the species' wide range, it can be seen in all other months. **NESTING** Nesting is recorded from August to March. The nest is built some 4–12m above the ground in small forking horizontal branches of a small tree, which may be in forest edge, savanna, open forest or even mangroves. Some of the nest material is entwined around the branches in order to anchor it more securely (F&B; Coates 1990). The nest itself is a deep bulky cup made from slender soft woody vines, tendrils and twigs, with many dead leaves and a large amount of dry rotten wood built into the base, as with the paradise-crows (*Lycocorax*) and the Crinkle-collared Manucode. A nest measured by Rand (1938) was 250mm in diameter and 140mm deep externally, with respective internal measurements of some 130mm and 100mm; the walls were asymmetric, with a thickness of 60mm at one side and just 20mm at the other, thin enough to see through. **EGGS** Only one clutch is known, the two eggs being of the usual slightly pointed oval shape and largely lacking gloss, coloured pale pinkish with small dark brown dots and larger lavender-grey oblongs, clustered most about the larger end; measurements 32.2 × 24.4mm and 31.5 × 24.4mm (Frith in F&B 1998). Rand observed one bird sitting and believed the incubation period to last 14–18 days, which seems rather brief. He thought that the female alone was incubating, but believed that the male may have been kept away by human disturbance. **NESTLINGS** Both sexes brood the chicks and both feed them by regurgitating fruit (as with paradise-crows and the Trumpet Manucode), as well as removing faecal sacs. Although this feeding behaviour is very poorly known for the rest of the genus *Manucodia*, it is likely to be widespread. Bergman (1961) noted one young still in the nest when at least 19 days old.

HYBRIDS The species is not recorded as hybridising, but such would be extremely difficult to detect within the genus.

MOULT This species appears to moult throughout the year, but with most such activity in December–June.

STATUS AND CONSERVATION

Classified as Least Concern by BirdLife International. The Glossy-mantled Manucode is a wide-ranging and often quite common, if somewhat inconspicuous, species. Once its advertising call is known, it is surprising how frequently it is observed; indeed, it is one of the commonest paradisaeids around Kiunga, with daily counts easily into double figures. The species is frequent in the savanna/rainforest ecotone at Varirata NP, in Central Province, PNG, and Bishop and Diamond (in F&B) recorded it as common and widespread in suitable habitat in the Aru Islands.

Glossy-mantled Manucode, adult with juvenile, October 2018 (*Huang Kuo-wei*).

TAGULA MANUCODE
Manucodia alter   Plate 2

Manucodia ater altera Rothschild and Hartert, 1903, *Novit. Zool.* **10**: 84. Tagula Island, Louisiade Archipelago, south-east Islands.
Etymology The Latin word *ater* means black or matt black, while *alter* means second, next or another.

FIELD IDENTIFICATION This is the only manucode species on the island of Tagula. Monotypic and sexually monomorphic, it is a large, heavy manucode with a heavy bill and a rather long tail. The male's tracheal loop is long, extending to the distal edge of the sternum (Beehler & Pratt 2016). **SIMILAR SPECIES** The only possible confusion would be with the male Eastern Koel (*Eudynamys orientalis*), which has a quite different shape, with longer tail, pointed wings, a whitish bill and much more slender build. Less likely confusion species include the Papuan Spangled Drongo (*Dicrurus carbonarius*), which is far smaller and has a distinctively forked tail, although it, too, has a red eye; and perhaps even the Metallic Starling (*Aplonis metallica*), which has a crimson eye, metallic plumage and a long diamond-shaped tail, besides being much smaller and with a rapid and direct flight.

RANGE This is an uncommon endemic resident of Tagula (Sudest) Island, in the Louisiade Archipelago of PNG. Tagula lies 175 miles (280km) south-east of mainland New Guinea and is the largest island of the

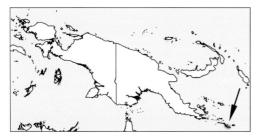

archipelago, measuring 50 by 15 miles (80 by 24km). It is a rugged island and rises to 806m at Mt Riu, with a total land area of 310 square miles (800km²). Much of it is still forested, but coconut plantations exist in the lowlands and lower slopes. **MOVEMENTS** Sedentary.

DESCRIPTION Sexes alike. *Adult* length 43cm, tail 165mm, bill 42mm. Large, with a heavy long bill and a much longer tail than the Glossy-mantled Manucode (latter has bill 39mm, tail 150mm). The bill length of adults shows no overlap with that of Glossy-mantled (F&B). The Tagula Manucode's all-black plumage has a purple gloss on the flanks and belly, with far more violet than is shown by the allopatric Glossy-mantled Manucode. The *iris* is dark, that of a captured female being red; *bill* black; *legs and feet* dark. The female lacked tracheal elongation and weighed over 300g. *Immature* and *juvenile* plumages are undescribed.

TAXONOMY AND GEOGRAPHICAL VARIATION Formerly synonymised within the Glossy-mantled Manucode, this isolated and almost unknown population on Tagula, which also has other little-known endemics, is vocally very distinctive (on a par with the two New Guinea riflebirds), and has some morphological differences. It is treated here as a separate species, as it seems rather distinct. A single-island endemic, it is monotypic.

VOCALISATIONS AND OTHER SOUNDS The calls are very distinctive, and quite unlike those of the Glossy-mantled Manucode. W. Golding has recorded them: there is a deep short sonorous *woo* hoot, and short raspy *kek* notes are given in quite a long series. The *woo* note is like the sound made by a low-pitched tuning fork, with some individual variation in pitch and some higher notes, too, but nothing like as high as those of mainland Glossy-mantled birds. There are no recordings on xeno-canto or the IBC at the time of writing (August 2017). The flight is reported as being noisy, which may refer to its making a rustling sound in flight; this requires confirmation.

HABITAT This species inhabits forest on Tagula.

HABITS W. Golding (*in litt.*) reports that this species flies noisily and hops on strong legs from perch to perch. It appears to have poor flying skills and is reluctant to cross forest gaps and open areas. Most are usually encountered while skulking through more shaded, thickly vegetated areas of contiguous forest and forested gullies on the island, including adjoining sago stands. It is typically located as it moves through the understorey, well-developed shrub layers and vine tangles. Generally found singly or in presumed pairs, it is often difficult to see clearly, and can be inconspicuous when wary. It frequently falls silent and observes keenly from a distance, peering through small gaps in the foliage. The noisy flight, in addition to agitated *kek* calls, usually confirms the close presence of this species. The metallic *woo* vocalisations are most commonly heard in the pre-dawn and early hours, in the evening and during overcast and rainy periods. Tagula Manucodes react to and often investigate the alarm calls of the endemic Tagula Butcherbird (*Cracticus louisiadensis*).

FOOD AND FORAGING This manucode is presumably largely frugivorous, as are its congeners, but as yet very little is known. It is often seen near figs, a favourite food for this genus.

BREEDING BEHAVIOUR A single female caught in mid-December had a brood-patch. No other information on breeding is available, with courtship/displays, nest and eggs all undescribed.

MOULT Almost unknown, but in mid-December a single female had the central rectrices, body and wing-coverts in moult.

STATUS AND CONSERVATION Likely to be Least Concern but not assessed by BirdLife. This isolated species has a restricted range and may be Data Deficient, but is currently not assessed as it has hitherto been regarded as a subspecies of the Glossy-mantled Manucode. The conservation status of this and of the three other Tagula Island endemics – Tagula Meliphaga (*Meliphaga vicina*), Tagula Butcherbird (*Cracticus louisiadensis*) and Tagula White-eye (*Zosterops meeki*) – is similarly little known. Forest clearance could conceivably be a problem on the island.

Tagula Manucode, female, Tagula Island, PNG (*Will Goulding*).

JOBI MANUCODE
Manucodia jobiensis   Plate 2

Manucodia jobiensis Salvadori, 1876, *Ann. Mus. Civ. Genova* (1875) **7**: 969. Wonapi, Yapen Island, Geelvink Bay.
Synonym: *Manucodia rubiensis* A. B. Meyer, 1885, *Zeitschrift Ges. Orn.* **2**, 374. Rubi, Geelvink Bay.
Other English names Allied Manucode
Etymology Named after Jobi (Yapen) Island, in Geelvink Bay, from where the type specimen came.

FIELD IDENTIFICATION This is a very tricky species to identify. It is a typical red-eyed, broad-winged manucode and is sympatric with three other manucode species over parts of its range. Identification problems mean that some data may not in fact be referable to this species, and the voice remains remarkably poorly known. **SIMILAR SPECIES** The Jobi Manucode is most similar to the Crinkle-collared Manucode, but much confusion exists and it is likely some individuals have been misidentified, particularly as regards calls. My field experience suggests that this is quite a small, compact species, neater-looking and shorter-tailed than either Glossy-mantled or Crinkle-collared Manucodes. It does seem to have some head bumps (supraorbital tufts), but these are not so obvious as those of the

Crinkle-collared Manucode. The crinkling of the neck and breast feathers is reduced, and it lacks the bronzy yellow-green iridescence evident on museum skins of Crinkle-collared, although this is of doubtful value in the field. The sympatric Trumpet Manucode is a shorter-tailed species with distinctive spiky head plumes (occipital tufts) and longer neck feathering. For other possibly confusing medium-sized to large black forest birds, see under Glossy-mantled Manucode.

RANGE The Jobi Manucode is endemic to the lowlands of northern New Guinea, and in south-western West Papua from Etna Bay eastwards to the Mimika and Setekwa Rivers. In the north it ranges from the south-western shore of Geelvink Bay east to the Ramu River, PNG (F&B). It is found also on Yapen Island, in Geelvink Bay, where it is the only manucode. It is

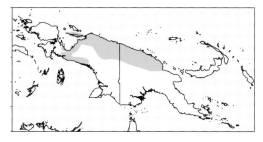

apparently absent from the Vogelkop and the Bird's Neck Peninsula. It has been reported from Kau FR, near Madang, but its presence there requires confirmation as field identification of this species is always somewhat problematic. **MOVEMENTS** Resident, but poorly known.

DESCRIPTION This species is sexually monomorphic, but males are somewhat larger than females. *Adult Male* 34cm, weight 212–257g; *Adult Female* 31cm, weight 150–205g. Its plumage appears less glossy than that of the Glossy-mantled Manucode. The head and neck are blue-black, glossed with purple, and tipped iridescent blue-green. The feathers of the head and neck appear crinkled and, like the underparts, have glossy greenish-blue bars, with velvety-black bars on the mantle. The head bumps may be less obvious than those of the Crinkle-collared Manucode, and the bird appears smaller and more compact, the last admittedly very hard to judge if one is unfamiliar with manucodes. Beehler & Pratt (2016) discovered that the best way of identifying the species in the hand is by the narrow and keeled proximal terminus of the culmen where it meets the skull: this structure in *chalybatus* is widened and rounded, with the culmen flattened and meeting the feathering of the forecrown in a 'U', rather than in the V-shape that typifies *jobiensis*. It is also possible that the structure of the trachea of the male Jobi Manucode, reported as being a single loop, may be distinct, but this requires further research. *Iris* is deep red to orange-yellow, while the *bill* and the *legs and feet* are black. *Immature* is dull sooty black, acquiring gloss as it moults into adult dress. *Juvenile* is downy brownish, the wings and tail becoming blacker.

TAXONOMY AND GEOGRAPHICAL VARIATION Monotypic, although birds from Yapen Island may average larger (Frith & Frith 1997d; Gilliard 1969), and have a more intense purplish suffusion (Cracraft 1992).

VOCALISATIONS AND OTHER SOUNDS This species' voice is remarkably poorly known owing to the identification problems. Beehler *et al.* (1986) noted a series of slowly delivered hollow *hoo* notes, much like that of the Crinkle-collared Manucode and presumably a display call, and a harsh *chig* or *bcheg* call note. C. Grant (in Ogilvie-Grant 1915a) heard a long-drawn moaning note, again perhaps a display song but possibly a misidentification of the Glossy-mantled call. Crinkle-collared-type manucodes of uncertain species at Karawari have head bumps and appear small, yet have a rising tuning-fork-type call very like that of Glossy-mantled Manucodes, thus emphasising the fact that so much remains to be discovered about the vocabulary of the present species. There is no mention in the literature of whether or not the birds make a rustling sound in flight as their congeners do, but this is likely. No recordings on xeno-canto or the IBC at time of writing (August 2017).

HABITAT Lowland rainforest, swamp forest and hill-forest edge. This species occurs from sea level up to about 750m, more usually to 500m, and is replaced at the higher elevations by the Crinkle-collared Manucode.

HABITS Very poorly known owing to identification difficulties. This manucode will join mixed-species foraging flocks along with other paradisaeids and the Rusty Shrike-thrush (Pitohui) (*Pseudorectes ferrugineus*) and/or Papuan Babblers (*Garritornis isidorei*) (Beehler & Beehler 1986). It has been seen perched out on tall dead branches like its congeners, and seems not to be particularly wary. A female was extremely nervous and difficult to approach near its nest (C. Grant in Ogilvie-Grant 1915a).

FOOD AND FORAGING Fruits have been reported in the diet, and this species has been observed to search for insects. It will join the brown-and-black mixed-species foraging flocks typical of its habitat and seems to occur primarily in the middle and lower stages. Beehler (in F&B) noted three individuals foraging for arthropods in the mid-stratum some 15m from the ground.

BREEDING BEHAVIOUR This remains almost completely unknown. It is to be expected that the species will be monogamous, like its congeners. The display behaviour is unknown. **NEST** One nest is recorded, a female with two eggs having been collected at her nest on the Mimika River, West Papua, on 28th December. She was extremely nervous and hard to observe near the nest, which was 2.4m above ground and suspended between two horizontal branches. It was a deep cup-shaped structure made of roots, creepers and leaves. It would be interesting to know if rotten wood was used in the base, as is the odd habit with this species' congeners. **EGGS** Clutch two eggs, pale pinkish, wreathed with small brown dots concentrated at the larger end, and with lavender-grey elongated spots. Size averaged 32 × 24mm.

MOULT Museum specimens indicate that moult occurs in all months of the year.

STATUS AND CONSERVATION Classified as Least Concern by BirdLife International, and with a large range, but this species could almost come into the category of Data Deficient as very little is known about it, and the difficulties of field identification compound the problem. It is reported as being locally common and is unlikely to be under any threat in the remote areas which it inhabits.

Jobi Manucode, sexes similar, PNG (*Tim Laman*).

CRINKLE-COLLARED MANUCODE
Manucodia chalybatus   Plate 2

Paradisaea Chalybata Pennant, 1781, *Spec. Faunula Indica*, in Forster's *Indian Zoology*, p.40. New Guinea, restricted to Arfak Mts. Type is an illustration.
Other English names Green-breasted Manucode, Green Manucode, Crinkle-breasted Manucode, Green Paradise Bird
Etymology Latin *chalybeatus/chalybatus* meaning 'steely', presumably a reference to the plumage gloss.

FIELD IDENTIFICATION A large black, red-eyed forest species with broad wings, a longish tail and a loose floppy flight. The distinctive head bumps above the eye are a good field character, giving a different profile from the much smoother head shape of the Glossy-mantled Manucode. The bumps (supraorbital tufts) can be erected to become very obvious, but even when not erect they are still distinct. The ruffled crinkly texture of the breast and neck feathers is visible at close range or through a telescope. Plumage gloss is highly variable in the genus, although this species has a bronzy yellow-green iridescence on the breast and neck feathers, whereas this is generally blue-green in the Glossy-mantled Manucode. Tail length may be shorter on average, but there is much overlap and sexual variation, making this criterion of little use in the field.
SIMILAR SPECIES The Jobi Manucode is a major field identification problem, being a poorly known similar species of northern and western New Guinea. Its head bumps are less obvious, and the bird appears smaller and more compact in the field, with a notably shorter tail. The crinkling on the neck and breast is more reduced, and the iridescence is more purple and lacks the bronzy yellow-green of the Crinkle-collared, although how useful this is under field conditions is debatable. For distinctions from other large and medium-sized forest species, see under Glossy-mantled Manucode.

RANGE The Crinkle-collared Manucode is endemic to New Guinea and Misool Island, where it is found mainly in hill forest. **MOVEMENTS** This species is resident.

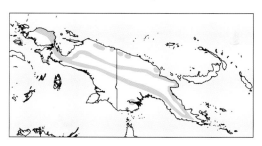

DESCRIPTION A large paradisaeid with a fairly long, graduated tail and prominent eyebrow tufts. *Adult Male* 36cm, 164–265g. The entire head to nape is blue-black, with blue-green to purplish iridescent feather tips forming inverted V-shapes; the hindneck feathers have glossy greenish-blue bars, while the blue-black of the mantle, back, rump, uppertail-coverts and uppertail has a violet-purple iridescence with magenta and/or blue

sheens, the mantle with velvety-black bars. The feathers of the throat and upper breast have glossy greenish-blue bars, more yellowish-green and crinkled across the lower throat and upper breast, and the belly has glossy bluish to purplish-blue bars. *Iris* deep red to orange-red, with a dark brown inner ring; *bill* dark grey to black; *legs* black. *Adult Female* 33cm, 160–255g. Averages smaller than male and duller below, with the underparts more green-blue, less purple. *Juvenile & Immature* are like the adult female, but with breast and belly even greener (lacking purple), the chin and throat matt blackish, and the iris greyish-brown to dark brown. The eyebrow tufts may be less obvious, making the young difficult to differentiate from those of the Glossy-mantled Manucode. *Subadult* is similar to the adult, but less purple and less barred, with the iridescent blue-green of the throat and breast duller, and the belly more blackish (Frith & Frith 2017).

TAXONOMY AND GEOGRAPHICAL VARIATION Monotypic. No racial variation is known across the wide range of the species. Synonym *Manucodia orientalis* Salvadori, 1896.

VOCALISATIONS AND OTHER SOUNDS This is a much less vocal species than the Glossy-mantled Manucode although, paradoxically, a greater variety of calls is known for it. The usual vocalisation in the Ok Tedi area is a rather querulous *took* or *chook*, deeper and richer in tone than a similar call of the Glossy-mantled Manucode. A similar but quiet *chig* call is on xeno-canto at XC152534 (by P. Gregory) and XC38115 (by F. Lambert). A presumed advertising call noted at Ok Tedi was a deep hooting series almost like that of a coucal (*Centropus*), not far-carrying, very seldom heard and quite distinct from any call of the Glossy-mantled Manucode; moreover, it was given from within deep bushy cover, and not out on songposts like those frequently used by the Glossy-mantled. Frith & Beehler (1998) give the song of the male as a slow series of up to eight haunting, low-pitched hollow *hoo* notes, which is probably the aforementioned coucal-like vocalisation. A response, presumably from a female, was a soft, ghostly whistled series of descending notes, *u-o-u-o-u-o-u-o*, perhaps likely to be sex-specific. Coates (1990) mentioned a short, sharp, high-pitched *kok*, a sharp *tchich*, *kick* or *chack* and a nervous, high-pitched, almost mammal-like trill of about two seconds' duration, the last occasionally heard from a presumed pair of these birds. He also noted short mammal-like whining cries, and a low pig-like grunting given by a male. A deep *ummmh* like that of an imperial-pigeon (*Ducula*) or a deep hollow *hoouw hooouw*, uttered in display, may be uttered once or repeated at intervals of many seconds. A captive individual at Baiyer River Sanctuary, Western Highlands, repeatedly gave an unusual-sounding two-note display call: the first note was a low hollow and humming sound like that of a frogmouth (Podargidae), this followed after a pause of about a second by a louder and higher-pitched note, *hmm-hooo*, duration more than three seconds; this call was repeated many times over some 14 seconds, sometimes with only the first note given. This species, like its congener the Glossy-mantled Manucode, makes a rustling sound in flight, but this is not so loud as the noise made by an airborne riflebird or astrapia.

HABITAT This species inhabits rainforest in the hills and occurs also in secondary habitat, including gardens with forest nearby. It is found also in lowland forest but seems to avoid swamp forest and monsoon forest, and we know of no records of this manucode from mangroves. It lives primarily in hill forest, extending up to about 1,500m and occasionally to 1,750m, but it does occur also at sea level in some areas. In the Ok Tedi area of Western Province, PNG, this is the typical manucode of the hill forests, replacing the Glossy-mantled above about 400m and here extending up to only about 800m. It is sympatric with the Glossy-mantled in the lowlands at Kiunga and Lake Murray, in Western Province, but is much less common there and seems to prefer the denser forest cover.

HABITS This rather shy species keeps within cover and, unlike the more confiding Glossy-mantled Manucode, perches out in the open on dead snags or branches only for brief periods as a rule. It is easily overlooked as it is a rather silent bird, and quite difficult to distinguish from the Glossy-mantled. Crinkle-collared Manucodes occur singly or in pairs or small groups, frequenting the lower and middle stages of the forest, but occasionally ascending to the subcanopy. Coates (1990) noted that they will drink and bathe in forest pools during dry weather. They are usually seen feeding on fruits in the shrubby layers some 2–6m above ground. When alarmed, they typically dip the front of the body downwards, which raises the tail, sometimes cocked, and they may make rapid jerky movements with tail-flicking. They are often seen flapping across clearings or roads, the flight being floppy and ungainly, dipping slightly, with the tail raised a little. They tend to bound about in bushes with powerful hops, the tail cocked, and Coates has seen them bounding along fallen logs.

FOOD AND FORAGING This species is primarily frugivorous, taking particularly figs (*Ficus*) and the small black berries of the same tree as that exploited by Magnificent Bird of Paradise in the Tabubil area of Western Province, where it is noted as taking also the red fruits of *Piper* species. Crinkle-collared Manucodes will join brown-and-black mixed-species foraging flocks, typical flocks in the Ok Tedi area containing Magnificent Bird of Paradise, Ochre-breasted (White-eared) Catbird (*Ailuroedus geislerorum*), Southern Variable Pitohui (*Pitohui uropygialis*), Grey Whistler (*Pachycephala simplex*), Northern Fantail (*Rhipidura rufiventris*), Frilled Monarch (*Arses telescopthalmus*) and Papuan Spangled Drongo (*Dicrurus carbonarius*). The Crinkle-collared Manucode can also be pugnacious at fruit sources, driving off other frugivores that approach (Mack & Wright 1996). It will also glean for insects. A radio-tracked individual at Mt Missim covered a range of about 45ha during one week (Beehler in F&B).

BREEDING BEHAVIOUR The Crinkle-collared Manucode is a monogamous species and seemingly non-territorial, with a wide foraging range. Bulmer (in Gilliard 1969) collected a pair at a nest, which suggests that both sexes attend the young. The breeding season is poorly known, but appears to be mainly from July to September. Coates (1990) suggested that breeding is

from the late dry season to the early wet season at least. **COURTSHIP/DISPLAY** This remains almost unknown, the above-mentioned observations of captive individuals vocalising being one of the main sources. Males will chase females through the forest, stopping occasionally to display or displaying repeatedly from one branch. The male expands his breast and mantle feathering as he leans forward to stretch his neck and raise his head as he calls (Coates 1990). **NEST** Almost unknown. Hartert (1910) noted a nest suspended between the forking branches of a tree limb, like that of an oriole (*Oriolus*). The nest was a basket-like structure made from brown wiry stalks, intermingled with leaves, and lined with finer stalks and fibres. It is not known if, as with some congeners, rotten wood is incorporated. Nests have been found in September and January, and nesting is recorded also in July–August (F&B). **EGGS** Little known. The clutch would appear to consist of either one or two eggs (Bulmer in Gilliard 1969; Hartert 1910). The eggs are rather rounded and chalky to creamy white, sometimes with a faint pinkish wash, and some brown and purplish markings. The average measurements of the four samples was 35.4 × 27.3mm (Frith in F&B 1998). There is no documented information on the incubation and fledging periods.

HYBRIDS There are no records of hybrids involving this manucode, which is hardly surprising for such a little-known species.

MOULT Museum specimens show evidence of moult during January–April and July–November, with most seeming to be during February–April. (F&B)

STATUS AND CONSERVATION Classified as Least Concern by BirdLife International. The Crinkle-collared Manucode is a rather uncommon species with a very wide range in the hill forests of New Guinea. Its biology is poorly known, particularly the intergeneric relations with sympatric species. Mayr (1941) linked it as a superspecies with the island endemic Curl-crested Manucode, but this seems unlikely, although it may be a sister form. Given its wide range and spread of habitats across New Guinea, the Crinkle-collared Manucode is unlikely to be at any real risk.

Crinkle-collared Manucode, sexes similar, Kiunga, PNG (*Jun Matsui*).

CURL-CRESTED MANUCODE
Manucodia comrii   **Plate 3**

Manucodia comrii P. L. Sclater, 1876. *Proceedings of the Zoological Society of London*, p.459, pl. 42. 'Huon Gulf', in error for Fergusson Island.
Other English names Curl-breasted Manucode, Comrie's Manucode
Etymology Named by P. L. Sclater for Dr P. Comrie, surgeon aboard HMS *Basilisk*, who was the original collector.

FIELD IDENTIFICATION Unlikely to be confused with any other species on the islands except the Trumpet Manucode, with which it is sympatric on Goodenough, Fergusson and Normanby. This is a much larger species which lacks the occipital tufts of the Trumpet Manucode, and has a mat of short curled feathers on the crown and heavily crinkled feathers on the throat and neck. The tips of the central tail feathers are also curled into a curious partial spiral, giving a somewhat singular tail shape. This is also a much more obvious and vocal species, less retiring than the Trumpet Manucode, and often sitting out in the open with the tail cocked. The flight is typical of the group, being loose and floppy, with the tail carried in an inverted V-shape, opened and closed in flight in synchrony with the wing movements (LeCroy *et al.* 1984). **SIMILAR SPECIES** The Eastern Koel (*Eudynamys orientalis*) male is quite different in shape and has a whitish bill. Trumpet Manucode (race *hunsteini*) is sympatric but far smaller, has head wires, and lacks the distinctive tail and head shape of this large species.

RANGE Endemic to the D'Entrecasteaux Archipelago islands of Goodenough, Fergusson and Normanby, and the smaller islands of Wagifa and Dobu, and also the Trobriand Islands of Kiriwina and Kaileuna. **MOVEMENTS** Resident.

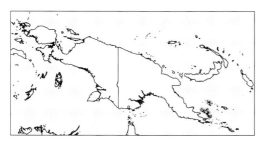

DESCRIPTION A large manucode with a large mop of short curled feathers on the crown and thickly crinkled and curled feathers on the neck; tail fairly long, with terminal third of central pair of feathers twisted and recurved to expose underside. Sexually monomorphic. *Adult Male* 43cm, one male 448g. Plumage is basically black, but with considerable and variable iridescence. Feathers of crown glossed purple and those of neck and breast glossed greenish-blue, with a purple-glossed area across lower breast. The mantle shows black barring, and the underparts have similar black bars with a coppery-purple gloss; the back feathers are glossed blue-purple

and the wings and tail have a violet gloss. *Iris* is red to red-orange; *bill* black; *legs and feet* black. **Adult Female** *c.*41cm, one female 418g. Smaller and slightly lighter in weight than male. *Newly fledged juvenile* is very odd-looking, for weeks after leaving the nest being bare-headed except for a fine central crown line of matt black feathers; it also has extensive bare blue-black facial skin. Plumage soft and fluffy, matt black flank feathers, wing feathers glossed blue. *Iris* greyish-brown, *bill* blackish, *legs* pale yellow. *Immature* is mainly matt black, much less iridescent than adult, with throat and breast not yet crinkled. *Iris* paler; mouth bright pink with small pink gape.

TAXONOMY AND GEOGRAPHICAL VARIATION Mayr (1941) considered this species to form a superspecies with the Crinkle-collared Manucode, but marked morphological and vocal differences would seem to make this very unlikely. Monotypic. Weakly differentiated form *M. c. trobriandi* Mayr, 1936, from Kiriwina and Kaileuna, in the Trobriands, was described on basis of smaller wing and tail measurements (wing 226mm *vs.* 244mm), but the wing measurements overlap with those of birds elsewhere in range by some 6mm, and the islands were likely joined to the D'Entrecasteaux Archipelago as recently as the last Ice Age. Trobriands form synonymised by Beehler & Pratt (2016), which treatment is followed here.

VOCALISATIONS AND OTHER SOUNDS This species is very vocal and has a considerable range of calls. It often calls well before dawn and may continue throughout the day. The presumed male utters a low haunting series, while presumed females have a prettier, more musical series which descends sharply. A common call is a beautiful, rather fluttery and far-carrying, mournful, tremulous *woodloodloodloodloodl* descending slightly in pitch and dropping in volume. This is on xeno-canto at XC70354 (by D. Gibbs), from Fergusson Island. A recording from Normanby Island at XC42029 (by J. Dumbacher) has a rather more throaty, deeper-toned version of this call. The species' local name of *mailulu* may be derived from the sound of this call. There is also a briefer more musical call, higher-pitched and rapidly descending in pitch (perhaps the female call), and a metallic croak (Beehler *et al.* 1986; Bell 1970); Bell also lists a low-pitched resonant *ko-ko-ko-ko-ko*. Gibbs (in F&B) recorded a low, almost pulse-like, even series of rapidly repeated soft and rounded *oo* notes, conceivably the same as the common call mentioned above. In addition, a harsh raspy *tek* note recorded on Fergusson Island is on xeno-canto at XC70355 (by D. Gibbs) and is rather reminiscent of a call of the more distantly related Halmahera Paradise-crow. Beehler noted that the sexes occasionally produce a rudimentary duet (as with the Crinkle-collared Manucode), the presumed female giving a very rapid descending series of piping notes, rather loud and with a whistled but throaty quality, and much higher-pitched than the presumed male call. Birds of this species on Goodenough Island emitted a ghostly monotone whistle, louder and then softer, followed by a very rapid series of notes all on the same pitch (Beehler in F&B). A harsh nasal-sounding *ench*, oft repeated, is used as a scolding call, and flying birds produce a wing noise (Beehler in F&B), although a rustling sound like that of congeners in flight is not confirmed. LeCroy & Peckover (2000) reported that the Curl-crested Manucodes on Goodenough made bubbling and echoing calls that were a dominant feature of the forest sounds. 'The common bubbling call seems to be a contact call, given as small flocks move together through the forest, but sometimes with individuals widely separated. It is one of the first calls heard in the morning, just as the sky begins to show light, and is one of the last calls heard in the evening, becoming fainter and fainter as the light fades.' Another common call here was termed 'clacking' by LeCroy *et al.* (1984), this being a peculiar low sound heard only when a small flock was nearby and moving as a group, and it may be a flock-cohesion call. On several occasions these observers also heard a long trill, more continuous and less bubbling than the common call, and possibly the vocalisation attributed by Beehler to the female.

HABITAT Curl-crested Manucodes inhabit forest at all elevations, and occur also in *Casuarina* groves, littoral woodland and gardens, from sea level to around 2,200m. They are sometimes found in mangroves (F&B).

HABITS This is an active, noisy and conspicuous species, and is both unwary and inquisitive, the males often perching in the open while calling and posturing oddly (Beehler *et al.* 1986). LeCroy & Peckover (2000) report that the perched posture of the species is peculiar: it sits like a chicken with its extravagant tail cocked up and constantly flicked. The body is held horizontal across the branch. This manucode may be seen singly or in pairs, and small aggregations may form at fruiting trees. Small flocks of four and another of six or seven on Goodenough at *c*.2,000m moved silently as tight groups, but individuals were spaced well apart when perched (Beehler in F&B). Males of this species are larger and have a more ornately crinkled plumage than females, and once a presumed male gave the bubbling call when directly overhead. His wings were held out from the body but were closed, and his breast swelled, inflating the coiled trachea just before the call began. The bubbling noise was then made as the air was released (LeCroy & Peckover 2000).

FOOD AND FORAGING The feeding behaviour of this species is little known, but it is, like the rest of the genus, a frugivore, recorded stomach contents being a large orange fruit and small cranberry-like fruit (Beehler in F&B). LeCroy *et al.* (1984) recorded it as feeding in a *Medusanthera laxiflora* tree with Goldie's Bird of Paradise at 1400m on Fergusson Island.

BREEDING BEHAVIOUR Almost nothing is known of the reproduction behaviour of this species, which is presumed to breed in monogamous pairs, as do other manucodes. Breeding has been recorded during June–November and is reported also for March, and is therefore likely at any time of the year; nest-building has been observed in mid-November (F&B; LeCroy *et al.* 2000). Fledglings from the D'Entrecasteaux group have been seen in October (Frith in Coates 1990), and in the Trobriand group LeCroy saw a fledgling on Kiriwina Island on 27th October. **COURTSHIP/DISPLAY** LeCroy and Peckover were on Goodenough in August, which

would seem not to be the courtship time, but they saw activities that may be courtship-related. Once, when a pair was feeding, the larger (presumed) male followed the other bird up a sloping limb and then perched on a small limb just above, leaning over until almost touching the (presumed) female. He then twice gave a version of the bubbling call, fanning the wings out as he did so and shaking them slightly. The pair then flew away. A second observation involved a flock of six moving through the treetops, at least one bird giving a low *clack* call: it had the wings out like a young bird begging and was vibrating them slightly, but did not appear to be a juvenile. Schram (2000) observed an intriguing group-display sequence on Fergusson Island on 23rd September 1999 when he found a small group of the manucodes in high thick forest on the edge of a clearing, having been attracted by their low burbling calls. There were four to six individuals and they were from six feet to 30 feet apart in thick foliage. Each bird would flutter forwards as if to fall bill first off the branch, but holding on with its feet while simultaneously half-raising the wings and fluttering them; the tail was raised and spread, with the curled and frilly feathers on the outer edge, the whole effect being of a large bird trying to regain its balance. The birds would stop and regain balance for a while, and then pitch forwards and repeat the performance, looking about between bouts. The iridescence of the upper body and curly crest was obvious. Occasionally the manicodes would move to another perch beneath the canopy, bounding about like big agile crows from branch to branch. The whole effect was curiously like that of a lek, with incessant calling, peering about and displaying, and the choosing of obvious perches from which to display. **NEST** Coates (1990) describes the nest as a basket-like structure of twigs and vines, lined with finer twigs and decorated externally with large, thick leaves; the bottom is very thick and includes many pieces of rotten wood (*cf.* Glossy-mantled Manucode and paradise-crows). It is suspended by its rim from the fork of a branch, rather like the nest of an oriole (*Oriolus*). An individual was seen by LeCroy to be carrying nest material *c*.20m up into a tree, while one nest was placed at a height of 7.6m at the extremity of a branch of a breadfruit tree (*Artocarpus altilis*) (Rickard in North 1892). **EGGS & NESTLINGS** The clutch is of one or two eggs, pale buffy-salmon with dark brown or rufous-brown blotches, underlying pale cinereous and pale purplish-brown blotches, and may be streaked like typical *Paradisaea* eggs (F&B; Coates 1990). In a sample of 39 eggs, average size was 44 × 29.8mm. Nestlings of this species are unusual in having a bare head except for a fine central crown line of black feathers; this is retained for weeks after they leave the nest and gives them an odd bare-faced appearance (Frith in F&B).

MOULT Museum specimens show evidence of moult in April–May and November–December (F&B).

STATUS AND CONSERVATION Classified as Least Concern by BirdLife International. The Curl-crested Manucode is a restricted-range island endemic, but the D'Entrecasteaux population seems fairly common to abundant overall, recent visitors seeing the species easily on Fergusson (Schram 2000), Normanby (pers.

obs. 2004) and Goodenough (LeCroy & Peckover 2000). The Trobriand form ('*trobriandi*') seems less secure, given the high human population density there, but it does occur in the overgrown gardens on Kiriwina, which is the entire available habitat on that island (LeCroy in F&B). It would be useful to know the current status of the population in the Trobriands. Gilliard (1969) was informed that the feathers of this species are used for cultural ceremonial adornment in the Trobriands, while Bell (1970) was told that the bird is sometimes eaten on Goodenough, where it remains tame and common.

Curl-crested Manucode, a nest with chick, Milne Bay Province, PNG (*Tim Laman*).

Genus *Paradigalla*

The genus *Paradigalla* consists of two long, slender-billed, montane species with black plumage and colourful yellow, red and blue wattles at the base of the bill and a striking yellow butterfly-shaped forehead wattle. They are not known to have elaborate displays and are presumed to be polygynous. *Paradigalla* was shown by Irestedt *et al.* (2009) to be sister to *Astrapia*, and was subsumed within *Astrapia* by Dickinson & Christidis (2014). Given the substantial morphological, behavioural and vocal distinctions, however, we feel it best to retain *Paradigalla* as a separate genus.

Etymology *Paradigalla* means literally 'paradise chicken' or 'paradise jungle fowl', derived from the Latin *paradisus* (= paradise) and *gallus* (= chicken), the connection being that both birds have facial wattles; the Latin *paradisus* derives from the Greek *paradeisos*, itself from an ancient Persian word *pairida za* meaning an enclosed park or garden, later interpreted as a pleasant place of happiness (or the Garden of Eden).

LONG-TAILED PARADIGALLA
Paradigalla carunculata   Plate 4

Paradigalla carunculata Lesson, 1835, *Hist. Nat. Ois. Parad.*, p. 242. Arfak Mts, Bird's Head.
Other English names Wattled Bird of Paradise, Carunculated Paradise Pie
Etymology *Carunculata* is the feminine form of the Latin for wattled (*carunculatus*), continuing the link with the wattles of chickens.

FIELD IDENTIFICATION A rare and remarkably little-known endemic of montane forests in the Vogelkop and perhaps the Fakfak Mts. This is a large black species with colourful facial wattles and a fairly long, graduated tail. Adult paradigallas seen in the Fakfak Mts in 1994 had a fairly short square-cut tail extending 3–4cm beyond the wingtips, with a pale yellow-white facial wattle, and more swollen paler blue malar wattles than those of the Long-tailed Paradigalla, and apparently lacked the red malar wattle (Gibbs 1994). The identity of these birds remains uncertain. **SIMILAR SPECIES** The smaller congeneric Short-tailed Paradigalla is allopatric, and has a very short tail. Western Lophorinas are smaller and lack the long tail, while the Arfak Astrapia and both the Brown and the Black Sicklebills are larger, with a much longer tail, and have no facial wattles.

171

RANGE Endemic to the Vogelkop, where it is found in the Arfak Mts. Another population, of uncertain taxonomic status, occurs in the Fakfak Mts of the Onin Peninsula (Gibbs 1994). The present species might be expected in the poorly known Tamrau and Wandammen Mts of the Vogelkop, and in the Kumawa Mts of the Onin Peninsula. **MOVEMENTS** Resident.

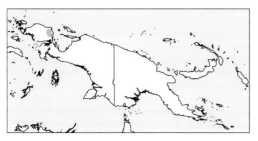

DESCRIPTION This is quite a large paradisaeid with prominent facial wattles and a fairly long and strongly graduated tail. *Adult Male* 37cm. The male has a bright yellow to greenish-yellow foreface wattle at the base of the upper mandible and a small swollen sky-blue horizontal malar wattle at the base of the lower mandible, with a small area of orange-red bare skin just below the blue wattle. There is also a small black pom-pom at the base of the upper mandible below the yellow facial wattle. The head and neck are a velvety jet-black with an oily bluish-green iridescence, more greenish-blue on the scale-like feathering of the crown and nape. The upperparts are a velvety jet-black from mantle to uppertail-coverts, while the upperwing-coverts, tertials, secondaries and central pair of tail feathers have a slight dark purple hue with a rich olive-green iridescent sheen; the primaries and outer (shorter) tail feathers are dark brownish-black. The entire underparts from throat to undertail-coverts are dark brownish-black with an iridescent coppery-purple sheen. *Iris* dark brown; *bill* shiny black; *legs* blackish. *Adult Female* 35cm; one female weighed 170g. Resembles the male but is paler, duller and smaller, with mean wing length notably shorter. *Immature* is like adult but duller-plumaged, with shorter tail. *Juvenile* undescribed.

TAXONOMY AND GEOGRAPHICAL VARIATION Monotypic. This is one of the least-known members of the family, seen by very few westerners; it is known from just a handful of sites and a few specimens. A paradigalla of indeterminate taxonomic status is reported from the Fakfak Mts.

VOCALISATIONS AND OTHER SOUNDS A long-drawn-out rising shrill *wheeeeeeeeeee* whistle, quite unlike any calls of the Short-tailed Paradigalla, was recorded by P. Hurrell (pers. comm. and tape). It has something of the quality of a shrill, upslurred jewel-babbler (*Ptilorrhoa*) call, and is repeated about every 30 seconds. This call is on xeno-canto XC26333 (by F. Lambert). In addition, some distinctive quite loud and far-carrying vocalisations, heard and recorded in the Arfak Mts above Mokwam at 1,250m in July 2015, seem not to have been previously reported: one was a hollow quite deep-toned *werr*, another was a mournful plaintive single whistle, followed by a disyllabic slightly

rising repetition, and a mournful, slightly raspy querulous rising and then falling and rather riflebird-like trisyllabic whistle, *whooeet weet wooo*. These are seemingly typical calls here, as they are well known to local guides. Whether or not this paradigalla produces a rustling sound in flight has not yet been recorded.

HABITAT This species inhabits mid-montane forests at 1,400–2,100m, but it is little known and may extend higher.

HABITS Probably similar to those of the Short-tailed Paradigalla, an unobtrusive species occurring at low density. The Long-tailed Paradigalla has been seen perched atop dead leafless trees (Beccari in Sharpe 1898).

FOOD AND FORAGING Very poorly known. Beccari (in Sharpe 1898) recorded one individual as feeding upon small fleshy nettle fruits. This species' diet is likely to be similar to that of its congener, suggesting that it forages for arthropods and insects, as well as fruit.

BREEDING BEHAVIOUR This species' breeding behaviour is almost unknown, but it is expected to be polygynous with promiscuous males, the females attending the nest alone. This paradigalla has hybridised with three sympatric species (see below), which supports the view that the males are promiscuous. Displays are unknown, but presumed males call from atop forest trees, or from within the forest canopy, where the loud ringing, almost *Ptiloris*-like call is given (pers. obs.). A woodpecker-like knocking is reported as being made as part of an as yet undescribed stamping display by a male on the forest floor (D. Hobcroft *in litt.*). As with most of the family, the female alone carries out the nest duties. In July 2010, above Mokwam, a female was sitting on a nest sited at mid-height in a bushy sapling, the nest exterior being a large mossy cup similar to that of the Short-tailed Paradigalla. In a series of videos by T. Laman on the Macaulay Library at Cornell Lab of Ornithology site, ML462849 shows the bird and nest nicely; this is only the second nest recorded to date (2018), the first having been in 2005 (see photo by M. Halouate on IBC). Two individuals seen together at this site in July 1994 seem likely to have been a female with a juvenile.

HYBRIDS The Long-tailed Paradigalla is recorded as having hybridised with Western Parotia, Western Lophorina and Black Sicklebill.

MOULT F&B reported moult in January, May and July specimens.

STATUS AND CONSERVATION Classified as Near Threatened by BirdLife International. One of the least well-known members of the family, with a restricted range and few sightings, this species is certainly somewhat data-deficient. Given the remote mid-montane habitat in which it lives, it is probably secure, and, not surprisingly, there are no records of its being used for decoration by local inhabitants. The paradigalla population discovered in the Fakfak Mts requires further research in order to establish its taxonomic, conservation and distributional status.

Long-tailed Paradigalla, Arfak Mts (*Gareth Knass*).

SHORT-TAILED PARADIGALLA
Paradigalla brevicauda   **Plate 4**

Paradigalla brevicauda Rothschild and Hartert, 1911, *Novit. Zool.* **18**: 159. Mt Goliath, Border Range.
Other English names Short-tailed Wattled Bird, Blue-and-yellow Wattled Bird of Paradise
Etymology *Brevicauda* is from the Latin, meaning 'short-tailed'.

FIELD IDENTIFICATION A highly distinctive, largely black species of the cool montane forests of western and central New Guinea. This is a typical but very short-tailed, stocky, stout-bodied member of the family with blunt rounded wings and a slender, slightly decurved bill. The sexes are very similar, with a distinctive pale yellow butterfly-shaped wattle at the base of the bill extending across the forehead, bordered laterally below by blue wattles. **SIMILAR SPECIES** The rare and larger congeneric Long-tailed Paradigalla is allopatric and that species has a long graduated tail, but note the anomalous and as yet unidentified paradigallas with intermediate-length tail reported from the Fakfak Mts (Gibbs 1994). Male Superb Lophorina and male Lawes's Parotia overlap only slightly at lower altitudes, and both lack the facial wattles, as well as having a longer tail. The Superb Lophorina has a triangular blue breast shield, while the parotia has distinctive cobalt-blue eyes and a white narial tuft. The male Loria's Satinbird, which does largely overlap in montane forest habitat, is smaller and shorter-billed and again lacks the distinctive wattles.

RANGE This paradigalla is found in the western and central sections of the cordillera, from the Weyland Mts east to the Karius and Bismarck ranges of PNG. It is not found in the Vogelkop, and the easternmost limits are not well known. **MOVEMENTS** Resident.

DESCRIPTION This is a rather large paradisaeid with prominent facial wattles and a very short tail. *Adult Male* 23cm, weight 160–184g. The male has a bright yellow foreface wattle originating at the base of the upper mandible, and a smaller swollen sky-blue malar wattle on the base of the lower mandible. The head, neck and entire upperparts are velvety jet-black, the head feathering has an oily yellowish-green iridescence (more apparent on the scale-like crown and nape), the mantle and back having a slight purple hue with a silky olive-green sheen; an oily olive-green sheen is apparent on the rump, uppertail-coverts, and the leading edges of the flight-feathers on the upperwing, and on the central pair of tail feathers (central pair 3mm shorter than others); primaries and tail feathers (except central pair) blackish-brown, rectrices with fine hair-like central points. The jet-black breast and remaining underparts are suffused with darkest brown and with the slightest iridescent sheen of dark coppery purple. *Iris* dark brown; *bill* black, mouth pale

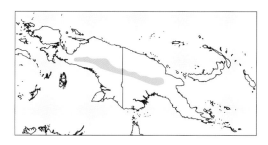

aqua-blue; *legs* purplish lead-grey. The tail is much shorter than that of the Long-tailed Paradigalla. **Adult Female** 22cm, weight 155–170g. Resembles the male but smaller, with a slightly longer tail, and duller, more brownish plumage, the black and iridescence being more subdued. *Immature* has a significantly longer tail than the adults, which appear virtually tailless. An immature at Tari Gap in August 2001 had a quite distinct tail, almost as long as that of a Superb Lophorina, and lacked the blue wattles, while the butterfly wattle was a pale greenish-yellow (pers. obs.). Beehler (in Coates 1990) states that the lower wattles are dull yellowish-grey with black contour lines below. Immature specimens were for many years described as the taxon *intermedia*, the longer tail causing the confusion (Frith & Frith 1997a). The immature of both sexes has a tail considerably longer than that of the adult, the tail becoming progressively shorter with age. *Juvenile* is similar to the female but uniformly duller (and quite unlike the barred young of most paradisaeids), lacking iridescence or sheen, and with facial wattles pale whitish-yellow.

TAXONOMY AND GEOGRAPHICAL VARIATION

Monotypic. The synonymised taxon *intermedia* (*Paradigalla intermedia* Ogilvie-Grant, 1913, *Bull. Brit. Orn. Club* **31**: 105. Otakwa River, Nassau Range, at 1,680m) was based on the longer-tailed immatures of this species, which were believed to be of a different race (Frith & Frith 1997a). The present species has been considered a short-tailed subspecies of the allopatric Long-tailed Paradigalla, but it differs substantially in morphology and call, with no evidence of clinal variation.

VOCALISATIONS AND OTHER SOUNDS
This is rather a silent bird, but in September 1993, at 1,500m near the Ok Tedi mine site in PNG, individuals gave a quite loud, far-carrying, rising bell-like *zheee* not unlike one call of the male Eclectus Parrot (*Eclectus roratus*). One bird would call and another reply, a loose group of four or five individuals forming a kind of dispersed cluster in the dense moss forest. Birds at Ambua have a similar call but seem not to form similar groupings, perhaps occurring at a lower density in the area. Several other vocalisations have been recorded. These include a melodious *hui*, which may be imitated to attract the birds (Stein 1936); a higher-pitched whistled series of very clear mournful notes, the last one prolonged and rising in pitch, the head being thrown back with each note (Cooper in Forshaw & Cooper 1977); a melodious *hooeee* and a harsh *churr churr churr*, the latter like a quiet version of the call of the male Superb Lophorina (Beehler in F&B); and a throaty croak, heard from birds at Wissel Lakes, West Papua (King 1979). Hicks & Hicks (1988a) noted a juvenile begging from the female with a faint high-pitched squeak, wings fluttering and head lowered and outstretched in a typical juvenile begging pose. Beehler (in F&B) heard a juvenile in the Baliem valley give a frog-like incessant interrogative upslurred *kwee*, presumably to attract its parent. A short, harsh flight call is also given by paradigallas at Ambua. There are no recordings of this species on xeno-canto at the time

of writing (2018). Short-tailed Paradigallas produce a distinctive loud rustling noise in flight, like that of *Astrapia* and riflebird species.

HABITAT This species inhabits montane forest from 1,400 to 2,580m, being found particularly in moss forest at 1,600–2,400m (Coates 1990; pers. obs.). It visits forest edge and gardens where fruiting trees occur, but avoids extensive cultivation and modified habitats.

HABITS A rather elusive species, the Short-tailed Paradigalla is generally observed in the middle or upper strata of the forest, visiting fruiting trees such as *Schefflera*. It calls from a perch high in dense moss forest, and may be seen perched atop tall dead tree branches beside roads and paths. Generally a rather quiet and unobtrusive species seen singly or in pairs, it is not particularly shy and is occasionally quite confiding, but normally is wary like many New Guinea species. In the Ok Tedi mine area, calling individuals may form loose groups of four or five individuals singing within earshot of one another. They forage on branches and at fruit clusters, with rather sluggish actions, flying away suddenly for no obvious reason.

FOOD AND FORAGING Food items recorded for this species include fruit, insects and small frogs (Rand 1942b). Gilliard (in F&B) noted an individual feeding on wild taro (*Colocasia esculenta*), and suggested that the wattles may help to keep the birds clean. Hicks & Hicks (1988a) observed a female with a well-grown youngster; the latter fed upon fruits of *Planchonella* species, *Sericolea pullei* and *Perrotetia alpestris*, along with grubs taken from dead leaves and from a hole in a tree. These paradigallas are rather sluggish and inactive feeders, climbing about on branches and in leaves or fruit clusters, and often probing with their long slender bill. They quite frequently perch on high dead branches for some minutes. Antagonistic intereactions have been observed with Princess Stephanie's Astrapia and Lawes's Parotia, these birds chasing the paradigallas out of a fruiting tree at Ambua Lodge (Beehler in F&B; pers. obs.). Yellow-browed Melidectes (*Melidectes rufocrissalis*) and Smoky Honeyeaters (*Melipotes fumigatus*) have also been seen to harass the paradigallas by pecking and flying at them.

BREEDING BEHAVIOUR Little is known of this species' breeding. It is believed to be polygynous, with promiscuous males which may display either solitarily or from songposts within earshot of one another. Only the females attend the nest. Courtship behaviour is unknown. There are no videos of courtship in the Macaulay Library on the Cornell Lab of Ornithology site, and no audio recordings on this site or xeno-canto at the time of writing (2018). In the Ambua area nesting has been recorded in all months except March and November, with egg-laying known in January, April–July, September, October and December (F&B). **NEST** The first known nest was collected by an expedition in 1915 but not described until 1970 (Frith 1970a). The second was found by a tour group in 1986 behind chalet five at Ambua Lodge, located in the middle section of a bushy sapling at the forest edge. This general area was used as a nest site for some seven years from 1986 to 1993 (Tano in F&B). Nests are sited in tree forks 5–11m (average 7m) above ground, in small bushy saplings or 15m-tall trees within the forest but close to gardens. The Ambua nest consisted of a bulky cup of leaves held together by creeper tendrils, the outside decorated or camouflaged with moss and fern leaves, epiphytes and some still live orchids. The cup was lined with dry leaves, and with fine dark stiff fibres that may be fern rootlets in the centre (F&B). **Eggs & Nestlings** Few nests have been found, but Frith & Frith (1992a) give the incubation period as about 19 days for the single egg, with a fledging time of around 25 days. The Friths (Frith & Frith 1992a) noted that about 65% of the nestling meals provided by a female at Ambua consisted of animal matter and 30% fruit, the rest being a mixture of the two; medium-sized frogs, skinks, insect larvae, crickets, beetles, a mantid, a katydid, a caterpillar and a large spider all figured in the diet. The parents removed all faeces from the nest and ate them. **Fledgling Care** The chick may stay with the parent for over three months after fledging.

HYBRIDS No hybrids have been recorded as yet. As the congeneric Long-tailed Paradigalla is known to hybridise with Western Parotia, Western Lophorina and Black Sicklebill, hybrids are doubtless possible.

MOULT Moult has been recorded in all months except April and June, with least activity possibly being in August–November (F&B).

STATUS AND CONSERVATION Classified as Least Concern by BirdLife International. This species has a wide range but is unobtrusive and patchily distributed, being generally uncommon over most parts. Its extensive range and montane-forest habitat mean that the species is likely to be secure. There appear not to be any examples of its being used for decoration by highland tribes, and, if such does occur, it must be rare and localised.

Short-tailed Paradigalla, immature, Ambua, PNG (*Nick Leseberg*).

Short-tailed Paradigalla, adult, PNG (*Markus Lilje*).

Genus *Astrapia*

The five species of the genus *Astrapia* are endemic to New Guinea in the mountains of the Vogelkop, the central ranges and the Huon Peninsula. Three species are allopatric and two have some overlap in the central highlands. This is a very distinctive genus, characterised by the small, slim bill and the greatly elongated central rectrices (including some of the longest such in the world, those of the Ribbon-tailed Astrapia reaching more than 1m in length). The males lack flank plumes and have specialised iridescent green feathering on the throat, breast, crown or nape, as well as producing a distinctive rustling wing noise in flight. Females, like many in the family, are heavily barred dark below but with a long tail. Molecular phylogenetic data suggest that the genus is monophyletic, roughly six million years old, and forms a sister-group with the two species in the genus *Paradigalla* (Irestedt *et al.* 2009).

Etymology *Astrapia* is derived from the Greek *Astrapios,* meaning a flash of lightning, presumably a reference to the iridescence of the plumage. Latin *astrapias* denotes a gem with flashes of light in it.

ARFAK ASTRAPIA
Astrapia nigra Plate 5

Paradisea nigra J. F. Gmelin, 1788, *Syst. Nat.* **1**(1): 401. Arfak Mts, east Bird's Head Peninsula.
Other English names Arfak Bird of Paradise, Arfak Astrapia Bird of Paradise, Long Tail, Incomparable Bird of Paradise
Etymology The epithet *nigra* is the feminine form of the Latin *niger,* meaning black or glossy black.
This was the first astrapia of this extraordinary family to be discovered, from a trade skin of unknown origin, but it remains one of the least-known and most seldom seen species.

FIELD IDENTIFICATION The Arfak Astrapia, with spectacular plumage and erectile neck adornments, is one of the most amazing but rarely seen of the genus. Sexually dimorphic, this is the only astrapia in the far western region; the very long, broad and blunt graduated tail of the male is distinctive, while females are blackish with a long blunt tail. The species bears a curious resemblance in some ways to the geographically far distant Huon Astrapia, which appears not to be its closest relative. The head is velvety black, complemented by spectacular iridescent green feathering on the crown, nape, throat and breast, and the wings, as with this species' congeners, make a loud rustling sound in flight. The sides of the neck and breast are of a rich metallic copper-bronze, narrowing to a pectoral band of the same colour fringing the breast shield. As is typical of the genus, this astrapia lacks flank plumes, and has a very distinctive nape–mantle cape. The female is mainly dark brown (for details, see DESCRIPTION, below). SIMILAR SPECIES May perhaps be confused with the larger Black Sicklebill, but that species has pointed tail feathers, a long decurved bill and a red (instead of dark brown) eye, the females also being brown and barred below but with a chestnut crown. The rare Long-tailed Paradigalla has a graduated tail and facial wattles, as well as being far smaller.

RANGE This species has a quite small range, being endemic to the Arfak and Tamrau Mts of the Vogelkop Peninsula, in West Papua. MOVEMENTS Resident.

DESCRIPTION Sexually dimorphic, as is usual with the genus. *Adult Male* 76cm, including central rectrices. A very large astrapia, one of the largest in the family, with an extremely long and markedly graduated tail. The plumage is patterned with highly iridescent and complex coloration which varies according to the light and the angle of view. The head is velvety jet-black with blue to purple iridescence, the crown and face side with a coppery-bronze wash, especially on the larger scale-like nape feathers. An unusual feature is an obvious cape, extending from the central nape to the mantle, consisting of large metallic yellowish-green scale-like feathers with a purplish-blue/mauve sheen, contrasting with elongate plush velvety jet-black feathers (with a dull blue to magenta sheen) on each side; the area from the back to the uppertail-coverts is sooty brownish-black with a dark coppery olive-green iridescence; the upperwing and uppertail are sooty black with a violet-purple sheen, with blue and/or magenta iridescence. The malar area is black, the chin and throat feathers with a velvety iridescent bluish-purple wash; the elongate dense feathers of the upper breast are velvety jet-black with dull coppery violet-purple iridescence, bordered below by a broad gorget of iridescent bronzed coppery/lime-green feathers, this extending up the side of the breast and malar area to beneath the eye; the remaining underparts are dull iridescent dark green, with large and broad scale-like iridescent sky-blue feather tips at the sides of the lower breast and belly, and with blackish-brown undertail-coverts. The *iris* is dark brown, the *bill* shiny black, and the *legs* brownish-black. *Adult Female* 50cm. Smaller than the male with shorter wings and tail. The black head and nape have an iridescent dark blue gloss, the iridescent blue of the nape grading into the upper mantle. The remainder of the plumage is

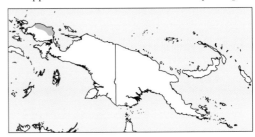

blackish-brown on the upperparts and chest, with the underparts dusky brown and narrowly barred buff. *Iris* dark brown; *bill* black; *legs and feet* blackish. **Juvenile** is fluffy, less blackish than the immature, the dorsal surface of the tail as that of older female-plumaged birds. **Immature Male** resembles adult female, but tail on average longer, after the first year with little to no paler barring below except on lower belly to undertail-coverts. **Subadult Male** varies, some being like adult female but with a few feathers of adult male plumage intruding, others like adult male but with a few feathers of female-like plumage remaining. The length of the central pair of rectrices increases progressively with age while that of other rectrices decreases, as is usual in the family (Frith *et al.* 2017).

TAXONOMY AND GEOGRAPHICAL VARIATION
Monotypic.

VOCALISATIONS AND OTHER SOUNDS
Remarkably little known, and the species seems to be rather silent. D. Gibbs (in F&B) mentions a simple double downslurred hollow *clu-uck* sound, like a person clicking the tongue, and a cut by A. Spencer on xeno-canto XC371175 has an incisive clicking *cluck* call, somewhat like the contact call of the Crinkle-collared Manucode. This call is also documented on the Macaulay Library at Cornell Lab of Ornithology site at ML163723 (by I. Fein). Scholes *et al.* (2017) liken this note to two pool balls knocking together. One non-vocal sound is known: a subadult in flight produced a *shhek-shhek-shhek* wing sound very similar to that of the Huon Astrapia, although probably not made with every wingbeat (Scholes *et al.* 2017). Birds seen in flight otherwise have appeared silent (pers. obs.).

HABITAT This species inhabits mid-montane and upper montane forests, 'in the highest and most difficult peaks of the Arfak mountains' (Beccari in Sharpe 1898). It is found from 1,700m to 2,250m.

HABITS Virtually unknown. Almost nothing has been published on this paradisaeid since Beccari's day, despite many observers having visited the Arfaks. It appears to be an unobtrusive, rather silent species which is not uncommon in its restricted range. Beccari noted that the neck feathers are erectile and can be expanded to form a magnificent collar around the head. The birds attend fruiting trees and this is the best way of locating them.

FOOD AND FORAGING The diet and foraging behaviour of this species are likewise almost unknown. Beccari mentioned its feeding on the fruits of *Freycinetia* pandans, as do other members of the genus, and the present author saw several at a fruiting tree in July 2015. It may be assumed that this astrapia is a gleaner of arthropods and insects and also frugivorous, like its congeners.

BREEDING BEHAVIOUR The breeding behaviour of this astrapia is virtually undescribed, but it may be assumed that it will prove to be a polygynous species, with promiscuous males, and that the females alone care for the young, as with Princess Stephanie's and Ribbon-tailed Astrapias. Whether or not the males form leks is not known. A female-plumaged bird was seen feeding a juvenile in July, and juveniles have been seen in the Arfaks in July–August (Poulson in F&B). Beccari (in Gilliard 1969) wrote that the adult males are rare, perhaps because they take some years to develop this plumage. **COURTSHIP/DISPLAY** Scholes *et al.* (2017) reported seeing one striking and undocumented behaviour at a practice display site, but no females were present and so the sighting stands alone. The behaviour seen was an *inverted display*, with the body feathers flattened and the facial and neck feathers extended forwards. As with the Huon Astrapia, the body is rotated backwards to drop, tail first, below the perch, with both bill and tail pointing skywards to form a rough C-shape from head to tail tip. When fully rotated backwards with the tail pointing up, the bird thrusts forwards and gives a *click* vocalisation, the tail dropping slightly backwards with the rectrices spread out in a fanning motion, again similar to that of the Huon Astrapia. This inverted display would presumably draw the female's attention to the rich iridescent greens and blues of the underparts, which otherwise tend to look blackish. **NEST** The nest remained undescribed until one was found in May 2017, 30m up in a tall evergreen tree. The nest was cup-shaped, approximately 30cm in diameter and 30cm deep, dark-coloured and composed of leaves, moss and twigs (D. Hobcroft *in litt.*). The female was sitting on the nest with the long tail elevated. She may have been feeding young, although she was not seen to carry food on three visits she made to the nest in the morning, so she may simply have been out foraging. No calls were heard in the vicinity of the nest. **EGGS** Undescribed.

MOULT Museum specimens showed evidence of moult in February, April–June and August.

STATUS AND CONSERVATION Classified as Least Concern by BirdLife International. Uncommon and hard to observe. This was the first of this extraor-dinary genus *Astrapia* to be described (in 1788), from a trade skin of unknown origin (specimens in Sir Joseph Bank's collection). The Italian collector Odoardo Beccari finally discovered it in its natural habitat in 1872 (in Sharpe 1898). It remains to this day very little known and is, indeed, one of the least-known members of the entire family. Although reported as common at higher elevations of the Arfaks (Eastwood pers. comm.), it seemed uncommon and at low density above Mokwam in July 2015. This species' status in the Tamrau Mts is uncertain, as it was only recently reported there. The remote montane habitat is lightly settled, and there are no records thus far of its plumage being used as decoration, but its restricted range could render it vulnerable to mining or large-scale logging despite conservation-area or national-park status for large tracts.

Arfak Astrapia, male, PNG (*Tim Laman*).

Arfak Astrapia, male displaying, PNG (*Tim Laman*).

SPLENDID ASTRAPIA
Astrapia splendidissima   Plate 5

Astrapia splendidissima Rothschild, 1895, *Novit. Zool.* **2**: 59, pl. 5. Type locality probably the Weyland Mts (*fide* Mayr 1941: 171).
Other English names Splendid Bird of Paradise, Splendid Astrapia Bird of Paradise, Most Splendid Long Tail
Etymology *Splendidissima* is the feminine form of the Latin word *splendidissimus*, meaning most magnificent, splendid or brilliant.

FIELD IDENTIFICATION The Splendid Astrapia is a rather distinct member of the genus, much smaller, notably shorter-tailed (though tail still long) and much greener above than its mainly black congeners. The male is dark iridescent green above but looks blackish in dull light, with a distinctive long club-shaped tail with largely white basal portion. The female has a shorter, graduated tail with whitish base, a greenish gloss to the entire head, nape and throat, and no narial tuft. **SIMILAR SPECIES** None within the range, although it may meet up with the Ribbon-tailed Astrapia at the eastern margin of its range. The female can be told by the white basal portion of the shorter and more rounded tail, the greenish iridescent gloss to the blackish-brown head, nape and throat, and the lack of a narial tuft. The Brown and Black Sicklebills can be distinguished by the long decurved bill and longer,

pointed tail shape plus the colour of the eye, which is dark in this species and pale blue or red in the Brown and Black Sicklebills, respectively.

RANGE Endemic to the western cordillera of New Guinea, from the Weyland, Snow and Nassau Mts of West Papua eastwards as far as the Star, Hindenburg and Victor Emanuel Ranges of westernmost PNG. It may come into contact with the Ribbon-tailed Astrapia in the easternmost part of its range, in the vicinity of the Strickland Gorge, but this is not yet established (see Coates 1990). Reports of Splendid Astrapia at Ambua are in error (Tolhurst 1989) and refer to hybrids between Ribbon-tailed and Princess Stephanie's Astrapias, which can show a lot of white in the tail (Barnes's Long-tailed Astrapia: see Frith 1995b). A report of two birds seen in flight from a helicopter at the remarkably low altitude of 600–700m in the Ok Tedi valley near Tabubil, PNG, is now considered erroneous (Johnston & Richards 1994). The species

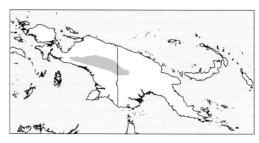

does, however, occur in the adjacent mountain ranges. **Movements** Resident.

DESCRIPTION This species, as is usual with the genus, is sexually dimorphic. *Adult Male* 39cm, weight 120–151g. Complex, unusually iridescent green plumage, with the crown and sides of the head to the short cape on the mantle an unusual brilliant iridescent yellowish to bluish-green, with black forehead and lores. The back is velvety black, faintly glossed purple, with wings, rump and uppertail matt black. The tail is distinctively long and club-shaped, the basal two-thirds of the central feathers white and contrasting with the black outer third and the black outer tail feathers; the underside of the tail appears black. The chin, throat and upper breast are a vivid iridescent blue-green, bordered by an iridescent coppery-red band extending from below the eye and widening out onto the upper breast, and this area can be erected as a shield (much as with congeners). The abdomen is a deep dark oil-green, with the lateral feathers tipped shiny blue-green; lower abdomen and undertail-coverts black. *Iris* blackish; *bill* black, with *mouth-lining* of pale yellow tinged aqua; *legs and feet* silvery blue-grey. *Adult Female* 37cm, weight 108–151g. Head is dull blackish to dark brown, with blackish-brown on the upperparts and down to the chest; the lower breast and belly are finely barred dusky brown and buff; white basal half of the central rectrices. The *iris* is grey-brown to dark brown; *bill* black; *legs and feet* leaden grey (Coates 1990). *Juvenile* is like female but with soft and fluffy plumage, duller and less black above, browner on underparts, more grey on chin, throat and upper breast, with tail feathers more pointed (Frith *et al.* 2017). *Immature* has a dark chestnut crown and hindneck. *Subadult Male* begins to acquire glossy green feathering on crown and throat, and tail becomes shorter with age. Some younger birds have female-type plumage with a few male-type feathers appearing, moulting to largely male plumage with some dark female-type feathers remaining. The timescale for attaining adult plumage is as yet undetermined but expected to be 5–6 years, as with congeners. **Leucism** This species is noteworthy in being one of the very few paradisaeids for which leucism has been noted. A female-plumaged individual from Lake Habbema (AMNH 342102) has the head, nape, throat and upper chest white with a few black flecks. The greater coverts, inner primaries and secondaries plus a tertial on one side are also white, and few odd white feathers are present on the breast, abdomen and vent; the legs and feet are piebald, with white claws (F&B 98). Such abnormalities are exceedingly rare in the family.

TAXONOMY AND GEOGRAPHICAL VARIATION Polytypic. Some authors accept three subspecies, but here we recognise two:

1. *A. s. splendidissima* Rothschild, 1895. Ranges from the far western end of the central cordillera, from the Weyland Mts, east to Wissel Lakes.
2. *A. s. helios* Mayr, 1936. (Synonymous with *A. s. elliotsmithi* Gilliard, 1961.) Extends from east of Wissel Lakes east to the Hindenburg and Victor Emanuel Range of PNG. Similar to the nominate, but averages larger, with crown, neck and dorsal

collar of adult male more bluish and less golden-green, spatulate tips of central tail feathers broader, and female-plumaged birds darker above. Both sexes differ from nominate in having extensive and unconcealed white bases on underside of the primaries excepting outermost two (F&B). Cracraft (1992) and F&B synonymise the slightly larger *elliotsmithi*, which Frith attributes to clinal variation, this form also having white primary bases of the *helios* type.

VOCALISATIONS AND OTHER SOUNDS This seems to be a fairly quiet species, but it is still poorly known. An adult gave a dry, frog-like *gree* every 20–30 seconds (Beehler *et al.* 1986). A raspy frog-like *shwik shwik shwik* call recorded by A. Mack is in the Macaulay Library at the Cornell Lab of Ornithology ML100385. E. Scholes and T. Laman document birds of this species in the Macaulay Library at the Cornell Lab of Ornithology, including a sharp, rapid, dry frog-like *wek wek* call at ML462765, and this is one of the most common vocalisations. Other calls include a *chick-chweee-chweee chweee*, similar in quality to the calls of the King Bird of Paradise (Palliser in Coates 1990). A nasal, insect-like *to-ki*, with the second note rising, is given by the male, and the female has a similar *teek teek* on an even pitch (King 1979). Bishop (in F&B) describes this latter double note as being 'dryish'. Males in the Hindenburg Range in March 1993 gave a strange and very distinctive *tch tch tch* clicking call, quite unlike calls made by any others of the genus and diagnostic of this species (Gregory 1995); a harsh growling *grrr grrr* was also heard here. D. Bishop (in F&B) has heard a distinctive yelping call at times, and Dennis (in F&B) mentions a mechanical, metallic whirring sound, apparently a vocalisation. In flight, males produce a rustling sound like that made by their congeners, while Beehler reports that the wings hiss, which may be the same thing.

HABITAT This species inhabits montane and sub-alpine forest and forest edge, up to the treeline on the higher summits, from 1,750m to 3,800m, mainly at 2,100–3,300m (Coates 1990). Occurs also in second growth.

HABITS This is an inconspicuous species, quite common in places but not very vocal. Frequents the canopy down to undergrowth shrubbery (Bishop in F&B). Adult males appear to be solitary and it is not known whether they hold leks, although female-plumaged birds will congregate in small groups at food sources (Stein 1936; pers. obs.) and pairs or trios are occasionally seen. The abundance of males varies from site to site (Gilliard 1969). Diamond and Bishop (in F&B) noted that, while the species was common at Okbap, Star Mts, West Papua, at *c.*2,600–3,200m, only female-plumaged birds were seen at the lower elevations to *c.*2,150m.

FOOD AND FORAGING Splendid Astrapias forage in a manner typical of the family, exploring moss and epiphytes along trunks and branches for arthropods and insects, and taking fruit such as *Freycinetia*, *Schefflera* and *Trema* spp. (Rand 1942b; Gilliard 1969; Beehler and Bishop in F&B). Known also to take frogs and skinks, as with other members of the genus. Beehler & Pruett-Jones (1983) estimated the diet to contain at least 75% fruit.

BREEDING BEHAVIOUR This species is likely to be polygynous, with promiscuous males displaying from arboreal courts, but, as with the Huon Astrapia, field data are very incomplete and, in common with that species, some observations have been of single males calling (Beehler & Pruett-Jones 1983). **COURTSHIP** Dennis (in F&B) makes some interesting observations, which may tend to support the normal lek-type behaviour for this species, but more observations are desirable. He found four adult males perched and calling some 40m apart around the forested edge of a natural meadow at 2,300m. They were perched on exposed branches near the tops of tall trees, and others were heard calling nearby; the males called at *c.*30-second to five-minute intervals over a period of 30–40 minutes, and regularly turned up to 180° on their perches, which emphasised their highly iridescent plumage. Two female-plumaged birds were seen in the area, but the males paid them no particular attention other than calling in their direction occasionally. The calling males appeared slightly fluffed up, but had no obvious plumage accentuations as if in display. Elsewhere, a male uttering the clicking *tch tch tch* call in the Hindenburg Range in March repeatedly depressed the tail at 90° to the body and flicked it from side to side two or three times, which had the effect of making the white bases flash; this behaviour lasted for about a minute, the tail being depressed and flicked every time the bird called. Scholes *et al.* (2017) document a **perch-hop display** similar to that of other members of the genus. In this, the male adopts a slightly hunched posture with the back feathers slightly expanded into a domed or hunched shape like that of Ribbon-tailed Astrapia, he expands the feathers of the face and neck to create a conspicuous 'feather beard' below the lower mandible, and he suddenly performs a flight hop through the tree canopy and alights on another branch, albeit only briefly, before jumping around to other branches around the crown of the tree, with little vertical movement; the wings are spread occasionally to assist the jumping, and the tail remains stiff. Sometimes the male returns to the initial branch, and sometimes not. This display concludes with a harsh frog-like call. It is not yet known whether this species may have some kind of inverted display like Arfak and Huon Astrapias, but the bright iridescent coloration of the underparts suggests that this may be a possibility. **NESTING & EGGS** Virtually unknown. Nest-building has been seen in March, a juvenile in August, egg in October and a nestling in November (Gilliard 1969; F&B). Eggs undescribed; the clutch is of a single egg, as with congeners.

MOULT Specimens showed moult in every month except June.

STATUS AND CONSERVATION Classified as Least Concern by BirdLife International. This is a little-known species but with a wide range in remote and lightly settled montane areas. The plumes are in little demand in the Ilaga valley of West Papua (Ripley 1964) and the Victor Emanuel Range (Gilliard 1969). During seven years in Tabubil, PNG, the skins of males of this species were seen in the local market only twice, selling for K20 (£4.85) in 1993. Lacking the long plumes of other astrapias, it is presumably deemed less desirable for personal decoration (*bilas*). The Splendid Astrapia is likely to be secure and is common in some parts of its range, as at Okbap, in the Star Mts (Diamond and Bishop in F&B), and the Hindenburg Range (pers. obs.), while it seemed uncommon and localised at Lake Habbema (Snow Mts) in 2015.

Splendid Astrapia, male, Jayawijaya Mountains, New Guinea (*Tim Laman*).

RIBBON-TAILED ASTRAPIA
Astrapia mayeri   Plate 6

Astrapia mayeri Stonor, 1939 (February), *Bull. Brit. Orn. Club* **59**: 57. Mt Hagen, Eastern Range.
Other English names Ribbon-tail, Ribbon-tailed Bird of Paradise, Ribbontail Astrapia, Shaw Mayer's Bird of Paradise
Etymology *Mayeri* is in honour of Fred Shaw Mayer, who sent the tail plumes to the British Museum (Natural History, now BMNH) for formal description of this remarkable species in 1939.
This spectacular species is a PNG endemic restricted to a small area of the central cordillera, and was the last of the family to be described. It occurs between the ranges of the Princess Stephanie's and Splendid Astrapias, overlapping with the former at the western edge of its range. The first westerner to see it was the famous Australian explorer Jack Hides, in 1936; he obtained tail feathers, which unfortunately were subsequently lost (Hides 1936). Two miners, the Fox Brothers, also reported seeing what were clearly birds of this then undescribed species some 80–100 miles (128–160km) west of Mt Hagen, even noting the black tip of the tail and seeing local people wearing the feathers in their hair. The naturalist and explorer Fred Shaw Mayer was given two central tail feathers of the species in August 1938 by a missionary, who had obtained them from a Mt Hagen tribesman who was wearing them in his head-dress! Shaw Mayer recognised their significance and sent them to the BMNH with the suggestion that they came from a new species of astrapia, which was duly described and named in his honour (Gilliard 1969). This was narrowly ahead of the Australian Museum, which received a whole specimen from the famous Taylor and Black Wahgi–Sepik patrol, and duly described it under the name *macnicolli* in honour of Sir Walter Macnicoll, administrator of the Territory of New Guinea, only to find that it had been beaten to it.

FIELD IDENTIFICATION The adult male of this mostly black species is unmistakable, as it possesses one of the longest tails of any bird, with the central white plumes sometimes more than 1m long. In addition, it has a striking small pom-pom at the base of the bill and an iridescent green breast and crown, the colours visible only in good light. Females and immatures are heavily barred black on cinnamon, with a fairly long tail and varying degrees of green iridescence on the head and breast. The species inhabits the higher levels of montane forest up to the treeline, generally above the habitat of the Princess Stephanie's Astrapia. **SIMILAR SPECIES** The only major confusion species sharing the same range is Princess Stephanie's Astrapia in female-type and immature plumages. The male Princess Stephanie's has long paddle-shaped tail feathers and lacks the pom-pom at the base of the bill. Females and immatures are more difficult, but those of Ribbon-tailed have narrower and more pointed tail feathers than the more rounded and broader rectrices of Princess Stephanie's Astrapia, with more white showing in the tail feathers; they also show traces of a pom-pom at the base of the bill, have dark brown (not blackish-brown) wings and tail, and have oily green iridescence on the head, while the barring of the underparts is narrower than that of Princess Stephanie's and they may appear paler. Hybrids with Princess Stephanie's Astrapia are not infrequent at the lower levels of Ribbon-tailed habitat, and these may be problematic; their outer tail feathers are intermediate in shape and the central feathers show quite extensive amounts of white around the vanes. The Splendid Astrapia is not as yet known to overlap geographically with the Ribbon-tailed, although it is possible that the two species do meet in the western extremity of the latter's range. Male Splendid Astrapias are a smaller and much greener species, with a shorter, spatulate tail having a different pattern of white, while females are smaller and have only a small amount of white at the base of the tail, which has more rounded feathers than those of the Ribbon-tailed. Reports of Splendid Astrapia at Ambua are in error (Tolhurst 1989), and refer to hybrids between Ribbon-tailed and Princess Stephanie's' Astrapias (known as Barnes's Long-tailed Astrapia: see Frith 1995b). The male Brown Sicklebill is of a similar size to the Ribbon-tailed, but has a strikingly decurved bill and a pale blue eye, and long black pointed tail feathers; females are brown above with a rufous cap, have a pale blue eye, and are heavily barred beneath. The Black Sicklebill occurs below the altitudinal range of the Ribbon-tailed, so the two species are unlikely to meet; again, the bill shape and red eye colour would readily distinguish the Black Sicklebill.

RANGE Restricted to the central cordillera of western PNG, ranging from Mt Hagen and Mt Giluwe west to both sides of the Strickland River (Coates 1990), including the southern Karius Range (Clapp in F&B). Replaced in the east by Princess Stephanie's Astrapia, but overlaps with that species around Mt Hagen and Mt Giluwe to near Wabag and the Tari Gap. Hybridises with that species in a narrow altitudinal band from about 2,200m to 2,600m, but maintains its specific identity and there is no evidence of genetic swamping as was at one time suggested. Replaced to the west by the Splendid Astrapia, but not as yet known to overlap and no hybrids are so far known. Local informants claim that the species occurs in the mountains around Ok Tedi (Gregory 1995), but this seems unlikely and is as yet unsubstantiated. **MOVEMENTS** Resident.

DESCRIPTION Among the most spectacular of the family, male and female-plumaged Ribbon-tailed

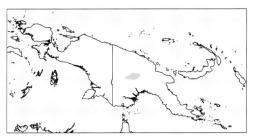

Astrapias are documented at the Macaulay Library at the Cornell Lab of Ornithology in a long series of videos by T. Laman and E. Scholes which show the astonishing plumage of this bird. Further examples are also on the Internet Bird Collection. Sexually dimorphic, as usual with the genus. **Adult Male** 32cm, but 125cm inclusive of tail, having the longest tail of any wild bird proportionate to body size, and presenting an extraordinary sight; weight 134–164g. There is a striking tuft or pom-pom of black feathers at the base of the bill, with the face, cap and a large breast shield brilliant iridescent emerald-green with bluish-violet highlights at some angles. Remainder of head, mantle, back and a broad breast-band velvety black, glossed purple on sides of head and bronzy elsewhere. Uppertail-coverts and tail are blackish-brown with two extraordinary elongated white central tail feathers, tipped with black on the outermost 3–4cm. The breast shield is bordered with a dark coppery-red edging, the remainder of underparts blackish-brown, tinged reddish on lower breast. *Iris* brownish-black; *bill* black, *mouth-lining* pale greenish-yellow; *legs and feet* dark grey. **Adult Female** 35cm, 53cm inclusive of central rectrices, and having no overlap in wing and tail measurements with the male; weight 102–157g. Similar to Princess Stephanie's Astrapia but with a narial tuft at the base of the bill, and narrower elongated central tail feathers which frequently show white along the vanes. The entire head and nape are jet-black with iridescent bronzy/blue colour (more so on older individuals), the blue iridescence more obvious on the sides of the face and lower throat. The mantle and back are velvety blackish with blue-violet sheen, the rest of the plumage overall a dull blackish-brown, with the abdomen rufous to buff-brown with fine blackish barring. **Juvenile** resembles female but with duller barred ventral plumage and has soft and fluffy plumage. **Immature** resembles female but the wings and tail are longer, while first-year birds resemble female but with duller iridescence. **Subadult Male** is variable depending on state of moult, from being like adult female with a few male-type feathers showing to being much as adult male but with some female-type feathers remaining. As usual with the genus, the central tail becomes progressively longer as other rectrices become shorter. Data on captive individuals suggest that these birds may attain full adult plumage only after at least six years (McGill 1951), suggesting a very long-lived species. A captive bird took at least six years to acquire the white tail, the tail being black for the first two moults and then gradually developing the white coloration. One adult-plumaged male at Kumul Lodge had a very short stumpy white tail about 30cm long in 2014, but this was much longer in the following year and getting close to normal length by 2016. Local New Guinea expert bird guide Joseph Tano reports an interesting observation at Tari Gap above Ambua, where on rare occasions the elongate white central tail feathers of this species, besides the more usual King of Saxony head plumes, are used by Archbold's Bowerbird (*Archboldia papuensis*) for bower ornamentation.

TAXONOMY AND GEOGRAPHICAL VARIATION
Monotypic, and endemic to PNG.
Synonym: *Taeniaparadisea macnicolli* Kinghorn, 1939 (Dec.), *Austral. Zool.* **9**: 295. west and north-west of Mt Hagen, between 142.4° and 142.4° E.
Synonym: *Astrapia recondita* Kuroda, 1943, *Bull. Biogeogr. Soc. Japan* **13**: 33. Locality unknown; probably from the region around Mt Hagen [not Morobe Province as guessed by the author].
Synonym: *Astrarchia barnesi* Iredale, 1948, *Austral. Zool.* **2**: 160. Mt Hagen [hybrid *A. mayeri* × *A. stephaniae*].
Hybrids with Princess Stephanie's Astrapia are quite frequent at the lower levels of the altitudinal range and are known as **Barnes's (Long-tailed) Astrapia** or Barnes's Long-tail. Males of this hybrid can be truly spectacular, having tails about 1.7m long, starting like Ribbon-tailed ribbons and then widening out into the paddle tail of Princess Stephanie's Astrapia, but they are rarely seen. The guide Joseph Tano at Ambua had seen them only three or four times in 30 years when we saw such a remarkable bird there in July 2014. The name Barnes's Long-tailed Astrapia was coined by the describer, Tom Iredale of the Sydney Museum, after his taxidermist William Barnes. It was first collected when the marvellously named Captain Neptune Blood collected a specimen during his wartime patrols in the Mt Hagen area. Captain Blood is well known also for being the man who collected the first specimens of the Sepik Blue Orchid (*Dendrobium lasianthera*), one of the major orchid discoveries from PNG, while fleeing from a Japanese patrol in the middle-Sepik.

VOCALISATIONS AND OTHER SOUNDS This can be quite a noisy species, especially in early morning around food sources, but, as with its congeners, it is often not very vocal. The typical call is a loud, harsh barking *waugh*, not dissimilar to one call of Archbold's Bowerbird, or a trisyllabic loud and quite piercing *wok-wik-wik* or *wauk wik* and variants (see xeno-canto XC176045, by P. Gregory). These latter calls have similar equivalents in Princess Stephanie's Astrapia. Kwapena (1985) lists various display calls, including a harsh *hisss-sss-ssh*, and a loud sharp call rendered *keaoo-ooo-ooo*. Taylor (in Iredale 1950) also notes a clicking or hammering sound something like a mechanical riveter at work and again given in display, and this is likewise noted by Kwapena. Beehler in F&B lists a *kenk!* call and a scolding *skaw skaw*, while Beehler (1986) notes a young male as giving plaintive single or double nasal frog-like notes. Gilliard (in Mayr & Gilliard 1954) mentions captives in display calling a raucous *grrrrow*, *grrr*, *grr* followed by hoarse cawing. Scholes *et al.* (2017) noted two upward-sweeping plaintive slow notes, *eert* or *weet*, the second higher-pitched than the first, given in practice courtship displays. The males of this species produce a distinctive rustling or swishing noise in flight (see XC116116, by M. Anderson, and XC279383, by J. Moore), similar to that of riflebirds and parotias, while Cooper (in Forshaw & Cooper 1977) noted a sound made by the lateral flicking of tail feathers just before an adult male left its perch.

HABITAT This species inhabits upper montane forest up to subalpine moss forest, from 1,800m (G. Clapp in Coates 1990) to 3,440m. Where sympatric with Princess Stephanie's Astrapia, however, it is usually not found below 2,450m, the lower elevations being occupied by that species (Coates 1990; pers. obs.). It occurs at some of the highest elevations inhabited by any member of the family.

HABITS Quite a conspicuous species, occupying forest edge and forest patches as well as disturbed, selectively logged and even regenerating burned forest. It occurs at all levels, more often in the upper canopy but also down to ground level. It regularly feeds on wild gingers (Zingiberaceae) less than a metre tall. Not shy, but distinctly wary, it is hunted for its plumes by the Huli people. A male in flight over the dark green forest canopy or across open grassland is one of the most astonishing sights in nature, the bird having a gently undulating flight, giving four or five wingbeats before each shallow dip, with the unique elongated dark-tipped white central streamers rippling out behind, and often spreading out in a scissor shape. The plumes are so long that they sometimes become draped at steep angles across or over branches in the forest, and it is a wonder that they do not often get broken or torn, although by October–November they can be looking very worn and bedraggled.

FOOD AND FORAGING Ribbon-tails forage at all levels of the forest but are seen mostly in the middle strata, although they do forage on the ground and come down low to feed on the red fruits of wild gingers. They forage on tree branches, and at fruit, leaves and flowers, with the umbrella tree (*Schefflera*) a particular favourite. Individuals will share fruiting spikes of these last-mentioned plants, spending much time at them. It is common to see two or three female-plumaged birds feeding together, although the adult males seem often to be solitary feeders. Pandanus crowns are frequently explored, the Ribbon-tail poking about with the bill and tearing into the dead leaf whorls, as with the Brown Sicklebill, and arthropods form a significant part of the diet. The birds obtain them by clinging to tree trunks and pecking into the epiphytes, mosses and lichens, or by hopping along branches and pecking similarly. Fruit is a major part of the diet, but gleaning for insects and arthropods is also very significant, and these paradisaeids may also probe into dead wood. Ribbon-tails in June at Tari were feeding avidly on the red fruit clusters of a ginger growing beside a track and, like the usually very shy Crested Satinbirds at that time, were very confiding. They also tend to work their way around tree trunks in a zigzag fashion, probing into the bark and moss cover. Ribbon-tailed Astrapias may displace Stella's (*Charmosyna stellae*) and Josephine's Lorikeets (*C. josefinae*) from *Schefflera* stalks, while Belford's Melidectes (*M. belfordi*) and Smoky Honeyeaters (*Melipotes fumigatus*) often show aggression towards them at such sources and may themselves be driven off, or drive the Ribbon-tail off. Adult males are often seen perched high up on dead branches in forest trees, although not often in the canopy, and they may remain there for minutes at a time.

BREEDING BEHAVIOUR Polygynous with promiscuous males, as is typical for the genus. **COURTSHIP/DISPLAY** Males do have favourite display areas and may display solitarily or in groups of up to five adult males. Late one afternoon in June 1999, at the Tari Gap, five such males were perched up in a large dead tree occupying the higher branches, all waving their distinctly arched-up tail plumes in a rather slow and graceful manner, and chasing about for some ten minutes. The breast shield can be partially raised to form a kind of iridescent green breast plate, bordered by the coppery maroon lower edge, and is very striking in good sunlit conditions which can be maximised in the bare display tree. No calls were heard, which was surprising, and female-plumaged individuals were in adjacent trees but not the primary display tree. A male Crested Satinbird also came and sat in the same tree for a few minutes while this was occurring. In August 2001, members of a group of about ten Ribbon-tails were calling and chasing about at the Tari Gap, in the vicinity of the aforementioned display tree but on this occasion not actually in it. There were some four adult males and various female-plumaged birds and subadult males, all chasing about and clearly very interested in what was occurring. Males kept displacing the females from their perches and repeatedly chasing them through the forest and across the road. Frith (in F&B) saw an adult male perform an extensive display flight with wing noise louder than normal: it made four or five wingbeats with a much deeper brief downward path and then an upward glide on closed wings before again wing-flapping, this causing the two white tail feathers to wave conspicuously. Individuals of this species are quite often seen flying over the forest canopy, and it is possible that this is also some form of aerial display rather than just normal flight. The wing noise and prominent tail feathers certainly make males very conspicuous, and suggest few, if any, predators in the area. This seems to be analogous to the slow, jerky flight over short distances described by Kwapena (1985), who states that the bird does not fly straight but as if travelling on an ocean wave.

Kwapena observed a display at Mt Giluwe on 25th August at 17:00 hours. Two adult males called and then joined up on tree branches 35m up in a *Schizomeria* tree, the longer-tailed bird slightly above the other. Three female-plumaged birds on adjacent branches then joined them. The longer-plumed male then jumped from branch to branch while making the rough harsh *hisss-sss-ssh* call followed by the loud, sharp *keaoo-ooo-ooo* call, before raising his breast plumage and tail feathers to jump from one branch to another with much calling. At the climax of the display he flew from branch to branch, trying to stimulate the three females looking on from below. He displayed for about ten minutes and then flew to a nearby branch, where he sat for some five minutes with a couple of the females. He then displayed again and continued to do so until darkness at 18:30 hours. Frith (in F&B) observed a male (with an onlooking subadult male) displaying by hopping gently from side to side on the same spot of his 18m-high perch, causing the central tail plumes to sway back and forth in a rhythmic arc. A similar display was seen which involved a subadult male with half-grown tail feathers displaying to a female some 6m above ground on a *Schefflera* plant; he jumped from side to side, with the central tail plumes held stiffly apart, and the plumes arched as they swung from side to side behind him. Scholes *et al.* (2017) made some valuable studies of videos archived at the Cornell Lab of Ornithology, although, as the filming was done at feeding sites where practice displays were performed, not much can be adduced about display sites and attributes or even order of activities. It is also lacking high-intensity displays directed at females, and this will no doubt be a focus for future research.

Four distinct **courtship and mating behaviours** were identified:

1. The *perch-hop* consisting of rapid hops from branch to branch. One non-vocal sound may accompany this display, a wing rattle which the bird produces in flight and when hopping back and forth between branches. This sound is louder than that of the Huon Astrapia and even more rattle-like.

2. The *hunchback-pivot*, a ritualised turning from side to side in a distinctive hunched posture with the mantle cape erect. This is shown in the Laman & Scholes videos at the Cornell Lab of Ornithology ML465288, ML465289 and ML465689. The rounded pom-pom forehead tuft is pushed forward over the outstretched and pointing bill. In this posture the iridescent green feathers atop the head are set up to accentuate the colour when seen head-on, with the plush black feathers around the neck also pushed forward to create a striking green-and-black pattern. This may be an introductory display analogous to the *flick-pivot* of the Huon Astrapia. Also in this display, the tail is lifted so that the black feathers align with the head and body and the long white plumes dangle behind the bird in a wide arc. The body pivots rapidly from side to side, often with small lateral hops, which cause the tail plumes to lift and swish up and to the side in an exaggerated way. The shorter the plumes, the less exaggerated is the swishing.

3. The *upright sleeked posture*, in which the ornamental plumage is held tight against the body in a ritualised posture. This is shown in the Laman & Scholes videos at the Cornell Lab of Ornithology ML465288 and ML465692. When seen from in front, the reddish belly feathering and the spectacular iridescent green of the throat and neck are highlighted. The long tail hangs directly below the bird in line with the body axis, and the bird appears to be looking intently at something in the distance, with the bill pointed and slightly upturned. This particular display has so far been noted only for this species, but it may be part of a more complex display and may be found also in other still poorly known astrapia species.

4. The *branch-sidle*, which is a lateral movement along a branch in a horizontal position with the body more or less parallel to the ground and bill pointed forward, the bird peering intently. This is another display behaviour so far known only for this species,

but it should be looked for in others of the genus, especially the closely related Princess Stephanie's Astrapia.

The Ribbon-tailed Astrapia has also been seen to practise nape-pecking, using a vertical knob on a branch as a surrogate female, initially crouching in a way similar to that in the *hunchback pivot*, then adopting a rigid posture resembling the *sleeked upright* but not so vertical, and then repeatedly lunging at a knob of wood in a ritualised motion, pecking at it with the bill. One individual actually copulated with the piece of wood then hopped off and performed hunchback pivoting before perch-hopping rituals.

NEST/NESTING This species nests in forest, including disturbed forest with secondary growth, and sometimes in remnant patches within subalpine grassland. It seems to prefer isolated small trees or saplings with no nearby canopy cover (F&B), and several old nest structures have been found in such sites (Kwapena 1985; Frith & Frith 1993b) adjacent to active nests in the same tree. This is known also for Loria's Satinbird and the Short-tailed Paradigalla, and it may reflect a climbing-predator avoidance strategy or simply a traditional favoured site which may be close to food sources or adult males (F&B). **Nests** are usually in near-vertical forks of trees or tree-ferns between 3m and 18m above ground (F&B), although one was on a tree stump some 2m high (Kwapena 1985). The nest itself is a firm, substantial cup of large leaves and pandanus-frond pieces, or thin rootlets and vines or orchid stems (some living), with a base of moss and containing leaf fragments and ferns (F&B; Coates 1990; Kwapena 1985; Gilliard 1969). The long green orchid stems are denser around the outer rim, and the egg cup is lined with finer dead straw-like leafless orchid stems (F&B). **EGGS, INCUBATION & FLEDGING** A single egg is laid, and incubation is as usual by the female only; the duration of incubation is uncertain. The fledging period is believed to be between 22 and 26 days (Boehm 1967). The Friths (Frith & Frith 1993b) noted a delayed fledging owing to adverse weather conditions, which must be common at the elevations occupied by this species, the chick emerging on to an adjacent old nest after 25 days and still there after 29 days following a period of wet weather. **TIMING OF BREEDING** Breeding is noted from May to March, with display seen in June–August and December, indicating that breeding may be feasible at any time of the year and may depend on the conditions in a particular year.

HYBRIDS Hybrids between Ribbon-tailed and Princess Stephanie's Astrapias are not infrequent at the lower levels of the Ribbon-tailed habitat, and these may be problematic; their outer tail feathers are intermediate in shape and the central tail feathers show quite extensive amounts of white around the vanes. See under TAXONOMY AND GEOGRAPHICAL VARIATION (above).

MOULT Individuals of this species have been seen moulting from July/August onwards at Tari Gap. Fully plumed males occur there from January to November, but the main moult seems to commence in July–August in most years and fully plumed males are scarce after those months. Individuals from August onwards often have very bedraggled-looking plumes, badly worn and frayed, and often lacking the black terminal tips. Museum specimens show evidence of moult almost throughout the year, with a peak in July–November, which is consistent with observations at Kumul Lodge, in Enga Province.

STATUS AND CONSERVATION Classified as Near Threatened by BirdLife International. A restricted-range PNG endemic, but quite common in much of this range, numbers of individuals seen being in double figures on many days, although adult males are much scarcer than female-plumaged birds. The plumes are used in tribal people's head-dresses in the Tari and Wahgi valleys; when the men are dancing the tall white plumes nod forward and bob, rather like the display by the birds themselves! This species remains common enough, though, and seems unlikely to be threatened in its remote and lightly settled montane habitat, although logging can be a local problem.

Ribbon-tailed Astrapia, immature male, Kumul Lodge, Enga Province, PNG (*Phil Gregory*).

Ribbon-tailed Astrapia, male, Kumul Lodge, Enga Province, PNG,18 March 2013 (*Phil Gregory*).

Ribbon-tailed Astrapia, male, Kumul Lodge, Enga Province, PNG, 17 April 2014 (*Phil Gregory*).

PRINCESS STEPHANIE'S ASTRAPIA
Astrapia stephaniae Plate 7

Astrarchia Stephaniae Finsch and Meyer, 1885, *Zeitschr. Ges. Orn.* **2**: 378. Mt Maguli, Southeast Peninsula.
Other English names Stephanie's Astrapia, Princess Stephanie's or Stephanie's Bird of Paradise, Princess Stephanie Bird of Paradise
Etymology *Stephaniae* is the Latinised genitive form of Stephanie, and is given for the tragic Princess Stephanie, Crown Princess of Austria, the wife of Crown Prince Rudolph (see Blue Bird of Paradise), perpetrator of the infamous murder–suicide of his

mistress and then himself at Mayerling in 1889. It was discovered by the collector Carl Hunstein (who sadly drowned off New Britain), in the Owen Stanley Mts in 1884.

This species is a PNG endemic found over much of the central cordillera, and overlapping with the Ribbon-tailed Astrapia in the west of its range.

FIELD IDENTIFICATION The adult male is a distinctive large astrapia with a very long black tail but with no pom-pom at the base of the bill. It bears a curious resemblance to a gigantic Broad-tailed Paradise-whydah (*Vidua obscura*) in flight. Females are heavily barred black on cinnamon below, have narrower blackish bars than the female Ribbon-tail, and show no or hardly any trace of a pom-pom. SIMILAR SPECIES Males, with their long paddle-shaped black tail, are distinctive on a good view. The only major confusion species within the range is the female and immature Ribbon-tailed Astrapia, and these plumages are more difficult. Princess Stephanie's Astrapia has more rounded and broader tail feathers than the narrower and more pointed ones of the Ribbon-tailed Astrapia, and they have a little white showing along the basal vanes of the central pair and usually none on the tail feathers themselves. They also show no traces of a pom-pom at the base of the bill and have more blackish (not dark brown) wings and tail, with less intense oily bluish-green iridescence on the head and narrower barring beneath. Hybrids are not infrequent at the lower levels of the Ribbon-tail habitat, and these may be problematic; their outer tail feathers are intermediate in shape and the central feathers have quite extensive amounts of white. The Splendid Astrapia does not overlap geographically. Reports of Splendid Astrapia at Ambua are in error (Tolhurst 1989), and refer to hybrids between Ribbon-tailed and Princess Stephanie's Astrapias (Barnes's Long-tailed Astrapia: see Frith 1995b). The male Brown Sicklebill is of a similar size, but has a strikingly decurved bill, a pale blue eye, and long black pointed tail feathers; females are brown above with a rufous cap, have a pale blue eye and pointed tail feathers, and are heavily barred beneath. The Black Sicklebill, occupying much the same habitat as Princess Stephanie's Astrapia, overlaps with it at the upper portions of its altitudinal range; again, the bill shape, tail shape and red eye colour would readily distinguish the Black Sicklebill.

RANGE Restricted to the central cordillera of western PNG, where it is the easternmost representative of the genus, ranging from the Tari Gap and the Bismarck, Schrader, Kubor, Tondon and Doma Peaks Ranges to Mt Hagen and Mt Giluwe and east to the Owen Stanley Mts. Overlaps with the Ribbon-tail around Mt Hagen and Mt Giluwe to near Wabag and the Tari Gap. The western limits are not well known. **MOVEMENTS** Resident.

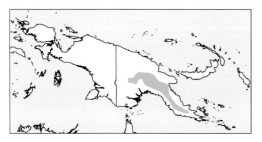

DESCRIPTION Princess Stephanie's Astrapia is another very striking member of a spectacular genus, the male with an unusually long and rather broad paddle-shaped tail. Some good videos by T. Laman (of this species feeding) are available at the Macaulay Library at the Cornell Lab of Ornithology, as at ML469270. Sexually dimorphic, as are congeners.

Adult Male 37cm, but 84cm with central rectrices included; weight 144–169g. Head is velvety black with blue/violet/magenta iridescence on crown, sides of face and nape, bluer on sides of face in some lights. There is a small tuft of black feathers at base of bill, not nearly so well developed as the pom-pom of the Ribbon-tail. The lores and the large broad breast shield are a vivid iridescent blue-green, appearing blue-sheened to purple-sheened at some angles. The upperparts are brownish-black, tinged glossy lime-green on the back, the flight-feathers tinged purplish. The central tail feathers are black with violet-purple iridescence and have a distinctive greatly elongated paddle-shape; the feather shafts often show as white, which can be quite striking when the bird is feeding. There is a broad velvet-black band across the lower breast, edged with rich coppery-red as with other members of the genus. *Iris* dark brown; *bill* black, *mouth-lining* pale green; *legs and feet* leaden-grey. *Adult Female* 53cm; weight 123–159g. The head, upperparts and chest are dull blackish-brown, with bluish iridescence on the head; rest of underparts below chest tawny to buff-brown, finely barred black. *Iris* dark brown or brownish-grey; *bill* black, *mouth-lining* greenish-yellow; *legs* dark greyish, sometimes with black *feet*. *Juvenile* is black above, with rich rufous nape feathering that may extend onto hindcrown. Plumage soft and fluffy, underparts buff, barred black. *Immature* resembles female; first-year male develops more iridescence on head, throat and upper breast and may retain rufous on nape. *Subadult Male* is variable, depending on state of moult: some like adult female with a few male-type feathers showing, others much as adult male but with some female-type feathers remaining. As usual with the genus, the central tail feathers become progressively longer as other rectrices become shorter. Time taken to reach adult plumage is uncertain but presumably much as with congeners, around 5–6 years, which is why female-plumaged birds greatly outnumber adult males.

TAXONOMY AND GEOGRAPHICAL VARIATION
Polytypic, different authors recognising two or three races. Beehler & Pratt (2016) recognise two subspecies, as below, but with the note that *feminina* is likely to be merged into the nominate:

1. *A. s. stephaniae* (Finsch and Meyer, 1885) (following Gilliard and LeCroy 1968, Cracraft 1992 and F&B), subsuming the poorly defined *ducalis* into the nominate form. This occupies most of the range from the south-eastern Owen Stanley Mts north-west to the central cordillera, including Mt Hagen, Mt Giluwe, Nipa and Doma Peaks and the Tari Gap area.

2. *A. s. feminina* Neumann, 1922. Occupies the Schrader Range, Bismarck Range and the Sepik–Wahgi divide, to the north of the nominate race. Shows less contrast between the crown and nape colour and that of the back, with crown and nape more bluish-black in adult males. Adult female has longer tail and shorter wings than the nominate (F&B).

Hybrids with Ribbon-tailed Astrapia are well known where the two species come into contact; see the text relating to **Barnes's (Long-tailed) Astrapia** under

Ribbon-tailed Astrapia TAXONOMY. This hybrid has a really quite extraordinary pedigree among the varied and colourful characters involved in the discovery of the various astrapias, with a complex and fascinating back-story. Princess Stephanie's Astrapia was named for Princess Stephanie of Austria, the wife of Crown Prince Rudolph (heir to the Hapsburg throne who committed suicide), which meant that the succession passed to the son of Archduke Franz-Ferdinand, whose assassination was one of the triggers for World War One. When the Prince was 30 he had a love affair with a 17-year old baroness, which ended in tragedy in a murder-cum-suicide at Mayerling in 1889, leaving Princess Stephanie bereft but immortalised with the wonderful astrapia named after her. Sadly, the epithet 'Princess' is these days often omitted for the much more prosaic Stephanie's Astrapia. It was common to name birds of paradise after royalty, as the collectors looked to gain possible sponsorship and an early kind of 'celebrity' endorsement. The tragic Crown Prince Rudolph himself has the rare Blue Bird of Paradise named after him, *Paradisornis rudolphi*. These links with German royalty also serve to remind us that, from the 19th century onwards until the First World War, the north-east of the island was a German colony (thwarted German colonial ambitions and jealousies also being among the causes of World War One).

VOCALISATIONS AND OTHER SOUNDS This is not a particularly vocal species much of the time, although it can be noisy near fruiting trees in the early morning. Males call a shrill, scolding, piercing *wok wik-wik* similar to the call of male Ribbon-tail at the display area, or a disyllabic *wauk wik* and variants. The calls on xeno-canto XC87571 and XC87572 (both by I. Woxvold) are typical, and those at the Macaulay Library at the Cornell Lab of Ornithology ML100595 (by A. Mack) present typical calls and the wing-rustling of the male. Shrill *quee quee quee* notes were given by a captive male, plus a weak cat-like *meow* by a female (Gilliard 1969), or an upward-inflected frog-like *whenh?* (Beehler *et al.* 1986). Various harsh, drawn-out scolding notes are made by female-plumaged birds, and a melidectes-like *hoo-hee-hoo-hee* is given by two or three birds at the same time (Gilliard 1969; Beehler *et al.* 1986; Coates 1990). Majnep & Bulmer (1977) refer to males making soft *ss, ss, ssw, ssw* sounds, females having a squeaking sound. An immature gives plaintive *weep*-type calls, probably soliciting attention from an adult female, as at XC24706 (by F. Lambert). The males produce a loud wing-rustling sound in flight; the Macaulay Library at the Cornell Lab of Ornithology has a good example at ML100582 (by A. Mack), where two males are chasing around in the late afternoon. This distinctive rustling or swishing sound is similar to that made by riflebirds and parotias, and gives rise to the male's vernacular name ('*ksks*') in Kalam country (Majnep & Bulmer 1977).

HABITAT This species lives in montane up to subalpine moss forest, from 1,280m at Efogi, in Central Province (Coates 1990), to 3,500m, but generally at 1,500–2,800m. It will inhabit disturbed and selectively logged areas, plus regenerating burned areas, and readily visits gardens and forest patches. Where sympatric with Ribbon-tailed Astrapia, it is usually not found above 2,450m, the higher elevations being occupied by that species (Coates 1990; pers. obs.). Hybridizes with Ribbon-tailed Astrapia in a narrow altitudinal band from *c.*2,200m to 2,600m, but maintains its specific identity. Those at Ambua are found at 1,900–2,200m, being replaced above that altitude by the Ribbon-tail in a very sharp transition there.

HABITS Princess Stephanie's Astrapia tends to keep higher in the habitat than the Ribbon-tail, more in the upper and middle strata of the forest, and is seldom found in the lower tiers. It also tends to be patchy in distribution in the forest, males keeping to taller, denser areas and having leks where they call and display. It is not shy but is often wary, though it can be confiding at fruiting trees. Males tend to be fairly silent unless displaying, and are hence unobtrusive; they perch high in forest trees, but not right on top. They have been seen bathing at forest pools, where they may be taken by native hunters concealed in hides (*cf.* King of Saxony) (Majnep & Bulmer 1977). The flight style is similar to that of the Ribbon-tail, being four or five flaps followed by a downward glide, giving a shallowly undulating flight.

FOOD AND FORAGING Princess Stephanie's Astrapia forages at all levels of the forest but is seen mostly in the middle strata. It seeks food on tree branches, and at fruit, leaves and flowers, and will tolerate other paradisaeid and frugivorous species feeding there. Fruit is a major part of the diet, and two large fruiting trees full of small black berries, which ripened consecutively in the garden at Ambua over a two-week period in August 2001, were visited daily by several female-plumaged individuals of this species. Female-plumaged individuals of Lawes's Parotia, Superb Lophorina and Blue Bird of Paradise also visited these trees, with at least one Princess Stephanie's Astrapia virtually constantly present as the other species came in and out, tolerating their presence; others feeding there at the same time included Crested Satinbird, Loria's Satinbird, an immature Brown Sicklebill and MacGregor's Bowerbird, along with Superb Fruit-dove (*Ptilinopus superbus*), Mountain Fruit-dove (*P. bellus*), Great Cuckoo-dove (*Reinwardtoena reinwardtii*) and Tit Berrypecker (*Oreocharis arfaki*) on regular occasions. Kwapena (1985) saw young being fed on the fruits of *Pittosporum* and *Rubus* species. Two or three (and occasionally up to six) female-plumaged Princess Stephanie's Astrapias may be seen feeding together, although the adult males frequently appear to be solitary feeders and do not often come to gardens, preferring the denser forest. Arthropods and small vertebrates such as skinks and frogs may form a significant part of the diet. The birds obtain them by clinging to tree trunks or by hopping along branches and pecking into epiphytes, mosses and lichens. Fruit was estimated by Beehler & Pruett-Jones (1983) to represent 85% of the diet, with the remainder animal and other prey, although this no doubt will vary locally. *Schefflera* species are a particular favourite, as with the Ribbon-tail, and the birds will share fruiting spikes of this plant with other individuals, spending much time there. Pandanus seems to figure much less in the preferences of Princess Stephanie's Astrapia compared with the Ribbon-tail, both as fruit and as a source of arthropods and insects (Kwapena 1985; pers. obs.). There seem to be no published observations of its digging into dead wood, either. Other species may displace the astrapias at fruit sources, both Belford's (*Melidectes belfordi*) and Yellow-browed Melidectes (*M. flavifrons*) being frequent harassers which sometimes succeed in driving the astrapias away. Opit (in F&B) saw a Great Cuckoo-dove displace an astrapia, while Beehler (in F&B) conversely saw an astrapia chase off a Short-tailed Paradigalla. Moreover, Laska (in F&B) saw two Princess Stephanie's Astrapias fly at and displace an adult male Ribbon-tail, and also witnessed them displacing two female-plumaged Ribbon-tails, an adult male Brown Sicklebill and an adult male King of Saxony. Princess Stephanie's Astrapias may join mixed-species foraging flocks with Brown Sicklebills, and two or three were once seen with five or six Crested Satinbirds, which foraged directly below the astrapias (Coates 1973a). Beehler (in F&B) saw an interesting association with a female-plumaged Brown Sicklebill, which was insect-hunting with a female-plumaged astrapia on the same branch, the two birds 3m apart and following one another as they moved about.

BREEDING BEHAVIOUR Polygynous, with promiscuous males, and the females alone responsible for nest duties. Breeding is known from May to December, but males display at any time of the year, especially during the drier months, and less so when in heavy moult in June–August. **COURTSHIP & DISPLAY** The males perform at **communal leks**. Healey (1978b) located four leks some 1.5–2km apart, each out of visual and auditory contact with the others, on the crest or side of steep-sided ridges. Display sites consisted of 4–7 trees about 25m tall with main bare limbs forking out from the trunk at heights of 17–18m. One or two trees are favoured, these typically having long, straight, bare branches beneath the canopy and sloping up at a fairly steep angle from the trunk, not horizontally (Majnep & Bulmer 1977; F&B). Leks may be occupied for many years. One in a partly cleared and regenerating site studied by Healey (1978b) had been used for 20 years, the adult males being hunted for their plumes there throughout (indicating low-level disturbance with traditional technology!). At another lek studied by Healey one tree had been in use for at least ten years, and a lek site at Ambua has been used for a similar period. Between two and five adult males gathered at dawn (06:00 hours), sometimes with female-plumaged birds present. Males would stay for about two hours, leaving only briefly to feed, preening and calling occasionally and hopping through to adjacent trees. The display of one would stimulate an immediate response from another, though only one at a time, other excited birds calling and hopping about or flying from tree to tree. Female-plumaged birds do not have to be present for displays to occur, as with *Paradisaea* species, but displays may not occur even if they are present. This accords well with observations above Ambua, where three or four males congregate similarly. Healey's sites were used

also in the late afternoon, which seems not to be the case at Ambua, while Kwapena (1985) noted them in use during mid-morning rain. Healey noted two levels of **display**. The *Low Intensity Display* consisted of the male hopping between two perches at the same level and sometimes briefly touching an intervening perch; he would pause at each main perch, the body held near vertical and the tail being swung forward in an inverted V under the perch, with a flicking sound audible (the tail striking the perch or the wings being rapidly opened and closed may have caused this). No other sounds were heard and the duration was about two minutes. This display, which was noted on some six occasions in August, seems analogous to the *perch-hopping* displays noted by Scholes *et al.* (2017) for other *Astrapia* species. The *High Intensity Display* was also silent except for the sound of wings or tail: the bird commenced by flying to the display tree, then hopping rapidly and briefly, for no more than four seconds, some nine or ten times between two perches about 1.5m apart; the tail streamed behind the fast-moving bird, not swinging below the perches, with the body held horizontal, shoulders hunched and bill cocked slightly upwards. This display was noted just three times in July. Both types of display ended with the bird flying from its perch. The tail can produce a loud churring sound, with ripples flowing along it, and this was noted both at the display site and away from it. Gilliard (1969) saw a male at Mt Hagen finish feeding and then fly to a horizontal limb 15m up beneath the canopy: 'It perched across the limb with the tail hanging in a wide inverted V, then displayed by lifting the wings in a most peculiar manner as though stretching, so that the primaries were held at right angles to the body and the wrists were held in a touching position over the back. The head was pulled down in a crook. This position was held for at least three seconds.' Kwapena (1985) was told by a man from Mt Giluwe that he had found a pair of birds which had fallen down from a tree while mating, still grasping each other by the thighs. He duly collected them in a *bilum* (string bag) and brought them home. Beehler (in F&B) witnessed a male flying at a female-plumaged bird perched by another such bird, and displacing it by locking claws and tumbling some 20m to near the ground. Captive individuals that have shown very violent pre-copulatory sequences also suggest an unusual mating strategy. These include grasping the thighs by the claws until the female is submissive, which has led to the deaths of males at times, and a pair in the throes of battle would have drowned had the two not been rescued from a pond (Boehm 1966). This has parallels with the so far unique remarkable *post-copulatory tumble* of the Huon Astrapia, one of the most extraordinary in the entire family, and it will be fascinating to see what is discovered by future field research. An excited male near Ambua in August was watched as it erected the iridescent breast shield, making the colours catch the light in stunning fashion, with the coppery-bronze lower border showing prominently, while the iridescent crown and ear-coverts also appeared flared and more prominent than usual. **NESTS** The few nests found have been sited from 3.8m to 10m above ground, in forked tree branches or in climbing bamboo within the forest (F&B; Kwapena

1985). The nest is a thick shallow cup of large leaves and creepers, lined with root fibres (Bulmer in Gilliard 1969), and sometimes partly concealed by surrounding vines (F&B). The nest collected by Kwapena was oval and lined with rootlets, small leaves and bark debris, the outside consisting of orchid and fern stems and leaves, like that of the Ribbon-tail (F&B). This seems to be a feature of the nests of these two species. Pratt (in F&B) gives a curious observation of a bird in adult male plumage that several times collected moss, epiphytic orchids and berries to add to a nest that contained shell fragments. **EGGS** Single-egg clutch. Egg colour appears normal, pale pinkish-buff with brown streaks, some red-brown spots, glossy and smooth-surfaced. Egg size measurements in the wild 39 × 27mm and 36.1 × 26.1mm (F&B). **INCUBATION & FLEDGING PERIODS** Data from the wild are very limited, but it would seem that the female alone incubates, as would be expected with a lekking species, and an incubation period in captivity was 22 days (F&B). Captive nestlings have fledged at 26–27 days (Yealland 1969; Everitt 1973).

HYBRIDS This species hybridises readily with Ribbon-tailed Astrapia where the two come into contact; see TAXONOMY AND GEOGRAPHICAL VARIATION above. See also the text relating to **Barnes's (Long-tailed) Astrapia** under Ribbon-tailed Astrapia TAXONOMY.

MOULT Adults are in heavy moult in June–August. Males at Ambua in November have usually lost their central tail plumes and are clearly moulting in October as well. Healey (1978b) had informants who gave June–August as the moult time, coinciding with waning display activity, while other informants gave the moult time as November–February near Jimi River. Museum specimens show evidence of moult in all months except December, while F&B give March–November, but especially March–June, as the peak time.

STATUS AND CONSERVATION Classified by BirdLife International as Least Concern. Quite a common species in some places, but tends to be patchy in occurrence even in areas of seemingly suitable habitat. The wide range and the existence of much remote lightly settled montane habitat mean that it is not threatened, despite the fact that its tail plumes are a very popular item for decoration (*bilas*) among highlanders, such as in the Tari and Wahgi valleys, where the black plumes form the centrepiece of many head-dresses. Some villages harvest males at leks, and have done so for many years without causing serious decline in the astrapia's population so long as traditional spears and bows and arrows are used (although the advent of home-made shotguns may locally present a different story). Logging can be a problem in some areas near highways, as witnessed near Ambua Lodge where road upgrades and a big increase in subsistence logging have led to recent declines in this species' numbers.

Barnes's **Long-tailed Astrapia**, male, Ambua, PNG (*Phil Gregory*).

Princess Stephanie's Astrapia, female plumage, Ambua, PNG (*Phil Gregory*).

Princess Stephanie's Astrapia, male (*Markus Lilje*).

HUON ASTRAPIA
Astrapia rothschildi   Plate 7

Astrapia rothschildi Foerster, 1906. In Foerster and Rothschild, *Two New Birds of Paradise: 2*. Rawlinson Mts, Huon Peninsula.

Other English names Rothschild's Astrapia, Huon Bird of Paradise, Huon Astrapia Bird of Paradise, Rothschild's Long Tail, Lord Rothschild's Bird of Paradise

Etymology Named *rothschildi* after the famous bird-collector Lord Walter Rothschild, collected by Carl Wahnes in the Rawlinson Ranges and sent to his lordship's private museum at Tring, in Hertfordshire, which is now the site of the Natural History Museum (NHMUK).

FIELD IDENTIFICATION This is a distinctive black species with a long, broad tail. It is endemic to the mountains of the Huon Peninsula and is the only astrapia in its range. The male is unlikely to be confused with any other species except perhaps male Wahnes's Parotia, which has a considerably shorter tail, six head wires and (both sexes) a cobalt-blue eye. Female Huon Astrapias also have a long, rather broad tail and are almost entirely sooty blackish, with some barring on the belly and undertail-coverts, again quite distinctive. **SIMILAR SPECIES** Neither Brown nor Black Sicklebill occurs in its range, and no other astrapia overlaps in range. Female Wahnes's Parotia is smaller and shorter-tailed, brown (not black) above and heavily barred below, with a dark head, prominent facial patterning and cobalt-blue eye.

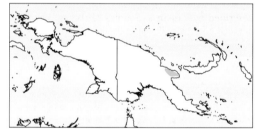

RANGE A restricted-range PNG endemic, found only in the mountains of the Huon Peninsula (the Saruwaged, Rawlinson, Cromwell and Finisterre Ranges), where still very little known. **MOVEMENTS** Resident.

DESCRIPTION Sexually dimorphic, as is normal with this genus. *Adult Male* 69cm; weight 186–205g. A distinctive blackish, long-tailed astrapia that bears some resemblance to the geographically distant and not very closely related Arfak Astrapia, rather than to any others of the genus. Head, foreneck and upper breast are velvety-black, with a bluish iridescence on the head and neck; the throat and breast feathers form a large dark shield that can be erected, and there is an erectile cape from the hindneck to the upper back with the feathers tipped glossy pinkish-violet and bordered blue-green subterminally. Remainder of upperparts black with a bronze-green tinge, and the very long broad tail (reminiscent of a less well-developed Princess Stephanie's Astrapia tail) has a violet sheen. The lower edge of the black upper breast is bordered with an iridescent coppery-orange band, a feature found in this entire genus although the colour varies slightly; the abdomen is deep oil-green, the rest of the underparts blackish. *Iris* brownish-black; *bill* black; *legs and feet* grey-brown. **Adult Female** 47cm; weight 143–200g. Mostly dull brownish-black, with narrow pale brown to whitish barring on the lower abdomen; hindneck on some females may be lightly barred pale brown. Tail very long and broad, but not so long as that of the male. *Juvenile* is undescribed. *Immature* is like adult female, but lacks pale nape barring and has abdomen barring much reduced; blue iridescence on head increases and abdomen barring decreases with age. *Subadult Male* variable, depending on state of moult: some are like adult female with a few male-type feathers showing, others much as adult male but with some female-type feathers remaining, the head developing increasing iridescence and the abdomen barring decreasing with age. As usual with the genus, central tail becomes progressively longer as other rectrices become shorter. Time taken to reach adult plumage uncertain, but presumably much as with congeners, around 5–6 years, which is why female-plumaged birds greatly outnumber adult males.

TAXONOMY AND GEOGRAPHICAL VARIATION
Monotypic.

VOCALISATIONS AND OTHER SOUNDS Little
known. Captives gave a thin scolding *kak kak kak kak!* as a call or alarm note (Crandall 1932). Xeno-canto XC324252 and XC324251 (by T. Mark) record a rather harsh tearing scold; this may be the throaty *rawk* note described by Laman & Scholes (2017), given when perched upright in display. Pratt in F&B mentions calls of wild birds as resembling some of those of Princess Stephanie's Astrapia, a muffled scolding *jj jj* or *jiw jiw*. In Macaulay Library at Cornell Lab of Ornithology, recording ML147993 (by T. Pratt) includes some odd, rather sibilant noises that may include wing noise. An adult male gave a five-note call that was more melodious than the scream of a parotia but not the yelp of a Princess Stephanie's Astrapia. The adult males produce a rustling sound in flight, like others of the genus, and there is also a non-vocal component in which the wings make a *shhek* noise during some displays.

HABITAT Montane forests from 1,460m to 3,500m.

HABITS This is an unobtrusive and fairly silent species
of the dense mid-montane and upper montane forests. Adult males are greatly outnumbered by females and immatures, as with other members of the family, and seem to be more frequent at higher altitudes. The flight is a gently undulating one typical of the genus, the wings after four or five flaps being closed in a gentle glide, before flapping again. Female-plumaged birds seem to be quite common and relatively unwary at some sites, whereas adult males are scarce and perhaps much shyer. This species frequents all stages of the forest except the topmost canopy, but feeds mainly in the middle and lower strata.

FOOD AND FORAGING This species is known to
eat *Pittosporum* seeds and the fruits of *Schefflera* and *Freycinetia* pandans, and its diet is no doubt similar to that of its congeners. It forages among mosses and

epiphytes, and has been seen to inspect knotholes in trees (Pratt in F&B). Lambley (1990) saw one feeding on the berries of the introduced Canadian elder (*Sambucus canadensis*). R. Donaghey (*in litt.*) reports that the nestling diet was mostly insects and some fruit, and once a small skink.

BREEDING BEHAVIOUR Very poorly known. Display has been recorded in January–February. Presumed to be polygynous with promiscuous males, and female alone undertaking nest care, as with congeners. **COURTSHIP/ DISPLAY** It is still not known whether or not the males form leks. Pratt (in F&B) heard an adult male make a loud whistling noise as it dived to chase a female-plumaged bird. C. Benjamin (*in litt.*) reports that female-plumaged individuals were plentiful, visiting fruit trees in a regular pattern in July 2017; males were much scarcer and harder to observe. A male was using a tall tree on the crest of a partly cleared ridge as its display tree, with a fruiting tree 15m away and also on the ridge; when a female came in to feed, a male would make several swooping passes with a distinct 'fluttering' noise coming from its tail (?wings) and then fly back to its display tree. As no mating display was observed, this behaviour may perhaps have been designed to entice the female. Videos by E. Scholes at the Macaulay Library at Cornell Lab of Ornithology depict this species well, ML458213 showing some display activity between a male with breast shield erect and another bird; ML458062 shows a male perched horizontally with cape and breast shield erect, the long tail spread out to appear really wide and being flicked and depressed at intervals, the black feather vanes being very shiny. ML456268 (by T. Laman) shows a male in display pose, again mainly horizontal, with the very long tail spread right out and being quite slowly waved about like a flag, reminiscent in shape of a giant Broad-tailed Paradise Whydah (*Vidua obtusa*). ML456250 (T. Laman) is of a male with breast shield partially erect, wing-flicking and waving the broad tail, as well as briefly turning upside-down on the perch, while ML456244 has the bird holding an upside-down pose with tail spread and breast shield erect for some 20 seconds, flicking the tail feathers in and out. Observations by Crandall (1932) on captive individuals describe two forms of display, both performed in silence. Performing a *Simple Display*, the adult male stood erect on the perch, tail pushed forward and slightly spread; the dark blue gorget was widely spread and flattened, its fiery golden margin glowing conspicuously and the green breast feathers laterally expanded; the bird stayed rigid for about ten seconds. Similar displays seen later involved the lateral tail feathers also being rapidly opened and closed. A *Complex Display* involved the male turning backwards under the perch at right angles to it, the body nearly horizontal with the anterior portion slightly lower than the posterior, the head and neck turned upwards at one end and the tail so at the other, so that the bird formed a kind of semicircle shape; the gorget and abdomen feathers were spread upwards around the head, with the gold margin again very conspicuous, and the ear-coverts were spread upwards likewise to join with the elevated ruff on the nape (a similar behaviour for captive

Princess Stephanie's Astrapia is briefly noted by Boehm 1966). The wings were pressed tight against the body, the back feathers expanded to cover them partly, and the tail was held upright and at first widely expanded, then the lateral feathers were rapidly opened and closed, the middle pair remaining stationary. This display was enacted four times at intervals of 4–5 minutes, each period lasting 10–15 seconds. This remarkable display is classified by Scholes *et al.* (2017) as the *inverted tail-fan display* (see below). Pratt (in F&B) saw solitary wild males perform the *Complex Display* between 06:50 and 09:50 hours in January and February: display trees were located on the broad crest of a ridge and often at the edge of a clearing. The throat and upper breast feathers seem to be expanded to form a disc with the raised and elongated nape feathers. Scholes & Laman (2017) made an exhaustive analysis of video materials in the Cornell Lab of Ornithology for the genus *Astrapia*, and made a number of significant discoveries, identifying some five distinctive **male courtship behaviours**:

1. The *perch-hop*, simply a series of short hops among branches, generally used both before and after other display activities. Can be used also to chase female-plumaged birds and other males, and can occur outside the actual display site as well.

2. The *flick-pivot*, repeatedly turning laterally from side to side while flicking the wings and tail open and shut. This is shown in the Laman & Scholes videos at the Cornell Lab of Ornithology (ML456250 and ML458238): the body is raised and lowered without the feet leaving the perch, and this is accompanied by a non-vocal *shhek* sound similar to that produced by the wings in flight; the bird may include a 180° rotation to face in the opposite direction. Female-plumaged birds are not on the display perch, this display occurring prior to the higher-intensity ones that lead to copulation. A variant action that can be classified here includes one in which the male holds the body, outstretched head and tail in a horizontal line, the bird appearing to be 'pointing' with the entire body. There is an acoustic component to some of the displays whereby the wings are flicked open and shut, producing a sound reminiscent of a baby's rattle and sounding like a repeated *shhek shhek shhek* with each wingbeat. (In the *flick-pivot* display the *shhek* notes may be given singly or as a two-note sound.) Vocal utterances are limited to a quiet throaty *rawk* given when the bird is perched upright and ranging from one to three *rawk* calls about a second apart; during calling the tail swings back and forth slightly, and is conspicuous as it accentuates the call by virtue of its length.

3. The *inverted tail-fan*, elaborate fanning of the tail while hanging upside-down, is routinely performed to females. This is shown in the Laman & Scholes videos at the Cornell Lab of Ornithology (ML458066 and ML458217). The male lowers himself tail-first backwards until he has become completely inverted, with the intense iridescent green belly feathers and the black underside of the tail facing skywards; the body feathers are sleeked down, and the plush ornamental breast and throat feathers erected outwards to form a black circular shape with an intense orange fringe that nearly encircles the head. The tail is moved repeatedly in an exaggerated rocking motion, which makes the tail wave up and down vertically, with the feathers expanded and sometimes creating a gap in the middle. The bird resembles an L-shape or C-shape depending on the attitude of the head, and he lunges repeatedly upwards at semi-regular intervals, which further emphasises the tail-waving and tail-fanning. The male tracks the position of the female and adjusts his own position accordingly to keep his bill pointed towards her. The display is high-intensity, but does not lead directly to copulation.

4. The *upright nape-peck*, a vigorous ritualised pecking at the nape of a female as a pre-copulatory display, and usually after a bout of *inverted tail-fanning*. This is shown in the Laman & Scholes videos at the Cornell Lab of Ornithology (ML458069 and ML458080). The female turns away from the male and wing-flutters, which signals the start of this nape-peck behaviour. The male rotates head-first upright from his inverted posture, with belly feathers sleeked and breast shield fully fanned into a distinct disc-shape. The tail remains cocked under the perch so the whole body forms an upright semicircle. The nape-pecking is a rhythmic plunge forward towards the female followed by a rigid rearing back with a slight pause; it can consist of two or three forward lunges after such a pause, and the pecking can become so intense that the male appears to be raking his bill along the female's neck and back; with each peck the female is uttering nasal cries and 'shrieks' reminiscent of some notes of a *Melidectes* honeyeater, and after several seconds of pecking and lunging the male will mount her, his wings open and flapping, standing on her back with bill towards the head and plumes fully expanded.

5. The *post-copulatory tumble*, a very unusual behaviour so far as is known (though something analogous is reported for Princess Stephanie's Astrapia), whereby the male and female spiral down towards the ground after copulation. This is shown in videos by Scholes and Laman at the Cornell Lab of Ornithology (ML458069 and ML458313); after mating, the male stands on the female's back, flapping his wings and leaning forward so that both tumble, entangled together, off the branch and towards the ground. This tumble can involve a dramatic twisting and spiralling downwards through the canopy for at least several metres. Sometimes the female may not release her grip on the branch, and both hang tangled together in a flapping mass under the perch before dropping down; in another variant, the male alone tumbles and the female simply flies off. This *post-copulatory tumble* is one of the most bizarre and extraordinary displays in the entire family.

Two **female behaviours** are a simple display interaction whereby the female actively engages with the courting male, and a wing-flutter, which is a solicitation behaviour. Courtship took place in the forest canopy at display sites, which are specific, small to mid-sized, largely horizontal branches (perches) from one or several adjacent trees used for display. These display sites appear fairly open and have multiple perches in close proximity that are routinely used for display, one such serving as a primary display perch. **NEST/NESTING** Nesting has been reported in October–November, but is still relatively little known. The nest is a shallow cup composed of roots, vines and creepers, sometimes with a lining of fine hair-like rootlets, built on a foundation of large, strong broad leaves and leaf pieces, leaf skeletons and pieces of moss. The outside of the structure also had odd pieces of moss, mostly on the rim (Frith 1971). R. Donaghey (*in litt.*) reports that a nest was built in an understorey tree with many of epiphytes, with another in the canopy of a tree-fern in a gully. Schmid (1993) reported that nests made from the stems of a *Bulbophyllum* orchid are known by the Nokopo people as 'house of the Huon Astrapias'. **EGGS & NESTLING** The eggs are pale pinkish-buff and variations thereof, with blotches and broad brush-like lavender-grey strokes, overlaid with some chestnut-brown. From a small sample of three single-egg clutches, presumably the norm for this species as with congeners (F&B), average egg size 35.3×27.4mm. At one nest studied by R. Donaghey (*in litt.*), the female brooded the young during the first week for an average of about 53% of the time per day, she regurgitated all food brought to the young, and swallowed almost all of the nestling's faecal sacs; the nestling period was 25–27 days.

HYBRIDS No hybrids with other species are known.

MOULT Museum specimens show evidence of moult in all months except May and September (F&B).

STATUS AND CONSERVATION Classified by BirdLife International as Least Concern. This restricted-range endemic is very little known, almost coming into the category of Data Deficient. The status of the species in the Finisterre Mts needs to be better understood, but elsewhere on the Huon it seems widespread. In January 1994, female-plumaged individuals were quite common in the forests above Satop village, with up to nine seen in a day, but males were exceedingly scarce. Given its remote and inaccessible habitat the species is likely to be secure, but it will experience local declines around areas of settlement. The skins and tail plumes are important cultural items for the Nokopo people (Schmid 1993). A large (760km²) community conservation project called YUS (after the Ypono, Uruwa and Som rivers that flow through it) is underway on the Huon Peninsula, with Matschie's tree-kangaroos (*Dendrolagus matschiei*) the flagship, but this should benefit the other Huon endemics as well.

Huon Astrapia, female feeding, PNG (*Tim Laman*).

Huon Astrapia, adult male, PNG (*Tim Laman*).

Genus *Parotia*

The genus *Parotia* is a distinct group of six montane New Guinea species, two widespread and the others restricted to relatively small areas. Most species have primarily black plumage, with striking iridescent scale-like breast shields and six wire-like head plumes with small spatulate tips. Some species have striking blue eyes, and one is known to be able to change the predominant colour of its iris. All make terrestrial display courts where they perform elaborate courtship dances, some of the most extraordinary rituals in the avian world. Diamond (1972) included this genus in his expanded *Lophorina*, a treatment which has not been widely followed. Two former subspecies have recently been recognised as being distinct enough to warrant full species status, as the Bronze Parotia and the Eastern Parotia. Beehler & Pratt (2016) mention that a local informant has reported a population of long-tailed *Parotia* from the Cyclops Mts, this necessitating further research (Beehler & Prawiradilaga 2010). When displaying, the species in this genus, as with the Superb Lophorina and the Black Sicklebill, can suddenly transform their everyday shapes into something bizarre and otherworldly, rearranging the feathers of the body into complex skirt-like metastructures. These are ornamented with iridescent patches made up of structural colours that are revealed only when arranged in a particular fashion, a remarkable example of evolution that is basically driven by female choice. Laman & Scholes made the remarkable discovery that, when the females of this genus view the display dance from their perches above it, their view is quite different from that when seen from the same level as the displaying male. When viewed from this perspective, the remarkable iridescence of the scale-like breast shield and nape patch would flash vividly as the male does the ballerina or 'wobbling ovoid' dance (Scholes 2008).

Etymology *Parotia* is from the Greek *parotis*, a lock or curl of hair by the ear, alluding to the head wires.

WESTERN PAROTIA
Parotia sefilata   Plate 8

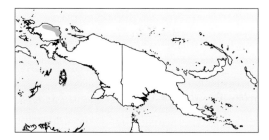

Paradisea sefilata J. R. Forster, 1781, *Spec. Fauna Ind., Ind. Zool.* 40. Arfak Mts, Bird's Head.
Other English names Six-plumed Bird of Paradise, Arfak Six-wired Bird of Paradise, Arfak Six-wired Parotia, Arfak Six-plumed Bird of Paradise, Greater Six-plumed Bird of Paradise
Etymology The specific name *sefilata* is derived from the Latin word *sex*, meaning six, and *filum*, a thread or filament, referring to the head wires. Given by Buffon in 1770.
A longer-tailed and isolated member of the genus, forming a superspecies with the other black parotias, *P. wahnesi* and *P. lawesii/P. helenae*. It is endemic to the Tamrau and Arfak Mts of the Vogelkop and the Wondiwoi Mts of the Wandammen Peninsula of West Papua.

FIELD IDENTIFICATION This is a distinctive, yet typical, species of the group which does not overlap with any congeners. It is one of the longer-tailed parotias, and is an inhabitant of the mid-montane forests. Males are relatively distinctive on a good view, with the characteristic head wires, bright blue iris and silvery-white narial tuft. **SIMILAR SPECIES** Female types might be confused with female-plumaged Arfak Astrapia or perhaps a Long-tailed Paradigalla, both of which have a much longer tail, or a female-plumaged Western Lophorina. The local nominate race of the Western has a largely black head with only a faint supercilium, lacking the stripey face and rusty wings of the eastern birds, so the cobalt-blue eye colour and curious flattened head shape of the female parotia are good field characters, the Western Lophorina also being smaller and paler and having a longer bill.

RANGE Confined to the Tamrau, Arfak and Wondiwoi

Mts of the Vogelkop and Wandammen Peninsulas, in extreme Western West Papua. **MOVEMENTS** None recorded.

DESCRIPTION Sexually dimorphic. *Adult Male* 33cm; weight 175–205g. A large parotia with a medium-length tail. The plumage is almost entirely velvety black, with an erectile triangular frontal crest of elongated finely pointed silver-white feathers atop the forecrown; the rest of the crown has a dark coppery-bronze sheen, with a broad nuchal bar of intensely iridescent blue scale-like feathers glossed purple to magenta depending on the light. There is an ear-tuft of pointed feathers, and three long, erectile, bare wire-like occipital plumes with near-circular spatulate tips extending from behind each eye (giving the six wires of one of the vernacular names). Chin and throat are velvety blackish with coppery-bronze to rich purple iridescence, grading into the breast shield of large intensely iridescent scale-like feathers; the remaining underparts are jet-black with a plum-purple sheen. *Iris* cobalt-blue with pale yellow outer ring; *bill* shiny black, *mouth-lining* greenish-yellow; *legs* purplish lead-grey. It is not known whether or not the iris can change colour, as is the case with Lawes's Parotia. *Adult Female* 30cm; weight 140–185g. Smaller than male, with almost no overlap in wing length, and lacking head plumes and iridescent plumage. The head is blackish, with a broad pale greyish submoustachial stripe flecked blackish-brown, a bold blackish malar stripe, and a very faint greyish-buff superciliary stripe. Upperparts and tail olive-brown, with paler buffish-white chin, throat and underparts all heavily barred blackish. *Juvenile* is very similar to adult female, but with rufous-red edgings on flight-feathers and upperwing-coverts. *Immature* male plumage like adult female. *Subadult Male* variable according to state of moult, from like adult female with a few adult male feathers to like adult male but with a few feathers of female-like plumage remaining; male tail length decreases slightly with age.

TAXONOMY AND GEOGRAPHICAL VARIATION
Monotypic, with no geographical variation recorded. The westernmost representative of the group, this species' relations with *P. carolae* are uncertain.

VOCALISATIONS AND OTHER SOUNDS
Harsh squawking notes typical of the genus are given, these rendered *gned gned* by D'Albertis (1880) and *gnaad gnaad* by Beehler *et al.* (1986), with cockatoo-like notes as well. Males at their courts squawk, but female-plumaged birds have only very quiet high-pitched mewing notes (Kirby in F&B). The court advertisement note is a harsh, somewhat nasal screech, *waugh* or *wengh*, again typical of the genus, and this is probably the harsh cawing note similar to that of a Eurasian Jay (*Garrulus glandarius*) described by Bergman (1957b). Crandall (1932) described captive males as giving a series of harsh squawks that were repeated so rapidly as to resemble the full call of one of the *Paradisaea* species. Good examples of this species' calls can be found on the xeno-canto site, as at XC163276 (a male, by F. Lambert) and XC40932 (a female, by P. Åberg). Like its congeners, this species makes a rustling sound in flight.

HABITAT
This is a species of mid-montane forests at around 1,100–1,900m, with a possible liking for old secondary forest and a fairly open canopy with many small saplings, much as with Lawes's Parotia. Males may prefer the thicker parts of the forest, with females and young at lower elevations.

HABITS
A quite shy and unobtrusive species, it inhabits the mid-montane forests and more open cleared areas. It may occur in the canopy at times, but more usually is seen in middle or lower strata. Little is known of its general behaviour.

FOOD AND FORAGING
Little known. Recorded as taking figs (*Ficus*) and nutmeg (*Myristica*). This parotia, like its congeners, forages primarily in the lower and middle strata of the forest. Pheasant Pigeons (*Otidiphaps nobilis*) and Cinnamon Ground-doves (*Gallicolumba rufigula*) have been seen repeatedly visiting display courts to feed upon seeds found in the droppings of the parotias (F&B).

BREEDING BEHAVIOUR
As with other parotias, males display at terrestrial courts, implying that this is a polygynous species. Females presumably attend to nesting duties alone. Gonad activity suggests breeding in July–January (F&B), but data are sparse. COURTSHIP/

DISPLAY Adult males are presumably promiscuous and may maintain exploded leks, but further study is needed. Courts are between 1m and *c.*2m in diameter in dense forest, and are carefully guarded, being kept clear of fallen twigs and leaves. Display perches on horizontal branches over the court are heavily utilised and may become stunted and worn, and vines and saplings around the court may form a part of the display area up to 3–4m above ground level (R. Kirby in F&B). The display sequence was filmed by R. Kirby in September–October 1994 for the BBC *Attenborough in Paradise* television series, and the following summarises the activities from this film and from Bergman (in Gyldenstope 1955). The adult male tolerates three or four female- or subadult-plumaged birds around the court, but none actually on or in it. No other adult males were seen at these courts, which seem to be the exclusive preserve of the male. These birds make way for the dominant male to enter the court unless they are soliciting him. Males displayed from 06:30 to 08:10 hours and again at 13:00–15:00, spending the first hour of daylight flying about in the forest canopy, advertising with squawking calls and seeming to 'shepherd' several female-plumaged individuals towards the court by means of chases moving progressively down through the canopy and subcanopy. These female-plumaged birds give soft mewing notes during this procedure. Once several female-plumaged birds are gathered, the male, as may be expected, commences to display. He uses a horizontal perch about 60cm above the court edge, and hops back and forth while repeatedly flicking the wings half open and fanning the tail. He pauses occasionally to wipe his bill on the perch, then hops down on to the court, continuing this flicking and fanning sequence, and begins pecking at the ground so as to flash his silver-white narial tuft before a little hop forward. The whole process lasts about 20–30 seconds, after which the male may preen and fly off with a squawk unless he goes into the hopping display. Here he adopts a stiff horizontal posture and hops back and forth in a bouncing gait across the court. The wings may be flicked half out with each hop, and he may adopt what F&B call the *Upright Sleeked Pose*, stretching steeply upwards with head wires flattened against the mantle. The hopping display may be repeated up to a dozen times, pausing beneath the perched female-plumaged birds, before stopping in the court centre to preen and then go into the astonishing 'ballerina' dance, like that of Lawes's Parotia one of the most amazing displays in the entire family. In this, the male adopts a stiff erect posture facing the audience, then makes a slow deep *Initial Display Bow*. He then stretches up with tail cocked to one side, and with the elongate flank plumes erected to form a skirt or tutu about him, the classic ballerina pose. He raises the six occipital plumes and brings them forward to a near-horizontal position while shaking his head rapidly from side to side, causing the plumes to wave and waggle. He moves alternately from side to side, often in a semicircle in front of his audience, side-stepping and using a curious mincing gait, and becoming faster after the first few seconds. He stands beneath and right in front of the assembled birds, with legs wide apart as the occipital plumes are raised and then fanned out; he slightly lowers his skirt and then raises it to near horizontal before suddenly snapping down to crouch on horizontal tarsi and perform intense head-and-neck-waggling, raising himself up on fully extended legs. He then makes a couple of large hops to each side, pointing his head forward and pumping it vigorously down into his shoulders, which causes the iridescent breast shield to flick up and down. The performance climaxes with the male flying up and on to a female for mating for 2–3 seconds, before both birds fly away. This display may stop at any point for mating attempts, the male approaching with wing-flicking or hopping directly on to the female's rump. The females solicit by squatting low on their perch with bill agape and gentle wing-fluttering, curiously reminiscent of juvenile behaviour. Mating may last for 2–3 seconds, the male holding the female's neck feathers in his bill and squawking loudly before flying off along with the other assembled birds. Some elements of the display may be omitted, and young males may copulate with a perch after a dancing display. Kirby saw three copulations during one morning display. The Macaulay Library at the Cornell Lab of Ornithology has a good collection of videos of the display and behaviour of this species by Laman & Scholes (e.g. ML469922 and ML468920). Males have been seen in display from August to October (D'Albertis 1880; Bergman in Gyldenstope 1955; Kirby in F&B), and captives in Sweden displayed all year except during their four-month moult period (Bergman 1958). A court at Mokwam, in the Arfak Mts, was in use for some six years, until the male was killed by hunters (Laman & Scholes 2012). **NESTING** No nests or eggs have yet been found and the breeding cycle remains unknown.

HYBRIDS Hybrids with Long-tailed Paradigalla and Western Lophorina have been recorded. The intergeneric hybrid with Long-tailed Paradigalla is confusingly named **Sharpe's Lobe-billed Riflebird** ('*Loborhamphus ptilorhis*') Sharpe, 1908, presumably because the bill is quite long and even though no riflebird is involved in the parentage. Only a single specimen is known, an adult male. The intergeneric hybrid with Western Lophorina is known from two adult male trade skins and is called **Duivenbode's Six-wired Bird of Paradise (Parotia)** (*Parotia sefilata × Lophorina superba*) Rothschild, 1900.

MOULT Recorded in January and April–July (F&B), and captives have shown a four-month moulting period (Crandall 1932; Bergman 1958). Fresh-plumaged individuals were collected in the Arfaks from mid-July to 20th August (Gyldenstope 1955).

STATUS AND CONSERVATION Classified as Least Concern by BirdLife International. Reported as common in the Arfak Mts in the historical literature and in August 1995 (Eastwood, pers. comm.), although seemed localised and quite scarce in July 2015 (pers. obs.), and historically common on the Wandammen Peninsula (Gyldenstope 1955). This species is likely to be relatively secure, despite its restricted range, as a result of the inaccessible nature of its mid-montane habitat and the low human population density. There are no reports of its being used for decoration (*bilas*) by local people, but this may well occur. Basic biological data concerning nesting and diet are scanty, although it is to be expected that they may be similar to those for Lawes's Parotia.

Western Parotia, female, Arfak Mts, West Papua (*Huang-Kuo-wei*).

Western Parotia, male in display pose, Arfak Mts, West Papua, 29th June 2018 (*Huang-Kuo-wei*).

WAHNES'S PAROTIA
Parotia wahnesi   Plate 9

Parotia wahnesi Rothschild, 1906, in Foerster and Rothschild, Two New Birds of Paradise: 2. Rawlinson Mts, Huon Peninsula.
Other English names Wahnes' Six-wired Bird of Paradise, Wahnes' Six-wired Parotia, Wahnes' Six-plumed Bird of Paradise, Huon Parotia
Etymology The specific epithet *wahnesi* alludes to the German collector and discoverer of the species, Carl Wahnes.

FIELD IDENTIFICATION A long-tailed, isolated member of the genus, this species is endemic to the mountains of the Huon Peninsula and Adelbert Range, PNG, where it is the only parotia. **SIMILAR SPECIES** The male is distinctive and not readily confused with any other species except, perhaps, the much longer-tailed and quite differently shaped male Huon Astrapia. Males of Wahnes's Parotia have six head wires, a long wedge-shaped tail, and a small pom-pom at the base of the bill giving a distinct head shape. Females could be confused with the smaller female-plumaged Superb Lophorina, but are larger, and have a dark head with only a short pale eye-stripe, a pale malar stripe and a long, wedge-shaped tail. Huon Astrapias are much longer-tailed and lack head markings or pom-poms at the base of the bill; the female-plumaged birds also show much less barring beneath. Superb Lophorinas here have distinct pale supercilia and the race concerned (*latipennis*) lacks the dark head of the female-plumaged parotia.

RANGE Endemic to the northern coastal ranges of PNG in Morobe and Madang Provinces alone: Cromwell, Rawlinson, Saruwaged and Finisterre Mts of the Huon Peninsula, and the Adelbert Mts of Madang. No recent records from Adelbert Mts. **MOVEMENTS** Nothing recorded; resident in the mid-montane forests of its limited range.

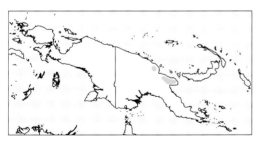

DESCRIPTION Sexually dimorphic. *Adult Male* 43cm; weight 170–172g. A large parotia with a long, wedge-shaped tail which is often cocked slightly upward or to the side. Plumage is entirely velvety jet-black, with a large erectile narial tuft of elongate coppery-bronzed feathers, and a narrow nuchal bar of intensely iridescent blue to pink-purple scale-like feathers. It has also a tuft of pointed feathers behind each eye, each with three long, bare erectile wire-like occipital plume shafts with black spatulate tips;

the black mantle has a faint plum-purple sheen; the velvety-black chin and throat have slight purple iridescence, grading into the breast shield of large scale-like feathers intensely iridescent and bronzed emerald-green to green-yellow with complex and variable purple/magenta to bluish-violet highlights; remaining underparts jet-black with a plum-purple sheen. *Iris* a striking cobalt-blue, with quite broad pale greenish-cream outer ring; *bill* shiny black; *legs* purplish lead-grey. It is not known if the iris can change colour as with Lawes's Parotia. *Adult Female* 36cm; weight 144–154g. Adult female is smaller than male, with no overlap in wing length, and lacking head plumes and iridescence; upperparts rich brown with brownish-blackish head, short off-white superciliary stripe behind eye, and broad submoustachial stripe flecked blackish-grey, bordered below by a narrow blackish malar stripe. Chin and throat whitish-grey, flecked and barred brownish-black, grading to brownish-cinnamon underparts closely barred brownish-black. *Juvenile* undescribed; likely to resemble female but with rufous edgings to flight-feathers and wing-coverts, as with congeners. *Immature Male* is like adult female, but iris cobalt-blue as with adult male; in first-year plumage has pointed rectrices, and male tail length decreases slightly with age.

TAXONOMY AND GEOGRAPHICAL VARIATION Forms a superspecies with the other black parotias, namely Western and Lawes's. Monotypic.

VOCALISATIONS AND OTHER SOUNDS Little known. Data here are taken from F&B and the Macaulay Library at the Cornell Lab of Ornithology. Adult males close to display courts give a harsh, rasping double *khh kaakk* with a cockatoo-like quality, seemingly a court-advertisement call. Notes of a similar harsh screeching quality are a double sharp, dry *wetch* or *snatch*, which may form part of a longer and more complex song based on them. A single loud, nasal *garr* note is also given (G. Opit in F&B). The scream of the adult male is described by T. Pratt in F&B as essentially similar to that of Lawes's Parotia but slightly higher-pitched, quieter and given more often. Macaulay Library ML167001 (by Ben Freeman) gives a good example of these calls. Thin, soft cheeping notes are given as the male is about to descend to the court floor from branches above; a low, nasal twittering may also be given in flight or just after landing, or when the bird is active on perches. There are no recordings on xeno-canto at time of writing. Adult males in flight often produce an audible rustling sound, but seem able to reduce or eliminate this. More observations are needed.

HABITAT Mid-montane forest between 1,100m and 1,700m on the Huon Peninsula and at 1,300–1,600m in the Adelbert Mts.

HABITS This is a quite shy and unobtrusive species that inhabits the dense mid-montane forests. May occur in the canopy at times, but more usually in middle or lower strata.

FOOD AND FORAGING Primarily frugivorous, but will readily take insects and arthropods. This species' feeding behaviour appears similar to that of Lawes's Parotia, foraging in the lower to middle stratum of the forest. Captive individuals have been seen to tear up and eat leaves (Frith & Frith 1979). Opit (in F&B) watched a female-plumaged bird in the Adelberts foraging for arthropods and then hanging head down on an epiphytic ginger to pluck and swallow some 15 of the small yellow fruits. Pratt (in F&B) saw a wild female-plumaged bird hold an insect to its perch and then bite it into pieces, being subsequently attacked and chased in flight by a female-plumaged Superb Lophorina; in the same area he observed two female-plumaged birds foraging below the treetops by probing into the moss and epiphytes for arthropods, reminding him of the feeding actions of astrapias. Schmid (1993) lists the following mid-montane foodplants as reported by local people: *Notocnide melastomatifolia*, *Freycinetia* sp., *Harpulia ramiflora*, *Harpulia* sp., *Elattostachys obliquinensis* and members of the Zingiberaceae (ginger) family.

BREEDING BEHAVIOUR Males display at terrestrial courts as do other parotia species, implying a polygynous species, again as with congeners. Adult males are promiscuous and may maintain exploded leks, but further study is needed. Females presumably attend to nesting duties unassisted. **COURTSHIP/DISPLAY** Two display courts (often called 'bowers' in earlier books) found near Mindik, in the Rawlinson Range, were some 5m apart in rather dense shrubbery under a break in the forest canopy. Both courts were flat or slightly depressed areas that had been cleared of all debris to expose the spongy mat of interwoven rootlet fibres, and included several small saplings from which most leaves had been removed. The courts were circular, with a diameter of 1.5m, or oblong, measuring 1.5m × 1m. A video by Scholes of a male dancing can be found at Macaulay Library (ML456819). Display behaviour in captivity is well described by Crandall (1940) and repeated in F&B, to whom the reader is referred for details. It appears similar to that of Lawes's Parotia, but is preceded by a little court-clearing, the narial tufts are opened but no crest extends forward to cover them, the occipital plumes are thrown forward before the bird stands erect, and the species uses the tail and wings when crouched (T. Pratt in F&B). Crandall describes an arboreal display very similar to the ground one, except that it begins with the male moving about on horizontal perches with its tail held to one side and its body stiffly horizontal. Opit (in F&B) describes how he was attracted by the loud calls to a young male in female-like plumage except for a bumpy, glossy feathered head and six flag-tipped wires emanating from the head above the eyes. This subadult was chasing and displaying to a female-plumaged bird: he faced the latter and spread his wings open slowly while calling a very loud, harsh, nasal parrot-like *garrr*, then twisted back and forth on his perch, causing his head wires to whip back and forth. A tree about 100m away held three immature males without head wires, all three of which twisted back and forth and leapt from branch to branch, calling noisily; others were calling in various directions a couple of hundred metres

away. T. Pratt (in F&B) wrote of a wild bird tolerating female-plumaged individuals and an adult male on his court, but once also attacking another adult male and knocking it off a court perch. Coates (1990) described display by a captive male on or near his court at Baiyer River Sanctuary: 'The male hops about with a bouncing motion, body slightly elevated showing the tibia, and tail cocked and sometimes angled to one side. Sometimes he hops about at speed, and sometimes, probably when in display mood, he hops swiftly across and about the court with a peculiar mammal-like appearance, hunched like a hopping rodent. Display starts when after moving around and about the court and tending it, he comes to its edge. The display includes an **Initial Bow Display** and a **Ballerina Pose** with the body plumage extended to form an amazing wide, circular skirt. When fully upright and just before beginning to dance, the head is flicked at right angles to one side and gives a fleeting glimpse of a bright jewel-like flash, which is either the eye or the iridescent nape patch. The occipital wires are then brought forward and, leaning forward and with the tail touching the ground, he begins the dance by walking forwards and to one side, head bobbing from side to side presenting a sideways view. He then stands still but the sideways head bobbing increases in tempo causing the flag-tipped wires to wave about wildly, and he may tilt his body from side to side. The tail hangs slightly to one side. This phase continues in silence for a minute or more then he stops suddenly, shuffles his plumage and resumes normal behaviour.' **NESTING** Nest and nest site unknown. **EGGS, INCUBATION & NESTLINGS** Two eggs laid on consecutive days by a captive female at Baiyer River were a dully glossed pale cream colour, with heavy streaking at the larger end becoming sparser halfway along and practically absent at the small end; the markings vary from small dots to elongate grey or broad tan streaks in about equal proportions, the tan marks overlying the grey in some places (F&B). There is no information on incubation and nestling care, but both are presumably by the female alone. Downy young undescribed.

HYBRIDS No hybrids between this species and others have been reported.

MOULT Coates (1990) suggested that this parotia moults during the wet season, based on captive birds at Baiyer River Sanctuary. Specimens examined by Frith (F&B) and involving all months but August showed moult in October–May. A captive individual in the New York Zoological Society (NYZS) collection had a four-month moult October–February (Crandall 1940).

STATUS AND CONSERVATION Classified as Near Threatened by BirdLife International. A rather sparse endemic, uncommon to rare on the Adelbert Range in March (Opit in F&B) and with no recent records there; locally common on the Cromwell Range (Lindgren in Coates 1990). In January 1994, near Satop, Huon Peninsula, I saw just a single male twice and one lone female over a three-day period; the species appeared rare, particularly when compared with the frequency of sightings of the Huon Astrapia. Habitat around villages is being lost, but much forest remains in the more remote terrain and the species should be secure despite its relatively restricted range. Although there are no recent records from the Adelbert Mts, where suitable habitat for this species is limited and perhaps even decreasing with climate change, the species' apparent absence may be due to the fact that few observers visit, and parotias can be elusive. The Tree Kangaroo Conservation Project is working in the region and has established with the local people in the northern Huon the large YUS conservation area (PNG's first conservation area, named after the three rivers – the Yopno, Uruwa and Som – which flow through it), which should benefit all the endemics. There is an increasing human population in the area as these mid-montane altitudes are favoured by local people for settlement and agriculture, resulting in more forest clearance around villages, so declines in numbers of this species

may be expected. While this region does not have a high population density, the human population is expanding rapidly and is clearing areas of forest within the species' range (I. Burrows *in litt.* 1994; W. Betz *in litt.* 1999); this forest loss, however, currently remains fairly minimal (B. Beehler *in litt.* 2012). This parotia is known to forage near active gardens and appears to be tolerant of human activities (W. Betz *in litt.* 1999), as is the better-known Lawes's Parotia, but these observations may represent simply feeding excursions from nearby undisturbed forest. There is no evidence that it is currently hunted for plumes or food (Frith & Beehler 1998; W. Betz *in litt.* 1999; B. Beehler *in litt.* 2012). There are, however, historical reports of the species being used for feather wheels and feather poles (in ceremonies to celebrate major stages in the life cycle) by the Nokopo people (Schmid 1993).

Wahnes's Parotia, dancing area, Huon Peninsula, PNG, July 2017 (*Jun Matsui*).

LAWES'S PAROTIA
Parotia lawesii   Plate 8

Parotia lawesii Ramsay, 1885, *Proc. Linn. Soc. NSW* **10**: 247. Mt Maguli, Southeast Peninsula.
Other English names Six-wired Bird of Paradise, Lawes's Six-wired Bird of Paradise
Original scientific name *Parotia lawesii* based on specimens secured by Carl Hunstein from the Astrolabe Mts, later amended to Mt Maguli in the Owen Stanley Range (Schodde & McKean 1973).
Etymology The name *lawesii* commemorates the British missionary Rev. William Lawes, who assisted in the obtaining of trade skins of this species.

FIELD IDENTIFICATION The male is a distinctive dumpy black bird, quite short-tailed and with broad, rather rounded wings. It has a bronzy-gold iridescent breast shield, six wire-like racket-tipped head plumes, short white supranarial tufts, a small and generally concealed frontal crest and dense flank plumes, with a bright blue eye. The female has a black head with chestnut-brown upperparts, paler black-barred underparts and a blue eye. The species is endemic to PNG, related to the congenerics Huon, Western, Bronze and Carola's Parotias in the Huon, Arfak Mts and western central hills and mountains, respectively. Eastern Parotia, an allospecies from the northern section of the South-east Peninsula of PNG, was previously treated as a subspecies of the present species. **SIMILAR SPECIES** Male Superb Lophorinas have a triangular blue breast shield and an extensive black cape, and lack the blue eye and white supranarial patch, which is a good field character for this species. Female Superb has an obvious pale supercilium, lacking in female Lawes's Parotia, and a much more sloping head shape, lacking a blue eye. Lawes's Parotia is sympatric with Carola's Parotia in the upper Jimi River–Western Bismarck Range–Schrader range (Schodde & McKean 1973; Healey 1976) and also in the Mt Giluwe–Kubor Range. Females of Carola's Parotia have a distinctive whitish superciliary stripe and forehead with a pale iris, and rufous in the wings, whereas male Carola's are readily told by the white flank plumes and yellow to creamy-white iris.

RANGE Endemic to the mountains of the central ranges of PNG, from the Nipa and Tari areas of Hela Province (formerly Southern Highlands) west as far as Oksapmin, in Western Province, and then through the Schrader and Bismarck Ranges, Baiyer and Jimi valleys, Mt Hagen, Mt Giluwe and Ubaigubi, in the Eastern Highlands. Eastern limits uncertain, and may come into contact with Eastern Parotia. **MOVEMENTS** Largely sedentary in montane forest and well-wooded cultivated areas. Female-plumaged birds and subadult males appear at Crater Mt Biological Research Station area in March–September, presumably moving from higher altitudes (Mack & Wright 1996). An adult female caught and ringed by T. Pratt on Mt Missim was recaptured at the same place eight years and one month later (Anon 1989).

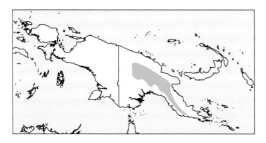

DESCRIPTION Sexually dimorphic. This is a relatively short-tailed parotia. *Adult Male* 27cm; weight 153–195g. Male is velvety jet-black, with an erectile silver-white narial tuft (prominent in display) and a frontal crest of coppery-brown feathers. There is a narrow nuchal bar of intensely iridescent blue to pink-purple scale-like feathers, which, like the breast shield, is an important feature during display dances. Likewise, an ear-tuft of pointed feathers behind each eye, each with three long, bare erectile wire-like black occipital plumes that have roughly circular spatulate tips (the six wires of one of the vernacular names), is also a prominent display feature. The black mantle, back and tail have a coppery-bronze and/or green sheen in some lights; the primaries are paler, more brownish-black, edged dull iridescent green in some lights. Chin and throat velvety jet-black with iridescent purple sheen, grading into the breast shield of large intensely iridescent scale-like feathers bronzed metallic

emerald-green to greenish-yellow with purple-magenta to bluish-violet in some lights; underparts-jet black with coppery sheen. *Iris* cobalt-blue with narrow pale cream-yellow outer ring (in-hand observations reveal that, amazingly, the bird can alter this colour from mostly blue to mostly yellow: it apparently has two parts to its iris, the inner blue and the outer yellow, and it can make one or the other predominate); *bill* shiny black, *mouth-lining* lime-yellow to lime-green; *legs* purplish lead-grey. *Adult Female* 25cm; weight 122–169g. Smaller than male, lacking head plumes and iridescence, with a distinctive brownish-black head, contrasting paler submoustachial stripe and a dark malar stripe. Upperparts mostly chestnut-brown, underparts closely barred blackish, bill brownish-black. *Juvenile* Resembles female, but with some more rufous edgings to wings and wing-coverts. *Immature* male is like adult female but iris duller, more greyish to brownish. *Subadult Male* variable, depending on state of moult, from like adult female with few adult male feathers to like adult male but with a few female-like feathers remaining. Adult head plumage is acquired first, when wings and tail similar to those of adult, and breast shield has coppery sheen, rather than clearer green of fully plumaged male; male tail length decreases slightly with age.

TAXONOMY AND GEOGRAPHICAL VARIATION Monotypic. Described forms *P. l. fuscior* Greenway, 1934, from Herzog Mts, and *P. l. exhibita* Iredale, 1948, from Mt Hagen area, are synonymised with *lawesii*. *Parotia helenae* is often treated as a subspecies of *P. lawesii*; see account for *P. helenae*.

VOCALISATIONS AND OTHER SOUNDS A loud, harsh sneezing call, often given in flight and seemingly an alarm call, is emitted by both sexes, although males also utter a similar but more raucous version from a perch. Also a harsh rasping four- or five-syllabled *weh weh weh weh*, which is not unlike some calls of Superb Lophorina. Xeno-canto has typical vocalisations recorded by Niels Krabbe at Ambua (at XC26627 and XC26626), and in Macaulay Library at the Cornell Lab of Ornithology there is a good sequence by Thane Pratt (ML22736), while Scott Connop has a loud ringing single *kweh* note also recorded at Ambua (ML65891). The loud squawks seem not to form part of a multi-note phrase as with other parotias (Laman & Scholes 2016). A faint whistled, downslurred, almost *Aplonis*-like or *Chrysococcyx*-like *tsiieeuw*, is given by the male and often as response to the same call from another male, but also uttered by the female from a perch above the display court. A faint repeated *sip* is given by the male as he hops about the court and adjacent area. Also soft guttural squawks, soft guttural cawing, a rapid soft trilling *treet treet* and a nasal twittering given at the court have been reported (Frith & Frith 1982; Beehler 1986; Coates 1990).

HABITAT Lower montane forest and well-wooded gardens from 500m to 2,300m, the main range being from 1,200m to 1,900m. Quite often seen along forest edges or around isolated stands of trees, and comes readily to fruiting trees.

HABITS This species is wary, but may permit a cautious approach, especially when in the forest. Males on the court are very wary and flush easily. Often seen flying across roads and tracks, when the short tail and rounded wings are very distinctive, and a flash of white from the supranarial tuft is often glimpsed. Female-plumaged birds seem to be less obvious, and are often found at fruiting trees or with mixed-species feeding flocks.

FOOD AND FORAGING Primarily frugivorous, fruits accounting for up to 95% of diet, but occasionally takes insects and small lizards (Schodde 1976; Bechler & Pruett-Jones 1983), and possibly snails (Majnep & Bulmer 1977; Healey 1980). Gleans for food and may forage creeper-like on tree limbs (Beehler *et al.* 1986). Important foodplants include *Schefflera*, *Ficus* and *Gastonia* (Beehler 1989a), also *Omalanthus*. Forages singly, in twos and in smallish groups. The largest concentration was reported by Eastwood (1987), who saw 25 individuals, including an adult male, along with single female Blue Bird of Paradise and Superb Lophorina at a fruiting tree. This parotia will readily associate with other paradisaeids at such sites, and has been seen feeding alongside Brown Sicklebill and Blue, Magnificent and Raggiana Birds of Paradise as well.

BREEDING BEHAVIOUR Polygynous; promiscuous males hold courts in a kind of **exploded lek** system. It is likely that only females build nests and rear the young, as with *Paradisaea* species. Some females show fidelity to a particular male by returning to mate with the same individual each year (Pruett-Jones & Pruett-Jones 1990). Courtship/Display The court of the male is an area completely cleared of debris on the forest floor, ranging in size from 0.5m^2 to 20m^2 of irregular shape, and with several perches running above it for the females to sit and watch. Courts are on level or sloping ground. Males clear the courts by using their bill, using the feet to hold tough plants as they tear at them. The male decorates his court with various items such as snakeskin, the droppings of mammals, pieces of chalk, fur, feathers and bone, which he places centrally on the court floor, often after rubbing them on the display perch and bare vertical saplings of the court; small snail shells, rootlets and clear plastic wrapping are also sometimes used. He may have an adjacent court as well, occasionally up to five such courts being maintained by one male. Males in visual contact (<15m) often interrupted each other's displays, or engaged in cooperative ones; those with courts 20–70m apart were in auditory contact alone, and these seldom interacted with one another. Male and female interactions were seen only at the courts (Pruett-Jones & Pruett-Jones 1988b). Males will give advertising calls from court perches or nearby, but are mostly silent when interacting with females. Up to six males have been seen at one court, but at the height of the display the resident chases the other males from his court. Females approach the court either singly or as groups of up to eight, sometimes with the male. Individual females visited the court for up to 12 weeks during the display season. At a study site on Mt Missim (Beehler & Pruett-Jones 1983), females had on average a choice of 17 courts to visit within their home range, each female visiting most or all of them. This study site of *c.*4km^2 had 28 courts of 25 males, and three other males called there but apparently lacked a court. Males, unlike male bowerbirds, do not use the organic objects which they collect in display, but they will steal them from each other's courts. The females usually remove them quickly, within 24 hours, throughout the nesting season (Pruett-Jones & Pruett-Jones 1988b). The snakeskin may be used for nest decoration, and the chalk perhaps eaten as a mineral supplement (F&B 98), while the mammal scats may deter predators. Males with more court objects did not enjoy greater mating success, but they did have longer visits from the females, which might have given them an indirect advantage by attracting more females (Pruett-Jones & Pruett-Jones 1990). The **display period** on Mt Missim extends from May until March, i.e. from the early dry season to the middle of the wet season there. Most activity occurs from September to February, the period when all matings were observed. Some males may be found displaying at any time of the year, as with other plumed species of paradisaeid (Coates 1990). Males remain close by their courts throughout the day (Schodde & McKean 1973) and most activity there is in early morning and late afternoon, as would be expected. *Details of Display* The following is largely adapted from Coates (1990) and Frith & Frith (1981) and concerns mostly captive individuals. The male preens during the day, paying much attention to the long, dense flank and belly plumage, which is sometimes extended skirt-like from the body. Displays are mainly silent, and may be performed solitarily or with another bird present. The display is directed at other males but seemingly not at females, and if another male is in view the behaviour of one influences the other, so the two will be performing the same activities, such as preening, displaying or bathing in the rain (captives at Baiyer River: Coates 1990). Laman & Scholes discovered that, when the females view the display dance from their perches above it, what they see is very different from that seen when on the same level as the perch. The remarkable iridescence of the scale-like breast shield and nape patch would flash vividly as the male performs the ballerina (or, as Scholes put it, the **wobbling ovoid dance** if viewed from above). The male adopts a hunched pose, crouching somewhat, with the head retracted, and he hops about and beyond the court and back again, repeatedly uttering a faint, short *sip*; occasionally he will bound across the court and beyond at speed, as if excited, and sometimes he walks slowly about the court in an upright posture with the flank plumage swept back. Suddenly, while attending the court, he performs the **Initial Display-bow** in which he crouches quickly, with the head and neck extended and bill angled towards the ground, the forecrest raised but the occipital plumes held flat over the back; the eyes have a peculiar glazed, distorted appearance. This pose is held for a couple of seconds, then he suddenly stands fully upright and, flicking his wings slightly from the body to free the flank plumes, he stretches upwards to his full extent on near-vertical tarsi while at the same time shaking out the flank plumes to form a complete skirt. This distinctive attitude is aptly called the **Ballerina Pose**. The occipital plumes are initially held behind the head, but they are brought upwards and forward as the **Dance Phase** commences, the male beginning to walk, bobbing his head from side to side. Frith & Frith (1982) describe a solitary captive male performing this dance: bobbing his head vigorously, the male steps sideways and then backwards, bowing slightly; he moves in a direct line or, more usually, a semicircle, finishing about 0.5m from and facing his starting point. After pausing momentarily at this second point, he then steps directly backwards for a similar distance, bowing slightly, to a third point, then forwards, returning to the second point. This is done three times in rapid succession, and this Dance Phase takes about 12 seconds, throughout which the head wires remain forwards. Then follows a period of bobbing up and down and swinging the head wires about while remaining on one spot, this being known as the **Stationary Phase**. Standing at the second point the male lowers himself slightly and remains momentarily motionless, but brings the head wires to a near-vertical position; then he draws himself stiffly erect and freezes, while bringing the head wires forward to an elevated angle of 45°. Suddenly he lowers himself again and snaps his head downwards several times, this accompanied by an outward and upward flicking of the breast shield, which results in an iridescent flash. Remaining crouched, he then bobs his head rapidly from side to side in a snake-like manner, three or four times on each side, and with each nod the breast shield is again flicked up to produce a flash of colour. The head wires remain at about 45°, and suddenly the bobbing stops and, after a pause, the male again draws himself up motionless, then swings the head wires backwards over the back and forwards again, just past the vertical. The finale is when he lowers himself and bobs the head, and flicks the breast shield more vigorously than before so that with each pump of the head a short hop is made. This Stationary Phase lasts for about 14 seconds. When the performance is over, the bird lowers all feathers to their normal position and resumes the usual behaviour. A **Dance Phase** given by a captive bird (Crandall 1931) differed from that described above. It began with 'slow short hops about two feet right and two feet left, and back again': the six wires are now seen at their best, as the head is moved rapidly from side to side in a rhythmic motion, causing the spatulate tips to bob madly; as a finale, the body is thrown forward into a horizontal position and the head-bobbing increases in speed until the movement of the wires becomes a blur. 'There is a flap of wings, the body feathers are lowered and the dance is over.' Captive males displaying at each other had another variation (from Coates 1990). The **Initial Display Bow** was given either sideways to or facing away from the neighbour, but always from the same point on the court. From this point the male would do an about-turn as he assumed the **Ballerina Pose** and, with head and neck arched slightly forward, immediately walk in the direction of the neighbour's court. Sometimes

he would pause on reaching the edge of his court, turn around and face the opposite direction, the **Dance Phase** continuing, but often he would bend his body forward with head and neck arching horizontally and head wires pointing to the ground ahead, giving the **Agonistic Pose** – which is reminiscent of a bull about to charge! Maintaining this pose, he would then move quickly towards the other male until the two met at the wire partition, where they would attempt to fight, wings fluttering and face to face, trying to ascend the partition. Our understanding of the displays of this species was greatly improved following the studies of Laman & Scholes and their remarkable documentaries on paradisaeids and their courtship (the Macaulay Library at Cornell Lab of Ornithology contains a superb collection of video and acoustic material). They discovered that the display courts of this parotia are often very close together, even contiguous, along ridgetops, such that the males have to interact with others far more than their congeners do. An amazing cooperative or coordinated display has evolved, where up to six males gather on one court without antagonism, and start hopping together across the court, first one way and then the other (video ML45942 at macaulaylibrary.org). This dance can last for several minutes with the birds in a rigid posture as if not quite at ease with each other, and gradually falling out of synchronisation until the whole procedure disintegrates and they depart in haste with loud angry-sounding calls. **NEST & NESTING** The nest of this species was described by Shaw Mayer as an 'open and rather shallow structure composed of fern tendrils and creeping fern stems woven together; lined with very fine tendrils. On the outside of the nest a few green fern fronds.' Interestingly, Majnep & Bulmer (1977) reported an active parotia's nest, probably of this species, having pig dung on its rim. This correlates well with the observations of mammal dung being used by males to decorate their courts, and being removed by females. Nest sites have been recorded high up in a large tree (Shaw Mayer in Harrison & Frith 1970), 11m above ground in the outer canopy foliage of a large tree (Frith in F&B 1998), 12m above ground in a forest tree (Coates *et al.* 1970), and 5–10m high in a thick vine tangle at the top of a *Macaranga* tree (T. Pratt in F&B 1998). **EGGS & INCUBATION** The clutch is said to consist of one egg (Gilliard 1969; Pruett-Jones & Pruett-Jones 1990). That in a single-egg clutch from Boneno, Mt Mura, in south-east PNG, measured 38.4 × 27.8mm and was a glossy light buff with broad longitudinal streaks of brown, grey and purple-grey and some small spots and blotches (Harrison & Frith 1970). It is likely that only the female incubates, as with other polygynous paradisaeid species, but the duration of incubation is not known. Little else is recorded about the nesting cycle, and the downy young remain undescribed.

HYBRIDS A unique and unlikely intergeneric hybrid combination known as **Schodde's Bird of Paradise** (Frith & Frith 1996a) is recorded: Lawes's Parotia × Blue Bird of Paradise of the western subspecies *margaritae*. After it was collected by R. Bulmer, in 1956, this unusual specimen was thought to be a female Lawes's Parotia, until recognized by R. Schodde (in Christidis & Schodde 1993). Hybridization is known only with the Blue Bird of Paradise but is to be expected with other species as well, since Lawes's Parotia will interact with other paradisaeids at or near their display sites. Healey observed a female-plumaged individual twice visiting the court of a Carola's Parotia, on one occasion carrying a snail, and a male Superb Lophorina has been seen to alight beside an active court of the present species. Both Lawes's and Carola's Parotias may be seen on the same courts at Crater Mt (Mack & Wright 1996) and both species have been seen together in the Wahgi valley (Bishop in F&B).

MOULT Seems to be throughout the year, but skins showed a predominance in January–June (F&B). Birds at Tari in November were moulting heavily.

STATUS AND CONSERVATION Classified as Least Concern by BirdLife International. Fairly common in the Tari valley, individuals coming up to the level of Ambua Lodge (2,000m) to feed but very seldom any higher at this locality. Reported as fairly common in the Owen Stanley Mts and at Mt Missim. The species inhabits a zone heavily populated by humans, but is under little immediate threat. The head wires are used as *bilas* (personal decoration) by the Huli people, but the skins very seldom appear in markets. The use of objects to decorate the court is also interesting, and it would be worth knowing whether the females do use

them for nest decoration and, if so, why they do. The exact relations with Carola's Parotia and whether the two do hybridise, as seems likely, would also be worth clarifying. Potential overlap with Eastern Parotia is also worthy of study.

Lawes's Parotia, female plumage, Ambua, PNG (*Markus Lilje*).

Lawes's Parotia, male, PNG (*Tim Laman*).

EASTERN PAROTIA
Parotia helenae   **Plate 8**

Parotia helenae De Vis, 1897, *Ibis* 390. Neneba, upper Mambare River, 1,220m, north of Mt Scratchley, Southeast Peninsula.

Other English names Helena's Parotia

Etymology The origin of the name *helenae* has two explanations, the first being that it was named after Helena Ford, daughter of Australian oologist A. Scott (Jobling 2010), and the alternative that it was named for Princess Helena Augusta Victoria, a daughter of Queen Victoria (Frith & Frith 2010). The common name Eastern refers to its geographical location relative to all other members of the genus *Parotia*, of which this is the most easterly representative.

FIELD IDENTIFICATION The male is a dumpy black bird, quite short-tailed and with broad, rather rounded wings. It has a bronzy-gold iridescent breast shield, six wire-like racquet-tipped head plumes, short bronzy-brown supranarial tufts, a small and generally concealed frontal crest and dense flank plumes, and a bright blue eye. The female has a black head, chestnut-brown upperparts, paler black-barred underparts and a blue eye. **SIMILAR SPECIES** Males of both Superb and Eastern Lophorinas have a triangular blue breast shield and an extensive black cape, and lack the blue eye and golden-brown supranarial patch (which are good field characters for Eastern Parotia). Females of Superb have an obvious pale supercilium, lacking in the female Eastern Parotia, and have a much more sloping head shape, again lacking a blue eye. Eastern Parotia is possibly sympatric with Lawes's Parotia in the western margins of the South-east Peninsula in the forests below the Herzog Mts, the upper reaches of the Waria River and Mt Scratchley. The male of Lawes's has somewhat longer, silvery-white (not bronzy golden-brown) narial tufts; females of the two species are very similar to each other but, when viewed from the side, the shape of the maxilla by the culmen appears slightly concave on *helenae* and convex on *lawesii* (Beehler & Pratt 2016).

RANGE Restricted to the eastern sector of the South-east Peninsula, west in the northern watershed to the Waria River, in the south probably only to the low, dry Keveri Hills, just west of Mt Suckling, which separate it from Lawes's Parotia there. The range is centred on Mt Suckling, Mt Dayman and Mt Simpson, at the eastern end of the Owen Stanley range, where it can be quite numerous in the mixed montane rainforest. It occurs eastwards into the peaks of Milne Bay Province. The

Eastern Parotia is seen on the Kokoda Trail above Popondetta, but remains remarkably little known and is seldom reported, and there are hitherto no well-known localities for seeing it. This taxon occupies a remote and inaccessible region and was largely ignored until its recent elevation to species status. **MOVEMENTS** Resident, so far as is known.

DESCRIPTION Almost all measurements are virtually identical to those of Lawes's Parotia. *Adult Male* 27cm; weight 170g. The plumage is primarily black, the velvety feathers of the upperparts tinged bronze to reddish-purple. This species has a very short, erect supranarial tuft of bronzy golden-brown (not silvery-white), a prominent frontal crest of dark blackish-brown, rather hair-like feathers, and a nuchal patch of variable iridescent green or bluish-purple (varying according to the light); six erectile wire-like spatulate-tipped head wires emerge from behind the eye, three on each side. The breast has an iridescent oil-green shield suffused greenish-gold to pinkish-purple. *Iris* ultramarine-blue with a yellowish-cream outer ring; *bill* short, quite broad and black in colour; *legs* blackish-brown. Field observations are scant, and what happens in the zone of overlap with Lawes's Parotia is unknown, it is possible they may be sympatric. *Adult Female* 25cm; weight 149g. Crown and sides of head are black (no frontal crest or head plumes), with an indistinct buff-grey malar stripe. The upperparts, including the tail, are a deep rufous-brown, with the wings more blackish-brown; the underparts are well-barred black on a pale rufous background, darker on the underwing-coverts and undertail-coverts. Eastern Parotia female is reported as being duller and more russet than female Lawes's Parotia, but this is seemingly not the case; the best diagnostic characters are the sharply keeled culmen and the more extensively exposed (unfeathered) distal portion of the culmen in *helenae* (18.3–19.5mm [sample size *n*=3], against 11.5–14.8mm [sample size *n*=6] for *lawesii*). When viewed from the side, the upper mandible by the culmen appears slightly concave on *helenae* and convex on *lawesii* (Beehler & Pratt 2016). Quite how useful this might be in the field, however, is debatable (although good photographs should show it), and range remains the best signifier. *Juvenile* unknown. *Immature* unknown, but presumably much as female initially.

TAXONOMY AND GEOGRAPHICAL VARIATION Monotypic. Eastern Parotia is found in a limited range of the northern watershed of peninsular PNG, from the Waria south-east to Milne Bay. This species has a chequered taxonomic history, having originally been described as a species, then subsequently regarded as a subspecies until Schodde & McKean (1973), Forshaw & Cooper (1977), Cracraft (1992), Sibley & Monroe (1993) and Clements (1996) upgraded it to species status, though the later edition of Clements (2000) again listed it as a subspecies. It was regarded as a subspecies by Gilliard (1969), Diamond (1972), Beehler & Finch (1985), Beehler *et al.* (1986), Coates (1990) and F&B (1998), but more recent treatments, such as the IOC and Beehler & Pratt (2015), regard it as a full species following improvements in our understanding of the complexity of speciation in

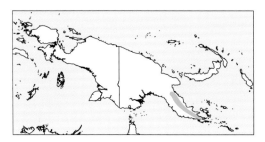

New Guinea. Nevertheless, this parotia is still one of the least-known birds of paradise, and the potential overlap with Lawes's Parotia remains in need of study.

VOCALISATIONS AND OTHER SOUNDS Beehler (in F&B 1998) reports a double harsh *kschack kschack* call for this taxon, but nothing else is recorded in the literature and there are no cuts on xeno-canto at present (2018). The Macaulay Library at the Cornell Lab of Ornithology, however, has a good series of videos of this species by E. Scholes, which include call notes. These sound very similar to those of Lawes's Parotia, a loud harsh, tearing *schraak schraak schraak* (as at ML460347).

HABITAT Inhabits primary forest from 500m to at least 1,500m, presumably venturing into nearby gardens and cultivation as with congeners. It is perhaps more numerous above 1,100m.

HABITS This species' behaviour appears to be much as for Lawes's Parotia, but few details as yet reported. It was said to be numerous in mixed montane forest of the eastern Owen Stanley Range, with loose groups of males and female-plumaged birds seen moving through the forest understorey in the early morning, giving their strident, rasping calls (Schodde & McKean 1973).

FOOD & FORAGING Undoubtedly primarily frugivorous and will no doubt also take arthropods, but details lacking.

BREEDING BEHAVIOUR Little is as yet recorded about this remarkably neglected species apart from some studies by Ed Scholes in 2002, and it remains therefore one of the least-known of the family.

DISPLAY Display behaviour appears similar to the sequences recorded for the sibling species Lawes's Parotia, the species being polygynous, with cleared terrestrial display courts and a variety of elaborate dances by the males. The male erects the 'ballerina skirt' metastructure and waves the head about while taking short steps backwards, forwards and sideways, the breast shield flashing but the narial tufts much less conspicuous, being golden-brown (as opposed to the silvery-white tufts of Lawes's Parotia). **NEST/ NESTING** Breeding season largely unknown. A female was incubating on a nest with an egg in early December, at 1,500m. The **nest** is an open shallow structure with a depth of 13mm, constructed from fern tendrils and creeping-fern fronds woven together and lined with very fine tendrils; the outside of the nest had a few green fronds.

HYBRIDS No hybrids have been identified, but this species remains one of the least-known birds of paradise, and the potential overlap with Lawes's Parotia is in need of detailed study.

MOULT Moult data are lacking.

STATUS AND CONSERVATION Classified as Least Concern by BirdLife International, and therefore considered not threatened. Although its geographical range is fairly limited, it appears common in some locations. This taxon occupies a remote and inaccessible region and was largely ignored until its recent elevation to species status. Field observations are scant, and what happens in the zone of overlap with Lawes's Parotia is unknown; it is possible they may be sympatric.

CAROLA'S PAROTIA
Parotia carolae   Plate 9

Parotia carolae A. B. Meyer, 1894, *Bull. Brit. Orn. Club* 4: 6. Weyland Mts, Western Range. Original type locality Amberno River = Mamberamo River, but in fact seemingly from the Weyland Mts (Mayr 1941).
Other English names Queen Carola's Parotia, Queen Carola's Six-wired Parotia, Queen Carola of Saxony's Six-plumed Bird of Paradise, Queen of Saxony's Bird of Paradise
Etymology The species was named *carolae* after Queen Carola of Saxony.

FIELD IDENTIFICATION Males of this species are quite distinct, with six head wires, white and golden-buff crown stripes, a pale eye, and puffy whitish throat feathering above a bronzy-iridescent breast shield, while the white flank plumes are not possessed by any of the congeners except the Bronze Parotia, a recently rediscovered related species from the Foya Mts. Female-plumaged Carola's Parotia have a distinctive striking face pattern with pale supercilium, blackish eyestripe, pale cheek stripe and dark malar stripe, and a pale yellowish eye, the underparts closely barred dark, and the flight-feathers reddish. SIMILAR SPECIES

Males of the Bronze Parotia are broadly similar to male Carola's, but readily told by their bluish eye and black throat; females of the two are very much alike, but the species' ranges do not overlap. Carola's Parotia is a distinctive and rather aberrant species, somewhat curiously occupying a range between members of the black parotia superspecies, and sympatric with Lawes's Parotia in some areas. Females are surprisingly similar to the much longer-billed and smaller Superb Lophorina, which may be a case of mimicry by the smaller species of the larger and more aggressive female Carola's. Female-plumaged Magnificent Birds of Paradise are much smaller and lack the striking face pattern; they also have a bluish bill. Magnificent Riflebirds are larger and much more brightly coloured above, and have a long decurved bill. Female-plumaged Lawes's Parotia is dark-headed, without the face pattern, and has a striking yellow (female) or violet-blue (male) eye.

RANGE This species is found in the western two-thirds of the central cordillera (but absent from the Vogelkop), from the Weyland Range eastwards to Mt Giluwe and the Hagen Range, the Sepik–Wahgi divide, Schrader and Bismarck Ranges and Crater Mt, in the Eastern Highlands. It is known also from

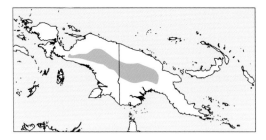

Mt Bosavi. The distribution appears rather patchy within this area. **MOVEMENTS** Primarily resident, but some altitudinal wandering in the Ok Tedi area, PNG, where the species is found down to 650m at certain times of the year, this probably related to food supply. In some years female and immature individuals are found during May–September at the lower levels, when it is slightly less wet at the higher altitudes. They are seemingly absent at times from these lower altitudes, and this seems to vary with the particular year.

DESCRIPTION A distinctive short-tailed parotia, sexually dimorphic. *Adult Male* 26cm; weight 205g (one of race *chrysenia*). A medium-sized, quite short-tailed, black-and-white parotia with six head wires. This striking bird is readily identified, having large white flank plumes, a silver and golden-bronze crest and a golden-bronze breast shield. Male of nominate race has a velvety jet-black head with coppery-bronze sheen, with an erectile black anterior crest tipped silvery white followed by gold-tipped crest feathers which fold backwards into a dish-shaped skull; eye broadly encircled by iridescent coppery-gold feathering; a narrow nuchal bar of highly iridescent scale-like feathers that appear blue-green to purple and/or magenta; an ear-tuft of elongate narrowly pointed feathers extends behind each eye, each tuft with three long, erectile bare wire-like occipital plumes with relatively small spatulate tips. Mantle to uppertail velvety jet-black with a coppery-bronze sheen, the primaries and their coverts more brownish-black. Chin is dusky olive-brown, smudged blackish, with paler tips of elongate 'whiskers' surrounded by malar area and buff throat with a golden sheen; the lower central throat is more whitish, flecked cinnamon, with fine long whiskers on each side, grading into a breast shield of large scale-like feathers which have an intense iridescence of bronzed yellow-green and/or magenta to pink. Remaining central underparts, to undertail-coverts, are blackish-brown with iridescent lustrous coppery sheen, becoming browner to dark reddish-brown adjacent to an extensive patch of elongate, recurved white flank plumes. *Iris* sulphur-yellow; *bill* black, *mouth colour* pale green; *legs* blackish-grey. It is not known whether or not the iris can change colour, as with Lawes's Parotia. *Adult Female* 25cm; weight 110–163g. Smaller than male (notably in wing length) but with tail longer; lacks head plumes and iridescence. The plumage is very different, with the head brownish to grey-brown with broad supercilium, the moustachial and submoustachial stripes dirty white, flecked olive-brown, with some paler flecks extending on to anterior ear-coverts; upperparts and tail brown, the upperwing with rufous flight-feathers. The malar area is olive-brown, chin faintly barred greyish-brown, throat paler; the underparts are buff with heavy blackish-brown barring. Females vary subtly with race, notably in the extent of their pale facial stripes and in overall colour saturation; further study required. *Iris* pale grey or cream to yellow (difference possibly age-related), but confirmation required. It is not known if this species can change the colour of the iris as Lawes's Parotia can. *Juvenile* is undescribed. *Immature* male is like adult female, with iris pale grey. *Subadult Male* variable with stage of age-related moult: some are like adult female but with few feathers of adult male plumage, initially on head, others like adult male with few feathers of female-like plumage remaining; some have a pale crown stripe and often partly grown head wires. Male tail length decreases considerably with age, as with congeners.

TAXONOMY AND GEOGRAPHICAL VARIATION
Polytypic, with four subspecies:
1. *P. c. carolae* A. B. Meyer, 1894. Weyland Mts east to Wissel Lakes area, West Papua.
2. *P. c. meeki* Rothschild, 1910. Snow Mts east of Wissel Lakes and south of the Doorman Mts, east to the PNG border. Like the nominate, but with slightly larger bill and with chin and sides of throat blackish.
3. *P. c. chalcothorax* Stresemann, 1934. Doorman Mts, just south of the lower Idenburg River. Like the nominate, but with longer occipital plumes, upperparts with a bright copper sheen, underparts similarly more intense copper, and long loral feathering brownish, less intensely black.
4. *P. c. chrysenia* Stresemann, 1934. Lordberg, Sepik Mts. The north scarp of the central ranges of PNG, including Hunstein Range and probably the Schrader and north scarp of the Bismarck Range. Form *P. c. clelandiorum* (= *clelandiae*) Gilliard, 1961, described from 'Telefolmin' (= Telefomin) in Victor Emanuel Mts, PNG, is now synonymised with *chrysenia*: range given as from the PNG border south-east probably as far as the southern watershed of the Eastern Highlands, including Crater Mt, north of *chrysenia* but exact boundaries unclear. Reportedly similar to nominate, but blacker above and larger, with slightly shorter bill, coppery sheen to loral feathers, copper-coloured eye-ring and longer occipital plumes. Both geographical range and physical characters of *clelandiorum* unclear, specimens being old and worn, and it seems best meanwhile to treat this form as a synonym.

VOCALISATIONS AND OTHER SOUNDS These are quite poorly known, and the western and north-western populations (of nominate subspecies, plus *meeki* and *chalcothorax* in particular) are very deficient in examples, with some marked differences suggested in the very limited data available. The usual call in the Ok Tedi area (of uncertain subspecies but which may be *meeki*, which is reported from the PNG border very close to this area) is a loud, husky, rather nasal ascending and then downslurred *wrenh wrenh*, a little reminiscent of the call of a *Melidectes* honeyeater, which may be repeated up to four times but is usually given as a double note. Laman & Scholes render this call as *kwoi-kweer* or *kwoi-kwoi-eeng*. It is a characteristic sound of the species, which is quite vocal on early mornings and middle to late afternoons. The call is often given three or four times before several minutes of silence. It is audible from several hundred metres and seems to be a contact or advertising call, which may help to keep small foraging groups together. Sometimes the final note breaks down, becoming upslurred and querulous. It appears to be given by males or subadult males. Given the vagaries of sound transcription, what is probably this same double call is transcribed as *prat prat* by Ogilvie-Grant (1915a), as *kwoi kwoi* by Cooper (in Forshaw & Cooper 1977) and as *Kuck Kurrk* by Bishop (in F&B). When an individual is disturbed or stimulated by other birds, this double call may become louder and more frantic-sounding, almost a shriek, described by Laman & Scholes as *kwa-a-a-a-ng*. The call described by King (1979) as being like that of a large parrot, loud and raspy and audible at great distance, is also probably this sequence. It seems likely that all these transcriptions are referable to race *chrysenia*. Healey (1980) records a loud song as *scree scree scree, oo-wit, oo-wi-oo*, the first three notes short and grating, the last two phrases powerful whistles and low-pitched; the two-note whistle rises and the three-note call is reminiscent of a human wolf-whistle (but was this correctly identified? The transcription accurately describes the Magnificent Riflebird call!). No vocalisations like this were ever heard during seven years in the Ok Tedi area and the identity must be in doubt. Healey also records a grating *chack* cry, a bell-like squeaking call and a loud grating *cor cor cor*. No sound cuts were available from xeno-canto at the time of writing (September 2018). Near the display court, Beehler (in Coates 1990) cites metallic *shre* noises and little pleading notes. A musical *whee o weet*, sometimes slurred into a brief *kwoieet*, is a characteristic of males, perhaps a group contact call (Beehler *et al.* 1986). Beehler (in F&B) recorded males (of subspecies *chrysenia*) duetting in late June at 1,800m at Mt Pugent, one bird giving a *kweer* and another responding with a loud and ventriloquial *kweer* note; a four-note ventriloquial whistle was also distinctive. Various notes, including a bleating whistle, a metallic note suggestive of Lawes's Parotia and some quiet musical notes, were also heard there. Birds from Crater Mt give a single frantic quavering whistle, this again referable to *chrysenia*. Adult males can produce a clacking sound with their wings, which younger males and females seem unable to do (Beehler in F&B).

HABITAT This species lives in both primary and secondary mid-montane forests, and will visit gardens and regrowth areas adjacent to such habitats. Tall trees seem to be an essential requirement. In the Ok Tedi area the species was recorded from 650m up to 1,500m, and in the rest of its range from 1,100m to 2,000m, mainly at 1,450–1,800m (Coates 1990).

HABITS This species is seen singly and in loose groups of up to half a dozen individuals, often including both adult males and a few subadult males. Often very vocal, which seems to be an important characteristic of this species and may help in maintaining flocks or causing groups to converge at feeding or display areas. Singles and small groups of female-plumaged individuals move through the foraging areas, calling occasionally, males often appearing briefly. They will often sit prominently atop tall trees for quite long periods when undisturbed. Males are not infrequently seen alone, as well. The birds forage in fruiting trees and along mossy limbs, clambering about and flying to nearby trees, one following another at short intervals. They seem to spend a few minutes in each tree, when they can be very unobtrusive, and then move off after uttering the loud, rather husky nasal *wrenh wrenh* contact call, which is usually the first indicator of their presence in an area. The birds are wary but not unduly shy, although a close approach is difficult.

FOOD AND FORAGING Primarily frugivorous, this species also forages for arthropods and insects by stripping mosses off branches. It also probes among beard lichens hanging from high branches. Rarely seen on the ground when feeding. One bird was seen to take a grasshopper from cushion moss and then hold it in its foot while tearing at it (T. Pratt in F&B). Foraging occurs mainly in the upper and middle storeys, the species tending to keep to itself in the Ok Tedi area, although Magnificent and Greater Birds of Paradise use the same regrowth. It has been seen in old gardens in the Bismarck Range, frequented also by the Trumpet Manucode and Superb Lophorina, feeding on the small red fruits of *Callicarpa pentandra* (Cooper in Forshaw & Cooper 1977). *Schefflera* fruits are a favourite food, as with many other paradisaeids; it also takes *Riedelia*. Gilliard remarks that, at a display court near Telefomin, 'all about were droppings containing many red and yellow seeds, and in places the clearing was so splattered it resembled an artist's palette' (Gilliard & LeCroy 1961). Captive individuals have been seen to eat green leaves (Coates 1990).

BREEDING BEHAVIOUR As with other parotia species, males display at terrestrial courts, implying a polygynous species. Adult males may maintain exploded leks and are promiscuous, and females attend to nesting duties alone. The courtship of captive birds is fully described in F&B, to which the interested reader is referred for details. **DISPLAY** Display courts are maintained in June, July and October (the dry season) in the Jimi River area (F&B); in the Herowana area of PNG, adult males court and mate October–December (D. Gillison in F&B). The displays seem similar to those of Lawes's Parotia, but the crest, flank and throat feathers also play a part. A court on the south slopes of the Bismarck Mts in July consisted of a cleared patch of forest floor about 1.5 × 1m, with five or more thin branches 30–75cm above it on a 20–25° slope; this was partly concealed by light undergrowth but with a canopy gap above permitting occasional sunlight to reach it (Cooper in Forshaw & Cooper 1977). Laman & Scholes made important new discoveries that shed much more light on the very elaborate displays of this species, which are among the most complex of the entire family. Each display area is the preserve of one male, and this individual, as with its congeners, maintains a cleared area on the forest floor, with a horizontal perch or perches for the females. Unusually, this and the Bronze Parotia are the only species known to use a prop, an object brandished in the display ceremony, with the present species usually a yellowish leaf. This is a kind of *modifier display* that may help to draw attention to the elaborate head markings and throat whiskers possessed by this species, which are wiggled during some displays. A good series of videos of the display and behaviour of this species by Laman & Scholes can be found at the Macaulay Library at the Cornell Lab of Ornithology, and these are highly recommended. One display court was described as near a large dead tree, cleared of debris, roughly circular (*c*.2m in diameter) and below a thick canopy of leaves (10–12m above) with a kind of 'jungle gym' consisting of dead sticks, vines and saplings with several near-horizontal perches up to two inches (5cm) in diameter. The favoured branch, *c*.3.5 feet (*c*.1m) above the court, was stripped of bark and moss and so worn from use as to be buff-coloured (Gilliard 1969). Another court observed in October near the Jimi valley had a half-buried log at its centre, which was used as a platform for performance (Healey 1980). A court on Mt Pugent, PNG, watched for three days in June, was attended by 3–5 individuals, including two males and some subadult males. All birds occasionally called excitedly together (Beehler in Coates 1990). A group of excitedly calling males could be heard approaching from a distance and they descended, still calling, into the vicinity of the court, this behaviour possibly designed to guide a female down to the court (Beehler in F&B). The adult male of subspecies *chrysenia* observed in the Jimi valley was seen to perform a display similar to that of Wahnes's Parotia. Just after 08:34, the male hopped on to a half-buried log at the centre of his court in an upright posture and fanned out his flank plumes to form the ballerina tutu. He briefly stretched his neck fully upwards before rapidly bobbing his head from side to side 3–4 times and then bobbing the body up and down several times on flexed legs. He then shuffled on his perch to turn 90° and extended his skirt farther forward to an angle of some 45° to the vertical, while raising the skirt sides to almost horizontal so that the two white flank plumes almost met over the back; he bobbed up and down again, while the now extended breast shield pulsed in and out so as to be almost at right angles to the breast at maximum extent. The skirt plumes were then returned to the normal position, but the breast shield remained expanded and was pulsed several times more. The bird was still in the upright position, and waggled the elongated throat feathers like a beard several times before returning all plumes to normal and flying to a low tree above the court, where he called loudly before flying off. Curiously, the six occipital plumes were not raised in this brief display, which lasted for 90–120

seconds and was carried out in silence. The silver and golden-bronze crest was noticeable and was probably manipulated by the bird (Healey 1980). At Mt Pugent, the male court-owner evicted other birds before performing high-intensity ballerina-like displays to a female-plumaged bird, one of these bouts lasting 30 minutes, with a few intervals of 15–60 seconds in which the female rebuffed attempted mountings (Beehler in Coates 1990). **NEST/NESTING, EGGS & NESTLINGS** There is no information on the nest, the nest site and the eggs, and the incubation and nestling periods remain unknown.

HYBRIDS Intergeneric hybridisation with *Lophorina superba* is recorded. The lone specimen, collected on Mt Hunstein in 1913, was identified as a female Carola's Parotia in 1923, and then named as a new subspecies, *pseudoparotia*, of Superb Lophorina by Stresemann in 1934. The Friths examined the specimen and realised that it is in fact a hybrid resembling an intermediate between the females of the two parent species; it is generally referred to as **Stresemann's Bird of Paradise**. Females of the two parent species exhibit parallel geographical variation, one explanation being that smaller *L. latipennis* mimics the larger and more aggressive present species. The partial sympatry of Carola's Parotia with the congeneric Lawes's Parotia also suggests that hybridisation between these two must be a strong possibility. Both species have been seen attending the same court at Crater Mt (Mack & Wright 1996), and Bishop (in F&B) saw a female-plumaged Carola's foraging with five Lawes's Parotias at the south-eastern edge of the Wahgi valley where it meets the north Kubor range.

MOULT Most moult activity appears to be in September and November–January, although, of the 22 specimens listed in F&B, none was moulting in April, May, July or November.

STATUS AND CONSERVATION Classified as Least Concern by BirdLife International. This is a reasonably common, albeit unobtrusive species, with a wide range across the mainland New Guinea mountains, and occupying a remote and fairly inaccessible habitat which is unlikely to undergo massive disturbance. Adult males are outnumbered at least 3:1 by female-plumaged birds in the Ok Tedi area, and the species may at times be quite common, if sparsely distributed, in the forests there. **FURTHER RESEARCH** The exact distribution of the constituent subspecies is poorly known, and there seem to be some striking differences in vocalisations, which require much further study. The nest and eggs are still unknown, as are details of its interactions with Lawes's Parotia in the limited areas of overlap. The species occupies a range between members of the black parotia superspecies, and it is something of a biogeographical puzzle as to how this evolved, and whether *carolae* is a member of the *Parotia* superspecies or a separate radiation. Carola's Parotia is known to have lived for 24 years in captivity (Gilliard 1969), but data on wild individuals are lacking.

Right: **Carola's Parotia**, female
plumage, Tabubil, PNG (*Markus Lilje*).

Carola's Parotia, male displaying to five females, PNG (*Tim Laman*).

BRONZE PAROTIA
Parotia berlepschi   **Plate 10**

Parotia berlepschi Kleinschmidt, 1897, *Orn. Monatsber.*
5: 46. New Guinea in Foja Mts, Mamberamo basin,
north-west New Guinea.

Other English names Foja Parotia, Foya Parotia,
Berlepsch's Parotia, Kleinschmidt's Parotia

Etymology The name *berlepschi* commemorates Hans
von Berlepsch, a German ornithologist and collector
who was the first to obtain specimens of this mysterious
taxon.

Scholes *et al.* (2017) give more details as follows:
'Kleinschmidt described a new species from the
two Berlepsch specimens, which he christened
Parotia berlepschi in honor of Count von Berlepsch

(Kleinschmidt 1897a,b). Kleinschmidt argued that *P.
berlepschi* was distinct from *P. carolae* in seven characters:
(1) shorter supranarial tufts (sensu Scholes 2006), (2)
loral tufts (sensu Scholes 2006) that lack white tips and
do not curve inward, (3) a thicker and more curved
bill, (4) black throat and cheeks, (5) smaller and
darker eye-ring, (6) darker crown, and (7) a golden-
brown (bronze) sheen on the hindneck (Kleinschmidt
1897b).' The study by Scholes *et al.* (2017) found all
the above points to be more or less accurate with the
single exception of the bill shape. 'Described from
trade-skins of unknown origins, *Parotia berlepschi*
Kleinschmidt, 1897 was the subject of a long-standing
ornithological mystery that remained unresolved for
well over a century. With few specimens and no known
wild population, most taxonomic assessments over the
last century have treated *P. berlepschi* as a subspecies of

Parotia carolae A. B. Meyer, 1894. Following discovery of its geographical home in 2005, most authorities returned to giving *P. berlepschi* full species status.' The type locality was formerly unknown (and the type specimen lost), but it was suggested as being the Van Rees Mts (Mayr & Greenway 1962) or perhaps the Foya Mts (Diamond 1985), where it was rediscovered in 2005.

FIELD IDENTIFICATION This species is a rather aberrant parotia resembling Carola's, the females of the two being very similar, but the males much more distinctive. No other *Parotia* occurs within the limited and isolated range in West Papua. **SIMILAR SPECIES** Male differs from that of similar but allopatric Carola's Parotia mainly in being much darker-headed, with a slightly shorter bill with upper mandible more decurved, the cheeks and chin to throat black, nape and hindneck with golden-brown sheen, elongate loral feathering without whitish terminal tips and less inwardly curled, occipital plumes slightly longer and spatulate tips smaller, and having iris pale blue (not yellow). Females also are distinguishable, that of Bronze Parotia showing more extensive white in the centre of the forehead, extending almost to the forecrown, the brown of the loral area also more extensive and with less white near the base of the upper mandible, and the supercilium reaches farther back onto the nape, giving this species a whiter nape than female Carola's Parotia. Eye colour, too, is distinct, being pale blue-grey and not yellow, and this is believed to be an important intersexual signalling character, the eyes being of focal importance as ornamental features (Scholes *et al.* 2017).

RANGE Confined to the uplands of the Foya Mts, east of the upper Mamberamo. **MOVEMENTS** Resident.

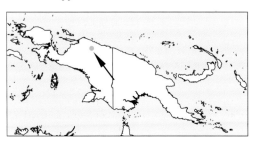

DESCRIPTION Sexually dimorphic. *Adult Male* 26cm. The adult male has the head and body blackish, with a heavy bronze sheen on nape and mantle, and a short erectile hindcrown crest tipped silver-white. The occipital plumes (or head wires) arise from above and behind the eyes, with ear-tufts of elongate narrowly pointed feathers extending behind each eye, and three long, erectile bare wire-like occipital plumes with relatively small spatulate tips on each side (12cm long). The loral and foreface feathering is slightly elongated, from nostril to above eye; the mid-crown has a lateral band of striking golden yellowish-brown iridescence with a white spot in the centre (which is very significant in display), and the nape is washed rich deep bronze. The chin, cheek and throat are black (with slight reddish-brown iridescence in some lights), the lower central throat slightly paler, grading into the breast shield of scale-like feathers with iridescence of bronzed yellow-green; rest of underparts blackish-brown, with striking predominantly white inwardly curving flank plumes. The throat lacks the prominent whiskers of Carola's Parotia and the head is altogether much plainer and darker. *Iris* is cerulean grey-blue; *bill* dark grey; *legs* blackish. It is not known whether or not the iris can change colour as with Lawes's Parotia. *Adult Female* 25cm. Smaller than male, lacking occipital plumes and iridescence. Upperparts brown, with brownish head and complex head pattern with a broad white supercilium which extends farther on to the nape than that of Carola's Parotia, and white moustachial and submoustachial stripes lightly flecked olive-brown; the brown of the lores is also more extensive than Carola's Parotia, and the malar area is olive-brown. The chin is faintly barred greyish-brown, the throat whitish with tiny brown markings. There is some rufous on the flight-feathers and greater coverts, and the underparts are whitish with fine dark barring. *Juvenile* undescribed. *Immature Male* has slightly more extensive whitish on elongate loral feathering; tail longer than that of adult male, as with congeners. *Subadult Male* This plumage was described by Scholes *et al.* (2017): 'As with *P. carolae*, subadult males range from primarily female-type plumage with partial adult male feathering appearing about the head (e.g., photo ML48065341) to nearly full adult male plumage with only small patches of female-type plumage remaining (e.g., photos ML48068171, ML48068161).'

TAXONOMY AND GEOGRAPHICAL VARIATION Until recently this taxon was treated as a race of the more widespread *P. carolae*. It was known only from six old specimens first collected in 1897 and found also in the Rothschild collection in 1908 (Rothschild 1908); this was one of the semi-legendary lost birds of paradise, one of the great ornithological mysteries and unknown in the field. Diamond saw a female parotia in the Foya Mts in 1985 which he suspected perhaps to be this then lost taxon, but it required confirmation by sightings of the male. He argued that this is a more distinct taxon than the eastern *helenae* (previously regarded as a subspecies of Lawes's Parotia but now elevated to species level by most). No adult male paradisaeid of any species is known to show differences in eye colour among subspecies, another feature in favour of specific status for *berlepschi*. Access is very problematic in the remote and inaccessible area inhabited by this parotia, and it was finally rediscovered in December 2005 (along with other little-known and new species) when an international team of eleven scientists (from USA, Australia and Indonesia), led by Dr Bruce Beehler, relocated it in the Foya Mts of West Papua. The species is still known from just seven specimens, including one recently recognised (previously unknown) in the Vienna Museum. The analysis by Scholes *et al.* (2017) confirms 'that *P. berlepschi* is phenotypically well-differentiated from *P. carolae* with regard external appearance and voice of males, which are important intersexual signals among polygynous birds-of-paradise and therefore justify treating *P. berlepschi* as being a distinct species from *P. carolae*. Evidence for differentiation in courtship behavior is inconclusive; however, we are confident that the full courtship repertoire of *P. berlepschi* has yet to be observed.'

VOCALISATIONS AND OTHER SOUNDS Primary song is quite distinct from that of Carola's Parotia, but remains little known, with no recordings on xeno-canto at time of writing (September 2019). The Macaulay Library at the Cornell Lab of Ornithology has a series of four recordings by Bruce Beehler; ML139542 has a series of harsh squawks and short, shriller, almost trilled notes, while ML139538 (listed as the song) has shrill rapid *pee-dee-deet* notes described as a tremulous whistle or whinny, and a querulous *weet*. The primary vocalisation is presumably the note described by Laman & Scholes as a shrill ascending two-note *wee deet*, reminiscent of a squeaky toy and unlike any call of Carola's Parotia. Deep harsh squawks are documented on Macaulay Library ML163704 (by E. Scholes). Scholes *et al.* (2017) described the following. When excited, for example on approach to the display court, males of the present species gave a harsh, raspy, 'white noise-like' scold ranging from one to more than six 'notes' (e.g. the single scold note in video ML457925 at 00:48, and also audio ML139631) and several irregular chattery squeaks and chortles; the scold notes are similar acoustically to the advertisement calls given by the 'all-black' blue-eyed parotia species, namely Lawes's, Eastern, Western and Wahnes's Parotias, but are not known from any of the Carola's Parotia complex of subspecies. The primary advertisement vocalisations of the Bronze Parotia are a shrill, ascending, two-note *whee-dee*, reminiscent of a plastic squeaky toy, and a similarly shrill, four-note tremulous whistle or whinny, *we-e-e-et*, in which the first three notes are ascending (audio recordings ML139538 and ML139631). The *whee-dee* vocalisation is occasionally truncated to just a single note, which sounds like *whee* or *wheep*. At other times it is modified such that the two notes descend slightly, with a more abrupt beginning and end to each note, so that it sounds like *chee-deep*, the *deep* note having a slightly lower frequency.

HABITAT This species inhabits the interior of montane forest (Beehler & Prawiradilaga 2010; Beehler *et al.* 2012). It is generally found at *c.*1,200–1,600m, occasionally to 1,700m.

HABITS Presumably much as for Carola's Parotia, but as yet very little known.

FOOD AND FORAGING Very little known. Undoubtedly largely frugivorous like its congeners, but will also take arthropods.

BREEDING BEHAVIOUR Bronze Parotia is a polygynous species, the males forming exploded leks. Males were attending display courts in late November and early December in 2005 and 2008; in June 2007, males were vocal but were not observed attending courts, although vocal activity was consistent with expectations of when courts are being actively maintained in the early part of breeding season (Scholes *et al.* 2017). **COURTSHIP/DISPLAY** Each male maintains a terrestrial display and courtship site with a floor clear of vegetation and a suitable horizontal perch therein. The courts of this species have been difficult to find and were seldom visited, but the bird does clear a court area and maintains it free of leaf debris by picking up and flicking away such items; a video sequence by Scholes of the male clearing the dance floor is available at Macaulay Library at the Cornell Lab of Ornithology (ML457925). There is also a loose carpet or mat of rootlets on part of the court, and the male was seen to bring a large clump of dark brown rootlets to the court. Carola's Parotia also has root mats on its court, but this is not known from other parotia species (Scholes *et al.* 2017). The court behaviour requires much more detailed observation, but broadly resembles that of Carola's Parotia. Females call loudly when coming to the site. Upon the arrival of a female, the male drops to the floor of the court and engages in complex dance sequences that incorporate movements of the head plumes and the fanning of the flank plumes into a 'skirt' when performing the remarkable dance display; clearly, the iridescent patch above the ear-coverts with its central white spot would be very significant in display, as would the scale-like iridescent breast shield. Two courtship behaviours documented by Scholes & Laman include (i) a **horizontal perch pivot**, in which the male holds a rigid horizontal body posture with head and tail held up and the white flank plumes flared out from the body to form two conspicuous white semicircular discs, the bird pivoting from side to side to show them off to maximum advantage; and (ii) a version of the **hop and shake display**, in which the male stands upright on the court and flutters the throat whiskers while opening the frontal crest. The male was also seen hopping in place, dipping the upper body and giving an exaggerated ruffling of the flank plumes and a ritualised shake of the upper body. No distinct hop was observed in the short time of observation, but the display here is very similar to the *hop and shake* of Carola's Parotia, although the skirt-shake element was not seen. Another behaviour documented is the **modifier display** (Scholes), the holding of a pale or yellow leaf while performing other displays, something that is done also by Carola's Parotia. Scholes *et al.* (2017) state: 'With approximately 20 elaborate behaviors used in the context of courtship, members of the *P. carolae* group have one of the most complex courtship repertoires of any bird-of-paradise (Scholes 2006, 2008). Field observations, photographs and video of *P. berlepschi* reveal that its courtship repertoire contains at least five courtship behaviors known from *P. carolae*, but to the extent discernable from the limited observations presented here, courtship behavior does not appear to differ from *P. carolae*. Nevertheless these observations are preliminary, and more field study is needed before differentiation in the behavioral components of the courtship phenotype can be fully assessed.' Those authors continued by noting that courtship displays not seen to be performed by the Bronze Parotia but likely to be part of its repertoire are the *hops-across-court display* and the *ballerina dance*, while less certain, but of particular interest, are the *head-tilting* and the *swaying bounce*. The first two of these are widespread within *Parotia* and therefore likely to be performed by the Bronze Parotia (with some species-level differentiation probable); the last two (head-tilting and swaying bounce) are derived within the *P. carolae* complex and may not exist (at least in the same form) in the repertoire of the Bronze Parotia, but more fieldwork is required. **NEST/NESTING, EGGS & NESTLINGS** There is no information on the nest, the nest site and the eggs, and the incubation and nestling periods remain unknown.

HYBRIDS No information.

MOULT Undescribed.

STATUS AND CONSERVATION The Bronze Parotia is currently classified as Least Concern by BirdLife International, but with a restricted range of about 820km², occurring in a narrow band of mid-montane forest in a small mountain range. It appears to be uncommon. Habitat within its very small range is currently secure, and the area is mostly uninhabited by humans. Nevertheless, in view of this species' remote, isolated and highly restricted range within western New Guinea, any major disturbance such as logging of its presently undisturbed habitat would potentially put it at risk. Scholes *et al.* (2017) propose that its conservation status be reconsidered even if the species is not decreasing, as the area which it inhabits is so small. The population is likely to be significantly below the level of 10,000 mature individuals, which is considered important for special conservation status. Extensive further fieldwork and surveys are required.

Bronze Parotia, female, Papua, Indonesia (*Tim Laman*).

Bronze Parotia, immature male, Papua, Indonesia (*Tim Laman*).

Bronze Parotia, male, Papua, Indonesia (*Tim Laman*).

Genus *Pteridophora*

The King of Saxony Bird of Paradise is classified in the monotypic genus *Pteridophora*, with the distinctive male and female plumages quite unlike those of others in the family. Gilliard (1969) considered it to be close to *Lophorina* and *Parotia* on account of the expandable dorsal cape, complex occipital plumes and the traces of a metallic breast shield. Diamond (1972) subsumed this genus into *Lophorina*, a treatment not subsequently followed (Frith & Beehler 1998; Dickinson & Christidis 2014). Irestedt *et al.* (2009) placed this genus as sister to *Parotia*. The occipital plumes are highly modified into plastic-like platelets and form a unique structure within the family, while the vocalisations and many elements of the display are also very distinct.

Etymology *Pteridophora* is derived from the Greek words *pteris*, *pteridos*, meaning a feathery fern, and *phoros*, which means 'carrying'. It refers to the curious head plumes of the male, which bear a resemblance to elongated fern fronds. The English name is also remarkable in being one of the longest vernacular names for a bird.

KING OF SAXONY BIRD OF PARADISE
Pteridophora alberti   Plate 10

Pteridophora alberti Meyer, 1894, *Bulletin of the British Ornithologist's Club* 4, 11. Type locality mountains on the Mamberamo River, but restricted to the Weyland Mts by Mayr (1941).

Other English names King of Saxony's Bird of Paradise, Enamelled Bird of Paradise, The Enamelled Bird

Etymology The epithet *alberti* was given in honour of Albert, King of Saxony.

FIELD IDENTIFICATION A medium-sized, fairly short-tailed, rather dumpy species of the higher montane forests. The male is black above with an orange-buff wing patch, and has a dark chin and throat and pale yellow underparts. The amazing uniquely modified enamelled head wires are sky-blue above and black or brownish beneath, and stream behind the bird in flight so that they may appear like tail-streamers on a brief flight view. The voice, a strange crackling, spitting, popping, buzzy combination like the song of a Corn Bunting (*Emberiza calandra*) crossed with a deep-fat fryer or intense radio static, is a distinctive sound of the forests in the species' range. Female-plumaged birds are more difficult to identify, being a rather featureless grey above and with heavy blackish scaling on the underparts, and distinctive cinnamon-buff undertail-coverts. **SIMILAR SPECIES** The adult male is unmistakable in a reasonable view, though the all-blackish female-plumaged Archbold's Bowerbird has a small buff patch in the wing and can be confused in a brief view despite the bowerbird's longer slightly notched tail. Female-plumaged birds could be confused with some kind of cuckoo initially, owing to the scaled underparts, but the short tail and cinnamon-buff undertail-coverts should soon eliminate any such possibilities. Loria's Satinbird or perhaps Crested Satinbird may have a rare grey-phase female (or immature?) plumage, but both lack any scaling and the cinnamon-buff undertail-coverts. Belford's Melidectes (*M. belfordi*) has similar undertail-coverts and is grey above, but it has a long decurved bill and striking face pattern. Some female-plumaged King of Saxony Birds (actually transitional-plumage young males) also show traces of head wires as two spikes or points in the occipital region.

RANGE Endemic to the montane forests of the western and central portions of the main cordillera of New Guinea. It extends from the Weyland Mts of West Papua eastwards perhaps as far as the Kratke Range of central PNG, including the Snow Mts, Star Mts, Hindenburg, Victor Emanuel, Schraderberg (Sepik Mts), Bismarck, Karius and Kubor Ranges, Mt Hagen and Giluwe, and the mountains of Ambua and the Tari Gap. Eastern limits as yet uncertain. **MOVEMENTS** None recorded.

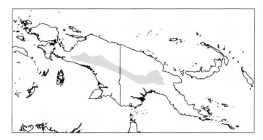

DESCRIPTION This is one of the more bizarre and remarkable species in a family noted for extreme plumage adaptations. Sexually dimorphic. *Adult Male* 22cm; 80–95g. Head, mantle and back are velvety black with an iridescent bronzy-green sheen, and an elongated 'cape' of feathers on the mantle. Behind each eye there is an elongate ear-tuft, with a striking and uniquely modified occipital plume up to 50cm long, each plume consisting of a bare central shaft with 40–50 plastic-looking 'flags' along the outer side only, each of these singular flags an enamel-looking glossy sky-blue above and dark brownish-black below. The rump, uppertail-coverts and uppertail are matt black, washed dark brown, the upperwing similar but with exposed bases and broad leading edges of secondaries and most primaries cinnamon. Chin and throat velvety black, but with scale-like feathers below black throat which are narrowly tipped iridescent green-blue to purple (suggesting a faint breast shield); underparts quite rich mid-yellow, brighter on breast, paler and duller on vent and undertail-coverts, with flanks more creamy. In flight the bird shows a broad buff wingband, and the head wires can be seen protruding beyond the tail. *Iris* dark brown; *bill* black, *mouth* pale to rich aqua-green; *legs and feet* dark brownish-grey. *Adult Female* 20cm; 68–88g. Smaller than the male, lacks occipital plumes, and has a very different and much more modest plumage. Head and upperparts greyish, darker and browner on lower back; chin, throat and sides of neck buffy grey with broad dark brownish-grey bars, giving scalloped appearance. Underparts distinctive, with breast to vent whitish, heavily marked with blackish-brown chevrons, and ochreish undertail-coverts, both good field characters. *Juvenile* undescribed. *Immature Male* resembles adult female, but upperparts paler, more uniformly brownish-grey to grey (less scalloped), underparts whiter (less dark barring and spotting). Some have the beginnings of head wires showing in this plumage. Iris red-brown. *Subadult Male* variable, from like adult female with few feathers of adult male plumage intruding to like adult male with few female-like feathers remaining; first sign of adult plumage is black nasal-tuft feathering and darker primaries with orange (not grey) concealed bases, followed by more black head plumage and some yellow on breast; male tail length decreases slightly with age.

TAXONOMY AND GEOGRAPHICAL VARIATION

Monotypic. Treated as monotypic by Diamond (1972) and Cracraft (1992), as well as by Beehler & Pratt (2016). Proposed subspecies *P. a. buergersi* Rothschild, 1931, from Schrader Mts, and *P. a. hallstromi* Mayr and Gilliard, 1951, from Mt Hagen, are treated as synonyms of *alberti*.

VOCALISATIONS AND OTHER SOUNDS

The voice of the male is one of the most distinctive in the entire family, being a strangely structured sibilant series of jumbled, confused notes with a curious jangling, hissing quality reminiscent of a Corn Bunting song, becoming louder as it progresses. It comes pouring out as a cascade of sound lasting perhaps ten seconds, the singer then going quiet and awaiting a response of some kind. It may be repeated every 2–3 minutes for up to half an hour, the male then going off and foraging, only to return later. Peak song activity is from early morning up to about 09:00 hours, with decreasing frequency thereafter, although sporadic singing can occur at any time. Gilliard (1969) described the male's song as beginning as a drawn-out hissing note (sounding like escaping steam) and terminating in an explosive rasp that can be heard for almost a mile over the crown of the cloud forest; *kiss-sa-ba*, the native name for this species in the Kubor Range, is based on the sound of its call. Xeno-canto has some good examples of the advertising song, such as XC38109 (by F. Lambert) and XC23174 (by M. de Boer). Diamond (1972) describes it as about three seconds of gradually increasing volume and of a 'weird unbird-like quality, a very dry rattling, a spitted jumble of insect-like notes poured out at machine gun pace and suggestive of bad static on the radio, which it briefly turns into a twittering at the… crescendo'. D. Frith (in F&B) saw a courtship display which involved a quiet subsong of hissing sound (*cf.* Gilliard 1969) within which can be heard soft high-pitched clucks, chatterings, mewings and squeakings, some like a soft form of the advertisement song, the whole somewhat reminiscent of the subsong of Common Starling (*Sturnus vulgaris*).

Some mimicry of other species may be involved here, but this requires confirmation as mimicry is unknown in the family. Another characteristic call is a loud melidectes-like harsh *grrrrreeeaaa*, downslurred and lasting just a couple of seconds, which seems to be a general sort of contact call, but this, too, may even be some kind of mimicry, as it is easily dismissed as just another odd melidectes call. It seems to be given by males or by those female-plumaged young males which have pointed wires starting from the crown. It may be given at the display post, but more often is given from within dense foliage in the mid-stratum. Xeno-canto has an example at XC176650 (by P. Gregory). Beehler (in F&B) notes young birds as having a jeering call reminiscent of a Hooded Cuckooshrike (*Coracina longicauda*), *chweer chweer chweer chweer*, each note harsh, rolling and downslurred.

HABITAT This species inhabits middle and upper montane forest, also forest edge and clearings, where it ranges from 1,400m to 2,850m, mainly 1,800–2,500m. It seems to avoid small forest islands in montane grassland, but can be found around logging areas and trails so long as good forest is nearby.

HABITS During the early part of the day the males perch on and sing prominently from branches in the middle to upper canopy of medium-sized to large trees. They may be out in full view, or tucked back in behind some epiphytes or foliage and difficult to see. They sing also in mid–late morning, but much less frequently, and generally from within cover and not on obvious song perches; song is very infrequent and sporadic during the afternoon. They are occasionally seen much lower down in the shrubbery, but at such times seem very shy and are easily disturbed. Female-plumaged individuals are much more catholic in their choice of niche, occurring from the lower shrubbery up to the canopy, probably mostly in the middle layers. They are also encountered much more frequently than the adult males (as would be expected with immatures included). They will join mixed-species foraging flocks consisting of scrubwrens, Grey-streaked Honeyeaters (*Ptiloprora perstriata*), Smoky Honeyeaters (*Melipotes fumigatus*), Belford's Melidectes (*Melidectes belfordi*) and Ribbon-tailed Astrapias. Agonistic reactions with the large, noisy and aggressive melidectes are frequent; in November 2001, at the Tari Gap, a Belford's Melidectes chased off a male King of Saxony from some fruiting vines, and was holding on to a head wire as the bird flew away (pers. obs.) – maybe this is one possible explanation of how Archbold's Bowerbirds obtain such plumes for their bowers! Smoky Honeyeaters also chase the birds and are chased by them, and Bishop (in F&B) saw a male Loria's Satinbird displace a male King of Saxony briefly. A **strange association** of this species with the rare and localised Archbold's Bowerbird occurs at the Tari Gap, PNG, and presumably elsewhere where the two are sympatric (Frith *et al.* 1995): the dominant male bowerbirds, those with the largest mat bowers, collect the occipital plumes of the male King of Saxony, with as many as six such plumes in a single bower. Joseph Tano (at Ambua), who has often collected them, said that he had taken nine sets from one particular bower over the years, this bower-owner apparently also getting the central tail feathers of Ribbon-tailed Astrapia for ornamentation. Other mat bowers may have none, and it is interesting to speculate on how and when the birds come by the plumes. Is the behaviour age-related? Is it a learned behaviour? Do all birds do it at some stage? This is presumably the alpha males showing off their prowess and status with these rare natural objects, but how they come by them is an intriguing and as yet unanswered question.

FOOD AND FORAGING This paradisaeid is primarily frugivorous, with a particular fondness for green fruit, but it takes some arthropods as well in varying proportions depending on local and perhaps seasonal circumstances. A male near the Tari Gap in November 2001 was eating the green acorn-like fruit out of the support cups in a vine some 3m up a tall tree, until driven off by the melidectes referred to previously (see HABITS, above). King of Saxony Birds will forage along mossy branches and in epiphytes, as well as at flowers and fruit; their foraging height varies from the shrubbery layer up to canopy level. They tend to be solitary, but small groups of three or four individuals are sometimes encountered at good feeding places. The Brown Sicklebill will often displace this species when the two chance to encounter each other; Ribbon-tailed Astrapias, on the other hand, seem more placid and less disposed to drive them away, although habitat-niche segregation seems to work to keep them separate anyway, the King of Saxony foraging in denser lower shrubby growth for much of the time.

BREEDING BEHAVIOUR Display has been noted from at least September to April/May and seems likely at all times of the year; breeding must be similar in timing, though the only known egg-laying was in January. Polygynous; promiscuous males sing solitarily from traditional perches. **COURTSHIP/DISPLAY** Males sing from various sites, from exposed bare limbs to more concealed positions among foliage. One such perch near the Tari Gap has been used constantly for at least five years, while Bishop (in F&B) reports a perch used for late-afternoon calling (sporadic in my experience) from 1985 to 1995. Calling is most frequent from about an hour after sunrise to about 09:00, tailing off rapidly thereafter but still heard occasionally at odd times during the day, least often at the hottest hours. Calls may be given about once every 1–3 minutes, and often within earshot of other calling males, creating small, widely spaced clusters of males at favoured sites, with large seemingly empty spaces in between. One area at the Tari Gap had about five males along a 300m walk, each singing solitarily but within earshot of at least one neighbour. The closest two males are about 70m apart, and they seem not unduly territorial, generally ignoring playback. Gilliard (1969) suggested that males gather in small areas to display and can be common there, but rare elsewhere in the habitat; they are spaced some 400 or more yards apart, and this seems broadly consistent with the Tari Gap observations. In the Kubor Range he found three groups of 3–7 males, the groups some 8–16km apart and each perhaps forming a kind of **exploded lek**. Some aspects of the display are fairly well documented, but others remain poorly known. The following is a composite based on Gilliard (1969), Coates (1990) and F&B together with my own observations, and I recommend the video and audio archive at the Macaulay Library at the Cornell Lab of Ornithology, where Laman & Scholes have some fine material. The males sing from perches, which may be 20–30m up in quite large trees, in the **Canopy Singing Display**. The perches often command a wide vista, and the male spends much of the day, but particularly the early morning, around his songpost. When he is singing the bright pale green gape is displayed, and the head wires can be moved around either synchronously or individually at almost any angle; the wires appear as semaphore signals in the sunlight, the blackish undersurface contrasting sharply with the sky-blue upperside and giving a twinkling appearance as the vane twists in the breeze, or is twisted by the bird's movements. The uniquely modified occipital plume has 40–50 flaglets along its outer side, these made of a kind of plastic-looking substance and nothing like a normal feather. The plumes may be held forward over the head, and the cape and breast shield erected, or only the cape is erected as the bird sings (when the bird is not excited the plumes hang down one to each side and can appear distinctly cumbersome). When the head is moved back and forth the plumes may wave and spring about, the tips at times nearly touching in a gentle lateral waving motion; the bright yolk-yellow of the breast can be prominent during such actions. During the course of peak singing, the male may turn completely around on the songpost several times, as if to broadcast the advertising song in all directions. When males leave their songposts it is notable that they frequently dive off the perch at a steep angle, rather than just flying directly away. When a female visits, the male will drop to a perch lower down among vines some 3–15m from the ground and perform a remarkable **Bouncing Display**. This is shown in the videos in the Macaulay Library at the Cornell Lab of Ornithology by Scholes (at ML465218) and by Laman (at ML456545), and also features in the *Attenborough in Paradise* film. Here the male begins violent bouncing motions causing the perch to bounce up and down, and sweeps his occipital plumes forward in slow, graceful, deep sweeping bows towards the female while expanding and pulsing his cape, which forms a semicircle behind the head, connecting with the black and iridescent breast shield. While bouncing some 40–60 times, the male kept up a constant hissing like the sound of escaping steam, and performed the bowing display six or more times. The female may approach with bill open and wings fluttering like a juvenile begging (Gilliard 1969), and will be bouncing on the vine through the actions of the male. When she was a bit closer, the male lost control during one of the upward bounces and let go of its perch, emitting a harsh gurgling hiss as the display terminated and the birds flew off, the male just behind the female. Frith & Frith (1997e) added significantly to the knowledge of this display. If the female turns and moves away, the male may stop bouncing and hold his closed wings barely away from his body, shivering them in juvenile fashion and holding the head low with the plumes just above his back, hissing all the while. The female may approach him again and the bouncing resumes, and the male's occipital plumes may be brought together and then held apart in front of him. If the female maintains interest, the bouncing slows and the plumes are brought around together in front of her more frequently. At this stage the male begins to rotate his upper body stiffly and to waggle his increasingly erect head and neck in a strange side-to-side motion. Mantle, cape, breast and head feathers are fully erect and the plumes held at 45°, wide apart and then brought around either side of the head to project forward and above him. He hops with increasing speed up the vertical perch in this pose to mount the female for about three seconds; during mating he points his bill into the female's nape, holds the head wires at 45°, flaps his wings to maintain position, and holds the fanned tail to one side. After copulation the female was seen to fly lower down the display perch while the male flew to an adjacent one and again directed his wing-shivering juvenile display at her until she left. **NEST/NESTING** Nest-building and subsequent duties are presumed to be undertaken by the female alone, as is usual with the family. Only one nest has been found, in December at Tari Gap (Frith & Frith 1990c), constructed *c.*11m above ground in a triple-forked upright branch of a tree close to a disused hut and disturbed forest, some 30m from where the forest edge met the subalpine grassland. The nest was a shallow cup about 170mm across externally and about 55mm deep, made from a loose accumulation of fine epiphytic-orchid stems and fresh green comb-tooth fern fronds; it was sparsely lined with fine epiphytic-orchid stems and some other plant material. **EGGS, INCUBATION & NESTLINGS** The single known nest held two eggs, smooth-surfaced with a slight gloss, buff-coloured with many dark streaks, spots and flecks, the streaks forming a band around the larger end, size 33.6 × 23.5mm (Frith in F&B). The incubation period was more than 22 days, and the female fed the young by regurgitation.

HYBRIDS No hybrids are known, despite sympatry with the Brown Sicklebill, Ribbon-tailed and Princess Stephanie's Astrapias and Short-tailed Paradigalla. It is not usually sympatric with the putative closest relatives Superb Lophorina and *Parotia* species, being altitudinally segregated, which may explain the lack of hybrids.

MOULT Museum specimens show evidence of moult throughout the year, and males shed their head plumes from November onwards in the Ambua region. Some adult male-plumaged birds have a different moult timing as they can have much shorter head wires than normal-plumaged males, maybe half the usual length, as seen at Rondon Ridge, near Mt Hagen, in June 2016 (pers. obs.).

STATUS AND CONSERVATION Classified as Least Concern by BirdLife International. The King of Saxony Bird of Paradise is a fairly common species over a wide range, but patchy in occurrence, and its density seems to vary considerably. It is quite common in the Tari Gap area from 2,100m to 2,300m and it was also quite frequent in disturbed habitat around the Ok Tedi mine site, Western Province, PNG, at about

2,000m, and common at 2,745m in the Okbap area of the Star Mts (Bishop & Diamond in F&B). Several of the long-established songposts above Ambua Lodge disappeared from 2014 onwards when local logging interfered with their habitat. Males are highly prized for their enamelled sky-blue and black plumes, which are still worn through the septum of fashion-conscious indigenes in the Wahgi valley, bent to form a facial rim between the nose and the crown (Gilliard 1969; pers. obs.). Local inhabitants in the Tari valley wear them as a part of the head-dress, the waving plume being a valued adornment. Central Aviation, a third-level airline in PNG, uses the plume motif as a striking logo on its aircraft. Max Mal, a guide from Enga Province, recounts a local story about how the King of Saxony acquired its head plumes. The story is as follows (Max Mal pers. comm.): In olden times these plumes adorned the Long-tailed Shrike (*Lanius schach*), who was friends with the King of Saxony, who greatly envied the plumes. One day, the two were working together in a garden and the King of Saxony suggested that the shrike remove the head plumes to avoid damage. Awaiting his moment, he then stole the plumes when the shrike was busy and disappeared into the forest. This is why the King of Saxony is now a forest bird, and the shrike haunts the grassy areas outside, still looking for his lost head plumes. Nowadays, there appear to be few threats facing the King of Saxony Bird of Paradise. The more accessible males close to major roads or tracks will suffer from human predation, but much of the species' range is remote and inaccessible, with low hunting pressure and only a small likelihood of major development or logging. The species should be secure over most of its range.

King of Saxony Bird of Paradise, female, Wahgi Valley, PNG (*William S Peckover*).

Genus *Lophorina*

This formerly monospecific genus was established for what was the Superb Bird of Paradise, but is now split into three allopatric species. Diamond (1972) linked it with a number of other genera, and it has similarities in some aspects of display and plumage both to *Ptiloris* riflebirds and to *Parotia* species. Indeed, some recent authors (Frith *et al.* 2017) subsume *Ptiloris* and combine both the Magnificent and the Growling Riflebird into the genus *Lophorina*, despite many points of difference. The extraordinary triangular iridescent blue breast shield is unique, as is the bizarre cape of elongate nape feathers, while aspects of the display and vocalisations are also very distinctive. *Lophorina* is widespread throughout the mainland New Guinea mountains and seems adaptable to disturbed habitats so long as some tree cover remains. This genus has hybridised with more species than any other member of the family, producing some bewildering intergeneric hybrids. Genetic studies have shown that the clade containing the two genera *Ptiloris* and *Lophorina* dates back over five million years, and that the two started to diverge in the mid-Pliocene around four mya, a little later than some earlier estimates. The dating of *Lophorina* speciation appears likely to be sequential, with Eastern (Lesser) Lophorina around two mya, Western (Curl-caped) Lophorina of the Vogelkop around 900,000 years ago, and the Superb (Greater) Lophorina populations of the central ranges and Huon around 850,000 years ago. Such changes may be driven by climatic fluctuations and the shifting of altitudinal habitat zones, with relatively low-altitude barriers to avian dispersal. The former Superb Bird of Paradise group is polytypic, with widespread variation that is still somewhat poorly known. The revision of Irestedt *et al.* (2017) has now been adopted by the major checklist authorities IOC and Clements, including a reallocation of the nominate, and three species are now recognised, with a further two subspecies. The reallocation of the type, however, has been controversial, and here, while we accept the split, we prefer to follow the traditional taxonomy in respect of the scientific names, and not to transfer *superba* (Western Lophorina) to the Superb Lophorina. Frith & Beehler (1998) and Beehler & Pratt (2016) treat the genus as monotypic, with six subspecies, but here we subsume *sphinx* (basically of unknown origin) into *L. minor*.

Etymology The genus name is derived from the Greek *lophos*, a crest, and *rhis, rhinos*, a nose or bill.

WESTERN LOPHORINA
Lophorina superba   Plate 11

Paradisea superba Pennant, 1781, *Specimen Faunulae Indicae* in Forster's *Zoolica Indica Selecta* p.40 (based on an illustration in Daubenton, 1774, *Table des Planches Enluminées d'Histoire Naturelle* pl. 632). New Guinea; restricted to Arfak Mts.

The identity of the bird in this drawing has recently been the subject of much controversy, and Irestedt *et al.* (2017) proposed that the taxon *superba* be reallocated to what is here called the taxon *latipennis* (Superb Lophorina).

Holotype The type on which the drawing was based is lost; it may have been destroyed during a fumigation of the cabinet where it was stored. Irestedt *et al.*'s reallocation sees AMNH 294594 (♀ adult, collected on 13 May 1928, by Ernst Mayr, no. 602 – type locality: Siwi, Arfak Mountains, Vogelkop, West Papua) proposed, though I do not agree with their suggested nomenclature.

Other English names Curl-caped Lophorina, Superb Bird of Paradise, Arfak or Vogelkop Superb Bird of Paradise

Etymology The specific epithet *superba* is from the Latin meaning superb, excellent or splendid; subspecific name *niedda* is from the onomatopoeic local vernacular name 'niedda', taken from the call (which is rendered *nied-nied*).

FIELD IDENTIFICATION The male is a medium-sized black paradisaeid with the usual dumpy body and broad wings, and a medium-length tail. It is a distinctive species, having an iridescent blue inverted-V breast shield lacking black central spots and which pokes out laterally like diminutive wings, a blue crown, and a curious black cape which is more curled at the margin than with congeners. The female is heavily barred blackish beneath on a paler more whitish background than that of congeners; it has a rather blackish head with little or no superciliary stripe, and dull dark brown upperparts. Males of all *Lophorina* taxa are broadly similar, although varying in the shape of the cape, and whether or not the blue inverted-V breast shield has black dots: all these males have a distinctive bizarre shape, the breast shield sticking out like diminutive wings, and the singular cape on the back. The females, however, are very distinct, so much so that, if species were described from females, this specific distinctiveness would have been picked up far earlier (a situation paralleled by some *Pachycephala* species, where very distinct females have been underrated or ignored as descriptors of specific rank). There are marked specific differences in the *Lophorina* group, with nominate *L. superba* and taxon *L. s. niedda* having a blackish head, paralleled again by the far south-east taxon *L. (s.) minor*. Populations in the Vogelkop (Arfak and Wandammen Mts) also have a proportionately longer tail, which is reflected in tail shape. Both sexes in the Vogelkop group have a moderately tapered tail with significantly longer central rectrices than the square-cut, semi-emarginated tails of cordillera and eastern-peninsula populations of genus *Lophorina*. **SIMILAR SPECIES** Western Parotia is chunkier, slightly larger, has a flat head shape and a violet eye, and is richer buff below than the black-barred quite pale whitish underside of the black-headed Western Lophorina. Magnificent

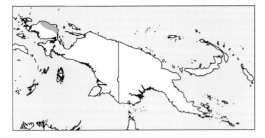

Bird of Paradise females are considerably smaller and shorter-tailed, with a bluish post-ocular stripe and bill, and blue legs. All the astrapias have a much longer tail.

RANGE The nominate race *superba* (synonym *inopinata*) occurs in the Arfak and Tamrau Mts of the Vogelkop, West Papua, while subspecies *niedda* is known from the Wondiwoi Mts of the Wandammen Peninsula. **MOVEMENTS** No movements recorded as yet, but some local altitudinal dispersal in response to food availability is quite likely.

DESCRIPTION Sexually dimorphic. *Adult Male* 26cm; weight 81–95g for Arfak and Tamrau individuals (nominate), 100–105g for those in Wondiwoi Mts (*niedda*). Plumage is mainly velvety black with olive-green to purplish sheens, head with a coppery-green sheen and scale-like metallic blue-green iridescent crown feathers; a short velvety-black tuft of erectile feathers is located above and behind the nostrils. The feathers of the nape are greatly elongated and form a distinctive large erectile cape, a striking feature shared only with the two other *Lophorina* species, though the present species has distinctive elongated downward-pointing sides to the cape giving it an extraordinary shape even at rest, and, as with congeners, the cape can be raised to form a fan and is very striking in the display. The chin and throat are velvety black with dark olive-green feathers, merging into the extraordinary inverted V-shaped, iridescent, intensely metallic blue breast shield (unique to this genus), which lacks the black dots of the Superb (Greater) Lophorina. This shield has the uppermost feathers small and rounded, becoming larger towards the sides, where the outermost are greatly elongated and narrow but square-ended. The rest of the underparts are black, with some olive-purple sheen on the belly. *Iris* dark brown; *bill* black with lemon-yellow to lime-green *mouth-lining*, very prominent when the male is calling; *legs* blackish. *Adult Female* 25cm; weight 68–85g (*niedda*), 71g (nominate). The head and sides of the throat are entirely blackish, lacking the superciliary stripes of Superb Lophorina. The upperparts are dark brown, with some rufous on the wing-coverts and secondaries; below, whitish-grey on the chin and throat, shading to buffy on underparts, and all closely barred brownish-black and appearing colder than the warmer-toned Superb Lophorina females. The tail is more tapered than that of the more squared-off tail of the species' congeners. Females of all *Lophorina* taxa have a longer tail than the respective males. *Iris* brown; *bill* black with pale greenish-yellow *mouth-lining*; *legs* blackish or dark grey. *Juvenile* as yet undescribed, but

likely to be similar to congeners. *Immature* resembles the female. *Subadult Male* begins to acquire the blackish plumage of the adult and becomes blotchy, the moult to adult plumage taking around five years, but this age progression still rather poorly known. Older birds may be like the female but gain an iridescent crown and moderately large iridescent breast shield.

TAXONOMY AND GEOGRAPHICAL VARIATION The species as now defined consists of two subspecies (with *inopinata* a junior subjective synonym of nominate):

1. ***Lophorina superba superba*** (J. R. Forster, 1781). Arfak Lophorina (Curl-caped Lophorina or Arfak Superb Bird of Paradise). Arfak and Tamrau Mts of the Vogelkop, West Papua. Head and sides of throat entirely blackish, lacking superciliary stripes. Males lack black central dots in the stiffened breast-shield feathers (Diamond 1972). Irestedt *et al.* (2017) controversially propose that this taxon be renamed *Lophorina niedda inopinata* subsp. nov., a change from what was the nominate. It also has a very distinct display. 'When expanded for courtship display, the western male's raised cape creates a completely different appearance, being crescent-shaped with pointed tips, rather than the oval shape of the widespread forms of the genus.' The way the western male dances for the female is also distinctive, 'being smooth instead of bouncy' (Scholes 2017).

2. ***L. s. niedda*** Mayr, 1930. *Ornithologische Monatsberichte* **38**, 179. Mt Wondiwoi of the Wandammen Peninsula, West Papua. Another black-headed race, the male lacks black central dots in the stiffened breast shield feathers (Diamond 1972). The female has darker and more ochraceous underparts than the nominate, with a pale cinnamon wash over the ventrum and clearer ochreish underwing-coverts in females (Mayr 1930; Cracraft 1992). Irestedt *et al.* (2017) propose this and the previous taxon as **Curl-caped Bird of Paradise**, *Lophorina niedda*, a course we have not adopted here. We have opted to use **Western Lophorina** for this group, having a neat parallel with the Eastern Lophorina (*L. minor*) and the Eastern and Western Parotias, and avoiding two not particularly helpful common names.

These two races are interesting in showing a curious parallel with the female plumage of the Western Parotia, with a black head, dark back, little or no superciliary stripe (sometimes a short whitish line just behind the eye, and a whitish chin) and reduced rufous edges to the wing feathers (Diamond 1972). The large black satin cape has the outer tips likely to curl much farther under the breastplate in Vogelkop populations than in congeners in the cordillera and the South-east Peninsula. This cape also exhibits significant differences in its shape and fine structure when it is erected in display (Irestedt *et al.* 2017) as now confirmed by Scholes (2017). This, along with the strikingly different female plumage and the very different vocalisations of at least the Arfak taxon, indicates differences at species level.

VOCALISATIONS AND OTHER SOUNDS There is very marked variation in call that must be significant at the species level, as suggested by Irestedt *et al.*

(2017). The loud rasping calls of both the Superb and the Eastern Lophorina are lacking in the Western Lophorina's vocabulary. Birds from the Arfaks, the nominate race of *L. superba* (now reclassified by some authorities as *L. niedda inopinata*), have entirely different calls from those of other *Lophorina*, the male giving a disyllabic slightly raspy, scolding *weep weep*, as at XC62469 (by A. Spencer), while female-plumaged birds give a loud and quite shrill repetitive *schwee* series very like the call of some *Meliphaga* honeyeaters, as at XC62468 (by A. Spencer). The calls of these Vogelkop birds were reported as sounding like the syllables *mjat-mjat* (Bergman in Gyldenstope 1955b), while the local name *nied-nied* also represents the call. Irestedt *et al.* (2017) note that some calls resembling those of Vogelkop birds have been recorded also from the western cordillera at Lake Habbema and Freeport (respectively XC26375, by F. Lambert, and XC140367, by B. van Balen); neither call function nor the age or sex of calling birds was established in these recordings, and until such is achieved evidence for vocal differentiation between Vogelkop and other populations will remain unclear.

HABITAT Inhabits mid-montane to upper montane forest, forest edge, and remnant patches in gardens in montane valleys. Altitudinal range is similar to that of Superb Lophorina, from around 750m up to about 2,300m, and it is most common at the middle ranges.

HABITS This distinctive species is not well known, and its behaviour was for long obscured by the lumping of it within what was then known as the 'Superb Bird of Paradise'. The males are quite vocal, but seem less obvious and are harder to find than the Superb (Greater) Lophorina. Males tend to be solitary, but female-plumaged birds will join mixed-species foraging flocks. They tend to keep to the lower and middle levels of the forest, but habitat-niche exploitation is still poorly known.

FOOD AND FORAGING Known to take primarily fruits, especially berries, supplemented by insect prey, but full details are currently lacking. It ranges from the canopy to the shrub level in search of fruit and insects, the proportions no doubt varying with the local conditions. Foraging behaviour is rather like that of riflebirds, exploring branches and trunks and probing with the bill among epiphytes, dead leaves and bark. Female-plumaged individuals will join mixed-species foraging flocks, including those with Western Parotia.

BREEDING BEHAVIOUR Polygynous; promiscuous males hold solitary display areas within auditory or visual contact of others. Males spend much time in the area of their calling posts, located in the middle or upper stratum 15–30m above ground in lightly leafed trees, and can often be readily observed there. **DISPLAY** The split of the genus into three species is now well supported by film of the distinct display of the Arfak taxon (Scholes & Laman *in litt.*). The display is broadly similar to that of the Superb (Greater) Lophorina, being performed low down in the forest on a fallen tree trunk, the male erecting the much longer and more pointed cape which hangs down in points on each side of the triangular iridescent blue breast shield. The cape is raised as a forward-pointing crescentic (not circular) structure, the blue 'eye spots' on either side of the narial tufts appearing whitish in full display, and the tail held at a shallow angle with the blue upperside quite prominent at times. The dancing is much smoother than and not so jerky as that of the Superb Lophorina, and mating at the display site may occur quite rapidly, the male mounting the female almost as soon as she lands, though copulation seems to be quite brief, albeit repeated if she stays for long enough. The male may call during the full display, a non-rasping *schur schur ip ip* call, with longer or shorter variations, reminiscent in structure of that of the Superb Lophorina but nothing like so raspy. There is also a sharp *click* occasionally made by wing-flicking, as with the Superb Lophorina. During calling, the greenish-yellow gape is quite prominent. There are some fine video examples of the display at the Macaulay Library of the Cornell Lab of Ornithology, as with ML481568 and a lengthy sequence at ML481540 (both by T. Laman). **NEST & EGGS** Little has been published on the nesting of this species, the details of which are thought to be very similar to those for the Superb Lophorina.

HYBRIDS Members of the genus *Lophorina* are known to have hybridised with more species than has any other paradisaeid, reflecting the wide range and extensive sympatry exhibited by the species complex. Generally only very small numbers of specimens exist, often just two or three examples, sometimes just a single specimen, all gathered during the heyday of the plume trade in the late 19th and early 20th centuries. Hybrids involving what is here now called the Western Lophorina are known as follows:
Duivenbode's Six-wired Bird of Paradise '*Parotia duivenbodei*' Rothschild, 1900. Western Parotia × Western (Curl-caped) Lophorina (Superb Bird of Paradise)
Putative hybrids, ideally needing genetic testing to confirm the parentage, are known:
Rothschild's Lobe-billed Bird of Paradise (or Noble Lobe-bill) '*Loborhamphus nobilis*' Rothschild, 1901. Western Lophorina (Superb Bird of Paradise) × Long-tailed Paradigalla
'**Mysterious Bird of Bobairo**' in Fuller (1995), probably a cross between Western Lophorina (Superb Bird of Paradise) and Black Sicklebill, from the Wissel Lakes, Weyland Mts.
For full details and discussion of these bizarre forms, see Fuller (1995) and Appendix One in F&B.

MOULT No details available, but moult presumably takes place immediately after breeding.

STATUS AND CONSERVATION This species appears fairly common within its remote range, notwithstanding some habitat loss caused by farming and some limited hunting for the breast shield as an ornament in traditional dress, or as food. Not assessed by BirdLife as the species was subsumed into a much wider *L. superba* ('Superb Bird of Paradise'), but it seems to be reasonably well distributed at low densities within its quite restricted geographical range and likely to be Least Concern.

Western Lophorina, female, Arfak Mts, West Papua (*Huang-Kuo-wei*).

Western Lophorina, male displaying, Arfak Mts, West Papua (*Huang-Kuo-wei*).

SUPERB LOPHORINA
Lophorina latipennis   Plate 11

Paradisaea latipennis Rothschild, 1907, *Bulletin of the British Ornithologists' Club* **19**, 92. Rawlinson Mts, Huon Peninsula.
Other English names Greater Lophorina, Superb Bird of Paradise, Weyland Superb Bird-of-Paradise (*feminina*), Herzog Superb Bird-of-Paradise (*latipennis*)
Etymology The species epithet *latipennis* is from the Latin *latus*, meaning broad, and *penna*, meaning plume or feather, thus with broad plumes.

FIELD IDENTIFICATION The male is a medium-sized black paradisaeid with the usual dumpy body and broad wings, and a medium-length tail. It is a distinctive species with an iridescent blue inverted-V breast shield (with small black dots in the central area) which pokes out laterally like diminutive wings, a blue crown and a curious black cape which gives the bird a very singular shape. The harsh squawking voice is reminiscent of that of a Eurasian Jay (*Garrulus glandarius*). The female is heavily barred blackish beneath, has a prominent whitish superciliary stripe and a broad black eyestripe, and is largely brown above with rather rusty flight-feathers. It has recently been found that the specialised structure of the black feathers of birds of paradise means that almost all direct incident light is absorbed, making the black colour the blackest of the black and greatly accentuating contrast with the iridescent parts of the plumage for display purposes, this no doubt quite significant for all the *Lophorina*.
SIMILAR SPECIES The male could be confused with male Loria's Satinbird, which is, however, entirely black and lacks the cape and breast shield, or perhaps with a male Lawes's Parotia, but that has head wires, a white narial tuft, no breast shield or cape, and a vivid violet eye. The male is inseparable from Eastern Lophorina in the field, but distributional overlap is marginal or non-existent. Female Eastern Lophorina is distinctive, as it has a primarily black head. Females are otherwise more problematic, and in some areas there exists a curious parallel between the females of the local parotia species and the females of the lophorina. The black barring on the underparts will eliminate just about every other species except other paradisaeids. Female Lawes's Parotia is somewhat larger, with a different, more elongated head profile, a violet eye, and no superciliary stripes. Female Wahnes's Parotia has a limited range, a longer tail, and much less well-marked head pattern. Magnificent Bird of Paradise females are considerably smaller and shorter-tailed, and have a bluish post-ocular stripe and bill and blue legs. All the astrapias have a much longer tail.

RANGE Endemic to the montane forests of New Guinea, where it has a very wide range through the entire central cordillera from the Weyland Range eastwards, including the Snow and Star Mts, also the Hindenburg and Victor Emanuel Mts, then through

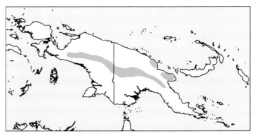

the Central Highlands as far east as the Herzog, Kuper and Ekuti Ranges of the Eastern Highlands. It is present also in the Hunstein Mts (A. Mack in F&B), Adelbert Mts, the Huon Peninsula, and Mt Bosavi. **MOVEMENTS** There is a suggestion of altitudinal wandering during the non-breeding season (Pruett-Jones & Pruett-Jones 1986), who found them much more commonly at the lower levels during this time, and also some sporadic records from Dablin Creek, near Tabubil, at 750m (pers. obs.).

DESCRIPTION Sexually dimorphic. *Adult Male* 26cm; weight 60–105g. Plumage is mainly velvety black with olive-green to purplish sheens, head with a bronzy sheen and scale-like iridescent metallic blue-green crown feathers. A short velvety-black tuft of erectile feathers is located above and behind the nostrils. The feathers of the nape are greatly elongated and form a distinctive large erectile cape (up to 29cm when spread out), a striking feature shared only with the two other *Lophorina* species, this giving the bird an extraordinary shape even at rest. This cape can be raised to form a fan and is very striking in the display. Chin and throat are velvety black with dark olive-green feathers, becoming purplish and merging into the extraordinary inverted V-shaped iridescent intensely metallic blue breast shield unique to this species. The uppermost feathers of this shield are small and rounded, becoming larger towards the sides, where the outermost are greatly elongated and narrow but square-ended, and can measure 20cm when spread. Both subspecies of the Superb Lophorina have small black dots in the central breast shield, the rest of the underparts being black with some olive and purple sheen on the belly. *Iris* dark brown; *bill* black with lemon-yellow to lime-green *mouth-lining*, the latter very prominent when the male is calling; *legs* blackish. *Adult Female* 25cm; weight 54–85g. Individuals of subspecies *feminina* are characterised by having a brown crown extensively spotted whitish, prominent white superciliary stripes that meet broadly over the frons and extend backwards around the nape to encircle the crown, the brown mantle frequently white-spotted, white malar stripes with black flecks, a mid-brown, often rufous-tinged back, and usually an ochreish moderately dark-barred ventral region (Irestedt *et al.* 2017). Females of nominate *latipennis* have a black crown (rarely spotted whitish) with prominent long, narrow pale white supercilia extending across the nape and over the forehead, giving a distinctive stripey-headed appearance; the dark blackish-brown eyestripe is also very striking. The upperparts are shades of mid-brown, rarely spotted white, with rufous more apparent on the wing-coverts and secondaries. The chin and throat are whitish-grey, shading to buffy on the underparts, and closely barred brownish-black; the ventral region is creamy white and moderately barred, without an ochre wash. This species has a more squared-off semi-emarginated tail like that of Lesser Lophorina, unlike the more tapering tail of the Western Lophorina of the far west. (The females of all *Lophorina* taxa have a longer tail than the males.) *Iris* brown; *bill* black with pale greenish-yellow *mouth-lining*;

legs blackish or dark grey. *Juvenile* has a darker crown, with ochraceous barring on the rear and sides of the neck. *Immatures* resemble the female, but *Subadult Male* begins to acquire the blackish plumage of the adult and becomes blotchy, the moult to adult plumage taking around five years; this age progression is still rather poorly known. Older birds may be like the female but gain an iridescent crown and moderately large iridescent breast shield.

TAXONOMY AND GEOGRAPHICAL VARIATION Irestedt *et al.* (2017) note that 'in considering the nomenclatural impact of a shift in application of *superba* Pennant, we submit, given the splitting of species advocated here, that transferring *superba* to the most widespread and familiar segregate species serves stability better than keeping it for a localised endemic in the Vogelkop'. This treatment remains subject to some debate, is not yet finalised, and is not adopted here. Following the split of *Lophorina* into three species, this species has two races:

1. *L. l. latipennis* (Rothschild, 1907). Synonyms *L. s. connectens* Mayr, 1930, and *L. s. addenda* Iredale, 1948. Mountains of the central cordillera of West Papua from the Weyland Range east to at least the Victor Emanuel and Hindenburg Ranges; eastern limits not yet known. Distinguished by the broad superciliary stripe, which extends across the nape to the other eye, female smaller in all measurements. Males have black central dots in the feathers of the stiffened iridescent blue breast shield (Diamond 1972). The female plumage again parallels the female of a parotia, the species this time being Carola's Parotia, having a brown crown, olive back, rufous wing feathers, similarly coloured underparts and a prominent pale superciliary stripe.

2. *L. l. feminina* Ogilvie-Grant, 1915, *Ibis Jubilee Supplement* **2**, 27. Utakwa River, Nassau Range. The central and eastern highlands of PNG, eastwards to the Herzog and Kuper Ranges and perhaps the Ekuti Range, where it presumably meets with *L. (s.) minor*. Found also on the Huon Peninsula and possibly in the Adelbert Range. Head of female dark brown with a broad whitish superciliary stripe, and white streaking on the forehead, crown and nape; chin and throat whitish. Smaller and lighter than the nominate form, both sexes slightly larger than *L. minor*. Males have black central dots in the feathers of the stiffened iridescent blue breast shield (Diamond 1972).

VOCALISATIONS AND OTHER SOUNDS The male has a distinctive advertising song consisting of loud, rather metallic, harsh, rasping upward-inflected *scheee* notes, which may be rendered as *scherr* or syllables sounding like *au-aggh*. There is also a slightly falling series of four notes in two seconds repeated up to twelve times, *schree-schree-schree-schree* (Coates 1990). This is a characteristic sound of the lower montane forest and gardens and may be given throughout the day, though less frequently in the afternoons. Examples are on xeno-canto at XC176171 and the IBC (by P. Gregory) and XC140367 (by S. van Balen). A distinctive higher and thinner call has been noted from captive females (Crandall 1932), and local people also claim that the birds may be sexed by call (Healey 1986), although female-plumaged individuals are problematic with regard to sexing. Males also have a harsh scolding note, and captive birds a soft mewing *pe-er, pe-er, pe-er* given by both sexes (Frith & Frith 1988). A rasping nasal note very similar to that of a female Blue Bird of Paradise near the nest has been heard from a subadult male and a female-plumaged individual (R. Whiteside in F&B). A piercing metallic *chee* note very like the contact call of female-plumaged Magnificent Bird of Paradise has been recorded near Mt Hagen; see IBC 21.04.14 (by P. Gregory). Males do make a rustling sound in flight, but it is much less obvious than that of riflebirds.

HABITAT Mid-montane to upper montane forest, forest edge, and remnant patches, particularly those with *Casuarina* trees, in gardens in montane valleys; ranges from 750m to 2,300m, mostly at 1,650–1,900m. Diamond (1972) noted that males are scarce at the lower altitudes, where female-plumaged birds predominate. In the Ok Tedi area the latter occurred from as low as 750m, but adult males were found only above 1,500m.

HABITS The males of this lophorina, one of the commoner species in many areas, are very vocal and often perch prominently high in trees in clearings, gardens or forest edge. Males tend to be solitary, but female-plumaged birds will join mixed-species foraging flocks including fantails, scrubwrens and whistlers. This species can be quite confiding, but becomes shy where hunting is rife. It tends to keep to the lower and middle levels of the forest, but will ascend to the canopy if food is available. Individuals will visit pools in the forest to drink and bathe.

FOOD AND FORAGING This species ranges from the canopy to the shrub levels in search of fruit and insects, the proportions no doubt varying with the local conditions. Schodde (1976) reported stomach contents as primarily fruit, much of it of the highly nutritious capsular type (Beehler 1983a). It is quite acrobatic when feeding, hanging head down (Coates 1990), and uses the feet to hold items. Foraging behaviour is rather like that of riflebirds, exploring branches and trunks and probing with the bill among epiphytes, dead leaves and bark. A fruiting tree at Ambua in June 2001 had female-plumaged individuals of Superb Lophorina, Princess Stephanie's Astrapia, Loria's Satinbird, Lawes's Parotia and a Blue Bird of Paradise feeding there at the same time, with a Crested Satinbird on occasion. Cooper (in Forshaw & Cooper 1977) noted a female-plumaged bird feeding on fruit of *Callicarpa pentandra* trees in old garden secondary growth with Lesser Birds of Paradise, Carola's Parotia and Trumpet Manucode. Birds at Rondon in July 2018 were feeding on small berries in trees near a *Schefflera* which was being visited by Princess Stephanie's Astrapia, but rarely by the present species, suggesting a particular food preference.

BREEDING BEHAVIOUR Polygynous; promiscuous males maintain solitary display areas within auditory or visual contact of others. Males spend much time in the area of their calling posts, located in the middle or upper stratum 15–30m above ground in lightly leafed trees, and can often be readily observed there. On Mt Missim, the average distance between territories was 140m (Beehler & Pruett-Jones 1983), with mean territory size of 1.5ha. Observations at Ambua suggest that the species is sparser in primary forest and commoner in old gardens and secondary growth. Pratt (in F&B) found that males at Mt Bosavi like to centre territories on ridge crests and to use the steep spur slopes for songposts. **DISPLAY/COURTSHIP** has been claimed by various authors to take place upon high perches, but most published observations suggest that it in fact occurs low down near the forest floor, on fallen trunks or small trees (Stein 1936; Crandall 1931; Frith & Frith 1988; Pratt in F&B 1998), where the birds may be easily trapped by local people. There are two basic kinds of display, an **Initial Display Activity** and a **High Intensity Display** as recognised by Frith & Frith (1998). In the *Initial Display* the bird briefly holds a sleeked pose, crouching with head and bill pointing up, the triangular breast shield and cape held tight against the body, and bifurcated narial tufts projecting conspicuously forward. The eyes remain fixed on the female, and then this pose changes to a repeated, sudden extension of the breast shield with all else the same. Interspersed with this breast-shield flashing, the cape is flicked well forward over and above the head, but not spread out as it is in the more intense display; the head is also held downwards to show off the short iridescent blue head feathers. These activities become faster as the *Initial Display* continues, and are often but not always performed in the presence of a female plumaged bird. This usually leads into the next phase, the *High Intensity Display*, which is performed on the display areas low in the forest and is a truly extraordinary spectacle, one of the most bizarre sights in nature. The breast shield is thrust forward and fully expanded, the bifurcated narial tufts erect, and the cape flicked forwards and spread laterally to form a complete semicircle over the head and down each side to at least the upper edge of the breast shield, if not below and behind it. Below the lower edge of the breast shield black feathers extend inwards and around to meet the body feathers and form a complete black circle, broken only by the blue iridescent triangular breast shield. Near the centre of the cape, just above the eyes and bill, are two very conspicuous iridescent blue-green eyespots, which are very striking when suitably lit; they may appear as a white patch, bisected by the raised bill and narial tufts to make the eye spots, and according to Frith (in F&B) formed by reflection from the elevated iridescent short forecrown feathers. The bill is kept closed, and in this remarkable circular pose the male dances around the female in short, sharp steps rather similar to those of a parotia in display. These remarkable displays were first shown in *Attenborough in Paradise*, and some incredible video footage is at the Macaulay Library of Cornell Lab of Ornithology, with some amazing sequences by E. Scholes as at ML458167 and ML458000. There is also a short sequence by K. Bostwick at ML487529 showing how the corners of the cape are rounded, not pointed like those of Western Lophorina. The bird may make a harsh cawing during this display, as did one from the Wahgi valley (nominate race *latipennis*), which showed the bright yellow mouth-lining as it gave harsh screams (Timmis 1968). Both the calls and the stepping dance are similar to features of parotia display, and the *Initial Display* pose is nearly identical to that of Magnificent Riflebird (Frith & Frith 1988). **Mating** occurs after the above displays, the female approaching in submissive posture with slightly drooped wings as the male dances around her, their bills almost touching and the female always facing the male. Mounting was for less than five seconds, the female then flying up as the male reverted to *Initial Display Activity* for a few seconds before he, too, flew off (Frith & Frith 1988). During courtship display the flicked wings produce a sharp *click* or *tick tick* sound, this is apparently emitted also in flight as with a similar sound made by Magnificent Birds of Paradise (F&B). A pair of captive Superb Lophorinas has also been observed to indulge in **allopreening** (Crandall 1932), and males have been seen to solicit preening from females and immature males (Frith & Frith 1988); this is curious behaviour atypical of a polygynous promiscuous species and more like a monogamous one. A captive male was also seen to present food to a nesting female, but did not assist with nesting duties and later killed the newly hatched chick (Timmis 1968). All of these observations may, however, be artefacts of captivity. Males also produce a rustling sound in flight, but this is much quieter than that of riflebirds. **NESTING** The species has a wide range, and nesting may be at any month somewhere across it, although Coates (1990) considered that most breeding activity was from the early dry season to the mid wet season. Displays in the wild are noted from September to December (F&B), though this is no doubt a very incomplete picture. The **nest site** may be up to 30ft (a little more than 9m) up in a forest tree (Harrison & Frith 1970), or at an average of 2.1m above ground on Mt Missim, where nests have been found in *Pandanus* crowns, *Calamus* species and a *Calyptrocalyx* palm (Pratt in F&B). Pratt (in Beehler & Pruett-Jones 1983) makes the interesting suggestion that females nest within the territories of males. The **nest** is a loosely made open cup of dark rootlets and fibres, with the addition of a few large dead leaves, small fern leaves and some creeper strands (Coates 1990). It is built by the female alone. Beehler (in F&B) records seeing a female-plumaged bird watching as another such – perhaps a young bird from the previous year? – which was testing out a nest site by crouching down into it. **EGGS, INCUBATION & FLEDGING** The normal clutch seems to be one or sometimes two eggs, although all eight nests found by Pratt (in F&B) on Mt Missim contained just a single egg or nestling. The eggs are pale creamy or brownish-buff, well marked with variable spots, dots and longitudinal streaks of grey, lavender, brownish or rufous colour. Captives incubated for 18 and 19 days, and fledging took 18 days (Timmis 1970). Nestling care is by the female only, as is usual in the family (Pratt in F&B).

HYBRIDS Lophorinas are known to have hybridised with more species than has any other paradisaeid, reflecting the wide range and extensive sympatry shown by the species complex, and its somewhat intermediate spread of plumage and display characters in relation to other genera. Generally only very small numbers of specimens exist, often no more than two or three examples, and sometimes just a single specimen, all gathered during the heyday of the plume trade in the late 19th and early 20th centuries. Hybrids involving what is now referred to as the Superb (Greater) Lophorina are known as follows:

Wilhelmina's Bird of Paradise or **Wilhelmina's Riflebird** '*Lamprothorax wilhelminae*' A. B. Meyer, 1894. Superb Lophorina (Superb Bird of Paradise) × Magnificent Bird of Paradise.

Stresemann's Bird of Paradise was first identified by Frith & Frith (1996b). It was originally identified as a female Carola's Parotia, and later as a new subspecies of Superb Bird of Paradise (described as *Lophorina superba pseudoparotia* Stresemann, 1934) from the Hunstein Mts. It appears to be intermediate between the females of Carola's Parotia and Superb Lophorina.

Duivenbode's Riflebird '*Paryphephorus (Craspedophora) duivenbodei*', Meyer, 1891. Superb Lophorina (Superb Bird of Paradise) × Magnificent Riflebird.

For full details and discussion of these bizarre forms, see Fuller (1995) and Appendix One in F&B.

MOULT has been recorded in all months, with most in January–March and very little over the breeding season. A captive male took about five months to complete his moult.

STATUS AND CONSERVATION Classified as Least Concern by BirdLife International. This species has a very large range, and hence does not approach the thresholds for Vulnerable under the range-size criterion (Extent of Occurrence <20,000 km^2 combined with a declining or fluctuating range size, habitat extent/quality, or population size and a small number of locations or severe fragmentation). The Superb Lophorina is a common and widespread species across the forests of the lower mountains of New Guinea. The iridescent blue breast shields of the males are in high demand as the centrepiece of head-dresses in the Tari and Wahgi valleys, PNG, but the species remains quite frequent in forest, gardens and secondary scrub in these areas, albeit much shyer in areas of heavy hunting pressure. Males are heavily outnumbered by female-plumaged birds, perhaps by as much as five to one, but there is some altitudinal and perhaps seasonal variation in the proportions as noted by Diamond (1972) and Pratt (1982). The species is not in any immediate danger, but it will suffer local declines as human populations and hunting pressure increase, along with land clearance. This lophorina's populations at both the western and eastern extremes of the range are poorly known and would repay further study, as would the mating system and the ways in which males maintain their territories. Affinities with other genera, such as *Ptiloris*, *Parotia* and *Cicinnurus*, await further molecular study.

Superb Lophorina, male plumage showing cape and breast shield, Rondon Ridge, Western Highlands Province PNG, July 2018 (*Steven Brumby*).

Superb Lophorina, female plumage (*feminina*), Rondon Ridge, Western Highlands Province PNG, July 2018 (*Steven Brumby*).

EASTERN LOPHORINA
Lophorina minor   Plate 11

Lophorina superba minor E. P. Ramsay, 1885, *Proc. Linn. Soc. NSW*, **10**, 242. Astrolabe Mts.

Other English names Lesser Lophorina, Superb Bird of Paradise, Rasping Bird of Paradise, Lesser Superb Bird of Paradise, Stanley Superb Bird-of-Paradise

Etymology The species epithet *minor* is Latin, meaning small/smaller or lesser, a reference to a marginal size difference from other *Lophorina* species.

FIELD IDENTIFICATION The male is very similar to that of the Superb Lophorina, being a medium-sized black paradisaeid with the usual dumpy body, broad wings and a medium-length tail. It has the distinctive *Lophorina* shape with the iridescent blue inverted-V breast shield poking out laterally like diminutive wings, with small black dots in the central area of the breast shield, and has a blue crown and a curious black cape, plus a harsh squawking voice reminiscent of that of a Eurasian Jay (*Garrulus glandarius*). The female is heavily barred blackish beneath, and has a rather blackish head with little or no superciliary stripe, and dull dark brown upperparts. The immature initially resembles the female, but gradually over four or five years acquires the adult's black plumage with the blue inverted-V chest shield. **SIMILAR SPECIES** The male could be confused with male Loria's Satinbird, which is entirely black and lacks the cape and breast shield, or perhaps a male Lawes's or Eastern Parotia, both of which have head wires, a white or bronze narial tuft, no breast shield or cape, and a vivid violet eye. Males are inseparable in the field from those of Superb (Greater) Lophorina, with which overlap is very limited or non-existent, but the black-headed females are very distinctive. Females are otherwise more problematic, and in some areas there exists

a curious parallel between the females of the local parotia species and the females of the *Lophorina*. The distinctive close black barring on the underparts will eliminate just about every species except other paradisaeids. The sympatric female-plumaged Lawes's and Eastern Parotias are somewhat larger, with a more elongated head profile and a violet eye, though also with a blackish head. Female-plumaged individuals of the Magnificent Bird of Paradise are considerably smaller and shorter-tailed, and have a bluish post-ocular stripe and bill and blue legs. All the astrapias have a much longer tail.

RANGE This species is confined to the mountains of the South-east Peninsula of Papua, probably from south-east of Wau and through the Owen Stanley Range. **MOVEMENTS** Although no movements have been recorded as yet, some local altitudinal dispersal in response to food availability is quite likely.

DESCRIPTION Sexually dimorphic. *Adult Male* 26cm; weight 77–93g. A medium-sized black paradisaeid with the usual *Lophorina* dumpy body and broad wings, and a medium-length tail. It is a member of a distinctive genus, males of which have an iridescent blue inverted-V breast shield which pokes out laterally like diminutive wings; it has small black dots in the central area, a blue crown and a curious black cape. *Iris* dark brown; *bill*

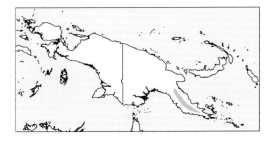

black with pale greenish-white *mouth-lining*, the latter very prominent when the male is calling; *legs* blackish. **Adult Female** 25cm; weight 56–68g. The females of all the *Lophorina* species are heavily barred blackish below. The Eastern (Lesser) Lophorina has little or no superciliary stripe (if present, only immediately behind the eye) and has a rather blackish head and chestnut-infused dull deep brown upperparts, similar in many ways to the geographically remote Western Lophorina. Female Eastern Lophorina is blacker-plumaged than Superb Lophorina of nominate race *latipennis*, with the head and throat blackish-brown and the underparts (including the ventral region) barred rich dark chestnut on a paler, more whitish background, another curious parallel with the Western Lophorina. As with the latter, the superciliary stripe is largely absent, being just a small post-ocular streak with little or no pale nape marking. The wings are shorter than those of all other *Lophorina* taxa, and the tail shorter than on all but Superb Lophorina of race *feminina*. **Juvenile** is little known. Likely to resemble that of Superb Lophorina, and with young males retaining female-like plumage for several years. **Immature** resembles the female, but **Subadult Males** begin to acquire the blackish plumage of the adult and become blotchy, the moult to adult plumage taking around five years, though this age progression is still poorly known. Older birds may resemble the female, but develop an iridescent crown and moderately large iridescent breast shield. Details for this species are almost lacking, but it is likely that the above progression is valid.

TAXONOMY AND GEOGRAPHICAL VARIATION

Monotypic. Described form *lehunti* Rothschild, 1932, is synonymised with *minor*. Type locality of *L. superba sphinx* Neumann, 1932, is unknown, but perhaps in far south-eastern PNG: the unique specimen (female or perhaps immature male) is said to be larger than *minor* (Frith & Frith 1997d), more reddish-brown on upperparts, with a less extensive eyestripe and lacking white flecks on forehead and neck (Gilliard 1969); validity of *sphinx*, however, is uncertain, and it is here synonymised with *minor* following Irestedt *et al.* (2017). Taxon *minor* was proposed as a distinct species, '**Rasping Bird of Paradise**' *Lophorina minor*, by Irestedt *et al.* (2017). We have opted to use the geographically descriptive name **Eastern Lophorina**, making a neat pair with Western Lophorina and similar to the situation with Western and Eastern Parotias, while also avoiding a couple of not unduly appropriate common names.

VOCALISATIONS AND OTHER SOUNDS

Very little known, as the core range of this species is quite remote and difficult of access, and is not on the major birding routes. There was no material on xeno-canto, the Internet Bird Collection (IBC) or the Macaulay Library at the Cornell Lab of Ornithology at time of writing (September 2018). Males give a scolding rasp almost like that of a Eurasian Jay (*Garrulus glandarius*), and an advertising song consisting of loud, rather metallic harsh, rasping, upward-inflected *scheee* notes, sometimes repeated and very similar to calls of the Superb Lophorina (pers. obs.).

HABITAT Mid-montane to upper montane forest, forest edge, and remnant patches in gardens in montane valleys. Altitudinal range similar to that of Superb Lophorina, from around 750m up to about 2,300m, and the species is most common at the middle ranges.

HABITS Not well known, and for long obscured by the lumping of this distinctive species within what was the Superb Bird of Paradise. The Eastern Lophorina appears to be a fairly common species, the males quite vocal and tending to be solitary. They usually keep to the lower and middle levels of the forest, but habitat-niche exploitation is still poorly known.

FOOD AND FORAGING Remarkably little documented owing to its remote and inaccessible range and the fact that the Eastern (Lesser) Lophorina was for long subsumed into the Superb Bird of Paradise. Its diet is presumably much as those of congeners, foraging primarily on fruits from middle to upper levels of the forest. Female-plumaged individuals will join mixed-species foraging flocks.

BREEDING BEHAVIOUR Polygynous; promiscuous males hold solitary display areas within auditory or visual contact of others. Males spend much time in the area of their calling posts, located in the middle or upper stratum 15–30m above ground in lightly leafed trees, and can often be readily observed there. **DISPLAY** Presumably much as with congeners, on a fallen trunk or sapling in the understorey of the forest, but as yet almost unknown in the wild and with no videos on the Macaulay Library at Cornell Lab of Ornithology. Display activity is likely to be during the less wet season, from August to January. Captive observations of this species describe a display very similar to that of Superb Lophorina, with the extraordinary cape erected to form a circle, but the bill may be kept open to show the pale greenish-white mouth-lining, and a clicking noise is made by the wings flicking as the tail is raised and lowered (Morrison-Scott 1936, taxon *minor*). Morrison-Scott also noted the staring green eye spots on each side of the bill created by erected head or narial feathers, which have a small black central spot, serving to heighten the impression of two green irides with black pupils (especially as the true eyes are hidden from view); during this display the bird made a harsh cawing, similar to that of the nominate race (*latipennis*) of the Superb Lophorina. **NESTING** At the time of writing (2018) no data on wild birds are available.

HYBRIDS No hybrids are known.

MOULT No information on wild birds is available.

STATUS AND CONSERVATION This species appears fairly common within its remote range. There has been some localised habitat loss due to traditional farming and some limited hunting for the breast shield as an ornament in traditional dress, or as food. Not assessed by BirdLife as the taxon was hitherto subsumed into 'Superb Bird of Paradise', but it seems to be reasonably distributed at low densities within its quite restricted geographical range, and likely to be Least Concern.

Genus *Ptiloris*

Traditionally the genus *Ptiloris* has been viewed as comprising three species, the Magnificent Riflebird of North Queensland and New Guinea and the two smaller Australian species, namely Victoria's Riflebird and the Paradise Riflebird. All three occur allopatrically in Australia. The two smaller Australian taxa are treated as *Ptiloris*, while the larger Magnificent Riflebird complex is sometimes separated at the genus level as *Craspedophora* on account of its elongated flank plumes, lacking in the smaller species. Recently, the Growling or Eastern Riflebird has been recognised as a distinct species, largely on the basis of its very distinctive vocalisations, and the Cape York taxon *alberti* would repay further analysis. The origin of the vernacular name 'riflebird' has been the subject of some debate, and it is now agreed that is has nothing to do with any of the calls of any of the species. The French naturalists Lesson and Garnot, who arrived at Sydney in 1824, recorded that many riflebirds (the Paradise Riflebird) were killed at Port Macquarie, near Sydney, a few months before their arrival, and that the bird was commonly known as a 'Rifleman' because a soldier had killed the first individuals. The more generally accepted version (Newton 1895) has it that soldiers of the British Army rifle regiments in the early 19th century had bright blue dress uniforms, which resembled the striking blue breast of the male riflebird. The flank plumes and short tail also bore a resemblance to the hanging pelisse (fur-lined cape) and the jacket worn by members of those corps! Another plausible explanation is that the name is derived from ruffle-bird because of the feathering on the culmen, and taken from the Greek generic name (Macdonald 1987).

Etymology *Ptiloris* is from the Greek *ptilon*, a feather, and *ris*, *rinos* = *rhis*, the nose. Riflebirds have the nostrils concealed by dense feathering.

PARADISE RIFLEBIRD
Ptiloris paradiseus   Plate 12

Ptiloris paradiseus Swainson, 1825. *Zoological Journal* **1**, 481. No type locality originally designated but accepted as northern New South Wales.

Other English names Paradise Rifle-bird, The Original Rifle Bird, Rifle-bird, Riflebird, Rifle bird, Velvet Bird, New South Wales Rifle Bird of Paradise.

Etymology *Paradiseus* means 'paradise', derived from the Latin *paradisus* and a reference to the beautiful iridescent plumage of the male. This was the first riflebird to be discovered, reportedly shot by a soldier or a convict in 1823.

FIELD IDENTIFICATION This is the most southerly bird of paradise, the first to be discovered by Europeans and the only one within its range, which makes identification straightforward. The medium size, strongly decurved bill, short tail and broad wings give it a distinctive shape. The jet-black male with his iridescent blue-green breast shield, crown and central tail is unmistakable, while the female, with brown upperparts, rusty wings, short pale supercilium and buff-coloured underparts with irregular blackish cross-barring, is almost equally obvious. **SIMILAR SPECIES** There is none within the range. Perhaps some of the larger honeyeaters with a decurved bill could be misidentified, but Blue-faced Honeyeaters (*Entomyzon cyanotis*) have pale wing patches, a black face with a vivid blue (adult) or green (immature) bare eye patch, yellow eyes, a black chest and whitish underparts. Friarbirds (*Philemon*) are dull brownish, with bare facial skin and bill knob. Spangled Drongo (*Dicrurus bracteatus*) lacks the decurved bill and has a quite long, distinctly notched outward-curving tail.

RANGE This species is the sole non-tropical representative of the family, endemic to the eastern coast of Australia from the Calliope Range south of Rockhampton, Queensland, to just north of Newcastle, New South Wales, in the Great Dividing Range, and some lowlands to the east. The southern limit is the northern Hunter valley near Barrington Tops. **MOVEMENTS** Mostly sedentary, but some very local dispersal into adjacent non-rainforest habitats occurs during the non-breeding season.

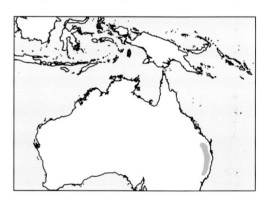

DESCRIPTION Sexually dimorphic. ***Adult Male*** 28–30cm; bill length 54mm; weight 134–155g. The forehead, crown and nape are iridescent metallic bluish-green, edged purplish at feather tips on nape and, as with breast gorget, varying in colour and intensity according to the angle and intensity of the light. The hindneck and sides of neck are black, tinged with a purplish sheen, the ear-coverts blackish, faintly tinged metallic purple, and the lores, eye-ring, malar stripe and chin blackish. The upper throat and sides of neck are blackish with metallic purple sheen, the centre of the throat scaly metallic iridescent bluish-green, this merging into the same colour on scaly breast gorget; this gorget is slightly erectile at the sides, and much larger than that of the Victoria's Riflebird.

The mid-breast is black with a purplish sheen, the lower breast black with metallic dark green scalloping, the flanks and belly dark green with patchy blackish mottling; the rear flank feathers are slightly elongated but do not form plumes. The undertail-coverts and vent are blackish, with thighs blackish-brown. The uppertail is a vivid iridescent metallic bluish-green with dark outer feathers (very striking in flight when seen from above), the undertail and underwings blackish. *Iris* dark brown; *bill* black, very long (but slightly shorter than bill of female), quite slender and decurved, *mouth-lining* bright yellow (prominent when calling in display); *legs and feet* grey-black. **Adult Female** *c.*27–29cm; bill length 59mm; weight 86–112g. The forehead, crown, nape, hindneck and sides of neck, ear-coverts and lores are dark olive-brown, with fine pale streaking on crown; a distinctive pale cream or off-white supercilium extends from lores to side of nape, becoming broader, more diffuse and flecked brown at the rear; the pale cream eye-ring is narrow and broken. The sides of the chin and throat are cream or off-white, shading to buff in the centre; there is a narrow dark brown malar stripe. The upperparts are entirely olive-brown with faint cinnamon-brown tinge, most marked on the wings (and noticeably colder and greyer than on Victoria's Riflebird). The underparts are buff, grading to creamy-buff on the lower body, with bold dark brown chevrons on breast, flanks and belly, and dark brown barring on rear flanks, vent, thighs and undertail-coverts (this species appears much paler, less buffy and much more heavily marked below than Victoria's Riflebird). The uppertail is brown, the undertail a slightly greyer brown. *Iris* dark brown; *bill* black to greyish-black, with paler pinkish base of lower mandible, very long, slender and decurved (slightly longer than bill of male); *legs and feet* grey-black. **Juvenile** Undescribed. **Immature Male** For the first couple of years resembles female in plumage, but gradually moults into male dress, some older birds appearing much as adult with some brown feathers on body, others like female but with patchy black on head and underparts. It is likely that this species acquires full adult plumage after four or five moults, taking four or five years.

TAXONOMY AND GEOGRAPHICAL VARIATION
Monotypic. Described form *Ptiloris paradisea queenslandica* Mathews, 1923, from the Blackall Range north of Brisbane, was later invalidated. The species has in the past been considered conspecific with allopatric Victoria's Riflebird, but significant differences in morphology and behaviour and its long isolation make this view untenable. No variation is apparent within the quite small range.

VOCALISATIONS AND OTHER SOUNDS The typical call of the male is a characteristic loud, scolding double note, transcribed as *scraarsh, scraarsh* or *yaaarss* by Morcombe (2000), while Pizzey & Knight (1997) describe it as *yaass yaass* or *y-a-a-a-ss*. The harsh call is very similar to that of its northern congener, Victoria's Riflebird, but usually given twice, whereas that species utters mainly single notes in the Cairns–Atherton region, though in the south of its range

they are also often given twice. Paradise Riflebirds also emit a long upward-inflected whistle, and give churring noises when in display (Morcombe 2004); a short whistle, a low chatter, and a loud *cluck* like that of a Noisy Friarbird (*Philemon corniculatus*) are also reported (G. Holmes in F&B). The species was called 'yass' by local aborigines, derived from the single harsh tearing call, which is a characteristic sound in the hill forests of its range, audible up to 800m away. This call is loud, slow, drawn out and very rasping, fading as it ends. It is usually given singly, but quite often doubled, Campbell (1901) suggesting that younger female-plumaged males give the single *yaass* note, older males the same note rapidly twice, and even older males leaving a measured interval between the two notes of the call. The *yaass* calls at Lamington in October seem to be given every couple of minutes, with a longer interval between them than is the case with its sister-species (pers. obs.). The duration of the single *yaass* call is probably about a second, although Morcombe (2004), when presumably referring to this call, notes it as being 3–4 seconds in duration. At Binna Burra, in Lamington NP, the period of least calling was found to be from March to August, with an increase in September and peak vocal activity during October–February (G. Holmes in F&B). Other, non-vocal sounds made by this species include a loud *woof-woof* produced by the alternately opened wings during high-intensity courtship display (Ramsay 1919), and this is presumably analogous to a swishing sound made by the wings during the more intense displays. Other observers have reported wing-claps audible from 60m away, but this requires verification, although Victoria's Riflebird certainly makes such a sound. In flight the males, like their congeners, make a loud rustling sound resembling the sound of a piece of thick stiff silk or taffeta being shaken.

HABITAT The species inhabits the subtropical and temperate rainforests of southern Queensland and northern New South Wales, but may be found also in nearby wet sclerophyll forest, and in dry sclerophyll forest during the winter months. Much of its former lowland habitat has disappeared, and nowadays it is found mainly in the hills. Generally this species is most numerous above 500m, extending up to about 1,100m; it occurs occasionally down to sea level in winter. It has been recorded during the non-breeding season from lantana (*Lantana*) on the edge of the forest and from sclerophyll forest dominated by eucalypts, and has been reported even from a Hoop Pine (*Araucaria cunninghamii*) plantation with lantana and vine-forest remnants.

HABITS The riflebirds are usually encountered singly, but may form flocks of six or seven individuals at fruiting trees (F&B). During the breeding season (mainly September–January), the males spend much of the day at their songposts in the middle stratum of the forest, where they display and devote much time to preening or calling, becoming active and vocal as the sunlight hits the treetops. Songposts are spaced about 180–200m apart in optimum habitat. The Paradise

Riflebird is often quite shy, and can be difficult to locate, tending to remain high in the forest. It is noteworthy that its congener, the Victoria's Riflebird, will readily visit feeders and fruit sources at viewing stations such as Ivy Cottage, at Paluma, and Cassowary House, at Kuranda, Queensland, and becomes quite confiding, whereas the present species seems to avoid such sites, not being an easy bird to find at such famous feeding sites as O'Reillys, at Lamington NP, near Brisbane. Lloyd Nielsen (*in litt.*) had a feeding station at his hut on rainforest edge at Lamington, and recorded this species as visiting about ten times in total over a three-year period, and always female-plumaged birds, never the males. A canopy food station about 13m above ground near the canopy walkway at O'Reillys did attract them more, and of both sexes, but not so regularly as it did other species. Nielsen records them as being much more wary than Victoria's Riflebirds, and they are certainly far more difficult to locate. They may come to fruiting trees with Regent Bowerbirds, Satin Bowerbirds and Green Catbirds. During the drought of 1991, a friend of Nielsen's had Paradise Riflebirds coming to bathe at his ground-level birdbath, surrounded by rainforest; they would arrive, bathe very quickly and then fly off. As soon as the rain returned, they ceased coming. For bathing they would usually use hollows containing water in the treetops.

FOOD AND FORAGING The diet of this species consists largely of arthropods, particularly those found in dead wood and dead foliage, or fruit (F&B). The birds behave like Australian treecreepers (*Climacteris*), climbing about branches and trunks, listening intently at times for insect noises within (Ramsay 1919). They spend much time in foraging for arthropods along branches, probing into crevices and behind bark, and inspecting leaf debris, arboreal ferns and epiphytes. They have powerful feet with sharp claws, and often use them to hold fruits or prey against a perch as they tear them up (G. Holmes in F&B). They can cling to the underside of branches for long periods when feeding, Ramsay (1919) recording one as holding such a position for some 40 minutes. They sometimes feed from the ground, and have been seen to drink water from tree cavities (F&B). In prime habitat at Lamington NP, Nielsen (*in litt.*) saw this species in eucalypt forest adjacent to rainforest only a few times over many years. During the drought year of 2002, however, Paradise Riflebirds were seen feeding out in the drier eucalypt woodlands on both visits to the Lamington area in September and November (Gregory pers. obs.): all sightings were of female-plumaged individuals, one in September foraging and probing under bark along dead eucalypt branches, while the November sighting consisted of three female-plumaged birds which were coming, usually singly but sometimes together, to the seed heads of grass trees (*Xanthorrhoea*) and seemingly foraging for insects by probing among the dense seed clusters. As with its sister-species Victoria's Riflebird, the smaller female has a larger bill than that of the male, which suggests exploitation of a slightly different food spectrum from the male's among the somewhat restricted guild of wood-boring

invertebrates (Moorhouse 1996; Frith 1997). The bill is also longer than that of Victoria's Riflebird, and the Paradise Riflebird perhaps may probe rather deeper into wood and cavities than that species (F&B), maybe reflecting a difference in the floristic composition of the habitat and the dependent insects.

BREEDING BEHAVIOUR Breeds mainly September–January. Males are promiscuous, solitary displaying birds that hold songposts in the middle or upper stratum of the forest and are likely to be territorial, as with congeners; they remain tied to the one area during the peak display months, although Beehler & Pruett-Jones (1983) suggested that they are non-territorial. Surprisingly, it is still not known for certain whether the female alone undertakes the nesting duties, as would be expected in both this genus and most of the family. There is an anomalous old report (Foster in Campbell 1901) of a pair of birds building a nest on 10th November, the male being immature as it was not entirely black; perhaps it was merely observing the activity and not taking part, but it is an interesting and curious observation. Pizzey & Doyle (1980) state that the Australian Koel (*Eudynamys orientalis*) is a brood parasite of this species; Higgins *et al.* (2006) are cautious about this record, though it is recorded once for the sister-species. **COURTSHIP/DISPLAY** This remains rather little known, despite the fact that the species occurs quite close to large human-population centres. One delightful early account from Jackson (1907) describes an adult male displaying on a thick horizontal limb of a Red Cedar tree (*Toona ciliata*): the wings would be fully opened, then brought over in front until their ends touched the limb, and with the head well thrown back he would walk majestically up and down the limb for a distance of about three feet, bobbing up and down and causing the wings to make an extraordinary noise resembling the rustling of a piece of new silk; then he would suddenly turn around and around, and every few seconds make quite an unusual sound resembling the faint croaking of a frog. Nielsen (*in litt.*) reports that Paradise Riflebirds seem to use a wide branch as a display perch, always high up close to or in the canopy. He never saw them displaying from broken-off tree trunks as Victoria's Riflebird does, and usually all that one would see would be the bill of the male poking over the edge of the branch, it being nearly impossible to get a view. Displays have been seen between at least August and December, Strange (in Iredale 1950) noting the pairing months as being November and December, at which time the male is easily found. Frith & Cooper (1996) have a full account of the display sequence, and report that the male may attend his display perch at any time of the day but especially early in the morning; when a female appears he commences display posturing, fully extending his wings and holding them out in front of him and parallel with the ground, while cocking the tail and progressively fluffing the flank and abdomen feathers. He will face the female throughout whether she is level with or above or below him, lowering the head and bill or looking up as appropriate. He may occasionally utter the *yaass* call, or gape to show the bright yellow mouth-lining. When the display intensifies, his wings are twisted back to become more vertical, their leading edges held at 45° and the tail lowered. If the female approaches he brings his wingtips forward to touch or nearly touch each other, the outer primaries contorted as he does so. The head and bill are raised vertically upwards, the head and neck plumage sleeked, and he begins to sway his head rhythmically from side to side with increasing rapidity. When the head reaches its rightmost extent his wings are swayed to the left, and vice versa; the female may now perch immediately in front of him if she is seriously intent on mating. The male's wings now all but encircle or embrace the female's head. He flicks his head rapidly from side to side as the female peers into the hole framed by his wings, occasionally seeming to pick or peck at his head or throat. The iridescent colours of the throat and breast are conspicuous, changing from blue to green with the light and angle of view, while the pale and now slightly iridescent bronzy-greenish lower breast and body feathers are puffed out to form a cushion; this cushion complements the iridescent metallic breast and throat coloration, and is edged by broken black curves formed by the bars on the flank feathers. The male now leans backwards and props himself upon his lowered tail, while increasing the speed of his movements until he suddenly lowers his wings and mounts the soliciting female. She may vigorously flutter her wings after mating and depart, if not already chased off by the male. Ramsay (1919) noted a loud *woof woof* sound made by the wings during **high-intensity display**. This is analogous to the swishing sound made by Victoria's Riflebird, and is well shown in a video by John Young in the television series *Australia Wild* (Series 1, episode 17: 'Bird Man of Paradise'): here the male rocks from side to side and the erect cupped wings make a loud swishing sound, intensifying as the female comes closer and gets buffeted by the male's wings, which often leads to mating. A very singular adult male display was noted at Mt Glorious, Brisbane, by A. Hiller (in F&B), in which the bird hung still and completely inverted from a broken bare display bough for some 2–3 minutes, silently gaping. Several other birds of paradise in different genera are noted to display by hanging upside-down, examples being the Blue Bird of Paradise and the King of Saxony. The Paradise Riflebird prefers more horizontal perches for display and calling than does Victoria's Riflebird, some forest-edge sites at Mt Glorious being just 3m or 4m above ground. Sites in the denser forest may be considerably higher, up to 20m, perhaps reflecting less risk of predation there, although it is a strange and perhaps paradoxical fact that rates of predation on displaying male birds of paradise seem very low. **NESTING** Jackson (1907) reported two nests being built on top of old ones of the same kind, and another was placed 18 inches (*c*.45cm) away from the new one; this proves that the birds certainly build year after year in the same tree. Bailey (in Campbell 1901) found an active nest with one egg adjacent to two old nests in the same tree. The Short-tailed Paradigalla can also nest in the same tree or its immediate area for years, but it would be useful to have more recent observations of this habit relating to this riflebird. The most frequently used **nesting site** seems to be in tangles of dense vines, sometimes the prickly lawyer vine (*Calamus*), some 5–40m above ground (F&B). Ramsay (1919) noted a nest 39m up in mistletoe on a eucalypt at the rainforest edge, with another 35m up in a eucalypt some 800m from rainforest. Nests have been recorded also in tree-fern crowns. Nielsen (*in litt.*) found a nest some 20m up in the very top of a small rainforest tree beside a road. **NEST** The nest is a shallow bowl, larger and bulkier than that of Victoria's Riflebird, made of vine stems lined with finer vines, fibres and rootlets. Fresh fern fronds are used to decorate the rim, as often are snakeskins (F&B). Large dead leaves are incorporated into the nest base. Campbell (1901) gives the dimensions as 203–228mm in diameter, with a total depth of 102mm, and the internal cup as 102mm in diameter and 51mm deep. The nest found by Nielsen in the very top of a small rainforest tree by a road had printed paper as a decoration around the edge, and was discovered when the female hopped up the tree and then into the nest to brood; Nielsen also reports that in the 1940s Mervyn Goddard found a nest at Dorrigo, in New South Wales, the rim of which had been decorated with a lunch wrapping which Goddard himself had discarded the week before. This species and Victoria's Riflebird have the strange habit of decorating their nests with sloughed snakeskins, which may perhaps help to deter predatory birds and mammals. Jackson (1907) recorded the skins of Black Snake (*Pseudechis porphyricus*), Carpet Python (*Morelia spilota*) and Death Adder (*Acanthophis antarcticus*) as being used in nests; he also noted that, in the 1–2 weeks between nest completion and egg-laying, one nest was filled with leaves from trees not of the species within the immediate area, and he was sure that the bird itself had placed the leaves there. **EGGS & INCUBATION PERIOD** Clutch size is usually two eggs, sometimes one, variably reddish-cream to pinkish-buff, marked with brush-like longitudinal streaks and a few irregular spots of reddish to brown colour, more numerous at the larger end; average size 32.9 × 23.8mm. Distinctly different from the eggs of Victoria's Riflebird, being darker and richer-coloured with more numerous smaller markings more evenly distributed over them (Frith in F&B). Incubation period not yet known; given as 15–16 days in Schodde & Tidemann (1988), but this seems unduly short as the sister-species takes 18–19 days (Frith & Frith 1998). **NESTLING** The nestling is naked and black-skinned initially (Campbell 1901), but very poorly known. Schodde & Tidemann (1988) give the nestling period as about four weeks, with nestlings in one nest for at least 21 days (F&B); for comparison, that of Victoria's Riflebird is 13–15 days (Frith & Frith 1995a, 1998a). Females with fledglings have been seen in February and March (F&B).

HYBRIDS No hybrids are known, as no other birds of paradise occur within the species' range.

MOULT This species' moult begins in late spring or early summer, and finishes around April. It is likely to take 4–5 years to reach full adult plumage, but there are few data, and none on longevity either.

STATUS AND CONSERVATION Classified as Least Concern by BirdLife International, with a range extending over some 128,000 km². This species is endemic to a fairly small tract of subtropical and temperate eastern coastal Australia, and may perhaps best be described as being sparse and elusive in the hill forests there, not occurring at high densities. The southern part of the range in New South Wales has seen much of its former lowland habitat destroyed, and it seems less frequent in the remaining areas of suitable habitat here. It may be seen in Lamington NP, near Brisbane, but again is often quite hard to find. It was present at Mt Tambourine, near Brisbane, until the early 1970s but had gone by 1976, and the surrounding area is now subdivided into housing zones. Vagrants were historically reported from sites near Rockhampton, a long way north of the species' current range, and may possibly reflect a previous outlying population whose habitat was lost early on. Estimates of numbers seem hard to come by in modern times, but Lucas (in North 1901–14) seldom found more than one male to every 400m of scrub, each male being sedentary. The species seems to avoid small forest patches (2.4ha or less) in northern New South Wales (Howe 1986). Holmes recorded six individuals of this paradisaeid in 112ha of subtropical rainforest, and 13 in 102ha of wet sclerophyll forest near Dorrigo, an average of one bird per 10–12ha. Holmes (in F&B) reported the disappearance of the species from forest remnants as large as 60ha in the Big Scrub. In late Victorian times this riflebird was in some demand for the millinery trade, large numbers being sent to London (Sharpe 1891). The species can become a nuisance in soft-fruit orchards such as pawpaw (papaya), and in Queensland damage-mitigation permits may technically be issued to kill birds that do extensive damage to fruit crops and/or cause severe economic hardship. Reassuringly, given the extensive habitat loss undergone by this riflebird species, it is not on Schedule 1 (common species for which permits may be given without undue restriction), which includes Rainbow Lorikeet (*Trichoglossus haematodus*), Eastern Grey Kangaroo (*Macropus giganteus*) and

Australasian Figbirds (*Sphecotheres vieilloti*). Any such permit application would, one hopes, undergo careful scrutiny, as the Queensland Parks and Wildlife Service has become more sensitive to potentially contentious situations involving such species as riflebirds and bowerbirds. The present species is, like its congeners, classified as common under Queensland legislation (an essentially meaningless categorisation), and is not listed in the rare or threatened categories. The Paradise Riflebird is not currently globally threatened, but because of its restricted range and poorly known biology it is one worthy of monitoring. The forest habitat is well protected in several extensive national parks, but state forest can be subject to fragmentation with corresponding loss of diversity in both plants and birds such as this riflebird. With development and the growth of the human population local losses may be expected, and the bird may in time come to be largely dependent on extensive protected areas. Much of the basic biology of the Paradise Riflebird is rather poorly known, with no recent data, despite the proximity of the bird's range to large universities and urban centres. Much further fieldwork is desirable.

Paradise Riflebird, male at blossoms, Queensland, Australia (*Todd Burrows*).

Paradise Riflebird, female plumage, New South Wales, Australia, June 2016 (*Lionel Hartley*).

Paradise Riflebird, female plumage, New South Wales, Australia, June 2016 (*Lionel Hartley*).

VICTORIA'S RIFLEBIRD
Ptiloris victoriae Gould, 1850   Plate 12

Ptiloris Victoriae Gould, 1850, *Proceedings of the Zoological Society of London* 1849, p.111, pl. 12. Barnard Island, North Queensland.
Synonym *P. paradisaea dyotti* Mathews, 1915. Cairns, North Queensland.
Other English names The Queen Victoria Rifle Bird, Queen Victoria's Rifle-Bird, Victoria Riflebird, Victorian Riflebird, Victoria Rifle Bird of Paradise, Lesser Rifle Bird, Barnard Island Riflebird.
Etymology Named in honour of Her Majesty Queen Victoria, and beginning a sequence whereby many of the family Paradisaeidae were named in honour of European royalty of varying degrees of importance.

FIELD IDENTIFICATION The only member of the genus in its area of distribution, and therefore relatively easy to identify with its typical stout body, short tail and broad rounded wings, plus the long decurved bill. The male is entirely black with an iridescent blue-green crown, chin, throat and breast, and uppertail surface. The lower breast is black, and the entire flanks and belly a curious bronzy-greenish colour. The female is grey-brown above with rusty wing edgings, and has a white superciliary stripe and whitish chin and throat with a dark malar stripe, and irregular dark mottling on the flanks and breast. **SIMILAR SPECIES** None within the range. The Helmeted Friarbird (*Philemon yorki*) is basically brown with a bill knob, while, of the medium-sized black birds in this species' rainforest habitat, Spangled Drongos (*Dicrurus bracteatus*) have red eyes, no blue-green iridescence and a distinctive notched 'fish-tail' and Metallic Starlings (*Aplonis metallica*) have a long, pointed graduated tail and a bright red eye, and both lack the decurved bill. Black Butcherbirds (*Melloria quoyi*) have a heavy, straight, blue-grey bill with a dark hooked tip. Female-plumaged riflebirds are sometimes initially misidentified as some kind of large honeyeater owing to the long decurved bill.

RANGE Endemic to Far North Queensland, where it is found in the wet tropics from just south of Cooktown southwards to Mt Elliot, just south of Townsville. **MOVEMENTS** Basically sedentary, with some very local dispersal into adjacent non-rainforest habitats in the non-breeding season.

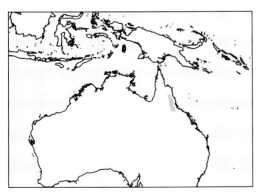

DESCRIPTION Sexually dimorphic. The sexual bill-size dimorphism of the Paradise Riflebird is found to a much lesser extent in the present species (some overlap between the sexes), though the female's bill averages longer than the male's. **Adult Male** 25cm; bill length 43mm; weight 91–119g. The forehead, crown and nape are iridescent metallic bluish-green, edged purplish at tips on nape and, as with breast gorget, varying in colour and intensity according to the angle and intensity of the light. The hindneck and sides of neck are black with a purplish sheen, and the ear-coverts blackish and faintly tinged metallic purple, with the lores, eye-ring, malar stripe and chin blackish. The upper throat and sides of neck are blackish with a metallic purple sheen, centre of throat scaly metallic iridescent bluish-green, merging into the same colour on the scaly breast gorget, which is slightly erectile at the sides and can appear like a shallow inverted V-shape. The breast is black with a faint metallic purple sheen, sharply demarcated from the dark olive-green upper belly, the front of flanks dark olive-green (and lacking mottling), the wings with a faint tinge of dark blue iridescence; the rear flanks and lower belly are bright metallic green with bold black mottling, the flank feathers slightly elongated but not forming plumes; undertail-coverts and vent are blackish, the thighs blackish-brown. The uppertail is an iridescent metallic bluish-green with dark outer rectrices, very striking in flight when seen from above, with the undertail and underwings blackish. *Iris* dark brown; *bill* long, slender and decurved, slightly shorter than bill of female, black, *mouth-lining* rich yellow and prominent when calling in display; *legs and feet* grey-black. **Adult Female** 23cm; bill length 45mm; weight 77–96g. The forehead, crown, nape, hindneck and sides of neck, ear-coverts and lores are rather greyish olive-brown. A distinctive narrow pale cream or off-white supercilium extends from lores to sides of nape, becoming broader, buffier and more diffuse at rear. The sides of the chin and throat are buff or orange-buff with a narrow dark brown malar stripe. The upperparts are entirely greyish olive-brown with faint cinnamon-brown tinge, most marked on wings. Underparts rich yellow-brown (much buffier than those of Paradise Riflebird) with many prominent small dark brown chevrons, spots or streaks on breast and flanks, sometimes with spotting on belly. Uppertail brown; undertail slightly greyer brown. *Iris* dark brown; *bill* black, very long, slender and decurved, slightly longer than bill of male, *mouth-lining* slightly paler yellow than male's; *legs and feet* grey-black. **Juvenile** Resembles female plumage but has less diffuse face pattern with broader whitish supercilium, and sometimes showing much more red on the secondaries, with paler and less well-defined mottling on underparts. **Immature male** Resembles female plumage for the first couple of years, but gradually moults into male dress, some birds appearing much as adult with some brown feathers on body, others like female but acquiring black patches on head and underparts. Some in the later **subadult stages** acquire a patchy blue iridescent crown, and have cheeks, chin and throat black, upperparts greyish with rufous secondary edgings, upperside of tail beginning to show some blue iridescence, and rich buff underparts heavily spotted with black on breast. Others presumably earlier in this subadult stage have the iridescent blue crown, but upperparts and underparts much as female, with just a few dark spots appearing on breast and a darker chin. Likely to acquire full plumage after four or five moults, taking 4–5 years to attain adult dress.

TAXONOMY AND GEOGRAPHICAL VARIATION

Monotypic. Southern birds have a double harsh call somewhat different from calls of northern individuals; furthermore, F&B suggest that females from north of the Bloomfield River to Big Tableland south of Cooktown are more uniform on the underparts than are others. Further study may be needed.

VOCALISATIONS AND OTHER SOUNDS The usual

call, given by the male only, is a loud, harsh, rasping, scolding *scaaarsh* or *yaaas*, sometimes repeated twice (in the southern portions of the range thus resembling the call of the Paradise Riflebird), but in the Cairns–Atherton Tablelands region more usually given singly. The call is similar to that of the Paradise Riflebird but perhaps slightly longer in duration. It is heard throughout the year, but the species is more vocal in spring and early summer, from about August to January. Calls may be spaced three or four minutes apart or be given every couple of minutes, the frequency being greater in the breeding season. The call is sometimes given in isolation, with no further ones heard, and is audible over a distance of several hundred metres. Individuals will respond to one another's calls almost immediately, much as with Eastern and Magnificent Riflebirds. Recorded examples of male calls can be found at xeno-canto XC199652 and XC139533 (by P. Gregory), and that of a female-plumaged bird at XC98590 (by P. Åberg). A scolding *krrsh*, quieter than the advertising call and uttered every 10–15 seconds, is given when a male is perched quietly in cover and is just announcing its presence, as at xeno-canto XC324600 (by P. Gregory); a much higher-pitched rapid chatter is heard when two individuals come into close contact, an example being available at XC32460 (by P. Gregory). A scolding *chuk chuk chuk* is emitted when two birds come into unduly close contact at feeders, and has been heard from both the adult male and female-plumaged birds. A very specific predator-warning call, a harsh scolding *kuh-kuh-kuh-kuk* series with an almost rattling quality, may be constantly repeated from low down or middle levels in dense thickets, and this attracts other species to mob the potential predator. An example of this vocalisation can be found at XC334913 (by P. Gregory), where a female-plumaged bird was mobbing an Amethystine Python (*Morelia amethistina*) emerging from under a roof. A sharp *kek* or *kek kek kek* is directed at other riflebirds or Spotted Catbirds when disputing a food source. Frith & Frith (1995a) record *kuk* or *kruk* notes given by an agitated female when humans were too close to a juvenile. Juveniles have a soft, lisping begging call, as on the Internet Bird Collection and at XC348005 (by P. Gregory). F&B note a soft low repeated *kuk* given by a female on the rim of a nest,

followed by a more musical and bubbling form of these notes as she flew and then called nearby, apparently trying to entice the juvenile out of the nest. Males of this riflebird, like those of their congeners, make a distinctive rustling sound in flight, recorded on xeno-canto XC324600 (by P. Gregory) at the end of the cut.

HABITAT An inhabitant of tropical rainforests in the lowlands and hills, this species is found also on some small offshore islands. It occurs also in adjacent wet sclerophyll eucalypt or *Melaleuca* woodland, and the landward edge of mangroves (F&B), particularly in winter when some local dispersal to such habitats occurs, probably from the higher altitudes as with a number of other species. Extends from sea level to about 1,200m, probably most frequent from 250m to 1,100m.

HABITS Victoria's Riflebird is typical of its genus, foraging for arthropods along tree trunks and branches somewhat like a giant Australian treecreeper (*Climacteris*), and clambering about fruiting trees. The flight is laboured and undulating, the wings partly closing after three or four strokes, the males producing a quite loud distinctive silken or taffeta-like rustling in flight. Individuals have been seen to use the bill as well as the feet to assist them in climbing a vine stem. They are quite dominant at food sources, displacing the heavier Spotted Catbird and even the large and potentially aggressive Black Butcherbird; in one case, a female-plumaged riflebird grappled with a butcherbird and the two tumbled off the feeder to land in a shrub below, the butcherbird then flying off and the riflebird coming up to resume feeding. Female-plumaged individuals have been noted bathing in a birdbath at Kuranda. Victoria's Riflebirds play a major role in the Kuranda area in alerting the bird community to the presence of predators, especially snakes and, more particularly, Amethystine Pythons and goannas (*Varanus*). They have a particular loud harsh vocalisation series which they use repeatedly, drawing in other species such as drongos, honeyeaters and catbirds, which harass the potential predator. They are frequently seen singly in the forest, but can form small groups at food sources with four or five birds sitting in the same tree. Males will at times even share food trays with other males, although female-plumaged birds seem to stay longer in such situations; on some occasions, however, they do not tolerate others of either sex. Males call loudly from songposts located in the middle to upper stratum of the forest, on the tips of dead sloping or vertical limbs or trunks, or often on the mud nests of termites that are sited on such trunks; songposts are usually between 8m and 20m above ground, sometimes lower. Individuals may be seen chasing through the middle levels of the forest, often a male and a female-plumaged bird, but sometimes two female-plumaged individuals. The young accompany the female for some time after fledging (see under BREEDING BEHAVIOUR, below).

FOOD AND FORAGING The diet consists primarily of arthropods gleaned from trees and dead leaves, and fruit plucked from the upper strata of the forest. The birds often forage at low levels, and sometimes even on the ground. The proportions of fruit to arthropods will no doubt vary according to what is available in the forest at a particular time. Riflebirds will hold prey in the feet to dissect it, and regurgitate items quite often, presumably the indigestible portions of both fruits and arthropods. The succulent bracts of bright red flowering pandans (*Freycinetia*) may be consumed, as are pandan fruits (Frith in F&B). Bird seed was noted by Griffin (in F&B) as being taken, but we have never seen this on feeders at Kuranda, Queensland. At Kuranda, flower-piercing has been noted rarely during the winter period, the birds piercing the base of the elongated pink bell-like flowers of *Tecamanthe* creepers, presumably in quest of nectar. Both sexes readily come to feeders, as at Ivy Cottage, Paluma, and Cassowary House, Kuranda, taking such fruit as bananas, paw-paw and particularly avocado, which seems to be a favourite. They will forage in the dead hanging leaves of banana trees, picking for insects, and make loud rustling noises in the vegetation by which they may be located; honey water is occasionally taken from feeding dishes by female-plumaged birds. These paradisaeids are especially fond of cheese, butter and cream, and have developed the not always endearing habit of raiding breakfast tables to acquire such supplies. This behaviour is mainly a winter phenomenon, presumably when food supplies in the forest may be scarce. The large inverted flower spikes of low-growing red and yellow *Heliconia* species are likewise often explored for food, the birds seeming to take water or nectar from some. Female-plumaged individuals have been seen to probe into arboreal termite nests near Kuranda (R. Gregory pers. comm.). The reverse sexual dimorphism in bill size of this species, whereby the smaller female may have a bill up to 4% longer than that of the male, is as yet unexplained (F&B). It may conceivably relate to dietary differences and is not really discernible in the field, unlike with the Paradise Riflebird.

BREEDING BEHAVIOUR Fledged young have been seen in November–February, and breeding has been noted from August to February, with display in July–December (F&B), though adult males at Kuranda do occasionally perform the *Circular Wings Display* (see below) in the austral autumn in March and May. The males are sedentary and promiscuous, displaying from songposts and holding territories based around them. They seem also to hold foraging territories and will drive off other males, although this may break down at rich food sources, which seem to be shared at times.

COURTSHIP/DISPLAY Female-plumaged individuals, presumably young males, will perform agonistic displays on feed trays, with wings fully spread and erect, stabbing vigorously downwards to repel the larger, heavier and usually dominant Spotted Catbirds, or sometimes other riflebirds. The bright yellow mouth may be a feature of such behaviour, acting as a kind of visual signal as the bill is held open wide. They will also drive off Hornbill Friarbirds and Macleay's (*Xanthotis macleayana*) and Yellow-spotted Honeyeaters (*Meliphaga notata*) by poking at them with the bill, and jumping towards them. Female-plumaged individuals

have also been seen to hang upside-down from feed trays, with wings extended, which seems to be some sort of agonistic display. The songposts are from about 3m to 20m above ground, usually in the middle or upper middle strata of the forest, on dead trunks or limbs that are steeply sloping or vertical. Bare limbs and live boughs may also be used, but dead substrates seem to be preferred, at least in the Kuranda area. Males in this area may call from the mud nests of arboreal termites, which are sometimes built atop vertical snags or broken side limbs. Females alone perform the nest duties, as with others of the genus. The **displays** are better known for this species than for any others of the genus, and for full details the reader is referred to Frith & Cooper (1996) and Frith & Beehler (1998). The bright yellow gape is a feature of the calling, and is directed at females when they approach. The ***Circular Wings and Gape Display*** occurs when a female is close, the male continuously facing her as she moves around him: during the high-intensity phase, the wings will meet or overlap above, or above and in front of, the male's head and be held still as the mouth gapes widely; the tail is cocked to the horizontal or just above, and the abdomen feathering is raised. As with the Growling (Eastern) Riflebird in New Guinea, he will now repeatedly, slowly and rhythmically raise and lower himself on his legs; the feathers of the lower breast, belly and flanks are fluffed out to appear as a pale bronzy, slightly iridescent cushion, with dark barring along the lower sides, complementing the enlarged metallic iridescent gorget, which flashes green or blue depending on the angle and the light. No call is given during this display, and there is no wing sound. A version seen at Kuranda involved a female very close to the male, who was standing with feet still but quite slowly swaying the cupped wings from side to side to create a loud swishing noise as he did so, and increasing in tempo as she showed interest. When seen from behind, the shiny shafts of the primaries are quite obvious, but these would not be seen from the female's front-view perspective. If the female moves away, the male may relax by closing his wings in the manner of shutting a fan, each feather sliding behind the next until the wing is shut, which causes a dry rustling sound. Sometimes a male will sit with his wings half-raised for several minutes, as if waiting for something to happen, and then subside to normal posture. When a female arrives on a display post, the male greets her by leaning back and erecting his wings in a ***Circular Wings Display***, with the head and closed bill hidden behind the leading edge of one wing. He may sway his body slowly to one side, and the wings are now held forward of vertical towards the female, forming a kind of shallow dish into which she peers. If the female is nervous the male holds this posture rigidly, but if she seems interested he begins the ***Alternate Wings Clap Display***. He raises and fully extends one wing, and brings it towards the female as he slightly closes and lowers the other wing and hides his upward-pointing bill behind the leading edge of the raised wing. During this rapid action he extends one leg to raise that side of the body while lowering the other.

He then suddenly raises his wing to hit the other above his head with a muted clap, and immediately swings his head across to hide behind its leading edge. He then repeats the wing and leg actions on the opposite side, and repeats the process with great rapidity, the raised head and bill swinging like an inverted pendulum between the alternately raised wings as the body sways from side to side and twists slightly. This display is very similar in many respects to that noted for captive Growling Riflebirds. During the high-intensity phase of this display, the tempo of the head and wing movements may become so fast as to be impossible to see clearly. The bill stays shut, and the iridescent throat shield appears 'as a bright shining green, sometimes blue, vertical line shot with electric white highlights as it is now jerked… from one side to the other' (F&B). The male leans towards the female and out of the vertical stage of the display, and this may become so vigorous that the perch sways and the female leaves! Assuming that she stays, he will then embrace the female with each wing, almost beating her with wings moving around her about twice per second so that she is continuously embraced; a receptive female will flutter her closed wings, and the male then sways to one side in order to hop on to her rump to copulate. Females may initially appear uninterested, pecking at the stump or preening, but as the tempo of the display mounts they peer intently at the centre of the male's alternately raised wings to watch his rigidly upright iridescent throat and bill flick rapidly from side to side, the enlarged iridescent pale bronzy cushion of the lower breast and belly feathers setting if off nicely. Sometimes these female-plumaged birds are presumably young males, as they then start to perform the *Alternate Wings Clap Display* themselves, almost as if imitating or learning from the adult male, which seems to tolerate their presence, at least in the early part of the breeding season. These female-plumaged, presumed immature males will try the *Circular Wings Display* at other female-plumaged birds, or at Spotted Catbirds, and the Friths saw one perform to a Musky Rat-kangaroo (*Hypsiprymnodon moschatus*) and even to a moth as it flew slowly past (Frith & Cooper 1996; F&B). They seem to like to practise the performance, with early efforts often very clumsy and uncoordinated, and groups of two to four such birds may gather to perform for each other. Female-plumaged birds displaying to each other or to an adult male are most likely immature males, rather than females (F&B). One presumed immature male riflebird at Kuranda in October 2016 was perched 5m up in the middle of an open waterberry (*Syzygium*) and made loud swishing noises with the cupped wings, performing at high intensity for several minutes to a watching female-plumaged bird. A male did exactly the same here in late May 2017. A lateral tail-flicking display was given by a male to a female-plumaged individual as both perched on a trunk, as was a downward-directed gaping display with raised tail as the bird leant forwards. See F&B for details. As with an intriguing report for the Paradise Riflebird, a male was once seen hanging upside-down from a perch with tail fanned and wings spread (Breeden & Breeden

1970). A male at Kuranda hung upside-down with wings closed for several minutes. **NESTING** Relatively well known in comparison with the others of the genus. The nest is built 1.5–20m above ground, often well concealed and invisible from below. Sites utilised include vine tangles, pandanus, *Cyathea* tree-fern crowns, fan palms and cordylines, and broken-off trunks that are shooting new foliage, while a banana tree-crown nest site was recorded once (F&B). Hislop (in North 1901) recorded three successive old nests plus a new one in a pandan, indicative of the use of a traditional site as with its congener the Paradise Riflebird, and also the Short-tailed Paradigalla. There is one record of an Eastern Koel (*Eudynamys orientalis cyanocephala*) parasitising a nest (HANZAB 2006). **NEST** The nest is a cup of fern stems and moss, the internal cup made from dead leaves and lined with vine-flower stems (Frith & Frith 1998a). Meek (1913) recorded the finding of many nests that had sloughed snakeskin as a part of them, including Black Snake (*Pseudechis porphyricus*) and Carpet Python (*Morelia spilota*). This curious behaviour is noted also for the Paradise Riflebird, and may perhaps help to deter predators, rather than being purely decorative. **EGGS & INCUBATION** The eggs are variably reddish-flesh, warm creamy-buff or pink-buff with a few broad brushstroke-like reddish-brown to purplish-brown streaks and sparse small spots. The **clutch** is of two eggs, sometimes just one, with a suggestion from an egg-collector with much experience of the species that two clutches could be raised in a season (Hislop in Iredale 1950). Incubation is solely by the female, as is usual in the family, and lasts 18–19 days (Frith & Frith 1998a). **NESTLING & CHICK CARE** Chick is naked at hatching, with skin dark grey-purple above and paler purplish-brown below, legs and bill pinkish-grey with darker tip of bill, gape pale yellow. Nestling care again is exclusively by the female, feeding the chick with fruit, larger insects and regurgitated items; the diet is largely of arthropods and fruit, about 11% being fruit in one study, and small skinks are also occasionally taken for the nestling (Frith & Frith 1998a). Females have been seen to shelter their offspring from the rain by standing in the nest with the wings open over it (Frith & Frith 1995a, 1998a). Nestling period is probably 13–15 days; nest-departure figures vary, Frith & Frith (1995a, 1998a) giving 13–15 days, though Schodde & Tidemann (1988) estimated about four weeks, which seems unduly protracted. **POST-FLEDGING CARE** Cooper (in F&B) has a record of a juvenile begging with quivering wingtips and soft calls and being fed by the female some 74 days after leaving the nest. Those around Kuranda seem to associate with the parent for about 10–30 days before disappearing. Juveniles with females at Lake Eacham and Kuranda have a distinct plumage, with a more diffuse face pattern with broader whitish supercilium and paler and less well-defined mottling on the underparts. Fledged young have been seen in November–February, and breeding has been noted from August–February, with display from July–December (F&B), though adult males at Kuranda do occasionally perform the *Circular Wings Display* in the austral autumn in March and May.

HYBRIDS No hybrids are known, as no other birds of paradise occur within the species' range.

MOULT Recorded in November–January at Kuranda, when individuals may develop partially bald pates, although an adult male over the course of one winter (May–September 2000) had a completely bald head, perhaps a result of old age or some skin disease. A male in 2016 had completed his moult over some four weeks in November–December, while one in December 2018 had a dull blackish head with very short pin-feathers regrowing, and with faint iridescence on the new feathers of the crown (pers. obs.).

STATUS AND CONSERVATION Classified as Least Concern by BirdLife International, albeit with a restricted range of around 31,600km². This is a reasonably common if low-density species in suitable habitat, much of which is now protected by the Wet Tropics World Heritage Area designation. In regrowth rainforest near Kuranda there is approximately one adult male bird per 2ha, plus assorted females and immature males. There is some winter dispersal to lowland sites in non-rainforest habitats, but the species is, like the rest of the family, relatively sedentary. It is known from some of the coastal Queensland islands, the only member of its genus to occur on such islands with the exception of the Magnificent (Cape York) Riflebird on Albany Island, in Torres Strait. Historically, Victoria's Riflebird was hunted for the millinery trade in the 19th century, and for museum collections. Campbell (1901) records shooting 10–17 of these birds per day on Barnard Island. The species is under no threat overall, but it will lose habitat to creeping suburbanisation at the lower-altitude sites. It has also historically lost a lot of habitat on the Atherton Tablelands owing to logging and clearance for farms. It can be a pest in soft-fruit orchards, where illegal killing is a possibility. The Queensland Parks and Wildlife Service can theoretically issue a damage-mitigation permit to kill birds that extensively damage fruit crops and/or cause severe economic hardship. This species, however, is not listed on Schedule 1 (common species for which permits may be given without undue restriction). Any such permit application would undergo careful scrutiny, as the Queensland Parks and Wildlife Service has become more sensitive to potentially contentious situations involving species such as black cockatoos, riflebirds and bowerbirds. The Victoria's Riflebird is, like its congeners, classified as common under Queensland legislation, and is not listed in the rare or threatened categories. It appears to be quite long-lived, as is typical of the family, with a ringed male known to have lived for 15 years in the wild, until killed by a cat.

Victoria's Riflebird, female plumage feeding on berries, Figtree Close, New South Wales, Australia (*Don Hadden*).

Victoria's Riflebird, male displaying, Queensland, Australia (*Jun Matsui*).

MAGNIFICENT RIFLEBIRD
Ptiloris magnificus   **Plate 13**

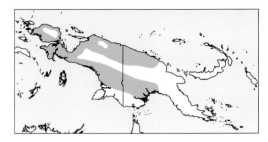

Falcinellus magnificus Vieillot, 1819, *Nouveau Dictionnaire d'Histoire Naturelle*, Nouvelle Edition 28, 167. La Nouvelle-Guinée, restricted to Dorey (Manokwari), Vogelkop, by Mayr 1941.

Other English names Magnificent Rifle Bird, Scale-breasted Paradise Bird

Etymology The Latin *magnificus* denotes splendid or magnificent, certainly appropriate in this case.

IDENTIFICATION A large, spectacular and distinctive species, not readily confused with any others except for its largely allopatric sister-taxon the Growling Riflebird, owing to the long decurved bill and distinctive broad rounded wings, stout body and rather short tail. Males have a bright iridescent blue breast, crown and uppertail with matt black plumage glossed with violet-purple, and make a characteristic loud rustling sound in flight. Females are quite bright orangey-rufous on the back, with a long decurved bill and prominent pale supercilium, and are heavily barred black on the underparts. **SIMILAR SPECIES** In a reasonable view the male Magnificent Riflebird is unlikely to be confused with anything else, and does not overlap in range with its congeners except for an ill-defined zone of abutment or sympatry in the Huon Peninsula (Coates 1990) and perhaps Eastern Highlands of PNG with the Growling Riflebird (de Silva in Mack & Wright 1996). Telling the two species apart would be difficult as field characters are subtle and of limited use, but fortunately the voice is diagnostic (see relevant VOCALISATIONS section). In a good view, though, the lack of feathering at the base of the culmen of the present species might be of some use for both sexes. Females could be confused with female Twelve-wired Bird of Paradise, which also has a long decurved bill and heavily barred black underparts, but that species lacks the supercilium and has a marked blackish cap and nape and a red eye. Female Superb Lophorina is smaller, much shorter-billed, and dark brown on the back with rusty wings, lacking the bright orangey coloration of the riflebird; it also has a bolder and longer pale supercilium and is absent from lowland forest, although the two do overlap in the hills. Female-plumaged Raggiana Bird of Paradise is duller and browner on the back, has a dark face with a bluish bill, and lacks the barred underparts and long bill, as does the chestnut-coloured female Greater Bird of Paradise. Manucodes all lack the long decurved bill and are entirely black, with varying degrees of gloss and, in the case of the Trumpet Manucode, head tufts.

RANGE Endemic to the western and central parts of the island of New Guinea, from the Vogelkop eastwards into PNG, where it ranges to the Wewak area of the Sepik drainage in the north and in the southern watershed as far as the Purari River, Gulf Province. The species seems to be absent from the forests and savannas of the lower and middle Trans-Fly, where the habitat is not optimal. It is absent also from all of the satellite islands. Eastern PNG is occupied by the allospecies the Growling Riflebird, while an outlying Australian riflebird population on Cape York, Queensland, bears more vocal resemblance to the western Magnificent Riflebird than to the eastern Growling Riflebird. This Australian population may be best treated as a distinct species, the Cape York Riflebird, which is detailed in a separate account below. **MOVEMENTS** Apparently sedentary.

DESCRIPTION *Adult Male* 34cm; bill length 60mm; weight 180–230g. Plumage entirely velvety jet-black with metallic iridescent green-blue and purple scaly feathering from lores and over crown to nape; the rest of the head, neck and upperparts show a dull metallic purple sheen in some lights. Most of the tail is rich metallic greenish-blue, with blackish outer feathers and narrow dark tip. The underparts show a distinctive large iridescent scaly-feathered greenish-blue gorget, with apex extending from chin and throat down to upper breast, and longer lateral feathers extending as erectile points on each side. This gorget can show a purple sheen in some lights, especially on the lower portion, which is bordered by a narrow black band across mid-breast, itself bordered by a yellow-green band across lower breast. The rest of the underparts are greyish-black, with long, thin, curved black filamentous plumes extending from the lower flanks to well past the tail; the undertail and underwing are glossy black. *Iris* blackish-brown; bill black, very long, slender and decurved, with pale yellow to greenish-yellow gape and *mouth-lining*, which is an important element in display; *legs and feet* greyish-black. An **aberrant adult male** leucistic specimen (in F&B), lacking iridescence except for a purple-blue trace on the lores, is a washed-out pale smoky brownish-grey above and slightly glossy brownish-grey on the breast shield. *Adult Female* 28cm; bill length 53mm; weight 120–185g. The upperparts are a warm olive-brown, the uppertail-coverts with cinnamon fringes, and the uppertail reddish-brown. The cap (forehead to hindneck) is olive-brown, flecked or streaked buffish-brown (except on hindneck). The long narrow white supercilium is washed buff at the sides of the nape, and a partial broken eye-ring forms narrow whitish arcs above and below eye. A broad dark brown facial stripe is bordered by whitish submoustachial stripe, which merges into off-white sides of neck and is bordered below by a narrow blackish-brown malar stripe. The wings are a quite bright rufous-brown. The underparts are creamy white, shading to pale buff in the centre, and closely barred blackish-brown on the breast, with sparser blackish-brown barring elsewhere. The undertail is olive-brown;

the underwing has a broad creamy white lining finely barred blackish-brown, with cinnamon-brown flight-feathers and dark trailing edge and tips. *Iris* blackish-brown; *bill* black to grey-black with pinkish base of lower mandible; *legs and feet* grey-black. **Juvenile** has upperparts much paler and more rufous, the soft and downy plumage having widely spaced sooty-brown bars on breast and belly. **Immature Male** resembles female at least in the first couple of years. **Subadult Male** shows a mixture of male and female plumages, becoming progressively more blotched with black on body, head and wings as it moults, and probably assuming adult plumage in 4–5 years.

TAXONOMY AND GEOGRAPHICAL VARIATION

Current taxonomy treats this species as polytypic, with two subspecies, the nominate of central and western New Guinea, and *P. magnificus alberti* of Cape York, Queensland. This latter taxon is here detailed in a separate account (below). The splitting of the Magnificent Riflebird assemblage into two species, as first suggested by Beehler & Swaby (1991), causes a problem with the Cape York population, the individuals of which, surprisingly, bear more resemblance vocally to the Magnificent Riflebird of the west than to the closest neighbour the Growling Riflebird. Again, the vocalisations are quite distinctive and morphological differences are of the same order of magnitude as the differences between the eastern and western taxa. Vocally, these Cape York birds fall into two groups, and this would be worthy of further studies. An argument can be made for separating out these birds from the Magnificent Riflebird, their inclusion within which only obscures their relationships and phylogeny, and, while genetic studies are desirable, a case can be made for separating the Cape York Riflebird as a third distinct species. The zone of abutment or sympatry between the Magnificent Rifleman and the Growling Riflebird in PNG would be a rewarding place to study the **interactions** between these two sibling species. Do they in fact hybridise, and is the transition zone abrupt or sharp? How frequent is any hybridisation? What is the exact relationship with the populations of Cape York, with which it has some vocal similarities? Much has been obscured by the treatment of the three taxa within one species, and there are many gaps in our knowledge of the basic biology of these beautiful species. Growling Riflebird *intercedens* is distinct from the nominate *P. magnificus*, but the Australian *alberti* is just as distinct in both morphology and vocalisations. The advertising songs of the three forms are different from one another, being more distinct than is the case with virtually any other taxa of the entire family (but *cf.* Buff-tailed Sicklebill and Black Sicklebill) except for the Trumpet Manucode assemblage, which may itself be polyphyletic. There are a number of relatively minor but significant **morphological differences** between the two New Guinea taxa and *alberti*:

· Males of the two New Guinea species have a blue gloss on the plumage and a less curved bill than the Australian taxon.
· Both sexes of *magnificus* have very little feathering at the base of the bill, as does *alberti*, whereas *intercedens*

has a largely feathered base of the culmen and a slightly shorter bill.

Bill length in mm
Adult male *magnificus* 60, immature male 59
adult female *magnificus* 53
Adult male *intercedens* 56, immature male 55
adult female *intercedens* 48
Adult male *alberti* 51–57
adult female *alberti* 46–56

Weight in g (from very small samples, but enough to suggest clear differences for *alberti*)
Adult male *magnificus* 207, immature male 180
adult female *magnificus* 142
Adult male *intercedens* 184, immature male 176
adult female *intercedens* 126
Adult male *alberti* 160, immature male 131
adult female *alberti* 104

Wing length in mm
Adult male *magnificus* 192, immature male 177
adult female *magnificus* 157
Adult male *intercedens* 193, immature male 175
adult female *intercedens* 152
Adult male *albert* 181, immature male 161
adult female *alberti* 149

Males of the Australian taxon have an oil-green gloss to the plumage and a more slender and curved bill than those of their northern relatives. They also have a broad greeny-gold breast band, and both sexes are notably smaller and lighter in weight. Their flank plumes project beyond the tail, as with *magnificus*, while the flank-plume projection of *intercedens* is only equal to or shorter than the tail. Given the sedentary nature of the group, it is evident that each group has been in isolation for a very long time indeed. The zones of overlap with *magnificus* in New Guinea are not well known, but each grouping maintains its identity, and a few records of one form within the range of the other, plus a possible hybrid along the border zone, need not invalidate the treatment as distinct species. As Schodde & McKean (1973) demonstrated with the genus *Parotia*, supranarial feathering may function as a subtle isolating mechanism in sister-species of some bird of paradise groups. Furthermore, there appear to be some significant display and perhaps nesting differences among the three populations which for long have been obscured by the traditional treatment as one species.

VOCALISATIONS AND OTHER SOUNDS The **advertising call** is a characteristic sound of lowland and hill forest, being an exceedingly loud and powerful, often (but not always) disyllabic wolf whistle, *WHOOOIT-WHOOOOO* or *WHOIIEET-WOIT* (Coates 1990). The first note is upslurred and the following note or notes are downslurred, the second phrase sometimes being repeated three or four times. The call is audible at a distance of well over a kilometre, and is given both by adult males and by birds in female-type plumage, which are presumably immature males. It is given throughout the day, though most often in the early morning and late afternoon. Birds in the Kiunga and Tabubil areas of Western Province, PNG, will typically allow two or three minutes to elapse between calls. Call frequency is much lower over the November–March wet season, when many birds are moulting. The species will occasionally be responsive to playback, although it will more usually ignore it (pers. obs.). Three loud, mellow whistles are also emitted at times by these Western Province birds, the first note slightly upward-inflected, the second slightly downward and the third falling away to a deeper tone altogether, a sort of *whiit-whooo-wh-hooo*. Examples of males calling can be found on xeno-canto at XC40382 (by P. Åberg) from Nimbokrang, West Papua – complete with New Guinea Harpy Eagle (*Harpyopsis novaeguineae*) in the background – and at XC38120 (by F. Lambert) from Kiunga, PNG. There are examples of vocalisations also in the Macaulay Library at the Cornell Lab of Ornithology site by T. Laman and E. Scholes, and on the Internet Bird Collection (IBC) site. Birds in the Kumawa Mts of West Papua utter lower, upward-inflected clear, hollow gibbon-like hoots, while those in the Cyclops Mts of West Papua and near Wewak call *KOIT-KOIT, KOIT KOIT* (F&B). All of these could conceivably be transcriptions of the basic whistled advertising calls. Ogilvie-Grant (1915a) reported the call from the Setakwa River, in southern West Papua, as a long drawn-out *ooû*, followed by two sharp and loud *wah wah* notes (which also sounds like a transcription of the regular call). He noted that the riflebirds would often call at night if disturbed by a falling tree or some other event. There are reports of this call being heard within the range of *intercedens*, as at Yalumet on the north side of the Huon at 760m (Coates 1990), while at Wasu it is *intercedens* in the lowlands (Coates 1990) and hills to the south (Peckover in Coates 1990). Reports of *intercedens*-type calls at Lake Kutubu are in error (Schodde in F&B), although Diamond (1972) reports *magnificus*-type calls at Karimui, Eastern Highlands. The cut-off between the two taxa is, however, quite sharp overall, although mixed pairings would be very hard to detect. The Magnificent Riflebird can also make strange guttural growling sounds, and Coates (1990) reports one as making growling sounds and then the advertising call. Pratt (in F&B) noted that *magnificus* in the Adelbert Mts made two or sometimes three frog-like croaks, less loud than *intercedens* calls near Wau (and conceivably not an advertising song?). The feathers of the male when in flight also make a loud and distinctive taffeta-like rustling, which can be quite startling at close range. Display sounds are different from any of the above, and include a *rush-rush-rush-rush rushrushrushrush* sound, audible from some distance, produced by the wings of an adult male. The calls of the taxon *alberti* from Cape York Peninsula are distinct, being variable and higher-pitched, and are detailed in a separate account of this taxon.

HABITAT Rainforests in the lowlands, hills and lower montane zones, also monsoon, swamp and gallery forest. Occasionally visits mangroves (Saenger *et al.* 1977, but which taxon?) and also timber plantations such as teak (Peckover 1990, but again which form?). Selectively logged forest is used, too, the birds preferring the taller trees and forest patches. The altitudinal range is from sea level to as high as 1,450m, and exceptionally to 1,740m (Coates 1990), but mainly to 700m; those in the Ok Tedi area range up to about 700m but drop out dramatically above that altitude, with a very sharp cut-off point (pers. obs.), and in other areas 600m seems to be the cut-off level, as at Tabubil, in Western Province (pers. obs.; Coates 1990).

HABITS The Magnificent Riflebird is a shy and rather wary species, this being particularly true of the adult males, which can be difficult to observe in its dense forest habitat. Males call from perches high up in large trees, usually from dead branches, and can then be out in full view, although more usually not fully exposed and viewable only from certain angles through the branches and foliage. Individuals at Kiunga will also perch and call fully exposed on the topmost growing points of Black Palms (*Borassus flabellifer*), which seem to be a favourite perch here. At food sources they will join with other paradisaeids such as Raggiana and Greater Birds of Paradise, as well as Great Cuckoo-doves (*Reinwardtoena reinwardtii*) and fruit-doves (*Ptilinopus*). Their flight is slightly undulating, the wings nearly closing after three or four beats, and can be reminiscent of a giant butterfly with the broad rounded wing shape and short tail.

FOOD AND FORAGING The diet consists of arthropods and fruit, much of the latter of the nutrient-rich capsular type (Beehler 1983a). This seems to be the case in Western Province, PNG, although the exact proportions of food types in the diet will no doubt vary according to what comes available at any particular time. Known food species include *Schefflera*, *Cyrtostachys* palm fruits (Beehler & Beehler 1986) and fruit of the aroid *Amorphophallus paeoniifolius* (Peckover 1985). Males forage by probing and working along branches, pecking at bark and prying into cavities and fern clusters; dead limbs are often explored, and the birds move actively through the forest, not staying for long at any one site. Males are usually seen alone or with up to three female-plumaged birds. They explore the upper and middle strata of the forest, and come quite readily to edge areas along roads, tracks and fields. Both sexes will join mixed-species foraging flocks, and female-plumaged individuals are a frequent component of the brown-and-black foraging flocks which characterise these forests. Rusty Pitohui (*Ornorectes ferrugineus*), Papuan Babbler (*Garritornis isidorei*), Black Cicadabird (Cuckooshrike) (*Edolisoma melas*), Papuan Spangled Drongo (*Dicrurus carbonarius*), Rufous-backed Fantail (*Rhipidura rufidorsa*) and Frilled Monarch (*Arses telescopthalmus*) are typical core members of such flocks (Coates 1990; F&B; pers. obs.). They are also sometimes seen on the ground, foraging in leaf litter. Magnificent Riflebirds readily consort with other paradisaeids such as Lesser (Hoogerwerf 1971), Raggiana and Greater Birds of Paradise and Glossy-mantled Manucodes when at fruiting trees (pers. obs.). Beehler & Beehler (1986) recorded an adult male Magnificent Riflebird along with both Twelve-wired and King Birds of Paradise in a mixed-species flock at Krissa, in PNG, and an adult male with a female-plumaged bird accompanying a Jobi

Manucode and King Bird in a similar flock at Puwani River, PNG (F&B).

BREEDING BEHAVIOUR There is a large collection of videos of the various taxa in this group in the Macaulay Library at the Cornell Lab of Ornithology site. ML455440 (by E. Scholes), taken in West Papua, shows a male of the nominate taxon displaying to a female and vocalising. This species is, like its relatives, polygynous, the promiscuous males singing from songposts in territories, and displaying to female-plumaged birds. The males at these songposts are usually in auditory contact with other males. Along the Ok Ma Road at Ok Tedi, in 1993, singing males were very vocal and were spaced at approximately 500m intervals along the road where the habitat was undisturbed or tall forest patches remained, up to 12 males being heard in an afternoon (Gregory 1995); there can be no doubt that these well-spaced sites represent territories for the males, as they will chase off other males when they come into contact. Females alone are likely to be responsible for the nesting duties, as with other members of the family (whether this has been documented for this species in New Guinea as yet is uncertain, most of the pertinent references relating to the Cape York Riflebird). The birds descend lower to feed, and when seen in sunlight the iridescence of the males is an amazing sight, constantly changing as the birds move: they will appear all black one minute, and then the throat and breast shield will show iridescent bright blue or purple with vivid yellow or green highlights. This is especially evident when the male is calling, as these colours are an integral part of the display and are emphasised when the head is thrown back to call or when displaying. A velvety-black pectoral band with violet-purple highlights, itself bordered by a broad bronzy yellow-green band, sets off the intensity of the breast coloration; moreover, the iridescent blue-green crown and iridescent metallic blue on the centre of the uppertail at times catch the light in striking fashion. **DISPLAY** Knowledge of the displays of all the species in the Magnificent Riflebird complex is still rather fragmentary, much remaining to be described, but observations on captive individuals help to give an idea. Judging by the male's vocalisations, active display occurs from March to September at least, with a lull during the wet season, when many birds are in moult. Diamond (1972) recorded an observation in August 1964 of what he thought to be probably an immature male and a female of this species alighting next to each other on a branch 4.6m above ground, at the top of a bush on the forest edge: one bird faced the other, reared up on its legs, spread out its wings and bent them backwards, threw back its head, and remained in this uncomfortable posture for some time; the second bird then adopted this posture, and finally both assumed it together before flying off. No calls were given. On a second occasion, in September 1965, Diamond saw two female-plumaged Magnificent Riflebirds facing each other on a branch 9.1m above the ground in the forest: one tilted its body until it was vertical, held the tail back at 90° to the body so that it was horizontal, opened the wings and bent them at the

shoulder so that they nearly met behind the back, and arched its breast towards the partner; it then moved up and down on its legs, and the pair flew off to another tree, where the performance was repeated. A possible explanation for this behaviour could be that these individuals, rather than being pairs, were immature males practising their displays, as Victoria's Riflebirds perform similarly (F&B; pers. obs.). Coates (1990) describes (presumed *magnificus*?) display from captive birds at Baiyer River sanctuary. Here, an immature male flew to a horizontal perch, where it assumed a normal perching posture but with the head and bill pointing vertically upwards, the throat pulsating in its centre; a female also flew to the perch about 18cm from the male and crouched with the head pointing vertically upwards, but without the pulsating throat. Both birds then hopped about the perch with wing-flicking and sudden jerky movements, but with throats facing each other. Suddenly the male adopted a rigid upright posture with fully outstretched wings, the head lying along the upper edge of one wing, and the tail slightly cocked, and he then began rhythmic up-and-down movements, produced by stretching and lowering the legs; with each upstroke the head jerked across to the other wing at such speed that the movement could hardly be seen, and at maximum height one leg was stretched more than the other, depending on which wing the head was lying along. This uneven stretching of the legs caused the body to rotate slightly around a vertical axis. No sounds were made apart from plumage noise from each rapid movement. The rhythmic movement increased in rapidity as the display climaxed, and then it was suddenly all over and both birds resumed normal posture. The entire sequence lasted about 30 seconds, and probably gives a good idea of what display might be like in the wild. Again at Baiyer River sanctuary, Mackay (1990) saw a captive male conclude his display by turning to the female with his wings still extended and clapping her between his open wings several times; this action is typical of Victoria's Riflebird courtship (F&B) and is an interesting parallel with a congeneric species. Wild adult males at Ok Tedi have been seen to spread the wings as described above and flick the head rapidly backwards and forwards, though without the up-and-down bobbing. The iridescent blue of the throat and chest seems to be expanded, and is bordered by a coruscant purple band, which catches the light and makes an amazing burst of colour. Such displays are brief, but the birds are shy and may perhaps be disturbed by observers. They seem also to display in the heat of the day rather than early morning, possibly to maximise the iridescent effect of the plumage in the stronger light; they appear also to be more vocal on clear days than on wet or misty ones, when calling is markedly reduced. Riflebirds in this area seem to call throughout the year, but less in November–March, the latter months coinciding with peak wet periods. Beehler (in F&B) observed a male Magnificent Riflebird which displayed within 50m of the display post of a male Twelve-wired Bird at Kakoro, PNG (presumably the present species and not Growling Riflebird, although Kakoro is right on the boundary of their respective ranges). The calling of one male seemed to stimulate a vocal response by the other. Not surprisingly, hybrids between these two species are known (see below). **NESTING** It is presumed that the female alone carries out the nest duties, as is usual for the sexually dimorphic species in this family. The nesting habits of the Magnificent Riflebird are remarkably little known, most published material referring to the two Australian sibling species. Gould & Sharpe (1875) state that Dr Beccari's nest was found by one of Mr Bruijn's hunters in the branches of a *Calophyllum inophyllum* tree. **EGGS & NESTLING** The clutch is usually of two eggs, creamy-brown with quite broad brushstroke-like longitudinal bands of reddish to brownish or grey, with scattered darker flecks, patterning being more obvious on the blunt end of the egg. **Nestling** Naked, dark blackish-purple above and paler below at 1–2 days of age.

HYBRIDS The Magnificent Riflebird is known to hybridise with the Twelve-wired Bird of Paradise.
Mantou's Riflebird (*Craspedophora mantoui* Oustalet, 1891) is a hybrid with Twelve-wired Bird of Paradise, and known from at least a dozen skins.
Bruijn's Riflebird (*Craspedophora bruyni* Büttikofer, 1895) refers also to this hybrid of Magnificent Riflebird × Twelve-wired Bird of Paradise.
Duivenbode's Riflebird (*Paryphephorus (Craspedophora) duivenbodei* A. B. Meyer, 1890) is a hybrid of Magnificent Riflebird × Superb Lophorina, and is known from three male specimens. Meyer named it after the Duivenbode father-and-son team of Ternate, who were plume-traders and supplied some skins of these rare hybrid combinations. It an intergeneric hybrid, though the two genera are closely related and sometimes *Ptiloris* is merged with *Lophorina*.
Bensbach's Riflebird or **Bensbach's Bird of Paradise**

(*Janthothorax bensbachi* Büttikofer, 1894) is a putative hybrid with the Lesser Bird of Paradise, and is known from just a single specimen.
The reader is referred to Iredale (1950), Fuller (1995) and Appendix 1 of F&B for detailed discussion of these curious forms.

MOULT Moult is believed to take place mainly over the wet season, December–March, but is likely to occur at any time of the year.

STATUS AND CONSERVATION Classified as Least Concern by BirdLife International. The Magnificent Riflebird is quite a common species where the habitat is undisturbed, and its voice is a characteristic sound of lowland forest in Western Province, where 12 males have been heard in an afternoon, spaced at about 500m intervals along a 7km stretch of road that was only in part prime habitat. The species is also quite common in selectively logged forest at Kiunga, occupying both swamp forest and forest along low rolling ridges. Recently there has been a big local decline in this area as the habitat has become fragmented and disturbed with improved road access. Bishop (in F&B) records calling males as being 100–200m apart. The species was uncommon in the Cyclops Mts of southern West Papua between 300m and 1,200m in July 1990 (Diamond and Bishop in F&B). This paradisaeid seems not to figure in native ornamentation in Western Province, perhaps surprisingly in view of its size and coloration. There will be local declines as the forest is cleared or logged and settlements spread, but much of the habitat is remote and with low-density human populations, so there are no obvious threats at this time. This species appears to be quite long-lived, as is typical of the family, but data on wild individuals are lacking.

Magnificient Riflebird, female plumage (*Marcus Lilje*).

CAPE YORK RIFLEBIRD
Ptiloris magnificus alberti   Plate 13

Ptiloris magnificus alberti Elliot, 1871, *Proceedings of the Zoological Society of London*, p.583. 'Cape York', Queensland, Australia.

Other English names Albert Riflebird, Prince Albert Riflebird, Prince Albert's Rifle-bird, Prince Albert's Rifle Bird; local vernacular name Jagoonya is an aboriginal name for this bird.

Etymology Named for the late Prince Albert, consort of Queen Victoria.

The paper by Irestedt *et al.* (2017) gives an estimated age of 710,000 years since this taxon diverged. Ancestral *alberti* reached the Cape York Peninsula via a Torres Strait land bridge during one of the cool 'glacial' cycles of the Pleistocene, probably along with many, if not all, of the other 18 rainforest-adapted New Guinean avian species limited in Australia today to the upper Cape York Peninsula, examples being Chestnut-breasted Cuckoo (*Cacomantis castaneiventris*), Yellow-billed Kingfisher (*Syma torotoro*) and Trumpet Manucode (*Phonygammus keraudrenii*).

FIELD IDENTIFICATION A large, spectacular and distinctive species, not readily confused with any others in its restricted range on account of its long, decurved bill and distinctive broad rounded wings, stout body and rather short tail. Male has a bright iridescent blue breast and uppertail, with most of remaining plumage matt black with oil-green gloss, and makes a loud rustling sound in flight. Female is a dull orangey-rufous on the back, has a prominent supercilium, and is heavily and narrowly barred black on the underparts. **SIMILAR SPECIES** None within range, the Victoria's Riflebird occurring no nearer than around Cooktown, several hundred kilometres to the south. Two other fairly large black species live in the rainforest here, but the vaguely similar Black Butcherbird (*Melloria quoyi*) has a stout hook-tipped blue-grey bill with a dark tip, and the Trumpet Manucode has a short straight bill, a floppy lateral crest or head plumes and a red eye.

RANGE North-east Australia. Endemic to lowland rainforest, riparian forest and vine thickets in northern part of the Cape York Peninsula, in the east extending from Silver Plains and Rocky River area of McIlwraith Range north through Iron Range NP to northern tip of the cape at Bamaga and Pajinka. An outlying population is found on the west coast near Weipa. Reported also from tropical monsoon vine forest on Albany Island, just off Somerset on northern tip of peninsula (Blakers *et al* 1984), where Trumpet Manucode also present. **MOVEMENTS** Sedentary.

DESCRIPTION Sexually dimorphic. *Adult Male* 33cm; bill length 51–57mm; weight 143–171g; notably smaller and lighter in weight than its New Guinean relatives. The plumage is very similar to that of nominate *magnificus*, but with wing length slightly shorter and the bill narrower and more decurved than on either nominate or Growling Riflebird. The feathering at base of culmen is intermediate in extent between *magnificus* and Growling Riflebird. The central rectrices are slightly shorter than the rest of the tail, slightly more so than in related taxa. Australian *alberti* males have an oil-green gloss to their plumage, with a broad greeny-gold breast band; their flank plumes project beyond the tail, as with *magnificus*, whereas the projected flank plumes of Growling Riflebird are only equal to or shorter than the tail. *Adult Female* 28cm; bill length 46–56mm; 94–112g; again, notably smaller and lighter than the northern relatives, but plumage otherwise much as for those taxa. Adult female plumage is attained by the second basic moult, after just over one year. *Immature* plumages much as for nominate *magnificus*. *Subadult Males* gradually moult into adult plumage over a period of 4–5 years, some having a mixture of male and female-type plumage; body moult seems to occur quite rapidly in the later stages, leaving feathers of head and neck in pin and giving a rather naked appearance (Higgins *et al.* 2006).

TAXONOMY AND GEOGRAPHICAL VARIATION This taxon (*alberti*) is usually treated as a subspecies of Magnificent Riflebird. Irestedt *et al.* (2017), however, suggest that the possibility that *alberti* merits treatment as a full species would be worth investigation: 'The outlier of the tropical riflebirds on Cape York Peninsula, *alberti*, may deserve species status, as suggested by its PP value of 0.54, above the bPTP threshold for molecular species delimitation… the critical patterns of narial feathering and voice are also alike, and the differences in morphology are largely quantitative. We thus include it in *magnificus*… pending clarification from more data.'

Synonym *Craspedophora magnifica claudia* Mathews, 1917. *The Australian Avian Record* 3, 72. Claudie River, North Queensland. This form was separated on the basis of its having the throat and upper breast greener, with the feathers more rounded and not so pointed; it is also smaller, wing 173mm. Not now recognised, but it is worth noting that this population has a call distinct from that of birds farther north in the cape.

VOCALISATIONS AND OTHER SOUNDS The voice is quite distinct from that of both congeners, and curiously enough bears more resemblance to that of the more distant Magnificent Riflebird than it does to the voice of the geographically closer Growling Riflebird in PNG. It is also interesting that the Cape York Riflebird is separated vocally into two distinct

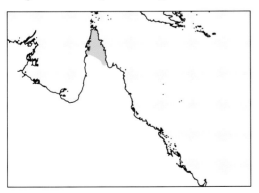

song types (which are presumably dialects): those at Pajinka utter two identical upward-inflected clear whistles, which may sometimes be given as a series; birds from the Claudie River, Iron Range, however, are somewhat different, the first note being shorter and lower-pitched and continuous with the second in a liquid downward inflection prior to the somewhat sharper last part. It would be interesting to know what the birds around Weipa sound like. MacGillivray (1918) describes this call as very striking, being a loud whistle, easily imitated by a human, and to which imitation it may often respond; he transcribes it as *wheeoo*, repeated three times and ending abruptly on a note like *who-o-o*. He also writes that both sexes make this note, the male being much the louder, which is in error and presumably due to mistaking immature males (in the female-type plumage) for adult females. He reports the adult males as calling from the tops of the highest trees, where they would call at intervals of 2–5 minutes. The call of the male at Iron Range is archived on xeno-canto at XC161222 (by H. Krajenbrink) and XC104961 (by E. Miller). A large collection of audio cuts of this taxon made by B. Clock can be found in the Macaulay Library at the Cornell Lab of Ornithology site. Australian field guides vary considerably in their transcriptions, Morcombe (2004) describing the voice (presumably at Iron Range) rather well as a deep powerful *awoo-arr-WHEET*, the sound swooping low through the *woo*, a growling *arr* and then rising to a powerful, sharp, drawn-out whip-crack, *WHEET!* Pizzey & Knight (1997) describe the call as a very loud mellow whistle with a dip in the middle, rising to a high *whiip*, as though whistling to attract attention. They then describe a breeding-season call as three loud clear whistles and a long-drawn diminishing note, each of the first three notes sounding like '*wheeoo*', and ending abruptly in a note like '*who-o-o*'. The males make a loud silken rustling in flight, like their congeners.

HABITAT This riflebird is restricted to relict rainforest patches, vine forest and riparian forest, and avoids eucalypt savanna. On Albany Island it is found in vine scrub forest (tropical monsoon vine forest), the same habitat as in the immediately adjacent mainland. Most of the peninsula is dry savanna country and thus unsuited to this species, which occupies the relict wetter forest habitats of the eastern coastal area and the tip of the cape. Recorded from sea level up to the lower hill forest.

HABITS This paradisaeid frequents primarily the middle to upper strata in dense rainforests, keeping to tree trunks and branches, where it probes vigorously for arthropods among the mosses, ferns and epiphytes that festoon the bark. It is shy and largely solitary, and can be hard to discern in its dense arboreal habitat. Several individuals may gather at fruiting trees, along with orioles, figbirds, Metallic Starlings (*Aplonis metallica*) and Trumpet Manucodes (Uhlenhaut *in litt*). They may sometimes be found by the noise of tearing bark or falling debris as they probe animatedly for insects among the massive clumps of epiphytic ferns, mosses and orchids. Calling birds are very hard to pick out, and are often, but not always, located high

in tall trees. They descend lower to feed, and when seen in sunlight the iridescence of the males is an amazing sight, constantly changing as the birds move. They will appear all black one minute, and then the throat and breast shield will show bright blue or purple iridescence with vivid yellow or green highlights, this being especially evident when the bird is calling as these colours are an integral part of the display and are emphasised when the head is thrown back to call or when displaying. A velvet-black pectoral band with violet-purple highlights, itself bordered by a broad bronzy yellow-green band, sets off the intensity of the breast coloration; an iridescent blue-green crown and iridescent metallic blue on the centre of the uppertail catch the light in striking fashion at times.

FOOD AND FORAGING This species takes a lot of fruit, including figs (*Ficus*) and the introduced custard apple *Annona muricata*, as well as seeds and flowers and the adults and larvae of insects, including ants and beetles (F&B). The birds will hammer powerfully into dead wood, cavities and rotten logs and stumps to reach beetles. The proportions of arthropods taken seem to be somewhat less than those of the two New Guinea congeners, no doubt reflecting habitat differences. A nestling taken by MacGillivray (the discoverer of the species in Australia) had eaten insects, grasshoppers and beetle remains. He saw a female-plumaged bird running up the trunk of a tree like a creeper, and its stomach was found to be full of nothing but insects, chiefly ants. A male shot at about the same time had just a few small round berries, the fruit of a tall tree, in its stomach (MacGillivray 1918). Significantly, there are no observations of these riflebirds associating with mixed-species foraging flocks in Cape York (pers. obs.; Uhlenhut and Nielsen *in litt*.), which is very different from the behaviour of the congeners in New Guinea, which are habitual elements of such flocks.

BREEDING BEHAVIOUR As is typical of the family, this riflebird is polygynous, the promiscuous solitary males calling from songposts to attract females, which carry out all the nesting duties alone (Barnard 1911; Frith & Frith 1997d). It seems likely that the males are territorial. The note by Thorpe (in Iredale 1950), writing of his stay on Cape York in 1867–68, bears out this assumption: 'Each male has a certain haunt of his own, averaging about two or three hundred yards in diameter… at frequent intervals they call as a sort of challenge to each other; but when they trespass on one another's domain, and meet, they have a pitched battle.' He writes also of their 'jealous nature' and of being able to bring them in by imitating the call, those that did not respond generally being accompanied by several females. DISPLAY/COURTSHIP There is a large collection of videos of the various taxa in the Magnificent Riflebird group in the Macaulay Library at the Cornell Lab of Ornithology site, B. Clock having a comprehensive selection taken at Iron Range. There are also two videos by J. Matsui on the Internet Bird Collection (IBC) site which show a male in display and calling. Frith (in F&B) reviewed videotapes of the **display**, but little else is published about it. Frith makes the following observations. The male perches on horizontal branches or vines, and upon seeing a female becomes stiffly sleeked and agitated. Stretching his head and bill in the direction of her, he makes repeated small side-to-side hops on the spot, sharply flicking his wings slightly, and occasionally running the bill down the underside of a wing as a kind of preening. When the female comes closer, he stretches upright with bill skywards and breast towards her, sometimes jerking his body and/or wings and continuously pulsing the bright blue breast shield, which causes the light to play upon it. He then snaps his wings fully out to each side of the body with an accompanying sharp rustling sound, and swings his neck and bill rhythmically back and forth between the wing edges with increasing speed. He also raises and lowers himself on his legs, at times with a side-to-side rocking motion, and raises and lowers his wings with a rustling sound which is audible from about 50m and is produced at a rising tempo. The tail is held out horizontally behind and the wings outstretched as he performs a bouncing dance along his display perch, hopping from a crouched position (which conceals his legs) and then up and forward for some 15cm before crouching again, before repeating the process. The male jumps off the perch and, with its extended head and bill, swings a graceful rapid arc from the leading edge of one wing to the other; when he reaches the end of his perch, he swings around and dances in the opposite direction. The wings may be held at 90° to the perch, or parallel to it as he hops along sideways. A male at Iron Range in October 1999 made a curious low-level flight with very slow measured wingbeats, almost like a roding display of the Eurasian Woodcock (*Scolopax rusticola*), but made at about 1.5m above the ground in quite thick rainforest (pers. obs.). This was quite different from the normal rather undulating flight, being far slower and with rhythmic deep wingbeats; the significance is unknown, but it was suspected that it might be part of a display sequence. NEST/NESTING Nesting details are much better known for this taxon than for its two New Guinean congeners, which remain virtually unknown. Barnard (1911) describes the nest as being composed of large leaves and vine tendrils very loosely put together. Unlike Victoria's and Paradise Riflebirds, the Cape York Riflebirds do not decorate the nests with snakeskins. Barnard (1911) examined about 50 nests and found no such decorations; he also remarked that if a single egg was examined, but not collected, it would be gone the next day, but if it was taken the second egg of the clutch would be laid! MacGillivray (1918) described a new but empty nest as constructed from broad leaves and twigs wound around with a green parasitic climbing plant; the lining was of fine midribs of leaves and fibres. Nests in monsoon vine forest at Iron Range in November were built between the bases of living fronds of pandanus tree crowns, and atop epiphytic basket ferns (*Drynaria*) on tree trunks at a mean height of 4m (F&B). EGGS & NESTLINGS The clutch is usually of two eggs but sometimes one, incubated solely by the female. Eggs more rounded than those of Australian congeners, dull white to buff or horn, boldly streaked dark and with sparse dark freckles, most streaks at larger end of egg. Size averages 33.1 × 23.9mm. Nestling still naked at 1–2 days old; female regurgitates

all meals to the nestlings (F&B). One nestling died after falling from a tree (MacGillivray 1918). Fledging period not known. **Juveniles** have been seen as early as 29th September (RAOU Nest Record Scheme).

HYBRIDS No hybrids involving this taxon are known.

MOULT Seems to moult mainly over the wet season, most museum specimens showing evidence of moult during December–March and to a lesser extent October–November and April–May, coincident with the onset and finish of the rains. Moult can, however, occur in all months. (HANZAB)

STATUS AND CONSERVATION Not assessed by BirdLife International, but would be Least Concern if split. This taxon is a restricted-range Cape York endemic, vulnerable to land clearance or degradation but with some populations within national parks on Cape York, as at Iron Range and Jardine NP. An adult female banded at Iron Range was recaptured there some eight years and 11 months after initial banding. Historically, the populations of central Cape York were much reduced 'mainly through the depredations of scientific collecting' (MacGillivray 1914); in 17 months one collector shot 106 adult males and 80 female-plumaged birds, with another said to have collected 70 skins (Higgins *et al.* 2006). Nowadays the riflebird is quite common where suitable habitat exists, and under no obvious threat beyond local land-use changes. It is listed in the category of 'common wildlife' under Queensland legislation (admittedly a fairly meaningless broad category), not having rare or threatened status. The taxon has been isolated from its congeners in New Guinea since the late Pleistocene and has, as might be expected, evolved some significant differences in morphology, vocalisations and behaviour. DNA studies would shed further light on the exact relationships with the sister-taxa, and it is to be hoped that further research will be stimulated by separating this form. Work on the two distinct song dialects within this taxon would also be interesting.

Cape York Riflebird, male displaying, Cape York Peninsula, Queensland, Australia (*Tim Laman*).

Cape York Riflebird, male and female, Iron Range National Park, Queensland, Australia (*Jun Matsui*).

GROWLING RIFLEBIRD
Ptiloris intercedens   Plate 13

Ptiloris intercedens Sharpe, 1882, *Journal of the Linnean Society London, Zoology* **16**, 444. Milne Bay and East Cape, Southeast Peninsula.
Other English names Eastern Riflebird
Etymology The species epithet *intercedens* is Latin and means 'interposed' or 'coming between', a reference to the fact that this species' eastern PNG range lies between those of the Magnificent Riflebird and the Cape York Peninsula (Queensland) taxon *alberti*.

IDENTIFICATION A large, spectacular and distinctive species, not readily confused with any others except for the sister-taxon, the Magnificent Riflebird, owing to the long decurved bill and distinctive broad rounded wings, stout body and rather short tail. Males have a bright iridescent blue breast, crown and uppertail, with rest of matt black plumage tinged with a blue wash, and the adult males produce a loud taffeta-like or silken rustling sound in flight. Females are quite bright orangey-rufous on the back, have a prominent pale supercilium, and are heavily barred black on the underparts. **SIMILAR SPECIES** In a reasonable view, the male Growling Riflebird is unlikely to be confused with anything else, and is allopatric with its congeners except for an ill-defined zone of abutment or sympatry on the Huon Peninsula and perhaps Eastern Highlands of PNG with the Magnificent Riflebird. Telling the two species apart would be difficult as field characters are subtle and of limited use, but fortunately the voice is diagnostic (see relevant VOCALISATIONS sction). The extensive feathering at the base of the culmen of the present species (both sexes) may, in a good view, be of some use. The flank plumes on males of this species when projected reach only level with or shorter than the tail, whereas the Magnificent Riflebird has

the flank plumes projecting beyond the tail. Females could be confused with the female Twelve-wired Bird, which also has a long decurved bill and heavily dark-barred underparts, but that species lacks the supercilium and has a marked blackish cap and nape and a red eye. The females of Superb and Eastern Lophorinas are smaller, much shorter-billed, and dark brown on the back with rusty wings, lacking the bright orangey coloration of the riflebird; Superb Lophorina also has a bolder and longer pale supercilium and is absent from lowland forest, though it does overlap with Growling Riflebird in the hills. Female-plumaged Raggiana Bird is duller and browner on the back, has a dark face with a bluish bill, and lacks the barred underparts and long bill.

RANGE Endemic to eastern PNG from the Adelbert Mts of the north slope and the Purari–Kikori River of the southern watershed eastwards to Milne Bay. It is absent from the satellite islands. Western PNG and West Papua are occupied by the sister-species, the Magnificent Riflebird. The Growling Riflebird is allopatric with its congeners except for an ill-defined zone of abutment or sympatry with the Magnificent Riflebird in the Huon Peninsula (Peckover in Coates 1990) and perhaps Eastern Highlands of PNG (de Silva in Mack & Wright 1996). An outlying riflebird population in Cape York, Queensland, bears more vocal resemblance to the western birds than to the

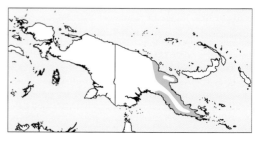

eastern and may be best treated as a distinct species itself, the Cape York Riflebird (see separate account, below). **MOVEMENTS** Sedentary.

DESCRIPTION Sexually dimorphic. *Adult Male* 34cm; bill length 56mm; weight 155–214g. Much like Magnificent Riflebird in plumage, but the bill averages about 4mm longer and is narrower, the base of the culmen with more feathering, and the flank plumes are shorter and do not extend beyond the tail. *Iris* blackish-brown; *bill* black, *mouth-lining* pale yellow to greenish-yellow; *legs and feet* dark slate to blackish. *Adult Female* 28cm; bill 48mm; 101–149g. Plumage much as for Magnificent Riflebird, but has largely feathered base of the culmen and a slightly shorter bill. *Juvenile* Undescribed. *Subadult Male* gradually assumes adult male plumage, becoming blotched with dark patches as it progressively moults. Presumed to take 4–5 years to attain full adult male plumage, as with congeners.

TAXONOMY AND GEOGRAPHICAL VARIATION Formerly treated as a subspecies of the Magnificent Riflebird, *intercedens* is, however, vocally very distinct from that species. Monotypic.

VOCALISATIONS AND OTHER SOUNDS This species has a remarkably loud and diagnostic call, a prominent aural feature of lowland forest in its range. It is a deep, guttural, rather growling and throaty-sounding disyllabic *HRAACK HRAOW*, the first syllable rising and the second falling, a kind of mutant reverse wolf whistle. The last syllable may be repeated up to seven times on occasion. The typical disyllabic call is variously rendered *Hraah-hraoou* (Finch 1983) or *CRRRAIY-CRRROW* (Coates 1990) or *uRAUow-urauow* (Beehler & Swaby 1991), reflecting the vagaries of voice transcriptions. Males above Wasu, on the Huon, in a zone of apparent sympatry with *P. magnificus* (or perhaps allopatry with latter in the hills at 760m: Coates 1990), give a variation of this call, which has a harsher, raspy quality with the second note either upward- or downward-inflected (Pratt in F&B). There are some good examples of this distinctive call on xeno-canto at XC24847 (by N. Krabbe) and XC279444 (by J. Moore), and also at the Macaulay Library of the Cornell Lab of Ornithology site. The call is audible at well over a kilometre, and is given both by adult males and by birds in female-type plumage, which are presumably immature males. It is given throughout the day, most often in the early morning and late afternoon. The birds at Varirata, PNG, will typically allow from 1–3 minutes to elapse between calls. Individuals of this species will occasionally be responsive to playback, although they will more usually ignore it (pers. obs.). Call frequency is much lower over the November–March wet season, when many birds are moulting. There are reports of *magnificus*-type calls being heard within the range of *intercedens*, as at Yalumet, on the north side of the Huon at 760m (Coates 1990), while at Wasu it is *intercedens* in the lowlands (Coates 1990) and hills to the south (Peckover in Coates 1990). Reports of *intercedens*-type calls at Lake Kutubu are in error (Schodde in F&B), though Diamond (1972) reports *magnificus*-type calls at Karimui, Eastern Highlands, well within the range

of *intercedens*. The cut-off between the two taxa is, however, quite sharp overall, although mixed pairings would be very hard to detect. Birds at Varirata make quiet growling noises when foraging, and when chasing through woodland can be quite noisy, emitting a variety of harsh, guttural single notes. A presumed immature male at 750m near Gare's Lookout, Varirata, in June 2000 was obviously practising the advertising calls, which were quieter and with the syllables less distinct than normal; the bird emitted an extraordinarily prolonged sequence, lasting more than 30 seconds, far longer than even the repeated series sometimes given by adult males. As with the Magnificent Riflebird, males of this species produce a loud silken rustling noise in flight, and loud blowing noises during some display behaviour have been described (Coates 1990); further, the wings have been described as opening with a rustling *plop* sound during one form of display given by a captive individual.

HABITAT This species inhabits rainforests in the lowlands, hills and lower montane zones; also monsoon, swamp and gallery forest. It occasionally visits mangroves, and also timber plantations such as teak (Peckover 1990), but which taxon (*intercedens* or *magnificus*) is involved is not clear. Selectively logged forest is also used, the birds preferring the taller trees and forest patches. The altitudinal range is from sea level up to as high as 1,450m and exceptionally to 1,740m (Coates 1990, but which form?), but mainly up to 700m, and in other areas 600m seems to be the cut-off point.

HABITS This is a shy and rather wary species, much like its sibling, and the adult males in particular can be difficult to observe in the dense forest habitat. Calling birds are very hard to pick out, and are often, but not always, located high in tall trees and calling from dead horizontal or slightly sloping branches, not from snags or vertical posts as the smaller Victoria's Riflebird does. Casuarina trees are a favourite perch in Varirata NP, near Port Moresby, and it is sometimes possible to see the riflebirds quite well in these more open trees. Males can be out in full view, although they are usually not fully exposed and are viewable only from certain angles through the branches and foliage. Coates (1990) saw Growling Riflebirds bathing and drinking at forest pools, probing the water with the long bill, perhaps to gauge the depth, before jumping in. Males are usually seen alone or with up to three female-plumaged individuals. They perhaps seem less willing than the Magnificent Riflebird to come to edge areas along roads and tracks.

FOOD AND FORAGING Bell (1982) thought that this species fed mostly on arthropods, though the exact proportions of the diet will no doubt vary according to what becomes available at any particular time. Stomachs of five birds from eastern PNG contained entirely fruits or arthropods of various groups (Schodde 1976). Food includes *Schefflera* species, and fruit of the aroid *Amorphophallus paeoniifolius* (Peckover 1985). Opit records that in October, at 1,500m on Mt Missim, the birds are often associated with pandanus palms, and are frequently seen hunting for insects among them (in F&B). When the pandans fruit these riflebirds are especially attracted and examine every one that they find. One individual was observed to eat seven orange pandanus fruits. This species was seen feeding on *Elmerrillia* fruits at Varirata, along with a Raggiana Bird, Brown Orioles (*Oriolus szalayi*) and Fawn-breasted Bowerbirds (*Chlamydera cerviniventris*) (Beehler in F&B). The diet at Mt Missim included *Chisocheton cf. weinlandii*, *Omalanthus novaeguineae* and *Gastonia spectabilis*. Capsular fruits seem significant, being nutrient-rich, drupes, berries and figs also featuring but much less importantly (Beehler in F&B). The riflebirds seem to avoid the topmost canopy, but feed mainly in the lower canopy 30–35m above ground, the subcanopy 8–25m up and the understorey at 0–8m (Bell 1982). Fruit-seeking was concentrated in the main canopy, where the main sources are, of course, located. Individuals forage by probing and working along branches, pecking at bark and prying into cavities and fern clusters. Dead limbs are often explored, and the birds move actively through the forest, not staying for long at any one site. Foraging for arthropods is undertaken with great enthusiasm, and the birds can be quite noisy as their actions cause wood and leaf debris to shower down. Frith (1990) aptly describes them as 'animated, acrobatic and vigorous', terms which apply to all members of the Magnificent group. These riflebirds will join with other paradisaeids such as Raggiana Birds at food sources, and are frequent components of the brown-and-black mixed-species foraging flocks in these forests. Hooded Pitohui (*Pitohui dichrous*) and Rusty Shrike-thrush (*Pitohui*) (*Pseudorectes ferrugineus*) are core species in the hills, with the latter and Papuan Babblers (*Garritornis isidorei*) in the lowlands. Such flocks at Varirata may contain female-plumaged Growling Riflebirds, Raggiana Birds, Hooded Pitohui and Rusty Shrike-thrush, Papuan Spangled Drongos (*Dicrurus carbonarius*), Chestnut-bellied Fantail (*Rhipidura hyperythra*), Frilled Monarch (*Arses telescopthalmus*), Yellow-bellied Gerygone (*Gerygone chrysogaster*) and Goldenface (*Pachycare flavogriseum*). Coates (1990) reports having seen them foraging on the ground in leaf litter with Piping Bellbird (formerly known as Crested Pitohui) (*Ornorectes cristatus*).

BREEDING BEHAVIOUR This species is polygynous, promiscuous males displaying solitarily at songposts and attracting female-plumaged birds. It is likely that the females alone are responsible for nest duties, as is typical of the family. Display posts in regrowth forest at Varirata are several hundred metres apart in two shallow valleys separated by a ridgeline. The males are, however, in clear auditory contact, at least four being audible along one short section of the track. Those here are regularly seen to chase both male and female-plumaged birds through the middle storey of the forest, and playback of males' advertising calls will sometimes induce another adult male to come in and investigate, which certainly suggests the maintenance of territories. There were no videos of the display in the Macaulay Library at the Cornell Lab of Ornithology site at the time of writing (2018). **DISPLAY/ COURTSHIP** For all the members of this Magnificent Riflebird species complex, our knowledge of the displays is still rather fragmentary, much remaining to be described, although observations on captive individuals help to give an idea. Courtship displays are performed on a horizontal or gently sloping perch, usually in the understorey of the forest 0–4m above ground, with some in the subcanopy at up to 25m and none in the canopy proper (Bell 1982c). Crandall & Lester (1937) made detailed observations of the displays by captives of this species, and noted two forms, one short and the other long. Their male was displaying to a stuffed female! The **Short Display** was usually preceded by a few abrupt, jerky movements of the head and neck, and then, with this slight warning, the wings are thrown slightly forwards and opened wide, while the neck is extended and the head moved to one side and brought to rest just behind the bend of the outstretched wing. The head is then moved from side to side, first over a period of about two seconds, but immediately and regularly at increased speed until brought to an abrupt stop in the middle after about a dozen movements, the last being so fast that it is virtually impossible to see them. This is analogous to some elements of a captive display noted for the Magnificent Riflebird, but lacks the bobbing motions (see the account for that species, above). In the **Long Display** the wings are opened suddenly with a rustling *plop*. While the head is being moved from side to side, the bird rises up on its perch and slightly elevates the extended and thereby relaxed wings. Simultaneously, as the head reaches the opposite side, the body is lowered on the perch and the wings snapped back to their fully extended position; this causes a sharp rustling, repeated with every similar wing movement during the performance. All of these movements are repeated in unison, again and again. The pose with wings extended and head to the side is held rigidly for about three seconds, then decreases to about three-quarters of a second until the end of the display. During the longest of these displays the head was moved from side to side some 35 times, punctuated by the rustling snap of the wings. No two displays were exactly alike, the rhythm of the movements varying. No vocalisations were heard, and the bright yellow lining of the mouth was not shown. Coates (1990) witnessed another form of display near Ower's Corner, PNG, in an apparently **agonistic** context. One male, followed by another, flew low through the forest understorey to alight on an almost horizontal twisted woody vine about 3m above ground. Here the two perched with jerky movements and faced each other about 60–90cm apart. The higher of the two then pointed its bill upwards, stretched its neck and expanded the gleaming blue breast and throat shield. Suddenly it stretched its wings fully and bent its head, the neck still stretched to one side to touch the outstretched wing. It then elevated itself to full height, at the same time emitting what sounded like a deep blowing which gradually increased in volume until the full height was reached, at which point the sound began to subside as the bird lowered itself, stopping when the original height was reached; just as it began to lower itself, the head was turned rapidly to the other outstretched wing. The complete motion lasted perhaps less than two seconds, and was repeated five or more times without pause, the effect being of a slow-motion up-and-down bobbing accompanied by a rising and falling blowing sound, the head being jerked swiftly from one wing to the other with each motion. The other male adopted a crouching posture, with head held low, and made quick, jerky movements. The first bird repeated this display several times, then the second bird gave an identical display while the first one crouched, also repeating the performance several times. Both males then flew off, one following the other, to a nearby perch, also low in the forest, which was out of sight. Sounds then indicated that the display was continuing, and a female-plumaged Raggiana Bird of Paradise was seen to fly rapidly towards them. Eventually the two males flew off in opposite directions. Frith & Beehler (1998) suggested that the deep blowing sound in this display was caused by the wing feathers. A different sort of display, involving **two female-plumaged birds** (one of which was presumably an immature male), was witnessed by Opit (1975b) at Varirata NP. The probable immature male was displaying on a horizontal casuarina branch some 23m above ground, at a site where the rainforest met the savanna. The bird had its back to the observer, but it rose upon its perch and extended its wings to about three-quarters open and began to rock from side to side. As one side was raised, so was the wing on that side lifted to its full extent over the head, which was held back with the bill pointing skywards; the head and neck were twisted from side to side towards the extended wing, and all of these movements were repeated in unison again and again, creating a perfectly rhythmic dance of great beauty. At times the extended wings almost touched each other over the head, forming an almost perfect disc with a central hole around the bird's head and shoulders. After dancing for some seven or eight seconds, the bird snapped its wings shut and spun 360° around on its perch, opened its wings, and then repeated the display. A displaying male Twelve-wired Bird has been observed to attract female-plumaged Growling Riflebirds (Coates 1990). **Nest/Nesting** Surprisingly little known. A nest with two eggs was found at Moroka, Central Province, in early June (Ogilvie-Grant 1915), and females carrying nest material have been seen in July (Coates 1990). Carl

Wahnes, who did some collecting on the Huon Peninsula, found a nest with two eggs, the nest being a deep cup composed of thin wire-like fibres surrounded by large dry leaves (Hartert 1910), much like that of Magnificent Riflebird but presumably belonging to Growling on basis of geographical range. There is a very short video by R. Currie on the Internet Bird Collection (IBC) site of a nest in Varirata NP, depicting an untidy ragged cup formed of sticks. **Eggs & Incubation** Undescribed; this taxon was for many years subsumed within Magnificent Riflebird, and published data seem to refer to that species or the Cape York taxon *alberti*. The incubation and nestling periods remain unknown. **Timing and Seasonality of Display** Display by this species is seen from middle to late morning, at midday and during the afternoon, and the season extends from the late wet season throughout the dry (Coates 1990), which accords well with the observations of Bell (1982), who also recorded display from April to September near Port Moresby.

HYBRIDS No hybrids involving this taxon are known.

MOULT Moult takes place in the early wet season (November–December), and display is not seen at these times.

STATUS AND CONSERVATION Classified as Least Concern by BirdLife International. The Growling Riflebird is quite common in the lowland and hill forests of eastern PNG. Bell (1982) calculated a population at Brown River near Port Moresby to be six birds per 10ha. One 2km ridge walk at Varirata NP has some half-a-dozen males audible from the valleys on each side. The species is not hunted for its plumes, and it is not currently threatened, although, as with the Magnificent Riflebird, it will decline locally as a result of logging, forest clearance and the expansion of settlements as the human population grows. Proposed massive oil-palm plantations in the Gulf of Papua drainage would also be highly detrimental to this and all other forest-dependent species.

Growling Riflebird, male, Madang Province, PNG (*Nick Garbutt*).

Genus *Epimachus*

The two species of the genus *Epimachus* are endemic to the mountainous uplands of New Guinea, where they have a wide range. They are the largest members of the family, and the genus is characterised by the very long lanceolate tail, the strongly decurved bill, and the iridescent green/purple coloration on the mantle and crown of the otherwise mainly blackish plumage of the adult male. This genus was merged with the much shorter-tailed *Drepanornis* by Diamond (1972), but separated by Frith & Beehler (1998). The phylogeny of Irestedt *et al.* (2009) placed this genus as sister to a clade that includes *Paradigalla* and *Astrapia*, implying that the long, curved bill has been independently acquired in *Epimachus*.

Etymology *Epi* is derived from the Greek, meaning over or upon, and *machaira*, a curved sword, referring to the scimitar-like bill, hence *epimakos*, a fighter.

BLACK SICKLEBILL
Epimachus fastosus   Plate 14

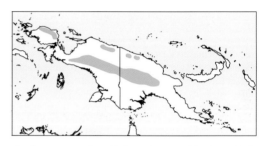

Promerops fastuosus [*sic*] Hermann, 1783, *Tabula Affinitatum Animalium (Argentorati)*, p.194. New Guinea, restricted to the Arfak Mts, Vogelkop, by Hartert (1930), *Novitates Zoologicae* **36**, 33. Spelling of the species epithet was emended by original author to *fastosus* on a later page (p.202) of the same publication, in which the author made his original intention clear (David *et al.* 2009).
Other English names Black Sickle-billed Bird of Paradise, Greater Sicklebill, Black Sabre-tailed Bird of Paradise
Etymology The species name *fastosus* is from the Latin noun *fastus*, and means proud or haughty.

FIELD IDENTIFICATION This species is the largest of the paradisaeid family, the distinctive male has a very long, slightly downward-curved black tail, and mainly black plumage. Both sexes have a prominent long decurved bill and a red eye. Females are brown above and heavily barred beneath, with a chestnut cap which is duller, less extensive and less well defined than that of the Brown Sicklebill. **SIMILAR SPECIES** The Brown Sicklebill is similar in shape, but the two species have only a narrow range of altitudinal overlap. Males have browner underparts and a pale blue eye. The females of the two species are similar, but the Black has a duller chestnut cap and a red eye. The calls of the males are diagnostic; see VOCALISATIONS section. The only other long-tailed montane species are astrapias, which are smaller, lack the decurved bill and have dark eyes, the males also with specifically distinct tail shapes. The rare Long-tailed Paradigalla in the Arfaks is very different and has a relatively short tail.

RANGE Endemic to the lower mountains of western and central New Guinea, from the Weyland Mts of the central ranges east to the Bismarck and Kratke Ranges, Mt Bosavi, and the Bewani and Torricelli Ranges of the north coast. Present also on the Vogelkop and Wandammen Peninsulas. **MOVEMENTS** Sedentary, with no significant movement recorded.

DESCRIPTION Sexually dimorphic. Following descriptions summarised from Frith & Frith (2010). *Adult Male* 63cm (110cm including central rectrices); bill length 70–86mm; weight 250–318g. This species,

the largest paradisaeid, has a long, sickle-shaped bill and a very long graduated tail with greatly elongated central feathers. Male of the nominate race has entire head black, with iridescent scale-like metallic green-blue feathers washed purplish; chin tuft and throat blackish. The entire upperparts are velvety black, the mantle to uppertail-coverts showing violet-purplish iridescence. The central back has striking large scale-like, highly metallic, vivid blue-green iridescent feathers, which feature prominently during display. The upperwing shows a variable blue to magenta sheen; primaries blackish-brown; uppertail blackish-brown with purple iridescence except on central feathers, which are more iridescent metallic blue/purple. Upper breast brownish-black, shading to paler and browner on lower breast and to sepia on the elongate filamentous flank plumes, vent and undertail-coverts (black of underparts can show a plum-purple sheen). The curious greatly enlarged axe-head-shaped pectoral plumes form an erectile fan, and are black with metallic green and purple iridescence; the shorter overlying feathers are broadly tipped iridescent metallic blue/violet, another major feature when erected in display; the elongate modified feathers on each side of belly and vent also have broad highly iridescent metallic deep green, violet and purple tips. *Iris* bright blood-red; *bill* black, *mouth* bright yellow; *legs* blackish-grey. *Adult Female* 55cm; bill length 66–89mm; weight 160–255g. Female is markedly smaller than male, with a shorter tail and quite different non-iridescent plumage. Dark brown above, with forehead, crown, nape, greater coverts and flight-feathers more reddish-brown; face, chin and throat to upper breast dark brown, rest of underparts pale buff with dark brown barring, bars on breast becoming broader down on to lower breast and belly, with still broader barring on flanks, vent and undertail-coverts. *Iris* brown; *bill* black, *mouth* bright yellow; *legs* blackish-grey. **Juvenile** resembles adult female, but crown and upperparts are more rust-red, chin and throat browner, underparts more buff. **Immature Male** is like adult female, plumage varying with state of moult and age; can take five and perhaps up to seven years to gain full adult plumage, but this process is still remarkably poorly known. Tail averages longer and primaries are more tapered and pointed. **Subadult Male** variable and can appear very blotchy: younger birds resemble adult female with a few adult male feathers showing, older birds are much like adult male with a few female-type feathers remaining. Subadults lack the elongated tail feathers of adult male, as the tail grows gradually longer with age, the central pair eventually more than doubling in length.

TAXONOMY AND GEOGRAPHICAL VARIATION

Two subspecies recognised here. F&B recognise three races, but Beehler & Pratt (2016) synonymise all of them, with bill length variable among all forms and the longer tail of the female *ultimus* only marginally different. There are, however, very distinct vocal differences between the far western population and the eastern birds, and we accept two taxa accordingly:
1. *E. f. fastosus* (Hermann, 1783). Confined to the Tamrau Mts and Arfak Mts of the Vogelkop. Synonym *E. f. ultimus* Diamond, 1969 (known from Mt Menawa of the Bewani Mts and Mt Somoro of the Torricelli Mts, now usually synonymised with nominate), having bill shorter than that of other populations (average 77mm) and female's tail longer than in nominate and most *atratus*; resembles *atratus* in being blacker, less brown, on underparts of adult males, and more olive, less rufous, on upper surface of adult female's tail.
2. *E. f. atratus* (Rothschild and Hartert, 1911). Mt Goliath, Border Ranges. Forms a cline with the synonymous *E. f. stresemanni* Hartert, 1930, of the Schraderberg, Sepik Mts (Gilliard & LeCroy 1961; Cracraft 1992; Diamond 1969). Ranges from Wandammen Peninsula and the central cordillera east to the Kratke Range. Now often synonymised with nominate, but does have a very distinct advertising call. Adult male of this otherwise weakly differentiated race has darker ventral plumage than nominate and a longer bill (average 80mm); female had average bill length 77mm, as opposed to 72mm for nominate female.

VOCALISATIONS AND OTHER SOUNDS

Males of eastern birds (subspecies *atratus*) have a very distinctive loud and far-carrying incisive *WHIK WHIK WHIK*, given from a songpost on bare branches atop tall trees or tall dead trunks. This can be heard on xeno-canto at XC64636 (by I. Woxvold). There is some marked **regional variation** in this call, birds at Ambua and in the Ok Tedi area giving this trisyllabic call, whereas those in West Papua utter a very powerful, far-carrying, rapid, loud and resonant metallic *WITIT WITIT*, sometimes broken up or given singly (presumably the 'du-dug du dug' call mentioned in F&B). Recordings are on xeno-canto XC215333 (by P. Åberg) and XC284600 (by P. Gregory), with a longer cut in this species account on the Internet Bird Collection (IBC). This call seems to be given only in the early morning at Ambua, from about 06:30 to about 07:00, but Gilliard (1969) states that it may be heard at intervals throughout the daylight hours elsewhere. The interval between the call series varies from about one minute to about 40 minutes (Bishop in F&B). This advertising call is audible from a distance of about 2km and is one of the loudest and most powerful calls in the forest. Birds bend forwards when calling, and then jerk upright as if forcing out the call as the bill opens, the whole body shaking as they do so. Diamond (1972) noted that what he designated as the **race *ultimus*** gives a very rapid, staccato machine-gun-like burst followed immediately by a single liquid note, as well as a two-note *whik*-type call; this call is somewhat intermediate between the usual calls of this species and those of the Brown Sicklebill. Pratt (in F&B) at Mt Bosavi heard songs like that of the Brown Sicklebill, and these may have been made by an immature male, such birds being prone to odd renditions of calls. Other calls are described as guttural notes, including a deep growled *grr-grrk grr K-WICK!* or *guck-er-ruk bl-whit!* The latter disyllabic note is sometimes repeated (F&B). Bell (1969) noted a displaying male which made a croaking noise, and Pratt noted peculiar frog-like or 'animal-like' guttural noises, interspersed with a soft honking and, more rarely, a song. Ripley (1957) mentioned a loud penetrating '*whick*' call given by an adult male in full display posture prior to a display flight. Frith (in F&B) notes a dry rustling sound coming from the plumage on a video of a bird giving hollow, dry knocking notes, described by Kirby (in F&B) as a rattle exactly like a distant machine gun. Bishop (in F&B) reports that this species will sometimes respond dramatically to playback of its calls, by flying past or perching close to the observer.

HABITAT Lower montane and mid-montane forest, sometimes at forest edge and in second growth or forest patches. Occurs from 1,280m up to 2,550m, mainly 1,800–2,150m. Kwapena (1985) states that in east of range, on Mt Giluwe, it is found at up to 2,860m.

HABITS This is a shy species, difficult to approach, and generally not so confiding as the Brown Sicklebill of higher altitudes. Peckover (1990) observed their ranges overlapping on a ridge at 2,120m at Ubaigubi, in Eastern Highlands Province of PNG, where display stations of the two species were within throwing distance of each other. A similar situation exists atop Rondon Ridge, near Mt Hagen. Males use tall bare tree branches sited below ridgetops as songposts, and can be conspicuous when calling, but any accessible males in inhabited areas are likely to be shot for their plumes. The flight is silent and rather slow and undulating, gliding after a few wingbeats. Foraging birds may bound from perch to perch like their congener the Brown Sicklebill, and spend much time at food sources such as *Freycinetia* pandans if undisturbed. This species is usually encountered in the middle or upper levels of the forest, unless displaying. It is sometimes seen very low down at or near ground level, presumably foraging. It will also join mixed-species flocks, but tends to be rather solitary. Bell (1969) observed a female-plumaged bird probing under the edge of strips of moss, which it had torn from a trunk to which it was clinging. This species and the Brown Sicklebill often replace each other altitudinally where they are sympatric, as at Ambua, where they rarely overlap at the higher levels for this species.

FOOD AND FORAGING The diet is not well known, but this sicklebill consumes fruit and arthropods, which it obtains by gleaning among dead leaves, mossy branches and epiphytes. Beehler & Pruett-Jones (1983) put the ratio of fruit to animal material in the diet at 50:50. A female-plumaged individual at Ambua spent many minutes in probing into the colourful fruit spike of a pandanus species growing high on a tree on the edge of a forest patch; it ate the orange fruit pulp, and would repeatedly return after a few minutes away. The yellow gape of the bird was conspicuous as it fed. Birds in female plumage have on occasion been seen to regurgitate several times, probably ejecting the seeds of some fruit as seen by Frith with captive birds (F&B). Beccari (in Gilliard 1969) noted that they liked to feed on pandans, especially those of the genus *Freycinetia*, which are epiphytic on the trunks of trees, and this is borne out by later observations such as shown in the BBC's *Attenborough in Paradise* film. Female-plumaged birds are sometimes seen poking around among the dead leaf clusters on pandans and palms near Ambua, presumably for arthropods. Males are usually solitary, but several females may congregate at food sources.

BREEDING BEHAVIOUR Polygynous, with promiscuous but solitary and sedentary adult males that occupy a home range with one or more prominent display perches (Gilliard 1969; Majnep & Bulmer 1979; pers. obs.). Whitney (in F&B) saw an adult male display on a fallen log on the ground at Ubaigubi. **DISPLAY/ COURTSHIP** The breeding behaviour remains little known as this is an elusive, rather sparsely distributed species, but this paradisaeid has one of the most astonishing display shape transformations in the family. Ripley (1957) witnessed a very rapid display sequence at 1,585m in the Tamrau Mts, Vogelkop, where an adult male was perched high in a huge *Agathis* tree, just above a female-plumaged bird, with the pectoral plumes spread out and upwards like two raised arms; the male called a loud penetrating *whik* as described above, retracted the pectoral plumes too swiftly to be seen, then turned and dived rapidly down about 30m to near the forest-floor shrubbery, before turning upwards with spread wings to sail back up to the original display perch almost as if on the rebound. Bell (1969) saw a male in the Ok Tedi area which started to display by quivering the pectoral plumes around its head, almost forming an arch, while it stood very erect and croaked. The *Attenborough in Paradise* video (1996) has a remarkable dawn sequence, taken during 05:20–0530 at 2,100m in West Papua in September–October 1994 (Kirby in F&B; Attenborough 1996), which shows the amazing shape transformation performed by this species, when he becomes something quite alien and other-worldly. This male was on a typical tall, vertical dead tree trunk in mid-canopy, surrounded by forest except where the ridge fell steeply away to the valley below. He would become active at first light, moving around the canopy and calling for a few minutes, and would then fly to the top of or just below his display perch and hop up, facing the downslope clearing when displaying; there

was always at least one female-plumaged bird nearby when he performed. The displays finished by 05:45, and no afternoon ones were noted. This compares well with the male near Ambua Lodge, although the timing is slightly later and no females were visible at that site (they may well have been present but out of view). The Macaulay Library at the Cornell Lab of Ornithology has a good series of videos by E. Scholes and T. Laman which shows these displays. The male begins with his loud ringing disyllabic or trisyllabic advertisement song and jerks his pectoral plumes back and forth until finally bringing them right up and meeting over his head, much as described by Bell (1969). He then stretches forwards and assumes an almost horizontal posture on the branch, swaying from side to side while giving a quiet rattle call like soft knocking on an empty box or a distant machine-gun sound (F&B). This posture is held for just a few seconds and he then becomes upright again and repeats the advertisement song, repeating this whole performance up to five times. While he is in this *Horizontal Display Pose*, a conspicuous iridescent electric-blue line is formed along the entire length of his shape by the broad iridescent tips of the pectoral tufts and flank plumes and by the iridescent tail feathers. The male spent no longer than ten minutes on his display perch. When the adult did not appear on two mornings, Kirby (in F&B) saw immature males use the vacant perch to practise their display. Gillison (in F&B) observed how adult males near Crater Mt, PNG, in July used a short vertical tree stump as a display perch for the aforementioned display, and he noted that on one occasion two adult males displayed facing each other, presumably where overlapping or adjoining territories met. In another case, an adult male on his display perch responded immediately to the appearance of a female-plumaged bird flying into an adjacent tree by adopting the *Horizontal Display Pose*; he then raised his pectoral plumes and remained static until she flew off. He also directed a head-down crouching pose, with the tail cocked >45° and the bill gaping widely, at a female-plumaged bird nearby (Frith & Cooper 1996). The mouth colour may be a significant visual signal here, as with the lophorinas and Victoria's Riflebird. At Ambua in July 2000 and July 2001, a male on a tall vertical dead tree snag performed a *Vertical Display Pose*, similar to the *Horizontal Display Pose* but done in an upright position, with the shoulders hunched up in an extraordinary, bizarre 'heraldic vampire' pose, like no other bird except for the Brown Sicklebill, which has a similar pose known as the *Leaning Display*. No female was visible, but she may have been concealed from view. On one occasion in August 2000, Tano (pers. comm.) observed a Black-billed Sicklebill which came on to the display branch just after the male Black Sicklebill had vacated it. Display by the present species has been observed in September, October, February and April. Peak display near Crater Mt is in September (Gillison in F&B), while at Ambua in 2000 and 2001 the birds were displaying in July and August but not in September–November. **NEST/NESTING** In November 1978, at 2,030m on Mt Giluwe, Kwapena (1985) found a nest which he attributed to this species and which was made entirely of orchid stems and moss at the base. The nest site is said by the Telefomin people to be in pandanus trees in the midstorey of the forest, active during February–April (Gilliard & LeCroy 1961). Wallace (1869), in his book on the Malay Archipelago, mentioned a bizarre local tale of this species nesting in a hole underground or in the rocks, always with two access holes! EGGS, INCUBATION PERIOD & NESTLING are unknown, as are details of the nestling cycle. TIMING OF BREEDING Meek collected a juvenile in February on Mt Goliath, Oranje Mts, West Papua (F&B), while Kwapena (1985) saw a juvenile on Mt Giluwe in February and suggested that the species bred there in the November–April wet season. Gregory (1995) heard singing males at Ok Tedi in September–January and June.

HYBRIDS Putative hybrids are known with Arfak Astrapia, Long-tailed Paradigalla and Superb Lophorina (see Fuller 1995, and F&B) but not with either Brown or Black-billed Sicklebills, which are sympatric in a limited altitudinal zone in a few areas. Genetic work is desirable in order to establish the exact parentage of some of the more problematic hybrid forms.

Elliot's Bird of Paradise (*Epimachus ellioti* Ward, 1873), also described as *Epimachus astrapoides* (Rothschild, 1897) and *Astrimapus ellioti* (Mayr, 1941), and known from two specimens, is a hybrid between Arfak Astrapia and Black Sicklebill, these two species inhabiting the same altitude band in the Arfaks. The male astrapia is the dominant, the hybrid having the broad elongate tail of that species, but the pectoral shields are of the sicklebill.

The False-lobed Long Tail or **False-lobed Astrapia** (*Pseudastrapia lobata* Rothschild, 1907), known from one specimen from the Arfaks, appears to be a cross between the Black Sicklebill and the Long-tailed Paradigalla, also sympatric in this region, and resembles the latter but with the tail of the sicklebill. A third possible hybrid bears the intriguing and evocative name of the **Mysterious Bird of Bobairo**, named by Junge in 1953, and may be a cross between Black Sicklebill and a Superb Lophorina, or just possibly a Black Sicklebill × Splendid Astrapia or some sort of hybrid throwback.

MOULT is recorded in almost every month, with a peak in June–February.

STATUS AND CONSERVATION Classified as Least Concern by BirdLife International; formerly listed as Vulnerable. This spectacular New Guinea endemic is the cause of some concern, even though much of its range still lies in remote and thinly settled lower montane regions, as these are also a favourite zone for human habitation in some areas, especially in PNG. Many such areas in PNG have been deforested, particularly affecting this species and the Blue Bird of Paradise, which overlaps in part of its altitudinal range. Many parts of its range are unsurveyed, and it is a low-density species even where locally quite common, as at Mt Bosavi. Reported to be common in the Arfaks (Gibbs in BirdLife International 2000). This species not only is a very large one, but also has the unfortunate habit of calling extremely loudly from prominent bare perches,

resulting in its rapid disappearance from settled areas or places with easy access. The skins are a very valuable cultural item and fetch very high prices. Gilliard (1969) reported on the numbers of tail plumes of this species worn in the head-dresses of the Wahgi valley men in 1950–51, and the situation is little different today, both there and in the Tari valley of PNG. The central plumes form the centrepiece of many a prize head-dress, and prices in the market at Tabubil, Western Province, reflected this demand as a good skin of a male Black Sicklebill cost 250 kina (K250) in 1995, as opposed to K5–K20 for other paradisaeid species (Gregory 1995). Gillison (in F&B) reports that this species was virtually extinct at Crater Mt prior to that site becoming a Wildlife Management Area, the decline being due to plume-trading and the giving of ritual gifts by the Gimi people of Ubaigubi. The birds seem to have made a strong recovery since 1996, when several men were fined for killing them. A similar situation may apply at Ok Tedi, where the species was also reported as virtually extinct by Coates & Lindgren (1978), being hunted out by highlanders during the construction phase of the copper and gold mine. Gregory (1995), however, found it at low density but not uncommon in the inaccessible areas of the lower montane slopes, this despite the persistence of some hunting with shotguns, which are technically banned. It has become markedly less frequent around Ambua as the forest patches there have dwindled and hunting continues. The species is vulnerable to the effects of logging and plume-hunting in all but the most remote and inaccessible sites.

Black Sicklebill, male display, Arfak Mts, West Papua (*Gareth Knass*).

BROWN SICKLEBILL
Epimachus meyeri   Plate 14

Epimachus Meyeri Finsch and Meyer, 1885, *Zeitschrift für die gesammte Ornithologie* **2**, 380. Hufeisengebirge, southern New Guinea (Mount Maguli, Owen Stanley Range, PNG).
Other English names Brown Sickle-billed Bird of Paradise, Meyer's Sickle-billed Bird of Paradise, Meyer's Sicklebill, Long-tailed Sicklebill, Long-tailed Bird of Paradise, Grey Sabre-tailed Bird of Paradise
Etymology The species is named after Adolf Meyer, Finsch's associate at the Dresden Museum, *meyeri* being derived from the Latinised form of Meyer; the German collector Carl Hunstein sent the first specimens there in 1884.

FIELD IDENTIFICATION The adult males of this species and its congener are large and distinctive, the long sabre-shaped tail plumes and decurved bill being unique to the genus. Females are barred beneath, and have a chestnut cap and a pale blue eye. Juveniles have a much shorter, blunter tail, but the chestnut cap and pale eye are still present and distinguish them from the smaller and shorter-tailed Black-billed Sicklebill, which also has a bare dark skin patch on the face. SIMILAR SPECIES The Black Sicklebill has a red or brown eye, but is otherwise quite similar and can be hard to separate if not seen well in the limited zone of overlap, but the songs of both species are very distinct (see below). The chestnut cap is less extensive on the female-plumaged Black Sicklebill. Black-billed Sicklebill is much smaller and has a much shorter and paler tail. Astrapias lack the decurved bill, and all the various females are darker in plumage than the female Brown Sicklebill and have dark eyes.

RANGE found in the central mountain ranges of New Guinea from the Weyland Mts of West Papua to the south-eastern Owen Stanley Mts, extending farther east than the Black Sicklebill but not occurring in the Vogelkop. Absent from the Bewani and Torricelli Mts of the Sepik. MOVEMENTS Sedentary within its habitat range, with no significant movements recorded.

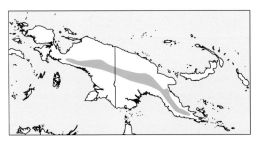

DESCRIPTION Sexually dimorphic. Following descriptions are summarised from Frith & Frith (2010). *Adult Male* 49cm (96cm including central rectrices); bill length 84–85mm; weight 144–310g. This is a large paradisaeid with a long, sickle-shaped bill and long, graduated tail with greatly elongated sabre-shaped central rectrices. The entire head is black, in some lights the scale-like feathers of crown and face appearing an iridescent metallic green-blue with purplish washes. Chin, throat and entire neck are black with iridescent magenta feather tipping; upperparts black, mantle and back with green-blue and/or magenta sheens and modified large scale-like central back feathers which are a highly iridescent metallic blue-green, the rump with a purple or plum gloss. Velvety-black upperwing with variable blue-green to plum gloss; brownish-black uppertail with blue iridescent sheen on outer webs, elongated central feather pair with metallic green-blue and/or magenta iridescence. Breast dark brown, washed plum-purple at side; curious greatly enlarged axe-head-shaped pectoral plumes black with metallic dark magenta iridescence, shorter overlying plumes broadly tipped iridescent metallic blue and purple/violet; elongate modified feathers on each side of breast, belly and vent have broad (but tapering) highly iridescent metallic purple and/or magenta tips; sparse filamentous flank plumes variable fawn-brown with paler, straw-coloured, central shafts; vent and undertail-coverts olive-brown. *Iris* pale chalk-blue; *bill* black, *mouth* bright yellow; *legs* dark greyish to blackish. *Adult Female* 52cm; bill length 81–83mm; 140–202g. Markedly smaller, less bulky, than male, with shorter tail, and very different plumage with no iridescence. Mainly dark brown above, more reddish-brown on forehead, crown and nape; face blackish, chin and throat dark sooty-brown, finely flecked dull buff; underparts greyish-white to light buff, barred blackish-brown, barring slightly paler towards rear. Bare parts as for male. *Juvenile* resembles adult female, but crown and mantle brighter and more rust-coloured. *Immature Male* like adult female, but tail longer. *Subadult Male* is variable with state of moult and age; can take 5–7 years to attain full adult plumage, but details still remarkably poorly known. Some younger males resemble adult female with a few dark feathers of adult male plumage, while older ones are like adult male but with a few female-type feathers remaining. Young male initially acquires darker crown than adult female, with blackish feathering around face, chin and upper throat, this followed by adult head plumage and then, with subsequent moults, an increasing proportion of adult plumage, Male acquires progressively longer tail with age, central feather pair more than doubling in length. (Avicultural records suggest that acquisition of adult male plumage may take at least 7.5 years [Aruah & Yaga 1992]: a female-plumaged bird thought to be a year old when caught in September 1978 was in adult plumage by May 1985. This bird first gave the machine-gun call in July 1982, when over five years of age.)

TAXONOMY AND GEOGRAPHICAL VARIATION Three races recognised by F&B and Coates (1990), but the species is regarded as monotypic by Beehler and Pratt (2016). We follow the latter treatment here. Variation appears to be clinal, the flank plumages of adult male becoming paler and whiter and body size smaller as one progresses westwards. Described form *E. m. bloodi* Mayr and Gilliard, 1951, from central section of the central ranges (probably from Kratke

Range to Mt Giluwe, Mt Hagen and most likely Tari–Doma Peaks area, westernmost edge as yet unknown), is considerably smaller than *meyeri*, with male flank plumes paler, rather more dirty whitish, than the dirty pale brownish of *meyeri*. Form *E. m. albicans* (van Oort, 1915), from Treub Mts of central New Guinea, is slightly smaller than *meyeri* and male has flank plumes whitish. *E. m. megarhynchus* Mayr and Gilliard, 1951, from Gebroeders Mts, in Weyland Range, slightly smaller than *meyeri*, with male flank plumes whitish, was erected on basis of a longer bill, but bill lengths of the three known specimens fall within the range of all other populations (F&B; Cracraft 1992).

VOCALISATIONS AND OTHER SOUNDS The male has one of the most extraordinary calls in the entire family, likened to the staccato stutter of a distant machine gun, and in popular mythology rumoured to have slowed the Japanese advance along the Kokoda Trail in WW2! This call is a characteristic and unforgettable sound of the montane moss forests and is given at any time of the day, but mostly in the mornings. It may be uttered just once, or repeated after a couple of minutes over the course of an hour, the male changing his calling site once or twice during this time. The males seem to occupy their own territories, within auditory contact of neighbouring ones, with interactions of two calling birds along the presumed boundaries of such territories. Whitney (in Beehler *et al.* 1986) reports that birds of south-east PNG give a *Tat-at, tat-at, tat-at* call, while those of the central highlands call *Tat-at-at-at, Tat-at-at-at-at-at*. Calling birds here (in Hela Province) lower and then jerk up the head and squeeze out the call, the body, wings and tail shaking with the force of the delivery (pers. obs.). The call is audible at long distances, probably up to 2km or so. Examples of this call can be found on xeno-canto at XC18880 from Kumul Lodge (by N. Athanas) and at XC40300 from the Baliem valley (by P. Åberg), and a call sequence at Ambua recorded by N. P. Dreyer is on the IBC. Birds of the Eastern Highlands of PNG have a call of three double notes lasting two seconds, repeated at intervals of two minutes (Diamond 1972). This call is audible at 2km, and the male pumps the throat and jerks the chest and wings as he delivers it. This vocalisation, when analysed, appears to be like a much speeded-up delivery of Black Sicklebill-like notes (F&B), and some Black Sicklebills have calls similar to those of the present species as well. These double notes have not been reported from the Ambua area. A **contact note** is listed as a nasal *nreh!* or *wahn?*, similar to a call note of the Blue Bird of Paradise (Beehler *et al.* 1986). Gilliard (1953) reports the machine-gun (jack-hammer) call being followed by 'drumming like a grouse, beating the wings against the sides and making loud, cracking reports which quite mystified us. It seems impossible that wings alone could make such a fuss.' The same author (in Gilliard & LeCroy 1961) heard a male in the Victor Emanuel Mts, over several successive days, making very loud snapping or cracking noises. These sounds were repeated with alarming suddenness about every half-hour during the mornings and afternoons.

A second male about 1.6km away also gave these calls. Gregory recorded a female-plumaged bird in July 2013 at Tari Gap giving a plaintive nasal *wheep* note, maybe a begging call or perhaps a contact call (see XC148732). Beehler (in F&B) reports a gurgling series from excited birds, rather like an underwater version of the usual machine-gun vocalisation. Pratt (in F&B) noted a recently fledged juvenile as making a quiet honking *ur ur*.

HABITAT Inhabits montane forest at middle and upper elevations, occurring above the range of the Black Sicklebill, the two being sympatric only in narrow altitude bands in a few areas. Peckover (1990) observed that their ranges overlapped on a ridge at 2,120m at Ubaigubi, in Eastern Highlands of PNG, where display stations of each species were 'within throwing distance'. The species will utilise disturbed forest but much of its range is above the main human habitation zones, in the moss forest. Brown Sicklebill is found from 1,500m to 3,200m, mainly 1,900–2,900m (Coates 1990), occurring at the lower altitudes in the eastern part of the range where the Black Sicklebill is absent (Gilliard 1969). This suggests that some sort of competitive exclusion is operating where these two sister-taxa occur on the same mountain range, and zones of sympatry are certainly narrow.

HABITS The Brown Sicklebill is a fairly common but low-density species of the moss forests, and seems to be often rather more confiding than its congener, though still wary. It is often, perhaps usually, observed singly, and sometimes as a pair, or a female with a juvenile, although loose groups of up to six individuals will form when foraging. Males call solitarily in the mid-stage of leafy tall forest trees, and calls may be heard at any time of the day, though there is a trend towards more calls in the mornings in the Ambua area. These sicklebills seem to favour the middle strata of the forest, and avoid the upper canopy, but they will come down to ground level on occasion. The flight is silent and rather slow and undulating, a few wingbeats followed by a glide.

FOOD AND FORAGING Shaw-Mayer (in Sims 1956) found berries, grasshoppers, other insects and various fruits in the stomachs of specimens. Gilliard (1969) found hard green fruits in the mouth of one specimen, and Rand noted the stomach contents of another individual to consist of 80% *Elaeocarpus* fruit and 20% insects. Fruits taken include cauliflorous figs, species of *Omalanthus*, *Garcinia*, *Freycinetia* and *Cucurbita*, *Pandanus brosimos*, and Urticaceae species (Kwapena 1985; F&B). Majnep & Bulmer write that many different fruits from trees and vines are taken, including taro (*Alocasia*) and wild raspberries (*Rubus*). During many days at Ambua we have never seen this species on the fruit spikes of *Schefflera* or of wild gingers (Zingiberaceae), both of which are favourites of Ribbon-tailed Astrapias in the area; Frith & Beehler (1999) also made the observation that in the central highlands the Brown Sicklebill is not seen at *Schefflera chimbuensis*, which is a major fruit source for these astrapias. Frogs have been recorded as prey of this sicklebill both in the wild and in captivity (F&B).

An interesting observation by D. Frith (in F&B) and P. Gregory (pers. obs.) at the Tari Gap involved two female-plumaged birds foraging on the ground at the roadside forest edge, probing into small plants and mud. This species is typically observed searching for food along moss- or epiphyte-laden branches and trunks, favouring the large epiphyte clusters especially, and at times being very hard to see when almost buried in the foliage. Gilliard (1969) saw one work its way up a trunk and nearby limbs from *c.*5m above ground, jumping along mossy limbs near the trunk. These sicklebills are often found working noisily through the dead leaf clusters that hang beneath pandanus fronds, probing for arthropods and probably spiders, the leaves rustling loudly as they probe into the bases. An immature bird came in to a fruiting tree at Ambua Lodge and fed in company with two female-plumaged Stephanie's Astrapias, a female-plumaged Loria's Satinbird, two female-plumaged Superb Lophorinas and a MacGregor's Bowerbird, plus a Great Cuckoo-dove (*Reinwardtoena reinwardtii*). Just occasionally here, they may be seen with a Black Sicklebill and Mountain Fruit-dove (*Ptilinopus bellus*) as well at the same fruit source. Brown Sicklebills will join with other species in mixed-species foraging flocks, with up to six female-plumaged birds in loose association. Belford's Melidectes (*Melidectes belfordi*) and Smoky Honeyeater (*Melipotes fumigatus*) are often involved, being common species of this habitat, sometimes feeding peacefully nearby, but sometimes antagonistic towards each other. Ribbon-tailed Astrapias can associate loosely with them at fruiting food sources, as does Stephanie's Astrapia and Crested Satinbird. Up to four sicklebills have been seen foraging in the same tree.

BREEDING BEHAVIOUR Polygynous; promiscuous males maintain large territories within auditory contact of other males, and sing from display sites therein. Territories are sometimes sited along ridges and regularly dispersed along them (Beehler & Pruett-Jones 1983). Although relevant data are scanty, females alone seem to be responsible for nesting duties, as would be expected with the male territory system in this family. DISPLAY/COURTSHIP in the wild remains still poorly known, which is perhaps surprising for what is a reasonably common and widespread species, the sibling Black Sicklebill being considerably better known. T. Laman and E. Scholes, who were able to witness several displays near Mt Hagen, discovered that the species does have an ovoid display like its congener, but, unlike Black Sicklebill, it also leans fully horizontal (and is not vertical) when performing. A series of videos of Brown Sicklebill by these scientists is in the Macaulay Library at the Cornell Lab of Ornithology site, but much remains to be discovered. Calling males near Ambua sometimes fluff up their flank plumes when giving the machine-gun rattle call. Crandall (1932, 1946) described three forms of display by captives. The first type is the **Pumping Display** (of F&B) and **Horizontal Display** (in Coates), in which a captive immature male was in an upright position, with breast feathers spread. The tail was jerked wide open and then tightly shut, the alternation being very

rapid; the wings, which were closed against the body, were moved up and down along its sides, the upward movements coinciding with the opening of the tail. The display was continued for a minute or more, but was seen only the once. Kwapena (1985) saw a form of this *Horizontal Display* performed by an adult male on Mt Giluwe in mid-February at 18:00 hours: the bird was calling every 2–3 minutes, audible a long way away, and was jumping up and down and raising its breastplate; two females came and sat near the display branch; the male stopped calling and was chasing the females from branch to branch, and he was still chasing them when Kwapena left at 18:55. When it displayed, this bird held its body in an upright position – the whole body was vibrating in constant motion when it was calling. When the male is fully adult, this display is more complex, the bird leaning backwards at an angle of about 45° (called the **Leaning Display** in F&B), the breast feathers spread and the flank plumes forming a fringe around the sides. After calling, the adult male then turns the breast upwards with breast feathers fully spread and the body flattened; the short feathers of the upper breast turned upwards about the head, circling the throat so closely that the iridescent black of the face and throat became very conspicuous. The wings were closed and the tail was slightly spread, though not vibrating or moving. The bill was closed. The pectoral shields were folded beneath the plumage and invisible. A similar *Leaning Display* has been noted for the Black Sicklebill. In another form of high-intensity display, the **Vertical Display**, a male from Deva Deva, upper Angabunga River, Central Province, PNG, was sitting and ostensibly preening the loosely extended breast feathers and pectoral shields. Suddenly, without calling, the body was drawn erect, with the tail very slightly opened and the wings closed; the breast feathers, encircled by the decorative flank plumes, widely spread, and the pectoral shields thrown straight upwards so that they extended far above the head and wrapped it closely. At the upper extremity the shields were narrow and compressed; at their bases they broaden gradually, to pick up the line of the spread breast feathers. The bill was widely open to show the bright yellow lining of the mouth. This position was held rigidly for about five seconds. Sometimes in this attitude the bird rotates his body in a series of short jerks, pausing for several seconds after each, until it is at right angles to the axis of the perch; he then jerks slowly in the opposite direction until he has come to a right angle, with the perch facing the other way. This movement may last for 2–5 minutes, and there is no movement of wings, tail or plumes, and no sound (adapted from Crandall 1932). Another captive male from the Waria River performed a variation of the Vertical Display, with the pectoral fan plumes raised to form a continuous vertical oval above the head; the outer tail feathers were simultaneously fanned rapidly and repeatedly. This suggests some regional variation in the display types, as with vocalisations. Frith (in F&B) notes a captive male displaying to a male Magnificent Riflebird with both the *Leaning Display* and the *Vertical Display*. The sicklebill's eyes were concealed behind the raised pectoral fan plumes, the bill protruding through the plumage during this Vertical Display; the body was slightly vibrated and slowly rotated from side to side. A different brief display involved the male bowing with his bill lowered almost to the perch between the feet, with the pectoral fan plumes semi-erect (Frith in F&B). The **display period** on Mt Missim extends from June to March, which is early dry season to the latter half of the wet season there (Pruett-Jones & Pruett-Jones 1986). Coates (1990) gave display period as from about February to September in the central highlands, which is mid wet season through to the second half of the drier season. The birds at Tari Gap–Ambua call in most, if not all, months, but seem much less vocal in November. **NEST/NESTING** The nest is a cup of stringy living moss and small vines, lined with slender rootlets and dried leaves, placed 3–12m above ground in the base of a pandanus crown, in the crotch of a thickly leaved slender tree, or in a fork of thin branches (Gilliard 1969; Coates 1990). Nest size seems to vary, one from near Kup, in the Kubor Mts, being noted as 175mm wide × 100mm deep externally and 95 × 45 internally (Gilliard 1969); another, from Mt Hagen, was 150 × 100mm outside and 80 × 45mm inside (Mayr & Gilliard 1954). Tree-fern crowns are another possible nest site (Kwapena 1985). A nest found by Kwapena and presumed to be of this species was oval-shaped, 200 × 100 × 57mm deep, with inner walls of orchid and fern stems and outer walls of moss and fern fronds. **EGGS & NESTLINGS** A captive laid only a single-egg clutch (Crandall 1941); one egg was laid in the Kubor Mts nest (Gilliard 1969), which the female alone was seen to visit three times before she was collected; a Mt Hagen nest contained a single nestling (Gilliard 1969). Egg-laying seems to be from at least April to October (F&B), but data sparse. **TIMING OF BREEDING** Breeding noted from April to January and is possible in all months (F&B). in PNG, juveniles were seen at Bulldog road, Wau, on 19th December 1974 (Pratt in F&B), on Mt Kaindi, Morobe, on 12th March 1975 (Beehler in F&B), and at Tari Gap and Kumul Lodge in August 2015.

HYBRIDS No hybrids with other species are known, despite limited sympatry with both Black and Black-billed Sicklebills and the wide range of this species.

MOULT Recorded in every month, with a peak in October–May; a captive male moulted over a five-month period (Crandall 1932).

STATUS AND CONSERVATION Classified as Least Concern by BirdLife International. This is quite a common species over much of its range, though patchy in parts and thought to have declined as a result of plume-hunting in some highland regions. Ripley (1964) found it rare in the Baliem and Ilaga valleys of West Papua and attributed this to traditional hunting for head-dresses. Majnep & Bulmer (1977) thought the species less evident in the Kaironk valley,

Brown Sicklebill, female types, Kumul Lodge, Enga Province, PNG, March 2016 (*Vincent van der Speck*).

Brown Sicklebill, male, Kumul Lodge, Enga Province, PNG, 2009 (*Phil Gregory*).

Schrader Range, in the 1970s than in the 1960s owing to plume-hunting. This species seldom came into the markets in Tabubil, PNG, during 1991–97, only a couple of poor-quality skins being seen and these selling for only K20, much less than the superb Black Sicklebill skin at K250. This probably reflects the distance over which the hunter had to travel, as the species' habitat here is remote and less accessible than that of the Black Sicklebill of the lower elevations. Hundreds of square kilometres of habitat throughout the highlands of PNG were burned and badly damaged during the severe El Niño drought conditions of 1997, and remained very marginal habitat for a long time thereafter. The species is under no obvious threat as much of its habitat is remote and inaccessible, although localised declines will occur as a consequence of clearance, logging and hunting.

Brown Sicklebill, immature male, Kumul Lodge, Enga Province, PNG, 2012 (*Phil Gregory*).

Genus *Drepanornis*

The two species of the genus *Drepanornis* are of quite restricted range and endemic to some scattered lowlands, hills and lower mountains of New Guinea. The genus is characterised by the unusually long, narrow, and decurved bill, the curious bare facial patch unique to this genus, the rather rounded brown or buffy tail, and the dull brown-and-grey plumage, lacking the striking black and iridescent metallic plumage of the two larger species in *Epimachus*, within which this genus was often formerly included (Diamond 1972). The phylogeny of Irestedt *et al.* (2009) placed *Drepanornis* as sister to a clade that includes *Ptiloris*, *Lophorina* and *Semioptera*.

Etymology *Drepanornis* is derived from the Greek words *drepanon*, a sickle, and *ornis*, a bird.

BLACK-BILLED SICKLEBILL
Drepanornis albertisi   Plate 15

Drepanephorus albertisi P. L. Sclater, 1873, *Nature* 8, 125. *Proceedings of the Zoological Society of London June 1873.* Mount Arfak; restricted to Hatam by Mayr (1941).
Other English names D'Alberti's Bird of Paradise, Black-billed Sicklebill Bird of Paradise, Buff-tailed Sicklebill, Short-tailed Sicklebill, Red Sicklebill
Etymology Luigi D'Albertis was a 19th-century Italian naturalist and explorer who was the first to collect this species in the Arfaks in 1872, recognising it as both a new species and a new genus. The Latinised *albertisi* commemorates him.
The well-known vernacular name of Buff-tailed Sicklebill was changed to avoid confusion with a species of South American hummingbird (*Eutoxeres condamini*) of the same name.

FIELD IDENTIFICATION A medium-sized sicklebill of mid-montane forest, where it is sedentary, shy, scarce and very elusive. This species has a pronounced sickle-shaped bill and a rather rounded tail, which is buff-coloured in the central populations and darker in the far west and Huon. It may be encountered with mixed-species foraging flocks as it probes among bark or on dead branches. **SIMILAR SPECIES** The Pale-

billed Sicklebill is a sister-taxon occurring in lowland forest and not sympatric in range. It has a pale sickle-shaped bill and lacks the buff tail. The two *Epimachus* sicklebills can overlap in range, but both are much larger and have long, pointed tail feathers, as well as quite different calls; females of these species would show a chestnut crown, as well as red or pale blue eye coloration. Immature *Epimachus* sicklebills do have a much shorter tail than the adults, but they also have a chestnut crown as well as different eye colour. *Melidectes* species also have a decurved bill, but are smaller, with a pale or bluish bill, have bare greenish-yellow or blue skin by the eye, lack barring below and are very vocal, with quite different vocalisations.

RANGE Endemic to the lower montane zone of New Guinea, this species shows a curiously disjunct range, and appears to be either absent or undiscovered in a huge stretch of the central cordillera from west of the Hindenburg Range, PNG, westwards to the Wissel Lakes. There are a couple of probable sight records, as yet unconfirmed, from the foothills of the Star Mts at 1500m near Tabubil. The species is known from the mountains of the Vogelkop, and the Fakfak, Kumawa and Foya Mts (Gauttier Mts) of West Papua. It is not yet recorded from the Adelbert Mts or the north coastal ranges of PNG. In the east, it is found in the mountains of south-eastern PNG, including the Herzog Range and

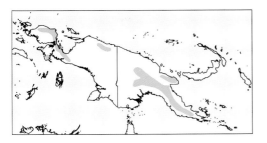

authors synonymise *geisleri* with nominate race, but that creates a bizarre and fragmented pattern, which does not accord well with the sedentary nature of the species and the ancient age of these mountain isolates. Pratt & Beehler (2016) recognise the following:

1. ***D. a. albertisi*** (P. L. Sclater, 1873). The nominate race is known from the Vogelkop and Bird's Neck in the Tamrau, Arfak and Wandammen Ranges. Presumably this race in the Fakfak, Foya and Kumawa Mts, but this is uncertain and in need of further elucidation. The upperparts are dark fuscous, with rump and tail strongly saturated with chestnut-brown, and the underparts are dusky grey-brown (Gilliard 1969).

2. ***D. a. geisleri*** Meyer, 1893. Confined to Huon Peninsula. Dark-tailed, resembling those of the Vogelkop and Bird's Neck, and quite distinct from the neighbouring pale-tailed taxon.

3. ***D. a. cervinicauda*** P. L. Sclater, 1883 (synonym *D. a. inversus* of Weyland Mts). Northern slopes of the central cordillera from the Weyland Mts to the Lordberg (Sepik Mts) and the Tari area, though there are huge tracts from which the species is as yet unrecorded, such as the Nassau Range, Van Rees Mts, Cyclops Range, Snow Mts and Star Mts (Diamond and Bishop in F&B). Uppertail-coverts and tail paler than on nominate, pale cinnamon, and barring of underparts of female paler brown; smaller than nominate, with proportionately shorter tail.

west as far as the Huon Gulf in the north and the Nipa/Tari areas in the south, extending to Mt Giluwe and the Hindenburg Range in the central axis. **MOVEMENTS** No significant movements have been recorded.

DESCRIPTION Slight sexual dimorphism. *Adult Male* 35cm; bill length 75–78mm; weight 103–125g. Head is cinnamon-brown, browner on crown, where feathers are tipped iridescent coppery-purple. The feathers of the forecrown are tipped purple and elongated, forming two small horns in front of the eyes; has pale maroon bare facial skin in a strip behind and just above the eye. Upperparts are shades of cinnamon-brown, with chin and throat feathers iridescent leaf-green, separated from lower breast by a band of feathers broadly tipped iridescent purple; the erectile pectoral plumes are iridescent bronze with a purplish wash, the flanks olive-brown and also broadly tipped purple, with rest of underparts whitish. Rump and tail strongly saturated with chestnut-brown in western (nominate) and Huon (*geisleri*) populations, paler cinnamon in central (*cervinicauda*) populations. *Iris* dark brown; *bill* shiny black, *mouth* pale green or pale yellow; *legs* brownish to greyish or blackish. *Adult Female* 33cm; mean bill length 78mm (nominate), 84mm (*cervinicauda*); weight 92–138g. Resembles male, but plumage lacks latter's iridescence, with chin and throat cinnamon-brown with pale buffy feather shafts; underparts barred dark brown, forming chevrons on lower breast. Has a strip of pale maroon bare facial skin behind and just above the eye, as on male. Bare parts the same as male's: *iris* dark brown; *bill* shiny black, *mouth* pale green/yellow; *legs* brownish to greyish/blackish. *Juvenile* undescribed. *Immature* resembles adult female, but has longer tail. *Subadult Male* shows a mixture of male and female plumages, with the horns, iridescent green throat and iridescent purple-tipped flank plumes being acquired first, along with an olive-brown central breast patch. Length of time taken to attain full adult dress as yet unknown, likely to be upwards of five years as the species is quite long-lived, one wild individual known to have reached more than 15 years.

TAXONOMY AND GEOGRAPHICAL VARIATION
The *Drepanornis* lineage is distinctly divergent from that of *Epimachus* with which it was formerly included, sufficient to warrant generic status (*contra* Diamond 1972). It seems unlikely that the well-differentiated Pale-billed Sicklebill forms a superspecies with the present species. Black-billed Sicklebill is polytypic, though the number of races is in some doubt and the whole arrangement is clearly in need of revision. Some

VOCALISATIONS AND OTHER SOUNDS The advertising song of the male, often delivered from a display perch, is the most frequently heard vocalisation. This is a powerful, musical, whistled piping series of downslurred liquid notes, *dyu dyu dyu dyu dyu dyu dyu...*, speeding up and rising in pitch, with an average of 18 notes per song, which lasts some 3.2 seconds (Coates 1990). It is also given occasionally throughout the day while the male is working his foraging grounds (Coates 1990). Arfak birds have a similar advertisement utterance, described by Gibbs (1994) as being somehow reminiscent of the flight call of a Whimbrel (*Numenius phaeopus*), of 3–4 seconds' duration, rising in volume and accelerating slightly. It is given in the very early morning or late afternoon from a traditional songpost high in the canopy, or from a perch that extends above the canopy, and is audible at some distance. Two examples of this call from the Hatam Mts of West Papua are on xeno-canto at XC23176 (by M. de Boer) and XC167207 (by F. Lambert). No vocalisations by the female have yet been noted (F&B). A calling bird is shown on the video by E. Scholes on the Macaulay Library at Cornell Lab of Ornithology site (ML456514). Gibbs (1994) noted that the song of the then newly discovered Fakfak Mts population differs from that in the Arfaks, being a series of downslurred whistles, slightly decelerating and much less rapidly delivered, rather reminiscent of a slowed-down yaffle of a Green Woodpecker (*Picus viridis*). Curiously enough, Stein (1936) described its call as a series of '*uë uë uë uë*' rather like that of the Green Woodpecker, which, when repeated, attracted the bird. The display song is infrequently heard and is given on the display grounds, being a prolonged and higher-pitched faster version of the advertisement song, and ending with a high hissing series, *tss tss tss tss tss tss* (Coates 1990); it is given from the lower part of the forest, sometimes within a metre of the ground. A *wrenh?* contact note similar to those of parotias or astrapias is given occasionally (Coates 1990), as also is a soft downslurred rasp, *ksp*, given once or twice (Diamond 1985). Opit (in F&B) mentions a loud *Yapp* note, while Pratt (in F&B) notes that a juvenile had a soft, barely audible piping call.

HABITAT Lower montane and mid-montane forest, occasionally in selectively logged areas, but rarely at the forest edge. Appears to be locally absent from much apparently suitable habitat. Occurs primarily from 1,100m to 1,900m in elevation, sometimes coming as low as 600m or ascending as high as 2,250m (Coates 1990). The species is sympatric with Brown Sicklebill on Mt Missim between 1,800m and 2,200m, with the lower limit of the Brown Sicklebill quite abrupt but the Black-billed upper limit being poorly defined (Pruett-Jones & Pruett-Jones 1986). The situation at Ambua is similar, with sympatry with Brown Sicklebill between about 2,000m and 2,050m, and with the Black Sicklebill from about 1,800m to 1,950m.

HABITS This is a shy and very elusive species, one of the hardest of all the paradisaeids to see. It forages in the middle strata of the forest, and female-plumaged birds sometimes join mixed-species foraging flocks (Diamond 1985), though apparently males do not (Coates 1990). It is usually seen singly, and appears to be territorial and sedentary. The species has been noted to rain-bathe at Mt Missim, sitting 10m up in a sapling and preening for some eight minutes in a heavy downpour (Beehler in F&B), while solitary males have been seen to bathe in small pools (Pratt in F&B). A male at a study site was preyed on by what was probably an *Accipiter* hawk in 1974 (Pratt in F&B), but within days it was replaced by a female-plumaged singing male. In 1977, the resident male there was in adult plumage, and was quite probably the same bird, which may give a clue to the ageing sequence and lifespan of what is likely to be quite a long-lived species.

FOOD AND FORAGING This species appears to be mainly insectivorous, and Schodde (1976) recorded the following arthropods from stomach contents: Orthoptera, Dermaptera, Coleoptera and the larvae of Lepidoptera. Beehler (1987a) observed it foraging for insects from 8m to 28m above ground. Fruit is also recorded as part of the diet, but these sicklebills seem seldom to forage for it, though the following are known as foodplants: *Cissus hypoglauca*, *Elmerrillia papuana*, *Chisocheton cf. weinlandii* and *Ficus* (Coates 1990). Beehler & Pruett-Jones (1983) estimated the diet as being 94% arthropod and 6% fruit. Interestingly, this species is rarely recorded as coming to fruiting trees at Ambua that were much favoured by other paradisaeids, which may reflect a dietary distinction and/or an avoidance of more open areas. Black-billed Sicklebills appear to be mainly insectivorous, and Schodde (1976) recorded

the following arthropods from stomach contents: Orthoptera, Dermaptera, Coleoptera and the larvae of Lepidoptera. Black-billed Sicklebills feed by gleaning and probing into moss, bark and dead leaves, also exploring knotholes and dead branches (Coates 1990; Beehler 1987a), while Coates (1990) was informed by local people that they will probe the large bulky nests of spiders. The bill is used like a forceps to seize prey or remove fruit from husks, but the entire length may be inserted into tree holes. Sometimes it is opened widely and either the maxilla or the entire mandible is used to probe into cavities for prey (Pratt in F&B). Goodfellow (1908) reported the bird as frequenting lower-growth forest and saw it foraging a short distance above the ground, while Diamond (1985) records it as foraging mainly in the lower canopy (with records from 3m to the canopy).

BREEDING BEHAVIOUR Polygynous and solitary, displaying from courts, presumably with the female alone responsible for nest duties. This species was studied by Beehler & Pruett-Jones (1983) over a 28-month period in a 200ha forest on Mt Missim, and is thus rather better known than other sicklebills despite its scarcity. They recorded five males singing from dispersed posts in the forest, in what appear to be exclusive territories, which are occupied throughout the year. Some behaviours of this species filmed by E. Scholes are shown on the Macaulay Library at the Cornell Lab of Ornithology site. **COURTSHIP/DISPLAY** Display sites at the Mt Missim site were on average 450m apart, with a colour-ringed male occupying a territory of about 14ha. This bird used his traditional site for at least four successive years, being at the centre of his home range in a patch of old-growth forest. A radio-tagged female moved over about 43ha in the eight days during which she was tracked. Males sing solitarily early in the morning from a high songpost, descend to saplings to display to females, and may also call and display from thickets near the ground. Opit (1975b and in F&B) heard a male calling from the canopy of tall trees and he saw him fly down to perch on a slightly sloping branch some 15–18m above the ground; this perch and all nearby branches had been cleared of foliage. The male whistled *to to to to to to to to* followed by a harsh but softer note, and began to raise his wings slightly, calling loudly, and dropping forwards to hang upside-down. The flank fan plumes were fully erected, then the pectoral fan plumes, and with a convulsive jerk both were made to form a complete feather disc around the body; the upper pectoral fan plumes were held behind the head and then sloped across to the front of the body to become contiguous with the flank fan plumes. This posture was held for about 30 seconds with the bird backlit, the sun showing through the feather disc but for a dark patch on each side of the head. The displaying male then resumed normal perching, and began pulling at and removing leaves from near his perch. This **inverted display** was repeated several times, calling each time before inverting, and followed each time by foliage removal. The entire display bout lasted about ten minutes. Beehler (1987a) found an advertisement-song perch in dead branches of the open crown of a 40m tall emergent tree (*Toona sureni*), with a display perch sited in a sapling just below the song perch. The bird also displayed near the ground at a site in thicket adjacent to the court of a Magnificent Bird of Paradise, some 45m from his advertisement perch. Beehler gives a composite of the display activities, as follows. The male arrives at his open canopy perch at 05:34 hours and gives 69 advertisement songs in 30 minutes, with the first display song at 06:03, interspersed with the advertisement song. At 06:07 he dropped to his display sapling in the leafy middle storey and three female-plumaged birds moved close, warily, to hop about the branches. The male chased these out of sight, then returned to the sapling area and perched 12m high to continue alternating the two song types. Suddenly he leaned backwards to 50°, flared his flank plumes and opened his bill, and held this pose for about 25 seconds. Becoming upright again, he continued singing for several minutes, then dropped back until entirely inverted. He flared all his plumes into a halo of wreath-like plumes around the head and chest, and held this pose while vibrating his body for 30 seconds, during which he gave one display song. He then turned upright again and began leaf-stripping (which seems to be an integral part of such displays). He returned to his canopy songpost at 06:20 and sang from it until 06:45, when he dropped once again to the display thicket to perch within 50cm of the ground and proceeded to give the display song for ten minutes. He then left to forage and sing in the forest canopy, before returning to his thicket site to sing from 07:20 to 07:50. LeCroy (in F&B) saw an adult male perform a curious display at the edge of a Lawes's Parotia court at 15:25m at Herowana, Eastern Highlands, PNG. The bird clung head uppermost to a vertical sapling 1m from the ground, and, with his body in line with the sapling, he reared backwards to about 20° from the vertical, and spread his plumes around until they formed a horizontal vibrating half-circle in front of the body, with the iridescent pectoral plumes prominent at the sides. The bird was silent and apparently alone, and used his legs to support this odd posture, which is reminiscent of a *Cicinnurus* type of display (see King and Magnificent Birds of Paradise). **REPRODUCTION & TERRITORY** This remains very little known, and mating has not been observed. Breeding is from September or December to March, with a male at Mt Missim vocal at his canopy perch annually from May to November but virtually silent for the remaining five months (F&B). Observations at Mt Missim suggest that mating occurs around October–November in central PNG, while Pratt (in F&B) reports a begging juvenile following its mother on 28th November 1974. A single vocal male held an exclusive, year-round territory of *c*.14ha. A single radio-tracked female travelled over a 43ha range in eight sample days, but spent most time in a 9ha core area abutting the male's territory. Although the female was silent, the territory-holding male sang daily for more than six months each year. The male performed an inverted nuptial display on a sapling near his regular song perch. This species exhibited a court-based mating system typical of many polygynous birds of paradise (Beehler (1987). **NEST** is a shallow basin about 25mm deep, set in the stems of a horizontal fork in a slim branch, and made of black wiry rootlets and wiry reddish-brown grass (Ramsay in Sharpe 1895). **EGGS** The egg is cream-coloured with a reddish tinge, spotted and dashed with brown and grey, most plentifully at the larger end (Ramsay in Sharpe 1895). The clutch is reported as being of one or two eggs (Gilliard 1969). No other details are known of the nesting cycle.

HYBRIDS No hybrids with other species have been reported, which is perhaps not surprising given the elusive nature and low density of this sicklebill.

MOULT Known to occur in almost every month, with a peak in December–March.

STATUS AND CONSERVATION Classified as Least Concern by BirdLife International. The Black-billed Sicklebill is a very sparsely distributed and elusive species, one of the most difficult to find of the widespread paradisaeids, and with a singularly disjunct range. It is nowhere common, and seems to be absent from vast expanses of the central ranges of West Papua, though it may yet be found in some of them and possibly also in the Adelbert Mts. It is not reported as being used anywhere for decorative purposes by local people, and any threats would seem to revolve around logging and land clearance as human populations grow. Subspecific limits have recently been revised, and the taxa of the outlying ranges in the Foya, Kumawa and Fakfak Mts require more study.

Black-billed Sicklebill, female plumage, Ambua, PNG (*Stephen Rannels*).

Black-billed Sicklebill, male displaying, PNG (*Tim Laman*).

PALE-BILLED SICKLEBILL
Drepanornis bruijnii   **Plate 15**

Drepanornis Bruijnii Oustalet, 1880, *Annales des Sciences Naturelles*, Ser. 6 9(5) p.1. Type locality is 'coast of Geelvink Bay between 136° 30' and 137° of longitude'. **Other English names** White-billed Sickle-billed Bird of Paradise, White-billed Sickle Bill, White-billed Bird of Paradise, Lowland Sicklebill, Bruijn's Bird of Paradise **Etymology** Named after the Dutch plume-trader Bruijn, who obtained the first specimens from one of his collectors.

FIELD IDENTIFICATION The Pale-billed Sicklebill is the only sicklebill known in the lowlands, and is not sympatric with any others. It is presumably a sister-species with the Black-billed Sicklebill of the lower mountain slopes, albeit well differentiated in morphology, voice, habitat and range. A rather shy and elusive species endemic to the lowland forests of the north-west coast and adjacent interior of New Guinea, it is best located by its loud and distinctive calls. When seen, however, it is readily identified by the sickle-shaped pale horn-coloured bill and the typical broad-winged, stout-bodied paradisaeid shape. **SIMILAR SPECIES** No congener is known to occur within the range, and no *Melidectes* occur at these altitudes. Long-billed Honeyeater (*Melilestes megarhynchus*) is superficially similar but considerably smaller, with a much less strikingly decurved bill, and the adult has a red eye.

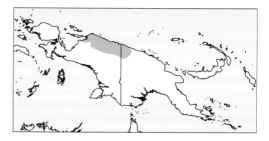

RANGE Endemic to the lowland rainforest of north-western New Guinea, from the east side of Geelvink Bay eastwards through the basin of the Meervlakte (Diamond 1981) and the north coastal lowlands to the vicinity of Vanimo, in PNG, and the north-western reaches of the Sepik drainage at Utai on the southern side of the Bewani Mts (Coates 1990). It is not yet known how widespread the species is in the Sepik drainage, or whether its range ever abuts or overlaps that of the Black-billed Sicklebill in the hills of the interior of West Papua. It is probably present also in the Idenburg drainage of West Papua, but not yet confirmed there. **MOVEMENTS** Sedentary.

DESCRIPTION Sexually dimorphic. This is a fairly large paradisaeid with a very striking long, sickle-shaped pale bill and mid-length tail. *Adult Male* 35cm; mean bill length 79mm; weight 160–164g. Has dark brown forehead, crown and anterior loral area. The elongated feathers above the eye form an erectile tuft that is iridescent blue-purple or red-purple in certain lights; chin, throat and ear-coverts darker, velvety blackish-brown, with leaf-green iridescence; extensive bare facial skin lead-grey with a slight purple hue, and a roughly circular small patch of strongly iridescent blue to purple fine scale-like feathers below the eye and over the base of the lower mandible. The mantle and upper back are dull brown, becoming cinnamon-brown on lower back, rump and uppertail-coverts; upperwing brown, with narrow slightly paler cinnamon leading edges to flight-feathers, paler on primaries; uppertail cinnamon-brown. Upper breast is dark olive-brown, with extensive tips of longer feathers iridescent olive-green; pectoral plumes dark greyish-brown, shorter row broadly tipped strong iridescent coppery-red, these overlying longer ones finely tipped iridescent purple/blue; grey feathers of side of lower breast have iridescent leaf-green broad tips, and beyond these is a line of jet-black feathers finely tipped with iridescent purple and/or blue; remaining parts of breast and belly are dark warm grey with dark lavender wash, the thighs, vent and undertail-coverts paler, more greyish-brown. *Iris* dark brown; *bill* ivory-whitish; *legs* purplish-brown. (C. B. Frith in HBW Vol. 14, 2009) **Adult Female** 34cm; mean bill length 75mm; weight 184–207g. Similar size to male but lighter in weight, and lacks iridescent feathering. Resembles adult male above but paler, more buff-coloured, and very different on underparts. Chin to upper breast buff, becoming darker on lower breast and paler cinnamon on belly, vent and undertail-coverts; chin and throat finely flecked blackish-brown, otherwise regularly barred blackish-brown below. *Iris* dark brown; *bill* ivory-whitish; *legs* purplish-brown. **Juvenile** undescribed. **Immature Male** like adult female, but tail longer than on adult of both sexes. **Subadult Male** plumage variable depending on age and state of moult; may take five or more years to reach adult plumage, still very poorly known. Some (presumably younger) birds are similar to adult female but with a few feathers of adult male plumage, whereas others (presumably older) are like adult male with a few feathers of female-like plumage remaining. The dark throat feathers and pectoral plumes are grown first, with much barring on warm grey of underparts, this area becoming greyer as barring diminishes with subsequent moults.

TAXONOMY AND GEOGRAPHICAL VARIATION

The *Drepanornis* lineage is distinctly divergent from that of *Epimachus* with which it was formerly included, sufficient to warrant generic status (*contra* Diamond 1972). The present species is monotypic.

VOCALISATIONS AND OTHER SOUNDS

Males have a very loud and variable, rather peculiar series of 4–10 hoarse, hollow whistles that rise and then drop in pitch, *wik-kew, KWEER KWEER kwer? kor kor kor*, lasting about five seconds (Coates 1990). Some songs are very similar to those of Lesser Birds of Paradise (Whitney 1987) or the Magnificent Riflebird. The song has a very striking wild, ringing quality with a plaintive mournful air, very distinctive and strange: examples can be heard on xeno-canto at XC209645 (by F. Lambert, at Nimbokrang) and XC140874 (by B. van Balen), and also on IBC (by P. Gregory). The song is given throughout the day, most frequently in early morning

from about 06:30 to after 08:00, declining after this during the heat of the day but with a resurgence in the late afternoon; it varies among individuals and even within individuals (Beehler & Beehler 1986). Ripley (1964), in what was then West Irian, noted the advertisement song of an adult male as being an extremely loud, not unmusical series of descending whistles reminiscent of that of Magnificent Riflebird: the first note starts at the tone and pitch of the second (descending) note of the riflebird's call, then goes into a series of descending whistles, repeated over again, several at a time; it also gave several gruff churring notes rather like those of a typical *Paradisaea* as it moved about in the tree. Songs from this sicklebill near Vanimo differed from those heard by Ripley. Whitney (1987) noted a pause of 2–3 minutes between songs, which consisted of one loud musical descending series of 4–10 whistles that sometimes ended with several lower, harsher whistles or *yuree* phrases; a typical song of about seven descending whistles lasted about five seconds. Males had one or two favoured trees from which they called every 1–3 minutes, mostly up to 09:30 but also again in late afternoon between 16:30 and 17:30. Whitney noted four males from one spot, singing along a linear road at about 200m intervals and within auditory contact of one another; playback response indicated that they were territorial. A call note given during foraging is a quiet *Whehn?* similar to that of some parotias and astrapias. Beehler & Beehler (1986) noted the male rattling his bill rapidly at the height of display, but making no vocalisation. No female-plumaged birds have been heard to call.

HABITAT The only sicklebill of lowland rainforests, inhabiting mainly the forest interior but also visiting the forest edges along roads and logging tracks. This species is reported as liking riverine and alluvial forest, and can be found within a few kilometres of the coast; in addition, it inhabits the limestone hills and the selectively logged forests around Vanimo. Ripley (1964) noted the birds in very tall *Agathis* trees. Recorded from sea level and just above, to 180m in the hills behind Vanimo.

HABITS This is a shy and inconspicuous species, easily overlooked and best found by listening for the amazing wild, loud and distinctive calls. It spends much time in the canopy, but descends to the lower strata to join mixed-species foraging flocks or to display (Beehler in F&B). The males may be seen perched up on prominent dead-limb songposts, and are sometimes observed atop palms rather like Magnificent Riflebirds. Usually seen singly, with habits similar to those of its sister-species the Black-billed Sicklebill (Beehler 1987a). When foraging, however, both males and female-plumaged birds have been seen in the same mixed flocks with other paradisaeids and other forest species.

FOOD AND FORAGING This species feeds on fruit and arthropods. It takes a lot more fruit than does the primarily insectivorous Black-billed Sicklebill, and its broader, heavier bill may be in accordance with this dietary preference (F&B). In West Sepik Province,

PNG, it was recorded as foraging on *Sloanea* species and an understorey *Cyrtostachys* palm (Beehler & Beehler 1986). It frequents the canopy and extends down to the shady middle and lower levels when in brown-and-black or mixed-species foraging flocks (Diamond 1987). Pale-billed Sicklebills clamber about on canopy limbs, peering under branches, poking the bill into cracks, knotholes, lichen and holes in dead wood, as well as searching dead leaves and checking bark (Beehler & Beehler 1986; Whitney 1987). Diamond (1981) saw them in the typical lowland-forest brown-and-black foraging flocks with Northern Variable Pitohui (*Pitohui uropygialis*), Rusty Shrike-thrush (Pitohui) (*Pseudorectes ferrugineus*) and Papuan Babblers (*Garritornis isidorei*), while Beehler & Beehler (1986) saw them foraging for fruit with these species plus Papuan Spangled Drongo (*Dicrurus carbonarius*), Jobi Manucode, Magnificent Riflebird, and Twelve-wired, King and Lesser Birds of Paradise. An adult male near Vanimo came into a huge fruiting tree full of Superb Fruit-doves (*Ptilinopus superbus*) and Claret-breasted Fruit-doves (*P. viridis*).

BREEDING BEHAVIOUR Polygynous, males holding courts and occupying a large display and foraging territory, similar to the Black-billed Sicklebill. Two adult males have been seen in a mixed-species flock (F&B), suggesting some sharing of food resources within territories. Males patrol their territory daily, singing at different points and countersinging and counter-displaying to rival males (Beehler & Beehler 1986). It is likely that, as with other paradisaeids, the female alone undertakes the nest duties, but field data are lacking. Some behaviours of this species, photographed by T. Laman and E. Scholes, are shown on the Macaulay Library at Cornell Lab of Ornithology site. **DISPLAYS** Display sites are in the lower middle strata

of the forest interior, about 7–10m above ground on small horizontal branches. Beehler & Beehler (1986) saw a posturing display in which the male sat erect on the perch, his short upper pectoral plumes erected into a wide skirt, and his tail fanned; the male held this pose for about ten seconds, and rattled the bill conspicuously without calling. No further displays have yet been noted, but it is likely that more elaborate ones occur. Display has been noted in August. **BREEDING STAGES**. No details of the nesting cycle are known.

HYBRIDS No hybrids with other paradisaeids are known.

MOULT Little known, but specimens showed evidence of active moult in March, August and October–November, with a peak August–October.

STATUS AND CONSERVATION Classified as Near Threatened by BirdLife International (2018), category C1 nearly met for Vulnerable listing. The Pale-billed Sicklebill is a restricted-range and very little-known endemic, and remains one of the least-known members of the family. May be not uncommon locally, as near Vanimo, PNG, where Whitney heard four males along one stretch of road and Gregory (in 1994 and 2000) recorded much the same. Beehler (1986) noted that a single male had a range over about 15ha at his Puwani river camp, PNG, and no others were heard there. Lowland-forest habitats are prone to logging and settlement, and one site at Nimbokrang, near Jayapura, has large refugee and resettlement camps nearby and much logging and hunting were obvious in 2018. The species is too sparse to figure in local decorative arts, and has not been noted in the ceremonial dress of local people. It is unlikely to be threatened owing to the remote and inaccessible nature of much of the habitat.

Pale-billed Sicklebill, female plumage, Remu, Nimbokrang, West Papua, June 2015 (*Phil Gregory*).

Pale-billed Sicklebill, female plumage, Nimbokrang, West Papua (*Gareth Knass*).

Genus *Cicinnurus*

I follow recent authorities in merging *Diphyllodes* into *Cicinnurus*; see Dickinson & Christidis (2014), Gill & Donsker (2016) and Beehler & Pratt (2016). The genus is endemic to New Guinea, ranging from the north-west Islands eastwards, including Aru Islands and Yapen Island, throughout the lowlands, hills and lower mountains. Most traditional treatments in the 20th century (e.g. Gilliard 1969, Forshaw & Cooper 1977) used *Cicinnurus* for a single species, the King Bird of Paradise, and *Diphyllodes* for the Magnificent Bird of Paradise and Wilson's Bird of Paradise. Diamond (1972) recommended folding these two genera into *Lophorina*. The phylogeny of Irestedt *et al.* (2009) combined the three species into a clade that is sister to the plumed *Paradisaea* birds of paradise. Characters defining the genus include the curled central tail wires, small bill, small body size and female plumage patterns, and also some vocal details.

Etymology *Cicinnurus* is from the Greek *kikinnos*, meaning a curled lock of hair, and *ouros*, meaning tail, referring to the coiled wire-like central tail feathers of these species.

KING BIRD OF PARADISE
Cicinnurus regius   Plate 17

Cicinnurus regia Linnaeus, 1758, *Systema Naturae* edition 10, p.110. East Indies; restricted to Aru Islands by Berlepsch (1911).
Other English names Little King Bird of Paradise
Etymology Species epithet *regius* is Latin, meaning royal or magnificent.
Named 'King's Bird' by the Dutch.

FIELD IDENTIFICATION The diminutive male, with its brilliant crimson-red and white coloration and emerald disc-tipped tail wires, is, with its congener Wilson's Bird of Paradise, the smallest member of the family and is unmistakable, albeit hard to see in the dull light of the dense vine tangles and forest which it inhabits. The female, too, is tiny, with a strange stubby, decurved yellow bill with feathered base, and has rusty wings, heavily barred underparts, and the typical quite broad, rounded wings, short tail and tubby body of the genus. The species has a curious head shape owing to the feathering on the culmen and the rather flat crown. The flight is gently undulating. **SIMILAR SPECIES** The congeneric Magnificent Bird of Paradise in female plumage is somewhat darker, without rusty wings, has a bluish bill and a pale streak behind the eye, and is not found in lowland forest, although the King Bird's range does extend up into its hill forest habitat in some areas. The bright violet-blue legs and feet are also a striking character, being duller purplish-blue in the Magnificent Bird.

RANGE Endemic to mainland New Guinea and the Aru Islands, and the West Papuan islands of Misool, Salawati, doubtfully Batanta, and Yapen Island in Geelvink Bay.

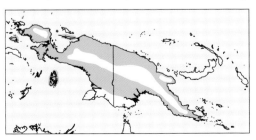

This is a widespread species, but absent from much of the lower Trans-Fly, where the eucalypt and *Melaleuca* savanna habitat is unsuitable. **MOVEMENTS** Sedentary.

DESCRIPTION Sexually dimorphic. This species and Wilson's Bird of Paradise are the smallest of the paradisaeid family. The culmen of both species is sharply keeled, unlike that of the Magnificent Bird of Paradise. **Adult Male** 16cm (31cm inclusive of central rectrices); weight 43–65g. Male of the nominate race has a vivid crimson head, chin, breast and upperparts, more orange on the forecrown, with a plush feather tuft over base of bill, and a discrete supraocular spot of iridescent dark green-black feathering. The upperparts, including the elongate mantle 'cape' feathers, rump, tertials and some upperwing-coverts, are glossy crimson-red, which flashes silvery white (due to refracted light) when the bird turns; the uppertail-coverts are duller, more orange and less glossy crimson; the upperwing is predominantly glossy crimson-red, flight-feathers and some coverts variably brown. The uppertail is dark brownish-olive, with brown-orange edges of the outer feathers; the two elongated central feathers are up to 15cm long and have thin, bare brownish central shafts, the latter tipped with extraordinary spiral circlets which are coloured a rich iridescent metallic emerald-green with a bronze-yellow sheen. There is a narrow breast shield (looking more like a broad pectoral band) which is iridescent dark green (appearing variably black to green-yellow depending on the light); on each side of the breast shield are several small elongate, erectile, fan-shaped olive-brown pectoral plumes, which are broadly tipped an iridescent bright metallic green; remainder of underparts are white. *Iris* pale to dark brown or greyish-brown; *bill* yellow, *mouth* pale aqua-green; *legs* violaceous cobalt-blue to blue-grey. **Adult Female** 19cm; 38–58g. Similar in size to male, but longer-tailed. Olive-brown above, with rusty margins on greater coverts, wing and tail feathers; often has a paler, more buff area above eye, with the entire underparts finely and uniformly barred dark brown on a somewhat variably buff breast, flanks and lower belly; *bill and legs* are duller-coloured (yellowish or blue) than those of the adult male. **Juvenile** The newly fledged young (11 days) has the upper head brown, tinged russet, with a pale superciliary stripe and a dark spot above. The upperparts and tail are grey-brown, the wing feathers darker brown, the greater coverts and outer primaries red-brown, and the chin grey-tinged brownish-yellow with fine streaks and points; underparts are light grey with dark barring. *Iris* is grey-brown, *bill* horn-coloured, and *legs* paler blue than those of the female. **Immature Male** is like adult, but with rectrices pointed at tips, younger (darker-billed) individuals having much orange-rufous on wing-coverts and outer edges of flight-feathers, and an orange-rufous wash on upper breast, but steadily losing this as they develop an increasingly paler bill and then the red of the adult plumage. **Subadult Male** variable; like adult female with few feathers of adult male plumage intruding to like adult male but with few feathers of female-like plumage remaining (Frith *et al.* 2017). Like all paradisaeids, this species will take several years to attain adult plumage, but the exact timing is still not known.

TAXONOMY AND GEOGRAPHICAL VARIATION

Polytypic, with two subspecies. Females exhibit no geographic variation. Previously a variety of races was recognised by different authorities, with six in Gilliard (1969).

1. ***C. r. regius*** (Linnaeus, 1758). Synonyms *C. r. rex* (Scopoli, 1786); *C. r. spinturix* Lesson, 1835; and *C. r. claudii* Ogilvie-Grant, 1915. Aru Islands, Misool and Salawati, the Vogelkop and all of southern New Guinea, extending on to north slope of south-eastern PNG from Huon Gulf to Milne Bay; doubtfully occurs on Batanta (Mees 1982). Aru Islands birds are slightly larger than those formerly known as subspecies *rex*, but variation is great and morphological differences slight (Cracraft 1992). Mees (1964, 1965) attributes a rounded green eyespot to this race, and a more elongate-shaped eyespot in the following race.

2. ***C. r. coccineifrons*** Rothschild, 1896, Yapen Island (Jobi). Synonyms *C. r. similis* Stresemann, 1922; *C. r. cryptorhynchus* Stresemann, 1922; and *C. r. gymnorhynchus* Stresemann, 1922. Range northern watershed of New Guinea, from east Geelvink Bay to Ramu. Birds of north coast of Huon unassigned. Measurements of all these forms overlap, with a cline in *similis* from larger in west to slightly smaller in east (F&B), though the bill and central rectrices are supposedly slightly longer in this taxon. The forehead of this race is red, not orange as with nominate, and the supraocular dark mark is a narrow slash as opposed to a spot (Beehler & Pratt 2016). This might be an artifact of skin preparation, the two forms are accepted by Beehler and Pratt (2016) with some reservation

VOCALISATIONS AND OTHER SOUNDS

Although, as with many other paradisaeids, the females of this species are not very vocal, the male has an extensive vocabulary. Indeed, this is one of the most vocal of the entire family, and the males will call throughout the day and stimulate each other to respond. They will also respond vocally to imitations or playback even if not coming in to investigate. Birds at Kiunga do not start calling very early, waiting until the sun is up at about 07:30, but they will call sporadically even in the heat of the day. They will break off at times to forage, then come back

a few minutes later, a rather different system from those of others of the family, but typical of the *Cicinnurus* group. A characteristic call in PNG is a disyllabic and rather nasal *whei-waaa*, or *whei-wa* or *hiia-haa* as transcribed by Coates (1990), the first note higher-pitched than the second and having a quality reminiscent of calls of the Raggiana Bird, albeit of a higher pitch and quieter. This call may be repeated every two or three minutes and seems to be a statement of occupancy of a territory post; sometimes only the first part is given, as a single call sounding like *kaar*. A short series of three notes can be made, *whei wher wha* or *wher-whei-wha* (Coates 1990); these notes can also be extended into long series of from 7–15 notes, given in any order, these long series being emitted just occasionally among the regular calls. Beehler (in F&B) has noted several males associating and calling much together, while at Kiunga, in June 2001, two males seemed to be overlapping in display areas and would come close together at one point with much loud calling, presumably a 'pair' of males (see Display section under BREEDING BEHAVIOUR, below). Examples of the typical calls can be found on xeno-canto at XC 40349 (by S. Connop), XC140736 (by B. van Balen) and XC89934 (by F. Lambert). Another advertising song is similar to that of the Magnificent Bird of Paradise, being a deeper-toned series of about seven rising notes, well transcribed by Coates (1990) as *cho-chow-chaw-chaw-chaw-chaw-chai*. There is also another Magnificent Bird-like call, a series of slower slightly rising interrogative notes, *kaw kaw kyaw kyaw*, described by Coates (1990) as notes of a uniform pitch, *chow-chow-cheouw-cheouw*. Gilliard (1969) heard the following calls from an adult male at his display tree in the Finisterre Mts, PNG: *waa waa waa waa waaa*, much like a Lesser Bird but higher-pitched; and *kii-kii-kii-kii-kii-kii-kii*, much like that species but again higher-pitched. He noted a male near Astrolabe Bay, PNG, as making a deep and raspy *quaa-quaa-quaa-qa-qa*, with the wings held slightly open and the head bent downwards; this may be the Magnificent Bird-like call mentioned above in a slightly different context. Gilliard also noted a kingfisher-like *kreea, kreea* and a drawn-out *kaa, kaa*, and harsh, sharp *quaa* notes. Hoogerwerf (1971) reported Vogelkop birds (race *regius*) as emitting a fairly inconspicuous call, a high-pitched unmelodious *kie-kwew-kwew-kwew* or *kwew-kwew-kwew*. In addition, the male makes a variety of quiet sub-vocals, audible only at close range and almost as if it is talking to itself, these being transcribed as *whiee-whee* in a high-pitched and whining tone, or *whiee-hee-hee*, like a subdued Raggiana Bird (Coates 1990). A low-pitched, hard and cat-like *maouw* may also be given a couple of times, as well as cat-like but becoming coarse *caaw* notes which are repeated in a series of three or four, or given two bursts of five notes each, or a series of ten (Coates 1990). These are not display calls. During display a weak song is uttered continuously, being a subdued chittering consisting of a series of short buzzing, twittering, churring and grating notes delivered rhythmically (Coates 1990). A harsh rasping buzzy call is given in threat, suspicion or alarm, while Gilliard (1969) heard one making a low, whirring, insect-like noise. In flight the wings produce a rattling sound, similar to that made by the congeneric Magnificent Bird.

HABITAT Found in rainforest, vine thicket, monsoon forest, gallery forest, tall secondary forest and selectively logged forest. This species is especially fond of tangles of vines and lianas hanging from great trees deep in the forest, from which it will call and display. Favoured trees may be used for many years, e.g. the well-known King Bird tree at Brown River, Central Province, PNG, which overhung the main coast road, and was used until it was blown down in 1993. Occurs from sea level up to a maximum of 1,150m (anthropologist Harriet Whitehead reports it to 1,500m in Western Province), but mostly to 300m, though regularly but sparsely to 420m in Western Province.

HABITS This tiny species is quite common in the lowland forests, males occupying display trees in an exploded-lek system with clans of up to five birds spaced 45–90m apart (Coates 1990). See also the Display section under BREEDING BEHAVIOUR (below). Female-plumaged birds often join mixed-species foraging flocks with pitohuis and Papuan Babblers (*Garritornis isidorei*) as core members, or form flocks of their own containing up to ten or more individuals (Coates 1990). Female-plumaged *Paradisaea* species often join such flocks, and the male King Bird may join up only as the flock passes through its territory. The species can be hard to locate although it does not seem unduly shy. Males advertise their presence by their distinctive calls, usually the first sign that King Birds are in the area, but despite their brilliant plumage they can be astonishingly hard to pick out in the deep shade of the vine thickets and forest foliage. They are frequently high up in tall trees, where they can be frustratingly hard to find, the white belly blending well with patches of light in the subcanopy. This species is a major cause of 'warbler-neck syndrome' among birders! These birds are quite acrobatic, clambering about vine tangles and hanging upside-down, or picking items off trunks and branches. In flight this is a typical paradisaeid, with rounded wings, dumpy body, short tail and an undulating flight. It will descend to the forest floor to drink or bathe, as evidenced by the marvellous photos in Coates (1990).

FOOD AND FORAGING The diet of this species is arthropods and small fruits, which may be swallowed whole or, if too large, consumed piece by piece. Bell (1983) considered the diet to consist mostly of arthropods, although Hoogerwerf (1971) reported the stomachs of nine Vogelkop individuals as containing entirely fruit; no doubt the diet will vary depending on local availability. Hicks (1988a) saw these birds taking the fruits of *Pipturus argenteus* and *Dysoxylum* species in an area 70km north-west of Port Moresby, while Simpson (1942) noted the crop of one bird to contain wild banana (*Musa*) pulp and seeds. Gilliard (1969) reported one as eating soft, triangular green tree fruits which were covered with little spines. The species will forage from the understorey up to the lower canopy, and at Kiunga seems to occupy mainly the middle stratum of the forest, 8–25m up, although it also investigates the lower canopy and sometimes comes lower to near ground level. Insects are obtained by gleaning from foliage, branches, vines and creepers, female-plumaged birds often joining mixed-species foraging flocks. Males near Kiunga will associate with such assemblages, but only when they move through the territory, when the component species seem to be tolerated there. Species involved at Kiunga include White-bellied Shrike-thrush (Pitohui) (*Pitohui incerta*), Papuan Babbler (*Garritornis isidorei*), Spot-winged Monarch (*Monarcha guttula*) and Papuan Spangled Drongo (*Dicrurus carbonarius*).

BREEDING BEHAVIOUR Promiscuous males attend courts in thick vines in the middle to upper strata, usually in shade but occasionally quite well lit. Hanging vines seem to be an important feature, and the birds will defoliate favoured perches. A kind of **exploded-lek system** seems to operate, but curiously enough with two males keeping close by before the next set of two, which may be 150–530m apart (Beehler & Pruett-Jones 1983; F&B). This seems to be a kind of intermediate mating system, between solitary males on the one hand and, on the other, an exploded lek in which males aggregate into larger groups. Males may be solitary or they may cluster, but the groups of two ('pairs') seem to space themselves at intervals of about 45–90m, although the odd group of two or more may collect and form effective leks (F&B). Males spend a great deal of time at the display perches, more so than any other member of the family except perhaps the Magnificent Bird of Paradise, starting early in the morning as the sun hits the treetops and continuing to late afternoon. Hot sunny days seem to generate less activity, cloud cover being optimal for display activity. Two males may come into the area at the same time without aggression, but such an event does seem to stimulate calling and the resident will monitor the visitor closely. Bishop (in Coates 1990) once saw two males (from a number of such gathered in a tree) fall to the ground locked in combat, so the visits are not always peaceful. **DISPLAYS** A singular possible display activity not so far subsequently noted was reported in south-west New Guinea by Goodfellow (in Gilliard 1969), who saw 'a small bird rise from the top of a tree and soar into the air like a Sky-Lark (*Alauda arvensis*). After it had risen about thirty feet, it seemed to suddenly collapse and dropped back into the trees as though it had been shot.' Gilliard (1969) also heard 'very weak grating, churring notes high overhead after the male disappeared, and thought they may have come from the male as it flew above the forest'. No subsequent observers, however, have reported any similar aerial activity. Gilliard saw a male suddenly drop from its display arena like a rock, falling four to ten or more feet (1.2–3m+) before it opened its wings. Males move up the vines or branches of the display tree by means of zigzagging hops, switching the body from side to side (Gilliard 1969). Coates (1990) notes two basic display types, one emphasising the pectoral fans and tail wires and the other the open, vibrating wings, and the reader is referred to that work for full details. To summarise, in the ***Wing-cupping Phase*** the wings are briefly partly spread and vibrated while the bird is perched more or less upright. This is followed by the ***Dancing Display***, given with the body held roughly parallel to the perch. The plumage is fluffed out as he squats, pectoral fans spread high, and the tail steeply cocked so that the wires are tilted forwards to reach above and beyond his head; he dances by vibrating his body and shaking the tail wires about, while singing a weak song. If a female is present, he faces away from her. The pectoral fans may be flexed higher than the head, and he may make little steps along the perch, repeatedly lunging at the branch and striking it with his bill. He pauses to look about, and then continues, and this display may last for many minutes. He then suddenly turns completely about and, still singing, performs for several seconds the *Tail-Swinging Phase*, whereby the tail is switched from side to side, causing the head wires to do the same in wide arcs. This often concludes the display, but the male may go on to the *Horizontal Open Wings Display* by spreading both wings out and forwards with body parallel to the branch; the wings vibrate as he rocks his body several times from side to side. This display is given in silence, and lasts for several seconds, then suddenly, holding the same pose, he flips under the branch to repeat it in the *Inverted Phase* of the *Open Wings Display*. The bill is held open (showing the bright light apple-green mouth-lining) and the head is turned from side to side. This display usually lasts for only a few seconds when, still inverted, he closes his wings and hangs suspended like a brightly coloured fruit or hanging-parrot (*Loriculus*). Looking about, he swings from side to side in the *Pendulum Display*, which is brief and also given in silence. He may then right himself on his perch or, as a finale, drop like a stone from the branch into clear space before flying off. During these displays the green-gold tips of the tail wires and the emerald-green breast band and pectoral-fan tips may catch shafts of sunlight in striking fashion, set off by the white belly and the crimson upperparts with a more luminous orangey colour on the head. Even in the shaded conditions of many display sites the colours are still remarkable, and may possibly have a colour component beyond human sight that is striking in such conditions. In a mating seen by Coates in October, both male and female perched close together on the branch, the male performing the *Dancing Display* in one spot for over a minute. Keeping the same pose and still singing, he began to quiver his slightly open wings. The male then turned swiftly about and, in the same posture but with the bill partly open, he rocked back and forth several times, each time striking the female about the head with the bill. She adopted a hunched posture, turned slightly away from the male, who briefly mounted and mated with her before flying away. A male with two female-plumaged brown birds was once seen climbing a steeply sloping limb and stretching one wing out (Coates 1990). Displays by captive individuals appear to be similar to those described above, but Bergman (1957a) observed that a captive male of the race '*rex*' (= *C. r. regius*) from western Irian Jaya, when with a female, raised his tail and wires at right angles to the body before performing the full set of displays listed previously. A variation of the *Tail Swinging Phase* saw this male usually holding the pectoral fans drawn in under the wings, with the body stationary. Finally, Beehler (in F&B) noted a display at 70m in PNG in September in which a male performed a display like that of a Raggiana Bird of Paradise, with

tail and head arched down, the wings out and held downwards and tail wires below him; he rocked violently from side to side, and then performed the *Inverted Open Wings Display* with wings folded tightly, legs straight, mouth wide open and rocking from side to side with the tail pointed skywards, all conducted in silence. **NEST/ NESTING** It is quite astonishing that for such a common and widespread species only one nest has been found in the wild, on the Aru Islands in March 1929 (Frost in Gilliard 1969). Equally surprising is that the species was nesting in a cavity some 18 inches (*c*.46cm) deep and filled with palm fibres, access being via a hole about one-and-a-half inches (3.8cm) in diameter. No other bird of paradise is known to be a hole nester, yet captive studies bear out the original finding (Bergman 1956, 1957a). In August 2007, however, a local informant Edmund Woram, from Watame (on Ketu Creek, Elevala River, in Western Province of PNG), who is very familiar with the species, reported a cup-like nest of twigs and small dry leaves sited high on a fern clump growing on a vine, and containing a single egg that was being incubated by the female. **EGGS, INCUBATION & NESTLING** The female alone performs the nest duties, as would be expected. Two eggs were found in the Aru Islands nest; these were creamy pink, with dark brown streaks on the larger end, and measured 27.5 × 21mm. A captive female incubated for 17 days, and fed her young as much arthropod food as possible while eating mainly fruit herself (Bergman 1968 in F&B). The nestling is devoid of feathers, with dark red skin and yellow gape, up to six days old. **TIMING OF BREEDING** Display activity has been noted from October to January in the Port Moresby area and in May–August at Kiunga. The species breeds from at least March to October and probably in all months somewhere across its extensive range. Juveniles have been seen in the wild in January and May, with nest material being carried in June and mating seen in October (Coates 1990).

HYBRIDS Hybrids between the present species and the congeneric Magnificent Bird of Paradise are widely known, the two being sympatric in some areas. They are, however, rare and as yet have not been seen in the wild, being known solely from trade skins; given that the King Bird has an arboreal display whereas the Magnificent uses saplings in a terrestrial court, contact between them must be infrequent.
King of Holland's Bird of Paradise or **King William III's Bird of Paradise** (*Diphyllodes gulielmitertii* A. B. Meyer, 1875). The Magnificent Bird of Paradise influence predominates over the King Bird in this hybrid.
Lyre-tailed King Bird of Paradise, **Lyre-tailed King**, **Lonely Little King** or **Crimson Bird of Paradise** (*Cicinnurus lyrogyrus* Currie, 1900, and *C. goodfellowi* Ogilvie-Grant, 1907). In this form the King Bird influence is stronger than that of the Magnificent. For details of these bizarre intrageneric hybrids, see Fuller (1995), Appendix 1 of F&B or Frith & Frith (2010).

MOULT The male's central pair of tail feathers grows progressively longer with age as with each moult he acquires progressively shorter outer rectrices. Captive specimens showed evidence of moult in all months, with most in October–January.

STATUS AND CONSERVATION Classified as Least Concern by BirdLife International. The King Bird of Paradise is a common and wide-ranging species in lowland forests throughout New Guinea and some of the western satellite islands. It is, however, inconspicuous and may be hard to locate unless calling. Some populations range quite high into the hills and lower mountains, and these are poorly known as yet. Estimates of six individuals per 10ha have been made near Port Moresby (Bell 1982a); Beehler (in F&B) found 22 adult males in 100ha at Kakoro, PNG, but only nine adult males in 150ha near Nomad River, Western Province, PNG. The species is patchy but quite common in the swamp forests of the upper Fly above Kiunga, in PNG. The species' biology remains little known, particularly the curious nesting habits and the exploded-lek type of

mating system with 'pairs' of males. This paradisaeid is, perhaps surprisingly in view of its wonderful coloration, seldom used by local people for decorative purposes, although Gilliard (1969) reported men of the Wahgi valley and Telefomin areas using its feathers in head-dresses. He also records that it was called 'money bird' in West Irian, perhaps because of the resemblance of the round feathers at the tail-wire tips to coins, or possibly because of the value of the skins in trade. In 1993 a male skin was seen for sale in Tabubil market for five kina (K5), and women were not allowed to step over it or handle it, there being some cultural traditions attached to the species locally. There are no obvious threats to this paradisaeid beyond localised losses resulting from habitat conversion to oil-palm plantations, and forest clearance for settlement and logging.

King Bird of Paradise, male and female, Elevala River, PNG (*Huang Kuo-wei*).

King Bird of Paradise, male displaying, Salawati Island, Irian Jaya, West Papua, Indonesia (*Konrad Wothe*).

MAGNIFICENT BIRD OF PARADISE
Cicinnurus magnificus   Plate 16

Paradisea Magnifica Pennant, 1781, *Specimen Faunulae Indicae*, in Forster's *Zoologica Indica Selecta* p. 40. New Guinea; restricted to Arfak Mts by Mayr (1941).
Other English names None recorded; birding vernacular, however, refers to the 'Mag BoP'.
Etymology *Magnificus* is a Latin word meaning magnificent, grand, splendid or distinguished.

FIELD IDENTIFICATION One of the small sickle-tailed group, the male has narrow loosely coiled, recurved and partly iridescent central tail wires, vivid golden-orange wings and cape, a glossy green breast (which usually looks black) and black underparts. The male plumage is highly complex and made up

of disparate elements, so much so that it is very hard to interpret what is actually being seen when in the field. The female is similar to the congeneric female King Bird of Paradise, but is larger, darker and more heavily barred beneath, lacking the rusty wings but with a pale post-ocular streak absent in that species, and with a distinctive bluish bill (not yellow or horn like that of the King Bird). The flight is gently undulating.
SIMILAR SPECIES In practice confusable only with the female-plumaged King Bird (see above) where the two happen to overlap. The bill has a feathered base, this and the rather flat crown giving a strange head shape, and the present species also lacks the rusty wings of the female King Bird and has more heavily barred underparts than the latter. Has the typical short tail, quite broad rounded wings and tubby body of the genus. The legs and feet are purplish-blue, less

intensely coloured than those of the King Bird. The Long-billed Honeyeater (*Melilestes megarhynchus*) is of a similar size and stout build, but lacks barring and has a very long decurved bill.

RANGE Endemic to the hill forests of New Guinea, from the West Papuan island of Salawati (doubtfully Misool), and Yapen, in Geelvink Bay, eastwards across the entire mainland in all mountain ranges. Absent from the unsuitable lower-altitude savanna habitats of the Trans-Fly, and replaced by the sister-species Wilson's Bird of Paradise on Waigeo and Batanta, in the West Papuan Islands. **MOVEMENTS** Sedentary.

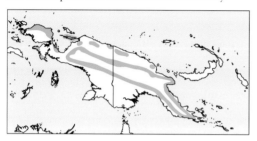

DESCRIPTION Sexually dimorphic. This species lacks the sharply keeled culmen of the King and Wilson's Birds of Paradise. Partially leucistic birds are known from museum specimens, a very rare occurrence in this family (F&B). *Adult Male* 19cm (26cm inclusive of 7cm central tail feathers); weight 75–119g. This is one of the stranger-looking members of the family, the plumage pattern very complex. Has rather short and bristly brownish head and nape feathers. A small patch of bare blue skin behind the eye flashes pale when turned in display; has a small iridescent dark green loral spot and a jet-black semicircle above the eye. The nape feathers are greatly elongated to form a glossy pale yellow cape, which flashes silvery white as the bird turns and can be erected to form a halo behind the head when in full display. This cape is usually obscured at rest by elongated dark brown feathers tipped brownish-buff at sides of neck and upper mantle; below the cape the back feathers form a semicircle of glossy dark carmine-red bordered with glossy black; the rump is a dull golden orange-brown, and the wings dark brown to blackish-brown with a large area of bright glossy golden orange-brown on secondaries and greater coverts. The upperside of the tail is dark brownish-olive, the central rectrices elongated into strange and striking metallic blue-violet iridescent sickles which cross near tip of tail and are sharply recurved. The chin and upper throat are dark brownish-black with a dark green iridescence, while the breast feathers form an erectile shield of glossy iridescent dark green, this becoming like an elongated heart when erected in display. Scale-like feathers extending in a line down the ventral mid-line are intensely iridescent turquoise, interspersed with dark green barring, but they are often invisible and show only when the shield is fully erect at a certain angle. The breast shield is edged with a band of elongate iridescent turquoise-blue feathers,

with the rest of the underparts blackish-brown. *Iris* dark brown; *bill* rich blue-grey; *legs* blue. **Adult Female** 19cm; weight 52–60g. Tail is slightly longer than male's, but lacks the elongated central wires. Has upperparts olive-brown to reddish-brown, richer on face and darkest on lores, giving a darker-headed appearance, with wings somewhat brighter-coloured than mantle; has a distinctive thin line of pale blue skin behind eye (as on male). Chin olive-brown and throat dirty white, flecked brown, with entire rest of underparts uniformly barred dark brown on a pale buffy-whitish background. *Iris* dark brown; *bill* dull bluish-grey; *legs* blue. *Juvenile* resembles female, but has dark bill and orange mouth-lining, and bright rufous wing-covert edges. *Immature Male* like adult female, but with bill blackish-brown; older birds develop orange on secondaries and wing-coverts, mouth of younger birds (less than two years or so) orange. Older birds in female-type plumage develop narrow, pointed and increasingly longer central rectrices prior to acquiring wire-like sickles of adult. *Subadult Male* variable, like adult female with few feathers of adult male plumage intruding, initially on head, to like adult male with few feathers of female-like plumage remaining. Full adult plumage not acquired until at least six years; with age, male acquires progressively shorter outer rectrices while simultaneously gaining longer central pair (Frith & Frith 2016). A wild-caught individual that was in female-like plumage in August 1969 was first seen to commence acquiring adult male plumage in September 1975, at an age of at least six years (Aruah & Yaga 1992).

TAXONOMY AND GEOGRAPHICAL VARIATION
Polytypic, with three races recognised in more recent texts, including Beehler & Pratt (2016).
1. *C. m. magnificus* (Pennant, 1781). Synonym *Diphyllodes rothschildi* Ogilvie-Grant, 1915. Western-most New Guinea, including Salawati Island, the Vogelkop, Wandammen Mts and the Onin Peninsula (F&B).
2. *C. m. chrysopterus* (Elliot, 1873). *Diphyllodes speciosus* var. *chrysopterus* Elliot, 1873 *Monograph of the Birds of Paradise*. Synonym *Diphyllodes m. intermedius* Hartert, 1930. Ranges from western and central New Guinea along the central cordillera from the Weyland Mts east to western PNG, and along north scarp to the Jimi River and Sepik–Wahgi divide. Also in the Adelbert Range. Measurements of *intermedius* are very similar to those of *chrysopterus* (F&B). Differs from nominate (both sexes) in secondary coverts and outer margins of flight-feathers being brighter, more dull orange, less yellow, and with crown darker.
3. *C. m. hunsteini* (Meyer, 1885). Synonym *Diphyllodes magnificus extra* Iredale, 1950. Eastern PNG, extending west to the Huon Peninsula, the Wahgi Highlands (Kubor, Bismarck, Hagen and Giluwe) region and along south scarp of central cordillera to upper Fly River region. Like nominate, but generally lighter on head, back and wings, a brighter orange on flight-feather margins and secondary coverts, and fractionally smaller (F&B).

VOCALISATIONS AND OTHER SOUNDS The advertising call of the male is a typical sound of the hill forests of New Guinea, and this is, like the King Bird, a decidedly vocal species. Around Port Moresby and in Western Province the advertising call is a quite loud sequence of some 4–6 downslurred harsh, throaty *chaw* notes, sometimes repeated about every three minutes and often within earshot of other calling males. Gilliard (1969) transcribes it as a hoarse or squalling *caar ca ca ca*. Occasionally this call may be repeated with one series after another, and it is audible over several hundred metres. Examples of male calls can be found on xeno-canto XC40374 (by P. Åberg) and XC163228 and 38112 (by F. Lambert). A shorter and quieter version of the advertising call may also be given by female-plumaged birds (Diamond 1972; Coates 1990), which may, I suspect, be immature males; this is perhaps a nasal *kew-kew kew kew kew* series that is quite often heard. Female-plumaged birds have a fairly quiet but distinctive somewhat rising, high-pitched interrogative metallic *tching* or *kying* contact note, often given during foraging and sometimes the first clue that the bird is in the area. Examples of this call are on xeno-canto at XC324603 and 324604 (by P. Gregory). A similar call may be given from the display court by the male, and repeated at intervals for minutes on end (Coates 1990), but gradually slowing to three calls in ten seconds from an initial rate of eight in ten seconds; this call becomes more pleading and mellow when a female is around (Coates 1990). A single hard, scolding *carr* is occasionally uttered, as is a low, harsh, hissing rasping note, given from in or near the court and made when the bird is suspicious or scolding (Coates 1990). Another note given at the court is a distinctive dry, scolding *ksss-kss-kss-kss-kss-kss*, downward-pitched towards the end and seeming to indicate some alarm or excitement; it is uttered in varying combinations of syllables, sometimes just *ksss-ksss-ksss*. Adult males disturbed at the court make low spitting and clicking notes and a scolding *char*, and in low-intensity display or when following a female about the court the male emits low, enticing, questioning *eek* or *eee* calls (Rand 1940b). Coates reports a peculiar low, rhythmic hard clicking, buzzing song with higher and lower notes when the male is performing the intense *Dancing Display*. Coates (1990) also notes a rattling sound seemingly made by the male's plumage in flight when in the vicinity of the display court, and sometimes the first indication of his presence. Beehler *et al.* (1986) mentioned adult males producing a clacking or rattling sound like two pebbles being struck together, but Coates (1990) also noted males flying about the courts silently, so this is not the equivalent of the taffeta rustling flight sound of astrapias or riflebirds. Female-plumaged individuals of the present species can also produce this flight noise (Coates 1990).

HABITAT This species inhabits primary and secondary hill forest and disturbed areas, often along the sides of quite steep or shallow ridges. It extends from about 400m to 1,200m, locally as high as 1,600m and exceptionally to 1,780m (Coates 1990). Found also at some low altitudes where the hill forest meet the foothills, and in adjacent lowlands, such as 75m at Utu at the base of the Adelbert Mts (Gilliard 1969), 115m at Nomad River, and 100m at Palmer Junction, Western Province (Rand 1942b), where a number of other montane species have outlying populations, and 160m at Aiome in the Ramu valley (Berggy 1978).

HABITS Female-plumaged birds are relatively easy to see but wary, often joining mixed-species foraging flocks with Hooded Pitohuis (*Pitohui dichrous*), drongos and Frilled (*Arses telescopthalmus*), Golden (*Carterornis chrysomela*) and Spot-winged Monarchs (*Symposiachrus guttula*). They come readily to fruiting trees, being fond of one bearing small black berries in the Ok Tedi area. The Amboyna or Slender-billed Cuckoo-doves (*Macropygia amboinensis*) share a preference for the same species of fruit tree, and are often seen feeding with the paradisaeids there. Males are more cryptic and solitary, and can be difficult to observe, but they, too, may come to fruiting trees, and are sometimes seen climbing high in clear view. The voice is a characteristic sound of the hill forests and the species is quite vocal, males seeming to call throughout the day, with a peak about two hours after sunrise. The males spend a lot of time at their display site, leaving for short bursts of foraging before coming back, behaviour much like that of the King Bird of Paradise. Magnificent Birds keep to the lower and middle stages of the forest, going up to about 30m in the subcanopy but generally being found in the 2–15m stage. Males are usually solitary, while female-plumaged birds may be seen singly or in small groups of up to six individuals at fruit sources. They can be seen with Carola's Parotias in the Ok Tedi area, as a rule keeping to lower levels in the tree than those preferred by the parotia, and they have been observed here also with female-plumaged Greater Birds of Paradise, Crinkle-collared Manucodes and the occasional female-plumaged Superb Bird of Paradise. Males will defend their display areas but seem to share intervening areas and fruiting trees amicably, not keeping home ranges (Beehler & Pruett-Jones 1983).

FOOD AND FORAGING This species takes a lot of fruit but also some arthropods. Beehler & Pruett-Jones (1983) estimated the diet to consist of 70% fruit, especially of the nutritious capsular type. Grant (in Ogilvie-Grant 1915) recorded these paradisaeids as apparently feeding upon flowers in large trees, in company with sunbirds and honeyeaters. Diamond (1972) found only fruit in some ten stomachs which he examined, while Schodde (1976) found that the stomach contents of nine individuals were 90–100% fruit and 0–10% arthropods, including crickets. Fruits taken in PNG include *Myristica*, *Psychotria*, *Trema orientalis*, *Pipturus argenteus*, *Sloanea*, *Harretia* and *Piper* species (Healey 1986; Hicks 1988a; Mack in F&B), *Omalanthus* and *Gastonia* (Beehler 1983a) and *Elmerillia papuana* (Pratt in F&B). Coates (1990) noted that males often regurgitate fruit seeds and that the courts are visited by Cinnamon Ground-doves (*Gallicolumba rufigula*) and Thick-billed Ground-doves (*Trugon terrestris*), which forage on them. This suggests that the species may be a significant seed-disperser in the forests. The Magnificent Bird of Paradise forages by gleaning along branches and trunks and poking about in mosses and epiphytes, climbing about acrobatically in twigs and foliage and often foraging upside-down rather in the manner of a nuthatch (*Sitta*). Males generally forage singly, but females often join mixed-species flocks.

BREEDING BEHAVIOUR Polygynous; promiscuous males attend display courts and the females alone are responsible for nest duties. The terrestrial courts, which were regrettably called 'bowers' in some of the early literature, inviting confusion with the rather different structures made by bowerbirds, are from several to many square metres in extent, and kept clear of all surface debris by the male. This species displays from small saplings within the court, and not on the ground. The **display courts** resemble small, neatly tended gardens hidden in the forest (Coates 1990) and are often located on the sides of ridges, in treefall or landslip areas (Gilliard 1969; pers. obs.) where the light shines down. The court area will have several vertical saplings on which the male spends much time perched, removing their leaves and wearing the stems smooth with his feet. One court studied by Gilliard (1969) had some 20 fish-pole-like saplings, and the leaves stripped from such and also cleared off the court may pile up around the edges of it. Saplings may be stripped of leaves to a height of about 8m (Gilliard 1969). Beehler (1983a) found one court maintained for three consecutive years, and some courts near Tabubil, Western Province, have been in use for some five years. Courts are spaced approximately 200m or so apart in suitable habitat in this area, while Beehler (in F&B) found nine males dispersed at intervals of 170–280m. Although the courts are maintained, the males seem not to be territorial (F&B), though in Varirata NP and in Western Province I have had males fly directly over me in response to playback, which does suggest some degree of territoriality. More usually, however, the birds are not particularly responsive. Males will sometimes maintain two courts, spaced some 60–70m apart (Beehler in F&B). Home ranges of males may overlap, as do those of females (Beehler in F&B). The display behaviour of the species is moderately well known, the observations of Rand (1940a) in West Irian (West Papua) and Coates (1990) in PNG being the major sources, along with some excellent videos by E. Scholes on the Macaulay Library at the Cornell Lab of Ornithology site (e.g. ML459291 shows a bird in full display pose, and ML460180 shows the breast shield pulsing). Some courts have large roots crossing them, but are otherwise swept clear of all leaves, twigs and small plants. The foliage is often stripped from overhanging trees, and the court saplings may have bark removed and even be killed because of the activities of the male bird. This all serves to accentuate the vivid colours of the male when the light hits the court. The male keeps the court clear by picking up objects in his bill and then flicking them out. Moss in the courts may be combed or groomed by the bird, but removing no objects. The male pecks at the ends of sticks and pick off bits of bark, the sound of its blows on the wood being audible some 3–5m away. The male spends much time on perches near the edge of the court, from two to fifteen feet (0.6–4.6m) off the ground, often for 30–40 minutes at a time. Much of this period is spent in preening, the bird stretching and putting the cape forwards over the head, but Rand never saw it preening while actually in the display area. The birds are vocal when perched near the court, giving the advertising calls here but only seldom within the court itself.

DISPLAYS Rand observed three types of display, the commonest being a **Pulsing Display** (*Breast Display* of Coates and F&B) of the breast shield when the male was perched low down near the display area, and performed regardless of whether or not a female was present. The breast shield expands and contracts so as to send shimmers or pulses of colour across the glossy iridescent green shield, and iridescent spots in front of the eyes may become conspicuous. The **Cape Display** was given when the male was clinging to the side of a sapling, usually a special display sapling, 30cm or so above the court, again regardless of female presence, but only after actively clearing the court if she was there. A **Pecking Display** was given, showing the inside of the mouth, from a normal perching position after copulation. Rand (1940b) gives a classic summary of the **display sequence** over a ten-minute period after a female appeared, following a period of court-clearing and calling from beside the court, and the following is adapted from his account. When she appeared the male immediately flew up and perched on a perpendicular sapling about 30cm above the ground. The female then came into the display ground and landed on the same sapling about a metre above the male. The male pulsed his breast shield towards her in the **Pulsing Display**, and continued to display his shield to her as she hopped from sapling to sapling, keeping about four feet (1.2m) above the ground. He was calling for much of the time, with low enticing questioning calls of *eek* or *eee*. The female was still and quiet while perched, then flew to the edge of the display ground and made as if to leave. The male at once turned his back and made as if to hop down to the ground and clean it. The female immediately came back directly above him and he again turned towards her while pulsing his shield. She flew away again and the whole ceremony was repeated. This time, however, while he was giving his breast-pulsing display, she began to hop towards him. He pressed his breast closer to the sapling and gave low, eager, single little calls. The female paused about a foot above the male, who was about the same distance off the ground, and the male went into the *Cape Display*, breast shield lengthened and flattened, the iridescent line down the centre clearly visible to Rand and the cape held straight out. The tail was in line with the body and was vibrating, but the bird was otherwise motionless. The male held this pose for 30 seconds and then the female hopped closer. He abandoned the pose and rather deliberately hopped up and mounted her, copulation ensuing. The male then dismounted, hopped to a perch just below the female, and gave the *Pecking Display*. The tail was erected at right angles to the back and he vigorously pecked the nape of the female. After each peck he drew back with widely opened mouth, plainly displaying the bright yellowish-green mouth-lining. Shortly thereafter the female flew from the court, the whole sequence having lasted for some ten minutes. Rand reported also that an immature male would visit the ground when the owner was away, and it performed the *Pecking Display* though without first cleaning the ground. Coates (1990) provides another classic description of the **display sequence** as witnessed at a court on the Sogeri Plateau, PNG. This series contained two new displays not witnessed by Rand in West Irian, and is illustrated with some amazing photographs in Coates's book. The adult male in this case appeared not long after first light and remained until the sun was well risen, leaving from 08:20 to 09:20 presumably to forage, before the sun hit the court. He would call to advertise his presence, and perform court-cleaning activities, hopping on the ground with tail plumes vibrating as he flicked away litter with sharp lateral movements of the head. He was heard calling in the general vicinity after leaving, but did not usually reappear until early afternoon. He would, however, come to the court if other large birds passed nearby, as when a party of Piping Bellbirds (formerly known as Crested Pitohui) (*Ornorectes cristatus*) passed through the ground. All displays occurred within 20–60cm of the ground. This male used two centrally located saplings in his court to perform the *Cape Display* after court-cleaning; if another Magnificent Bird was around, he would perform the *Pulsing Display* from a vertical sapling. Coates observed a new element of the display sequence, too, the **Back Display**, which is performed when the male first alights on the display sapling. This is an inverted display, given in silence with tail up and head uptilted, showing off the brightly coloured upperparts and the shape of the tail plumes. This is usually held momentarily before jerking upright, but is sometimes maintained for longer periods with the bird jerking from side to side, holding the pose at each new position. Following this *Back Display*, he assumes a normal sideways perching position and then pulsates the breast plumage in the *Pulsing Display*, accompanied by low pleading calls. Coates notes that this is the commonest form of display and may be given without other elements of the display sequence and/or without other birds being present. During the *Cape Display*, an intense part of the sequence given only when a female or females are present, the body is held in a rigid near-horizontal line with the head slightly uptilted and the plumage compressed except for the expanded pulsing breast; the pleading calls become more urgent, directed to the female, and the breast is expanded with pectoral fans extended to become heart-shaped. Viewed from the side or back, there appear to be fans twitching at the sides of the breast. The **silent** *Cape Display* is performed as a solo 'practice' display at least once and sometimes up to eight times during the course of a morning, and can be given in the late afternoon, too; it is used also to attract when a female is close by. When performing it, the male leans back from his display pole with neck stretched, bill pointing up, and cape fully expanded and erected over the head like a brilliant yellow ruff; the breast shield is flattened and elongated, and the tail wires do not figure in this display; the bill is sometimes opened to show the bright yellow-green mouth-lining, but this was not noted when a female was present. An extraordinary and previously unrecorded *Dancing Display* is given, presumably as a **prelude to mating**. The male faces up the pole, his body puffed out

with neck withdrawn, the breast expanded and pulsated, and the tail cocked at an obtuse angle to the body to show the shining metallic blue upper surface of the loosely coiled tail wires; the head is jerked from side to side to show the brilliant yellow-green mouth-lining, and the tail is quivered from side to side as the bird dances up and down the pole in short jerky movements; the cape is flashed but not fully opened as the head is jerked from side to side. This intense display, which lasts just a few seconds, is accompanied by a peculiar low, rhythmic, hard, clicking buzzing song with higher and lower notes. Coates did not actually witness mating after this as the female flew off each time. The male also sometimes gave this *Dancing Display* alone in his court, and the full display sequence was given after an unsuccessful mating. **Visits to other males' courts** Adult males may visit the courts of other males without eliciting an aggressive response, the resident male remaining tense and motionless on his pole as the visitor inspected the court (Coates 1990). An adult male was seen to come with four female-plumaged birds to another such court, again without aggression but just a performance of the *Breast Display* from the owner. This male also performed the *Cape Display* from the ground to a female 15cm above him on a vertical sapling (Thair & Thair 1977). **NEST/NESTING** The nest is an open cup of green moss with a little fur or dry leaves and weeds, lined with rootlets and fibres. The four known examples have been found 0.8–4.3m up in the thick leaf cluster of a parasitic plant in a tree in a fallow garden, in a low bush in kunai grass, and in a *Pandanus* tree crown at the edge of a garden (Bulmer in Gilliard 1969), and once on a ridge slope (Campbell 1977). This is totally unlike the nest site of the King Bird, which appears, from very scanty data, to be in a hole or other cavity. Bulmer (in Gilliard 1969) found three nests in the wild, on the northern watershed of Mt Hagen, two in September and one in December. **EGGS & INCUBATION** The eggs are pale horn-coloured with a slight pinkish wash, and boldly marked with broad brush-like streaks and blotches of purplish-grey overlaid with darker brown ones from around the larger end down most of egg length. Average size (sample of 13 eggs) 31.2×22.6mm (Frith in F&B). The clutch is of one or two eggs. As expected, the female alone attends the nest. Incubation period in captivity is 18–19 days. **NESTLING** Naked, with bright pink skin and yellow gape, darkening to blue-grey skin over a couple of days. Campbell (1977) saw a female regurgitating some large red and numerous small black berries at a nest. The fledging period in captivity was noted as being 16–17 days, young becoming independent at 38 days (Everitt 1965, 1973). **TIMING OF BREEDING** Display and breeding seem likely in any month over the extensive range of the species, but with a peak from July to February. Rand noted that immature males in female plumage were often found to have enlarged gonads.

HYBRIDS Hybrids with the closely related King Bird of Paradise are known, the latter species having a similar song and certain display aspects to those of the Magnificent Bird. Such hybrids, however, are rare and as yet not seen in the wild, being known only from trade skins; furthermore, given that the King Bird has an arboreal display whereas the Magnificent uses saplings in a terrestrial court, contact between the two must be infrequent.

King of Holland's Bird of Paradise or **King William III's Bird of Paradise** (*Diphyllodes gulielmitertii* A. B. Meyer, 1875). The Magnificent Bird of Paradise influence predominates over the King Bird in this hybrid.

Lyre-tailed King Bird of Paradise, **Lyre-tailed King**, **Lonely Little King** or **Crimson Bird of Paradise** (*Cicinnurus lyrogyrus* Currie, 1900, and *C. goodfellowi* Ogilvie-Grant, 1907). In this form the King Bird influence is stronger than that of the Magnificent. For details of these bizarre intrageneric hybrids, see Fuller (1995), Appendix 1 of F&B or Frith & Frith (2010).

Wilhelmina's Bird of Paradise or **Wilhelmina's Riflebird** (A. B. Meyer 1894, and named for his wife) is an intergeneric hybrid with the partly sympatric Superb Bird, known from three specimens (and with no relationship at all to riflebirds).

Ruys's Bird of Paradise (van Oort, 1906) is an intergeneric hybrid with the partly sympatric Lesser Bird of Paradise and known from a unique specimen.

MOULT Museum specimens show evidence of moult in all months, greatest in October–January.

STATUS AND CONSERVATION Classified as Least Concern by BirdLife International. This is a common and widespread species in suitable habitat throughout New Guinea. The birds are sometimes used for personal decoration, mainly in the Wahgi region, where the tail wires are used in head-dresses (Gilliard 1969). As with the King Bird of Paradise, I rarely found the Magnificent Bird for sale in the markets at Tabubil, seeing it offered just a couple of times over a seven-year period. It is a common enough bird in the Ok Tedi area, one 2km length of track having at least five males audible and with female-plumaged birds common members of mixed-species foraging flocks. There have been some losses of display sites here as a result of clearance for bush farms, growing bananas and cassava on amazingly steep and rocky slopes, and the species is retreating from the lines of roads as the habitat becomes degraded, although it does seem tolerant of disturbed habitats so long as some bushes and small trees remain. Diamond (in 1972) found this species to be one of the most abundant birds of the hill forest of eastern PNG, his figures suggesting that immature males tended to occur more commonly at the lower altitudes where adult males were scarce. The altitudinal range adjoins that of the Superb Bird of Paradise, which replaces it at the higher elevations. Local losses due to logging and clearance for farms or plantations will occur, but the species is secure across the bulk of its extensive range.

Magnificent Bird of Paradise, female plumage, Mokwam, West Papua, 19 June 2015 (*Phil Gregory*).

Magnificent Bird of Paradise, male, Mokwam, West Papua (*Gareth Knass*).

Magnificent Bird of Paradise, male, Mokwam, West Papua (*Gareth Knass*).

WILSON'S BIRD OF PARADISE
Cicinnurus respublica   Plate 16

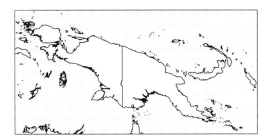

Lophorina respublica Bonaparte, 1850, *Comptes Rendus des Séances de l'Académie des Sciences*, Paris **30**, 131, 291. Waigeo, north-west Islands.

Other English names Waigeo (Waigeu) Bird of Paradise, Bare-headed Little King Bird of Paradise

Etymology A complex and ethically dubious provenance is attached to the scientific name. Edward Wilson had purchased the skin to be sent to the Academy of Natural Sciences in Philadelphia, but Prince Bonaparte briefly examined the first specimen in Paris and was so charmed by this little beauty that he (somewhat controversially) formally named it at a meeting of the Academy of Sciences. He compared it to the Superb Bird of Paradise, and corrected his mistake only a few months later, by which time Cassin, in the USA, had also described it as a new species named in honour of Wilson, unaware of the pre-emptive description. Bonaparte named it *respublica*, from the Latin for the republic, as up to this time it had been customary to name most of the paradisaeid family after royalty and he had not the slightest regard for all the rulers of the world. He named the new species for the French Republic, 'which might have been a Paradise had not the ambition of Republicans, unworthy of the name, made it by their evil actions more like a Hell' (Iredale 1950). The bird was finally discovered in the wild on Waigeo Island in 1863, by Bernstein, and again described as a new species as Bernstein was unaware of the controversy regarding the previously unique skin of the species.

FIELD IDENTIFICATION This is, along with its congener the King Bird, the smallest member of the family and is the only member of its genus on the two islands where it occurs. It should be unmistakable, being a typical stubby-bodied, broad-winged and short-tailed paradisaeid with a unique cerulean-blue tonsure, resembling a sort of hot-cross-bun pattern but divided into six segments by a fine black tracery of lines in both sexes, the blue being duller in the female. The head pattern resembles an exposed brightly coloured blue brain, a very distinctive pattern. The male has sickle-shaped violet-purple central tail feathers, a glossy green throat and breast shield, a bright yellow cape and crimson back and crimson-edged flight-feathers. The female is heavily barred dark beneath, dark brown above, with wings and tail washed with reddish. **SIMILAR SPECIES** None; should be unmistakable within the restricted, island range.

RANGE Endemic to the West Papuan islands of Waigeo and Batanta. **MOVEMENTS** Sedentary.

DESCRIPTION Sexually dimorphic. *Adult Male* 16cm (21cm inclusive of central rectrices); 53–67g. The male has a sharply keeled culmen like that of King Bird of Paradise (but unlike Magnificent). The bill base has a tuft of velvety-black feathering, with the crown and nape forming a skullcap of bare skin coloured a striking bright intense cobalt-blue and divided into

six smaller segments by crisscross lines of fine velvety-black feathering. The upper mantle has a semicircular brilliant yellow cape contrasting with the crimson lower mantle and bordered by a broad black line, showing as a shallow black U from above. The mantle (except uppermost part), upperwing-coverts, secondaries and the outer margins of the dark brown primaries are glossy crimson-red; the upper back and a narrow line across the lower back are velvety black, with rump, uppertail-coverts and uppertail blackish-brown. The central tail feathers are elongated into two iridescent violet-blue sickle-shaped spiralled wire-like plumes or sickles (these are more strongly recurved and both shorter and broader than the plumes of the Magnificent Bird of Paradise, sometimes almost forming a complete circle when curving back). The chin and upper throat are velvety black with a coppery-bronze to purple gloss, and an extensive lower throat and upper breast shield of smooth, oily glossy emerald-green, variably marked on the central throat with small iridescent turquoise-blue feathers. The lower breast shield is broadly tipped iridescent turquoise-green, with the rest of the underparts sepia-brown with a violet to purple sheen. *Iris* dark brown; *bill* blackish, with *gape* narrowly pale yellow (often concealed) and *mouth* bright yellow to green; *legs* a violaceous cobalt-blue to blue-grey. *Adult Female* 16cm, 52–60g. Similar in size to the male, but with a longer tail lacking central tail wires. The crown and nape have less extensive and darker, more lilac-blue bare skin, again divided into six areas by narrow black-feathered lines. The upperparts are dark olive-brown to reddish-brown, rustier on uppertail-coverts, with the tail warm brown and upperwing dull brown. All remiges have narrow rusty margins, with the primaries duller. The underparts are pale buff, finely and uniformly barred brown-black, with pale buff chin finely flecked greyish. Bare parts as for male, but *legs* duller blue to blue-grey than those of adult male. *Juvenile* undescribed. *Immature Male* like adult female, except that rectrices have pointed tips. *Subadult Male* variable, some being like adult female but with a few feathers of adult male plumage intruding, initially about head and breast, to like adult male but with few feathers of female-like plumage remaining. Tail of male becomes progressively shorter as it moults while simultaneously gaining progressively longer central pair.

TAXONOMY AND GEOGRAPHICAL VARIATION Monotypic, with no distinctions between the populations of the two islands where it occurs.

VOCALISATIONS AND OTHER SOUNDS The advertising song is a quite loud and incisive series of 8–13 *tchow tchow tchow* notes, closely spaced and repeated 10–12 times with about 30 seconds to two minutes between songs. It is akin to the song of the Magnificent Bird of Paradise but higher-pitched, richer and more incisive, with less of an interval between the notes. A good example is on xeno-canto at XC140737 (by B. van Balen). The birds start to call early, from 06:10, from perches up to 9m tall near the court (Bergman in Gyldenstope 1955), and can be heard through the morning with gaps for presumed foraging. Harsh, churring scolding notes are also made. A call series may also be given as sharper, more metallic clicking notes followed by three more distinct and even sharper click notes (F&B). Displaying males produce a whisper song of complex squeaky twitterings, mixed with guttural bowerbird-like notes which bear some resemblance to notes made by displaying Red Birds of Paradise, which are partially sympatric here (Frith in F&B). Males in flight about the court also produce a rapidly repeated loud, sharp, dry tick note, reminiscent of some tropical manakins (Pipridae), and also a dry rattling, perhaps an equivalent to the rattling sound made by the Magnificent Bird of Paradise. A recording by A. Spencer at XC62531 shows this nicely. The whirring noise of the flapping wings is also noticeable at times at the court, often accompanied by a harsh *tchaw* call, as at xeno-canto XC23182 (by M. de Boer). Captive individuals have been heard to emit a sharp, explosive *keeetch* call and a pleasant, soft high-pitched whistled *teel*, the latter repeated five or six times (F&B). A stronger and louder call is a *too-too-too-too-too-too-wit* series, the last note raised in pitch to make a whip-crack type of ending. A louder and more penetrating form of the latter call ended with a *zeet* note, with the head outstretched on each note, possibly some form of advertising call (F&B). The vocabulary of the species is not yet well known apart from singing males, with much remaining to be discovered. Female-plumaged birds have apparently not yet been noted as calling, with no examples on xeno-canto (2016). It seems odd that an equivalent of the *ching* contact call of Magnificent Bird has not yet been reported, although the sharp explosive single *keetch* may be it.

HABITAT Inhabits hill forest and mid-montane forest from about 300m up to the summits of the hills at 1,000–1,200m, often occurring on steep ridges. Occasionally reported from around 60m in lowland rainforest of the interior.

HABITS Wilson's Bird of Paradise is quite vocal during the day, much like its congeners, its voice being a characteristic sound on mid-altitude ridges on Batanta and Waigeo. It occupies the lower and middle strata of the forest and will join mixed-species foraging flocks. In general habits it is likely to resemble the Magnificent Bird.

FOOD AND FORAGING Field observations are very limited, but this species has been seen to take small orange fruits from the lower crown of a tree (Bishop in F&B). Pheasant Pigeons (*Otidiphaps nobilis*) and

Cinnamon Ground-doves (*Gallicolumba rufigula*) have been seen to come to display grounds to take seeds voided by this species (F&B; Dutson *in litt*).

BREEDING BEHAVIOUR Promiscuous males occupy terrestrial courts and the species is polygynous, much as with its two congeners. **COURTSHIP/DISPLAY** Courtship has seldom been seen in the wild, although captive studies have been made. Bergman (in Gyldenstope 1955) observed the display in the field, and noted that the display courts are about a half to one metre in diameter, situated in dense forest and preferably in a small clearing around or near a fallen tree (note a similar preference by the Magnificent Bird). The clearing is surrounded by dense tangled vegetation which almost completely conceals it except at very close quarters. The display ground is reminiscent of a stage illuminated at certain times by sunrays. The court is carefully cleaned by its owner, who removes anything that may have dropped in since his last visit, exposing the bare soil. Small bushes and some slender low trees are always left intact, although they are stripped of their foliage. The following details have been noted (pers. obs.; Gilliard in Greenway 1966): the species uses roughly circular or oval display courts between 1.8m and 3m in diameter and open to the sky, and these may be on gently sloping ground or on slopes of up to 45°, and are kept bare by the male, who also strips saplings and vines within the court of their foliage; the court limits may be marked by fallen trees or dense foliage. Males tend to be wary but not unduly shy at the courts. Stripped saplings at the centre of the court seem to be the preferred vantage points of males, which perch some 15–20cm from the ground. Advertising songs may be given from the court or nearby. **Displays** are poorly known and the following details are primarily from studies of captives; it is likely that elements of them are missing or still unknown. The apparent absence of a *Pulsing Display* and a *Dancing Display* like those of the Magnificent Bird may perhaps be due to lack of observations to date. The Macaulay Library of the Cornell Lab of Ornithology has a large collection of videos of this species by Laman and Scholes, some of which show elements of the display; for example, E. Scholes has ML459872 showing the breast shield erect and ML459852 documents the wings snaps and clicks made by the male. The display filmed by the BBC for Attenborough (1996) shows an adult male responding to a female-plumaged bird by freezing on the base of a vertical sapling; the female perched above him and he points his head parallel with the trunk towards her, sleeking the body plumage and with the tail held normally; he sharply flicks his head and neck a short distance each side while making a soft whisper song of ticking and buzzing sounds; as the female approaches, he suddenly pulls back the head and neck into the body, while continuing to look up the perch and to flick out and expand his flattened breast shield and cock the tail up to 90° to show off the sickle wires to maximum effect. A film by T. Schultze-Westrum described in Frith & Beehler (1998) had a male and an immature male stretching the head, neck and bill towards a female with the tail cocked but the breast shield modified (see illustration in F&B). They were less than 1m apart, and pointed at the female wherever she moved, moving like two compass needles remaining stiffly parallel to each other. If she returns to the perch above him, the male reverts to the head-and-neck posture, similar to initial courtship of the Magnificent Bird (F&B). Captives have been seen to perform a ***Back Display*** similar to the one reported for the Magnificent Bird of Paradise (F&B). A display by a captive male recorded in Iredale (1950) has elements of both the above displays plus a few extra details. The bird jerked his head from side to side while emitting a low, faint whistling noise, and raised the neck feathers on one occasion. The notes became louder, the jerking of the head ceased and the bill was opened and closed several times. The feathers of the head were moved to assume their normal glossy intense black, which contrasts vividly with the light green of the mouth-lining. Suddenly the bird retracted its head and neck, elevating the breast and expanding the shield which was now seen to be bright green; in the bay at the upper margin of the shield, which is roughly bean-shaped, was the head, the blue crown just showing and two green spots appearing behind and above the base of the bill. The bird remained motionless and silent for a brief time, then thrust forwards again into almost the first position, which it resumed forthwith. Twice it was seen to open its bill when in this position. **NEST, EGGS & NESTLING** No information. Nothing is known about the nest, clutch, or incubation and fledging periods. **TIMING OF BREEDING** Birds with enlarged gonads have been collected in May–June and October (F&B). Advertising song has been recorded in most months.

HYBRIDS None. Glossy-mantled Manucode and the Red Bird of Paradise are the only other paradisaeids on the islands, but both are substantially bigger than Wilson's Bird.

MOULT The moult cycle is little known, but specimens show evidence of moult in May–June.

STATUS AND CONSERVATION Classified as Near Threatened and potentially Vulnerable by BirdLife International owing to its restricted range and the threats therein. Wilson's Bird of Paradise was fairly common in the hill forests of Batanta until the mid-1990s, and apparently much less common on less well-known Waigeo. It is a vulnerable species as a result of its restricted range, and disturbing recent reports have come in about extensive logging on Batanta (Tindige *in litt.*), where a lot of potential habitat has already been lost through clearance in the lowlands. This species is reported to be still fairly common in selectively logged forest and may, it is hoped, be an adaptable species like the Magnificent Bird of Paradise. The extent of habitat for the species on Batanta is estimated at about 100km², and it is presumed that it will be safe at higher altitudes. The situation on Waigeo is less clear; a nominal reserve which was established seems to be under some threat and may in any case have no practical reality, judging by the parlous state of some other Indonesian national parks and protected areas. Some logging is under way in the north and a cobalt-mining concession is another potential threat, although some ecotourism has now commenced and the income generated by these activities may help landowners to protect and maintain their forests.

Wilson's Bird of Paradise, male, Waigeo (*Gareth Knass*).

Wilson's Bird of Paradise, male and two females, Batanta Island, Papua, Indonesia (*Tim Laman*).

Genus *Semioptera*

This monotypic genus was erected by G. R. Gray in 1859 to accommodate the unusual new species which he had described. Recent genetic dating reveals this species as having arisen around 11.5 mya, during the later Miocene period, considerably after the evolution of the sympatric paradise-crow (*Lycocorax*), which is a basal member of the paradisaeids. Phylogenetic analyses (Irestedt *et al.* 2009) indicate that it is a member of a clade which includes also *Ptiloris*, *Lophorina* and *Drepanornis*. Earlier studies suggested a link with *Cicinnurus* as there are some similarities with the nest and egg of that genus. The sexual dimorphism is quite marked with this species, but both sexes are much less colourful than other dimorphic species. Males have a unique parachute-type display flight.

Etymology The name *Semioptera* is derived from the Greek and means standard wing or flag wing.

WALLACE'S STANDARDWING
Semioptera wallacii   Plate 18

Paradisaea (Semioptera) wallacii G. R. Gray, 1859, *Proceedings of the Zoological Society of London*, p.130. Near Labuha Village, Batchian (Bacan) Island.
Other English names Wallace's Standardwing Bird of Paradise, Wallace's Bird of Paradise, Standardwing Bird of Paradise, Standard Wing, Standard-winged Bird of Paradise
Etymology The specific epithet *wallacii* commemorates the great naturalist, anthropologist and explorer Alfred Russell Wallace, who discovered the species on Bacan in 1858. We conserve his name in the vernacular to perpetuate the cultural and historical link.

FIELD IDENTIFICATION A unique isolated species, one of the three most westerly of the family and the only plumed paradisaeid in its range. Readily identified by the long, sharply keeled (a character shared with King and Wilson's Birds of Paradise) and slightly decurved pale bill, which gives a curious rather friarbird-like head shape, plus the unique white standards on the wings and the iridescent green breast shield of the male. Both sexes have distinctive rather pale, washed-out plumage, with broad wings. Wallace wrote a marvellous description of the male: 'the general colour of this bird is a delicate olive-brown, deepening to a… bronzy olive in the middle of the back, and changing to a delicate ashy violet with a metallic gloss, on the crown of the head. The feathers, which cover the nostrils and extend halfway down the bill, are loose and curved upwards. Beneath, it is much more beautiful. The scale-like feathers of the breast are margined with rich metallic blue-green, which colour entirely covers the throat and sides of the neck, as well as the long pointed plumes which spring from the sides of the breast, and extend nearly as far as the end of the wings. The most curious feature of the bird, however, and one altogether unique in the whole family, is found in the pair of long narrow delicate white feathers which spring from each wing close to the bend. On lifting the wing-coverts they are seen to arise from two tubular horny sheaths, which diverge from near the point of junction of the carpal bones. They are erectile, and when the bird is excited are spread out at right angles to the wing and slightly divergent.' Individuals in the field are an odd pale biscuit colour, with their bright orange legs very noticeable. Bishop

(1984) also notes the females as having 'a beautiful metallic violet patch on their wings'. Wallace's original description of the male notes that it 'had a mass of splendid green feathers on its breast, elongated into two glittering green tufts; but what I could not understand was a pair of long white feathers that stuck straight out from each shoulder'. Gooddie (2010) has an unusually entertaining perception of this species, likening the extended metallic green breast shield to Salvador Dali's moustache drenched in absinthe, while the snow-white elongated shoulder plumes are spindly at the base but spatulate at the tip, splaying outwards in four directions and held aloft but slightly drooping, like helicopter blades at rest. He likens the bird to 'the ridiculously flamboyant lovechild of ET and Liberace!'. **SIMILAR SPECIES** Basically none, although the Dusky-brown Oriole (*Oriolus phaeochromus*) may cause momentary confusion, but that is a much smaller and darker bird with a pink eye patch and dark legs.

RANGE The Indonesian islands of Halmahera, Kasiruta and Bacan, in the North Moluccas (Maluku). **MOVEMENTS** Nothing significant recorded; a sedentary species of the hill forests.

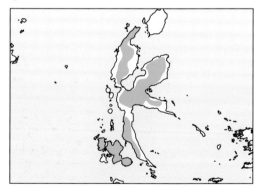

DESCRIPTION Sexually dimorphic. See Frith & Beehler (1998) for a thorough series of comparative measurements. *Adult Male* 26cm; weight 152–174g. The buff feathers forming a tuft covering the nostrils and extending halfway down the bill are loose and curved upwards, grading into the rather flattened crown of dull iridescent lavender-grey, paler on the sides of the face. The upperparts are a pale biscuit-tinged greyish-brown, with the upperwing-coverts greatly elongated to form the remarkable narrow white standards erected in the aerial display. Chin and throat dark buffy. The breast feathers are a highly iridescent bronzed yellow-green that can appear greenish-blue, entirely covering the throat and sides of the neck to form an extensive iridescent emerald-green breast shield; this shield merges into the feathering of the throat and olive-brown of the lower breast and belly, which have a yellow-green sheen. *Iris* deep olive to dark brown; *bill* horn-olive; mouth colour not noted; *legs and feet* bright orange to yellow-orange. **Adult Female** 23cm; weight 126–143g. Female is remarkably plain, being entirely dull pale earthy brown, similar to male but smaller and lacking the breast shield and wing-covert standards, with smaller buff tuft above nostrils, and the crown with only a slight ashy-violet tinge. The central tail feathers are also longer than those of adult male. *Iris* deep olive to dark brown; *bill* horn-olive; the quite colourful bright orange to yellow-orange *legs and feet* are one of the most striking features. **Juvenile** unknown. **Immature Male** resembles female but has slightly longer tail; gradually moults into subadult plumage and acquires shorter tail. **Subadult Male** has mixture of male and female plumages, the green breast feathering gradually appearing. It is likely that adult plumage is acquired over 3–5 or more years (accounting for the preponderance of female-plumaged birds first noted by Wallace).

TAXONOMY AND GEOGRAPHICAL VARIATION
Polytypic, with two subspecies.
1. *S. w. wallacii* Gray, 1859, Bacan Island, and probably this taxon also on Kasiruta.
2. *S. w. halmaherae* Salvadori, 1881, Halmahera Island. Wallace (1869) wrote that these have the green breast shield rather longer, the crown darker violet, and the lower parts of the body rather more strongly scaled with green. More recent descriptions stress the rich pinkish coppery-purple iridescence on crown and nape (sometimes also mantle), and a slightly longer tail; other measurements slightly smaller bar mean wing length, which is the same as the nominate.

VOCALISATIONS AND OTHER SOUNDS
Wallace wrote that 'it continually utters a harsh, creaking note, somewhat intermediate between that of *Paradisaea apoda*, and the more musical cry of *Cicinnurus regius*', and indeed the voice has similarities to both *Cicinnurus* and *Paradisaea* species. The first pre-dawn advertisement calls are single, loud, nasal upslurred *bark* notes resembling the calls given away from the lek in daylight. These calls gradually increase in intensity, volume and complexity, the main ones including a series of six or seven of these barking notes that are increasingly loud and distinctly pulsed, each note forming a complex sound (F&B). Other males often answer such calls, and males can give an individually distinct greeting which may be age-related or status-related (Bishop 1992). The strident *bark* or *wark* (F&B) notes are given also in far less intense form as a contact call by foraging individuals, and can be heard over distances of up to about 300m (Bishop 1992). A faster series followed by three distinctive friarbird-like *waa-kuck* notes, descending as the call fades out, has also been recorded (F&B), and Bishop (1984) noted a *cheung cheung cheung* given during display. In the *Aerial Display* the intensity of the calling of the adult males

rises to almost cacophonous levels as they leap up off their perches; the barking calls are given, while other males give a more musical chatter lasting 5–10 seconds and often indicating that a female has arrived at the lek (F&B). Coates (in Frith 1992) noted males hopping about their lek while uttering repetitive, variable, harsh, loud churring *waughh* notes, and also a clearer *wau-wau-wau* resembling a call of the Twelve-wired Bird and also of *Paradisaea* species. Bishop (1992) noted males landing on their display perch after the Aerial Display to utter a series of high-pitched sweet, rapidly repeated twittering notes quite distinct from their other calls, given when the males were displaying with shield and standards extended. This would also be given in a less intense fashion prior to the Aerial Display. Goodfellow (1927) mentioned a similar series that reminded him of a Common Starling (*Sturnus vulgaris*), bubbling, gurgling and producing occasional explosive sounds. Daily vocal activity of Wallace's Standardwing peaks at around 08:00 and the species is quietest around midday. There are some lengthy recordings on xeno-canto which were made at a lek: see XC214261 and XC214260 (by P. Åberg). The alarm call is recorded by F. Lambert at XC193813, and some harsh calls at XC18813 by M. Catsis. Other sounds made by this species include clicking of mandibles in unison at the lek, while individuals in flight produce a distinctive wing rustling which is not so loud as that made by Magnificent Riflebirds or the Twelve-wired Bird (F&B). Standardwings are reported as cracking their wings in display, as well as flapping and rustling them (Bishop 1992), although Friedmann (1934) noted the cracking to be of vocal origin.

HABITAT This is a hill-forest species, patchily distributed on the steep and hilly terrain, especially where limestone is found, and apparently absent from the lowland forests. It occurs from about 200m up to 1,150m on Bacan (Lambert 1994). On Halmahera it is found from 350m (occasionally 250m) to at least 900m and probably up to 1,300m, being most common in the elevational range 485–750m (Bishop 1994). It inhabits forest dominated by *Canarium*, *Eugenia*, *Vitex*, *Diospyros* and *Agathis*, with a diverse understorey dominated by palms with *Calamus* (rattan) thickets (F&B). Rarely found in secondary woodland (Coates & Bishop 1997).

HABITS This species frequents the lower forest canopy and the shaded mid-storey, ranging from 2m to 12m, where it is inconspicuous, shy and hard to observe away from leks. Standardwings usually forage in dense forest canopy (Coates & Bishop 1997). Goodfellow (1927) found them curious and bold, coming down to about 5m to investigate. They tend to be found singly or in twos, but groups of three or four may be found and, like other paradisaeids, they join mixed-species flocks at times. Wallace (1869) wrote: 'this bird frequents the lower trees of the forests, and, like most Paradise Birds, is in constant motion – flying from branch to branch, clinging to the twigs and even to the smooth and vertical trunks almost as easily as a woodpecker'. He also noted that, as with most paradisaeids, female-plumaged birds far outnumbered males. The males at short intervals open and flutter their wings, erect the long shoulder feathers, and spread out the elegant green breast shield (Wallace 1869). Bishop (1992) believed that the males roosted overnight at the leks, many still being present after sunset. Earlier (Bishop 1984) he recorded them as being less shy in undisturbed forest, moving rapidly from perch to perch in short low-level glides, and also noted their woodpecker-like habit of clinging to the side of large trees, followed by hopping upwards while using their feet to grasp vines. Large leks of up to 100 individuals have been reported (Anon in F&B), quite unlike any other paradisaeid in terms of numbers. The species performs a unique *Aerial Display* at lek, and the birds are very vocal at such places from well before dawn.

FOOD AND FORAGING Primarily frugivorous, but occasionally taking some insect food. Goodfellow (1927) noting his captive birds as having a fondness for green-coloured insects, with 'a pronounced partiality for the large soft bodied grasshoppers (? phasmids), nearly five or six inches [*c.*125–150mm] long, which live on the branches of the coconut palms'. They will forage on trunks in a manner similar to that of sicklebills in New Guinea (Bishop 1992; Wallace 1869), presumably looking for arthropods, although no one has yet published any observations of these actually being taken in the wild. They will join mixed-species foraging flocks as paradisaeids in New Guinea do (Poulsen in F&B).

BREEDING BEHAVIOUR Wallace's Standardwing is polygynous, the promiscuous males displaying at leks and nest duties likely to be done by the females alone, as with other paradisaeids that lek. Display is believed to take place from April to December i.e. from early in the dry season, when birds are in fresh plumage, and finishing with the onset of the rains. **DISPLAY** Goodfellow (1927) gave a vivid account of his visit to a lek on Halmahera, visiting a stunted kind of jungle with a few enormous trees here and there, but most being saplings 30–40 feet [*c.*9–12m] tall, straight and with few branches; very little undergrowth was present, and in some places none at all. There were at least 30 standardwings here, with just a couple in female plumage (unlike Wallace's experience, where there were few males, this perhaps related to time of year.) The birds were all constantly on the move, flying backwards and forwards from tree to tree with a great fluttering of their wings, and at times hanging in all sorts of positions from the slender branches, some turning around and around like a cartwheel, and all the time making a variety of squawks and calls. Some birds had the green breast shield elevated perpendicularly, so the head was visible only through the deep V-base of green feathers. The long shoulder plumes stand out at right angles below the shield, and are constantly raised and depressed. Goodfellow also noted captive individuals with their silvery-purple short scale-like feathers on the top of the head seeming to undulate with a curious effect, visible only from certain positions. He also noted that at midday they all disappeared, a few probably remaining in the dense top of one of the huge trees nearby, coming out again for a short time later in the afternoon. The birds avoided display during full sun but would continue if overcast. The Macaulay Library at Cornell Lab of

Ornithology has good series of videos of this species by E. Scholes. On 25th August Bishop (1984) watched two males displaying to about 12 female-plumaged birds, the males leaning forwards and constantly flicking their partially extended wings, with the iridescent green breast shield fully extended and catching the light, similar to the displaying Superb Lophorina. The two wing standards moved independently of each other like antennae, being one moment vertical, and then horizontal and at right angles to the body. The males called excitedly throughout with growling calls and the *cheung* series noted above; they never stayed at a perch for more than 30 seconds, but moved through a series of what appeared to be an established sequence of perches, closely attended by the female-plumaged birds, which often passed them and jumped around them on nearby branches. Bishop (1992) recorded display perches as being on thin, near-vertical twigs or branches in the lower crown of tall canopy trees some 25–30m above ground. Lek activity commenced about 05:35, about 20 minutes before sunrise, with intermittent, tentative calls and birds rustling, flapping and cracking their wings as well as bill-clicking. The calling became more complex and intense, and reached a crescendo at about 06:15, intense display and calling continuing until about 07:30; activity continued less intensely until after 08:00, to about 08:30, and sometimes to 10:30 if the weather was overcast. In June 1990, Coates (in F&B) saw a single male display low in the understorey just a metre off the ground. Standardwings pluck or tear foliage from lek perches, as do other paradisaeids. A lek at Tanah Batu Putih, on Halmahera, was estimated to hold up to 100 birds, although Poulsen (in F&B) saw no more than 30–40 at the lek itself at any one time. Bishop (1992) was also the first to record a communal *Aerial Display* unique to this species, although the individual elements do correspond to the major elements in the general displays of other male paradisaeids (F&B). In the initial **Convergence Display** the males charge about while beating their partly opened wings and giving rapid repeated sharp call notes; the breast shield is not used, and the standards are only slightly raised as the birds flutter from perch to perch. This may lead into other forms of display, including the spectacular *Aerial Display*, in which males perform one after the other and an individual may perform at least twice in succession. It begins with a male perching upright with shut wings, the breast shield and wing standards relaxed. He leans forward to call, stretching the head and neck up and opening the breast shield slightly. Duetting and countersinging against other males increase the excitement until he leaps vertically up from the perch by thrusting his legs down and beating the wings; he rises to about 7m, stops flapping and holds the wings open with bill pointing down, then 'floats' downwards while rapidly vibrating the white primaries so that they show as a blur at each wingtip. The descent lasts for several seconds and the tail may be fanned as a brake. The wing standards seem to be held vertically upright and are blown back. Once this display is over, the male adopts the *Wing Standard Display*, which may lead to copulation. The male usually displays his wing standards from a steep perch, with body held rigidly upright and bill slightly raised as he changes the position of his vibrating wings from partly open to half or fully open, the carpal joints moving from wide apart to close together. This alternation causes the raised white wing standards to gyrate in striking fashion. A crack sound may be made vocally (Friedmann 1934, from captive birds), and the vibrating pale primary tips become conspicuous, while the mandibles may click together. The crown feathers are raised and lowered in the perch display (see Goodfellow, above), with the iridescent lilac-purple colour prominent, and the tips of the upper breast shield meet behind the neck. Bishop noted copulation at this lek on 10th May, a female-plumaged bird approaching seven displaying males as they performed the *Wing Standard Display*, landing on the most frequently used perch. The occupying male immediately increased his calling and bounced from perch to perch with horizontally extended quivering wings, snapping his breast shield open and holding his wing standards alternately horizontally and then vertically. The other males performed frenzied calling and displays at the same time, and the female leant forward on a branch attached to the display perch as the male mounted her to copulate. During the mating, the wing standards were held vertical and the wings were extended out inclined backwards, and the breast shield was held erect. The other males performed the *Aerial Display* as the two birds mated. Bishop also noted two males grappling with one another and tumbling to the forest floor, where they still fought until they were caught and disentangled. **NEST/ NESTING** Almost nothing is known. A single nest, found in May, contained one egg (Bagali in F&B). The nest, an open cup with dried leaves at the base, was located some 10m above ground between epiphytic orchid leaves and a palm tree where the frond bases emerge. **EGGS, INCUBATION AND NESTLING** A single oval egg was in the sole described nest. Nothing is known about incubation and fledging periods.

HYBRIDS None; no other paradisaeids except the much larger and very different Halmahera Paradise-crow occur in this species' range.

MOULT Poorly known. Likely to be mostly in the wet season (non-breeding), December–March, most specimens showing evidence of moult in November–June.

STATUS AND CONSERVATION Classified as Near Threatened by BirdLife International. The conservation status of Wallace's Standardwing has become the focus of some concern owing to extensive logging and settlement within its range. Political and religious ferment in the Moluccas prevented surveys and observations there from 1999 to 2006. Lambert (1994) noted this species as being moderately common on Bacan in both primary and logged forest habitat from 70m to 1150m, and scarce in flat areas of logged forest. It was seen on Kasiruta (Lambert 1994) but nothing is known of the conservation or taxonomic status of this population. Bishop (1992) noted it as being absent from many lowland areas on the west coast of Halmahera and on the east coast of the eastern peninsula, but as this seems to be a hill-forest species this is perhaps

not surprising. It has been described as common but local on Halmahera (Coates & Bishop 1997). The total forest area on Halmahera and Bacan is estimated as being 19,000km^2, with less than 10% on limestone, which Bishop considered to be a key factor. Surveys in 1994 found 0.2–0.3 Wallace's Standardwings per hectare on limestone and non-calcareous sedimentary-rock soil formations, while few were seen on igneous rock or in montane areas (Mackinnon 1995); this contrasts with a survey by BirdLife International workers in 1994–95, which found 0.4–0.8 birds/ha, with higher numbers on the volcanic substrates (Frith & Poulsen 1999). Bishop considers the species to be at serious risk as a result of logging and clearance, which were exacerbated by social upheavals around the turn of the century. This spectacular and unique species must be gravely threatened over the medium to long term without substantial effective reserves being established. BirdLife International identifies the primary threat to the species as being habitat loss through commercial logging for timber, and clearance for shifting agriculture, mining, settlements, and plantations of oil palm (*Elaeis guineensis*), coffee, rubber and timber species (Vetter 2009; Burung Indonesia *in litt*. 2014). Another potential threat is posed by wildfires, which have devastated areas on other Indonesian islands, the chances of such fires being increased by the conversion of forest to scrub and grassland and the opening-up of forests for road construction, as well as selective logging and fragmentation (Vetter 2009). The species occurs in Aketajawe Lolobata NP (Halmahera) (Burung Indonesia *in litt*. 2014). It is vital that surveys are carried out to assess the current population size of this species and that regular surveys be conducted in order to monitor the population trend. Rates of habitat loss should be tracked through regular studies of satellite images. The area of suitable habitat with protected status should be increased.

Wallace's Standardwing, male, North Maluku (*James Eaton*).

Wallace's Standardwing, male with standards erect, North Maluku (*James Eaton*).

Genus *Seleucidis*

Seleucidis is a curious monotypic genus erected by Lesson in 1835. The Twelve-wired Bird of Paradise exhibits characters shown by a number of other genera, so diverse in fact that, if known from just a few specimens, it would probably be considered a hybrid. It has flank plumes and breast cushion like a *Paradisaea*, a long decurved bill and a noisy plumage-swishing flight like *Ptiloris*, pectoral fan plumes like *Epimachus* or *Cicinnurus*, a black-headed female like *Parotia* or *Lophorina*, and a red eye like *Epimachus fastosus*. It also has bright yellow flank plumes, and some vocalisations quite similar to those of Raggiana Bird of Paradise, but the twelve sharply recurved wire-like tail feathers are unique. Irestedt *et al.* (2009) placed this monotypic genus as sister to a clade that includes *Ptiloris*, *Lophorina*, *Semioptera* and *Drepanornis*.

Etymology *Seleucidis* is from the Latin *seleucides*, migratory birds that were believed to be sent by the gods from paradise to destroy locusts (Greek *seleukis*, a locust-eating bird).

TWELVE-WIRED BIRD OF PARADISE
Seleucidis melanoleucus   Plate 19

Seleucidis Lesson, 1835. *Hist. Nat. Ois. Parad., Synopsis:* 28, pl. 35. Type by monotypy *Seleucidis acanthilis* Lesson = *Paradisea melanoleuca* Daudin, 1800. *Traité d'Ornithologie* 2, p.278. Type locality Salawati or Vogelkop. Synonym *Paradisaea ignota* Forster, 1781.

Other English names Twelve-wired Paradise Bird

Etymology *Melanoleucus* is from the Greek words *melos*, black, and *leukos*, white. The bird has no white in the plumage, but this seeming contradiction is explained by the fact that the vivid yellow pigmentation is lost after death, and the first specimens showed white flank plumes. A Kiunga specimen hanging from a car-driver's window mirror retained the slowly fading yellow for about a year, but had totally bleached to white a few months later.

FIELD IDENTIFICATION In a reasonable view the male Twelve-wired Bird of Paradise is unmistakable, with its long decurved bill, bright yellow flank plumes, black underparts, and bizarre bare wires emanating from the flanks and bending sharply forwards along the body. It bears a passing resemblance to the Magnificent and Growling Riflebirds, but the colour scheme is quite different. The wires may break off, and it is common to see males with ten or 11 wires rather than 12. The female resembles the female of these two riflebirds, but has a blackish head and a red eye. The shape also is very odd, with very broad rounded 'butterfly' wings, a short tail and a long, decurved bill. **SIMILAR SPECIES** Greater, Lesser and Raggiana Birds may be sympatric, but all have a much shorter bluish bill and lack the flank wires. Female-plumaged riflebirds lack the black head, red eye and reddish legs of female-plumaged Twelve-wired.

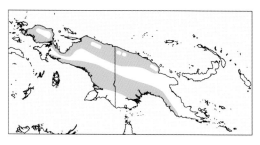

RANGE This is a species endemic to the coastal lowland plains, lowland forests and great river systems of New Guinea. It ranges from Salawati Island eastwards along the southern Vogelkop and throughout the lowland forests of New Guinea, in the south as far as the Port Moresby area and in the north to the Ramu River valley. It is absent east of these areas, where the coastal plain is lacking or narrow. Occurrence in the Onin Peninsula and southern Trans-Fly is uncertain as yet, but the species is known from Lake Murray. **MOVEMENTS** Sedentary.

DESCRIPTION Sexually dimorphic. *Adult Male* 33cm; weight 170–217g. Entire head is velvety black with an iridescent coppery olive-green sheen, the crown showing purple iridescence. The head has a bumpy profile with erectile tufts above the ear-coverts, and a long, rather riflebird-like slightly decurved bill; unusually, there is a black post-ocular stripe of bare skin. The upperparts are velvet-black with an iridescent oily sheen of coppery olive-green; flight-feathers and uppertail iridescent violet-purple and/or magenta, primaries black. The chin, throat and entire breast are velvety black with a slight iridescent sheen, the breast forming an extensive elongate and dense 'cushion' which can be flared out into a fan shape in display; the lower breast has a gorget of broadly tipped iridescent emerald-green feathers, with violet-purple adjacent bases. The rest of the underparts, including the greatly elongated, inwardly curving filamentous flank feathers, are brilliant yellow; six flank plumes on each side have a white central shaft that is extraordinarily and uniquely elongated into sharply recurved black 'wires' which become white again after they bend upwards. *Iris* bright blood-red; *bill* shiny black, the *mouth-lining* aqua-green (very prominent in display); *legs and feet* (including bare thighs) pink. *Adult Female* 35cm; weight 160–188g. Smaller than the male, but with the tail much longer and the plumage very different. Has an unusual bare black post-ocular stripe, hard to see in the field. Head and upper mantle are sooty black with a dull iridescent purple sheen, lower mantle and rest of upperparts rich chestnut-brown; underparts greyish-white, malar area and throat flecked/spotted dark greyish-white, upper breast and rest of underparts uniformly barred blackish. *Iris* red; bill black; *legs and feet* bright pink. *Juvenile* undescribed. *Immature Male* like adult female (tail longer than adult male), but some individuals are paler, washed with sandy orange-yellow, and sometimes with a brown base to lower mandible, but the time taken to reach adult plumage

remains poorly known, likely to be anywhere between five and eight years. The iris is pale brown, turning yellow with age. *Subadult Male* variable depending on state of moult and age: younger birds resemble a blotchy adult female, with a few feathers of adult male plumage intruding, while older ones resemble the adult male but with a few remaining feathers of female-like plumage; the iris changes from yellow to almost red with age, and the male progressively acquires the shorter tail.

TAXONOMY AND GEOGRAPHICAL VARIATION
Monotypic. Population from the Mamberamo River, West Papua, eastwards to the Ramu River, PNG, sometimes separated as subspecies *S. m. auripennis* Schlüter, 1911, smaller with shorter bill than birds in rest of range, and underparts of female darker and more heavily barred, but differences minor and *auripennis* seems best synonymised.

VOCALISATIONS AND OTHER SOUNDS The advertising call of the male is a far-carrying, nasal, resonant interrogative *Haw?*, dropping slightly in pitch at the end. It is reminiscent of a Raggiana Bird of Paradise call, but more reedy and nasal. This is often given singly, particularly when the male has just arrived at the display area in the early morning, but may also be repeated in series up to eight times. It is audible over several hundred metres, and males will respond vocally to imitations of the call or to other males calling. The single *Haw?* note may be followed by a soft, descending whistled *twi* note in some areas (F&B). When other males are present a five-note call may be given, *houw-wah-wah-wah-wah*, the *wah* notes being slightly shorter and upslurred (Coates 1990 from Port Moresby area), while Frith (in F&B) noted a similar *Paradisaea*-like series from a male on the Karawari River, PNG, followed at times by *Wauk wauk wauk*. Males call early in the morning, not long after first light, with greatest frequency from about 06:30 to not long after 07:00. Occasional calls may be given during the day, and there is a brief resurgence in the late afternoon on some days, but the peak calling period is quite limited. Good examples of this species' calls can be found at xeno-canto XC18686 (by M. Catsis) and XC20960 (by N. Athanas), and also at the Macaulay Library at Cornell Lab of Ornithology site ML20960 (by T. Laman). In flight the males, and some female-plumaged birds (Beehler in F&B), perhaps immature males, produce a swishing *wish wish* sound with the wingbeats, which Finch (1983) likened to the song of the Black Sunbird (*Leptocoma aspasia*)! The sound is akin to the rustling produced by riflebirds, but higher-pitched.

HABITAT Widespread in swamp, monsoon, riparian and lowland forest, extending to uplifted karst hill forest in the West Sepik. This paradisaeid likes seasonally flooded forest such as that along the Fly and Sepik Rivers, and around Vanapa and Brown Rivers near Port Moresby. Much of its habitat may be very difficult of access, being swampy and low-lying, with sago swamp and pandanus thickets. It occurs from sea level up to about 180m in the karst hill forest of the West Sepik, PNG (Beehler in F&B).

HABITS Once the distinctive call is known, the species may be found to be widespread but elusive in its swamp- and monsoon-forest habitat. The males are vocal early in the morning from about 40 minutes after sunrise up to about 07:00, but frequency of calling diminishes rapidly as the day heats up, with a resurgence late in the afternoon. The birds are quite wary, but males may show nicely when calling atop display posts on dead trunks and palms. They are often seen flying across rivers and wet areas, their flight being distinctive and rather shallowly undulating on broad 'butterfly' wings, dipping after every three or four wingbeats. They forage quietly in the lower and middle strata, sometimes in mixed-species flocks. Males at display posts are often in auditory contact with those at other posts, and the calling of one bird will set off the next. The birds are reported as using traditional nocturnal roosts some 3–4m above the ground in low bushy trees, this being one of the few paradisaeid species for which roosting information is available (Wallace 1869; Bergman 1968).

FOOD AND FORAGING The diet of the Twelve-wired Bird of Paradise appears to consist of fruit and arthropods as with many of the family, Beehler & Pruett-Jones (1983) estimating the proportions to be about 50:50, though doubtless varying according to local circumstance. Pandanus fruits are a favourite, and the birds may be seen probing for arthropods in dead leaf clusters and on palms and sago palms. Wallace (1869) saw them frequenting flowering trees such as sago palms and pandans, sucking the flowers and (as later discovered) having stomachs full of brown sweet liquid which was probably nectar, a curious habit not reported for other members of the family, although Victoria's Riflebird is known to pierce some blossoms, presumably to reach nectar, and female-plumaged birds are known to take sugar water. These birds forage quietly in the lower and middle strata, poking around in palm and pandan fronds. They can also be quite acrobatic, clambering about and hanging upside-down to probe into crevices or bark or to feed at fruit sources. They will join mixed-species foraging flocks along with other paradisaeids, and Beehler & Beehler (1986) reported an adult male in the West Sepik foraging with a male Pale-billed Sicklebill, a female-plumaged Lesser Bird of Paradise and an adult male Magnificent Riflebird. They are sometimes members of mixed flocks with pitohuis, Papuan Babblers (*Garritornis isidorei*) and monarchs.

BREEDING BEHAVIOUR The Twelve-wired Bird has a polygynous mating system with solitary, promiscuous adult males dispersed every few hundred metres in good habitat. Beehler & Pruett-Jones (1983) reported the distance between neighbours at Kakoro and Nomad River, PNG, as from 420m to 940m. Those along the Elevala and Fly Rivers may be adjacent but separated by the river, although still in auditory contact. One display site along the Fly River above Kiunga was in regular use from 1986 to 2001, although the male shifted his site slightly when his display post fell down in 1997 (Bishop in F&B; pers. obs.), and a male was utilising the general area up to 2017 at least. This site is used from April to

September at least. The female alone performs the nest duties, as is usual for the family. **DISPLAY** The male uses a bare upright snag or dead trunk as his display post, usually in clear view above the surrounding vegetation, although often with a backdrop of taller trees. The topmost decurved growing point of the *limbum* or Black Palm (*Borassus flabellifer*) is also sometimes used as a call post, though not for display, a feature shared with the Magnificent Riflebird. From this exposed position, probably the single most exposed true display perch of any member of the entire family, the male calls loudly, from at least April to September in the Kiunga area. Those wishing to see the bird must make an early start, for one has to be on site not long after first light, and calling will cease as the sun begins to gain strength, a relatively brief window of activity which is strangely reminiscent of the behaviour of the Black Sicklebill. Various observers have described the display, and the following composite account of the male's behaviour in the wild is taken from Gilliard (1969), Coates (1990) and F&B, plus my own observations. The Macaulay Library at Cornell Lab of Ornithology site also has videos by T. Laman of some display behaviours. Calling may be done with the wings half-opened above the back, to show off the vivid yellow flank plumes. The bird may circle around the display post, sometimes with wings raised, and broadcast the call through 360°, and he may call horizontally across the display post, or in a head-down posture with the head below the level of the feet (F&B). When a female arrives, the male points his bill at her and erects the breast shield while splaying out his flank plumes and accentuating the yellow (rather than black) of his plumage The iridescent green and purple edging of the breast shield can be quite apparent in the early morning sun. Frith (in F&B) notes a curious and unique feature of the display of this species, where the colourful deep pink bare thighs are shown off as the belly plumage is sleeked: the birds touch bill tips, and may spiral or back down the perch in this position; the female then flies over the male and back up the song post, while the male will then about-face and come back up to meet her in a series of stiff, jerky hops until the two can again touch bills. The breast shield remains erected and may be pulsed, and the bare thighs remain conspicuous. The male is seemingly aiming to peck at the ventral surface of the female before pecking at her nape, and he may rotate around the perch as she keeps facing him (F&B). These actions may be repeated a number of times. The **Wire-wipe Display** is an additional phase, performed with the male atop the snag and the female perched beneath. The male sways slowly from side to side, flank plumes splayed and breast shield erect, moving his head from side to side while watching the female and directing his swaying plumes and wires at her for maximum effect. The flank wires brush across the female's upper body and the flank plumes are optimally presented, the rhythmic swaying pausing briefly at the end of each arc. The female may probe at his flank plumes and move backwards down the perch, the male continuing the *Wire-wipe Display* and backing down after her. Copulation may take place with just a brief pursuit, after much backing up and down the perch, or after the *Wire-wipe Display*, and begins with the female perching atop the songpost. The male approaches and touches her bill tip, his breast shield fully erect as a kind of partial ruff, wings held closed, and flank plumes splayed and exposing the colourful thighs. The male may flick his head up, and down, and attempt to probe at the lower flanks of the female as he rotates around the perch. She fences him off with her bill until she is ready to remain still, and he then probes with his bill at her breast, flanks, wings, and each side of her undertail-coverts. He stretches up over her to bill her nape, lowering his breast shield, and she lowers her head and vibrates her wings a little to solicit the male. The male then hops up and mounts her to copulate for several seconds, flapping his wings to maintain position and pecking at her nape. After mating he may perch below the female or nearby, and she may remain for a minute or so before flying off. The male then resumes his position on his songpost and commences calling again. The displaying male has been observed to attract female-plumaged Growling Riflebirds (Coates 1990), and Beehler (in F&B) observed a male Growling Riflebird display within 50m of the display post of a male Twelve-wired Bird at Kakoro, PNG; the calling of one male seemed to stimulate a vocal response by the other. Not surprisingly, hybrids between these two species are known (see below). Males will defend their songpost from other males (Coates 1990), but they range far from them during the day and seem not to be unduly territorial, although they will respond vocally to playback or imitations of their call during the early-morning peak display time. **NEST/ NESTING** The nest is shallow cup in a bulky structure of pandan bark and vines, on a foundation of sticks and leaves. It is lined with plant fibres and rootlets, the egg-cup lining being of vine or flower stems, and is carefully hidden in the leaf-base stems (Gilliard 1969). One nest was 200mm wide and 90mm deep externally, with internal measurements of 90mm wide and 25mm deep. Pandanus seems to be favoured for nest sites, three of the four known nests having been found there, the other being located in a Black Palm, both species typical of the preferred habitat (Gilliard 1969: Beehler in F&B). Nests were placed at heights of between 3m and 14m. **EGGS, INCUBATION & NESTLING** Clutch size appears to be one (from limited data on wild birds), although a captive has been known to lay two eggs. The female alone is responsible for the nest duties and is shy and secretive, coming in among deep shadows about a metre above the forest floor before flying steeply up to the nest (Gilliard 1969). A captive nestling left the nest after 21 days, but there are no records of nestling period in the wild. **TIMING OF BREEDING** Display has been noted from April to January, with breeding near Port Moresby from January to late October/early November, avoiding the November–January moult period here (Coates 1990), which coincides with the start of the rains.

HYBRIDS Two types of hybrid are known:
Wonderful Bird of Paradise (*Paradisaea mirabilis* Reichenow, 1901), the Twelve-wired × Lesser Bird of Paradise.
Mantou's Riflebird or **Bruijn's Riflebird** (*Craspedophora mantoui* Oustalet, 1891; also *Craspedophora bruyni* Büttikofer, 1895). Twelve-wired × Magnificent Riflebird, one of the more frequent examples of a hybrid, with at least a dozen specimens known. See Fuller (1995) and Appendix 1 of F&B for further details of these bizarre forms.

MOULT Moult seems to have been recorded in all months. Crandall (1937a, 1937b) reported a male as taking seven years to go from female plumage to fully adult, and it lived for nearly 23 years. A bird at Karawari in early October 2017 had moulted all the tail wires but still had the vivid yellow plumes and was actively displaying to a female, hopping jerkily up a slender emergent palm stalk and erecting the breast shield and ear-tufts.

STATUS AND CONSERVATION Classified as Least Concern by BirdLife International on account of the extensive range and no obvious evidence of population declines. The Twelve-wired Bird of Paradise has a wide distribution in inaccessible country, and can be quite common where the habitat is suitable. In Western Province, PNG, these birds occur along some 40km of the Fly and Elevala Rivers north of Kiunga in widely separated territories, with some clusters in favoured areas (Bishop in F&B; pers. obs.). Beehler & Pruett-Jones (1983) recorded at least three males in both their PNG study sites at Nomad River, Western Province, and Kakoro, Gulf Province, of 150ha and 100ha respectively. The species is also widespread along the Sepik River and in the lowland river systems around Port Moresby. It is not widely used for decoration by local people, perhaps a result of its fondness for relatively inaccessible habitat. Nevertheless, skins of males are occasionally seen hanging from car mirrors in Kiunga, where they soon lose their yellow pigmentation and turn white. There are no obvious threats to this unique and curious member of the family given its wide range, although localised losses through logging, forest conversion to oil-palm plantations and village settlement will increasingly occur.

Twelve-wired Bird of Paradise, female, Cenderawasih Bay, Indonesia (*Dicky Asmoro*).

Twelve-wired Bird of Paradise, pair on display post, Nimbokrang, West Papua (*Gareth Knass*).

Twelve-wired Bird of Paradise, male, Nimbokrang, West Papua (*Gareth Knass*).

Genus *Paradisaea*

The species of this genus are perhaps the iconic birds of paradise in the minds of most people, and are endemic to the New Guinea region, ranging throughout the lowland forests and hills of mainland New Guinea as well as to the West Papuan Islands, Yapen Island and the D'Entrecasteaux Archipelago. *Paradisaea* species are characterised by the males' iridescent green throat patch, the two long extended central tail wires and the extraordinarily colourful and erectile lace-like filamentous pectoral plumes (red, orange, yellow, or white), the brown wing feathers (male and female), and also the extensive yellow on the head. Despite clear affinities to *Paradisaea*, and formerly placed within it, the Blue Bird of Paradise of the monotypic genus *Paradisornis* is split from this compact group because of its many unique features. This treatment is supported by the phylogeny of Irestedt *et al.* (2009). The spelling *Paradisaea* (rather than *Paradisea*) was conserved (ICZN 2012, Opinion 2294, Case 3500).

Etymology *Paradisaea* is derived from the Latin *paradisus*, meaning paradise.

GREATER BIRD OF PARADISE
Paradisaea apoda   Plate 20

Paradisaea apoda Linnaeus, 1758. 'India' = Aru Islands, West Papua.
Other English names Great Bird of Paradise, Fly River Bird of Paradise
Etymology Latin *apoda*, from the Greek *apous*, *apodos*, 'footless', a reference to the belief that the bird nested in the heavens and never came to earth, hence the lack of feet.

The Sultan of Bacan presented some skins of this species to the survivors of the famous Magellan circumnavigation expedition in 1521, while they were anchored off Tidore, in the Moluccas. The birds lacked feet owing to the method of preparation then used, and caused a sensation when taken to Europe and presented to the King of Spain in 1522, being the first species of the family to become known in the west. In the older literature this is often given as being the Lesser Bird of Paradise, but the scant information surviving suggests that the Greater is the species involved: Portuguese apothecary Tomé Pires saw *Paradisaea* skins from the Aru Islands around 1512, which would be those of the Greater Bird of Paradise, and this is probably the First European report. This species remains to this day an important item of trade among local people for customary celebrations, where the plumes are a major part of headpieces of the traditional costumes. Recent genetic work indicates that this species evolved relatively recently, along with its sibling the Raggiana

Bird, about 0.7 mya in the late Pleistocene period. This is the largest of the plumed birds of paradise, but still relatively little known. It forms a superspecies with *P. raggiana* and *P. minor*, with the other infrageneric species less closely related.

FIELD IDENTIFICATION This is a large paradisaeid of the south-west of New Guinea, just extending into PNG in the Trans-Fly and Tabubil areas, adults of both sexes being unlikely to be confused with any other. The male has striking yellow or apricot-orange plumes and very long tail wires. The female is very distinctive, being uniformly dark chestnut-brown and lacking yellow on the nape, though subadult males may cause confusion. This species hybridises with the Raggiana in the narrow zone of overlap in Western Province of PNG, where hybrid swarms are known to exist among pure-bred examples of both species. These may be the origin of the distinctive apricot-orange-plumed males often seen here, although these birds appear not to show any other Raggiana traits such as the yellow throat collar or lack of such a distinct breast shield, and they are associated with typical female-plumaged Greater Birds. **SIMILAR SPECIES** Males are similar to male Lesser Birds of Paradise but larger, with a blackish breast cushion (poorly developed and maroon-brown in Lesser) and maroon back (yellowish in Lesser). The ranges of these two siblings do not overlap in PNG, although they come into close contact in the Ok Tedi region, with the Greater in the hills up to 850m and the Lesser Bird coming as close as Telefomin in the mountains to the north. Overlap and hybridisation with the Lesser Bird is to be expected in the south-west of New Guinea between Timika and Etna Bay, a poorly known area. Hybrids with the Raggiana Bird of Paradise at Kiunga may show a yellow breast-band while having yellow or apricot-orange flank plumes. Salvadori (1880–82) recorded variation in the flank plumes of birds along the middle Fly, which he regarded as hybrids with the Raggiana, ranging from yellow to orange to bright orange-red ('Lupton's Bird of Paradise'). Papuan Babbler (*Garritornis isidorei*) is smaller, and has a decurved yellow bill and entirely rusty plumage, while Rusty Shrike-thrush (Pitohui) (*Pseudorectes ferrugineus*) has pale eyes, is much smaller and has primarily rusty plumage; both species have entirely different calls from the raucous sounds made by the birds of paradise.

RANGE Aru Islands and southern New Guinea from Timika east to the Fly/Strickland drainage. The eastern and western extremities of the range are not well known; it may extend west to Etna Bay, and was reported by Mackay (1966) from its eastern extremity at Nomad River, where Beehler failed to find it in 1980. It is quite common in the foothills around Tabubil, though the Ok Tedi mine itself is well above the species' altitudinal range, and also at Ok Ma, Ningerum, Rumgenai, Kiunga and Lake Daviumbu. Coates reports a hybrid population near Bensbach. This species does not range nearly so far into the uplands as either the Raggiana or the Lesser Birds of Paradise. Curiously, a small introduced population existed on Little Tobago Island, for some 75 years, the result of

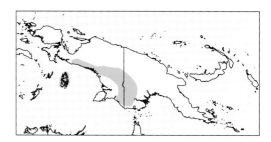

an Edwardian-era naturalisation scheme in 1909–1912 for what was then perceived as a rare and endangered bird. The population gradually dwindled in numbers owing to habitat stress and hurricanes, dying out in about 1984. **MOVEMENTS** No significant observations. A sedentary forest species, but female-plumaged birds do wander up into the hills in search of fruit, appearing occasionally at Dablin Creek (850m), near Tabubil, when trees are fruiting.

DESCRIPTION Sexually dimorphic. *Adult Male* 43cm, central tail wires 48–85.4cm; weight 222–225g. Resembles Lesser Bird, but the yellow on crown to hindneck does not extend onto the mantle, so the entire upperparts are rich chestnut-brown, the wings lacking a shoulder bar. The upper breast cushion is blackish-brown, shading to dark brown and then mid-brown to pinkish-brown on the lower belly and ventral regions. The central tail feathers are long bare wires (like those of Lesser and Raggiana Birds) up to 85.4cm in length, and the striking spray of lace-like filamentous ornamental flank plumes up to 50cm long and bright yellow or apricot-orange, shading to paler, whitish, and more attenuated towards the outer third of the cascade, with dark maroon-red streaks along the upper segment. *Iris* lemon-yellow; *bill* blue-grey; *legs and feet* grey-brown, tinged purplish. *Adult Female* 35cm; weight 170–173g. Distinctive, being entirely rich chestnut-brown, darker on face, chin, throat and upper breast. *Iris* lemon-yellow; *bill* blue-grey; *legs and feet* purplish grey-brown. *Juvenile* undescribed. *Immature* like female but with paler hindcrown and tinged orange on hindneck. *Subadult* Another plumage similar to the immature, but has a chin tuft developing and some scattered iridescent green feathers on lores and throat. A third type resembles the male, but without flank plumes and much brownish feathering on hindcrown and nape, rest of underparts as female. *Subadult Male* has bright yellow crown and nape, short buffish flank plumes, and the two bare wire-like rectrices extending from a few cm to well beyond the tail. The flank plumes gradually lengthen and change colour to yellow as the bird gets older. *Iris* yellow, *bill* blue-grey, *legs and feet* as adults. **PLUMAGE DEVELOPMENT** The birds take a long time to assume adult plumage, possibly as long as eight years to judge by avicultural experience (Seth-Smith 1923a), although this may be an artefact of an unfamiliar diet, and a period of 5–6 years (as with the sibling species) is considered more likely. The BOU Expedition of 1910–13 collected a long series from the Mimika and Utakwa Rivers and other sites, and these were recorded by Ogilvie-Grant (1915b) as follows:

'Adult males in full plumage were obtained between July 18 and January 7. The moult evidently extends over a considerable period, for birds with moulting head and half grown side plumes were killed in September, October and November. It would seem that in the wild state the males take at least five years to assume the adult plumage –

1st plumage brown like the female; no trace of metallic plumage on the head, middle pair of tail feathers not longer than the outer pairs.

2nd plumage similar to the above.

3rd plumage head and neck as the adult, but the chest scarcely darker than birds in the first or second years' plumages; middle tail feathers similarly coloured, but half as long again as the outer pairs, webbed on both sides to the tip, but with the vanes much narrower along the middle third.

4th plumage as in third plumage but the chest is very deep chocolate brown glossed with purplish and the lengthened middle pair of tail feathers is replaced by wires, as in the adult; there is still no trace of the ornamental side plumes.

5th plumage as in the fourth plumage but with long ornamental side plumes. The bird is now in adult plumage, but the side plumes are not so long or as fully developed as in the sixth and subsequent years. As might be expected, these changes are sometimes retarded in captivity when the bird is not taking exercise or receiving its proper diet.'

Birds seen on Tobago by E. T. Gilliard in 1958 were believed to be of the third generation since their introduction in 1909–12, giving some idea of just how long-lived birds of this family can be. Ogilvie-Grant adds the note that the species was seen only in the Middle Fly, not near the coast, and extended only into the very first foothills of the mountain range (Grant in Iredale 1950).

TAXONOMY AND GEOGRAPHICAL VARIATION
Monotypic. Proposed subspecies *P. a. novaeguineae* D'Albertis and Salvadori, 1879, described from middle Fly and said to apply to entire southern New Guinea population (from Timika, West Irian, eastwards to the Fly/Strickland watershed), differs from Aru Islands population in its paler upper breast, more maroon coloration and smaller size, but differences deemed insufficient to warrant recognition. Interestingly, sexual size dimorphism also differs between the two populations (Frith in F&B). The **Raggiana Bird of Paradise** was for long treated as a subspecies of the Greater Bird, despite the evident differences in size and coloration of both sexes. The two do hybridise in the Kiunga area, where mixed leks that include hybrid males occur, and hybrids are reported also from the Bensbach area (Coates 1990). It may be that introgression of Raggianas is becoming more frequent as the forest is cleared; one lek near Kiunga which consisted entirely of male Greaters up to about 1994 now has hybrids present (pers. obs.).

VOCALISATIONS AND OTHER SOUNDS Similar to those of the Raggiana Bird, perhaps slower and more discordant, a typical call being a loud ringing, resonant *wawk wawk wawk wok wok wok*-type series, the latter section shriller and more high-pitched and audible to well over a km, with various combinations of the call frequently heard. There is an example of

the sounds of males at the lek in Western Province, PNG, at xeno-canto XC149709 (by P. Gregory) and some calls from Freeport, in West Papua, at XC105044 (by B. van Balen). The male also has a louder, more emphatic *KA-KOW*, repeated several times, and birds at lek make quiet deep-toned buzzy *curr* calls and ringing *kow kow kow* and *wok wok wok* sequences, plus a plaintive rising single *kweh?*.

HABITAT Lowland and hill forest, from sea level to 850m in the Ok Tedi region and at least 950m in West Irian. Female-plumaged birds range higher than adult males in the Tabubil area, extending to 850m, whereas adult males were never found above about 650m. The species seems prone to disturbance, and several leks along the Ok Ma Road at Tabubil were abandoned as hunting and clearance encroached upon them. Birds at a lek near Kiunga became notably shyer when an access track was put in, probably permitting hunting to occur. The continuing clearance in Western Province will be of interest, as there is a hint that the species may be less adaptable than its siblings, preferring forest as opposed to cultivation. The Raggiana may profit from this sensitivity in this area, replacing the Greater or hybridising extensively with it.

HABITS Seems not to be so confiding as the Lesser and Raggiana Birds, but can be approached cautiously and quietly. The Greater Bird of Paradise frequents the lower canopy and the mid-stratum of tall forest, but seems to descend to the lower levels less often than the Raggiana. Males away from the lek are usually seen singly, but female-plumaged individuals and subadult males often form small bands of 2–6 birds. Birds at lek are as usual very noisy and may be heard at a distance of about 1km, but they tend to slip away if approached too closely, this perhaps a response to high hunting pressure in Western Province.

FOOD AND FEEDING Largely frugivorous, but readily takes insects, showing a liking for grasshoppers. Has a preference for *Schefflera* fruits. Wallace (in Elliot 1873) found birds' stomachs full of fruit, noted that they seek orthopterans, and found one of the largest phasmids almost entire in the gut of an adult male. This species will join mixed flocks of brown-and-black birds around Tabubil, with Rusty Shrike-thrush (Pitohui) (*Pseudorectes ferrugineus*) and Southern Variable Pitohuis (*Pitohui uropygialis*), Papuan Spangled Drongos (*Dicrurus carbonarius*), Crinkle-collared Manucodes, Magnificent Riflebird and Magnificent Bird of Paradise. Around Kiunga sometimes seen within flocks including Papuan Babbler (*Garritornis isidorei*), Glossy-mantled Manucode and King Bird of Paradise, as well. Gleans from branches and the underside of leaves, hanging almost upside-down at times, and has the usual generic habit of peering about. Explores dead trunks and opportunistically probes for prey in crevices.

BREEDING BEHAVIOUR Polygynous; males are promiscuous, while females nest-build, incubate and raise their young unaided. A superb and extensive archive of videos by T. Laman, showing this species in display, is lodged at the Macaulay Library of the Cornell

Lab of Ornithology site. The Internet Bird Collection (IBC) site also has some good video archive. **COURTSHIP/DISPLAY** Leks may be abandoned quite suddenly, as happened along the Ok Ma Road in the 1990s, but local guides know of some that have been in use for 40 years (S. Kepuknai pers. comm.). Leks along the Ok Ma held up to eight plumed males, but there have been up to 15 at sites around Kiunga, and Wallace recorded 20 in the Aru Islands in 1857. The birds choose large trees for their lek sites and show a liking for the big subcanopy limbs. Little has been recorded of their display and breeding since Wallace's time, apart from Dinsmore's observations on the introduced population on Little Tobago, which may be anomalous, although Attenborough published an account in 1996 and his film of one of the Kiunga leks is outstanding, as are those of Tim Laman and Ed Scholes in the *National Geographic* project on the birds of paradise. Surprisingly, the nest and nest site of the Greater Bird of Paradise remain little known. Alfred Russell Wallace, in his classic *The Malay Archipelago*, provides a vivid description of the lek of the species in the Aru Islands, from which the following extract is taken:

'*The birds had now commenced what the people here call their "sacaleli," or dancing-parties, in certain trees in the forest, which are not fruit trees as I at first imagined, but which have an immense tread of spreading branches and large but scattered leaves, giving a clear space for the birds to play and exhibit their plumes. On one of these trees a dozen or twenty full-plumaged male birds assemble together, raise up their wings, stretch out their necks, and elevate their exquisite plumes, keeping them in a continual vibration. Between whiles they fly across from branch to branch in great excitement, so that the whole tree is filled with waving plumes in every variety of attitude and motion. The bird itself is nearly as large as a crow, and is of a rich coffee brown colour. The head and neck is of a pure straw yellow above and rich metallic green beneath. The long plumy tufts of golden orange feathers spring from the sides beneath each wing, and when the bird is in repose are partly concealed by them. At the time of its excitement, however, the wings are raised vertically over the back, the head is bent down and stretched out, and the long plumes are raised up and expanded till they form two magnificent golden fans, striped with deep red at the base, and fading off into the pale brown tint of the finely divided and softly waving points. The whole bird is then overshadowed by them, the crouching body, yellow head, and emerald green throat forming but the foundation and setting to the golden glory which waves above. When seen in this attitude, the Bird of Paradise really deserves its name, and must be ranked as one of the most beautiful and most wonderful of living things.*'

With the exception of the Raggiana Bird of Paradise, other paradisaeids are not known to visit the lek of this species. The display appears quite similar to those of Raggiana and Lesser Birds, with three main elements distinguishable. The ***Convergence Display***, when males move into the lek, calling loudly and holding the wings in a rigid position in front of the body for few seconds, seems to attract females to the lek area. Then follows the ***Static or Flower Display***, when the plumes are held erect over the back, the tail tucked forwards under the perch and the body perpendicular

to it, the bird sometimes inverting in spectacular fashion, with the plume masses shimmering. When a female approaches, wing-flapping along with loud *wauk* calls may occur. The male may adopt the same wing pose, or may not erect the plumes but use a more upright pose. This cycle may be repeated every 10–20 seconds without interruption, some bouts lasting for more than 30 minutes, and sometimes leading to the ***Copulatory Sequence***. When a female is directly above the inverted male, he quickly bounces upright and starts to bounce rhythmically, shuffling back and forth or up and down on the branch and giving a click call. The male then cups the female between his open wings, cuffing her as he bills her nape or bill and swaying laterally at an increasing tempo, the swaying, wing-beating and bouncing increasing until he mounts the crouching female to copulate for a few seconds with wings vigorously flapping. Many females break off contact prior to the intense phase of the display, and the dominant males perform most copulations. **NEST/NESTING** As is usual with this family, females build and attend the nest alone. The nest is as yet remarkably poorly known. It is an open shallow cup, with a base of large leaves, and the bowl consisting of epiphytic orchid or vine stems and some fern fronds, without sticks, and lined with fine vine tendrils. It is placed in the fork of a large tree. **EGGS** The eggs are pale pinkish-buff with longitudinal purplish and purplish-grey streaks overlaid with similar markings of chestnut and browns radiating from the larger end; size 35.2–42.8mm × 26.5–27.3mm (F&B). Seven single-egg clutches, all from the Aru Islands, are housed in the NHMUK. Incubation period is not known. **NESTLING** Undescribed. **TIMING OF BREEDING** Food-carrying by a female has been observed at Tabubil in September (A. Murray in Coates 1990; pers. obs.), and a female with an enlarged ovary was collected at Lake Daviumbu in late August (Rand 1942a). Breeding has been recorded during March and May and August–December in mainland New Guinea.

HYBRIDS Lupton's Bird of Paradise, a hybrid between Greater Bird of Paradise and Raggiana Bird of Paradise, was first described by Percy Lowe in 1923 as a subspecies of Greater Bird of Paradise (noting that it may be a hybrid), from specimens seized by HM Customs and Excise and sent to the British Museum. Lowe named it for the officer concerned in the seizure.

MOULT Many males shed their plumes in September–October in the Kiunga area, and are fully plumed again by March. Lek activity is mainly from March to November in the Kiunga–Tabubil region of PNG. Wallace (1869) reported birds in the Aru Islands as moulting in January–February, showing new flank plumes by April and being in full plumage by May–June. Data from Little Tobago, where the species was introduced, appear to show a northern-hemisphere adjustment and are probably not very relevant, as 45% of displays were in December–February, which is the peak of the moult in PNG, although the 40% in March–May is consistent with PNG observations.

STATUS AND CONSERVATION Classified as Least Concern by BirdLife International, with a range of around 100,000km². This is still quite a common bird where good forest remains, which is extensive at present since the species inhabits the remoter and less populated regions. It may not be so adaptable as its congeners, and is under heavy hunting pressure for its plumes in the Aru Islands and Western Province of PNG at least, plumes at the latter site being sold for K20 each in 1999. Female-plumaged individuals greatly outnumber males, reflecting the lengthy period taken to attain adult dress. The species is

common around Kiunga at 30m above sea level, but it is the Raggiana and not this species that is seen along the Fly and Elevala Rivers to the north of Kiunga, suggesting an avoidance of the swamp–monsoon riparian-forest zone. Its status in 1988 on the Aru Islands, where Wallace first met the species, is that it was more common in the north, but no plumed males were seen and hunting pressure was high as the plumes were being sold in Dobo (Diamond & Bishop 1994). It is unfortunately all too likely that the situation has worsened, as demand for status objects in Asia is rising, and plumes regrettably fall into this category.

Greater Bird of Paradise, female plumage, Kiunga, PNG (*Phil Gregory*).

Greater Bird of Paradise, male, Kiunga, PNG, 24 July 2016 (*Phil Gregory*).

RAGGIANA BIRD OF PARADISE
Paradisaea raggiana   Plate 22

Paradisea raggiana P. L. Sclater, 1873, *Proc. Zool. Soc., London*, p.559. Orangerie Bay, south-east New Guinea.
Other English names Count Raggi's Bird of Paradise, Raggi's Bird of Paradise, Red-plumed Bird of Paradise, Empress of Germany's Bird of Paradise (*augustaevictoriae*), Marquis Raggi's Bird of Paradise, Grant's Bird of Paradise ('*granti*'). Red Kumul (Bird of Paradise) in Tok Pisin.
Etymology Named *raggiana* after the Marquis (or Count) Raggi of Genoa, a friend of the ornithologist Salvadori.

Modern genetic work gives a relatively recent origin of this species and its sibling, *P. apoda*, dated to around

0.7 mya in the mid-Pleistocene for this species. This is one of the best-known and most widespread *Paradisaea* species, being common in much of PNG and a very significant bird in the local cultures and national identity.

FIELD IDENTIFICATION Sexually dimorphic, all members of *Paradisaea*, the quintessential 'Bird of Paradise', have a rusty plumage with variably coloured red or orangey flank plumes in the males, yellow on the head and iridescent green on the throat, with yellow eyes and a large, stout pale grey-blue bill. The Raggiana is almost endemic to PNG, having only very recently been found to extend just across the border into West Papua in the Trans-Fly zone (M. Halouate pers. comm. 2017). It is a medium-sized bird, with distinctive floppy flight and the broad rounded wing shape typical of the genus, somewhat reminiscent of a Eurasian Jay (*Garrulus glandarius*) in shape. It is sometimes quite confiding, noisy and conspicuous, but more often wary and unapproachable. The males are maroon or vinaceous brown with splendid red or orange flank plumes extending up to 50cm, the shade varying with the race, rich golden-yellow crown and nape, and an iridescent oily emerald-green throat which often appears dark. Male Greater Bird has yellow or orange flank plumes and lacks the yellow throat collar, while females are distinctive in being entirely rich chestnut-brown. The male Lesser Bird has yellow plumes, and the female of that species has characteristic white underparts, both therefore being readily told from the Raggiana Bird. Adult female Raggiana is slightly smaller than the male and considerably lighter in mass, with forehead blackish-brown, crown to hindneck dull yellow, upperparts maroon-brown with faint yellowish traces, and face, throat and breast blackish-brown, sometimes with an indistinct yellowish collar on the foreneck; underparts pale buffy vinaceous brown, paler on the central belly. Both sexes have the iris yellow, bill pale blue-grey, and legs and feet greyish-brown. Immatures are similar to female but dark-eyed, males gradually developing bright golden-yellow head markings, bill and legs duller and darker. **SIMILAR SPECIES** Forms a superspecies on the mainland of New Guinea with Greater and Lesser Birds, the three replacing one another geographically. The island isolates Goldie's Bird of Paradise and the Red Bird of Paradise appear to be more distantly related, as is the other member of the genus the Emperor Bird of Paradise. Greater, Lesser and Emperor Birds of Paradise all have marginal overlap and are essentially the geographical replacements for the Raggiana. Lesser Bird of Paradise males have yellow plumes, the yellow of the nape extending more onto the mantle, and lack the dark breast cushion and yellow breast strap of the Raggiana, while male Greater similarly lacks the breast strap and has a dark chestnut mantle, with apricot or yellow flank plumes. Male Emperor is distinctive, with whitish flank plumes and extensive green on the head and upper breast. Females and immature Raggiana Birds lack the whitish underparts of the Lesser Bird of Paradise, and are not dark chestnut-brown beneath like female and immature Greater Bird of Paradise. Female and immature Emperor Bird of Paradise are quite similar but have a dark forecrown. Confusion with Rusty Shrike-thrush (Pitohui) (*Pseudorectes ferrugineus*) is unlikely, as that species is smaller, has a whitish eye, a shorter dark or yellow bill, and rusty underparts, and has a quite different voice.

RANGE This species is a fairly common endemic in the lowland and hill forest of southern and south-east PNG, barely extending into West Papua in the Trans-Fly (where some hybrids with Greater Bird of Paradise are now known). It is replaced by the closely related sibling species the Greater Bird of Paradise in the far west of PNG and the Lesser Bird of Paradise in the north-west Sepik–Ramu area. **MOVEMENTS** Some wandering by immatures occurs within local areas, visiting atypical

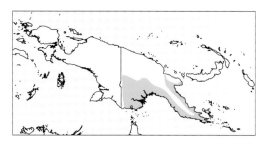

habits such as mangroves, and individuals will forage in savanna adjacent to rainforest, but no major movements are known.

DESCRIPTION Sexually dimorphic. Nominate race described (see TAXONOMY AND GEOGRAPHICAL VARIATION for other races). *Adult Male* 33–34cm excluding tail wires; weight 234–300g. Has black forehead and lores glossed with green, and a black chin tuft. The throat is glossy rich iridescent green but often appears dark; rest of head, neck and collar/strap on lower throat rich golden-yellow. Upperparts are maroon or vinaceous brown with a pale orangey-yellow bar on the wing-coverts. The central rectrices are greatly elongated and wire-like, up to 493mm in length and extending up to 360mm beyond the tip of the tail. Upper breast blackish maroon-brown, cushion-like with a velvety texture; rest of underparts pale vinaceous brown, with a striking spray of lace-like ornamental flank plumes up to 50cm long with deep scarlet undersides. *Iris* yellow; *bill* pale blue-grey, *mouth* dull pinkish flesh; *legs and feet* lavender-brown. *Adult Female* Averages 33cm, weight 133–220g (F&B). Forehead blackish-brown, the crown to hindneck dull yellow, rest of upperparts maroon-brown with faint yellowish traces; face, throat and breast blackish-brown, sometimes with an indistinct yellowish collar on the foreneck. Underparts pale buffy vinaceous-brown, paler on the central belly. *Iris* yellow; *bill* pale blue-grey; *legs and feet* greyish-brown. **Juvenile** undescribed. **Immature** initially much as adult female, but the *Subadult Male* has brighter yellow developing on head and the central rectrices often protruding beyond tail; but no flank plumes until nearly adult, which may take 5–6+ years (captives acquire first traces of adult plumage at five years); bill and legs duller and darker. One plumage stage lacks tail wires and has short, soft greyish-maroon flank plumes, the tail wires developing progressively, as do the flank plumes. Captive individuals were dark-eyed and took three years to acquire a yellow iris.

TAXONOMY AND GEOGRAPHICAL VARIATION Polytypic, with four subspecies (Beehler & Pratt 2016).
1. *P. r. raggiana* P. L. Sclater, 1873. Southern watershed of eastern part of South-east Peninsula from Cloudy Bay to East Cape; thence on north coast from East Cape north-west to Cape Vogel (Frith & Beehler 1998). Red-plumed and with a yellow upper back.
2. *P. r. salvadorii* Mayr & Rand, 1935. *Paradisaea apoda salvadorii* Mayr & Rand, 1935, *Amer. Mus. Novit.* 814: 11. Vanumai, Central Province, southern watershed of South-east Peninsula. Ranges in southern New Guinea from Oriomo River and Fly River east to

Cloudy Bay (east of Kupiano), extending also to upper Purari River and Wahgi valley, from where it sometimes wanders, via deforested habitat, across 'hybrid gap' into the Baiyer valley, northern watershed (where *P. minor* dominates). Adult male lacks yellow on the mantle, and has flank plumes less deep red (more 'brick-red') and slightly shorter. This is the taxon at Varirata NP, near Port Moresby.

3. **P. r. augustaevictoriae** Cabanis, 1888. *Paradisea Augustae Victoriae* Cabanis, 1888, *J. Orn.* 36: 119. Finschhafen, Huon Peninsula. Synonym: *Paradisea granti* North, 1906, *Victorian Naturalist* 22: 156, 'German New Guinea?'; apparently an intergrade between *augustaevictoriae* and *intermedia*. Ranges from coast of Huon Gulf and Markham River valley north-west up to the Uria River (a tributary of the upper Ramu), where it hybridises with *P. minor*; extends south-east to Wafa, Bulolo and Wau valleys, thence into the Waria and coastally to lower Mambare River (Frith & Beehler 1998); presumably also meets and hybridises with *P. minor* on north coast of Huon on northern flank of Saruwaged Mts. Adult male is quite distinctive, with much yellow on back and with orange flank plumes, very little or no yellow neck collar, extensive yellow wingbar, and underparts washed yellowish, with more extensive dark breast cushion, flank plumes bright apricot-orange on undersides. Hybridises with *P. r. intermedia* and with Lesser and Emperor Birds where ranges abut.

4. **P. r. intermedia** De Vis, 1894. *Paradisea intermedia* De Vis, 1894, *Ann. Rept. Brit. New Guinea* 1893–94: 105, Kumusi River, South-east Peninsula. Ranges from Collingwood Bay (plumes are a little more reddish) to Holnicote Bay, Kumusi River, and the lower Mambare River (Mayr 1941). Both sexes have a yellow back and yellow streaking down to uppertail-coverts. Hybridises with *P. r. augustaevictoriae* where ranges abut.

VOCALISATIONS AND OTHER SOUNDS A characteristic and loud sound of PNG lowland and hill forests, the advertisement call being a raucous far-carrying (over 1km) and quite variable series of a dozen or more scolding raucous *kwee kwee* and *kwok* notes, very loud ringing *ko ko ko ka-kow* series, rising in pitch and then slowing, a ringing *ko ko ko ko kwah kwah*, and quiet scolding notes. Recordings of subspecies *salvadorii* are available at xeno-canto XC87556 (by I. Woxvold) and of nominate race at XC23950 (by K. Blomerley) and XC149712 (by P. Gregory). The primary display call is a long series of excited high-pitched notes, which can continue unbroken for several minutes and often prompts nearby birds also to begin vocalising. In addition, this species, when feeding with Papuan Babblers (*Garritornis isidorei*), emits curious quiet growling guttural notes quite similar to some calls of the latter species. Unlike the Lesser Bird of Paradise, this species in display may strike its carpal joints together to produce thudding sounds, as well as making quiet clicks and growls.

HABITAT Occurs in lowland, hill and lower montane forest, from sea level to at least 1,400m, reaching about 1,800m in some highland valleys. Inhabits rainforest, riparian forest and forest edges, wandering into adjacent savanna at times, as well as disturbed and degraded habitats such as native gardens in the highlands; *Casuarina* trees are frequently utilised. Found also in second growth, and occasionally even visits mangroves (Beehler in Frith & Beehler 1998). Frequently uses large forest trees for leks, but may also use quite small trees with dense vine cover, which permit concealment when not performing. Trees may be live or sometimes dead, and may be deep within forest, in disturbed ground or at the forest edge, with no obvious preference for flat or sloping ground, although some sites are on ridgetops. Plumed males tend to keep within forest cover. Immatures occur within local areas, visiting atypical habits such as mangroves, and individuals will forage in savanna adjacent to rainforest.

HABITS Raggiana Birds of Paradise can be noisy and conspicuous at lek, but even then they tend to be cautious and may slip away if disturbed. Otherwise they are fairly shy, but can be bold when intent on feeding or foraging with a group of other species. This is mainly a species of the mid-storey to the canopy but quite frequently comes lower, almost to ground level at times when associating with Papuan Babblers, which frequent the lower storey. Raggianas have an inquisitive look owing to the large pale eye, and have a habit of peering at objects with the head held to one side. When suspicious, they crouch and jerk the tail from side to side, flicking the wings simultaneously. Known to bite off leaves from around perches, perhaps to improve sight-lines (pers. obs.).

FOOD AND FORAGING Primarily frugivorous, but also takes insects and arthropods from branches, bark and leaves. Raggiana Birds of Paradise will glean for insects and occasionally may sally for swarming insects. The long periods spent in display (see BREEDING BEHAVIOUR) suggest that food is readily available for the species. Joins brown-and-black bird flocks with core members the Rusty Shrike-thrush (Pitohui) (*Pseudorectes ferrugineus*), Hooded Pitohui (*P. dichrous*) and Papuan Babbler, and is frequently a member of such assemblages (Diamond 1987; Bell 1983). Foraging parties of female-plumaged individuals may be joined by King Birds of Paradise and occasionally a Growling Riflebird or manucode species. Raggiana Birds may congregate at fruiting trees such as *Ficus*, or make a series of individual visits. Two species of fruiting tree at Varirata were visited only by birds of paradise (Beehler & Dumbacher 1996) and no other species, an extreme specialisation of bird and foodplant if this is regular behaviour. Frequents the canopy and upper storey, but may forage low down near ground level with Papuan Babblers (pers. obs.) or come to pools for drinking during dry weather (Coates 1990).

BREEDING BEHAVIOUR Polygynous; males are highly promiscuous and form very vocal communal leks, which are indeed one of the wonders of the natural world. Females nest and raise young alone, as is usual in this family. Leks with half a dozen plumed males are quite regular, but others may have just a single or two, while leks with up to ten have been recorded (Coates 1990). Dominant males seem to attend more than others, defending their own perches, and some males may attend only sporadically (Forshaw & Cooper 1977; Beehler 1988). Early-morning and late-afternoon sunlight is important for these display sites, and leks sometimes move to nearby trees for no obvious reason (perhaps related to disturbance). The birds sometimes use adjoining trees as part of the lek site, as at Kiunga Km 17, although the alpha perches, the ones that attract the most females for mating, are in one main tree, with opportunistic matings nearby. **COURTSHIP/DISPLAY** The display sequences are complex and variable and have been described in various ways by many authors. The display has three main phases (see Coates 1990), and we adopt the standard terminology from LeCroy (1981), Frith (1981) and Beehler (1988) to describe these behaviours among most members of this genus. Display is generally in the early morning from about 06:30 to 08:30 and late afternoon from about 15:00 to 17:00. Rainy periods depress activity, which is greatest on clear sunny days. The birds often remain in the general area and may initiate calling at any time of day, although this does not lead to intense displays. Males may perch within 250m of the lek and call, moving in when females approach as they may do singly, in twos or in small parties. *Convergence Display* Males move into the lek tree and give loud display calls while fluttering about between perches and partly raising the flank plumes, one bird commencing and others rapidly joining, with up to 20 birds involved, including males and female-plumaged ones. The half-open wings are beaten and the head and chest are raised and lowered, causing the flank plumes to wave about as the birds call loudly. During this phase males may appear to 'moonwalk' along branches, giving the appearance of walking without actually changing position, a spectacular and bizarre sight. This initial warm-up phase may continue for 20 minutes or so, and sometimes goes no further, birds abandoning the lek perhaps because of disturbance, a predator, or simply lack of a receptive audience. The females move about the periphery of the lek and often seem to select a particular male for their attentions, associating with him for the first stage of the display. *Static or Flower Display* When females come close, many birds may adopt their next form of display on their vertical or steeply inclined perch, calling rising to a crescendo as the visitors appear and then dying away as the birds move into the next stage of their show, which is almost silent by comparison with the noise beforehand. They may hang almost upside-down and beat the wings together over the back before opening them; they then turn the slightly lowered head from side to side while rhythmically and repeatedly striking the carpal area of the wings sharply together. Quiet clicks and growls may be given during this performance, which differs from the display of the Lesser Bird of Paradise (which does not strike the carpals together). The flank plumes cascade out and give the appearance of a giant exotic flower (aptly named the *Flower Display* by LeCroy 1981, a sequence

of the static display that may well lead to copulation). *Copulatory Sequence* This intense inverted *Flower Display* may lead to copulation if the female is sufficiently enticed, the male half opening the wings and literally bouncing or hopping along the branch, swaying the body laterally as he does so. Given female presence, he then performs the *Frontal Display* (of F&B 1998), in which the throat and breast shield are puffed out and the head almost disappears into the expanded ball of feathers so that only the crown and bill protrude (Cooper 1977), the male hopping towards the female, swaying from side to side and pecking at her bill. Swaying, wing-beating and bouncing increase until he mounts the crouching female to copulate for a few seconds with wings vigorously flapping. Many females break off contact prior to the intense phase of the display, and successful matings seemed to be entirely in the mornings at Mt Missim (Beehler 1988), but were recorded also in the afternoon at a lek near Kiunga. The dominant males perform most copulations, and, of 305 visits by females to one lek on Mt Missim, 35 (11%) resulted in mating (Beehler 1988). **Other Detail Regarding Leks** Immature males, with yellow nape and tail wires present, often attend these leks and watch the proceedings as if learning by example; they may also call and initiate partial displays of their own on the periphery of the lek, as if practising. The long tail wires may play a tactile role in mating when the female is investigating the male closely, as suggested by Frith (in F&B) and as with the Twelve-wired Bird of Paradise. These wires develop long before the flank plumes and may be very significant where mating involves immature males. Other paradisaeid species may be attracted to the astonishing spectacle of a lek 'in full cry', such as the Twelve-wired, Magnificent Riflebird, Growling Riflebird and Lesser Bird, which no doubt explains the various hybrids known with closely related species. Leks may be used for many years, more than 20 years in some cases, or may be temporary to exploit nearby food sources. Females may breed with immature males, perhaps when the adults are unavailable. Some leks are objects of pride for the local village clan and the plumed male birds are sustainably harvested, enough being left so that the site is not abandoned. **NEST/NESTING** A flimsy open bowl-shaped nest of leaves, vine stems, orchids or ferns, moss and rootlets, lacking sticks (as usual with the genus), and with a lining of horsehair-like material (which is probably a fungus), built in trees at heights ranging from 1m to 18m. **EGGS & INCUBATION PERIOD** Clutch is usually of either one or two eggs, pale pinkish-buff to pale salmon and marked with brush-like brown and purple-grey strokes, denser about the larger end; average size 36.4 × 25.2mm (Frith in F& B 1998). Captive females have laid three single-egg clutches over a 12-month period, then two two-egg clutches over a subsequent 12-month timespan. Incubation period 18–20 days. **NESTLING** The hatchling is pink and naked, quickly becoming dark-skinned (data on captives). Nestlings seem to be fed with arthropods for the first five days, after which fruit becomes a component (Frith 2010), with capsular fruits and figs significant items. In captivity, the nestling period was 17–20 days (three breeding events). **TIMING OF BREEDING** Plumed males are seen in April–November and lek display is similarly noted from April to November, the peak at Varirata being in May–August (nominate) but that in Wau in June–October (race *augustaevictoriae*). Nesting is recorded from April to December, with a peak in September–November, which avoids the main wet season.

HYBRIDS Hybrids with Greater Bird of Paradise are seen at Kiunga, where the two species often visit the same leks. Fresh-looking Raggiana-type bird skins with almost whitish (pale candy-floss pink) plumes have been seen at markets in Mt Hagen (pers. obs.), and the Raggiana has been known to hybridise also with Lesser, Emperor and Blue Birds of Paradise (the last of those involving intergeneric hybridisation following the generic reclassification of Blue). Intraspecific hybrids between races of Raggiana also occurs. The following summary is from Frith & Frith (2010) and Fuller (1995).

Grant's Bird of Paradise ('*P. granti*' North, 1906) is a Raggiana Bird of Paradise hybrid between *P. r. augustaevictoriae* and *P. r. intermedia*.

Captain Blood's or **Blood's Bird of Paradise** ('*Paradisaea bloodi*' Iredale, 1948), is a very rare intergeneric hybrid between Raggiana and Blue Birds of Paradise, known from just one specimen collected in 1944.

Frau Reichenow's or **Maria's Bird of Paradise** ('*Paradisaea maria*' Reichenow, 1897), an interspecific hybrid between Raggiana and Emperor Birds of Paradise, is known from eight specimens.

Rothschild's Bird of Paradise ('*P. mixta*' Rothschild, 1921), an interspecific Raggiana (race *augustaevictoriae*) × Lesser Bird of Paradise (race *finschi*) hybrid, is known from at least four trade skins.

Gilliard's Bird of Paradise (Frith & Beehler, 1998), Raggiana (race *salvadorii*) × Lesser Bird of Paradise (race *finschi*), occurs in the upper Baiyer River valley.

Lupton's Bird of Paradise (Lowe, 1923), a hybrid of Raggiana (race *salvadorii*) × Greater Bird of Paradise (form '*novaeguineae*'), is quite frequent in the Kiunga area. Males are variable and intermediate between the parent species, ranging from individuals with vivid mixed red, orange and reddish-orange flank plumes, with or without yellow neck strap and shoulder bar, to those with orange flank plumes tinged pinkish at the extremities and likewise with or without yellow neck strap or shoulder bar. This was first described in 1923 as a subspecies of Greater Bird of Paradise (noting that it may be a hybrid), from specimens seized by HM Customs and Excise and sent to the British Museum, and was named for the officer concerned in the seizure.

MOULT Moult has been recorded in most months, with a peak in November–March. Birds at Kiunga and Varirata are certainly moulting by October and are in good plumage by April. Moult occurs over a period of 18–19 weeks in captivity, in the wild coinciding with the onset of the wet season in many areas, with moulting birds seen from October onwards, and the main moult season from then until March.

STATUS AND CONSERVATION Classified as Least Concern by BirdLife International, having a range encompassing around 200,000km². This is a common and quite adaptable species over much of its range, well-known sites including Varirata NP near Port Moresby, where several long-established leks exist, and also the Tari valley and upriver from Kiunga, in Western Province. Although it is reasonably adaptable, and quite tolerant of disturbance and modified habitats, some leks being very close to busy roads or settlements, just how long it can survive with continuing degradation and hunting owing to rising population pressure remains to be seen. Brown River forest was estimated to support one bird per hectare (Bell 1986), and in 200ha on Mt Missim ten adult males were resident and formed two leks (Beehler 1988). There are estimates of 16 individuals per km² of primary forest at 432–650m and 43 at 651–935m (Frith & Frith 2010). One of the best-known and most widespread *Paradisaea* species, the Raggiana is a very significant bird in the local cultures and national identity. It features on the national flag, the official crest, the uniform badges of the Customs and Police services, the logo of Air Niugini, many displays and signs in urban areas, and countless paintings and designs. It is also depicted on many PNG stamps, banknotes and coins, as well as giving its local-language (Tok Pisin) name to the national rugby side, the *kumuls*. It remains an important item of trade among local people for customary celebrations, in which the plumes are an essential component of the *bilas* or traditional costumes, where they are the centrepieces of the head-dresses. Novotny (2009) quotes some century-old records of bride price from the Wahgi valley, where the going rate for obtaining a bride was 36 skins of Raggiana, 30 of Lesser Birds, three of Princess Stephanie's Astrapia, 24 of the gold-lipped kina shells (which later gave their name to the currency), seven giant baler shells (used as chest ornaments in ceremonial dances), 48 pigs, 71 stone axes and three bundles of salt. The plumes of this and the Lesser Bird of Paradise are still the most widely traded of the various paradisaeid plumes. Although this species may be hunted legally only by traditional means, in practice this is often ignored; it is often hunted out for its plumes. Only citizens of PNG may keep the skins and plumes, which could be bought for K20 each at Mt Hagen market in 1999, the same price as in Tabubil in 1991, but in 2014 prices had risen to K50. Given the cultural significance of this bird for the people and nation, its long-term survival prospects are considered to be good despite local losses. The Raggiana Bird of Paradise is quite long-lived. An immature male (weight 147g) ringed on Mt Missim in 1980 was retrapped in 1997 in full male plumage at a minimum age of 16 years ten months (Dumbacher in Frith & Beehler 1998), giving an indication of **longevity** in the wild; individuals have lived for up to about 33 years in captivity (Frith & Beehler 1998).

Raggiana Bird of Paradise, male, Varirata, PNG (*Stephen Rannels*).

Raggiana Bird of Paradise, male, Varirata, PNG (*Stephen Rannels*).

LESSER BIRD OF PARADISE
Paradisaea minor   Plate 20

Paradisea minor Shaw, 1809, *Gen. Zool.* **7**, part 2, p.486. 'New Guinea' = Dorey (Manokwari), Vogelkop, West Papua.

Other English name Little Emerald Bird of Paradise
Etymology The species epithet *minor* is Latin and means smaller or lesser.

This was one of the first species to be reported in the west, perhaps as early as 1512, and one of the first specimens of the family to reach Europe. The specimens brought back by the remnants of Magellan's Expedition in 1522 were often reported as being of this species, but Greater Bird seems the more likely from the scant information surviving. The Lesser Bird was also the very first of the family to be seen in the wild by a European, René Lesson, naturalist on the barque *Coquille*, having seen a plumed male in 1824: he was too astonished to fire his gun at it, as it flew 'like a meteor whose body, cutting through the air, leaves a long trail of light'. This species is very significant culturally, its plumes still being an important part of the display costumes of the local people; its plumes are among the most frequently seen to be used as '*bilas*' (decoration) and are traded widely away from the species' main centres of distribution. Recent genetic work indicates that this species evolved quite recently, about 1.4 mya in the mid-Pleistocene epoch.

FIELD IDENTIFICATION This is a familiar species over the northern watershed and far west of New Guinea, forming a superspecies with the Raggiana and Greater Birds, with which it is known to hybridise. The males

have yellow plumes, similar to male Greater Birds in the south-west, but they lack such well-defined breast cushions, and the male Lesser Bird has a yellowish back. Raggiana over much of the range is red-plumed, but the distinctive race *augustaevictoriae* of that species does have much orange in the plumes, albeit with much red in the basal portion, and the usual Raggiana yellow neck strap below the green throat. The female Lesser Bird is very different, the dark head contrasting with the white underparts. In flight, the shape is again reminiscent of that of a Eurasian Jay (*Garrulax glandarius*). **SIMILAR SPECIES** Greater and Raggiana Birds are distinguished as above, and Lesser is, significantly, the only member of its genus over much of its range. MacGregor's Bowerbird is a smaller, dumpier species with a mainly uniform brown plumage and black bill, while Rusty Shrike-thrush (Pitohui) (*Pseudorectes ferrugineus*) also is smaller, with mainly rusty plumage, a different musical call and shape, and a pale eye.

RANGE Found on Misool Island and the Vogelkop of West Papua, extending across the northern watershed of New Guinea to the mouth of the Gogol and Ramu Rivers, and the north-west Huon Peninsula in the Finisterre Range. Penetrates into the interior along river valleys such as those of the Baiyer and Jimi. Replaced by the Raggiana Bird of Paradise in the east and by the Greater Bird of Paradise in the south; hybridises with the former in the upper Ramu valley and Finisterre Range and occasionally in the Baiyer valley. The incidence of hybridisation with Greater is unknown (likely where it presumably meets that species somewhere between Etna Bay and the Mimika River). **MOVEMENTS** No significant information; appears sedentary.

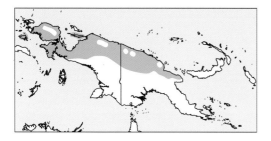

DESCRIPTION Sexually dimorphic. *Adult Male* 32cm; weight 183–300g. Forehead is iridescent green, orange-yellow of rest of head extending onto nape and mantle and then blending into mid-brown of back and wings; rest of upperparts paler brown, with an obvious orange-yellow shoulder bar and outer edges of the greater coverts. Central tail feathers greatly elongated into two long bare wires up to 641mm long. Chin and throat are iridescent emerald-green, which contrasts with reddish-brown breast. The underparts below breast are somewhat paler, with spectacular elongated filamentous flank plumes bright vivid yellow at base and shading to white with variable beige tinge. Local people in the Schrader Range of PNG report that, rarely, a very old male occurs which is really dark on the face, neck and nape, but this remains as yet undocumented. *Iris* yellow; *bill* pale blue-grey, *mouth-lining* pale aqua-bluish; *legs and feet* grey-brown with a purplish tinge. *Adult Female* Slightly smaller than male, weight 141–210g. Distinctive, having the head entirely dark brown with paler buff-yellow nape and mantle, and rest of upperparts mid to dark brown; throat dark brown, grading into striking white underparts washed light pinkish-brown. *Iris* yellow; *bill* pale blue-grey; *legs and feet* grey-brown with purplish tinge. *Juvenile* undescribed. *Immature Male* First-year is much as adult female in plumage but larger, sometimes with variably intensive yellow nape and hindneck, chin glossed green, with green feathering on lores, and flank plumes slightly elongated and white. *Subadult Male* Later stages include individuals resembling the adult male but with flank plumes not developed and maroon-brown on lower neck shading to white on mid-breast, and others similar to this but with more white on underparts and longer flank plumes varying from buffy to white, with central tail feathers long, narrower and more pointed. A captive individual took 8–9 years to attain full plumage, starting its transition at four years of age.

TAXONOMY AND GEOGRAPHICAL VARIATION Polytypic; two subspecies accepted by Beehler & Pratt (2016), although F&B (1998) recognised three, noting that *pulchra* requires assessment.
1. *P. m. minor* Shaw, 1809. Misool Island and western and northern New Guinea east to Gogol and Upper Ramu Rivers, in the southern watershed to the base of the Vogelkop at Etna Bay. Synonyms *P. m. pulchra* Mayr & Meyer de Schauensee, 1939, from Misool, and *P. m. finschi* A. B. Meyer, 1885, of northern PNG from just east of the West Papua border to the Gogol and Upper Ramu Rivers, presumably intergrading with *minor* in the Vanimo area. Similar in size to

nominate *minor*, but flank plumes slightly shorter and of a brighter orangey yellow; yellow shoulder marking less extensive than on the other races, but differences relatively minor and now synonymised.
2. *P. m. jobiensis* Rothschild, 1897. Yapen (Jobi) Island, West Papua. Larger than *minor*, with longer tarsus and flank plumes.

VOCALISATIONS AND OTHER SOUNDS Has shriller, higher-pitched and more raucous voice than Raggiana or Greater. Typical calls are ringing *waugh waugh waugh wauk* notes and a shrill *kwee kwee kwee* series, occasionally with single harsh screeches and squawks interspersed when displaying. A piercing, rather shrill *wook* note seems unique to this species, but deeper-toned equivalents are given by the Greater and Raggiana Birds. The calls vary greatly in length, volume and intensity, and depend on the degree of activity at the lek (when given during the display sequence). The calls of such birds carry long distances across the valleys, more than 1km in some areas. Recording ML169819 (by T. Pratt) at the Macaulay Library at the Cornell Lab of Ornithology provides some typical sounds, as do xeno-canto at XC209651 and XC267855 (by F. Lambert) and XC187345 (by P. Gregory).

HABITAT A common species in lowland and hill forest from sea level to 1,560m, living at the forest edge and in second growth, and able to adapt to modified environments, as has the Raggiana. Often found in parties in tall forest, also in the middle stratum and close to the ground, and in cultivation. Males tend to be seen more in deep forest, females and immatures being more catholic in their habitats and extending higher. Healey (1978a) found the species to be more forest-dependent than the Raggiana Bird, less tolerant of disturbed habitats, although it does occur in *Casuarina* groves in the highlands and near habitation there.

HABITS This species is reasonably confiding, but can be very wary when hunted and is heard more often than seen in places closer to some more developed areas such as Madang and Mt Hagen. Many leks are in disturbed areas or close to villages, one well-known one being at Baiyer River Sanctuary, in Western Highlands Province, and another at Kama, in the Wahgi valley.

FOOD AND FORAGING Predominantly frugivorous, this species will also take grasshoppers and other arthropods. Recorded fruits include those of *Ficus* species, *Elmerilla papuana*, *Trema orientalis* and *Omalanthus novaeguineensis*. It gleans arthropods from limbs, trunks and vines. Some fruits are held under one foot as the contents are extracted. Foraging individuals choose stable perches close to the trunk, from where they peer about. They may be quite acrobatic when feeding, even hanging upside-down (Forshaw & Cooper 1977). Female-plumaged birds often join mixed-species feeding flocks with pitohuis (*Pitohui*), shrike-thrushes (*Colluricincla megarhyncha* and *Pseudorectes ferrugineus*) and Papuan Babblers (*Garritornis isidorei*), one such flock in West Sepik including King Bird, Jobi Manucode and Pale-billed Sicklebill (Beehler & Beehler 1986).

BREEDING BEHAVIOUR A superb and extensive archive of videos by T. Laman, showing this species in display, is located at the Macaulay Library of the Cornell Lab of Ornithology website. The Internet Bird Collection (IBC) site also has some good video archive. Polygynous; males are promiscuous and attend communal leks, while females nest and raise young alone, as is usual in this family. Males display communally in leks in canopy trees, where from one to 20 plumed males may assemble; each one occupies a section of often quite open limbs, which is his court. These courts may be just 50cm apart (Beehler 1983c). Adult males come to the lek daily (except when moulting), usually early in the morning 06:00–09:00 and again in the late afternoon 14:30–17:30, dominant males attending more than others (Beehler 1983c). Leks may be used for many years, one at Baiyer River having been in use for 15 years before it was damaged and the birds moved to a nearby site, though some remained (Coates 1990), and some trees may have been in use for 60 years (Healey 1978a). Leks are active for at least seven months of the year, although many birds seem to moult during October–March and the Baiyer River lek was utilised mainly from March to November (Coates 1990). **COURTSHIP/DISPLAY** Adult males will display whether or not a female is present, and subadult males often visit the lek but without holding court there. The arrival of a female signals a crescendo of activity, the birds becoming agitated and calling loudly, with flank plumes raised and shivered, with much wing-flapping and hopping about on limbs. After this brief beginning, the males assume their courts as the female moves around the lek, inspecting various males. The following display accounts are taken from Coates (1990). The females initiate copulation by approaching the males of their choice, and mating occurs only at the courts. Beehler (1983c) spent 18 days watching one lek of eight plumed males, recording 99 visits by females and some 26 matings involving just two centrally located males. One male performed 25 copulations and the other male just a single one, while a third male failed to mount a soliciting female. The *Convergence Display* occurs when males come into the lek from the surrounding area, usually because females have arrived in the vicinity. Loud advertisement singing follows and males shake their flank plumes and hop from side to side, and lower the head with the body near horizontal and the neck crooked. The wings are opened fully and held out at 45° on each side above the back; the tail is lowered and pulled forwards even in front of the feet, and the flank plumes are spread and erected to arch above the back. After some 10–20 seconds during which the wings are slightly quivered, the male hops excitedly back and forth along the display limb with head lowered, wings fully opened horizontally or lower, and flank plumes spread and erected. The males of this species project their white filamentous flank plumes conspicuously forwards on each side of and over the head, giving a rather different display appearance from that of the congeners in similar mode. At the climax of the Convergence Display, the male gives the **Static Display**, in which he remains perched and still, but rubs or wipes his bill on the perch for several seconds (Ogilvie-Grant 1905). Gilliard describes this posturing in greater detail (in LeCroy 1981): 'The shoulders are drawn close to the body and the primaries are extended out… they are held still for a while and the bird makes no noise except for the occasional low "*graaaa*". Then the wings begin to move up and down, moving up slowly about three inches at their tops, then snappily to a point just below shoulders… as the performance continues [the flank plumes] begin to rise up behind in a splendid cascade. But the distinctive thing is that certain of the shorter yellow flank plumes are lofted through the opening normally covered by the scapulars and they stick up in random places like little golden fountains.' At this point the bird may almost turn under the perch, but no actual inversion has yet been reported for this *Flower Display* (LeCroy 1981), unlike the Raggiana. *Copulatory Sequence* The female approaches the male and crouches submissively in front of him with her back to his breast. Once she has signalled her readiness, the male rhythmically beats his open wings over and about the female, rocking from side to side for about 20 seconds while pecking at her (and/or she pecking at him), before mounting to copulate (Peckover 1973; Forshaw & Cooper 1977; Coates 1990). Gilliard (in Gilliard & LeCroy 1961) reports that this species does not thump the carpal-joint area of the wing over the back as do displaying male Raggianas. Males will perform without the presence of female-plumaged birds, and subadults will go through the display motions of adult males in the lek when the dominant males have left. The display of this species appears to be considerably simpler than that of the congeneric Raggiana and Greater Birds of Paradise. Female-plumaged individuals sometimes perform displays, and it is reported that, in captivity, females will often perform male-type displays among themselves (Boehm 1967) with both this species and the Raggiana. Presumably these birds are actually young males. **NEST/NESTING** Females undertake all the nest duties unaided by the males. The nest is reported as an insubstantial cup-shaped structure of twigs, lined with black wire-like fibres or rootlets, and partially covered by dead leaves. It is generally located 6m or more above ground in a slender tree. **EGGS** Poorly known. The eggs are a rich pinkish-cream with rufous-brown, dark brown and greyish longitudinal streaks, most plentiful at the larger end; average size of 12 eggs 38 × 27mm. A captive laid a creamy-pink egg peppered and spotted with fine purplish-grey, reddish and brownish (Bishop & Frith 1979), but this seems atypical. Clutch size usually one egg, sometimes two. **NESTLING** Undescribed. **TIMING OF BREEDING** Breeding has been reported in August and a nestling in October from Baiyer River (Forshaw & Cooper 1977).

HYBRIDS Hybrids with the Raggiana Bird of Paradise are known from the Upper Ramu, Baiyer valley and Finisterre Range, and this species is likely to hybridise also with the Greater Bird between Etna Bay and the Mimika River.

Rothschild's Bird of Paradise ('*P. mixta*' Rothschild, 1921) is an interspecific Raggiana (race *augustaevictoriae*) × Lesser Bird of Paradise (race *finschi*) hybrid known from at least four trade skins. Hybrids and birds typical of the parent forms indicate that this hybrid occurs in a band about 35km wide between the parent forms. Lesser Bird genes predominate in the north-west, with Raggiana in the south-east. The main hybrid zone seems to lie in the Gusap Pass between the Markham and Ramu Rivers. A second contact zone could be expected on the northern side of the Finisterre Range, to the south and east of Madang. There appears to be significant introgression, with much variation in individuals of hybrid origin. Those closest to being pure Lesser of the nominate race (synonym *finschi*) are predominantly yellow-plumed and lack the yellow throat collar, while those closest to parental Raggiana are predominantly orange-plumed and have a narrow yellow throat collar.

Gilliard's Bird of Paradise (Frith & Beehler, 1998), Raggiana (race *salvadorii*) × Lesser Bird of Paradise (race *finschi*), is found in the upper Baiyer River valley. Much like the preceding hybrid but with more reddish flank plumes, as would be expected with a red-plumed race of the Raggiana as a parent.

Intergeneric hybrids

Wonderful Bird of Paradise ('*P. mirabilis*' of Reichenow, 1901): Twelve-wired × Lesser Bird of Paradise; five adult male specimens known.

Bensbach's Bird of Paradise or **Bensbach's Riflebird** ('*Janthothorax bensbachi*' Büttikofer, 1894): Magnificent Riflebird (*magnificus*) × Lesser (nominate race), known from a single adult male trade skin.

Ruys' Bird of Paradise ('*Neoparadisea ruysi*' van Oort, 1906): Magnificent Bird of Paradise (race *magnificus*) × Lesser (race *minor*); one adult male trade skin specimen known.

Duivenbode's Bird of Paradise ('*Paradisea duivenbodei*' Menegaux, 1913): Emperor × Lesser Bird of Paradise, known from a single trade skin.

MOULT Museum specimens show evidence of moult mostly in September–February, and none in June–July (which would be peak display season). Birds seen in April and May in the Anji valley near Kumul Lodge, in Enga Province, appeared to be in fresh plumage, and a very early account by Valentijn (1724–26) stated that they moult during the eastern monsoon period December–April, and take four months to complete the cycle (Frith in F&B; Iredale 1950). Moult is basically known in most months, with a peak in November–January.

STATUS AND CONSERVATION Classified as Least Concern by BirdLife International, having a global range of around 300,000km². This is a common species in lowland and hill forest, and tolerant of disturbed areas provided that hunting is not excessive. The plumes are highly valued in traditional cultures and could be bought for K20 each in 1999 at Mt Hagen, the same price as in Tabubil in 1991. Hunters report having taken plumed males from leks over three generations, or about 60–100 years (F&B), under traditional hunting regimes, which are now breaking down. Nevertheless, given its wide geographical range and adaptable behaviour, the species is thought not to be under any major threat at present.

Lesser Bird of Paradise, male in flight, Kama lek, Enga Province PNG, July 2016 (*Stephen Rannels*).

Lesser Bird of Paradise, female, Cenderawasih Bay, Indonesia (*Dicky Asmoro*).

Lesser Bird of Paradise, male, Cenderawasih Bay, Indonesia (*Dicky Asmoro*).

GOLDIE'S BIRD OF PARADISE
Paradisaea decora   **Plate 21**

Paradisea decora Salvin and Godman, 1883. Fergusson Island.

Etymology The specific name is the feminine form of the Latin *decorus*, meaning beautiful or becoming. Named after Andrew Goldie, a Scottish plant-collector who discovered the species in 1872 (*cf.* Goldie's Lorikeet).

Recent genetic work shows this species as having evolved about 2.8 mya, in the late Pliocene epoch. Endemic to the hill forests of Fergusson and Normanby islands in the D'Entrecasteaux Archipelago, where it is the only member of the genus and is presumably a sister-form of the *P. apoda* superspecies, as it has some similarities in male plumes, head pattern and tail wires.

FIELD IDENTIFICATION This is a distinctive restricted-range species, the male with crimson-red plumes resembling a Raggiana, but with a lavender-grey breast lacking a breast cushion, while females are paler than those of Raggiana and finely but indistinctly barred with dark brown below. This is the only *Paradisaea* species in which the female shows diffuse ventral barring, and it is the only *Paradisaea* on the islands, where it is unlikely to be confused with any other species. **SIMILAR SPECIES** None in the range.

RANGE Endemic to the hill forest of Normanby and Fergusson Islands, in the D'Entrecasteaux Archipelago, south-east PNG. **MOVEMENTS** No significant movements recorded. This is a sedentary species of the hill forests.

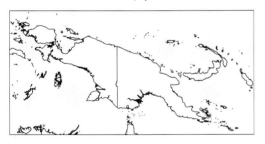

DESCRIPTION A distinctive island species, sexually dimorphic. **Adult Male** 33cm, plus tail wires of 368–536mm; weight 237g. The head, nape and mantle are shiny orange-yellow, the upperwing-coverts paler orange-yellow, rest of wing olive-brown with paler yellowy fringes; the back and tail are dark brown. Has a striking iridescent green frons and throat, with blackish chin edged by a very narrow yellow band before the unique lavender-mauve breast, which merges into vinaceous-beige underparts. The elongated blackish central tail wires have iridescent dark green vanes, and the filamentous flank plumes are crimson-red with paler greyish-buff tips. *Iris* yellow; *bill* blue-grey; *legs and feet* browny grey with purplish tinge. **Adult Female** 29cm. Distinctive: excluding the more distantly related Blue Bird of Paradise (now in genus *Paradisornis*), this is the only species of *Paradisaea* in which the female

has barred underparts. The yellowish-buff crown and nape shade into the brownish-olive upperparts, while the forehead, chin and throat are dark brown; the underparts are cinnamon, darker on the breast, and with indistinct narrow greyish barring. Bare parts as for male. **Immature Male** resembles adult female at first and gradually attains full male plumage, developing a brighter orange-yellow on the head and nape. Some males retain the barred underparts and have shorter tail wires, while another stage is like the adult male but with the flank plumes short and cinnamon-brown, and with traces of barring on the sides of the breast, with lower breast and belly chestnut-brown and barred as on the female. **Juvenile** undescribed.

TAXONOMY AND GEOGRAPHICAL VARIATION Monotypic. No significant variation recorded.

VOCALISATIONS AND OTHER SOUNDS Voice described as loud, varied and unmusical, and some calls are rather different from those of its congeners. LeCroy (1981) is one of the few observers to have seen this restricted-range species, and she reports that the voice at the lek is a ringing *wok-wok* male-to-male contact call, usually given when no female is present, and *whick-whick*, a low-intensity male contact version of the next call, which is a very loud, liquid ringing *WHICK-WHICK*, usually given when a female is present but occasionally when dropping out of the tree; a low growling call, given at the start of a display bout with a female present, and a loud duet by two males displaying together, often in the presence of a female. The duet commences with a loud, ringing metallic *WAAK*, given alternately by the males, and becomes progressively more rapid until a continuous metallic rattling or 'gargling' is emitted by both birds. This vocalisation, the characteristic call at an active lek, carries some distance through the forest and involves only two plumed males perched 1.5–3.5m apart, usually facing each other and often with plumes raised. The two may run up and down their perches while duetting, and any other males present do not take part. This species also utters the typical *waugh waugh* calls of this genus but rather guttural and drawn out (pers. obs.); see xeno-canto XC42106 (by J. Dumbacher). David Gibbs recorded the distinctive loud, clear explosive Black Sicklebill-like *whit whit whit* calls, plus a softer *wuk wuk whik-whik-whik-whik* series in November 1993; see xeno-canto XC70356. There is also a good series of this species by E. Scholes and T. Pratt at the Macaulay Library of the Cornell Lab of Ornithology site.

HABITAT Inhabits hill forest, forest edges and second growth in the hills of Normanby and Fergusson, sparsely in lowlands but mainly 350m to at least 600m and probably higher.

HABITS Little information available. The species is reported as visiting tree knotholes containing water, for drinking and bathing purposes.

FOOD AND FORAGING Presumably largely frugivorous, but there are few relevant data. Seen to feed close to leks on the capsular fruit of *Medusanthera laxiflora*; observed also to forage in the forest-floor leaf litter, which local inhabitants report as a regular occurrence, but unusual for the genus.

BREEDING BEHAVIOUR Very poorly known, with nest and eggs undescribed, although local people informed LeCroy *et al.* (1980, 1984) that the birds make a hole in a bird's-nest fern (*Asplenium*) to nest, which seems possible but unlikely. Presumed to be polygynous, the males displaying in loosely formed arboreal leks. These leks are formed in adjacent trees, the one studied by LeCroy *et al.* (1980) consisting of four main trees, with a fifth used occasionally, on a south-facing slope just below a ridge crest at 400m in mature rainforest. The canopy was about 30–33m high, and the display trees included at least three species with tall straight trunks and shallow crowns, not densely foliaged. This lek was spread out over a rectangle of about 92 × 46m, and is unusual for the genus (*Paradisaea* leks are mainly in single trees or several trees near one another). Display sites were in the mid-canopy 20–23m above the ground on larger open limbs near the tree trunk; the outer branches were used only briefly for displays. Some 8–10 plumed males performed here, usually two to each tree; no more than two females and six unplumed males were seen at any one time on the lek. Leks appear to be about a mile (1.6km) apart, with many birds able to hear other birds at least sometimes along ridges. Lekking males regularly pluck leaves near their display limbs.

COURTSHIP/DISPLAY A superb and extensive archive of videos by T. Laman, showing this species in display, is located at the Macaulay Library of the Cornell Lab of Ornithology site; ML469205 has a very nice sequence of the males at the lek. This species differs significantly from its congeners in various aspects of the display, with a more dispersed lek site, frequent duetting by plumed males, the presence of unplumed males which mate with the female during the displays, and the lack of several elements shown by congeners, such as the Flower Display. The following details are taken from LeCroy *et al.* (1980). *Convergence Display* At 06:10 hours on 11th November 1978 three plumed males were present at the lek, two of them duetting as the third left the area, together with up to six unplumed birds of which two were thought to be female because of their smaller size and quiet behaviour. The four unplumed presumed males were actively hopping about perches and 'rowing' their wings. Calling and display by the duetting males increased as they now and then chased unplumed males, but they mostly ignored the latter. Display postures were a wing pose similar to Greater Bird, but with the wings held not so far forwards and straighter down at the sides of the body and moved with a 'rowing' action, with the tail and body held in a more horizontal position. *Static Display* An adult male stopped posturing at the peak of the display and moved away from the main display area, perching quietly during the rest of the courtship. The remaining male continued his display, his movements becoming slow and the display almost static with no audible vocalisation. The unplumed males moved in to surround the female closely, and two or three of them started rowing movements of their wings; the remaining plumed male did not attempt to drive them away. The female perched quietly near the displaying plumed male, and left the main area several times but returned almost immediately. The plumed male retained his plumes in display position throughout this, but accompanied her a few times. The female began to solicit the displaying male by quivering her wings, held slightly out from the body, while the male continued his slow and rhythmic display; the female solicited for five minutes, during which time one unplumed male moved in and copulated with her, while two unplumed males in succession also copulated with her. These copulations were brief, lasting only a few seconds, and there was no preliminary neck-rubbing, although the unplumed male did once put his wings down around the female's body. The female did not stop soliciting after these copulations, remaining near the displaying plumed male. *Copulatory Sequence* After about 30 minutes of display, the plumed male began hopping stiffly up and down near the soliciting female. He edged over to her, put his neck and breast on her back and rubbed back and forth. Then he mounted her, brought his wings down around her body, and the two copulated. While still mounted he rubbed his neck and breast on her back again, and they copulated again. The copulatory sequence lasted for about 30 seconds in total.

'*After the plumed male had copulated with the female, all of the birds remained in the tree and the entire sequence of events was repeated, starting with the duetting and joint displays by the two plumed males. We had no way of knowing whether or not the male that now displayed alone was the one that had displayed alone during the previous display bout. The unplumed males were around as before, were chased away on several occasions during the duetting and joint displays by the two plumed males, but were tolerated during the period of intense display by the single adult male. Once again they copulated briefly with the soliciting female. The displaying plumed male copulated with the female as before, dismounted and moved immediately to the second female and copulated with her – this copulation lasting only a few seconds. The first female moved away about 3.5m, after the male dismounted, and began preening. After the second copulation all the birds left the tree. It was then 0715.*'

Another display sequence ending in copulation when no unplumed birds were present was observed on 8th November during 12:00–12:20 hours. When the single female appeared in the lek tree, two plumed males were present. She was greeted with low growling by at least one male. For the most part the second male sat quietly in the same tree, but twice approached and displayed briefly near the female. The main display perch in this tree was a sharply sloping limb. The female was on a horizontal limb branching from it. The male hopped slowly up and down the sloping limb, zigzagging his body back and forth as he did so, so that he presented alternately back, front and side views to the female. She watched him continuously, sometimes facing him, sometimes peering sideways at him with head cocked. This female appeared quite wary and frequently flew to the outer branches, where the male followed her and displayed. They always returned quickly to the main display perch. Movements by the male around the female were always slow and deliberate; several times he gave high-intensity *whick whick* calls in front of the female, but generally no calls accompanied the display. After 25 minutes of display, the female began soliciting and the male gradually moved closer to her, hopped up and down on the branch next to her, and mounted her with his wings down around her body. Copulation lasted about 15 seconds before both birds flew from the tree. No neck- or breast-rubbing was observed during this episode. **NEST, EGGS & NESTLING** Unknown. Nestling undescribed. **TIMING OF BREEDING** Lek displays and mating have been seen in the second week of November (LeCroy 1980). One male collected in September had much enlarged gonads.

HYBRIDS No hybrids known, as no other close relatives are found within this species' range.

MOULT Not well known, but specimens show moult in January, March–April, May, July and December (F&B). Fully plumed males seen in April.

STATUS AND CONSERVATION Classified as Vulnerable by BirdLife International. Recent research, however, indicates that the total population numbers as few as *c.*650 individuals, with a maximum of 500 on Fergusson Island (at Maybole Mt, Oya Tabu Mt, Edagwaba Mt, Sebutuia Bay lowlands, Lavu Lowlands and Lamonai) and 150 individuals on Normanby Island (at Lomitawa, Mount Solomonai, inland Sewa, Lonana and Mount Hobia) (D. Mitchell *in litt.* 2008). Surveys on Fergusson and Normanby indicate declines of *c.*20% from *c.*1997 to 2007 (D. Mitchell *in litt.* to BirdLife 2008). On the basis of these figures,

the total number of mature individuals is estimated at 450, with *c.*350 on Fergusson Island. The BirdLife classification may thus require revision. Habitat loss and degradation resulting from logging, mining and clearance for agriculture are ongoing threats. In 2013, the resumption of logging in the East Fergusson Timber Rights Purchase was in the second of its five years of logging (D. Mitchell *in litt.*). On Normanby Island, mineral exploration is taking place in areas close to populations of this paradisaeid. Elsewhere on Normanby, the expansion of subsistence agriculture has resulted in the replacement of previously occupied habitat with gardens (D. Mitchell *in litt.* 2013). It is recommended that the monitoring of this species in recent years be continued, and that further research on its tolerance of degraded forest be conducted. It is vital that significant areas of remaining primary forest be protected and that the large-scale development of forested areas on the islands be reduced or, better, prohibited. Logging and mining operations should be limited by means of agreements with government and the private sector. Any large-scale agricultural expansion for cash crops could be extremely detrimental to this species. Goldie's Bird of Paradise was reported as fairly common by early collectors, but there have been few modern-day visits to these islands and continuous assessment of the potential threat to population levels is desirable. In 1897, A. S. Meek (in Gilliard 1969) found it not rare but by no means very numerous on the hills of interior south Fergusson, from about 1,500 feet (*c.*460m) upwards, which just about sums it up. There are no data on this species' longevity in the wild.

Goldie's Bird of Paradise, two males displaying, Fergusson Island, PNG (*Tim Laman*).

Goldie's Bird of Paradise, female, Fergusson Island, PNG (*Tim Laman*).

Goldie's Bird of Paradise, a male and female, Fergusson Island, PNG (*Tim Laman*).

RED BIRD OF PARADISE
Paradisaea rubra   Plate 21

Paradisea rubra Daudin, 1800, *Traité Orn.* **2**: 271. Waigeo I, north-west Islands.

Etymology The name *rubra* is the feminine form of the Latin *ruber*, meaning red.

Recent genetic work shows this species as having evolved about four mya in the mid-Pliocene period. It is an island endemic found on Batanta, Waigeo and Gam, in the Raja Ampat Islands of West Papua, where it is the sole member of its genus.

FIELD IDENTIFICATION Correct identification is straightforward, as this species is the only member of the genus in the islands. The bright crimson-red plumes of the male are very distinctive, as are the very long and twisted black central rectrices. Females are typical of the group, with extensive yellow hindcrown and nape reaching in a broad yellow band across the chest. Rusty Shrike-thrush (Pitohui) (*Pseudorectes ferrugineus*) of the local subspecies *leucorhynchus* (Waigeo) is notably smaller, has a pale eye with pale orbital ring and striking pale yellow bill, and utters a loud musical call quite unlike the harsh voice of this paradisaeid. **SIMILAR SPECIES** None.

RANGE Confined to the islands of Batanta, Waigeo and Gam, and the two small islets of Gemien and Saonek (off Waigeo), in the West Papuan Islands.

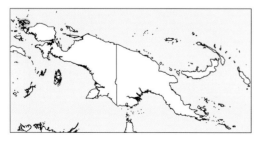

MOVEMENTS Nothing significant recorded.

DESCRIPTION Sexually dimorphic. *Adult Male* 33cm, with two unique long shiny black plastic-like tail wires that hang down in a spiral twist, each being 3–4mm wide with an average length of 56cm (these lack the branches off the central shaft that characterise a typical feather, resembling more the strange plastic-like modification shown by the platelet head wires of the King of Saxony Bird). A distinctive species, the face and throat being iridescent metallic dark green with a black chin, the green extending over the front of the head to behind the eye. The head has a curious bumped shape owing to a distinctive erectile iridescent scaly-feathered green cushion above each eye which can be raised to form blunt bumps. Back of head, nape and mantle shiny iridescent orange-yellow, merging into brown of remainder of upperparts and wings, with orange-yellow bend of wing and wing-coverts. The upper breast feathers are orange-yellow, somewhat elongated at sides and framing a kind of breast shield; the lower breast is dark brown, merging into the paler belly and undertail-coverts. The filamentous lace-like flank plumes are not so long as those of congeners, and are of a striking bright crimson, with whitish filament tips extending 10–12cm beyond the tail; these flank plumes are stiffer than normal and have a compact curved shape that is retained even when the bird is inverted, not forming a flower-like cascade as with some congeners. *Iris* reddish-brown (not yellow as with most congeners); *bill* yellow (not the usual blue-grey typical of the genus); *legs* bluish-grey. *Adult Female* 30cm. Face and throat are dark brown, with buffy-yellow rear crown, nape grading into dark brown of upperparts, and upperwing-coverts broadly edged buffy yellow. Upper breast forms a broad yellow band, and rest of underparts are brownish, paler on flanks. Central pair of tail feathers is narrower and more pointed than the rest. *Juvenile* No data on wild individuals, but a captive began to acquire a yellow breast at 60 days of age. *Immature Male* resembles adult female but is larger.

Subadult Male exhibits a variable mixture of male and female characters depending on age and state of moult, with central tail feathers gradually lengthening. Some resemble the adult but lack flank plumes, have the tail wires partially developed, and have an orange-brown (not yellow) back and brownish breast and belly. Adult plumage is acquired in 5–6 years.

TAXONOMY AND GEOGRAPHICAL VARIATION
Monotypic. No clear variation within the relatively circumscribed range.

VOCALISATIONS AND OTHER SOUNDS These birds are very vocal early in the morning and again in the late afternoon as is typical of the group, but they may call at intervals throughout the day. A common call on Batanta is a loud *wak* note, leading to a *wok wau wau wau wau wau* series typical of the group (Burrows and Gregory pers. obs.). The following vocal data are from F&B (1998). Captive individuals gave a loud clear *wak* note, building up to this with a throaty guttural *work-wok, wak wak, wak wak, wak* which leads into the clear and loud *wok-wau-wau-wau* sequence at about 1–4 notes per second. A similar loud but higher-pitched call was a *ca-ca-ca-ca-ca-ca* delivered at 4–5 notes per second. A commonly given crow-like call consisted of a coarse and guttural *kaw kaw kaw* at 1–2 notes per second, and what sounds like this note is often heard on Batanta during the day (pers. obs.). Captives often emit a soft pathetic-sounding *weep* note, and a very high-pitched, mewing *meew* was not uncommon, as was a single snap of the mandibles, *tick* (F&B). A loud resonant *whit-chow* or *whit-chow chow chow* was a frequent call on Waigeo in June. Many variations of these calls were heard, but the crow-like one was the most frequent and varied in tone, volume and duration. A series of long recordings by T. Laman and E. Scholes made at display sites of this species is at the Macaulay Library at the Cornell Lab of Ornithology site. Displaying adult males, prior to intense display, often give an odd bill-click 'call' by rapidly and repeatedly snapping the mandibles together, this often ending with a brief vocalisation. The individual *tick* notes, repeated four or five times, are preceded by a high-pitched *beep* and end with a guttural *book* note, thus *beep-t-t-t-t-book*, a mechanical sound that lasts for just over a second. This bill-click call is a display sound, given with stylised body movements and most often before intense displays. A continuous bill-clicking without other vocalisations or body movements was also made by one male chasing another male in flight. The soft mewing note and the single snap of the mandibles are infrequently given by displaying males.

HABITAT Inhabits lowland and hill forest, with an apparent liking for tall trees, from sea level to about 600m. Found at the base of steep ridges on some parts of Batanta (pers. obs.) and in tall forest on steep limestone on Waigeo.

HABITS On Batanta this species is quite noisy in tall forest, though plumed males could be surprisingly inconspicuous perched high atop trees (pers. obs.). Captives have been observed sunning, with one wing drooped and the body leaning away from the sun, with head held high and the bill pointing upwards or tilted to one side (see illustration in F&B). Birds at lek are quite wary and keep largely within cover until satisfied that there is no threat, while in the forest the species likewise keeps to dense cover and vine tangles high in the tall trees, seldom giving a clear view. Individuals have been observed to break off leafy twigs with the bill, presumably to give a clear view of the display area (C. Eastwood *in litt.*).

FOOD AND FORAGING Wallace (1869) observed birds of this species on Waigeo 'quite low down, running along a bough searching for insects, almost like a Woodpecker; and the long black riband-like filaments of his tail hung down in the most graceful double curve imaginable… but the gun missed fire and he was off in an instant among the thickest jungle'. Wallace also observed them eating the fruits of a *Ficus* species and shot two, before they ceased to visit the tree 'either owing to the fruit becoming scarce or that they were wise enough to know there was danger'. Captive Red Birds of Paradise (Frith 1976) were often seen to seek insects, either by pecking off the bark of dead twigs to probe the soft wood beneath or by carrying pieces of dead bamboo or leaves to a perch, where they tore apart these items of vegetation. Individuals also clambered about the fine outer twigs and foliage of shrubs to snatch insects from the leaves, and one bird hung upside-down to swallow small buds.

BREEDING BEHAVIOUR Polygynous, with promiscuous communally displaying males. Females presumably nest-build and rear the young unaided, as with other species in the genus. Seems rather shy and wary. An archive of videos by T. Laman showing this species in display is at the Macaulay Library of the Cornell Lab of Ornithology site, where ML465503 shows some of the male display. The lek trees are tall, these birds preferring near-vertical or sloping bare branches for their displays; these may be dead branches or branches defoliated by the birds (Eastwood 1996), most likely the latter as they seem still pliable and able to support the weight of a dancing bird, even amplifying his movements as he dances. From the scant data available, up to ten plumed males have been seen at lek (Bergman in Gyldenstope 1955), as have two, three or four individuals. **COURTSHIP/DISPLAY** A male on Waigeo in June 2015 displayed at intervals during 06:40–07:00 hours on a calm day, the birds avoiding display when it is too windy. It is possible that more than one individual came to dance, but only one bird at a time occupied the central branch. One display sequence lasted nearly three minutes; others were shorter, the male flying off with tail ribbons rippling without going into the full sequence. The peculiar long twisted ribbons of the male hang down and at times seem to frame the body in a flattened M-shape of wires to each side, rippling and swaying as the bird dances, he progressively moving up the slender bare display branch and stopping to sway rhythmically from side to side as he does so. When in the more intense display he spreads out the wings and quivers them rapidly, the red coloration of these together with the flank plumes and the body making an extraordinary vivid red spectacle. The standardised *Paradisaea* display sequence is detailed below, with data from Frith (1976) based on limited observations of ten adult males and a single female over six days in a Singapore aviary, which will not be the full picture. *Convergence Display* Excited males hop quickly from branch to branch with fluttering, flicking and extended wings, then land on their individual display perches. Males usually begin displaying on near-vertical branches, and Crandall (1937a) made the interesting observation that displays by captive birds did not commence until a downward-sloping branch at about 45° was made available, leading him to conclude that such a perch may be necessary for the full performance. The male gives the bill-click 'call' frequently, flicking the outer primaries back and forth, and occasionally bill-wiping. Once on the vertical branch, he perches diagonally with bill pointing upwards, and continues to flick the primaries. The wings are slightly spread and quivered as he becomes more excited, and he again wipes his bill and sways gently from side to side. Swaying stops, and he leans increasingly to one side, the head becoming lower, and spreads and shakes his wings while still flicking the primaries, until he is completely inverted. He then spreads his wings still further, shaking them very rapidly so that they vibrate, while he sways from side to side, making the flank plumes conspicuous but not raised or spread. He may bill-wipe occasionally, and move the head about rather stiffly in a peering motion. *Static Display* The male may have given the soft *meew* call, or more often a single snap of the mandibles. When completely inverted, he shakes his wings, fully extended at each side of the body, but does not raise them above the back. He vibrates them rapidly but so slightly that it is only just discernible as he hangs momentarily motionless, except perhaps for peering as if looking intently at the perch and giving an occasional mandible snap. The flank plumes are raised very slightly but not spread, and the tail is held slightly downwards so that the plumes are a little above the tip of the tail. The *Copulatory Sequence* is uncertain, but the following may be a component of it, taken from F&B and based on aviary birds (where only a single female was present and the perches were very low down). Suddenly the bird reverts to the normal non-display position, hops back down the perch, sometimes in a spiral fashion, until almost at the base, which may be on the ground in captivity. He then hops once around the branch and back up to the original spot or higher, sometimes in a spiral fashion but usually by hopping to face alternate sides. He hops down and up the perch in a slow and deliberate ritualised gait. Having ascended, the bird then performs a hopping-on-the-spot display, slowly and deliberately moving the head and body in a peculiar mechanical fashion. Hopping on the spot may be uninterrupted, or may be interspersed with bouts of head peering or bill-wiping. Suddenly he again opens his wings and becomes almost motionless except for a slight rapid vibration of the wings and continuous peering. He then begins to sway slightly from side to side, pointing the bill upwards much of the time, continuing to head-peer and occasionally giving a single bill click. Finally he augments the

head peering by tapping the sides of the utmost tip of his bill sharply on the perch every few seconds while rotating the head with its green plumage conspicuously erected. The bill-tapping produces a tick sound, which is clearly audible at 10m. This completes the entire captive display, lasting 45–120 seconds, the bird then hopping from the display perch to preen or feed (Frith 1976). Copulation was not observed, suggesting the possibility that sequences were misplaced, aberrant or missing, and data on wild-living individuals are highly desirable in order to give the full picture of the display behaviours of this species. The bill-click sound is accompanied by stylised movements that seem to be some sort of display. When the initial *beeb* note is uttered the head and bill are thrust downwards to the breast; during the bill-clicking the bill is thrust upwards, and during the concluding *book* note the head is again thrust downwards and then raised to the normal position. During his performance the male jerks his body very rapidly but almost imperceptibly, while hopping a short way along the perch and back again. The whole action appears mechanical and the sudden jerky body movements are emphasised by the stiff flank plumes. Poulson (in F&B) reports having seen wild adult males hanging upside-down while quivering their wings and stropping the bill from side to side against a branch, with the red plumes in close contact with the face of a female perched above. The male then climbed down the vertical branch for a short distance, which sounds like a part of the captive-bird behaviour described above. NEST & EGGS Data only on captive individuals. The female alone builds the nest and incubates, as would be expected (Isenberg 1961, 1962; Todd & Berry 1980; Hundgen *et al.* 1991; Worth *et al.* 1991). The clutch consists of one or two eggs, size 34.8–39.1 × 23.8–25mm from seven examples in the NHMUK (F&B). The eggs are incubated for around 14–17 days. NESTLING Undescribed. The young chicks leave the nest between 15 and 20 days after hatching.

HYBRIDS None, as no other close relatives occur within the species' range.

MOULT Data from F&B (1998), referring to captive individuals; no information on wild-living birds. Twelve specimens revealed active moult in all months except February, April and August. Fully plumaged males have been seen in June. A captive immature male in the New York Zoological Society collection took about three months to moult and a subsequent moult took four months (Crandall 1932). Captive females moulted two months before males (Isenberg 1961).

STATUS AND CONSERVATION Classified as Near Threatened by BirdLife International, and occupying an area of around 3,900km², with the population believed to be declining as a result of habitat damage and hunting. The species was fairly common on Batanta in 1994 (pers. obs.), and patchy on Waigeo, although it was reported to be common in the central parts of Waigeo in 1948 (Bergman in Gyldenstope1955b). Parts of both islands are designated forest reserves, but how meaningful this really is may be open to question. Trade in plumes was historically and up to recent times quite significant (Severin 1997; Swadling *et al.* 1996), and having such a small natural range means that the species is potentially vulnerable. Destruction of forest habitat is the primary threat affecting the survival of the Red Bird of Paradise. On Batanta in particular, a lack of protected areas means that major habitat loss has occurred through logging, while on Waigeo selective logging and a cobalt-mining concession pose further threats to the remaining suitable habitat. Although the Red Bird of Paradise is known to inhabit areas of the protected Pulau Waigeo Nature Reserve on the island of Waigeo, in the Raja Ampat Conservation Area, there are concerns that the reserve has been vastly reduced in size in recent years. Observations concerning the state of the habitat, numbers of individuals of this species in the wild and details of the display cycle in the wild are desirable.

Red Bird of Paradise, female plumage, Waigeo Island, West Papua (*Gareth Knass*).

Red Bird of Paradise, male, Waigeo Island, West Papua (*Gareth Knass*).

Red Bird of Paradise, male, Waigeo Island, West Papua (*Gareth Knass*).

EMPEROR BIRD OF PARADISE
Paradisaea guilielmi   Plate 23

Paradisea guilielmi Cabanis, 1888. Sattelberg, Huon Peninsula.
Other English name Emperor of Germany's Bird of Paradise
Etymology The Medieval Latin *guilielmus*, meaning William (genitive *guilielmi*, of William or William's), after Willem II, Emperor of Germany and King of Prussia, perhaps better known as Kaiser Wilhelm.
Recent genetic work indicates that this species evolved about six mya in the late-Miocene period.

FIELD IDENTIFICATION The white flank plumes, pale straw-yellow nape, mantle and shoulders, sides of breast and an incomplete pectoral band of the same colour, along with extensive green on the head and upper breast readily identify the male Emperor. In addition, the flank plumes are shorter and more lace-like than those of its congeners. The female is similar to the Raggiana female, but has a dark crown as well as more extensive dark on the face and chest, and, with dark eyes and brighter richer yellow on the hindneck and mantle, is more richly coloured than the local Raggiana race *augustaevictoriae*. **SIMILAR SPECIES** Raggiana and Lesser Birds of Paradise differ in the colour of the flank plumes and the amount of green on the head and chest, and lack the pale straw-yellow nape, mantle and shoulders of the Emperor. Female Lesser Birds are white beneath, while female-plumaged Raggiana have much less dark on the face and crown than the female-plumaged Emperor, although care should be taken in overlap zones and with subadults. Other large, rusty forest birds within the range are quite limited: Huon Bowerbird has a stout blackish bill and is notably smaller, with quite different proportions, and Rusty Shrike-thrush (Pitohui) (*Pseudorectes ferrugineus*) has a pale eye and also is much smaller.

RANGE Confined to the lower mountains and hills of the Huon Peninsula in Morobe Province, north PNG. Found on the Sattelberg, where originally collected

Emperor Bird of Paradise, male, Huon Peninsula, PNG, 2017 (*Jun Matsui*).

Genus *Paradisornis*

The single, striking and distinctive species in the genus *Paradisornis* is endemic to the mid-montane zone of the central cordillera of eastern and east-central PNG. The genus is characterised by the distinctive patterning of the black and unique cobalt-blue plumage, and the likewise unique blue pectoral plumes of the male, as well as the white eye-crescents present in all plumages. Interestingly, when first described in 1885 this species was placed in the genus *Paradisornis* and, although it is closely allied to the genus *Paradisaea*, it is distinct from the latter in more than a dozen plumage, behavioural and vocal characters as detailed below and in Frith & Beehler (1998: 61–62). The phylogeny of Irestedt *et al.* (2009) places *rudolphi* as sister to the *Paradisaea* clade (Pratt & Beehler 2016), but not part of that genus.

Etymology The genus name is from the Latin *paradisus*, meaning paradise, and the old Greek *ornis*, meaning bird.

BLUE BIRD OF PARADISE
Paradisornis rudolphi   Plate 23

Paradisornis Rudolphi Finsch and Meyer, 1885, *Zeitschr. Ges. Orn.* 2: 385, pl. 20. Mt Maguli, Southeast Peninsula. Discovered and collected by Carl Hunstein in 1884.
Other English names Prince Rudolph's Blue Bird of Paradise, Archduke Rudolph's Blue Bird of Paradise, Blue Kumul
Etymology The species epithet *rudolphi* honours Archduke Rudolph (1857–89), Crown Prince of Austria–Hungary, who tragically committed a murder–suicide with his mistress at Mayerling in 1889. The dedication was to 'His Imperial and Royal Highness Rudolph, Crown Prince of Austria, the high and mighty protector of ornithological researches over the whole world'; his wife was the Princess Stephanie of Stephanie's Astrapia fame.
Recent genetic work dates this species as having evolved about 9.4 mya in the late Miocene period.

An uncommon to rather rare endemic of the central cordillera of eastern and central PNG, it is one of the most beautiful, spectacular and bizarre birds in the world, with less marked sexual dimorphism than that found in the genus *Paradisaea*. Indeed, it is a highly distinctive species that is now placed back in its original genus, *Paradisornis*, as it is so divergent from the members of *Paradisaea*, within which genus it was for long subsumed.

FIELD IDENTIFICATION Very different from all the *Paradisaea* and should, given a reasonable view, be readily distinguished in any plumage. Adult male has the entire head, neck and mantle glossy black with various other sheens, the back and rump blackish, and the wings and uppertail bright cerulean-blue. There are conspicuous white crescents above and below the eye, and a large pale blue-grey bill. The flank plumes are blue below and rusty above, and the central rectrices are greatly elongated into dark strap-like ribbons with a blob on the end. The female is less different from the male than is the case with the other plumed paradisaeids: she lacks the flank plumes and elongated rectrices but otherwise resembles the male, although she has distinctive rusty-buff underparts which are barred or just slightly barred depending on the race. **SIMILAR SPECIES** None. Raggiana Bird of Paradise has a similar shape but always has the chestnut plumage tones, without any blue feathering. Other paradisaeid species which are locally sympatric are Lawes's Parotia, males of which are blackish with six head plumes and lacking any blue feathers; Carola's Parotia, which has head wires and large white wing patches; Superb Lophorina, which is smaller with an obvious blue triangular chest adornment, quite unlike the Blue Bird of Paradise; and the much smaller and quite differently shaped Magnificent Bird. Females of all of these have brown and barred plumage patterns with superciliary stripes, and should not be confusable with the Blue Bird, which always shows the distinctive blue on the wings and the striking white eye-crescents.

RANGE A PNG endemic of the central and south-east main mountain chain, ranging in the east from the Owen Stanley Range north-west to the Herzog Range and Okapa in the Eastern Highlands, and westwards to Mt Hagen, Kompiam and the Tari valley of the Southern Highlands. **MOVEMENTS** Sedentary within its habitat range, with no significant movements recorded.

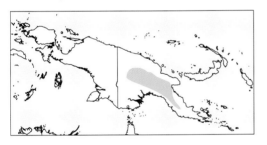

DESCRIPTION Sexually dimorphic. *Adult Male* 30cm, excluding tail wires of 341–458mm; weight 158-189g. The plumage is unusually complex even for this family, with the head, nape and mantle glossy black with an iridescent bronzy-green sheen, the rear crown and nape having a dark carmine-red sheen. The lower back to uppertail-coverts are black with a bluish sheen, the tail a darker blue than the wings. The wings are a muted cerulean powder-blue, with the primaries matt black with blue outer edges to all except the outer three. The central tail feathers are greatly elongated to form ribbons up to 458mm long, bluish-black in colour and with small oval spatulate tips that show iridescent centres in good light; these ribbons wave gracefully in flight, rippling behind the bird. Striking white half-circles above and below the eye form a broken eye-ring, which stands out very prominently in the black head. The breast is matt black with slight hints of dark cobalt-blue, more intense towards the belly, which is jet-black with blackish-brown thighs and undertail-coverts. The long filamentous flank plumes appear cinnamon-amber from above but are rich

dark purple-blue at the base, shading to cobalt-blue to mauve distally; a discrete flank patch on each side of the belly is coloured black and crimson and forms a continuous line of those colours when seen in display. The underwing is dark greyish, with a blue wash on all but the outermost feathers, and the underwing-coverts are variably cinnamon to blackish. *Iris* dark brown; *bill* pale bluish-white, *mouth-lining* yellow-green or pale green; *legs and feet* purplish-grey. *Adult Female* 30cm; weight 124–166g. Head and upperparts resemble those of adult male but duller, more brownish-black, with only faint iridescence and no carmine on the head. The underparts are chestnut-brown, blackish on throat grading to dark cinnamon, with blackish-brown thighs; some variable black barring also occurs, reduced or absent on the central belly and sometimes seeming entirely lacking. *Juvenile* Like adult female but wings darker and bill dark greyish, with whitish abdomen. The body is dark sooty black above and rich rufous below, with dull blue on the wings, and the mantle and back dull blackish; lores and forecrown unfeathered, the head otherwise black with a brownish tinge. *Iris* dark brown and *legs and feet* grey. *Immature* Much as adult female, but male with central tail wires progressively appearing. *Subadult Male* Much as female but with some male-type feathering showing, gradually assuming full male plumage.

TAXONOMY AND GEOGRAPHICAL VARIATION Polytypic, although Frith & Frith (1997d) merge previously recognised but weakly differentiated form *ampla* (Greenway, 1934) with *rudolphi*, and *P. r. hunti* Le Souef, 1907, is likewise synonymised. Pratt & Beehler (2016) recognise two marginally differentiated races.
1. *P. r. rudolphi* Finsch, 1885. South-east PNG, from the Owen Stanley Range north-west to the Herzog Range and to Okapa in the Eastern Highlands.
2. *P. r. margaritae* Mayr and Gilliard, 1951. Central PNG west of the range of nominate form, in the Bismarck Range, Sepik–Wahgi divide, Kubor Range, Mt Karimui west as far as Tari, Mt Giluwe, Mt Hagen and the highlands of Enga Province; western limits not yet fully known. Like the nominate form, but females have underparts uniformly and narrowly barred blackish and tail and tarsus on average shorter.

VOCALISATIONS AND OTHER SOUNDS The male's advertising call is a loud, nasal, bugling *KRO KRO KRO* series, sometimes given quite rapidly, at other times more drawn out, and a plaintive slightly rising nasal *kuwah*, delivered from high in a tree or dead-branch songpost during the early morning, from about 06:30 to 08:30, and seldom heard at other times. It is vaguely reminiscent of some Raggiana Bird calls but is slower and deeper, with a more bugling quality, and carries more than 1km across the highland valleys. A recording on xeno-canto at XC148731 (by P. Gregory) features this advertising call, as does ML200901 (by E. Enbody) at the Macaulay Library at Cornell Lab of Ornithology. Also given is a strange raspy, rather metallic-sounding *kraar kraar kraar*, quieter than and quite different in tone from the loud advertising calls. Harsh scolds reminiscent of those of Superb Lophorina are also emitted, and scolding *skss* notes are heard from juveniles. Displaying birds when hanging upside-down give a constant ventriloquial nasal chatter, with low *kaw* notes mixed in and various clucking, chittering and chattering notes. Other display calls start with guttural *kwaah* and quiet throaty single *ka* notes, and then in the intense phase, given when the female is close by, a unique throbbing, pulsing, vibrating electronic-sounding series lasting up to 25–30 seconds, followed by quiet *kar* and *kwaah* notes. This bizarre electronic-sounding unearthly call is recorded by E. Scholes on ML163699 at the Macaulay Library at Cornell Lab of Ornithology site, and is surely one of the strangest of all avian sounds. It is rarely heard by casual observers, as the birds display low down in out-of-the-way places.

HABITAT Occurs in lower montane forest, particularly *Lithocarpus* and *Castanopsis* oak forest, which seem to be important for males to hold territory, and also denser and more overgrown gardens. Found at 1,100–2,000m, but mainly 1,400–1,800m. This species inhabits a heavily populated mosaic of gardens and copses of tall trees at Kikita, near Tari, and below Ambua Lodge; also near Anjiwalya, in the Wahgi valley. It does not like very disturbed areas, and tall trees may be important for it, with Raggiana Birds in the less pristine environments. In the Wahgi and Anji valleys it is still found on steep slopes and ridges, but is now relatively uncommon or even rare there as a result of clearance and hunting. Pruett-Jones & Pruett-Jones (1986) found that on Mt Missim the adult males of this species inhabited the centre of the altitudinal range, young males tending to occupy the upper and lower extremes.

HABITS Males are generally solitary, though sometimes in twos, and can be difficult to observe; conversely, they may at times be quite viewable if not exactly confiding. Female-plumaged birds can form small groups at fruiting trees. Gillison (in F&B) noted a male displaying near a court of Lawes's Parotia, and territories of this species are often near Blue Bird territories in the Tari valley (with one hybrid known). Males choose tall trees from which to call, often using bare branches near the top as songposts and with the sound carrying far across the valley.

FOOD AND FORAGING The species is frugivorous but, as with the *Paradisaea* species, will readily take insects and arthropods. *Omalanthus novoguineensis*, *Trema orientalis*, *Schefflera*, *Piper* and *Musa* (wild bananas) are significant foodplants (Hicks & Hicks 1988a; Hopkins 1988), while field data from Pratt showed *Gastonia spectabilis* to be very important (Beehler 1989). Beehler (1983a) calculated the diet of birds on Mt Missim as being 85% fruit, although crickets, grasshoppers, a spider and a wasp were also taken as prey, while Schodde (1976) found that cockroaches were taken. Opit (1975b) saw one Blue Bird tearing open a hollow bamboo to get at a large spider. The species will glean bark and search vines and creepers at most forest levels for arthropods (Coates 1990; Beehler in F&B), and a plumed male has been seen moving up a mossy trunk in the manner of a treecreeper (Smyth 1970). Female-plumaged

individuals may congregate in small parties in fruiting trees, but have been seen also to defend fruiting sources, driving away Lawes's Parotia, Superb Lophorinas and Magnificent Birds (Pratt 1984). Such species are regular at Ambua, where they share the fruiting resources such as *Schefflera* with other species such as Superb Lophorina, Lawes's Parotia, Brown Sicklebill, Princess Stephanie's Astrapia and Short-tailed Paradigalla, as well as Loria's Satinbird (*Cnemophilus loriae*), Tit Berrypecker (*Oreocharis arfaki*) and Mountain Fruit-dove (*Ptilinopus bellus*). They may partition such resources temporally, not defending the resource exclusively but often dominating it when present. Feeding individuals may range from the canopy down to near ground level. Pruett-Jones & Pruett-Jones (1986) found that on Mt Missim the adult males of this species inhabited the central areas, young males tending to be more towards the edges.

BREEDING BEHAVIOUR Polygynous; promiscuous males establish a solitary calling station. Females alone nest-build and carry out all duties at the nest, as with other plumed paradisaeids. This is one of the most extraordinary members of this remarkable family, with an astonishing and very rarely seen intense display accompanied by some unique and strange sounds. Remarkable photographs of displaying males by B. Coates and F. Pekus are in the Internet Bird Collection (IBC) site, and the Macaulay Library at the Cornell Lab of Ornithology has some terrific videos of this species by T. Laman and E. Scholes. **CALLING BEHAVIOUR** A male at Kikita, near Tari, had several calling stations in and around a copse of tall trees, from which it would call between 06:30 and at latest 08:30 from a near-horizontal subcanopy branch some 15–20m off the ground. Calls were heard in May, June, July, August and November in 1998, although the male was moulting in November and barely called at all (pers. obs.). The head is inclined slightly and the pale green gape becomes visible, but calling is often erratic, with 5–10 minutes between bouts, although Beehler noted a bird at Trauna Ridge calling every two minutes in June (Beehler in F&B). Whiteside (in F&B) noted adult males, with numerous advertising perches within a territory, singing while preening and foraging during the early morning at Kompiam, as did LeCroy (in F&B) at Ubaigubi, and this accords well with my own observations near Tari. Whiteside (in F&B) also recorded both Carola's Parotia and Magnificent Birds as possibly associating with Blue Birds at their low display perches, no doubt attracted by the activity. Hybrids with these species must be a possibility. **COURTSHIP/DISPLAY** Males perform solitary and relatively static courtship displays, often only a few metres above ground. Like *Paradisaea* species, they remove leaves from the immediate area of their display perches, and these perches are usually well shaded. Some males may display at sites of up to about 20m in the vegetation, but typically they choose places 1–3m off the forest floor on a slim, gently to steeply sloping branch, bamboo, grass or vine stem (F&B). Smyth (1970) was able to get within a few metres of an inverted, displaying male, but Whiteside (in F&B)

found the males very nervous and unapproachable which may reflect local hunting pressure. Pruett-Jones & Pruett-Jones (1988a) found that the home range on Mt Missim was just under 5ha, and the advertisement area was in a central core of about 1ha. Advertisement areas seemed not to overlap, whereas home ranges did so sometimes. The distance between males on two ridges of Mt Missim varied between 160m and 520m, with 15 males along 2km of one ridgeline and 12 males along 3.5km of the other ridge. Below Ambua it is possible to hear three or four males in auditory contact with each other, as the sound travels long distances in the clear mountain air. Displays are most frequently performed in the mornings between 06:00 and 09:30, but may extend to midday, and again but less frequently in the afternoon. Captive individuals studied by Coates displayed 12:30–17:30, but this may be an artefact of captivity. Whiteside saw up to four female-plumaged birds attending a display area, but only one at any time perched on the display stem (Whiteside in F&B). The courtship display is unlike that of species in the genus *Paradisaea*, and lacks a *Convergence Display*, as might be expected from the solitary habits of the species, consisting solely of a **Static Display** with apparently no ritualised *Copulatory Sequence* (F&B). The following display description from Coates (1990) is based on captive birds at Baiyer River:

The male lowers himself backwards off his perch, to hang upside-down below it, and then spreads his flank plumes in the form of an exquisite broad triangular fan with the apex at the bird's head. The wings are closed and the long tail plumes raised to half their length and then dropped down on each side behind the plumes. The head is turned to either side and the male begins to deliver a continuous rhythmic song, a mixture of low nasal chittering and chattering notes interspersed with low caws. While singing he jerks his head and constantly moves his plumes to send shimmering waves of blue and violet over them. The focal point is the black band with its purple trimmed upper margin in the centre of the abdomen, which becomes ovate in shape and expands and contracts as he at the same time throws the body forwards from the hips. The whole effect is stunningly beautiful (and one of the most outstanding bird displays ever). The head is turned to one side or the other, but sometimes points up. This display continues for minutes, and after a while the bird seems to go into a trance, with eyes almost closed and making the broken white eye-crescents even more conspicuous. He always knows the location of the female and twists his body to aim the display at her as she moves from perch to perch.

A bird filmed by E. Scholes at ML456531 on the Macaulay Library at Cornell Lab of Ornithology site can be seen hanging upside-down for over a minute but, while throbbing and waving the flank plumes, he does not enter the intense form of the display. A longer sequence is featured at ML456794. The display may cease, or the bird may give a couple of caws and enter a new and more intense phase. This is when an incredible sound is produced, described as a fast rhythmic buzzing but sounding like no other bird noise in the world, an unbelievable pulsing, throbbing, mechanical wavering noise like something from a science-fiction film. The Tok Pisin phrase *kisim pawa* is very apt,

meaning 'to go electric', and it does indeed sound like an extract from some bizarre electronic experimental music, utterly astonishing and never really described to do it full justice, the dry text of science prosaically reducing a wonderful unique event almost to the near-mundane. If a female is nearby, this will attract her to perch close to or above him on the same perch, where she peers down at him. His plumes are spread wide, the black mark on the abdomen becomes crescentic in shape, the head is still turned to one side, and he continually moves his plumes in pace with the pulsing rhythm of the song. After about a minute of this intense phase, the plumage is held more stiffly, the body drawn up somewhat, and the black patch becomes ovate. His head points to the centre of his chest and he swings his tail quickly from side to side, causing the long tail plumes to lash to and fro (something similar also happens with the Red Bird, the *Paradisaea* species with greatly elongated central rectrices). He may step jerkily towards the female if she is to one side of him, maybe attempting to position closer to her. The tail-swinging phase lasts for about a dozen seconds and then the bird stops. He may revert briefly to low-intensity display before righting himself and flying to another perch. R. Whiteside and M. Feignon (in F&B) watched many inverted courtship displays at Kompiam in which the males never swung the tail, though sometimes a male would suddenly right itself breast first towards the female, to displace the startled female before inverting and starting again or chasing her. There are vague reports of more than one male displaying together but they lack detail and context. This would be well worth further research, especially in view of the distinctive solitary nature of the known display behaviours of this fascinating species. *Captive females* have been seen to give a modified but similar display to the male, complete with inversion and vibrating of the body, though without calls (Crandall 1932). The **Copulatory Sequence** in captivity at Taronga Zoo, Sydney, consisted of the inverted male regaining the upright position, and the female turning from him with head lowered in a submissive posture to solicit by flicking her tail several times, before the male mounted her to copulate (Hiller in F&B). Observations in the wild of this part of the display are lacking. NEST/NESTING The nest is made of strips of pandanus leaves and fibre from the leaves of some palm, with no lining (Goodfellow 1926). Hadden (1975) records an open flattish structure composed of sticks, which Frith supposes to have been a nest lining of woody fine tendrils (F&B); Mack (1992) records an oblong-shaped cup consisting mostly of needle-like *Casuarina* twigs, with small pieces of vine, and with a lining of finer examples of the external material; Whiteside (in F&B) records a sparse deep circular bowl of long, green, supple epiphytic-orchid stems that encircled the structure, and involved few or no leaves. The **nest site** was recorded as being <4m above ground in a low tree in thick bush near the top of a ridge (Goodfellow 1926a). Another was in the thickest part of low dense scrub on the ridge, and dense masses of bamboos, with larger trees on the slope (Simpson 1942), or 19m above ground in a subcanopy tree 21m tall (Pruett-Jones & Pruett-Jones 1988a). One was found *c*.9m above ground in the crotch of a *Casuarina* on a steep slope in secondary growth at the edge of temporarily disused gardens, the nearest mature forest being *c*.25m upslope (Mack 1992). **EGGS & INCUBATION PERIOD** Eggs are variably rich cream to pale pinkish-buff or pale salmon-coloured, with brush-like strokes of lavender-grey and dark brown or cinnamon-rufous and tawny, or blotched and spotted with these colours, most concentrated about the larger end, and generally resembling the eggs of the Raggiana Bird, size 36.3–38.6 × 23–27.3mm (Hartert 1910; Bishop & Frith 1979; F&B 1998). Clutch is usually of one egg, but may be two on occasion. Incubation known to be <18 days at one wild nest, and undertaken by the female alone as would be expected (Pruett-Jones & Pruett-Jones 1988a); the female at this nest was aggressive towards other Blue Birds, as well as to Trumpet Manucode and Superb Lophorina that came close to the nest. **NESTLING** Two nestlings looked 'grotesquely spiny owing to the abnormally long grey pin feathers which covered them all over'; the grey-down tip dropped off as they grew, the orbital region and much of the head remaining bare long after the rest of the body was feathered (Goodfellow 1926). A nestling had a white tip to the blackish bill and a pale yellow mouth-lining and gape, with white semicircles already developed around the eyes. **TIMING OF BREEDING** Breeding may be at any time of the year depending on locality, with a general peak in July–February. Single young were at Tari in February, August and October (Smyth 1970; Hadden 1975).

HYBRIDS Genetic work is desirable in order to establish the exact parentage of the problematic and very rare hybrid forms.

Captain Blood's or **Blood's Bird of Paradise** (*Paradisea bloodi* Iredale, 1948) is a hybrid of the Raggiana Bird *P. raggiana salvadorii* × *P. rudolphi margaritae*. This spectacular intergeneric hybrid may be due to human encroachment on the Blue Bird's range causing a meeting between it and the Raggiana (Gilliard 1969), but the two do overlap on Mt Missim and come into close contact in the Tari valley, so that a natural origin is equally likely.

Schodde's Bird of Paradise (Frith & Frith, 1996a) is another unlikely hybrid combination: Lawes's Parotia *Parotia l. lawesi* × Blue Bird of Paradise *P. r. margaritae*. This was not surprisingly thought to be a female Lawes's Parotia after being collected by R. Bulmer in 1956, until it was recognised by R. Schodde (in Christidis & Schodde 1993), but the parentage is far from obvious.

MOULT Males have been seen moulting in November (pers. obs.). Museum specimens show evidence of moult in January–June and September–October, mostly in January–April (Frith in F&B). Captive individuals of this species take about 17–19 weeks to complete a moult cycle (Crandall 1932).

STATUS AND CONSERVATION Classified as Vulnerable by BirdLife International. This fascinating and spectacular species is certainly rare or absent in some seemingly suitable areas, while it has lost range in others owing to habitat clearance. This is particularly so in some of the highland valleys, where the human need for cultivation has removed great swathes of habitat from formerly suitable areas, such as the Wahgi valley. The species is hunted for its plumes, which are used by a few of the highland tribes as ceremonial *bilas*, or costume. Plumes from a male Blue Bird of Paradise cost K20 at Mt Hagen in 1998, rather surprisingly the same as for the far more widespread Lesser and Raggiana Birds (pers. obs.). Much of the Blue Bird's range coincides with a high density of human population. It still exists in the heavily populated but fairly well-wooded Tari valley, but it is rather rare there, seen much less frequently than it was 20 years ago, and likely to decline for reasons given earlier. The Blue Bird may be quite plentiful where suitable habitat remains, as at Kompiam (Whiteside in F&B), but this may be a temporary situation. It may decline as the habitat becomes degraded, much as has happened in the Jimi, Anji and Wahgi valleys, where it is now pushed to the steeper slopes and less accessible remnant forest patches. Careful monitoring of the population status is desirable; moreover, the subspecific status and exact extent of occurrence of the western birds remain to be discovered. **CURRENT THREATS** BirdLife summarises the threats as follows. The major threat is the hunting of the species for its pectoral and tail feathers (Beehler 1985; Coates 1990; Frith & Beehler 1998). Although hunting is carried out mainly for collection of feathers for traditional customary practices, whole birds or feathers are occasionally sold to tourists (van den Bergh *et al.* 2013), even though it is illegal to take them out of the country. Despite a law designed to prevent the killing of birds by non-traditional means (i.e. shotguns), there are many more children than there were 40 years ago and they kill fairly significant numbers of birds on the nest by using slingshots (B. Beehler *in litt.* 2012). Research has also revealed that these laws and regulations are frequently not enforced, or are routinely misunderstood, and they have therefore had little influence on hunting pressure and trade (M. van den Bergh *in litt.* 2014). Even though the species is hunted for its plumes, these are not worn so commonly as are plumes of other species and are not frequently sold (particularly in the highlands). Nevertheless, a few tribal groups still use the plumes of this species and so hunting is likely to be concentrated in certain areas (B. Beehler *in litt.* 2012; M. Supuma *in litt.* 2012). Furthermore, although Christian priests forbid the hunting of birds of paradise, the increasing celebration of Christmas may add pressure for plume collection (M. van den Bergh *in litt.* 2014). Remaining forest, including that in the favoured elevational zone of this paradisaeid, is under significant pressure from clearance for agriculture by the increasing human population. Agriculture-related habitat alteration, however, does not necessarily preclude the species from these areas, as it has been found to occur in mosaics of highly degraded forest remnants and gardens, and can (so far) survive in human-dominated ecosystems (B. Beehler *in litt.* 2012; G. Dutson *in litt.* 2012).

There are still significant areas of its range which are inaccessible and largely uninhabited (Coates 1990; Frith & Beehler 1998; B. Beehler *in litt.* 2012). **CONSERVATION ACTIONS UNDERWAY** This species is listed in CITES Appendix II. It is protected by law in PNG (Fauna Act of 1966–73), although, as mentioned above, this is rarely, if ever, enforced. It is officially illegal for non-citizens to take birds of paradise without a permit from the Department of Environment & Conservation, and illegal also to kill birds of paradise by any means other than traditional ones (Beehler *in litt.* to van den Bergh 2009; Sekhran & Miller 1996). While all birds of paradise are protected by the PNG Fauna Act (1968), the enforcement of this protection is challenging, the more so when it is realised that more than 93% of land ownership rests with traditional custodians (M. Supuma *in litt.* 2012). In addition, there is a very marked lack of the funds required to support enforcement officers to monitor the trade of this and other threatened species. **CONSERVATION ACTIONS PROPOSED** Proposals have been made to survey the western boundary of the range, and also to survey historical sites in the north and east. In addition, population densities and sizes at known sites must be assessed, and investigation be undertaken of the species' tolerance of secondary forest and degraded areas both for foraging and for breeding, this to include the mapping and monitoring of male song perches in populous mid-montane valleys such as those of the Wahgi and Tari. The rates of forest loss in the Blue Bird's preferred altitudinal range should be determined through further research. Importantly, numbers of this species present at the most accessible sites, such as Ambua Lodge, should be regularly monitored, as also should the numbers of individuals traded and the prices involved; alongside this, the levels of hunting and the attitudes among hunters to controls require investigation. Finally, there is a real need for the creation of large, locally managed forest reserves with enforced hunting bans, combined with awareness and education programmes for both landowners and other highland inhabitants; enforcement of existing legislation is also a proposed priority. This unique species' conservation status should be emphasised to tourists, by utilising its well-known image as a flagship species for ecotourism and conservation ventures.

Blue Bird of Paradise, female, Tari Valley, PNG (*Phil Gregory*).

Blue Bird of Paradise, male, PNG (*Tim Laman*).

FORMER BIRDS OF PARADISE

1: FAMILY CNEMOPHILIDAE, SATINBIRDS

Genus *Cnemophilus* (2 species)

The species of this small, compact and enigmatic lineage can be found only in the central ranges of New Guinea. Formerly considered a distinct subfamily Cnemophilinae within the birds of paradise, it was shown by Cracraft & Feinstein (2000) to be only distantly related, despite the superficial similarity of the species. Barker *et al.* (2004) provided a phylogeny which indicated that this group is sister to Callaeidae (the New Zealand wattlebirds), and these two are sister to the Melanocharitidae (berrypeckers). This is confirmed by Jønsson *et al.* (2011), but see Irestedt & Ohlson (2008) and Aggerbeck *et al.* (2014) for conflicting placements. These three sexually dimorphic cnemophiline species are shy inhabitants of montane forest, where they feed on a diet largely or entirely of fruit. The most comprehensive treatment of the group is found in Frith & Beehler (1998), where they are dealt with as anomalous members of the Paradisaeidae, the cnemophilines. The family name of Satinbirds, incidentally, is derived from the silky texture of the plumage of these species.

Etymology *Cnemophilus* is derived from the Greek *knemos*, meaning mountain slope, while *philos* means loving or fond of, hence mountain-lover in reference to the preferred habitat.

LORIA'S SATINBIRD
Cnemophilus loriae   Plate 40

Loria loriae Salvadori, 1894, *Annale del Museo Civico di Storia Naturale di Genova* Ser. **2**, 14. Moroka, Owen Stanley Range.
Synonym: *Cnemophilus mariae* De Vis, 1894, *Annual Report for British New Guinea* 1893–96.
Other English names Loria's Cnemophilus, Loria's Bird, Lady Macgregor's Bowerbird, Crow Bird-of-Paradise
Etymology The species epithet *loriae* is thought to commemorate the wife of the discoverer, Dr Loria.

FIELD IDENTIFICATION Loria's Satinbird is an unobtrusive medium-sized dumpy inhabitant of mountain forest, having rounded wings, a quite short tail and no barring in the plumage. The male is deep velvety black with a greenish gape flange, whereas the female is entirely dull olive-green with faint dusky edges on crown, back and upper breast, and cinnamon edges to the flight-feathers. **SIMILAR SPECIES** Crested Satinbird is very similar in shape, and separating the females and immatures of the two species is sometimes difficult, although female-type Crested lacks the olive-green coloration of Loria's Satinbird and is rather more yellowish on the underparts. MacGregor's Bowerbird is another pitfall, but this is a much darker brown bird with no hint of green, and has a more robust black bill. The male Black Shrike-thrush (Pitohui) (*Melanorectes nigrescens*) is smaller and not so dumpy, and not likely to be seen perched high in a tree, being a shy dweller of the lower strata of the forest.

RANGE Montane forests of the central ranges of New Guinea, from the Weyland Mts of West Papua east to the Owen Stanley Range. Absent from the Vogelkop and the Huon Peninsula. **MOVEMENTS** Sedentary.

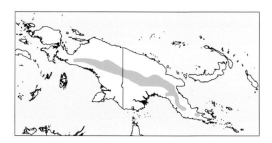

DESCRIPTION Sexually dimorphic. *Adult Male* 22cm; weight 75–101g. Has velvety-black plumage with purple and/or magenta sheens; head iridescent metallic green-blue, washed purplish; tertials iridescent green-blue to blue, washed purple to magenta. *Iris* dark brown; *bill* shiny black, with sharply keeled basal upper ridge of culmen, also has a wide gape with conspicuous bare gape flanges, and *mouth* yellowish-white (in the field can appear as lime-green); *legs* dark olive-brown to blackish. *Adult Female* Slightly smaller than male, weight 60–96g. Very different from male in having rather drab olive-green plumage, browner on wings and tail and more yellowish on the belly; lacks gape flanges. *Juvenile* Nondescript grey-coloured, this plumage worn only briefly. Reports of a grey morph presumably refer to individuals in this plumage. *Immature Male* Similar in plumage to adult female, but with more pointed, longer rectrices than adult male; acquires a shorter tail progressively with age. *Subadult Male* Varies progressively from being like adult female with a few feathers of adult male plumage intruding to being like adult male with a few feathers of female-like plumage remaining.

TAXONOMY AND GEOGRAPHICAL VARIATION Now classified as monotypic. Until recently this species was regarded as polytypic, with three subspecies, but

Beehler & Pratt (2016) state that variation within the populations (e.g. the colour of the gloss on the inner secondaries) is such that recognition of these subspecies is not merited. The races formerly recognised are as follows (with gender emended after David & Gosselin 2002):

1. *C. l. loriae* (Salvadori, 1894). Mountains of southeast PNG from the Herzog and Kuper Ranges through the Owen Stanleys to Mt Dayman (Coates 1990; F&B).
2. *C. l. amethystinus* (Stresemann, 1934). The Western, Southern and Eastern Highlands of PNG from Mt Hagen to Mt Karimui and the Bismarck and Schrader Ranges. Perhaps this form on Mt Bosavi, too (F&B). Differs from nominate in that male has violet-purple iridescent upper tertials, female slightly longer tail. Now synonymised with *loriae*.
3. *C. l. inexpectatus* (Junge, 1939). Central ranges of West Papua from the Weyland, Nassau and Oranje Mts east to the Hindenburg and Victor Emanuel Mts (Coates 1990; F&B). Male has much more green iridescence on tertials, female has slightly shorter tail. Likewise now synonymised with *loriae*.

VOCALISATIONS AND OTHER SOUNDS The male has a distinctive **advertising note** which is repeated monotonously, for long periods, from a subcanopy perch where the bird has a good viewpoint. This note has been described in a variety of ways, but is basically a quiet, unobtrusive rising and upward-inflected *zheee*, remarkably similar to a call of the Marsh Whydah (*Euplectes hartlaubi*). There is a curious far-carrying quality to the call, which is audible over several hundred metres and is surprisingly ventriloquial. Males will call during drizzle or light rain, and avoid direct sunlight. They call with the bill opened wide to display the lime-green interior of the mouth, often facing in different directions, including turning about, presumably to broadcast the call as widely as possible. Calling has been recorded in all months, but at Tari seems much less frequent in the drier period of April–June whereas the birds are much more vocal during July–November. Coates (1990) transcribes this advertising call as *kyerring!*, Frith (F&B) as *weep* or *peep*, and Beehler (F&B) as *kerrng!* or *Herrng!* given every five seconds at the Baliem Gorge in West Papua. There seems to be some variation in the frequency with which it is given, as Frith reports the species as calling every nine seconds from the upper Tari valley in December–January, whereas in July–September the call is given less often, more like once every 30 or so seconds. Other calls are much less frequent, as this is not a vocal species except for the advertising call of the male. An incubating female disturbed from her nest repeated a low harsh scold four or five times (Frith in F&B), and two chasing males on Mt Bosavi uttered *ch-ch-ch* notes (T. Pratt in F&B). Stein (1936) reported an adult as giving a screaming cry as it flew away. A clicking sound during display has been reported (Hicks & Hicks 1988).

HABITAT Inhabits montane forest, forest edge and second growth, being found from 1,500m to 3,000m but mostly at 2,000–2,400m.

HABITS This is an inconspicuous and readily overlooked species, most easily seen at fruiting trees or when the male is calling from a perch high in a tree. It occurs throughout the forest strata but is often seen higher in the forest than the Crested Satinbird, being not such a frequent skulker of the shrub layer. It is reported also as keeping quite close to the ground and having a preference for large mossy branches, along which it hops in a thrush-like manner (Stein 1936; Melville 1979). In addition, it seems also more confiding and can be quite approachable, not so likely as the Crested Satinbird to dive into thick cover and vanish. Loria's Satinbirds are often seen singly, but do occur in small groups of three or four, regularly including an adult male, and they are frequently seen feeding with other paradisaeids. A fruiting tree in a garden at Ambua had Loria's Satinbird as the most constantly present species. They will ascend high into fruiting trees on occasion, feeding alongside paradisaeids such as Princess Stephanie's Astrapia, Superb Lophorina, Blue Bird of Paradise and Lawes's Parotia (pers. obs.), with a Crested Satinbird also present. Stein (1936) reported female-plumaged birds as occurring in groups of up to ten birds, eating in fruit bushes close to the ground. Bishop (in F&B) reports this species and a King of Saxony as calling briefly from the same perch, the Loria's displacing the latter. Belford's Melidectes (*Melidectes belfordi*) has also been seen to displace Loria's from advertising perches by flying aggressively at them (pers. obs.).

FOOD AND FORAGING The diet is recorded as consisting entirely of fruit from drupes to berries of small to medium size (Schodde 1976; Beehler & Pruett-Jones 1983), gathered from the middle and lower levels of the forest, which are simply swallowed whole. F&B record *Pittosporum, Acronychia, Timonius, Drimys, Symplocos, Riedelia, Xanthomyrtus, Elaeocarpus* and *Psychotria* species as among the foodplants. Loria's Satinbirds often feed singly, but will associate with paradisaeids at food sources, where they will tolerate Stephanie's Astrapia, Superb and Blue Birds, Lawes's Parotia, Short-tailed Paradigalla, Brown and Black Sicklebills and occasionally Crested Satinbirds in the same tree, albeit not too close. Tit Berrypeckers (*Melidectes belfordi*), Superb Fruit-doves (*Ptilinopus superbus*) and Great Cuckoo-doves (*Reinwardtoena reinwardtii*) are other species that may feed in the same tree at the same time. Bishop (in F&B) reports these satinbirds as attempting to chase off astrapias, but I suspect that this is only when the two find themselves in very close proximity as my observations indicate peaceful sharing of food sources. Majnep & Bulmer (1977) claimed that Loria's Satinbird breaks open and eats large snails, but this seems unlikely given this species' rather weak bill morphology.

BREEDING BEHAVIOUR Adult males sing from songposts, and from limited data it seems that females alone attend the nest, which indicates a polygynous mating system with promiscuous males, akin to many of the sexually dimorphic paradisaeids (F&B). **DISPLAY** Little known. Adult males spend long periods at traditional calling sites, often on the forest edge, with early-morning and late-afternoon peak activity periods, although during cloudy conditions they will often call at other times of the day as well. The mouth colour of pale lime-green or whitish is prominently shown when the male is calling, and this may be a feature in a rather static form of display, along with the very inconspicuous iridescent lores and pale gape flanges. Males have been seen to 'yawn' occasionally, perhaps to expose the mouth colour (Frith in F&B). J. & R. Hicks (1988a) observed a brief form of presumed intense display at Ambua, where a lone male hung upside-down from a subcanopy perch about 6m high, his half-opened wings quivering for less than ten seconds before he flew off to feed. He made 'a regular clicking noise' at the same time, probably with his bill. No male–female display interactions have been noted as yet. **NEST/NESTING** Frith & Frith (1994) were the first to find the nest of this species, a report by Sims (1956) and Loke (1957) being in fact referable to the Crested Satinbird. They found nine nests in an area of 0.5km², two between 2,150m and 2,000m, at an average of 1.5m above ground. Eight of these nests were on richly vegetated rock faces and one on a moss-covered tree trunk on the sides of a narrow ravine, and all were extremely well camouflaged and cryptic. Five of these nests were close to others used in previous seasons, as happens with *Paradigalla* and *Ptiloris* species and some bowerbirds. The nest itself is a substantial globular domed structure, as suggested originally by Majnep & Bulmer (1977). It is made from fresh moss and filmy-fern fronds, with an aperture in front, and the nest chamber is lined with the stems of epiphytic orchids. As with the Crested Satinbird, comb-tooth fern fronds are also used to camouflage the exterior of the nest, and sticks are used in the nest entrance perch and within the moss below it. Average height

of five nests was 240mm, with average width 234mm and with a depth from front entrance to back wall of 178mm (Frith & Frith 1994). **EGGS, INCUBATION & CARE OF YOUNG** The eggs are pale pinkish-buff, spotted and blotched with russet to purple-grey markings and with a denser band around the larger end; size of one egg was 36.8 × 24.5mm (F&B). The incubation period at one nest was 26 days (Frith & Frith 1994), and care is believed to be by the female only (*contra* Majnep & Bulmer 1977). **TIMING OF BREEDING** Breeding is known from November to February, the single known egg being laid in January.

HYBRIDS No hybridisation is known, but it is perhaps possible with its fellow cnemophilines.

MOULT Seems to moult in all months, but specimens show a peak in active moult from February to October (F&B).

STATUS AND CONSERVATION Classified as Least Concern by BirdLife International. Loria's Satinbird is a fairly common species of the montane forests, albeit easily overlooked except when calling or when visiting fruiting trees. Ripley (1964) claimed that adult males were rare in the Ilaga forests of West Papua owing to hunting for the plume trade, but this seems unlikely and we have no records of the species being used as decoration in either the Southern Highlands or the Western Highlands. Its exact position in the paradisaeids was puzzling for a long time until the genetic work (Cracraft *et al.* 2000) which identified the anomalous cnemophilines as a separate family. The breeding biology of Loria's Satinbird remains very little known, especially regarding displays (if any), courtship and mating. There are no obvious threats beyond local clearance for logging and farming, and the species must be considered reasonably secure.

Loria's Satinbird, female, Kumul Lodge, Enga Province, PNG (*Markus Lilje*).

Loria's Satinbird, male, Kumul Lodge, Enga Province, PNG (*Markus Lilje*).

CRESTED SATINBIRD
Cnemophilus macgregorii   **Plate 41**

Cnemophilus macgregorii De Vis, 1890, *Annual Report of British New Guinea* 1888–89. Mt Knutsford, Owen Stanley Mts.
Other English names Sickle-crested, Multi-crested, MacGregor's or Black and Gold Bird-of-Paradise, Crested Golden Bird, Crested Cnemophilus; Yellow Satinbird (nominate race), Red Satinbird (race *sanguinensis*)
Etymology Named in honour of Sir William MacGregor, Administrator of British New Guinea, who was the first to collect the bird.

FIELD IDENTIFICATION A medium-sized, dumpy and somewhat lethargic species of montane forest and shrub, often shy and elusive but on occasion tame and confiding. Short-billed, with rounded wings and a fairly short tail. The male is one of the most striking species in New Guinea, and is readily identified by his bright orange or orange-yellow upperparts (depending on the subspecies) and black underparts. The female and immature are more problematic, being dumpy unbarred brown birds, readily confused with similar plumages of Loria's Satinbird or MacGregor's Bowerbird. Despite the name, the crest is not a useful field character and is in fact remarkably difficult to see at all. **SIMILAR SPECIES** Males are unmistakable given a good view, but care is needed to separate females and immatures from similar plumages of Loria's Satinbird, or even more so from female-plumaged MacGregor's Bowerbird. Female Loria's Satinbird is a much greener bird than the Crested, but often has a browny-yellow patch on the secondaries and can be quite yellowy brown beneath, leading to a tricky identification problem unless the bird is seen well. The bowerbird is even more problematic, but tends to feed higher in trees and has a slightly different head shape with a stouter blackish bill, and duller and darker brown plumage, especially on the underparts, which lack the yellowish tinge often found with Crested Satinbirds. Bill coloration is not a useful field character, as both Loria's and Crested Satinbirds have a dark bill, but bill shape is different with the bowerbird. A grey-phase individual would be very difficult to tell, but such birds are rare in both the *Cnemophilus* species.

RANGE Endemic to the highest mountains of the eastern half of the central ranges, extending into West Papua in the Star Mts with sightings from north of Lake Habbema in the upper Ibele valley, an offshoot of the Baliem valley (B. Poulson in F&B). This species

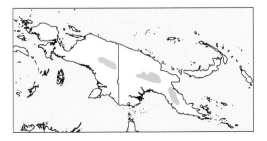

extends patchily eastwards to the Tari Gap, Mt Giluwe, Mt Hagen, the Kubor Range, Bismarck Range, Eastern Highlands, and the Wharton and Owen Stanley Ranges as far east as Mt Knutsford. Easily overlooked and may extend even farther westwards into the Weyland Mts. **MOVEMENTS** Resident; possibly some seasonal altitudinal movement.

DESCRIPTION Sexually dimorphic. *Adult Male* 24–25cm. **Nominate race** of south-east PNG ('Yellow Satinbird' of HBW–BirdLife, who split this taxon: del Hoyo & Collar 2016) has forehead, crown and sides of head to just below the eye, and the entire upperparts a vivid flame-yellow with a silky appearance, duller on the back and more cinnamon on the wings and uppertail. There is a small (and usually invisible) erectile crest of 4–6 sickle-shaped dark buff feathers concealed in a shallow groove in the feathering; lores, lower sides of head and underparts are brownish-black with a coppery-bronze dull sheen, usually appearing black in the field; some cinnamon feathers on thighs and flanks. *Iris* dark brown to bluish-grey; *bill* dark brownish-black, *inside of mouth* pinkish; *legs* purplish-brown to brown-black. **Race sanguineus** ('Red Satinbird' of HBW–BirdLife) differs from nominate in having very slightly longer wing and shorter tail, flame-red dorsal plumage, brownish-purple (not dark-edged buffy) crown plumes, a black bill, and darker, more rufous-tinged tail and wing edgings. *Adult Female* Slightly smaller than and very different from the male, basically fairly uniform brown-olive, the underparts slightly paler and buffier (often appearing yellowish), especially on belly and undertail-coverts. *Iris* dark brown-grey to dark bluish-grey; *bill* brownish-black, *mouth* pale green; *legs* dark brownish to brownish-black. Female *sanguinensis* much the same as nominate. *Juvenile* is for a short period grey in colour. A juvenile at Mt Hagen had a smoke-grey head and breast and the wings a darker grey; the gape and the inside of the mouth were white in colour (Loke in Sims 1956). *Immature Male* Resembles adult female but with a paler bill, and legs and iris brownish to brown-grey. *Subadult Male* Varies from being like adult female with a few feathers of adult male plumage showing to being like adult male with a few feathers of female-like plumage remaining; tail longer than that of adult male.

TAXONOMY AND GEOGRAPHICAL VARIATION Polytypic, with two well-marked subspecies, these treated as two species in the HBW–BirdLife Checklist (2016):
1. **C. m. macgregorii** De Vis, 1890. Mountains of south-east PNG from Mt Knutsford west and north to English Peaks, Mt Albert Edward and the Ekuti Divide (Bulldog Road) (Coates 1990). Synonym *Xantholmelus macgregori*, Goodwin 1890 *Ibis* **2**, series **6**, p.153. Mt Musgrave, Owen Stanley Range. This taxon was originally described by de Vis as a bowerbird, but then placed in what was at that time the same genus as that of Regent Bowerbird as there is a superficial resemblance. It is interesting that the Kalam people consider it as being a bowerbird (Majnep & Bulmer 1977).
2. **C. m. sanguineus** Iredale 1948. Red Satinbird: the central and eastern highlands of PNG in the Doma

Peaks, Mt Hagen, Mt Giluwe, Mt Karimui, Kubor and Bismarck Ranges. Diamond (1972) synonymised *kuboriensis* of the Kubor Range with *sanguineus*. Birds in the Kratke Range and west of Tari into West Papua are presently of uncertain taxon, as there is a lack of good specimens (F&B). A leucistic fawn-coloured form is known from two specimens (Frith & Frith 1998).

VOCALISATIONS AND OTHER SOUNDS The Crested Satinbird is a rather quiet species, seldom very vocal. It does have a quite loud and harsh muffled bark, *haah*, rather like that of Archbold's Bowerbird, but appears not to repeat it so often as does that species, usually calling only once or twice. Diamond (1972) and Beehler *et al.* (1996) report this as an explosive muffled bark similar to that of MacGregor's Bowerbird, and repeated at long intervals. Mack & Wright (2000) describe the call of the male as a loud *grwhaa* given once every few minutes, shorter, louder and more emphatic than some similar calls of the female; the female alarm call near the nest is described as being like the sound made by scraping or like heavy material being torn. They also report a loud snapping from a female-plumaged bird. Rand (in Mayr & Rand 1937) noted a low, harsh, hissing call, a loud clicking call repeated a number of times, and a loud call similar to two timbers rubbing together under stress. Cooper (in Forshaw & Cooper 1977) reports a prolonged squeak like the sound of a rusty gate, probably the former call, and a rasping *aa-aah* or *haah*. A female attending a nest with young gave a soft *wark, wark* as she approached (Loke 1957). Frith & Frith (1993c) noted that a female approaching her nest and young gave a repeated soft, sharp *whit*, and when disturbed there made a soft churring growl. The male of this species, unlike its congener Loria's Satinbird, seems not to have an advertising call. The male's wings produce a whirring noise in flight.

HABITAT A characteristic but rather unobtrusive species of upper montane forest, forest edge and subalpine shrubbery, occurring also in disturbed habitat (Coates 1990), from 2,000m to 3,650m, mainly at 2,600–3,500m.

HABITS Frequents mainly the middle and lower strata of the forest up to 12m, many sightings being from forest-edge habitat, which forms a significant vegetation type within the species' range (Frith & Frith 1993c). It is a frugivore, and can often be found at fruiting trees, bushes or gingers, sometimes in small groups but more usually singly or in pairs. It will on occasion ascend high into fruiting trees, along with paradisaeids such as Princess Stephanie's Astrapia, Lawes's Parotia, Superb Lophorina and Blue Bird of Paradise, and also Loria's Satinbird (pers. obs.). Known occasionally to visit the bowers of MacGregor's Bowerbird, but this species does not itself construct any kind of bower. Crested Satinbirds are usually shy and wary, often just glimpsed as they fly past below eye-level, but they can at times be confiding, particularly when feeding. They are an inconspicuous species and easily missed, which is surprising given the coloration of the male, which somehow manages to blend into the foliage. The red

fruits of certain gingers (Zingiberaceae) are a favourite food at Ambua, and these birds may be found feeding close to ground level along tracks where they grow.

FOOD AND FORAGING The diet is recorded as being almost entirely fruit from drupes or berries 3–12mm in diameter (Diamond 1972), which are simply swallowed whole, although larger fruits such as gingers may be dissected and eaten. Tiny shelled molluscs, perhaps accidentally ingested, have also been found in faecal samples (F&B), with *Garcinia, Riedelia, Rapanea, Schefflera, Xanthomyrtus, Elaeocarpus, Alpinia* and *Ficus* species recorded among the foodplants. This satinbird often feeds singly, but will associate with paradisaeids at food sources; Tano (in F&B) reports seeing some 20+ Crested Satinbirds at a fruiting tree, along with nine male Loria's, Brown Sicklebills and Ribbon-tailed Astrapias. They have also been seen foraging in leaf litter (Hoyle 1975) and often seem to fly low, diving into the foliage and disappearing.

BREEDING BEHAVIOUR Believed to be polygynous, with the female alone attending the nest, as may be expected with such striking sexual dimorphism. The species was for a long time thought to be a bowerbird, and there have been instances of males visiting the bowers of MacGregor's Bowerbird (F&B), while Clapp (1976) saw a male carrying a stick. No bowers have ever been found for this species, and recent genetic work (Cracraft *et al.* 2000) separates out the cnemophilines as a whole from the paradisaeid lineage, neatly explaining a number of anomalous features of the group such as the wide gape, lack of complex displays and domed nests. **COURTSHIP/DISPLAY** Courtship behaviour remains unknown, although Opit (in F&B) records an adult male on Mt Giluwe as having a territory a couple of hundred yards in diameter, using the same trees in succession as he patrols it; this is an intriguing hint at territoriality or even display areas, but the matter is still unresolved. A pair of Crested Satinbirds in July 1998 below the Tari Gap at 2,200m was engaging in some form of display activity, as the male would circle around a tree trunk with wings flicking, following the female (pers. obs.). **NEST** The nest is a highly cryptic globular domed structure of mosses and (usually) ferns, quite unlike those of paradisaeids, and has a meagre woody stick support; these sticks may form a kind of ramp in front of the nest, overlaid with fern fronds. The nest chamber is lined with long, fine supple stems of an epiphytic orchid (probably *Glossorhyncha*), and the egg cup is similarly lined but with finer stems (F&B). The nest site is atop a decayed mossy tree stump, on the side of a mossy trunk or within the branches of a tree, at an average height of 2.6m above ground (Frith & Frith 1993c). Mack & Wright (2000) described a very well-camouflaged nest found in January 1990 embedded in thick moss around a 20cm-thick tree trunk some 2.2m above ground on the downhill side of a fairly steep slope; a few epiphytic orchid stems grew from the nest and appeared to help in supporting it, and some plant stems protruded from the side entrance. **EGGS & INCUBATION** A nest with eggs collected by A. S. Anthony was described by Rothschild (1898) as being from this species, but Hartert (1910) expressed doubts as both nest and eggs differed from those of other paradisaeids (as this species was then regarded). The nest found by Mack & Wright contained a single egg, seemingly recently laid and measuring 38.9 × 25.8mm: the basal colour was a uniform pale salmon mottled with dull flesh-ochre markings *c.*0.5–1.5mm long, these sparser on the narrow end and forming an indistinct ring on the broader end; a few mahogany-red markings were superimposed over the other marks and helped to form part of the darker ring. The egg resembled that of Loria's Satinbird, but with the purple-grey markings of that egg replaced by the narrower mahogany-red ones. The female alone incubated (the male was never seen to attend), and was still sitting after 19 days, when the observers left the area. Clutch size is thought likely to be a single egg. **NESTLING** The chick hatches devoid of feathering and becomes darker-skinned, as with the paradisaeids. The nestling is cared for by the female alone, and is fed with regurgitated fruit; one nestling, shortly after being fed, ejected fruit stones from its crop with a force sufficient to shoot them out through the mouth of the nest (Loke in Sims 1956). The nestling period is around 30 days, similar to that of Loria's Satinbird and quite protracted for a bird of this size; the duration perhaps related to a limitation of food supply caused by the cool and wet conditions frequent at the altitudes where this species lives (F&B). **TIMING OF BREEDING** The known laying period, based on only three known nests, is between August and December in the PNG highlands (F&B).

HYBRIDS No hybridisation is known, but it is perhaps possible with its fellow frugivorous cnemophilines.

MOULT July–November seems to be the peak moult period (F&B).

STATUS AND CONSERVATION Classified as Least Concern by BirdLife International. The Crested Satinbird remains a poorly known species, with even its taxonomic position in a state of flux until the recent genetic work, which has made its somewhat anomalous position much clearer. It is now known to lie outside the paradisaeid lineage and to constitute, with the two other satinbirds, a separate family, the Cnemophilidae. The mating system, display (if any), diet, vocalisations and westward extent of its range remain very little known. The species seems quite common but is often secretive and easily overlooked. Gilliard collected 20 individuals between 2,400m and 3,600m on Mt Hagen; at Ambua it is possible to see ten and more in a day if fruit conditions are right. Kwapena (1985) suggested that numbers of this species above 2,600m increased during the dry season, as birds moved to the higher levels. Around Ambua, numbers certainly vary depending on which trees are in fruit, favoured trees causing local increases in numbers (Tano pers. comm.; pers. obs.), but birds can generally be found at the higher levels around 2,200m. We have no records of this species being used as ceremonial adornment, but Gilliard (1969) did see a male skin being utilised as a head-dress near Goroka. There are no obvious threats to the species beyond local habitat damage resulting from logging.

Crested Satinbird, female, Kumul Lodge, Enga Province, PNG July 2011 (*Phil Gregory*).

Crested Satinbird, male, nominate race, Kumul Lodge, Enga Province, PNG (*Jun Matsui*).

Genus *Loboparadisaea* (1 species)

The single species of this enigmatic and little-known genus is found only in the central ranges of New Guinea. It is sister genus to *Cnemophilus*, but the male has distinctive pale lime-green, blue or yellowish bulbous and bifurcated maxillary wattles, and the female and juveniles have mottled ventral plumage. As with *Cnemophilus*, it was formerly considered to belong in a distinct subfamily Cnemophilinae within the birds of paradise, but genetic studies have now shown it to be only distantly related to the latter.

Etymology *Loboparadisaea* derives from the Greek *lobos*, a lobe, and Latin *paradisaea*, a bird of paradise.

YELLOW-BREASTED SATINBIRD
Loboparadisaea sericea   Plate 40

Loboparadisea sericea Rothschild, 1896, *Bulletin of the British Ornithologist's Club* **6**,16. Dutch New Guinea, the type purchased at Kurudu Island, off Yapen Island, and probably originating from the Weyland Mts (Mayr 1941).

Other English names Yellow-breasted Bird of Paradise, Wattle-billed Bird of Paradise, Wattled Bird of Paradise, Yellow-chested Bird-of-Paradise, the Shield-bill, Shield-billed Bower-bird, Shield-billed Bird of Paradise

Etymology The species name *sericea* is the feminine form of *sericeus*, meaning silky, and the family name Satinbirds is derived from the silky texture of the plumage of these species.

The exact position of Yellow-breasted Satinbird within the paradisaeids was for a long time puzzling until recent genetic work which separated the anomalous cnemophilines as a distinct family (Cracraft *et al.* 2000).

FIELD IDENTIFICATION A typical dumpy, broad-winged cnemophiline without plumes or barring. The male is russet-brown above, with pale silky yellow lower back, rump and underparts. The female is dull brown rather than russet above, and has paler, more yellowish-buff underparts with darker streaking, especially on the breast. The face is blackish, as is the forecrown and forehead. The male has a bulbous bifurcate pale lime-green wattle atop the base of the upper mandible, the female lacking the wattle. Like other members of the family, this is a rather inactive species and is easily overlooked, with a strange patchy distribution. It appears to be generally a rather rare bird, seldom seen and poorly known, but can be common in a few localities. **SIMILAR SPECIES** The Yellow-breasted Satinbird is a unique species, rather distinct from any other. Crested and Loria's Satinbirds lack both the enlarged wattle of the male and the yellow underparts, rump and lower back, and are quite differently coloured above.

RANGE Endemic to the mountains of New Guinea, from the Weyland Mts through the Snow Mts, Victor Emanuel and Kubor Range eastwards to the Kuper Range north of Wau and Bulolo, but very patchy within this range. Known localities include the Ok Tedi area above 1,500m (pers. obs.), Telefomin, the north slope of the Wahgi Divide at Mt Pugent, Soliabeda, Jimi Ridge and Mt Karimui; the south slope of the Bismarck Range in the Aseki area; Dawong; and Mt Missim, Mt Bosavi and Crater Mt (Coates 1990; F&B). Appears to be

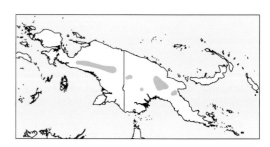

absent from the Vogelkop, Huon Peninsula and south-east PNG. **MOVEMENTS** Presumed sedentary.

DESCRIPTION Sexually dimorphic. *Adult Male* 17cm; weight 50–75g. **Nominate race** has a distinctive pair of bulbous bifurcated maxillary wattles and bare skin over the mandible bases a pale, chalky turquoise-green or yellow. The lores and side of face are dark brown, the crown similar but with a coppery-green sheen. The nape, mantle and upper back are honey-brown with darker feather tips, washed iridescent coppery yellow, the lower back and rump pale silky iridescent sulphur-yellow; the tail and upperwing and wing-coverts also are honey-brown, with darker tips on primaries, secondaries and rectrices. The malar area and the entire underparts are silky sulphur-yellow with brown thigh feathers. *Iris* dark brown; *bill* blackish, with basal upper ridge of culmen broadly flattened, the wide gape typical of the family, *mouth* dull and lacking bright coloration; *legs* blackish. **Race *aurora*** male is fractionally larger than nominate, has longer tail, and upperparts significantly brighter (paler), more brownish-yellow, the crown much paler and more greenish, and the narial wattle pale blue. *Adult Female* Slightly larger than male (uniquely within family), weight 60–77g. Lacks wattles, and has plumage rather different from that of male, being dark olive-brown above with cinnamon-brown on wings, and paler, more yellowish-buff, underparts with darker streaking, especially on the breast. *Juvenile* undescribed; possibly briefly grey, as with Loria's and Crested Satinbirds. *Immature* First-year plumage is dark olive-brown above, washed amber on the wings and tail, and lacks yellow; underparts are cinnamon, with broad greyish streaking formed by the dark feather edgings on the breast, and the belly pale greyish. Second-year plumage of both sexes is like adult female, but darker below and less yellow above; tail of immature male longer than adult's, becomes progressively shorter with age. Male's narial wattles require at least a year to develop fully, changing from all black to black mottled with turquoise-green, finally to adult male colour.

TAXONOMY AND GEOGRAPHICAL VARIATION

There is considerable variation within the species, and the exact nature of the geographical differences is not as yet understood. One option is to treat the species as monotypic and varying clinally, with some minor differences between the western and eastern birds. Two subspecies are recognised by F&B and Beehler & Pratt (2016):

1. *L. s. sericea* Rothschild, 1896. Western and central New Guinea at isolated localities in the central ranges, from the Weyland Mts eastwards to Soliabeda and Mt Karimui, in Chimbu Province. Narial wattles pale lime green or yellow.
2. *L. s. aurora* Mayr, 1930. Herzog Range east of the Watut River drainage in Morobe Province at Dawong (Upper Snake River near Mt Shungol), and at Wau at 1,150m. Birds from the upper Jimi valley have also been placed here (Gilliard 1969). This form is described by Beehler & Pratt (2016) as poorly differentiated, being slightly larger and brighter-coloured above, with a paler, more greenish crown. Mayr (1930) noted the narial wattle as being pale blue, but this may fade on museum specimens.

Frith noted a possibly distinct subspecies obtained by van Duivenbode's collectors, undoubtedly from somewhere in western New Guinea; the single adult male specimen, purchased in December 1910, exhibits a rich and bright mantle unlike that of the described forms; it would be useful to know the exact provenance of this specimen. Interestingly, a much darker and more chestnut adult male specimen is held in the NHMUK, and may represent an undescribed race from West Papua. The species remains very little known.

VOCALISATIONS AND OTHER SOUNDS

This species is vocally the least well-known of all the cnemophilines and paradisaeids. The only known call was described by Bishop (1987), who heard an adult male giving 'a series of loud, harsh, grating notes, slightly upslurred, "*ssh ssh ssh*", usually two notes followed by a brief pause, and again two to three notes'. Both sexes would respond to this call, and he remarked on its being not unlike some notes of the Superb Lophorina, but as the series continues the notes become slower and stronger but lower in pitch than those of that species.

HABITAT The species is confined to lower montane and mid-montane forest from 625m to 2,000m, mainly above 1,200m, and seemingly (unlike Crested and Loria's Satinbirds) mostly in the interior of the forest and not at the edges.

HABITS An often quiet and easily overlooked species, the Yellow-breasted Satinbird has been found from lower levels up to the canopy, where it may move slowly from branch to branch or just sit quietly for long periods. Usually seen singly, but recorded also in groups of six (Rand & Gilliard 1967) or ten (Stein 1936) in the few places where it is common. Cooper (in Forshaw & Cooper 1977) reported one as foraging over moss-covered vines and tree trunks some 3m from the ground in forest on the south slopes of the Bismarck Ranges, and Stein (1936) noted that those in the Weyland Mts

were never high in the forest. Gregory (1995) saw one in the mid-stratum near Ok Tedi about 5m up, while Bishop (in F&B) saw and heard them commonly in the subcanopy at Trauna Ridge, Mt Pugent. Beehler saw the species here in mossy canopy branches, and netted three males on a steep ridgetop (F&B). Pratt (in F&B) at Mt Bosavi noted this as a fast and unsystematic forager dashing through the canopy.

FOOD AND FEEDING Diet is almost unknown, but this species is believed to be primarily, if not entirely, frugivorous, Pratt suspecting that it might select only the ripest fruits. The few specimens examined were eating mainly fruits, drupes and berries plucked and swallowed whole, as with Crested and Loria's Satinbirds. One individual of six examined had arthropods in its stomach (F&B). Female-plumaged birds have been seen swallowing 8mm-sized *Ficus* from a canopy vine (Beehler in F&B).

BREEDING BEHAVIOUR Almost totally unknown, with nothing known of this species' display, courtship, mating system and nesting. For long presumed to be polygynous, but perhaps monogamous, unlike the other members of this family. It has been stated that the small amount of sexual dimorphism has led to speculation that the species may be monogamous (Schodde 1976; LeCroy 1981), but the sexual dimorphism in fact appears quite well marked. The male has swollen narial wattles which may vary in colour geographically, and which are likely to be of some sexual significance. Natives of the Weyland Mts thought that this satinbird nested in branches close to the ground, and laid a single egg in an open moss structure (Stein 1936).

HYBRIDS None known.

MOULT No information.

STATUS AND CONSERVATION Classified as Near Threatened by BirdLife International. Secretive, poorly known and apparently absent from large areas of seemingly suitable habitat, this species appears to have a moderately small population which is declining owing to, in particular, mining, subsistence gardening, and logging activities. It is consequently classified as Near Threatened. There are no known records of this species being used by local people for ceremonial adornment.

Yellow-breasted Satinbird, male nominate race, Crater Mountain, PNG (*W. S. Peckover*).

FORMER BIRDS OF PARADISE

2: FAMILY MELIPHAGIDAE

MACGREGOR'S HONEYEATER
Macgregoria pulchra   **Plate 41**

Macgregoria De Vis, 1897, *Ibis* p.251, pl.7. Mount Scratchley, south-east New Guinea. Type, by monotypy, *Macgregoria pulchra* De Vis.
Other English names Macgregor's Bird, Macgregor's Bird of Paradise, Orange-wattled Bird-of-Paradise, Macgregor's Lappetface; Giant Wattled Honeyeater (Beehler & Pratt 2016) is unfortunate as it invites confusion with two similarly named Fijian species (*Gymnomyza*); 'Big Mac' in birding vernacular. Apparently, Sir William signed his name as MacGregor so the long-standing usage of 'Macgregor' is incorrect.
Etymology Named after Lady Mary MacGregor, wife of the governor of British New Guinea; *pulchra* is the feminine form of the Latin *pulcher*, meaning beautiful or lovely.

FIELD IDENTIFICATION A large, black, rounded-winged, crow-like bird of the subalpine vegetation zone, having bright yellow-orange facial wattles and conspicuous ochre-yellow wing patches. The wings often make a loud and distinctive rushing whistle in flight. SIMILAR SPECIES This species might be confused with the far smaller and longer-billed Smoky Honeyeater (*Melipotes fumigatus*), which also has orange facial wattles and lives at the same high altitudes, but that species lacks the prominent ochre-yellow wing patches.

RANGE MacGregor's Honeyeater is a relict species of the highest altitudes of New Guinea, absent from huge tracts of seemingly suitable country. It is found in small remnant populations from the western Snow and Oranje Mts (Mt Carstenz, Mt Wilhelmina, Carstenz Meadow, Kemabu Plateau, Lake Habbema) and the Star Mts at Mt Capella and Dokfuma, then disjunctly again in south-eastern PNG in the Wharton (Mt Albert Edward, Mt Strong/Chapman and Batchelor) and Owen Stanley Ranges (Mt Scratchley–English Peaks, Mt Victoria). MOVEMENTS Some nomadic movements and short-distance altitudinal movements occur, both being related to food supply.

DESCRIPTION Sexes alike in plumage, male larger than female and has wattle slightly more extensive.

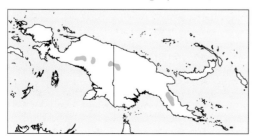

Male 40cm, 242–257g; *Female* 35–40cm, 190–230g. A very distinctive large black honeyeater, with large yellow wattle covering most of the sides of the head. The short erectile feathers of the forehead and lores are dense, soft and silky, and the body beneath the contour feathers covered with a thick layer of down, which is assumed to represent insulation for high altitudes (Frith *et al.* 2017). **Nominate race** is mainly velvety jet-black to sooty black, slightly duller on remiges, and with orange-ochre edges of primaries (forming a conspicuous panel on the folded wing, and shows as a large ochre-yellow patch on the primaries in flight; faint brownish tinge on belly, vent and undertail-coverts, brownish-black undertail and underwing-coverts, latter contrasting slightly with black remiges. A striking large, conspicuous fleshy yellow wattle forms a semicircle around the eye, broken in front and with a small notch in the upper edge. The colour of the wattle can change from yellow to more orange-yellow, similar to but not so intensely coloured as that of Smoky Honeyeater (Gregory & Johnstone 1993). *Iris* reddish-brown to red; *bill* glossy black; *legs* blue-grey. **Race *carolinae*** has smaller wing and tail than nominate. **Juvenile** is duller and browner than adult, with iris dark brown (not reddish).

TAXONOMY AND GEOGRAPHICAL VARIATION Polytypic, with two marginally distinct races; tail length may be clinal and the species could easily be regarded as monotypic.
1. *M. p. pulchra* De Vis, 1897. Eastern parts of the range, in the Wharton and Owen Stanley Ranges, at a few localities only.
2. *M. p. carolinae* Junge, 1939. Western parts of the range, from the Snow Mts east to the Star Mts along the West Papua–PNG border.
This species was for long considered an aberrant member of the paradisaeid family, although Iredale (1950) presciently treated it as a honeyeater. The molecular studies of Cracraft & Feinstein (2000) have convincingly shown that the bird is indeed a honeyeater, which explains many of the disparities involved when treating it as a paradisaeid. The species and its habitat merit additional detailed study (see Beehler 1991: 221–243). Beehler & Pratt (2016) state that the weight measurements provided in Frith & Beehler (1998) for the subspecies *carolinae* (340–357g) are almost certainly in error (more likely 240–257g). A single specimen showing leucism, having a few white feathers on the abdomen, is known (Frith & Frith 1998).

VOCALISATIONS AND OTHER SOUNDS This is quite a vocal species, the commonest call being a high-pitched musical whistle, *jeet!*, which seems to serve as a contact note and also as an alarm note when these birds are flushed. There is also an infrequently heard softer, longer *pseer* note, which seems to be given as a greeting when a bird returns to a roost (Beehler in Coates 1990), the bird flicking its wings and cocking its tail as it calls. Other notes reported by Rand are uttered chiefly during the courtship chases and include a low, sharp *click........click ....click* or *click-click.......click-click.........click-click*, a low, plaintive *quee-ee*, both of these sets of calls described as not being very loud, and a wheezy, unmelodic *cheu* (Mayr & Rand 1937). Beehler (in F&B) notes a pair as giving a nasal, slurred, quiet and lisping *chiff*, while a disturbed pair gave a loud frightening *krahh*! (from the larger presumed male) and a *jeet* (from the presumed female). A *fwooip*! was given by a bird separated from two others which were emitting *scheet* and *schweet* notes, and Beehler also mentions a *schweet schweet* call. The **wing noise** in flight is quite distinctive, something like the swishing wings of a diminutive hornbill. Rand (1940a) described the sound of the fully spread gliding wing as a loud, ripping '*zing*'; Beehler (in Coates 1990) described a musical ripping or whining sound or a remarkable rustling or zipping sound (Beehler 1991). As first suggested by Rand (1940a), the birds can seemingly control these sounds, which figure prominently in the display chases. The primaries are spread when the sound is made, but the birds can fly silently as well. Rand describes the wingbeats as a heavy rustling or a low, loud, hollow 'thopping' sound.

HABITAT MacGregor's Honeyeater is found primarily in subalpine woodland, descending at certain periods to the slightly lower levels of the upper montane forest. Podocarp woodlands seem to be essential to it, especially *Dacrycarpus compactus*. It occurs also in forest patches among subalpine meadow, as at Dokfuma, in Star Mts. Occurs from 2,450m up to 4,000m, but most frequently in the range 3,200–3,700m (Coates 1990).

HABITS The species is fairly common and unwary where not hunted, but local numbers vary depending on what is in fruit, and it can be absent from previously occupied sites if fruit is lacking or unsuitable at that time. These honeyeaters are found in pairs or twos, or in small family groups, and are usually sedentary, sometimes having a favoured roosting site in a *Dacrycarpus* grove (Coates 1990). They spend much time foraging in the canopy, but when fruits are unripe they can be found feeding low down in the shrub and bush layers. They have been reported as feeding on flowers (Ripley 1964), and will descend to the ground and feed upon fruiting cushion shrubs (Beehler 1991). The flight is heavy with noisy wingbeats (see VOCALISATIONS AND OTHER SOUNDS), followed by a short glide on outstretched wings. The birds will drop out of tall trees by gliding steeply (Bishop in F&B).

FOOD AND FORAGING *Dacrycarpus* fruits are an important part of the diet and may form the major part when there is a local abundance, although other fruits and berries are taken and both insects and small vertebrate prey may well form a part of the diet. *Dacrycarpus* fruits are high in aromatic compounds and may require special digestive adaptation (F&B). The fruiting of these species is also erratic, not annual, and

may require the birds to undertake local wanderings when the supply is finished or unripe. Other fruits eaten include *Eurya brassii*, *Cladomyza acroscela* and the terrestrial cushion plants *Styphelia sauveolens* and *Coprosma divergens*, *Rapanea* sp. and *Symplocos cochinchinensis* (F&B). The birds will feed together quite amicably, but can be aggressive to other large honeyeaters and Crested Berrypeckers (*Paramythia montium*) if they come near the nest site (Coates 1990), and Beehler (1983b) recorded them as driving a Painted Tiger-parrot (*Psittacella picta*) from a feeding site. They often forage in the canopy, but will use all levels of the forest if the main *Dacrycarpus* fruit is not ripe. Rand (1940) noted that they behave as if they are much smaller than they really are, perching out in the open, calling frequently and cocking the tail as they hop about. This is the largest frugivore in the subalpine zone and may be a significant seed-disperser, especially for *Dacrycarpus*. The birds will chase through the trees at times, sometimes as mated pairs but occasionally as three to six individuals, and may descend into the undergrowth while chasing (Rand 1940). Feeding on the ground and from flowers has been noted (see HABITS). MacGregor's Honeyeaters have been observed also to forage in an astrapia-like manner as if seeking arthropods among moss and epiphytes (Clapp 1986; Safford & Smart 1996; Beehler in Coates 1990).

BREEDING BEHAVIOUR Monogamous and non-territorial, with overlapping foraging ranges, and the home range of a radio-tracked pair being 12ha (Beehler in F&B). The breeding behaviour of this species is highly atypical of a paradisaeid, but becomes much more typical when the species is considered as being part of the honeyeater family instead, as initially suggested by Iredale (1950). **COURTSHIP** Courtship flight and hopping chases are made in the vicinity of a nest being constructed, and may involve pairs or sometimes up to six additional birds. The wing noise features prominently in these chases, which may lead to mating, and seems to be controlled by the birds, which fly back and forth across clearings, hop through treetops and glide across openings, with descents into the undergrowth at times. The large bare facial wattles can blush a darker shade of orange (pers. obs.), as happens with *Melipotes* honeyeater species which have a similar facial adornment, and this may be related to sexual excitement. **NEST** Just three nests are known; most data are from Rand (1940), who found two nests plus four old nests, which were probably of this species, within 50m of a courtship-chase area. One nest was at 17m in the upright fork of a small branch of an isolated podocarp rising above the surrounding moss-forest vegetation, while another was on a lateral bough about 11m above ground at the edge of a small clump of *Dacrycarpus*-type trees near the forest beside an extensive grass area. Pratt and Brown (in F&B) saw a nest high in the top of a canopy tree. The nest itself is a bulky cup-shaped structure covered externally with moss, mixed with herbaceous and woody-vine or orchid stems and lichens; only one or two short sticks may be incorporated. Inside one nest was a firm cup of slender herbaceous and woody stems, thinner at the bottom and lined with slender woody stems; the cup contained a number of broad leaves of the coniferous *Phyllocladus hypophyllus* and numerous leaves of *Nothofagus* beech. The external diameter was 240mm and depth 190mm, with the internal cup 130mm across and 90mm deep. The second active nest found by Rand was similar, but had a bulkier foundation of moss and was lined with semi-woody stems and small oval leaves of a shrub, measuring 315mm externally with a depth of 190mm, internally 135mm in diameter and 90mm deep (Rand 1940a). **EGGS & INCUBATION PERIOD** Only a single egg is known, earthy pink in colour, lightly spotted with brown, 39.9mm × 28.6mm (Coates 1990: F&B). The female alone incubates, the incubation period likely to be 26–30 days (F&B). **NESTLING AND ITS CARE** Both members of the pair look after the nestling and both feed the young by regurgitation; both also evict other frugivorous species from the nest area (T. Pratt & E. Brown in F&B). Nestling development appears to be slow, as may be expected from a frugivorous species living at high altitudes, the sole known one being about 12 days old and still very undeveloped (Rand 1940a). **TIMING OF BREEDING** Display flight chases have been seen in August at Lake Habbema and in November at Dokfuma (pers. obs.). The single known egg was collected in August. Beehler (in F&B) suggested that at Lake Omha, in south-east PNG, breeding may be synchronised with the fruiting of *Dacrycarpus*, which is not an annual event, but a nest with a small chick was found during a period when almost no ripe fruits were available (F&B). Rand (1940a) saw a male accompany a female as she built a nest and when she was feeding during the incubation period.

HYBRIDS Hybridisation with other species is unknown and since this species is a honeyeater, considered unlikely.

MOULT No information.

STATUS AND CONSERVATION BirdLife International (2018) lists this species as Vulnerable, with a small and declining population estimated at between 2,500 and 9,999 mature individuals, and the species may be due for revised threat status if the decline is severe. MacGregor's Honeyeater is a rather rare relict species which may be locally common at times in a few remote sites, and which is subject to partial and poorly understood nomadism related to *Dacrycarpus* fruit supplies. It is a specialist frugivore which could suffer severely from ongoing climate change. Its diet is primarily, or perhaps entirely, of fruit. Its optimal subalpine habitat is itself restricted and scattered, and this honeyeater's large size and unwary nature make it very vulnerable to hunting, as evidenced by recent declines around the now much more accessible site at Lake Habbema. It is quite possible that hunting has extirpated this species from the densely settled highland provinces of PNG, where much apparently suitable habitat remains but without any records of this species. These birds may become sparse or disappear for lengthy periods, as at Lake Omha where they can disappear for up to a year (F&B), and at Woitape, PNG, which is also sporadically occupied (Safford & Smart 1996). This phenomenon appears to be related to the production of the essential *Dacrycarpus* fruits on which this species is largely reliant. Where MacGregor's Honeyeater is not hunted it may be quite common and confiding, as at Okbap, in the Star Mts of West Papua, where the species is apparently revered and not hunted (Diamond & Bishop in F&B), and at Dokfuma (pers. obs.). The unpredictable nature of the food supplies means that some nomadism is required, and given the restricted habitat combined with hunting pressure, along with the impacts of climate change, the species is certainly vulnerable.

MacGregor's Honeyeater, sexes similar, Snow Mts, West Papua (*Mark Sutton*).

BOWERBIRDS

FAMILY PTILONORHYNCHIDAE

Human interactions with bowerbirds are nothing like so extensive as they are with the birds of paradise. A few bowerbird crests are used in some local areas as decoration, such as the crests from MacGregor's Bowerbird sometimes seen on sale at Mt Hagen market, and the vividly coloured skins of the male Flame Bowerbird have been used as car ornaments at Kiunga, in Western Province of PNG, and have also featured as part of the altar decorations of the local Catholic church. The widespread and highly significant cultural use of the head, tail and flank plumes of the paradisaeids in traditional costumes and masks, however, has no equivalent with regard to the bowerbirds. The remarkable bowers of some species are often well known in both New Guinea and Australia, where children sometimes vandalise them. In Australia, however, they are sometimes a source of civic pride and become local tourist attractions, as with some bowers of the Great Bowerbird at Mareeba Golf Course and at Mount Molloy Primary School, in North Queensland. The bowers of the Satin Bowerbird, with its remarkable affinity for blue items, are also frequently popular tourist attractions, as at O'Reillys Guest House at Lamington NP. The catbirds, genus *Ailuroedus*, do not construct bowers, while the Tooth-billed Bowerbird makes just a simple upturned leaf mat on the forest floor, a sort of intermediate stage towards a bower proper.

Genus *Ailuroedus* (Catbirds)

Etymology The name *Ailuroedus* is from the Greek *ailouros*, meaning cat, and *eidos*, referring to form (or perhaps from *oaidos*, meaning singer).

The *Ailuroedus* catbirds (the Spotted and White-eared Catbird assemblages) are now classified as some ten species rather than the traditional three (Green, Spotted and White-eared Catbirds), after recent genetic research revealed more of the history of this cryptic group. This group and the Tooth-billed Bowerbird (*Scenopoeetes*) are the basal members of the family, but somewhat divergent in their breeding behaviour. The ten species of the green *Ailuroedus* catbird group all lack marked sexual dimorphism, form pair bonds in which the male helps to build the nest, and have rather simple arboreal chasing displays without the use of bowers or stages. In contrast, the brown-and-black, sexually monomorphic but polygynous Tooth-billed Bowerbird has a terrestrial display and builds a simple stage of large leaves on the forest floor as a display arena; the male does not form pair bonds and does not help to build the nests or raise the young.

All the catbird species are most obviously more closely related to one another than they are to any other bowerbirds; recent biomolecular research indicates that they diverged from other bowerbird genera about 24 mya, and that they are quite distinct from the builders of avenue bowers. The Tooth-billed Bowerbird is, however, as distinct from the catbirds as are other bowerbird subgroups, and it appears to be genetically closer to the New Guinea *Amblyornis* gardener bowerbirds than it is to catbirds. The genus *Ailuroedus* is a basal member of the bowerbird family, and is now classified by the IOC as consisting of ten rather than three species following the genetic studies of Irestedt *et al.* (2015). To quote the latter's introduction:

'*The fauna in New Guinea is characterised by complex distribution patterns where species and subspecies often are replaced by sister-taxa in adjacent regions. One such pattern is that widely distributed lowland species are morphologically variable with distinct, geographically disjunct populations, but in spite of that there are few apparent barriers in the lowland. Today we know that New Guinea has a complicated geological history and that these patterns may reflect past geographical disjunctions.*

'*The avifauna of New Guinea is renowned for its diversity and includes families such as the birds-of-paradise and the bowerbirds. In this study we use catbirds of the bowerbird genus* Ailuroedus *to investigate past vicariance patterns in the lowlands and lower mountains. Catbirds were chosen as model group since the two species inhabiting New Guinea are found at slightly different altitudes and may thus have responded differently to past vicariance events.*

'*By examining molecular data within a spatio-temporal framework we found three deep genetically divergent populations in the lowland species that corroborate the presence of ancient lowland barriers. In the lower mountain taxa we found a different and more complex population structure. The result supports the hypothesis that taxa inhabiting lowland and lower mountains, respectively, have responded differently to past geological and ecological events in New Guinea. By using integrative taxonomy based on both molecular and morphological data we argue that several subspecies of* Ailuroedus *catbirds should be upgraded to species rank.*'

All the *Ailuroedus* species lack sexual dimorphism, do not construct bowers or stages, and are monogamous pair-bonding species, the males helping with nest-building but not incubating, although they do help to feed the young. Vocalisations of all the taxa within what was formerly the Spotted Catbird/Black-eared Catbird appear to be very similar so far as is known, both the cat-like squalling call and the *zik* contact note being given, and the calls of the White-eared Catbird group members also appear similar to one another.

An earlier treatment by Schodde & Mason (1999) defined some four groups within what was Spotted/Black-eared Catbird, each with its own zoogeographical integrity. They suggested that, if the Arafura Basin Black-eared Catbird *melanotis* and the Green Catbird *crassirostris* are separated specifically, so also should the other two groups be: black-headed ochre-breasted Arfak Catbird *arfakianus* of montane New Guinea and diffusely spotted Spotted Catbird *maculosus* of north-east Queensland. The four groups recognised are:

1. **Arfakianus** of montane New Guinea and Misool (and thus presumably synonymising *jobiensis*, *guttaticollis*, *astigmaticus* of outlying montane areas). Large; top of head black, fine sparse ochraceous mottling on crown changing to coarse mottling on upper mantle; face dull blackish without contrasting lores and brow, but with clear black cheek patch and chin, no pale bar on side of neck, whitish on throat reduced and heavily scalloped black; and underparts plain tawny-ochre with thick blackish scalloping restricted to upper breast. Birds seen at Kiunga and Tabubil, however, resemble this form with the rich ochre underparts (pers. obs.).

2. **Melanotis** of the Arafura Basin, including Aru Islands, and Trans-Fly. Large; head contrastingly marked in black and buffy white, with dense clear cream-white spotting evenly from crown to mantle; clearly patterned face with whitish brow and lores freckled black, clear black cheek patch and whitish to faintly black chin; no pale bar on side of neck, extensive whitish throat finely scalloped black; breast and belly plain greenish cream-buff, scalloped blackish all over breast and upper flanks.

3. **Maculosus**, the Spotted Catbird (*sensu strictu*) of north-east Queensland humid tropics (excluding Cape York). Small; head dully patterned dusky and off-white, with dense off-white spotting over crown and nape diminishing to dull whitish streaking on mantle; face dully patterned with dusky-freckled whitish lores and brow, dusky cheek patch and dusky chin; trace of a pale bar on side of neck, greyish throat mottled dully off-white; breast and belly greenish-cream, heavily and extensively chevroned mid-olive, washed ochraceous over breast, flanks and most of belly. Curiously, however, the birds on the Atherton Tablelands have white spots all over the underparts and no ochraceous on underparts (pers. obs.).

4. **Crassirostris**, the Green Catbird, disjunct in central east-coast Australia. Large; head plain green, plain greenish face freckled faintly dusky, clear whitish bar on side of neck; greenish-grey throat freckled lightly off-white, breast and belly extensively mid olive-green with variable light V-shaped white spotting.

Schodde & Mason (1999) made the point that intermediate-type forms link some of the taxa, although the geographical extremes at *arfakianus* and *crassirostris* are very distinct. The greening of the ventrum, reduction of head spotting and disappearance of black on the face can be traced from *arfakianus–melanotis* to *crassirostris* through the diffusely marked *maculosus*. *Maculosus* also has traces of the neck bar of *crassirostris* and has a short and similarly white-tipped tail; it could therefore be considered the vicariant of *crassirostris* rather than of *arfakianus–melanotis*, in accordance with Bergmann's Rule and contrary to current opinion (Gilliard 1969; Schodde 1975; Sibley & Monroe 1990; Christidis & Schodde 1992). The taxon *joanae* on Cape York compounds the matter, having a short white-tipped tail like the southern forms, but resembling *melanotis* in colour pattern and tones (and behaviour: this form is really shy!). Schodde & Mason state that it is unlikely that such similarities are due to convergence or are ecophenotypic, given that the Arafura Basin is so close. Cued by geographical trends, *joanae* can hardly be combined with *melanotis* at species level without including *maculosus*, too; and for a balanced taxonomy *maculosus* cannot be combined with *melanotis* without bringing in *crassirostris* as well (Ford 1977a).

For decades, the above argument weighed against making the split of the group, with *joanae* as the obstacle, but presumably the answer lies in Ice Age land bridges connecting Cape York to New Guinea allowing a variety of New Guinea species to colonise a small area of Cape York. The taxon *joanae* appears to be a relatively recent colonist from *melanotis* stock in New Guinea, like the Cape York Magnificent Riflebird (*Ptiloris magnificus alberti*) and the Frill-necked Monarch (*Arses lorealis*) that are likewise derived from their New Guinea representatives.

WHITE-EARED CATBIRD
Ailuroedus buccoides
Plate 24

Kitta buccoides Temminck, 1836, *Planch. Col. Ois.*, livr. **97** (1835): pl. 575. Triton Bay, Bird's Neck, New Guinea.
Other English names White-throated Catbird, Barbet-like Catbird, Least Catbird
Etymology The specific epithet *buccoides* means barbet-like and refers to the heavy bill and green plumage, from the modern Latin *bucco*, a barbet and the Greek *oides*, meaning to resemble.

FIELD IDENTIFICATION A typical large green catbird, this is a medium-sized very shy stocky green bird of dense forest, with ochraceous-buff black-spotted underparts, a blackish-brown crown and whitish sides of head and throat, and a broad blackish collar across the nape. The stout pale bill is also obvious. Usually located by the distinctive sibilant hissing calls, it is otherwise extremely difficult to see.
SIMILAR SPECIES Arfak (Spotted/Black-eared) Catbird has a much paler head, a large black patch on the ear-coverts, the underparts much less buff with no black spotting, and a whitish tip to the tail, as well as a completely different voice.

RANGE Endemic to West Papua on islands of Waigeo, Batanta and Salawati, also in the Vogelkop and Bird's Neck, and in lowland north-west New Guinea east to the Siriwo River in the north and the Mimika River in the south. **MOVEMENTS** Resident.

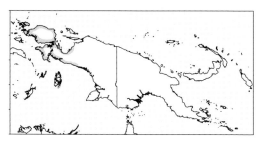

DESCRIPTION 25cm. No marked sexual dimorphism; female slightly smaller than male, crown and perhaps leg colour may be paler. *Adult* A heavily built, stout catbird. Crown dark blackish-brown, often tinged greenish, ear-coverts pure white, white extending onto lower lores. Upperparts including tail dark green, with small pale tips to the tertials and inner secondaries. Neck buff-coloured and feathers broadly tipped black, forming a dark blackish collar across hindneck. Underparts pale cinnamon-ochre with large black spotting (most pronounced on breast), lower belly and undertail-coverts unmarked and maybe paler. Underwing-coverts buffish-white. *Iris* dark red; *bill* pale bluish-grey; *legs and feet* grey. *Immature* Presumably a duller version of the adult.

TAXONOMY AND GEOGRAPHICAL VARIATION
Monotypic. Formerly regarded as a polytypic species incorporating *A. geislerorum* and *A. stonii* as subspecies, but recent genetic work suggests that these taxa are better treated as three separate species (Irestedt *et al.* 2015). Described form *Ailuroedus buccoides oorti* Rothschild and Hartert, 1913, *Novit. Zool.* 20: 526, from Waigeo Island, is synonymised following Mees (1964c) and Beehler & Pratt (2016). Those authors also consider the form *cinnamomeus* best combined with what was the nominate, which tends towards being paler ventrally in the west and darker and richer ventrally in the east, but this is not followed by Irestedt *et al.* (2015) or the IOC, who place *cinnamomeus* with the southern Ochre-breasted Catbird. Clearly much remains to be discovered.

VOCALISATIONS AND OTHER SOUNDS This species emits an extraordinary low, sibilant but harsh, rather wavering hissing, *kssssssssh*, which can continue for minutes on end and which does not sound bird-like, dropping in pitch as the call continues. The species will sometimes respond to tape playback of this call. In addition, a weak, short, high-pitched whistled *tseep* or *tsip* is given as an alarm call, analogous to a similar call of Spotted Catbird. There is no cat-like squealing *errrow*-type call made by this species, which is heard far more often than it is seen. The voice is not known to differ significantly from those of the two other members of what was White-eared Catbird, but this particular species is the least known of the newly split species group and there were no recordings on xeno-canto at the time of writing (2018).

HABITAT Dense tropical rainforest at up to about 800m. Sympatric with Arfak Catbird and in some areas may be sympatric with Black-eared Catbird, but this not as yet known for this taxon.

HABITS These catbirds keep to the dense middle and understorey of the forest, being well camouflaged and concealed among the vines, ferns and leaves. The species is very shy, heard more often than seen, but quite widespread in dense forest. Usually observed singly, but sometimes in pairs or family groups of three birds. It tends to keep quite low down and is rarely seen in the canopy. Other habits of this species are as yet unknown.

FOOD AND FORAGING Diet includes fruits and berries, also seeds and insects and possibly bird eggs and nestlings, but specific diet for the members of the newly split species is poorly known, especially for this taxon. The birds feed singly or in pairs, moving through the foliage in search of food.

BREEDING BEHAVIOUR Breeding behaviour of this taxon is basically unknown, but unlikely to differ significantly from that of the other members of this complex. The present species does not make any kind of stage or bower.

MOULT No details available.

STATUS AND CONSERVATION Not as yet formally assessed by BirdLife as this species group is not split by del Hoyo & Collar (2016). This species would be Least Concern, having a fairly wide range, mostly in lightly inhabited country. It is a fairly common but very unobtrusive resident.

TAN-CAPPED CATBIRD
Ailuroedus geislerorum   Plate 24

Aeluroedus geislerorum A. B. Meyer, 1891, *Abh. Ber. Zool. Mus. Dresden* (1890–91) **3(4)**: 12. Near Madang, north-east New Guinea.
Other English names White-eared Catbird, White-throated Catbird, Northern White-eared Catbird, Least Catbird
Etymology The species was named *geislerorum* for the Geisler brothers, German taxidermists who collected in New Guinea.

FIELD IDENTIFICATION A typical large green catbird, this medium-sized very shy, stocky green bird of dense forest has a pale chin and upper chest merging into pale ochraceous-buff black-spotted underparts, a distinctive tan crown, whitish sides of head and throat, and a chequered blackish and white collar across the nape. The stout pale bill is also obvious. Usually located by the distinctive sibilant hissing calls, being otherwise extremely difficult to see. **SIMILAR SPECIES** Huon Catbird and Northern Catbird might have marginal contact, though they inhabit primarily montane and hill forest and are not lowland taxa, but they lack the tan cap and white ear-coverts. Fruit-doves (*Ptilinopus*) seen very poorly may possibly be confusable.

RANGE Yapen Island and north-western lowlands, Sepik–Ramu, Huon Peninsula and northern watershed of the South-east Peninsula (Mayr 1941; Rand 1942b; Gilliard & LeCroy 1967). **MOVEMENTS** Resident.

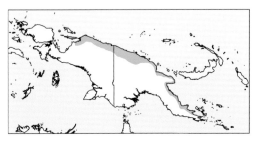

DESCRIPTION 25cm. No marked sexual dimorphism; female slightly smaller, crown may be paler and perhaps leg colour also paler. *Adult* A heavily built stout catbird having the crown a distinctive pale tan-brown. The pure white of the ear-coverts extends forwards onto the lower lores. Upperparts, including tail, dark green, the tertials and inner secondaries with small pale tips. Neck buff-coloured and broadly tipped black, this forming a dark blackish collar across hindneck. Underparts pale cinnamon-ochre with large black spotting, most pronounced on the breast, unmarked on lower belly and undertail-coverts. Underwing-coverts buffish-white. *Iris* dark red; *bill* pale bluish-grey; *legs and feet* grey. Photographs of this form are in Coates (1990: 382) and Coates & Peckover (2001: 206). *Immature* resembles adult, but is duller-plumaged.

TAXONOMY AND GEOGRAPHICAL VARIATION Polytypic, with two very similar subspecies, the nominate of north-central New Guinea and *molestus* (Rothschild & Hartert, 1929) of eastern New Guinea, the latter a recently reactivated designation (Irestedt *et al.* 2015). Formerly regarded as part of the polytypic White-eared Catbird, but recent work suggests that latter may be better treated as three distinct species (Irestedt *et al.* 2015); *geislerorum* was originally described as a species, and it has therefore reverted to the original designation.

VOCALISATIONS AND OTHER SOUNDS Shares with its close relatives the extraordinary non-birdlike low, sibilant, and harsh, rather wavering hissing, *ksssssssh*. This can continue for minutes on end and drops in pitch as the call continues. The species will sometimes respond to playback of this call. In addition, a weak, short high-pitched whistled *tseep* or *tsip* is given as an alarm call (Coates 1990), and is analogous to a similar call of Spotted Catbird. Coates also notes an occasional light chittering series of nasal notes which rise, fall and then rise again over 2–3 seconds. No cat-like squealing *errrow*-type call is known for this species, which is heard far more often than it is seen. Its voice is not known to differ significantly from those of the two other members of what was the White-eared Catbird *sensu lato*. Xeno-canto has a number of examples of the typical hissing call, such as XC62496 (by A. Spencer) and XC209032 (by F. Lambert).

HABITAT Rainforest, hill forest and monsoon forest in the lowlands, occurring mostly from sea level up to about 800m, but locally to higher levels. It is sympatric with the Huon and Northern Catbirds in some areas.

HABITS These catbirds keep to the dense middle and understorey of the forest, being well camouflaged and concealed among the vines, ferns and leaves. The species is very shy, heard more often than it is seen, but quite widespread in dense forest. Usually seen singly, but sometimes in pairs or family groups of three individuals. These birds tend to keep quite low and are rarely seen in the canopy. They seem to associate on the periphery of mixed-species foraging flocks, not being a core element but often found on the margins when such flocks are passing through. They call at intervals throughout the day, and calling sessions may continue for 20 minutes or more before they fall quiet again.

FOOD AND FORAGING The specific diet for the former members of the newly split species (*A. buccoides*) is poorly known, especially for this taxon. Tan-capped Catbirds forage singly or in pairs, moving through the foliage to take fruits and berries, also seeds and insects and maybe bird's eggs and nestlings, but details are little known. The present species has been known to attack birds in mist-nets, the head of the trapped bird always being eaten and sometimes other parts as well.

BREEDING BEHAVIOUR The breeding behaviour of this taxon is almost unknown, but is unlikely to differ significantly from that of the other members of this complex. During April at Mt Missim, near Wau,

Morobe Province, three or four very active individuals of this species were calling to one another high in the forest canopy, and flying from tree to tree, where they swapped perches in an excited fashion. Individuals with enlarged gonads have been collected in March or April in the Adelbert Mts (Gilliard & LeCroy 1967). The species maintains a **territory** during the nesting period, but does not make any kind of stage or bower. In March 1958, Gilliard noted one in a forest understorey space that was about 6m (20 feet) high and at least 13m (40 feet) in diameter and 'thickly cluttered with small moss-covered trees, tree-ferns, vines and brush, and the trunks of a few medium and large trees'. Two birds were often in this area and violent chases would then occur, the owner attacking and flying after the interloper much as Spotted Catbirds will do. Calling birds would bow towards the presumed mate, lifting the head and then lowering it below the level of the perch as it called (Gilliard 1969). Situated within this same general area was the court of a Magnificent Bird of Paradise, though no mention is made of any attention being paid to it by the catbird. The catbirds called throughout the day, often at 10–20-second intervals, with violent but strangely quiet chasing accompanied by wing-rustling. While defending this area the catbird made darting attacks on small birds that passed through the site, once also passing like an arrow right above the observer's head when he sat outside the blind. NEST & EGGS The nest appears to be undescribed. The eggs are described as being plain horn-coloured and unmarked. No other information is available, and the nestling apparently is not known.

MOULT No details available.

STATUS AND CONSERVATION Not as yet formally assessed by BirdLife as this member of the *A. buccoides* species group is not treated as a separate species by del Hoyo & Collar (2016). The present species would be Least Concern, having a fairly wide range, mostly in lightly inhabited country.

Tan-capped Catbird, Keki Lodge, Adelbert Range, PNG (*Lars Petersson*).

OCHRE-BREASTED CATBIRD
Ailuroedus stonii   Plate 24

Aeluraedus [sic] *stonii* Sharpe, 1876, *Nature* **14**: 339. Laloki River, Southeast Peninsula.
Other English names White-eared Catbird, Southern White-eared Catbird
Etymology The species epithet *stonii* commemorates the explorer and ethnographer Octavius Stone, who was in New Guinea in 1875–76.

FIELD IDENTIFICATION A typical large green catbird, being a medium-sized very shy stocky green bird of dense forest, with rich ochraceous-buff underparts lightly spotted with black, a blackish-brown crown, heavy black marks on the rear of the ear-coverts, whitish sides of head and throat, and a broad blackish and white collar across the nape. The stout pale bill is also prominent. Usually located by the distinctive sibilant hissing calls, it is otherwise extremely difficult to see. SIMILAR SPECIES Black-capped Catbird has a distinctive blackish cap and dark cheek patch, and lacks extensive white on the ear-coverts. Black-eared Catbird, which has limited range overlap with Ochre-breasted, lacks the white ear-coverts and has the crown heavily spotted with white. Fruit-doves (*Ptilinopus*) might be confusable if very poor views obtained.

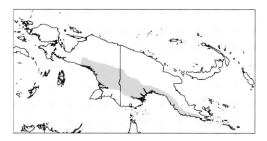

RANGE Southern watershed of the eastern ranges and South-east Peninsula east to Amazon Bay with a break in the centre, as it is not recorded from the Trans-Fly (Bishop 2005a). MOVEMENTS Resident.

DESCRIPTION 25cm. No marked sexual dimorphism; female slightly smaller than male, crown and perhaps leg colour may be paler. *Adult* A heavily built stout catbird with the crown dark blackish-brown, tinged olive-green. The pure white of the ear-coverts extends forwards onto the lower lores. The upperparts, including the tail, are dark green, the tertials and inner secondaries with small pale tips. Neck buff-coloured and broadly tipped black, forming a dark, blackish collar across hindneck. Underparts heavily washed rich cinnamon-ochre, with small black spotting most pronounced on the breast, with lower belly and undertail-coverts unmarked; underwing-coverts buffish-white. *Iris* dark red; *bill* pale bluish-grey; *legs and feet* grey. Photographs of this form are in Coates (1990: 383) and Coates & Peckover (2001: 206). **Immature** is paler and less rich ochre-buff beneath, with a more brownish cap on head and a paler greyish (not red) iris.

TAXONOMY AND GEOGRAPHICAL VARIATION

Formerly regarded as a subspecies of the White-eared Catbird, but recent work suggests that the latter may be better treated as three distinct species (Irestedt *et al.* 2015); *stonii* was originally described as a species, and has therefore been returned to the original designation. *A. stonii* is polytypic, with two subspecies: the nominate race in the southern coastal regions of south-east PNG from Amazon Bay west to the Upper Purari River (Karimui Basin), with larger black spotting and richer ventral area; and *A. s. cinnamomeus* in the south-west east to the Fly River basin and Lake Kutubu, with a gap in the Trans-Fly lowlands. Note that Beehler & Pratt (2016) consider that *cinnamomeus* is best combined with what was the nominate form of White-eared Catbird (*A. buccoides*, now treated as monotypic), which tends towards being paler ventrally in the west and darker and richer ventrally in the east.

VOCALISATIONS AND OTHER SOUNDS This

species sometimes responds to a recording of the call which it shares with its close congeners: the long, low, sibilant, harsh yet rather wavering hissing, *kssssssssh* (dropping in pitch). A weak, short, high-pitched whistled *tseep* or *tsip*, given as an alarm call (Coates 1990), is analogous to a similar call of Spotted Catbird. Coates also notes an occasional light chittering series of nasal notes which rise, fall and then rise over 2–3 seconds. No cat-like squealing *errrow*-type call is known for this species, which is heard far more often than seen. The voice is not known to differ significantly from those of the two other members of what was the White-eared Catbird *sensu lato*. Xeno-canto XC87583 (by I. Woxvold) has a short example.

HABITAT Rainforest, hill forest and monsoon forest in the lowlands, occurring mostly from sea level up to about 800m, but locally to higher levels as at Karimui (1,100m) and Baiyer River (1,200m). It is sympatric with the Black-eared Catbird in some areas, as at Kiunga and Tabubil, and with the Black-capped Catbird at the Sogeri Plateau.

HABITS These catbirds keep to the dense middle storey and understorey of the forest, being well camouflaged and concealed among the vines, ferns and leaves. The species is very shy, heard more often than seen, but quite widespread in dense forest. Usually seen singly, but sometimes in pairs or family groups of three individuals, they tend to keep quite low down and are rarely seen in the canopy. They call at intervals throughout the day, calling sessions often lasting for 20 minutes or more before they go quiet again. Coates (1990) has some wonderful photos of them coming to pools to bathe and drink during the dry season near Port Moresby.

FOOD AND FORAGING The Ochre-breasted Catbird is primarily a frugivore, but will take arthropods and seeds and very possibly nestlings and eggs, as its congeners are believed to do. The birds glean fruit, and are not very active, often remaining in one place for long periods. They can hop quickly and may bound along the ground, branches or logs in a similar manner to Spotted Catbirds. They seem to associate with the periphery of mixed-species foraging flocks, not being a core element but often found on the margins when such flocks are passing through.

BREEDING BEHAVIOUR The pair maintains a **territory** when breeding and, as with the Spotted Catbird, is aggressive to conspecifics and small birds that pass through (Gilliard 1969). Violent but strangely quiet chases accompanied by wing-rustling occur when another catbird enters the territory (Gilliard). When **calling**, the bird stretches upright with tail hanging down, then bows down to call, or it may adopt a horizontal pose and throw the head forwards (Coates 1990). This species shows an interest in the display courts of the Magnificent Bird of Paradise, sometimes visiting them. Cooper (in Forshaw & Cooper 1977) saw one alight by such a court and give a soft, low rasping *aah-aah-aah* call which could almost be described as hissing noises (and thus like the usual call?): it called for about five minutes and then flew to the main display perch above the arena, where it remained for about two minutes, intently surveying the cleared area below, before flying off. **Nest** The nest is a cup-shaped structure, often placed low down in *Pandanus* and accessible by hand without climbing (Simson 1907). Coates (1990) noted a platform of sturdy forked twigs, with a deep central depression lined with a few large leaves and with a mass of short sections of strong slender stems above it, making the central cup only a slight one. He reported the nest as being in the vertical fork at the top of a slender sapling some 1.8–3.4m above ground, and not concealed from above. **Eggs, Nestling & Fledgling** The clutch size is just a single egg, colour seeming to vary with the race concerned, but an egg of the nominate race from Crater Mt is described as 'coffee with cream' and unmarked (Mack in F&F 2004). No information on nestling. The fledgling has been noted as remaining with the parents for some 3.5 months (Coates 1990). **Timing of Breeding** Breeding seems to occur mainly during the wet season, with eggs or nestlings known in May, June and January (Coates 1990). Adults with enlarged gonads have been collected in August at Karimui, in Chimbu Province (Diamond 1972).

MOULT No details available.

STATUS AND CONSERVATION Not as yet assessed by BirdLife as this member of the *A. buccoides* species group is not treated as a full species by del Hoyo & Collar (2016). The present species would be Least Concern, having a fairly wide range, mostly in lightly inhabited country.

Ochre-breasted (White-eared) Catbird, Brown River, PNG (*Brian J. Coates*).

GREEN CATBIRD
Ailuroedus crassirostris   Plate 25

Lanius crassirostris Paykull, 1815, *Nov. Act. Reg. Soc. Sci.*
Uppsala **7**, 283. Nova Hollandia = Sydney, New South
Wales, Australia.
Other English names Spotted Catbird
Etymology The name *crassirostris* means thick-billed,
from the Latin *crassus*, thick, and *rostrum*, bill.

FIELD IDENTIFICATION A medium-sized, stocky
green bird of thick forest, this species has prominent
white tail tips and wingbars, a stout pale bill and
conspicuous large white spotting on the underparts.
There is a whitish bar on the side of the neck, and it
lacks black ear-coverts. **SIMILAR SPECIES** The allopatric
Spotted Catbird is similar but has an obvious blackish
patch on the ear-coverts, and lacks an obvious whitish
neck bar; its head pattern is much more marked, the
crown to upper nape being patterned in dusky and
spotted with whitish (not grass-green with small white
spots), giving a darker and much more contrasted
appearance to the Spotted Catbird. Superb (*Ptilinopus
superba*) and Rose-crowned Fruit-doves (*P. regina*) have
a different shape, show a colourful cap and underparts
and lack the spotted plumage.

RANGE Endemic to the lowland and hill rainforests
of south-eastern Australia from the Shoalhaven River,
in south-east NSW, north to the Bunya Mts and the
Kroombit Plateau of the Dawes Range in the Gladstone
area of South East Queensland. Not found inland of the
Great Dividing Range. **MOVEMENTS** Resident.

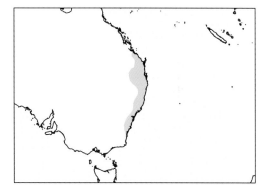

DESCRIPTION 31cm; male 167–289g, female
169–211g. Sexes alike in plumage, but female slightly
smaller than male. *Adult* The head and upperparts are
mainly green with blackish feather tipping, giving a
finely spotted appearance to crown, ear-coverts, malar
area and chin; a thin line of white feathers, finely
flecked blackish, encircles the eye, which appears far
more prominent in a paler face than with the Spotted
Catbird. The nape, mantle and side of neck are finely
streaked whitish, with a conspicuous white patch
on lower neck; primary-coverts and secondaries are
broadly tipped whitish, forming two spotted wingbars,
and the tertials also are tipped white. The chin and
throat are dirty whitish with a mottled appearance
owing to blackish feather bases and greenish tips. The
breast is paler green than the upperparts, becoming
washed with pale straw-yellow on the central abdomen;
the lower throat and breast are finely streaked and

329

spotted white, these markings larger on the breast and giving a spotted appearance, and more streaked white on rest of underparts; undertail-coverts washed very pale green. *Iris* deep red; *bill* pearly whitish, with variable greenish wash; *legs* pale greyish. **Juvenile** Appears fuzzy-headed with russet down on crown, upperparts duller green than adult, underparts dirty whitish with faint green wash on centre of belly. **Immature** Similar to adult but with finer, narrower, pale nape markings, reddish-brown iris, greyer bill. (Frith *et al.* 2017)

TAXONOMY AND GEOGRAPHICAL VARIATION

Monotypic. The Green Catbird is the isolated southernmost representative of the complex of taxa that comprise the Spotted Catbird group, and has at times been treated by some authorities as a subspecies of that bird (Schodde & Mason 1999; Dickinson & Christidis 2014). A recent revision of the whole complex using molecular data has confirmed its status as a species (Irestedt *et al.* 2015). Synonyms *Coracina viridis* Vieillot, 1817; *Ailuroedus crassirostris blaauwi* Mathews, 1912.

VOCALISATIONS AND OTHER SOUNDS

The loud, squalling, cat-like *err-row err-row err-row* gives the species its vernacular name, this call being not quite so harsh as the similar call of the Spotted Catbird in North Queensland. It may be given as a single *err-row* call or be repeated in a short series of up to three, and is delivered with a rising volume and pitch, dropping slightly as it terminates. This call is given most frequently at dawn and in late afternoon, but can be made at any time of day. It can be transcribed as *Wheere are ya?*, often with a squalling *Here I am* as a response as presumed pair members call to each other (G. Threlfo pers. comm.). I treat the usual vocalisations as calls rather than songs, as they seem too simplistic to fit a song designation, but these notes are sometimes described as songs. Cooper (in Forshaw & Cooper 1977) recorded a short *chuck* or *churt* note given by an individual when foraging in leaf litter. A quiet but incisive *zik* is the common alarm or curiosity note, similar to that made by Spotted Catbird but perhaps quieter and less incisive. Xeno-canto has various examples, with the call and the *zik* note at XC97339 (by P. Åberg) and the typical call at XC155114 (by F. Deroussen). An unusual squeaky rising note is on XC94549 (by N. Jackett). Marshall (1954) described a dawn vocal display given from a vantage point low in the forest as one or more clicking notes followed by three guttural cries, the first two drawn out and the last brief and sharp.

HABITAT

Inhabits rainforest of the temperate and subtropical lowlands; found also at rainforest edges and in adjacent eucalypt sclerophyll forest and woodland, where it may go to forage. Occasionally visits gardens and orchards, and can very rarely wander to more open habitats such as farmland. Occurs on densely forested ridges and hills, from sea level up to about 1,200m.

HABITS

A shy and secretive but very active, primarily frugivorous bird, which is hard to see in the dense dark forests, and is often best observed at fruiting trees. This species frequents mostly the lower and middle stages of the forest and is heard far more often than seen. It may be found along with Satin and Regent Bowerbirds at fruiting trees or fruit sources. The feeding tables at O'Reillys, in the Green Mts at Lamington NP, commonly host Satin and Regent Bowerbirds, but the wary Green Catbird is only an occasional visitor and then usually singly rather than in groups, and generally not staying long; this is in marked contrast to the behaviour of Spotted Catbirds in North Queensland, which readily come to such food sources, often in pairs. In some areas, however, Green Catbirds may forage also at picnic tables or feeders; they are inquisitive and will respond to tapes or pishing at times, albeit warily. For drinking and bathing this catbird exploits water found in crevices in mature trees, and also uses wet foliage in these damp forests, where the birds shuffle and shake their feathers among the leaves (Donaghey 1996).

FOOD AND FORAGING

Primarily a frugivorous species, eating wild figs, native cherries, berries and the seeds of various palms. It is known to take beetles and other insects, too, and is likely to be a predator of eggs and nestlings. Green Catbirds have been known to follow human observers, who may disturb nesting birds, thereby leaving the nests and chicks vulnerable to the catbirds (Slater in Forshaw & Cooper 1977). Fruits form a high percentage of the diet, especially *Ficus*, *Pandanus* fruits, fan-palm fruits and the fruits of umbrella trees (*Schefflera*). This catbird feeds alone or in pairs, and often with other frugivorous birds, including Regent and Satin Bowerbird. During the winter months it may form foraging flocks of up to 20 individuals (Forshaw & Cooper 1977), although this may simply be aggregations of birds at fruiting trees rather than flocks as such. Green Catbirds have been known to attack commercial fruit crops.

BREEDING BEHAVIOUR

Like the Spotted Catbird, the Green Catbird forms monogamous pairs, the female alone building the nest and incubating, but both sexes feeding the young. The species is territorial during the breeding season: the nesting territory is often about 0.6ha, the parents generally foraging not far from the nest (>50m), while at other times foraging seems confined mostly to a home range of over 1ha. Pair members will chase through the forest, calling loudly, this followed by silent preening. They will perform **distraction displays** near nests and/or fledglings, and will mob intruders by flying at them and fluttering along the ground (Grimes in Campbell 1901). This species has been seen to perform a 'rodent-run' distraction display if a human is near the nest, the bird fluttering to the ground and hopping for about 10m with head and tail lowered and wings drooped (Woodall 1994). It has also been noted to **feign injury** when a captured nestling gave an alarm call, the parents flapping about within a metre of the observers, and has been seen to feign a broken wing or leg. Predators near the nest, such as Lace Monitors (*Varanus varius*) and snakes, are mobbed by the catbirds, which swoop down and give harsh alarm calls. Miller (in Marshall 1954) reported that nestlings are fed mainly with insects up to fledging time, which is consistent with other bowerbirds at a time when the nutritional requirements are very high. An intriguing observation by Phipps (in Forshaw & Cooper 1977) describes how a captive pair in an aviary maintained a rudimentary display area decorated with leaves stripped from a lemon tree and placed face down in a rough circle on ground exposed to sunlight; the leaves had pale undersides uppermost and were always kept turned that way, being replaced once they withered, a habit very reminiscent of the Tooth-billed Bowerbird bower. There are as yet no observations of this behaviour in the wild, and it is presumably simply an artefact of captivity. **COURTSHIP** involves intense and often silent chasing by partners in tree canopies; the male will feed his mate throughout the year. **NEST** The nest is a fairly large bulky, open but thick-walled cup with a foundation of sticks and twigs, with dried vines, tendrils and twigs, interwoven with many broad leaves and lined with tendrils and fine rootlets; there is an inner layer of decaying wood, and sometimes earthy epiphytic-fern matter. External measurements are 20–25cm across and 15–20cm deep, with internal dimensions of about 13cm width and about 8cm depth (F&F 2004). It is well concealed in forked branches among the dense foliage atop a small sapling, shrub or epiphytic ferns some 2–18m from the ground. The nest site is often reused the following year. **EGGS** Clutch one to three, usually two, buff-coloured or rich cream glossy ovate eggs, size $c.43 \times 31$mm and laid on alternate days. They are very similar to the eggs of the Spotted Catbird despite statements to the contrary in Gilliard (1969). Incubation is by the female alone, for 23–24 days, she being fed on the nest by the male. **NESTLING** The nestlings are brooded by the female, brooding declining in frequency after day 13 but continuing at low intensity until fledging. They are fed by both sexes,

the nestling period being $c.21$ days. The parents will perform distraction displays in the face of predators, flying at intruders and fluttering along the ground. One juvenile remained partially dependent on its parents for more than 72 days, two others were independent $c.80$ days after fledging. Overall nest success 65% (25 nests), each pair averaging 1.1 fledged young per season; pairs may nest again if the clutch or brood is lost early in the season. Oldest marked individuals lived for at least 13 years. (Frith & Frith 2016) **TIMING OF BREEDING** Breeds in spring–summer from mid-September to February/March, the peak being October–December, with a season duration of $c.$four months. Nests have been found in the austral spring in October, November and December.

MOULT Moult of the secondaries and tail feathers begins around the time when the young fledge, when primary moult is about a third completed, the head and body moult ongoing at the same time (F&F).

STATUS AND CONSERVATION Classified as Least Concern by BirdLife International. The Green Catbird is fairly common where suitable habitat remains, but much has been lost to development, as in the clearing of vine forests at Nanango, in south-east Queensland, and particularly in increasingly heavily populated coastal New South Wales. It may be scarcer towards the southern extremity of the range, and rainforest fragments there may themselves be under threat from fires. It was historically considered good eating and was often shot. The species is under no obvious threat beyond continued loss of and degradation of habitat by feral plants and animals. It is found in a number of national parks and reserves and should be secure within these core protected areas.

Green Catbird, sexes similar, Coolendel campground, Shoalhaven River, New South Wales, Australia, November 2017 (*Sue Stanley*).

BLACK-EARED CATBIRD
Ailuroedus melanotis   **Plate 25**

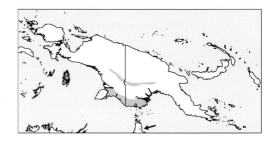

Ptilonorhynchus melanotis G. R. Gray, 1858, *Proc. Zool. Soc. London*, **9** p.181. Aru Islands.

Other English names Spotted Catbird

Etymology The species epithet *melanotis* is derived from the Greek *melas*, black, and *otis*, eared, hence black-eared.

The Spotted Catbird group underwent a recent revision of the whole complex following molecular analyses, which revealed the presence of six genetically rather distinct groups which could be regarded as species (Irestedt *et al.* 2015). These are detailed below in specific accounts and are summarised as follows:

· Spotted Catbird *Ailuroedus maculosus* in North Queensland; monotypic.

· Black-eared Catbird *A. melanotis* at Iron Range, in Far North Queensland (subspecies *joanae*), and then in the Trans-Fly (*melanotis*) and southern montane watershed (*facialis*) of central New Guinea. Note that the Friths (2004) used the name Black-eared Catbird for the whole group, including the Spotted Catbird of Queensland.

· Arfak Catbird *A. arfakianus* from the Vogelkop; monotypic (described form *misoliensis*, from Misool, in West Papuan Islands, now synonymised with *arfakianus*).

· Northern Catbird *A. jobiensis* from the northern slopes of west-central New Guinea and the northern mountains; monotypic.

· Huon Catbird *A. astigmaticus* from the Huon Peninsula montane region; monotypic.

· Black-capped Catbird *A. melanocephalus* from the montane South-east Peninsula of PNG; monotypic.

There are some morphological distinctions in the colour of the head, the throat and underpart markings and the head pattern, but the voices of at least five of the six suggested species are quite similar, with squalling cat-like notes as the primary vocalisation. There are some variations in how harsh they sound, and how drawn out they are, and most of the taxa also seem to have the quiet *sip* or *zik* contact note, although much remains to be learned and there are as yet few recordings of the New Guinea taxa.

FIELD IDENTIFICATION A typical shy and elusive, chunky green catbird but with striking black ear-coverts with white surround, a dark crown with variable pale spotting, and underparts heavily spotted buffish-white, with blackish feather edgings on breast. All three races are seldom seen and remarkably poorly known. **SIMILAR SPECIES** White-eared Catbird has much white on the ear-coverts, and a very different call. *Ptilinopus* fruit-doves might perhaps be confusable if seen very poorly.

RANGE This species is endemic to southern New Guinea and the Cape York Peninsula of Queensland, north Australia. It occurs in the Aru Islands and in the Trans-Fly lowlands east to the Oriomo River (nominate race); in the west-central mountains of New Guinea in hill forest at 600–1,700m (*facialis*); and in the lowland rainforest of Iron Range, Cape York Peninsula,

Queensland (*joanae*). **MOVEMENTS** Sedentary; some possible seasonal altitudinal movements in Australia.

DESCRIPTION 29cm. Sexes alike. *Adult* Has a large black patch on the ear-coverts with a variable whitish to buffy patch around it, a dark crown with variable pale spotting, and a tawny-buff to tawny-brown hindneck scaled with black edgings and large medium-buff spots on nape. The upperparts are rich green, lacking pale wingbars, and the tail has white tips on all but the central feathers. The throat and breast appear as a darkish green broad band with large buffy spots, and there is a greenish wash on the belly; the lower underparts are variably greenish to buffish, heavily spotted white, and contrasting with the darker breast. *Iris* deep red-brown; *bill* stout and whitish-grey; *legs and feet* blue-grey. *Immature* Differs from adult in having less mottled head and a brown (not red-brown) iris.

TAXONOMY AND GEOGRAPHICAL VARIATION Polytypic, with three subspecies now assigned to this species.

1. *Ailuroedus m. melanotis* (G. R. Gray, 1858). Aru Islands. Nominate race of Aru Islands and Trans-Fly lowlands, extending east to the Kiunga–Elevala Rivers and the Oriomo River; the adult male has a very large bill compared with that of other races (Ogilvie-Grant 1915a; Frith & Frith 2004). Mees (1982) reviewed this form in detail, comparing birds from the Aru Islands with those from the adjacent mainland; he found mainland birds to be slightly smaller, but considered the difference insufficient to warrant subspecific distinction.

2. *Ailuroedus m. facialis* (originally described as *Ailuroedus crassirostris facialis* Mayr, 1936, *Amer. Mus. Novit.* **869**: 4. Otakwa River, in western ranges). Similar to nominate subspecies, but throat darker, less white, more buff; spots on crown and upper back much darker and more numerous; more green in the smudgy breast-band (Frith & Frith 2004). Occurs on the southern slopes of western ranges east to where it presumably meets with the form *A. melanocephalus*, known to occur at least as far west as the southern slopes of the eastern ranges. It is uncertain which taxon occurs on the southern slopes of central ranges (between the Baliem River and Karimui), but the gap will presumably be filled by *facialis* from the west and Black-capped Catbird from the east (Beehler & Pratt 2016).

3. *Ailuroedus m. joanae* Mathews 1941, *Emu* **40**: 384. The name *joanae* is a female eponym but of unknown derivation, as a dedication was not given; perhaps

after a relative of the collector, Dr G. Scott? The taxon *joanae* is a restricted-range endemic of north-central Cape York Peninsula around Iron Range. Slightly smaller than Spotted Catbird, blacker on crown, nape and mantle, with less black on face and chin, a whiter throat, underparts more yellowish and less marked than Spotted Catbird, especially on belly and flanks, but with a distinct darker gorget heavily spotted pale. The underwing-coverts are largely white (dark grey on Spotted). The blackish scalloping on the breast is darker than on Spotted Catbird, but paler than on *melanotis* in New Guinea.

VOCALISATIONS AND OTHER SOUNDS

Voice varies slightly with taxon, but poorly known and in the case of some newly elevated taxa virtually unknown. Both *melanotis* and *joanae* have the harsh, wailing, scolding *err rur rurr rur rurr* call like the sound of an angry cat, and a quiet *sip* alarm call, and *facialis* is likely to be similar. Birds of taxon *joanae* from Iron Range sound slightly higher-pitched and more drawn out than Spotted Catbird, but very similar, as with the *sip* alarm call. Xeno-canto XC104924 and XC104925 (by E. Miller) give the squalling typical catbird-type call and the *sip* note of *joanae*. A recording of the cat-like sounds of *melanotis* from Lake Murray, in the middle Fly, is at XC388883 (by P. Gregory) and also at XC38132 from the Kiunga area (by F. Lambert). This species is not known to mimic other species or sounds. I treat the usual vocalisations as calls rather than songs, as they seem too simplistic to fit a song designation, but these notes are sometimes described as songs.

HABITAT

Inhabits lowland rainforest and hill forest in New Guinea, and monsoon forest and woodland/vine forest in Cape York. Recorded at around 650m at Tabubil and at up to 1,700m in hill forest in the Snow Mts. Sympatric with Ochre-breasted Catbird along the upper Fly and Elevala Rivers.

HABITS

A shy and secretive but very active bird, very hard to see in the dense dark forests and best observed at fruiting trees. Frequents mostly the lower and middle stages and is heard far more often than seen.

FOOD AND FORAGING

Omnivorous, but with fruits forming a high percentage of the diet. This species consumes especially *Ficus*, *Pandanus* fruits, fan-palm fruits and the fruits of umbrella trees (*Schefflera*). One ate fruits of a *Dianella ensifolia* that were placed as decorations on the bower of a Flame Bowerbird (Mackay 1989). Other, larger frugivores, such as Pacific Koel (*Eudynamys orientalis*) and Great Cuckoo-dove (*Reinwardtoena reinwardtii*), have been seen to chase catbirds away from fruit sources (Pratt 1984).

BREEDING BEHAVIOUR

The breeding behaviour of all the New Guinea taxa of the former Spotted Catbird complex is very poorly known or unknown. **NEST** The nest is seemingly a bulky cup of sticks, with a neat central cup of large dead leaves lined with woody stems and creepers, much as with Spotted Catbird. Nest sites recorded for nominate *melanotis* include 3m above ground in a slender tree in light rainforest and *c.*2.4m high in the main fork of a slender understorey tree in rainforest with a dense understorey canopy of vines, pandans and palms. Nests have been found in September, November and December and nests with a single fledgling seen in September and January (Frith & Frith 2004). **EGGS** The eggs are unmarked creamy-white, buff or light olive-brown, and much like those of the Green Catbird despite statements to the contrary in Gilliard (1969); size *c.* 43 × 30mm. Clutch size is one or two; clutches of two eggs have been recorded for *melanotis* in December, with a single fledgling in September in the Trans-Fly, and two eggs were collected from a nest of *facialis* in the Snow Mts in August.

MOULT

Museum specimens from New Guinea show wing moult mostly during December–March (F&F 2004).

STATUS AND CONSERVATION

Not assessed by BirdLife International as *A. melanotis* is not treated as a full species by del Hoyo & Collar (2016), but would be Least Concern. Fairly common, albeit not easy to see at times, with much of its habitat in remote and lightly settled regions and the taxon *joanae* occurring in Iron Range NP.

Black-eared Catbird, race *joanae*, Iron Range National Park, Queensland, Australia (*Jun Matsui*).

Black-eared Catbird, race *joanae*, Iron Range National Park, Queensland, Australia (*Jun Matsui*).

ARFAK CATBIRD
Ailuroedus arfakianus   Plate 26

Ailuroedus arfakianus A. B. Meyer, 1874, *Sitzungsber. Akad. Wiss. Wien* **69**: 82. Arfak Mts, Vogelkop.
Other English names Spotted Catbird, Black-eared Catbird
Etymology *Arfakianus* means from the Arfaks.

FIELD IDENTIFICATION A typical shy and elusive, chunky green catbird but with black ear-coverts, and a dark crown with variable pale (often whitish) spotting. This species is seldom seen and very little known. SIMILAR SPECIES White-eared Catbird may be locally sympatric, but has no black on ear-coverts and has a very different call.

RANGE Misool Island and Vogelkop (Bird's Head). The population in the Kumawa Mts is apparently referable to this form (see Diamond 1985: 81). The populations in the Fakfak Mts and mountains of Wandammen Peninsula (Diamond 1985), currently of unknown affinity, may likewise belong to this form. MOVEMENTS Sedentary.

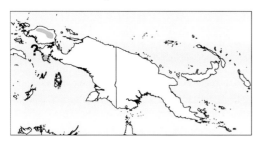

DESCRIPTION 29cm. Sexes alike. *Adult* has black ear surrounded by a larger white post-ocular patch, and a dark crown with variable pale, often whitish spotting, paler than in allopatric Black-eared Catbird. The dark throat with black spotting on a dirty-white to buff background is also distinct from the pattern of Black-eared Catbird; chest darker green with narrow pale buff feather centres, and not marked with blackish (Frith & Frith 2004). *Iris* red; *bill* stout and pale; *legs and feet* greyish. *Immature* No information; presumably much as that of Black-eared Catbird, with minor differences from adult.

TAXONOMY AND GEOGRAPHICAL VARIATION Formerly a part of the Spotted or Black-eared Catbird complex, which is now split into six components based on genetic work by Irestedt *et al.* (2015). Synonym: *Ailuroedus crassirostris misoliensis* Mayr and Meyer de Schauensee, 1939, *Proc. Acad. Nat. Sci. Philadelphia* **91**: 152. Tip, Misool, West Papuan Islands. The validity of described race *misoliensis* from Misool (with minor size differences and a blacker throat) was long ago doubted by Mees (1965: 195–196), and Beehler & Pratt (2016) subsume it into nominate. Monotypic.

VOCALISATIONS AND OTHER SOUNDS Typical squalling catbird-type vocalisations, similar to those of Spotted Catbird but perhaps slightly deeper-toned.

See xeno-canto XC163133 and XC163127 (by F. Lambert). Still poorly known.

HABITAT Hill forest on Misool and in the Arfak and Kumawa Mts, to at least 1,700m.

HABITS A shy and secretive but very active bird, very hard to see in the dense dark forests and best looked for at fruiting trees. It frequents mostly the lower and middle stages and, as with congeners, is heard far more often than it is seen. Visits the large red pandanus fruits left out at a couple of lodges in the Arfak Mts for the benefit of visiting photographers and birders.

FOOD AND FORAGING This species is omnivorous, but with fruits forming a high percentage of the diet, presumably much as with other members of the genus. One was observed in a feeding flock with Western Parotia.

BREEDING BEHAVIOUR No information. For all the New Guinea taxa of the former Spotted Catbird complex, breeding details are very poorly known or unknown.

MOULT No information, but wing moult likely to be during December–March as for Black-eared Catbird.

STATUS AND CONSERVATION Not assessed by BirdLife International as *A. arfakianus* is not treated as a full species by del Hoyo & Collar (2016), but would be Least Concern. Fairly common, albeit not easy to see, with much of its habitat in remote and lightly settled regions.

Arfak Catbird, sexes similar, Arfak Mts, West Papua (*Tony Palliser*).

NORTHERN CATBIRD
Ailuroedus (melanotis) jobiensis   **Plate 26**

Aeluroedus m. jobiensis Rothschild, 1895, *Bull. Brit. Orn. Club* **4**: 26. 'Jobi Island'; error = probably from mainland east of Geelvink Bay (Mayr 1962).
Other English names Spotted Catbird, Black-eared Catbird
Etymology The species name *jobiensis* means of or from Jobi Island, in Geelvink Bay, although it does not actually occur there.

FIELD IDENTIFICATION A typical shy and elusive, chunky green catbird but with black ear-coverts, blackish crown with variable pale spotting and a blackish upper breast finely spotted buff. This species is seldom seen and very little known. **SIMILAR SPECIES** White-eared Catbird has no black on ear-coverts, and a different call. *Ptilinopus* fruit-doves might be confusable if seen very poorly.

RANGE The northern slopes of the western, border and eastern ranges; Foya, Cyclops, Bewani, Prince Alexander and Adelbert Mts (*cf.* Diamond 1985); also the mountains of the Sepik and Jimi Rivers (Rand 1942b; Coates 1990). **MOVEMENTS** Sedentary.

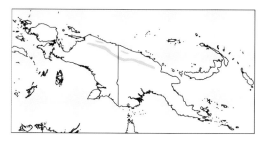

DESCRIPTION 29cm. Sexes alike. *Adult* Similar to Black-eared Catbird, but blackish crown has the pale spotting more buff, and chin, throat and upper breast are blackish with fine buff spotting; remaining underparts darker than those of *melanotis* (Frith & Frith 2004). See photographs of this form in Coates (1990: 384) and Coates & Peckover (2001: 206). *Iris* red; *bill* stout and pale; *legs and feet* greyish. *Immature* Presumably much as that of Black-eared Catbird.

TAXONOMY AND GEOGRAPHICAL VARIATION
Formerly a part of the Spotted or Black-eared Catbird complex, which is now split into six components based on genetic work by Irestedt *et al.* (2015). Synonym *Ailuroedus melanotis guttaticollis* Stresemann, 1922, *Orn. Monatsber.* **30**: 35. Hunstein Mts, middle Sepik, north-central New Guinea. Reported as having buffer spotting above, underparts more rufous, throat and chin darker, but differences considered minor. Monotypic.

VOCALISATIONS AND OTHER SOUNDS Voice virtually unknown. At time of writing (2018) there is one recording on xeno-canto at XC120528 (by K. Tvardikova), originally misidentified as Tan-capped (White-eared) Catbird until it was realised that that species group does not make the loud wailing calls.

HABITAT Occurs in hill forest and lower montane forest.

HABITS A shy and secretive but very active bird, always very hard to see in the dense dark forests, and best looked for at fruiting trees. Frequents mostly the lower and middle stages and is heard far more often than seen. Appears to be sympatric with Tan-capped Catbird in the hill forests around Madang.

FOOD AND FORAGING Omnivorous, but fruits form a high percentage of the diet. Details presumably much as for other members of the genus.

BREEDING BEHAVIOUR Males with enlarged testes in August and September and in breeding condition in March and April. Nothing else reported, but presumably much as for Black-eared Catbird.

MOULT No information, but wing moult likely to be during December–March as for Black-eared Catbird.

STATUS AND CONSERVATION Not assessed by BirdLife International as *A. jobiensis* is not treated as a full species by del Hoyo & Collar (2016), but would be Least Concern. Fairly common, albeit not easy to see, much of its range and habitat being in remote and lightly settled regions.

Northern Catbird, sexes similar, Prince Alexander Mts, Sepik-Ramu, PNG (*W. S. Peckover*).

HUON CATBIRD
Ailuroedus (melanotis) astigmaticus   Plate 26

Ailuroedus melanotis astigmaticus Mayr, 1931, *Mitt. Zool. Mus. Berlin* **17**: 647. Ogeramnang, Huon Peninsula.
Other English names Spotted Catbird, Black-eared Catbird
Etymology The name *astigmaticus* is derived from Greek and means without a brand, mark or spot.

FIELD IDENTIFICATION A typical shy and elusive, chunky green catbird but with black ear-coverts, the crown blackish with a few small and narrow white lines rather than spots, and a dark blackish collar with abundant pale buff spotting. This species is seldom seen and very little known. SIMILAR SPECIES Tan-capped Catbird has a tan cap, extensive white on ear-coverts, and a very different call. *Ptilinopus* fruit-doves could perhaps be confusable if seen very poorly.

RANGE This taxon is found only in the rainforests of the mountains of the Huon Peninsula.

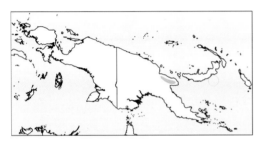

DESCRIPTION 29cm. Sexes alike. *Adult* This species has the crown blackish with a few small and narrow thin white lines (rather than spots), and a dark blackish collar with abundant pale buff spotting. Otherwise plumage much as that of congeners, with green upperparts, whitish tips to the tail, and whitish spotting on the buffish underparts. *Iris* red; *bill* stout and pale; *legs and feet* blue-grey. *Immature* No information.

TAXONOMY AND GEOGRAPHICAL VARIATION Formerly treated as a part of the Spotted or Black-eared Catbird complex, which is now split into six components based on genetic work by Irestedt *et al.* (2015). Monotypic.

VOCALISATIONS AND OTHER SOUNDS Voice virtually unknown. At time of writing (2018) there is one recording on xeno-canto at XC384384 (by J. Matsui), the call perhaps slightly more drawn out and slower than those of others of the group but overall very similar in sound.

HABITAT Inhabits montane forest up to at least 1,800m, the lower limits as yet uncertain.

HABITS A shy and secretive but very active bird, very hard to see in the dense dark forests and best observed at fruiting trees. Frequents mostly the lower and middle stages and is heard far more often than seen, as with its congeners.

FOOD AND FORAGING Omnivorous, but fruits form a high percentage of the diet, presumably much as with other members of the genus.

BREEDING BEHAVIOUR Details of breeding are as yet very poorly known or unknown for all the New Guinea taxa of the former Spotted/Black-eared Catbird complex. Males with enlarged testes have been recorded in October, but nothing else is known. Presumably much as Black-eared Catbird in its nesting behaviour.

MOULT No information, but wing moult likely to be during December–March as for Black-eared Catbird.

STATUS AND CONSERVATION Not assessed by BirdLife International as *A. astigmaticus* is not treated as a species by del Hoyo & Collar, but would be Least Concern. Fairly common, though not easy to see, with much of its habitat in remote and lightly settled regions. Endemic to montane forest in the Huon Peninsula, where the YUS Conservation Area should, it is hoped, protect a good part of the range.

BLACK-CAPPED CATBIRD
Ailuroedus melanocephalus   Plate 26

Aeluraedus melanocephalus Ramsay, 1882, *Proc. Linn. Soc. NSW* **8**: 25. Astrolabe Mts, Southeast Peninsula.
Other English names Spotted Catbird, Black-eared Catbird
Etymology The word *melanocephalus* is from the Greek and means black-headed.

FIELD IDENTIFICATION A typical shy and elusive, chunky green catbird with a stout pale bill and a rather blackish crown spotted with buff (not white). It lacks the white ear-coverts of Ochre-breasted Catbird and is much greener below. Appears rather dark-chested and lacks any pale spots on the wing-coverts. SIMILAR SPECIES There may be some marginal overlap with Ochre-breasted Catbird, but the lack of white ear-coverts and the blackish cap should distinguish it easily; in addition, the call is very different. Might be confusable with *Ptilinopus* fruit-doves in very poor views.

RANGE Mountains of the South-east Peninsula of PNG west to Mt Karimui and, in the northern watershed, to the Herzog Mts and head of Huon Gulf. MOVEMENTS Presumably sedentary.

DESCRIPTION 29cm. Sexes alike. *Adult* Similar to Black-eared Catbird (*melanotis*), but with the underparts generally darker, notably blacker on the chest and throat, and more ochre below. The crown is blacker, and the buff spots and streaks slightly smaller and sparser; the belly has a greenish wash, and this taxon, unlike others in the complex, lacks pale spots on the wing-coverts. Differs from *arfakianus* in having

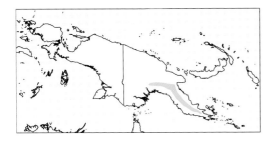

buff (not white) crown spots (Frith & Frith 2004). See photograph of this form in Coates & Peckover (2001: 206). *Iris* red; *bill* stout and pale; legs and feet blue-grey. **Immature** No information.

TAXONOMY AND GEOGRAPHICAL VARIATION Formerly treated as a part of the Spotted or Black-eared Catbird complex, which is now split into six component species based on genetic work by Irestedt *et al.* (2015). Monotypic.

VOCALISATIONS AND OTHER SOUNDS Voice virtually unknown. There are no examples on xeno-canto at time of writing (2018). It is presumed to be the usual cat-like raucous sounds, but requires documentation.

HABITAT A montane taxon, found in forests in the middle and upper levels of the mountains from about 600m to 1,700m, occasionally higher.

HABITS This is a shy and secretive but very active bird, very hard to see in the dense dark forests and best located at fruiting trees. It frequents mostly the lower and middle stages and, as with its congeners, is heard far more often than it is seen.

FOOD AND FORAGING Omnivorous, but fruits form a high percentage of the diet, presumably much as with other members of the genus.

BREEDING BEHAVIOUR The breeding details of this species are not known; presumably much as for Spotted or Black-eared Catbird. Males with enlarged testes have been collected in August, September, October and November, and nests with a single egg found in October and November.

MOULT No information, but wing moult likely to be during December–March as for Black-eared Catbird.

STATUS AND CONSERVATION Not assessed by BirdLife International as *A. melanocephalus* is not treated as a species by del Hoyo & Collar, but would be Least Concern. This species is fairly common, though not easy to see, with much of its habitat in remote and lightly settled regions and thus not in any danger.

Black-capped Catbird, sexes similar, Crater Mt, Eastern Highlands, PNG (*W. S. Peckover*).

SPOTTED CATBIRD
Ailuroedus maculosus   Plate 25

Aeluroedus maculosus E. P. Ramsay, 1875, *Proceedings of the Zoological Society of London* 601. Cardwell, Rockingham Bay, Queensland.
Other English names Black-eared Catbird, Grey-throated Catbird
Etymology The species name *maculosus* is Latin and means spotted.

FIELD IDENTIFICATION A medium-sized, often shy, stocky green bird of thick forest in North Queensland, this species has distinctive black ear-coverts and heavily pale-spotted plumage, as well as prominent white tail tips and wingbars and a stout pale bill. **SIMILAR SPECIES** The Green Catbird is allopatric, has a whitish bar on the side of the neck and lacks the black ear-coverts. Superb Fruit-dove (*Ptilinopus superbus*) and Rose-crowned Fruit-dove (*P. regia*) are of similar size and green above, but have colourful underparts or cap and a different shape.

RANGE Endemic to tropical north-east Queensland from just south of Cooktown south as far as the Seaview and Paluma Range just north of Townsville. The range is very similar in extent to that of Tooth-billed Bowerbird, and also the race *minor* of Satin Bowerbird. **MOVEMENTS** Resident; possibly some seasonal altitudinal movements.

DESCRIPTION 29cm; male 145–205g, female 140–199g. Sexes alike, female on average smaller than male. **Adult** A stocky green bird with distinctive black ear-coverts, small pale spots on the head, and conspicuous large white spotting on the underparts, which have broad browny-green feather edgings, the chest less heavily marked. There are prominent white tail tips and wingbars. *Iris* dull red; *bill* pale and rather stout; *legs and feet* blue-grey. **Juvenile** Has downy crown dark chestnut, back and wings as adult, ventral plumage uniformly pale greyish-white with green wash, iris mid bluish-grey, bill pale dirty whitish. **Immature** Like adult, with buff crown spots, pale throat spots, whiter spotting on underparts, broad greener feather edgings.

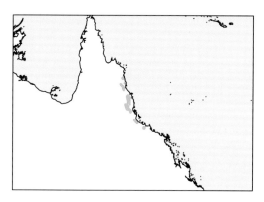

TAXONOMY AND GEOGRAPHICAL VARIATION

Formerly classified as part of a polytypic assemblage of taxa which is now broken up into six component species, as outlined above, following recent genetic evidence (Irestedt *et al.* 2015). Synonym *A. m. fairfaxi* Mathews, 1915, *Austral Avian Record* **2**: 132. Monotypic.

VOCALISATIONS AND OTHER SOUNDS

I treat the usual vocalisations as calls rather than songs, as they seem too simplistic to fit a song designation, but these notes are sometimes described as songs. The call around Kuranda is a loud, raucous, wailing, bawling and remarkably Siamese cat-like *wow er wow er wow*, upward-inflected and trailing off slightly as the notes finish. It can be given as a single rasping *err row* call or in short series, often followed by a quiet but piercing high-pitched *zik* note. These notes may be run together as a squeaky slow stuttering trill, *sisisi*, or be given as a double or triple *zik* or *sip* series, often as counterpoint following the cat-like wailing call. Xeno-canto XC45077 and XC161726 (by P. Gregory) give good examples of both calls. This *zik* note serves as a contact call and also as a mild alarm, and similar calls are given by the Black-eared Catbird taxa *joanae* (at Iron Range) and *melanotis* (in Western Province of PNG). Calls are most frequent over the breeding season and much less so during the winter. Adults utter a soft *grrk* call to nestlings, and larger chicks may reply with loud begging notes. Fledged young may give soft begging calls when following the adults. In addition, harsh scolding notes are uttered by adults when alarmed or mobbing a predator.

HABITAT

Found primarily in rainforest and tall secondary forest. Occurs from sea level in a few areas, but is a species mainly of hill-forest and tablelands rainforest, ranging from 350m to 1,700m.

HABITS

Shy and wary over much of the range, but at some well-known sites much more confiding and will visit feeders, becoming quite tame. This species frequents the understorey and middle levels, occasionally ascending to the canopy, and can be extremely hard to see. It will also forage on the ground. It occurs singly and in pairs, or in small groups of three to five individuals, some of which may be family groups.

FOOD AND FORAGING

The Spotted Catbird is primarily frugivorous, but it will take flowers, buds, invertebrates and vertebrates on occasion. An individual near Kuranda was seen to catch and kill a northern barred frog (*Mixophyes schevelli*), but did not eat much except for a small portion of the belly. It is an opportunist predator of nestlings and eggs, seen to smash and eat the eggs of Wompoo Fruit-doves (*Megaloprepia magnifica*) and also prey on nestling Grey-headed Robins (*Heteromyias albiscapularis*) while being mobbed by the parent birds. One was driven off from attacking the nest of a Noisy Pitta (*Pitta versicolor*), and another was seen to tear apart an empty nest of Red-browed Finch (*Neochmia temporalis*). Large fruits may be torn apart and eaten where found, or be plucked and carried in the bill to a nearby branch or hollow to be eaten or cached. Larger food items are taken apart and carried in pieces to the nestlings. These catbirds will forage at all levels, though least often on the ground, where they explore leaf litter for invertebrates or fallen fruits. They pluck fruit from trees, and glean invertebrates, also taking insects in flight by sallying if the chance arises. They are reported as displacing Tooth-billed and Satin Bowerbirds at fruiting trees, but may themselves be driven off by the larger Wompoo Fruit-dove (*Ptilinopus magnificus*). Some fruits such as figs may be cached atop epiphytic basket ferns (*Drynaria*) or in crevices in trees, to be retrieved later. Has been seen 'mouthing' cached fruits, presumably testing for ripeness, and then carefully covering with leaves, again for later retrieval. At a feeding site at Kuranda, in Far North Queensland, this catbird devours fruits from feeders, with banana and avocado great favourites, also grapes, the birds tearing off chunks and flying off with them, or taking two or three small green grapes at a time. They will also feed at the trays, gulping down mouthfuls and squabbling with conspecifics, also with Victoria's Riflebirds, which win about half the disputes for possession of the food, and Hornbill (Helmeted) Friarbirds (*Philemon yorki*), which they usually dominate; they often dominate the smaller honeyeaters, such as Macleay's (*Xanthotis macleayanus*) and Yellow-spotted (*Meliphaga notata*), and drive them off, but sometimes allow them to share the tray space. Spotted Catbirds have also been seen occasionally to drink from sugar-water bowls, and occasionally to take bread, and meat scraps off bones; one individual was seen to drink a quarter cup of sugar-water in about ten minutes. Young birds are fed primarily with fruit at Kuranda. The Friths (2004) saw this species taking sweet liquid from an exposed root that was exuding sap, and actively defending the source against other catbirds. These catbirds are pugnacious, flying at each other and chasing with much vociferous squalling unless paired, and presumed pairs will often dispute ownership of a food source, one moving in as soon as a vacancy arises. They can become tame, in sharp contrast to the sibling Green Catbird and the New Guinea and Cape York taxa. They begin calling at dawn and are at the food site as soon as it is light, coming in throughout the day (though with a lull in the hottest periods), and finishing off in the late afternoon. Juveniles are not tolerated at the feeder, being driven off smartly and having to sneak in when the adults are absent.

The birds seem to loiter in the vicinity and arrive very soon after fruit is put out.

BREEDING BEHAVIOUR This is a monogamous species, like the entire genus so far as is known. The female alone builds the nest and incubates, but both sexes feed the young. The parents will perform distraction displays, flying at intruders and fluttering along the ground. NEST/NESTING Nest is a large loose, bulky structure of dead sticks with a neat central cup, the exterior of the cup made entirely from large dead leaves or pandanus frond scraps, scantily lined with thinner woody stems and creepers, thicker about the rim. One nest 2m up in a small pandan at Kuranda in 2017 was a large, rather flat, shallow bowl of coarse sticks and stems, with large leaves used as a kind of lining; the adult brought what appeared to be earth into the nest and dropped it into the base, then landed inside the bowl and fidgeted about before settling down, as if adjusting the lining to suit. The nest site is often about 2m up in a slender forked sapling or pandan in rainforest, with a dense shady understorey of vines, palms, pandans and moss-covered litter. EGGS & INCUBATION Clutch is of one or two eggs, laid on alternate days. The eggs are unmarked creamy white or buff (and much like those of the Green Catbird, despite statements to the contrary in Gilliard 1969), size *c.*40 × 28mm. Incubation is by female, period 22–23 days. NESTLING Brooded by the female for up to 15 days, fed by both sexes. Adults at Kuranda have the habit of coming in to feeders, grabbing fruit and then departing quickly, always dropping down from the tray and flying off rapidly in a straight line back towards the nest. Nestling period 19–20 days. CARE OF FLEDGLING

Adults will feed the young once it has fledged, the latter shivering the wings and gaping to solicit food, but they have been seen to drive it away from food sources (perhaps a different pair being in possession of the food resource at that time?). Fledged young may remain with the parents for quite long periods, with up to 49 days recorded (Frith & Frith 2004). NEST SUCCESS Overall nest success (for 63 nests) was 57%, and average of one fledged young produced per pair per season. This species will nest again if the clutch or brood is lost early in the season. TIMING OF BREEDING Birds at Kuranda nest in October–November, with juveniles seen in December, breeding during the austral spring. LONGEVITY The oldest recorded ringed individual was more than 19 years of age (Frith & Frith 2016).

MOULT Primary and wing-covert moult begins on the Paluma Range at the end of November, at the start of the rains, and lasts until early May, with a peak in January–March. Secondary and tertial moults occur in February–March/early April, while head and body feathers are moulted in August–June, with fresh plumage in July (F&F).

STATUS AND CONSERVATION Not assessed by BirdLife International as *A. maculosus* is not treated as a full species by del Hoyo & Collar (2016). Would be Least Concern. Fairly common, albeit not easy to see at times, but in some places comes regularly to artificial feeders and can then be readily observed. Much of its habitat lies within national parks or World Heritage forest, which should, one hopes, afford good protection.

Spotted Catbird, sexes similar, North Queensland, Australia (*Dominic Chaplin*).

Spotted Catbird, juvenile, Cassowary House, Kuranda, Queensland, Australia, October 2017 (*Jun Matsui*).

Genus *Scenopoeetes*

This is a monotypic genus of uncertain affinities despite the striking resemblance to a brown *Ailuroedus* catbird in shape, and one of the older names was Tooth-billed Catbird. Recent genetic studies have allied it more towards *Amblyornis*, the gardener bowerbirds, although it does not build any kind of bower and it may be one of the more ancient members of the family.

Etymology The genus name *Scenopoeetes* is derived from the Greek *skene*, meaning a tent or stage, and *poietes*, a maker, in reference to the species' habit of making a stage.

TOOTH-BILLED BOWERBIRD
Scenopoeetes dentirostris   Plate 27

Scenopoeetes dentirostris Ramsay 1876, *Proc. Zool. Soc. London* 591. Bellenden Ker Range, 3,000–4,000 feet [*c.*915–1,220m], North Queensland, Australia.
Other English names Stagemaker, Stagemaker Bower Bird; Tooth-billed Catbird
Etymology The specific name *dentirostris* means tooth-billed, from the Latin *dens, dentis*, a tooth, and *rostrum, rostris*, beak or bill.

FIELD IDENTIFICATION A stout, catbird-shaped, medium-sized bird of the tropical rainforests of Far North Queensland, brown above and heavily streaked dark below, with a stout black bill. The voice is very distinctive during spring, when the males sing by their stages, being highly ventriloquial and mimetic. **SIMILAR SPECIES** The Spotted Catbird is of a similar size and shape, but is green, with white spots beneath, and has a pale bill. Female and immature Australasian Figbirds (*Sphecotheres vieilloti*) are less densely streaked below, and have a slimmer, less stocky shape, with a longer slender bill and a pale eye-ring.

RANGE The Tooth-billed Bowerbird is a restricted-range endemic of the uplands of north-east Queensland, occurring from the Mt Lewis massif and Bloomfield River headwaters, north-west of Cairns, southwards to Paluma and the Seaview Range and Mt Elliot, near Townsville. **MOVEMENTS** Largely sedentary, males living in their home ranges and during the display season not moving more than *c.*390m; has been known to wander to the coast during times of drought. Some winter dispersal to lower altitudes around 350m.

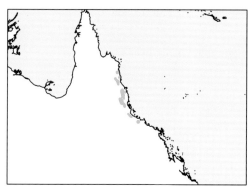

DESCRIPTION 27cm. Sexes almost alike, differing only in mouth coloration. *Adult* The upperparts are dark brown, with a narrow pale ring around the eye and no wingbars. Throat buff; underparts paler, almost whitish-buff, prominently streaked with dark brown, paler on the lower underparts; undertail-coverts barred blackish. *Iris* dark blackish-brown; *bill* blackish, paler at tips of mandibles and cutting edges, two or three notches neatly fitting into each other near tip when bill closed (hence the name Tooth-billed), *mouth-lining* blackish in male, yellowish to pinkish-flesh in female; *legs and feet* greyish. *Juvenile* Dark olive-grey above, upperwing-coverts olive-brown with rufous edges, primaries grey with outer webs edged brown, breast and throat buff with grey markings; iris grey, bill grey with darker tip, mouth pale orange, legs light grey. *Immature* Has pale mouth coloration, pale ochraceous edgings on flight-feathers, and more pointed tail feathers. It is likely that it can take *c.*four years to attain adult bare-part coloration. *Subadult* Resembles adult, but male has mouth becoming blackish to black.

TAXONOMY AND GEOGRAPHICAL VARIATION There is some uncertainty about which branch of bowerbirds the Tooth-billed Bowerbird belongs to or whether it is basal in the bowerbird subfamily; see Christidis *et al.* (1996) and Kusmierski *et al.* (1997). The latter study included Tooth-billed, Golden and Archbold's Bowerbirds within an enlarged *Amblyornis*, despite vast differences in morphology, bower type and vocalisations; such treatment is now not widely followed, although this species is closer to *Amblyornis* than to *Ailuroedus*. It has been called variously Tooth-billed Catbird and Tooth-billed Bowerbird, the latter despite not building a bower but closer to the taxonomic position. This species is monotypic. Described form *Scenopoeetes dentirostris minor* Mathews, 1915, another short-lived Mathews creation, is synonymised.

VOCALISATIONS AND OTHER SOUNDS This species is one of the most remarkable in the family in terms of vocalisations. Males begin calling around late August with the start of the austral spring and resumption of activity with their stages, with peak calling in October–December at the height of the breeding season, as might be expected; vocalising tails off with the coming of the rains in February. The birds are amazingly ventriloquial (it is sometimes almost impossible to work out where the sound is coming from. In addition, they can produce two sounds simultaneously, with an undercurrent of one noise and a different vocal over the top of it, and they are **skilled mimics**. Examples of the songs are on xeno-canto, such as XC174156 (by M. Anderson) and XC198676 (by P. Gregory). The list of bird species recorded as being mimicked is quite long, with at least 38 species, and is dependent on what may be heard locally, but frequently heard calls include those of Bower's Shrike-thrush (*Colluricincla boweri*), Little Shrike-thrush (*C. megarhyncha*), Spangled Drongo (*Dicrurus bracteatus*), Grey-headed Robin (*Heteromyias albiscapularis*), Golden Whistler (*Pachycephala pectoralis*), Lewin's Honeyeater (*Meliphaga lewinii*), King Parrot (*Alisterus scapularis*) and Crimson Rosella (*Platycercus elegans*). Mimicry seems to be used in the primary advertising song, and also as a sort of quiet subsong, which is given only during courtship. One commonly given call is a *chuck* very like the contact call of Bower's Shrike-thrush; it is thought that this may be used to indicate the establishment of presence at the songpost as it is often the initial series given, and it is suggested that it carries farther in the forest than does the song. Cicadas and crickets are commonly mimicked, as are frogs, and one individual on Mt Lewis picked up a squeaky 'pishing' call from the guide and immediately imitated it back in response (pers. obs.). A harsh rattling alarm call may be given in flight. Individuals are generally in auditory contact with those in nearby territories and seem to listen for a response to their song before replying. The voice is loud and quite piercing, at close range almost painfully so, and varies considerably, harsh raspy notes being interspersed with melodious whistles and mimicry, with certain phrases repeated at intervals. The male throws his head back, with throat fluffed out and the blackish mouth-lining exposed as he sings.

HABITAT This species inhabits sub-montane tropical rainforest with a thick understorey of vines, rattans and palms, mainly from 600m to 900m (rarely to 1,200m), occurring from the understorey to the middle levels and occasionally to the subcanopy. Recorded also from exotic pine plantations in the ecotone with rainforest (with rattans and shrubs), and orchards and isolated fruiting trees adjacent to rainforest. A large degree of range and habitat overlap with Spotted Catbird and with the Satin Bowerbird race *minor* is apparent.

HABITS This bowerbird is rather shy and inconspicuous, but males can be seen at quite close range when singing above their courts if approached carefully, although they will disappear immediately if any undue noise or movement occurs and they then take some time to return. Usually seen singly, sometimes in twos or threes or in small aggregations (not flocks as such) of 4–8 at fruiting trees, often with other frugivores such as pigeons or doves. They leave their songposts in short fast, direct flights. Birds fly in and out of fruiting trees and will fly out to isolated trees up to 100m from forest, or to visit gardens and orchards nearby. They roost above their courts, and anting has been noted (F&F 2004).

FOOD AND FORAGING Tooth-billed Bowerbirds are primarily frugivorous, but will on occasion take flowers and various invertebrates, including beetles as well as larvae, spiders and termites. They are not known to take snails, the shells of which are sometimes found near

their stages but result from the activity of Noisy Pittas (*Pitta versicolor*) or other snail predators nearby. Peak foraging times are early mornings and late afternoons, and activity is much reduced on wet days. They tend to forage in the upper and middle levels of the forest, sometimes lower down or on the ground, where they make short hops or take one or two steps. They will come to fruiting figs *Ficus* and *Schefflera* species along with Double-eyed Fig-parrots (*Cyclopsitta diophthalma*), Australasian Figbirds (*Sphecotheres vieilloti*), and Barred Cuckooshrikes (*Coracina lineata*). The notched bill is presumably an adaptation for the eating of leaves, up to ten small leaves being consumed during a foraging bout. Larger leaf pieces are folded and then chewed or masticated before being swallowed. Leaves are a main food source over the winter months when fruits are scarce. The distinctive series of two or three notches near the tip of the bill was described in the older literature as serving as a special modification to help in the severing of leaves used in the court. This overlooks the fact that most, if not all, of the bowerbirds would be capable of severing leaves, and that the females as well as the males have this odd bill structure, though only the male makes the stage. It seems likely that the bill adaptation is for a diet that includes lots of leaves, the unusual dietary habit of winter folivory being well known for this species.

BREEDING BEHAVIOUR This species is thought to be polygynous, with promiscuous males, and the females alone attending to the nest duties. **COURTSHIP/DISPLAY** The **display court** or **stage**, often called a bower in the older literature, is on flat or gently sloping ground under bushes, saplings and vines, roughly oval in shape and 1–2.5m in diameter. The male clears all debris from this space, then decorates it by laying the fresh green ovate leaves of favoured tree species; on the Atherton Tablelands these are laid upside-down to show the rather silvery underside, and are replaced by fresh leaves when they wither. The number of leaves utilised varies with the individual, but can be anything from 40 to over 100. The tree species used will vary with the locality: at Paluma Range, near Townsville, Ivory Basswood (*Polyscias australiana*) was the favoured species, while on the Atherton Tablelands *Schefflera actinophylla* was often used, as well as Brown Tamarind (*Castanophora alphandii*), Rusty Laurel (*Cryptocaria mackinnoniana*) and Anchor Vine (*Palmeria scandens*). Fruits or snail shells are not used as decoration (such items may simply be food dropped by other species), and no painting of the stage is recorded. When the male is not singing or is absent, the nearby rival males often pilfer or attempt to poach leaves from the court. Courts are occupied from August to December, in abundant fruit years maybe from late July, with some activity into February and the commencement of the rains (F&F 2004). Males spend much of their daily time by the court, around 62% in one study, reflecting the dual needs to advertise to the females and to minimise the risk of theft of leaves by rivals. It is very seldom that another bird is seen at the stage (except when other males come to steal leaves), and the females seem to be very circumspect about showing themselves near the court. The male's display or **song tree** is near his own court, often from 50cm to 6m just above it, on horizontal or gently sloping branches, and he spends far more time on the perches than on the courts. The male can sing for up to half an hour at a time without shifting position, and song bouts can last for up to an hour, often with five or six other males in auditory contact in a kind of clan arena or **exploded lek**. The mean size of the **home range** of four males attending courts was of 9.5ha, reflecting the need for a diversity of fruit sources (F&F 2004). Some sites have been recorded as active for at least 20 years, and are coveted by younger males, which may make simple temporary stages nearby and may eventually take over the prime location if the dominant male disappears. **Nest** The nest, built solely by the female, is a frail dish-shaped structure of thin twigs, hidden in thick vegetation among leaves or vines 4–25m above the ground. **Eggs, Incubation Period & Fledging Period** Clutch consists of one or two oval eggs, clear creamy-brown, buff, dark cream or deep yellowish-brown in colour, and measuring *c*.41 × 28mm. It is likely that only the female incubates and performs nest duties, but this is still not well known. The nestling period is around 22 days. **Fledgling** has some grey down on the head, the neck bare and some down on thighs, rump and chin. **Longevity** This species appears quite long-lived, with two individuals ringed at their courts retrapped ten and 12 years later; other ringed birds have lived for >13, 19 and >20 years, the last of those perhaps being as old as 24 as he was an adult holding a court when first marked. The mean annual survival of 24 adult males over 19 seasons from 1978 to 1997 was 90% (F&F 2004).

MOULT Wing moult occurs over the wet-season months of November–March, with head and body moult from September to June.

STATUS AND CONSERVATION Classified as Least Concern by BirdLife International. Despite its relatively restricted range, this species is fairly common within its habitat, much of which is in national parks or World Heritage areas and thus notionally safe from development.

Tooth-billed Bowerbird, head of male showing toothed bill serrations, Lake Barrine, Queensland, Australia (*Jun Matsui*).

Tooth-billed Bowerbird, with flower, Lake Barrine, Queensland, Australia (*Jun Matsui*).

Tooth-billed Bowerbird, upturned leaf stage at Lake Barrine, Queensland, Australia (*Jun Matsui*).

Genus *Archboldia*

This highly distinctive monotypic genus was the last of the family to be discovered and is endemic to the highlands of New Guinea. It is characterised by the distinctive black plumage, with a peculiar flat crown evident in both sexes, along with a relatively long tail and the adult male's erect golden forehead crest. The fern-mat bower construction is also unique and quite distinct from the bowers of *Amblyornis*. The molecular studies of Kusmierski *et al.* (1997) indicated that the single species is very close to those in the genus *Amblyornis*. There are, however, so many distinctions in behaviour, plumage, morphology and bower construction that we feel retention of this genus to be justified, much as with the Golden Bowerbird *Prionodura* (see Frith & Frith 2004).

Etymology *Archboldia* is named in honour of Richard Archbold (1907–1976), who sponsored and ran three famous collecting expeditions to New Guinea in the 1930s for the American Museum of Natural History.

ARCHBOLD'S BOWERBIRD
Archboldia papuensis   Plate 28

Archboldia papuensis Rand, 1940, Bele River, 2,200m, 18km north of Lake Habbema, Snow Mts, western New Guinea.
Other English names Black Bowerbird, Gold-crested Black Bowerbird; Sanford's Bowerbird, Sanford's Golden-crested Bowerbird, Tomba Bowerbird (all *sanfordi*)
Etymology The name *papuensis* is the Latinised version of Papuan.

FIELD IDENTIFICATION The male Archbold's Bowerbird is a very distinctive dark brownish or blackish species with a golden-yellow crest. Females lack the golden crest, but have a small but distinct ochraceous spot on the bend of the closed wing. The shape also is quite distinctive, this species appearing rather flat-headed, with a fairly long, slightly notched tail, and lacking the rotund shape of *Amblyornis* bowerbirds. Sɪᴍɪʟᴀʀ Sᴘᴇᴄɪᴇs Archbold's Bowerbird, with its longer tail (slightly notched) and curiously flattened head shape, appears much more elongated and less rotund than *Amblyornis* species. The relatively short bill and smaller size readily distinguish it from *Melidectes* species, sicklebills and astrapias, while the black male Loria's Satinbird is quite differently shaped, being smaller and shorter-tailed.

RANGE West Papua in Wissel Lakes region, Nassau Mts, Lake Habbema and Oranje Mts, and PNG in south Karius Range, Ambua Range, Mt Giluwe and Mt Hagen. Mᴏᴠᴇᴍᴇɴᴛs Sedentary.

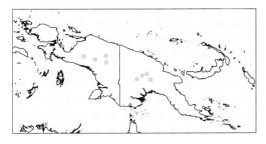

DESCRIPTION A strangely shaped species, appearing longer-tailed than usual in the family, the tail noticeably notched, and the head with a distinctly flattened crown. ***Adult Male*** 37cm; weight 170–195g. Nominate race is entirely sooty blackish, appearing almost spangled on lower breast and belly in some lights, with an elongated feather tuft on forehead forming part of a prominent black-streaked golden-yellow crest from forehead to nape, which tapers to a point on hindcrown; some tapering yellow crest feathers have black tips and the crest has a mottled appearance. Flight-feathers dark grey-brown, tinged olive or blackish; upperwing blackish, primary-coverts sooty brown or jet-black, sometimes with traces of cinnamon. *Iris* dark brown; *bill* black; *legs* blue-grey. ***Adult Female*** 35cm; weight 163–185g. Smaller than male, particularly in wing and tail lengths, but shape similar, with flattened crown and a quite long notched tail; lacks crest. Sooty-black to brownish-black plumage is browner than male's, with a small ochraceous patch on leading edge of wing. ***Juvenile*** Newly fledged young has downy crown, blackish bill, purple-grey legs and dark brown-grey iris. ***Immature Male*** Resembles female, but blacker and lacking paler chin/forecrown, and ochraceous wing patch may be darker. ***Subadult Male*** Like adult male, but with just a few crest feathers intruding into black crown (Frith *et al.* 2017).

TAXONOMY AND GEOGRAPHICAL VARIATION
Two subspecies recognised.
1. *A. p. papuensis* Rand, 1940. Mountains of west-central New Guinea (Wissel Lakes, in Weyland Mts area; Nassau Range and Oranje Mts; Bele River and Lake Habbema region in the Snow Mts).
2. *A. p. sanfordi* Mayr and Gilliard, 1950. Sanford's Bowerbird. Mountains of east-central New Guinea (Mt Hagen, Mt Giluwe, Tari Gap and south Karius Range). May occur also in Kubor Range, and status in Star Mts and Border Ranges unknown. Slightly larger and longer-tailed than the nominate, with the bill proportionately smaller; plumage blacker and less sooty-blackish (especially in immature and subadult stages).

The two races were for long thought by some to be two separate species, supposedly because *sanfordi* was crested and built a 'mat' bower and the male nominate was uncrested and possibly built an avenue bower. Gilliard was convinced that he had found a new species, as at that time the West Papuan birds were thought to lack the yellow crest and there was a report that that form built a stick-type bower (Ripley in Gilliard 1969), which was presumably based on an aberrant bower. Males of both races do, however, have a crest and build similar bowers, so they are now generally treated as conspecific.

VOCALISATIONS AND OTHER SOUNDS A harsh tearing *kshaak* or *kraaaa* is a frequent call heard within the forest near feeding sites at fruiting trees, and is often the first sign of the species' presence. It resembles a call of Crested Satinbird but is louder and harsher. The call is sometimes doubled, and is given sporadically. The male at bower gives loud advertisement vocalisations with a ventriloquial quality, including whistles, buzzing, snapping, tearing, and harsh grating or churring sounds (Frith *et al.* 2017). One at a bower in June 1992 kept up an odd chattering call, a rapid *wirry wirry wirry wip* with the *wip* note sometimes repeated, plus quiet scolding background notes (pers. obs.). Frith & Frith (2004) report vocal mimicry of at least ten sympatric bird species as well as ambient sounds from within the habitat. Subsong given during display includes bleating calls and mimicry of other bird calls, as well as the imitation of whirring bird wings and fluttering vegetation. An individual near the bower gave a very loud and long-drawn-out 'mighty whistle', *pheeuw*, with two sets of two notes, each with about three seconds between the notes. This species also emits ventriloquial *kee* notes, which are very hard to pin down and sound as if more than one bird is involved. The harsh *kra-kraaa* is very loud and can be followed by a snap and a kind of hiss (Gilliard 1969). *Quee* notes are given, and a constant churring series is uttered as the male displays. Xeno-canto has just two examples, some rather nasal scolds from a female *sanfordi* probably with an immature at XC38122 (by F. Lambert), and a typical harsh call from a nominate bird at XC23189 (by M. de Boer).

HABITAT Archbold's Bowerbirds favour frost-pockets in high montane mossy Antarctic beech (*Nothofagus*) forest with coniferous (*Podocarpus*) tree species in the canopy, and an understorey including pandanus (*Pandanus*), umbrella trees (*Schefflera*) and scrambling bamboo. It is found at 1,750–3,660m, mainly 2,300–2,900m. The nominate race of West Papua occurs in lower subalpine habitat in the Ilaga valley at 2,850–3,660m, this being the highest altitude for any bowerbird. Traditional bower sites are evenly dispersed within forest patches interspersed by grassland.

HABITS This is a rather wary species occurring at low density in the high-altitude forests, often first located by its loud harsh calls. Groups of two or three individuals may occur at fruiting trees along with Brown Sicklebills and Ribbon-tailed Astrapias in PNG, as well as Belford's Melidectes and Crested Satinbird.

FOOD AND FORAGING Primarily frugivorous, often seen at fruiting trees within the forest, and females have attended a feeder at Kumul Lodge, in Enga Province, to take fruit, especially paw-paw. Nestlings noted by Frith & Frith (2004) as being fed with skinks, insects and dismembered nestlings of birds, as well as fruit.

BREEDING BEHAVIOUR The polygynous, promiscuous male seasonally decorates the unique-style mat bower, which may be used over repeated seasons.

The female alone builds and attends nest. The species is non-territorial except for the defence of the bower by the male. For 16 traditional bower sites across 1000ha at Tari Gap (PNG), the mean nearest-neighbour distance was 370m. One traditional bower site was used for more than 15 years, and one adult male attended the same site for more than six seasons (Frith *et al.* 1996). Gilliard obtained from one of his Papuan hunters a series of 11 specimens (seven adult males, one subadult male, three females or female types), all from a single bower site or 'sing-sing' ground, and all trapped between 12th July and 22nd July, in addition to which another male had escaped from a snare (Gilliard 1969). This is an astonishing density, but has parallels with collections of similar numbers of MacGregor's Bowerbird and Golden Bowerbird from single bowers. **BOWER** The bower of this species is of a unique type, consisting of a cleared area (up to 6 × 5m) on which a mat of fern fronds is accumulated; perches up to 2.6m above this mat are draped with stems of epiphytic orchids or ferns, the orchids sometimes in flower but not used as decorations *per se*. Average male bower attendance was more than 50% of daylight hours at Tari Gap sites. Bower decorations include large snail shells, beetle wingcases, fruits, fungus, tree resin and charcoal, which are placed on the mat in discrete collections, also (and amazingly) the occipital nuptial plumes or head wires of the adult male King of Saxony Bird of Paradise (Frith *et al.* 2017). A bower at Tari Gap (in June 1992) was located just a few metres inside the forest, an extraordinary structure of dead ferns draped over saplings and branches, reaching up to 1m in height. This bower was subdivided by a branch in one corner, the whole thing appearing as dead leaves draped over sticks with only the avenue and the piles of shells and beetle wings giving it away; the length was about 2m, width about 1m, and it was roughly rectangular in shape. An avenue or pathway about 1m long and 70cm wide led to a gateway of branches with an apron of black snail shells just outside it, while just inside the structure was a small pile of beetle elytra. The floor consisted of dead leaves and ferns, with two head plumes from a male King of Saxony Bird of Paradise laid upon the centre, one plume with the colourful sky-blue side uppermost, the other with dark side up and much harder to see. Joseph Tano collected them and said that he had taken nine sets from this particular bower over the years, this bird also apparently acquiring the central tail feathers of Ribbon-tailed Astrapia for ornamentation. **DISPLAYS** The male's display involves two main elements, the ***Prostrate posture*** and the ***Grovel Display***, and he performs various other postures on the bower mat while giving a subsong, and also ***Displacement-chases*** the female from the bower mat until she succumbs to his advances. At Tari in 1992 (pers. obs.), I could hear the bowerbird scolding and it flew low over me twice before eventually coming into the bower and displaying, creeping about low to the ground with the golden-feathered crest erect, and flicking the tail occasionally. The male kept up an odd chattering call along with quiet scolding background notes (see VOCALISATIONS, above); the performance lasted about 15 minutes and the bird was often out of view. The displaying male attracted an immature male Archbold's Bowerbird and an adult male Ribbon-tailed Astrapia, the immature bowerbird coming right down on to the bower and the male astrapia peering in from nearby shrubs. A Belford's Melidectes also seemed interested in the performance. Gilliard (1969), who discovered this taxon in 1950, gives a graphic account of a complete display which lasted about 25 minutes, the female coming in as the male was churring on the fern mat. She changed perches every minute or two, and flew very close to the male, 'whipping her wings with such rapidity that they sounded as though they would be torn. This ripping, tearing sound, like stiff cardboard being torn, was in part delivered over the male and gave the appearance of whipping.' The male meanwhile appeared submissive and begging in the extreme, as he lay flat on the fern stage for 21 minutes, 'crawling about like a wounded animal. Body pressed to the ferns, wings half-open, their under surfaces against the ferns… so flattened was the bird that it resembled a reptile more than a bird.' The crest was folded flat against the head and only the yellow tuft at the forehead stuck up; the bill was usually wide open and the mandibles flexed as if gasping, with some of the time a shaft of bamboo or fern held crosswise in the mouth. 'The movements of the mouth parts gave the appearance of chewing. The direction of the crawl (which resembled a whipped dog towards its master) was always towards the female. Progress was slow. With open, elevated bill the bird crawled perhaps a foot in one or two minutes.' The female was nearby on horizontal perches around the edge of the fern stage, her head usually towards the male. She would fly over him as he got close, often hovering to whip her wings over his back, then fly across the bower to another low perch. The male would immediately turn and crawl slowly towards her again, quickly hopping at times from a prostrate position and keeping up an incessant churring before resuming the 'whipped dog' attitude. The male made a complete circuit of the stage in about five minutes, following the female as she moved around its edge, and never touching any of the shell, insect or resin adornments in its path, crawling among and even over the adornments, still with a small strand of vine in its mouth. **NEST** The nest is an untidy large deep, bulky bowl with a stick foundation, having a deep and substantial cup of large dried leaves (the uppermost ones still fresh and green) and a lining of curved twiglets. It is constructed 3–7m above ground in the leafy crown of a sapling within or adjacent to a small gap in the forest (no canopy directly above). The average distance of seven active nests from the nearest active bower was some 250m. The female may nest in the same area, sometimes in the same tree, for years in succession. **EGGS, INCUBATION PERIOD & NESTLING PERIOD** The clutch consists of a single egg; incubation 26–27 days. Average nestling period is 30 days. **NEST SUCCESS** The overall success rate for eight nests at Tari Gap was 88%; mean number of fledged offspring per female 0.88 per season. **TIMING OF BREEDING** The breeding season appears to be from September to February, with eggs in November–December; display season September–December, but Mt Hagen bowers are

attended as early as July. **LONGEVITY** The oldest marked adult male lived for at least six years (Frith *et al.* 2017).

MOULT Little known, and sample sizes are small, but wing-feather moult peaks in October–November at Tari Gap and in April–May at Mt Hagen and Western Highlands.

STATUS AND CONSERVATION Classified as Near-Threatened by BirdLife International. Restricted-range species: present in Central Papuan Mountains EBA. Although Archbold's Bowerbird can be reasonably common locally, it is generally rare throughout its patchy range. Global population thought to be rather small, and may be declining as a result of habitat loss. The nominate race appears to have the larger population and is considered probably secure, at least in the short term; eastern race, *sanfordi*, has a total range of no more than *c*.800km², and habitat on two mountains within its range is threatened by logging activities. Habitat disturbance and destruction resulting from an increasing human population, along with associated logging and habitat degradation, represent potential threats. Fire damage following a severe drought in 1997 badly damaged much of this bowerbird's prime habitat in Hela Province, and this, coupled with human activity, led to the abandonment of at least six traditional bower sites below Tari Gap. Climate change may also have an adverse effect, with reduced or different rainfall patterns, more fire potential, warmer temperatures and changing composition of the vegetation. Frost pockets within the forest, which appear to be a microhabitat niche favoured by this species, may become less frequent.

Archbold's Bowerbird, male, race *sanfordi*, Tomba Pass, PNG (*William S. Peckover*).

Archbold's Bowerbird, female, race *sanfordi*, Kumul Lodge, Enga Province, PNG, 18 March 2013 (*Phil Gregory*).

Archbold's Bowerbird, race *sanfordi*, Kumul Lodge, Enga Province, PNG (*Jun Matsui*).

Genus *Amblyornis*

A comparatively homogeneous grouping of five morphologically and structurally very similar species, all mainly or entirely dark brown above and buffy brown below, stocky and quite stubby-billed. Each species builds a remarkable and diverse variant on the single maypole-type bower. These birds are known also as gardener bowerbirds because of their use of fruits, flowers, fungi and sundry other materials in the constructing and decorating of their elaborate bowers. Some of these are the most astonishing architectural structures in the avian world, the bowers of the Vogelkop and Streaked Bowerbirds being the most elaborate and complex. A genetic study of mtDNA (Kusmierski *et al.* (1997) generated a phylogenetic tree that placed *Prionodura* and *Archboldia* among members of this *Amblyornis* lineage, despite those genera having dramatic differences in structure, coloration, bower design and vocalisations, and neatly illustrating the pitfalls of relying solely on one small suite of characters to erect a phylogeny without reference to the wider picture (Beehler & Pratt 2016).

Etymology The scientific name *Amblyornis* means 'dull bird', from the Greek *amblus*, meaning dull, and *ornis*, a bird.

MACGREGOR'S BOWERBIRD
Amblyornis macgregoriae   **Plate 29**

Amblyornis macgregoriae De Vis, 1890, *Ann. Rept. Brit. New Guinea*, 1888–1889: 61. Musgrave Range, 7,000–9,000 ft [*c.*2,130–2,740m], Southeast Peninsula.
Other English names Crested Bowerbird, Gardener Bowerbird, Mocha-breasted Bowerbird, Crested Gardenerbird, MacGregor's Gardenerbird
Etymology The specific epithet *macgregoriae* commemorates Lady Mary MacGregor, the wife of Sir William MacGregor the explorer, colonial administrator and later governor of British New Guinea 1887–1898. This species was discovered on the MacGregor Expedition in the Owen Stanley Range in 1889.

FIELD IDENTIFICATION A stocky, thrush-sized, mainly olive-brown bird with a short, stout black bill, the male with a large erectile yellowish-orange crown which forms a kind of fan when spread in display at the bower, but which is usually seen as a stripe of yellowish-orange or orange on the hindcrown extending over the nape.
Similar Species MacGregor's Bowerbird is allopatric with

the congeneric and very similar Vogelkop and Huon Bowerbirds. It is, however, sympatric with **Streaked Bowerbird** in a narrow zone of overlap in south-east PNG. The latter species generally occurs between 670m and 1,200m, mostly below the range of MacGregor's, but a hybrid with intermediate features has been reported. Both sexes of Streaked Bowerbird have buffy-ochraceous streaking on the throat and breast, but would be hard to separate in bad light or when seen high up in a tall tree. The bower morphology is also totally distinct, and if seen at or near the bower this is a good clue to the bird's identity. A more problematic confusion species is the female or immature **Crested Satinbird**, which has a similar stocky structure and quite short bill, and is olive-brown above: the underparts of this species, however, have a much more pronounced yellowish cast, the crown is more olive, and it is slightly smaller with a smaller, thinner bill which is brownish or bluish-grey, not black. **Black Pitohui** (Shrike-thrush) (*Melanorectes nigrescens*) has a larger and heavier, slightly hooked bill, and has much more rufous coloration below, often with a much greyer head.

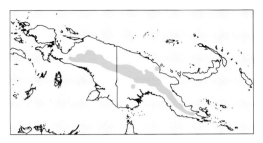

RANGE Mountain forests throughout most of New Guinea (except the Vogelkop and Wandammen Mts) from the western ranges and Border Range east to the Hindenburg Mts of extreme west PNG (Schodde & McKean 1973) and presumably farther east to the Strickland River gorge, thence to eastern ranges and mountains of South-east Peninsula east to Milne Bay, with outliers in the Adelbert Mts and Mt Bosavi. **Movements** Resident, although immatures and females may descend to lower elevations in winter months.

DESCRIPTION Sexually dimorphic. *Adult Male* 26cm; weight 100–145g. Nominate race is basically dark brownish-olive above, slightly paler on face and neck and slightly greyer on lores, with faint orange-reddish wash on crown, and darker on mantle and tail, some specimens showing an amber or orange suffusion. The crest is composed of narrow, rather filamentous or decomposed feathers and can be up to 78mm long; when erect and spread sideways it is quite large and striking, being broad and elongate, a glossy deep yellowish-orange in colour and with some feathers tipped brown. Underparts variable umber, darkest on side of breast and outer lower flanks, paler on chin, throat and belly, with undertail browner; axillaries and underwing-coverts are ochraceous to pale orange-yellow. *Iris* dark brown; *bill* short, stout and blackish (specimens with paler and bluish base); *legs* dark bluish-grey or dark grey. *Adult Female* 26cm; weight 104–140g. Similar to male but lacks crest, and has upperparts slightly paler and washed umber. *Juvenile* All feathers are very soft; upperparts dark mouse-brown with a faint maroon wash, and underparts mouse-grey, darker on sides and paler on abdomen. This is the only known description and was of a fledgling perhaps ten days out of the nest (Gilliard 1969). *Immature* (immature male weight 110–139g) Resembles female, but with upperparts slightly richer rufous, underwing-coverts and axillaries slightly deeper orange, lower underparts slightly warmer brownish and with a more distinct dark breast. *Subadult Male* Looks like adult female but with darker body coloration, and first signs of crest developing, with a few long orange crest plumes projecting like hatpins from midway between the eyes (Gilliard 1969).

TAXONOMY AND GEOGRAPHICAL VARIATION Polytypic. Now treated as having two subspecies (nominate and *mayri*), but formerly considered to have up to seven, with *germanus* now elevated to species status and five others now synonymised with nominate.

1. *Amblyornis m. macgregoriae* De Vis, 1890, Musgrave Range, south-east New Guinea. Occupies central and eastern parts of Central Ranges from Strickland River gorge east to Milne Bay, and also Adelbert Mts.

2. *Amblyornis m. mayri* E. Hartert, 1930 (synonym *Amblyornis inornatus mayri* E. Hartert, 1930), *Novit. Zool.* **36**: 30. Weyland Mts. This is the larger subspecies, with a longer crest: male tail length 87–98mm and crest length 62–74mm (Frith & Frith 2004). From Weyland Mts east to Hindenburg Mts and presumably to Strickland River gorge.

The following are now treated as synonyms of nominate.

1. *Amblyornis inornatus aedificans* Mayr, 1931, *Mitt. Zool. Mus. Berlin* **17**: 648. Dawong, Herzog Mts, Southeast Peninsula.

2. *Amblyornis macgregoriae kombok* Schodde & McKean, 1973, *Emu* **73**: 53. Minj–Nona Divide, Kubor Range, 2,140m, Eastern Ranges. Similar in size and proportions to nominate, but crest rather densely feathered, throat and upper breast pale brownish-olive, lower breast, abdomen and undertail-coverts bright light buffy brown. (*Kombok* was the local name for this taxon.)

3. *Amblyornis macgregoriae nubicola* Schodde & McKean, 1973, *Emu* **73**: 55. Mt Wadimana, Milne Bay Province, Southeast Peninsula. Underparts uniformly dull coffee-brown, and crest rather densely feathered.

4. *Amblyornis macgregoriae amati* Pratt, 1983, *Emu* (1982) **82**: 121. Mt Mengam, also Adelbert Mts, north-east New Guinea. The smallest form (but known from only two adult male specimens), with short wing (like *germana*) and tail (like preceding race), and with chin to upper breast dark olive-brown, only slightly paler than side of head and forehead.

5. *Amblyornis macgregoriae lecroyae* Frith & Frith, 1997, *Bull. Brit. Orn. Club* **117**: 201. Mt Bosavi, 1400m, south-central New Guinea. Eastern ranges and mts of Southeast Peninsula, presumably from Strickland River gorge east to Milne Bay; also the Adelbert Mts. Small and short-tailed, differing from previous in being smaller and darker, and more strongly suffused with orange on head and upperparts. Marginally smaller than *mayri*: tail (male) 78–93mm; male crest 49–69mm (Gilliard & LeCroy 1961; Frith & Frith 2004). Photographs of this now synonymised form are in Coates (1990: 392–396).

Gilliard & LeCroy (1961: 73) studied the available museum material and concluded: '…it therefore seems best to use tail length and crest length alone to differentiate races.' Beehler & Pratt (2016) agreed that plumage colour is not useful in determining subspecies, measurements alone being valid. Other authors, such as Schodde & McKean (1973b) and Frith & Frith (2004), differ, the latter recognising at that time seven races based on size and coloration of body plumage. Beehler & Pratt (2016) recognise two subspecies, and treat a third taxon, the Huon Bowerbird, as a distinct species, this arrangement being followed here.

VOCALISATIONS AND OTHER SOUNDS Male at bower site gives loud and often ventriloquial advertisement calls such as a plaintive repetitive *whoip* note with short trills, and sharp loud *wheet* or *wheep* notes, which are often the first clue to its presence. Also utters harsh tearing sounds, growls, thudding and tapping noises, and ventriloquial hollow whistles, as well as mimicry of Brown Sicklebill (*Epimachus meyeri*) and Sclater's Whistler (*Pachycephala soror*) among others, and other vocal mimicry including frogs calling, trees creaking in the wind, an axe chopping wood and a barking dog. Gilliard relates that at one time, when in his blind awaiting a chance to photograph the male, he heard a band of native people walking right up and was infuriated as they would scare away the bird, only to discover that the noise was in fact the bird mimicking! Male courtship subsong includes high-quality avian mimicry and, as with Tooth-billed Bowerbird (*Scenopoeetes dentirostris*), it often sounds as if more than one bird is calling at the same time. Bulmer in Gilliard (1969) likened the song to 'a piece of stiff greaseproof paper being vigorously crumpled and torn, as a background to a whole variety of whistles and screams'. The flight of an adult male about the bower produces a whirring noise, which it can also mimic vocally. Few recordings available, but xeno-canto has XC24862 (by N. Krabbe), which has a harsh scolding series, and XC357831 (by P. Gregory), which has a loud querulous rising *wiiip* note, and XC357832 (also by P. Gregory), with a rising much shriller *shwee* note.

HABITAT Primary tall mixed montane and southern beech (*Nothofagus*) rainforest, and rarely moss forest or cloud forest from 1,050m to 3,300m, mainly 1,600–2,300m. Traditional bower sites regularly and linearly spaced along forested ridges with appropriate slope and width, closure of canopy, and density of saplings adjacent to bower sites. Tree species important to diet are common along ridges and slopes where bowers are present.

HABITS A shy and wary species that inhabits the interior of montane forest, where it is fairly common but is heard far more often than seen. This species characteristically forages high in the canopy and displays on the ground at its distinctive ridge-crest bower. Females and immature males maintain relatively large, overlapping home ranges throughout year, foraging in the same areas as adult males.

FOOD AND FORAGING The diet is mostly fruits of medium to large size from many species of tree, shrub and vines, but also some flowers and especially arthropods. This species feeds singly or in twos or in small parties, eating the small fruits of the second-storey trees as well as visiting the tops of trees (Rand in Mayr & Rand 1937). It often forages high in the canopy, depending on the availability of fruit. Males will often cache fruit in the vicinity of the bower, a habit known also for the Golden Bowerbird (*Prionodura newtoniana*). Nestling diet is fruits and arthropods, including cicadas (Cicadidae) and ants (Formicidae). Some of the large trees whose fruits are important in the diet are often found along ridges and slopes near where the bowers are located, and it is feasible that males may place bowers in these sites

to have better access to fruits and to be in the best position to attract females using them (Pruett-Jones & Pruett-Jones 1982).

BREEDING BEHAVIOUR The display season extends for up to 9–10 months (May–February) with a peak in early October, when males can attend bowers for on average more than 50% of daylight time. The polygynous, promiscuous male seasonally decorates a terrestrial bower of simple maypole type, and this may be used over a period of years. The female builds and attends the nest alone. The species is non-territorial except for defence of the bower site; median distance which adult male travels from bower is *c*.88m, but up to 800m in order to steal decorations. The home range at some sites is *c*.150–200m in diameter, and generally elliptical about the bower site. What appears to be an immature male MacGregor's Bowerbird has been recorded as visiting the display court of Carola's Parotia (*Parotia carolae*). **BOWER** Gilliard (1969) provides some detailed descriptions of the extraordinary bowers of this species, located in thick wet mossy forest under trees up to 25m tall. They often have slender vertical saplings nearby, which are habitually used by the male when approaching and departing from the bower; these saplings have bare perches which contrast with the neighbouring moss-covered saplings. The bower is a conical column of sticks up to 3m tall erected around a thin sapling or tree-fern trunk within a nearly level, roughly circular moss-covered clearing platform (*c*.1–1.2m in diameter) raised at its circumference into an elevated rim. The moss can be up to 25–30cm deep, a surprising thickness, and the birds level this out into a densely packed platform, so tightly compressed that it may be rolled up like a carpet. At the base of the maypole the twigs are broken off to a height of 10–15cm and this section is covered with banked-up moss, forming a central rise not unlike the hub of a wheel with the rim of moss. The moss appears to be often derived from trees, as pieces of bark are often found still attached. Gilliard dissected a bower and found it to consist of 816 dead twigs and thin rootlets, mostly about 190mm in length, with a few longer ones up to 375mm. Most bower sticks were about 2mm thick and many had two or three short slender twigs attached. Mixed among the sticks were a few fern shafts and slender bamboo leaves, one very small orchid, one thin root, and a thicker longer stick with a leathery white lichenous covering located near the top of the bower. Bower decorations seem not to be present at the start of the season, but are evident later, often as small bundles of silk-like insect-derived material known as 'frass' produced by wood-boring caterpillars, with sawdust or woody pieces stuck to it. This frass dangles from the end of many of the maypole twigs at a height that can be reached by the bird from the platform, and the bundles sway in the breeze like tassles. Some bowers have dried fern fronds attached in similar fashion to this lower level of the maypole. Other decorations include black or white objects placed on the circular platform in loose groups and derived from charcoal, fungus, animal droppings, small black berries, dry fruit, lichen or vegetable material. There may also be clusters of small orange-coloured seeds, berries or pieces of fruit on the rim; pandanus leaves are often placed radially around the border, and bunches of small brightly coloured fruits may be hung in surrounding vegetation, along with lichens. The central maypole can also have decorations of beetle wings, pieces of charcoal and blue-grey lichen, as well as grass and fern fronds. Besides the central maypole, a small sapling by the rim may also have a column of twigs, or a stem bending over to connect with the maypole may be covered with twigs, and up to six auxiliary columns have been reported. In 2016, a bower on Mt Rondon, above Mt Hagen, consisted of a **double group**, with two distinct maypole circles about 1.5m apart, one relatively small, not a well-formed circle and about 1m in diameter with the typical maypole structure, the other larger and much better formed but with a maypole in three sections totalling about 1.5m in height and separated by bare sections of sapling. This bower had been in use since 2006, but it became a double formation in 2012 (J. Ando pers. comm.). Both towers were decorated with dried pale fern fronds around the lowest level. The study on Mt Missim found that about 10% of males there maintain two bowers simultaneously. A bower on Mt Missim had a pedestal platform up to 76cm tall and had possibly been in use for over 20 years, though probably not by the same male throughout. Simple or **rudimentary bowers**, being a simple clearing around a sapling with or without a short pile of sticks, are built by one or more (up to five reported) immature males. As the males mature, which probably takes at least four years, they build increasingly complex structures culminating in the fully formed version. The painting of bowers is not recorded for this species, although a type of whitish fungus, which resembles 'paint', has been found in some bowers. Mt Missim bowers are *c*.180m apart; bower destruction and theft of decorations by other males commonly occurs and this inveterate raiding is frequent throughout the range. Similar behaviour is well known for other bowerbird genera, including *Chlamydera* and *Prionodura* as well as *Sericulus*. Adult males spent about 54% of the daytime close to their bowers, which they defend against intrusion by other individuals. About 71% of time was spent perching quietly above the bower, around 14% vocalising (though this seems to vary and may be much longer), 12% in bower maintenance and 3% in interacting with visitors/intruders. Time spent at the bower varies but was around 4.6 minutes per visit at Mt Missim and 1.4 visits per hour, with no strong diurnal cycle. Neighbouring males interrupted about 39% of matings at the sample of five bowers observed. Mating has been observed very rarely, suggesting that females repeatedly visit numerous males throughout the long display period before choosing a mate. A male can take several weeks to several months to complete a bower, depending on whether it is a refurbishment or an entirely new one. At Mt Missim, decorations began to appear in June–July and slowly accumulate, with the main display and breeding period there in October-January, coincident with the start of the wet season. MacGregor's bowers on Mt Hagen are more ornate than those of the Kubor Range, and this may reflect the need for stronger mechanisms of isolation on Mt Hagen, where this species is sympatric with Archbold's

Bowerbird, which does not occur in the Kubor Range (F&F 2004). COURTSHIP The male displays to one female at a time, though the females can visit in groups of 2–4 individuals. The male is usually in the canopy above the bower when he detects the approach of a female. Courtship involves a display near the bower, the male leaping rapidly between vertical sapling trunks, after which he moves to the bower and gives a subsong with a 'hide-and-seek' on the bower mat, with the maypole kept between the birds for up to 30 minutes. The wings are flicked and the orange crest flashed, this followed by a sudden rush at the female in a rapid semicircular dance of mincing steps. Often the female flies up to the canopy after this and the display is terminated, but she may return and the dance then continues. The male's orange crest is erected intermittently and fanned out sideways, forming a striking visual display. NEST/NESTING The nest is a sparse stick foundation, with a bulky leafy cup, and the inner egg-cup lined with supple twiglets and rootlets, which provide a springy and dry foundation for the egg. The nest is placed 2–3m above ground in a pandanus (*Pandanus*) tree crown, a sapling, or a tree-fern crown; the mean distance of six active nests from the nearest active bower was 116m (Mt Missim). EGGS & INCUBATION Clutch a single egg, plain creamy-white or buffy-white, mean size 42.9×28.4mm (n=10). Incubation period almost unknown, but in excess of 17 days. NESTLING No information on nestling period. TIMING OF BREEDING The breeding season is September–February. In east New Guinea, eggs from October to January on Mt Missim and a well-incubated egg in early July on Mt Hagen.

HYBRIDS Has apparently hybridised with Streaked Bowerbird, one adult male reportedly showing intermediate plumage (Schodde & McKean 1973). This individual has a crest intermediate between that of MacGregor's and that of Streaked, with very slight throat and chest streaking, and a paler lower mandible than usual. The very limited range overlap between the two species, however, suggests that such hybrids will be both rare and difficult to detect.

MOULT Little known, but moult appears to be mostly during March–May (F&F 2004).

STATUS AND CONSERVATION Classified as Least Concern by BirdLife International, with a range of more than 140,000km², albeit with a trend of some decrease. Common and widespread throughout range, although some outlying populations are very little known. In certain areas of the highlands, walkabout sawmill logging combined with disturbance and habitat damage has caused abandonment of traditional bower sites, while it is known that children will destroy bowers. The crests of the adult males have been highly valued by some of the mountain peoples for many centuries, and are still being used locally in head-dresses and also in trade, the crest sometimes being seen for sale at markets in Mt Hagen. The species is traditionally caught in deadfall traps or shot by four-pronged arrows from blinds built close to the bowers.

MacGregor's Bowerbird, two males at maypole bower, Mt Giluwe, PNG (*Brian J. Coates*).

Below: **MacGregor's Bowerbird**, maypole bower with second smaller bower alongside, Rondon Ridge, PNG, 22 July 2017 (*Phil Gregory*).

HUON BOWERBIRD
Amblyornis germana   Plate 29

Amblyornis subalaris germanus Rothschild, 1910, *Bull. Brit. Orn. Club* **27**: 13. Rawlinson Mts, Huon Peninsula. The original description used *subalaris*, from the Latin meaning under the wing or underwing, a reference to the brightly coloured rusty-orange underwing of this bird, but this epithet now refers to the Streaked Bowerbird.

Etymology The word *germanus* is Latin, meaning brother (its feminine form being *germana*, meaning sister), referring to the close relationship with *Amblyornis subalaris* and *A. macgregoriae*.

FIELD IDENTIFICATION A stocky, thrush-sized, mainly olive-brown bird with a short, stout black bill. The male has large erectile yellowish-orange crown feathers which form a kind of fan when spread, but they are usually seen as a stripe of yellowish-orange or orange on the hindcrown and extending over the nape (except when in display at the bower). This is the only montane bowerbird on the Huon Peninsula, but the two *Chlamydera* species, the Fawn-breasted and Yellow-breasted, are both known from the lowlands. **SIMILAR SPECIES** No other *Amblyornis* occurs within its range on the Huon Peninsula. The Black Pitohui (Shrike-thrush) (*Melanorectes nigrescens*) has a larger and heavier, slightly hooked bill and has much more rufous coloration below, often with a much greyer head.

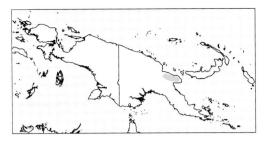

RANGE The mountains of the Huon Peninsula, including the Rawlinson Range and Saruwaged Mts. **MOVEMENTS** Resident; some may descend to lower elevations in winter months.

DESCRIPTION *Adult Male* 26cm. Very like the allopatric MacGregor's Bowerbird but short-crested and short-winged, the crest around 43mm long. *Adult Female* Very similar to MacGregor's Bowerbird but slightly smaller. The original description of a female stated 'agrees with *A. s. subalaris* (Streaked Bowerbird) from British New Guinea in the small size of the bill and wings and in the bright colour of the under wing-coverts and wing-lining, but they are even brighter, being almost rusty orange. It differs, however, in the darker and more rufescent upperparts, darker breast and flanks, and uniform throat.' *Juvenile* Unknown. *Immature* Very similar to MacGregor's Bowerbird, but slightly smaller.

TAXONOMY AND GEOGRAPHICAL VARIATION A member of the maypole or gardener group of bowerbirds and formerly considered conspecific with MacGregor's Bowerbird, but now treated as a distinct species on basis primarily of the very different bower morphology and location, smaller size and shorter crest and wings, and apparently some quite significant genetic variation divergent from that species (Benz 2011 in Frith *et al.* 2017). Originally described from a female and allocated to the Streaked Bowerbird, which was prescient given the recent discovery that in bower structure and genetics it is closer to that species than to MacGregor's Bowerbird. Monotypic.

VOCALISATIONS AND OTHER SOUNDS Very poorly known; data from Baylis (2015). Vocalisation bouts lasted anywhere from one second to ten min 45 seconds, many of the bouts consisting of assorted whistles, clicks, rasps, snaps and croaks. What seemed to be a territorial call was often heard at or near the bower and was well known to the local guides. The birds mimicked several species, including Lesser Melampitta (*Melampitta lugubris*), New Guinea Vulturine Parrot (*Psittrichas fulgidus*), Rufous-backed Honeyeater (*Ptilorrhoa rufescens*), and also what sounded like a frog and a bleating goat. Xeno-canto has four recordings by T. Mark of this taxon: XC324218 has a *chup chup chup chup* sequence and some harsh scolds, plus mimicry of both Papuan King Parrot (*Alisterus chloropterus*) and Regent Whistler (*Pachycephala schlegelii*), and XC324221 shows a nasal, querulous *oowup*. The recordist also noted mimicry of a small hunting dog which had barked on the night before his visit to the bower.

HABITAT Montane forests of the Huon Peninsula, from 1660m to 2940m.

HABITS Little known, but presumably much as for MacGregor's Bowerbird, a shy species that inhabits the forest canopy and opportunistically visits fruiting trees. From the limited observations, behaviour at the bower also appears similar to that of MacGregor's.

FOOD AND FORAGING A primarily frugivorous species of the montane forests. Its behaviour is little known, but assumed to be like that of congeners.

BREEDING BEHAVIOUR BOWER The bowers of this species are not on or near ridgetops like those of its sibling *A. macgregoriae*, but are built well downslope. Two examples cited by Baylis (2015) were at 2271m and 2305m, respectively, and at least 150m lower and some 500m away from the ridgetops, on sloping ground located above gullies, with the front of the bower facing downslope and with a decorated bower 'face' (lacking from the circular bower of *A. macgregoriae*). The bower of *A. germana* has a perimeter of sticks (and not moss), with a mushroom-shaped head of sticks atop the central moss wall, which is decorated extensively with small coloured items. There is a broader lower tower section and a circular bower mat, which comprises fibres like those of tree-ferns or rootlets (unlike the deep moss platform of MacGregor's Bowerbird). The bower, with its decorated wall, is most similar to that of the Streaked Bowerbird, rather than the platform of the bower of *A. macgregoriae* (which is constructed from moss and defined by a low moss perimeter). One bower had the maypole as a sapling about 2.2m tall, with the interlocked stick tower about 70cm high, with a flattened but roughly circular base some 1.8m across and 1.45m from front to back. Distinctively, the outer rim of the circle is also made of sticks (quite different from that of MacGregor's Bowerbird, which uses moss for the rim), this bordering an area of fibrous black material akin to rootlets about 85cm across and extending around the tower, with a wall of these fibrous rootlets-type materials about 50cm wide and 18cm high below the tower and more extensive in front. It is decorated with tiny fragments of blue and yellow, the blue being remnants of a tarpaulin and the rest maybe flowers and fresh green vegetation. This coloured decoration was separated into blue on the left and yellow on the right of the front face of the wall. The front rim base was decorated with many orange fruits and fresh green plant material, while the top of the wall below the tower was decorated with torn or chewed green leaves. An old decrepit bower was nearby. The second bower was similar but more rudimentary, the tower being less substantial and the decoration less elaborate. **COURTSHIP** The following is based on observations from dawn to dusk over five days, split between two bowers. The male spent relatively little time on bower maintenance given the size and complexity of the structure, but presumably the time constraints are much more demanding during the construction phase. The bird would typically spend a couple of minutes on the bower, moving quietly around as if studying the decoration, then adjusting or adding to it. Once the male arrived with a whole green leaf, which it then tore up and pieces of which it placed on the bower. It usually perched 5–10m directly above the bower, or farther away but in front of the bower with a good view of it. There was never more than a single bird in attendance, but a nest and eggs were discovered. **NEST** The nest is of a cup-shaped form and resembles that of MacGregor's Bowerbird. R. Donaghey (*in litt.*) reports that the foundation and external walls of the bulky cup-shaped nest consisted of sticks and the inner walls were composed of large dry leaves and twigs, the cup being lined with small twigs. The external diameter of the nest was 160mm and the outer depth 100mm; the internal diameter was 120mm and the interior depth 60mm. Prior to Donaghey's study in 2014, the nest site of the Huon Bowerbird was unknown; the nest which he observed was placed 1m above ground in a dense tangle of brambles and ferns, being well concealed from above, at an altitude of 2470m. This site is different from that reported for MacGregor's Bowerbird, but further observations are needed to determine if this is a typical site. **EGGS & INCUBATION** Clutch consists of a single egg, as is usual with this genus. The egg measured 42mm × 28.8mm and was uniformly beige or buffy-white (R. Donaghey *in litt.*). Incubation is by female only; the incubation period was some 19+ days, with a warm egg present on the last day the nest was checked. **FLEDGLING** Almost unknown. Rollo Beck collected a fledgling in March 1929 at Sevia, in Cromwell Mts of eastern Huon Peninsula.

HYBRIDS None recorded, with no congeneric siblings in range.

MOULT Nothing currently known.

STATUS AND CONSERVATION Not assessed by BirdLife as yet, as this taxon has only recently been elevated to species rank, having previously been treated as a subspecies of MacGregor's Bowerbird. Likely to be Least Concern, as its habitat is mainly in good condition and there is a large conservation area designated primarily for tree-kangaroos.

Huon Bowerbird, on nest, Huon Peninsula, PNG (*Donna Belder*).

Huon Bowerbird, maypole bower, Camp 12, Huon Peninsula, PNG (*Richard Donaghey*).

STREAKED BOWERBIRD
Amblyornis subalaris   Plate 31

Amblyornis subalaris Sharpe, 1884, *J. Linn. Soc. London, Zool.* **17**: 408. Astrolabe Mts, Southeast Peninsula.
Other common names Gardener Bowerbird, Striped Bowerbird, Striped Gardener-bowerbird, Orange-crested Bowerbird, Eastern Bowerbird, Eastern Gardener-bowerbird, Eastern Gardenerbird
Etymology The species epithet *subalaris* is Latin, meaning under the wings, a reference to the rich ochraceous colour of the underwing.

FIELD IDENTIFICATION Shy, wary and little known owing to its remote and generally inaccessible montane habitat, although walking the Kokoda Trail provides a transect through some of the range of this species. Its bowers are one of the most remarkable constructs in the avian kingdom, with some similarities to those of the equally remarkable Vogelkop Bowerbird. **SIMILAR SPECIES** Very similar to **MacGregor's Bowerbird**, with which it has very limited local sympatry (MacGregor's is usually at a higher elevation); Streaked Bowerbird, however, is slightly smaller, the crest of the male is shorter, and both sexes show indistinct streaking on the throat and breast. Another potential confusion species is the female or immature **Crested Satinbird**, which has a similar stocky structure and quite short bill, and is olive-brown above; the underparts of Streaked Bowerbird have a much more pronounced yellowish cast, the crown is more olive, and it is slightly smaller with a stouter bill which is brownish or bluish-grey, not black. **Black Pitohui (Shrike-thrush)** (*Melanorectes nigrescens*) has a larger and heavier slightly hooked bill, and much more rufous coloration below, often with a much greyer head.

RANGE Relatively restricted range in the lower montane forest of the eastern sector of the South-east Peninsula of PNG: on the southern watershed from the Angabanga River (north-west of Port Moresby) east to Milne Bay, and on the northern watershed from Mt Suckling east to the mountains of Milne Bay Province. Exact range still undetermined, with westernmost limits uncertain. **MOVEMENTS** Resident.

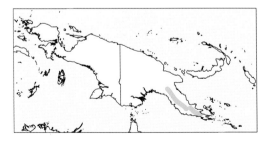

DESCRIPTION Sexually dimorphic. *Adult Male* 24cm; weight 96–107g. Has entire face and upperparts warm olive-brown, with lores and the feathers on each side of the crest a darker and richer brown. The rich glossy orange crest has variable dark brown tipping and is extensive, erectile and laterally spreadable

to form a spectacular display adornment. The crest colour appears to be variable, from deep orange to yellowish-orange, sometimes with much dark brown tipping occluding the coloration (?subadult); crest size also is variable, ranging from 34mm to 47mm. Underparts paler, with centre of chin, throat and upper chest feathers palest, almost tawny-olive, creating indistinct streaking; axillaries and underwing-coverts pale apricot or bright ochraceous orange, the undertail-coverts slightly darker; underside of tail olive-brown with whitish feather shafts. *Iris* dark brown; *bill* mostly bluish-grey, blackish basally and along culmen; *legs* blue-grey. *Adult Female* 24cm; weight 95–122g. Similar to male but lacks the crest. *Juvenile* Undescribed. *Immature Male* Like adult female, some with a darker, more chocolate-brown head with traces of orange at bases of some central crown plumes. *Subadult Male* Has crest smaller, probably with more brown feathering visible within it. Some individuals resemble adult male, but have just a few slender orange plumes showing.

TAXONOMY AND GEOGRAPHICAL VARIATION Monotypic.

VOCALISATIONS AND OTHER SOUNDS Reported as being similar to MacGregor's Bowerbird. Very poorly known, with no recordings on xeno-canto and nothing in the Macaulay Library at the Cornell Lab of Ornithology at the time of writing (2018).

HABITAT Primary forest and taller secondary forest. Inhabits the interior of lower montane forest dominated by oaks (*Lithocarpus* and *Castanopsis*) from 670m (rarely 650m) to 1200m, occasionally to 1500m.

HABITS Usually seen singly, ranging from understorey to canopy, where, like congeners, it seeks out fruiting trees and shrubs. Shy, wary and little known.

FOOD AND FORAGING A primarily frugivorous species of the lower montane forests. Behaviour little known, but assumed to be like that of congeners.

BREEDING BEHAVIOUR Polygynous, promiscuous male seasonally decorates complex maypole bower; female builds and attends nest alone. Non-territorial except for defence of bower sites. Average male attendance is more than 40% of daylight hours, but precise extent of timing of activities is little-known. **BOWER** Sited on the slope of a hill and well shaded by trees, usually 5–30m below the summit of a ridge. Distance between bowers is typically 50–75m, but sometimes up to 200m. Highly distinctive bower is rather more complex than that of the Vogelkop Bowerbird (which also makes an amazing hut-like structure), and is reported as being built from the ground upwards. Streaked Bowerbird constructs a hut-like or teepee-shaped structure around a central maypole formed by a column of sticks or tree-fern fibres built around a thin sapling. It has a roughly dome-shaped or teepee-like roof of sticks with back and side walls, and an opening at the front like a forecourt or front garden which leads in to a semicircular tunnel-like passageway the openings of which face the front of the bower on each side of the central maypole. The bower can be *c.*80cm high, 120cm wide and 100cm in diameter. Goodwin (1890) memorably described a bower as looking like a cartload of sticks with a rounded top when seen from behind, but a beautifully decorated structure in front. **Bower decorations** include small fruits and berries of various colours (many bright blue or scarlet), flowers (often bright yellow) and leaves (many yellowish-green), which are confined to the base of the central column; larger items include also beetle elytra (often mauve-coloured), tree resin and pieces of fungi, which are placed on the outer mat or garden and on top of the stick parapet surrounding, often in discrete piles. One bower studied by Opit (1975a) had only many brilliant blue berries decorating the central column, with pieces of yellowish-green resin in the forecourt. Pieces of blue plastic wrapping and blue paper have been recorded, too. Simson (in Gilliard 1969) reported that one bower had yellow flowers on one side and blue berries on the other, and that it was common to see piles of rotting scarlet fruits lying a metre or so away from the bower where they have been thrown by the birds. Bower decoration varies seasonally and perhaps locally, or just at the discretion of the individual male, as with Vogelkop Bowerbird. Also like other bowerbirds, bower destruction and theft of decorations occur. Birds at the bower are exceedingly shy. The display period around Efogi, in Central Province, seems to be centred on the dry season, bower use presumably starting at the end of the rains; bowers there are in prime condition in September and have fallen into disuse by late December. **Courtship** Courtship not fully described, and no videos currently available, but apparently similar to that of MacGregor's Bowerbird, the males uttering subsong at the bower and displaying the crest to the female, and mating occurring at the bower. **Nest** One nest was a large bowl-shaped structure with long, brown dead leaves and few sticks, lined with twigs. Five nests, each with a single egg, were found in December at Boneno, in far south-east PNG, by the famous naturalist F. Shaw-Mayer (Harrison & Frith 1970). No other information. **Eggs** Clutch one egg, yellowish-white in colour, 40.6 × 27.7mm. **Timing of Breeding** Few data. Display season September–December. Nests with eggs known in December–January, and a few eggs collected in those months.

HYBRIDS Has apparently hybridised with *A. macgregoriae*, one adult male reportedly showing inter-mediate plumage (Schodde & McKean 1973). This individual has a crest intermediate between the present species and MacGregor's, with very slight throat and chest streaking, and a paler lower mandible than usual. The very limited range overlap between the two species, however, indicates that such hybrids will be both rare and hard to detect.

MOULT Very little known, but specimens show evidence of moult in February–May, August and October (F&F 2004).

STATUS AND CONSERVATION Classified as Least Concern by BirdLife International. This is a restricted-range species, with an area of occurrence of around 18,500km² in Central Papuan Mts EBA, where it is widespread and locally common. Crests of adult males were in the past used by local people for head-dresses and also in trade, and this may still continue on a small scale.

Streaked Bowerbird, hut-style bower, Efogi, PNG (*William S. Peckover*).

VOGELKOP BOWERBIRD
Amblyornis inornata   **Plate 30**

Ptilorhynchus inornatus Schlegel, 1871, *Ned. Tijdschr. Dierk.* (1873) **4**: 51. Hatam, Arfak Mts, Vogelkop, north-west New Guinea.

Other common names Crestless Bowerbird, Plain Bowerbird, Brown Bowerbird, Crestless Gardener-bowerbird, Plain Gardener-bowerbird, Brown Gardener-bowerbird, Gardener Bowerbird, Vogelkop Gardener Bowerbird, Brown Gardener, Brown Gardenerbird, Crestless Gardener

Etymology The species epithet *inornata* means plain or unadorned, both this and the genus name (*Amblyornis* = dull bird) being a very apt description for this species.

FIELD IDENTIFICATION A typical plain brown *Amblyornis* gardener bowerbird with stocky body and quite short stout bill, the only member of the genus in north-west New Guinea (but see TAXONOMY AND GEOGRAPHICAL VARIATION). The bowers of this species are among the most remarkable constructs in the avian kingdom, with some similarities to those of the equally remarkable but much less well-known Streaked Bowerbird. SIMILAR SPECIES Apart from female Black Pitohui (Shrike-thrush) (*Melanorectes nigrescens*), no other major confusion species occurs here. The female of the local race of the latter species has a heavy hook-tipped bill and greyish cap.

RANGE Mountains of Vogelkop (Tamrau, Arfak), Onin Peninsula (Fakfak), Bomberai Peninsula (Kumawa) and Wandammen Mts, in north-west New Guinea. MOVEMENTS Resident.

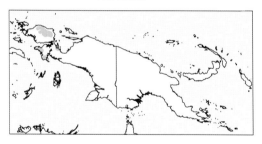

DESCRIPTION Sexes alike in plumage. *Adult Male* 25cm; 105–155g. Plain rich brown above with no coloured crest, underparts buff to pale cinnamon (Fakfak birds slightly darker above and brownish-yellow below). Chin indistinctly scalloped with greyish feather edging, axillaries ochraceous; tail feathers brownish-olive, undertail-coverts buff. *Iris* dark brown; *bill* blackish, with bluish base of lower mandible; *legs and feet* rich blue. *Adult Female* 25cm; weight 105–146g. On average fractionally smaller than male, with legs darker and less blue. *Juvenile* Undescribed. *Immature Male* Like adult female. *Subadult Male* Like adult male.

TAXONOMY AND GEOGRAPHICAL VARIATION Monotypic. Plain-plumaged and sexually monochromatic *Amblyornis* populations from the Kumawa and Fakfak Mts, in southern Bird's Neck region (Diamond 1987a; Gibbs 1994), are, however, of uncertain taxonomic status; they are slightly darker above and more brownish-yellow below than Arfak birds, but with very different bowers. These bowers are built from sticks loosely interwoven around a central maypole and can be over 2m tall. Males use exclusively drab-coloured decorations such as bamboo bark, rocks, and snail shells placed on the periphery of the circular court; they also typically prop dried palm (*Pandanus* sp.) leaves against the maypole tower (Uy & Borgia 2000). The maypole bowers of these populations are more like the bower of MacGregor's Bowerbird than like those of the Vogelkop Bowerbird, but they lack the mossy rim or parapet. Gibbs argues that these should be treated as a distinct species, but more detailed molecular work is needed to resolve this matter, as well as detailed studies of the bower behaviour. Genetic material thus far reveals little differentiation, but more study is required and it seems very likely that two species are involved. Further evidence of **geographical variation in bower style** over small distances within the same range is given in Diamond & Bishop (2015): bowers observed by F. Sadsuitubun in Kumawa in 1975, *c.*8km from J. Diamond's 1983 South Kumawa transect, differed from the latter bowers in being decorated with fruits of four colours, much more often decorated with white stones, and having dark stones scattered over the mat rather than assembled in a pile (Diamond 1987a). The eight bowers that Diamond and K. D. Bishop observed in their 2013 Central Kumawa transect between 1,026m and 1,654m, *c.*11km from JD's South Kumawa transect, differed in that two were decorated with a neatly rectangular row, 1.2m long by 13cm wide, of 5cm pieces of buff-coloured clay. These eight bowers consisted of 1–2 towers of sticks up to 1.5m tall, glued together with an unidentified white substance around a sapling, on a circular moss mat 0.9–2.1m in diameter. Decorations besides the clay rectangle were *Pandanus* leaves propped against the stick tower, pale snail shells, black beetle elytrae, grey or buff stones, long thin black sticks, whitish limestone and black palm seeds. Gilliard (1969) and Gilliard & LeCroy (1970) previously noted wide individual variation in Tamrau bowers, similar to the observation of individual variation in the presence or absence of a clay rectangle in Central Kumawa bowers. Like Diamond (1986, 1988) in South Kumawa and Wandammen, and Uy & Borgia (2000) in Fakfak and Arfak, JD & KDB in 2013 in Central Kumawa placed coloured poker chips at bowers to test whether the absence of coloured natural decorations at Central Kumawa bowers reflects a lack of available natural objects in the environment or, instead, the birds' preferences. Confirming the latter interpretation, when they placed poker chips on two of the bowers – chips uniformly coloured red, orange, yellow, blue, purple or violet, and green, blue or black chips with white spots – most or all of them disappeared, presumably discarded by the bower-owner. Similarly, South Kumawa bower-owners, which do not use coloured natural decorations, discarded coloured poker chips placed at their bowers, while Wandammen bower-owners, which do use coloured natural decorations, used coloured poker chips placed nearby to decorate their bowers.

VOCALISATIONS AND OTHER SOUNDS A plaintive, rather variable, either whistled or harsh *tu-ee* is given at intervals, is a good indication of the presence of the species; see xeno-canto XC163224 (by F. Lambert). Also given are quavering whistles, loud harsh scolds, and a variety of chugging, wheezing, spitting and ratcheting sounds, sometimes making two sounds at the same time as with Tooth-billed Bowerbird; see XC96178 (by R. Silva), recorded at the bower. Other vocalisations are a sharp click, a repetitive *kah kah* repeated, and a series of *keu keu keu keu* notes. Recordings of the maypole-building form from the Kumawa Mts (see TAXONOMY AND GEOGRAPHICAL VARIATION), by F. Rheindt, have some unusual deep mewing and squalling noises; see XC92511 and XC92510. Birds here also make a variety of trills, harsh rasps, coughs, gurgles, creaks, whistles, bleats, clucks, neighs, rattles and soft ventriloquial growls (Frith *et al.* 2017). This species is also a good mimic of other birds, including parrots, and of sundry human-related sounds such as the chopping of wood, a tarpaulin flapping, squeaking tripod legs, generators and barking dogs. Like its congeners, the male also has a courtship subsong which can involve mimicry, with Western Parotia and Friendly Fantail (*Rhipidura albolimbata*) copied frequently. Most calling seems to be in early morning and late afternoon, and voice is much more subdued on wet days. It is reported that the wingbeats make a loud sound when the birds are flying about the bowers, presumably as part of the advertising display. It is also said that the bowers may act as resonance chambers during the intense displays, when singing may continue for up to 70 minutes (Kirby in F&F 2004).

HABITAT Rainforest with canopy 25–30m tall and emergents of *Agathis labillarderi* and *Araucaria cunninghamii*; at 1,000–2,075m, mainly 1,200–2,000m. Traditional bower sites are located on ridge spines and slopes.

HABITS A shy species of the montane forests, most easily seen from a blind sited near a bower.

FOOD AND FORAGING Primarily frugivorous; some insects also taken. This species is poorly known, but its behaviour appears to be much like that of its congeners, appearing at fruit sources when they become ripe.

BREEDING BEHAVIOUR The polygynous, promiscuous male seasonally decorates either a complex or a simple maypole bower, depending on which population is involved. The female builds and attends the nest alone, as is usual for the genus. The species does not hold territories, except for the male's defence of his bower site. The male spends in the region of 50% of his day near the bower, and devotes a lot of time to its care and maintenance; Bailey (1996) noted that the male adjusts items and then flies out to make a visual inspection of his handiwork before returning for more work. **BOWER** This species as currently defined exhibits two distinctive maypole bower types, both based on a central column of sticks or other material placed around a sapling trunk. The type from the **Tamrau, Arfak and Wandammen Mts** is one of the most complex structures made by a bird, an extraordinary piece of avian architecture resembling the hut-like bower of the Streaked Bowerbird, with a roof of epiphytic-orchid stems or sometimes sticks or fern fronds. The first naturalist to see this species was Odoardo Beccari, who had heard rumours of a strange mountain bird with extraordinary architectural capabilities. He first saw it in 1872, on a spur of Mt Arfak in thick virgin forest at 4,800ft (*c*.1,460m). His description is worth quoting in detail:

'The *Amblyornis selects a flat even place around the trunk of a small tree that is about as thick and high as a walking stick… It begins by constructing at the base of the tree a kind of cone, chiefly of moss, the size of a man's hand. The trunk of the tree becomes the central pillar; another whole building is supported by it. On the top of the central pillar twigs are then methodically placed in a radiating manner, resting on the ground, leaving an aperture for the entrance. Thus is obtained a conical and very regular hut. All of the stems used… are the thin stems of an orchid (Dendrobium), an epiphyte forming large tufts on the mossy branches of great trees, easily bent like straw and generally about 20 inches [c.50cm] long. The stalks had the leaves, which are small and straight, still fresh and living on them – which leads me to conclude that this plant was selected by the bird to prevent rotting and mould in the building, since it keeps alive for a long time.'*

Beccari (1878) describes the garden of this bird: 'Before the cottage there is a meadow of moss. This is brought to the spot and kept free from grass, stones or anything that would offend the eye. On this green flowers and fruits of pretty colour are placed so as to form an elegant little garden.' He saw some small apple-like fruits, some rosy fruits, rose-coloured flowers, fungi and some mottled insects on the turf. There is considerable individual variation in what is used for decoration in the gardens, depending on the bird's preferences and what is available locally. In 1938, Ripley noted that the gardens which he saw in the Tamrau Mts were carefully segregated as to colour, although sometimes a small blue plum-like fruit had one or two yellow fruits admixed. In the entranceway to the bower were large brown decayed fruits or particularly large flowers. One peculiar 'bed' was found to consist of jelly-like white material which was the sap of a small tree-fern. Ripley (1944) saw another bower that was on gently sloping ground and looked to have been swept with a broom. The hut itself was about three feet high by five feet broad (*c*.91 × 152cm), with a front opening about one foot (*c*.30cm) high, fronting on a cleared area like a front lawn, with several small beds of flowers or fruit. Just under the door there was a neat bed of yellow fruit, farther out a bed of blue fruit and at the bottom of the lawn a large squarish bed with pieces of black charcoal and small black stones. A few brownish fruits lay there, some rather decayed. Off to one side were several big mushrooms in a heap, and near them ten freshly picked flowers. Gilliard (1969), in July and August 1964 in the Tamrau Mts, saw some rather differently styled bowers, differing in proportions and decorations – some with blacks, browns and blues predominating in the piles of ornaments, others with orange or red flowers and red, blue and green tree fruits. The floor was carpeted with moss and the base of the central column surrounded by a cone of soft moss; the roof was very evenly curved, and the end of the roof sticks sheared off to form a level ceiling that was aligned with the top of the doorway. On the mossy floor there were a red ornament and a few black beetle exoskeletons. There has recently been a significant change in bower decoration around and above Syoubri village, in the Arfak Mts, where the arrival of groups of visitors has meant that there is now rubbish discarded in the forest. In July 2015, many bowers were seen that utilised man-made objects, often the colourful wrappers from biscuit or sweet packets, as part of the decoration. One remarkable bower at 2,000m had a one-litre clear plastic bottle placed in the hut itself, and our guide assured us that the bird itself had done this! Discarded camera batteries have also been utilised as part of the bower ornamentation. Some bowers, especially the ones farthest away from campsites, used only natural flowers and fruits, whereas others had a mixture, and some were decorated mostly with human-derived detritus. The large brown decaying fruit piles were still in evidence, this being a decoration type first noted in 1939. One interesting behaviour that occurs widely among bowerbirds (and also the court-holding birds of paradise) is the meticulous keeping of the floor of the display area clean of extraneous material. It is well known that, if you place a leaf or piece of paper on such a bower area, the bird will remove it as soon as it is noticed, sometimes almost immediately if it is nearby. In July 2015, at a bower with a newly constructed blind above Syoubri, the owner was being very recalcitrant about coming in until the guide dropped a large red-and-white biscuit packet out through the viewing slot: the bowerbird, which had been perched largely hidden not far away, immediately flew in and began pecking at the package, calling harshly, and when it discovered that it was too big to shift it called even more emphatically and flew off (pers. obs.). One of the **mixed-object bowers** had the following highly complex display of materials in the front garden: in the entranceway to the hut were a red, white and yellow '*Twisties*' corn-chip packet, some small yellow vegetable objects, a large pile of orange fruit covers, and some large brown fungi plus a few small, round brown fruits; located centrally in front of these were three sticks about 10cm long and one large ovate green leaf, with to one side a big spread of large round brown fruits and a red flower, and to the opposite side a large pile of quite big pieces of charcoal up 8cm long, separated in the centre by a spread of discarded biscuit-wrappers and grey plastic wrapping. At the outermost front edge were a large orange leaf and some small green fruits. Each pile was quite discrete and formed a clearly defined pattern in front of the hut structure. Another bower nearby was very different and also anomalous in having another quite large opening midway up the sidewall (apparently quite regular in some Arfak bowers), and a mound of small sticks piled up by a small tree-fern that was growing on the edge of the forecourt, reminiscent of some MacGregor's Bowerbird bowers which have ancillary columns nearby. This main court was a particularly large and diverse one, with a dark-coloured flat area of dead moss forming an apron in front. Within the hut opening and on this dark moss area were two large plastic bottles

and many small blue berries scattered across, and a heap of charcoal to one side, with a large greyish pile of unidentified organic material in the centre of the opening. Some large red flowers formed a pile at the very front side of the moss apron, and in front of this was a veritable garden of long strips of blue plastic in an untidy pile, small orange fruit coverings, an empty red-and-silver tuna tin, two small tin cans, some biscuit wrappings and a pile of red and yellow flowers plus a large segment of orange peel. There was a third, smaller bower nearby, built around a small tree-fern and with another tree-fern bisecting the front entrance of the hut, so that effectively there were two openings on each side, with a mossy mound around the tree-fern base. Each frontal opening had a pile in the centre, consisting of charcoal to the left and blue and white plastic discards on the right. There was a mossy interior to the hut, and outside in the garden proper ten long dead pandanus leaves arranged to point away from the hut, a neat pile of orange fruit coverings and a pile of blue plastic (pers. obs.). **Feathers** are occasionally used and have included tail feathers from New Guinea Harpy Eagle (*Harpyopsis novaeguineae*), Black Sicklebill and Moluccan King Parrot (*Alisterus amboinensis*) (McConnell in F&F 2004; Betz in F&F 2004). One Arfak bower had the same sardine can as one of the focal mat decorations for at least seven years. In one area, the left of the central maypole base is covered by black fern fronds leaning at 45° and the right with moss covered by a shiny dried tar-like glue which was thought to be saliva and charcoal, and not excrement (as the birds typically defecate away from the bower). Hard and shiny pig dung is also used as decoration, and one bower had 19 species of mushroom placed centrally beneath the canopy. The author remarks that a small blue bioluminescent mushroom is typically placed similarly (McConnell in F&F 2004). **Dimensions** of the bowers vary according to location, with the hut-type bower towers from the Wandammen being much lower, 0.4–0.8m as opposed to 2.4–2.7m for Arfak and Tamrau structures. Wandammen huts have the central column base covered with moss, which extends out about 45cm from the opening to form a frontal mossy mat or lawn. The entrance height is about 20–28cm and the width 18–58cm; the height of the bower hut is 0.43–0.76cm and the diameter is between 0.94m and 2.07m. Colour and quantity of display items vary greatly from one site to another, as detailed previously. Bower destruction and theft of decorations occur, much as with congeners. The other bower type, from the **Fakfak Mts and Kumawa Mts**, is somewhat different and far less architecturally advanced, being a simple roofless structure built around saplings. It resembles that of MacGregor's Bowerbird, but lacks the raised mossy perimeter rim; the tower can be up to 2m tall and the basal circular mat up to 2m in diameter, decorated with discrete piles of colourful fruits and flowers, insect exoskeletons, pieces of charcoal, fungi and other items. Fakfak bowers had only discrete piles of snail shells, black bamboo leaf sheaths and small white limestone rocks, which are seemingly hard to come by (Gibbs 1994) and may be status objects; other Fakfak bowers also used only drab-coloured objects, even when white was available and some coloured items were left by the observers to see if they would be chosen, in sharp contrast to the colourful objects used in the Arfaks at the hut-type bowers. In the Kumawa Mts, Diamond (1987a) noted that the moss mat, sticks and brown stones were all coated with a black glossy substance, and since the excrement of birds which he netted was oily and black it was suggested that they used this to paint the bower mat and some decorations; it would be good to confirm this painting behaviour, since this is not yet certain for any of this genus. A whitish sticky substance is recorded as being utilised for fusing sticks together, and this may be of fungal origin (F&F 2004). Diamond (1987a) recorded white sap, fungus and eggshells on Tamrau Mts bowers but not on those in the Wandammen Mts, where the birds were much tamer than the very shy Kumawa individuals, which had probably not seen humans before. Kirby (in F&F 2004) writes that the species collects the beetle elytra by digging out the creatures from rotting wood, eating the body and then carrying the elytra back to the bower, where they are placed shiny side uppermost. **Younger males** are reported as building aberrant structures and use fewer decorations and of the 'wrong' colours (Diamond 1987a). **Courtship** In courtship, the male responds to the arrival of a female by rushing into the back of the bower behind the maypole to crouch and hide, while uttering a continuous echoing and very varied subsong, then standing stiff and erect and occasionally running out from concealment to the bower entrance, head briefly cocked to one side, and then dashing back inside again. A study by Uy (2002) of 16 bowers in the Arfaks noted that only half of the birds actually mated, and that the three most successful males performed 66% of the matings, these particular individuals being the ones with

the largest bowers and with the most items of a blue coloration. Conspecifics are driven away, as are itinerant visitors such as the Black Fantail (*Rhipidura atra*), Grey-streaked Honeyeater (*Ptiloprora perstriata*) and Mountain Fruit-dove (*Ptilinopus bellus*), although a Black Sicklebill was tolerated (Diamond 1987). **Nest/Nesting** Two nests found, each built into forking branch of a sparsely foliaged sapling, respectively 1m and 2.5m above ground; one nest was an untidy structure of sticks with a leaf lining, like that of MacGregor's Bowerbird but with fewer sticks in base. **Eggs, Incubation & Fledging** Clutch a single whitish egg. No information on incubation and nestling periods. **Timing of Breeding** The display season is from July to February, the months of peak activity varying across the range. Two nests each had a whitish single egg in early October and in May, respectively.

HYBRIDS None recorded, as no congeners occur in the area.

MOULT Very little known, but specimens show evidence of wing moult during April–July (Frith & Frith 2004).

STATUS AND CONSERVATION Classified as Least Concern by BirdLife International. This is a restricted-range species: present in West Papuan Highlands EBA. It is widespread and common throughout its range, which is centred on some remote West Papuan mountain areas, some of which are part of designated national parks. People near Hatam, in the Arfak Mts, avoid destroying the bowers of this species, and a local tourism initiative has this species and its amazing bower as one of the key targets for visitors.

Vogelkop Bowerbird, sexes similar, Arfak Mts, West Papua (*Gareth Knass*).

Vogelkop Bowerbird, hut-style bower, Arfak Mts, West Papua, 21st June 2015 (*Phil Gregory*).

YELLOW-FRONTED BOWERBIRD
Amblyornis flavifrons   **Plate 31**

Amblyornis flavifrons Rothschild, 1895, *Novit. Zool.* **2**: 480 (Fig. *Novit. Zool.* 3, pl. 1, figs 3 and 4). Foja Mts, north-west lowlands.

Other common names Golden-fronted Bowerbird, Yellow-fronted Gardenerbird, Golden-fronted Gardener, Gold-maned Gardener

Etymology The name *flavifrons* is from the Latin *flavus*, meaning yellow or golden-yellow, and *frons*, meaning forehead or brow.

The English name is somewhat confusing and can lead to assumptions that the bird has a yellow breast as opposed to a yellow forehead, but it is long established. The colour is golden-yellow or rich yellow rather than golden, despite a renaming using the last colour descriptive by Beehler & Finch (1985) and several recent authors. An alternative name that avoids possible confusion would be Foya Bowerbird, since the species is restricted to that area.

This taxon was originally known only from trade skins, purchased by Rothschild from the Dutch traders Duivenbode of Ternate, who had bought them from hunters who obtained them from an unknown location. About a dozen attempts were made over many years to discover the bird in life, but it remained one of the great ornithological mysteries until 1979, when it was found by J. Diamond in the remote Foya Mts of what was then Irian Jaya (Diamond 1982a).

FIELD IDENTIFICATION The male is a distinctive dull rufous-brown bowerbird with an unusually extensive rich golden-yellow forecrown and crest. The female is similar but lacks the colourful crest. SIMILAR SPECIES No other *Amblyornis* bowerbird occurs in the range, and the male of this species should be unmistakable. Females may be confused with Little Shrike-thrush (*Colluricincla megarhyncha*), which is paler and more slender in build and has a more slender bill. Crested Satinbird does not occur here.

RANGE Foya (Gauttier) Mts of north-west West Papua. MOVEMENTS Sedentary.

DESCRIPTION Sexually dimorphic. *Adult Male* 24cm. This species is a little smaller than MacGregor's and Vogelkop Bowerbirds with an extensive, broad and brilliant glossy deep golden-yellow crest, and a few brown-tipped feathers. The crest extends from bill to mantle and is erectile and can also be spread laterally. The upperparts are dark rufous-brown, the upperwing

and uppertail brownish-olive. The sides of the face, lores, chin, throat and upper chest are similar in colour to the back, but paler and less reddish. The breast and abdomen, undertail-coverts and underwing-coverts are rich buff, variably washed with deep cinnamon. *Iris* dark brown; *bill* blackish; *legs* bluish-grey. **Adult Female** Resembles male but without crest, and in some videos appears lighter on belly. **Juvenile** Undescribed. **Immature Male** Probably like female. **Subadult Male** Acquires orange-yellow underside to crest before the more yellow superficial ones (Diamond 1982a).

TAXONOMY AND GEOGRAPHICAL VARIATION Monotypic.

VOCALISATIONS AND OTHER SOUNDS The male at the bower gives loud and varied advertisement vocalisations, including short high, quite shrill, nasal repetitive screeches, a deeper almost barking note, repeated deep rasps and yelps, rolling gravelly sounds, clucks, wheezes, croaks, crackling and whip-like sounds. This species is also a mimic, with imitation of the calls of the high-altitude Cinnamon-browed Melidectes (*Melidectes ochromelas*) and Smoky Robin (*Peneothello cryptoleuca*) and the lower-altitude Sulphur-crested Cockatoo (*Cacatua galerita*) and Grey Crow (*Corvus tristis*). Other noises include sounds like sudden exhalation, the striking of a tree trunk, and clicking the tongue against the palate. During courtship, the male gives a weak high-pitched two-note whistled *who-chik*. The vocalisations are generally deeper than and rather distinct from those of MacGregor's Bowerbird. A series of videos by E. Scholes is at the Macaulay Library of the Cornell Lab of Ornithology site, and the sounds made by the male during his display at the bower can be heard on ML459746.

HABITAT Montane moss forest dominated by *Araucaria*, southern beech (*Nothofagus*), *Podocarpus*, and *Lithocarpus* oaks, from 940m to 2,000m, mainly 1,100–1,800m. Adult males rarely below 1,600m; females, subadult males and immatures tend to inhabit the lower altitudes (Diamond 1982a).

HABITS Frequents mostly the middle storey and lower canopy, but sometimes descends to the forest floor.

FOOD AND FORAGING This species is frugivorous, as are its congeners. Along with pigeons it is one of the commonest arboreal frugivores in the moss forests.

BREEDING BEHAVIOUR A series of videos by E. Scholes at the Macaulay Library of the Cornell Lab of Ornithology site (ref. ML459746) shows the display of this species at the bower. It appears to be non-territorial except for defence of bower sites, and, as with congeners, the male is polygynous. BOWER There are excellent videos by E. Scholes at the Macaulay Library site, which show the bower of this species (ML459699). The male builds a simple maypole bower, with a stick tower 0.5–1.2m high, and rather resembling a single tower of the Golden Bowerbird. There is a basal circular moss mat, about 1m in diameter, which lacks the elevated perimeter rim of the bower mat of Macgregor's Bowerbird, while the adjacent forest floor is kept clear of debris. The bowers are often built around the trunk of a live tree-fern. Bower decorations comprise discrete piles of blue, green and yellow fruits, the last a species of *Ficus*, each laid on a distinct part of the bower mat. The blue shows well in the dull light of the forest interior, and also contrasts well with the crest of the male; some blue berries are often laid on the mossy base of the maypole. Bower-painting is not recorded for this species. Traditional bower sites are *c*.0.5km apart on ridge crests at 1,650–1,800m, and other males are not usually audible from any one site. Courtship The male plays hide-and-seek at the bower with the female, as is usual with the gardener bowerbirds, both birds circling around and the male peeking out at intervals, but with the crest closed. In the more intense phase he frequently spreads the crest out laterally to show the rich golden-yellow colour, dipping and rising with the tail wagging, and then shaking the tail rapidly from side to side, which causes the fan-like crest to vibrate rapidly (ML459746, by E. Scholes). An adult male was seen displaying by making an odd noise, like the sound of a large mammal walking on loose gravel, and holding a small blue berry (which contrasted well against the golden-yellow crest) in the tip of its bill; this was held towards a presumed female as she moved about the bower site, the fruit being thus held for about 20 minutes (Diamond 1982a). When the female perched above the male he stretched up vertically, and when she was low down he stretched out horizontally towards her. This particular display was outside the bower on a small horizontally bent sapling some 1m high about 6m away, and the female moved around the male at heights of 2m to 10m and keeping a distance of about 10m. Nest Unrecorded. Eggs & Nestling Unrecorded.

HYBRIDS None recorded.

MOULT No details available.

STATUS AND CONSERVATION Classified as Least Concern by BirdLife International. This is a restricted-range species, present in the North Papuan Mts EBA (Endemic Bird Area). It has a very small global range of around 980km² within the Foya Mts Natural Reserve. The total population is estimated at between 1,500 and 7,000 individuals at most, but this is a remote area and remains very sparsely inhabited by humans.

Yellow-fronted Bowerbird, female or young male, Foya Mts, Papua, Indonesia (*Tim Laman*).

Yellow-fronted Bowerbird, two males at bower, Foya Mts, Papua, Indonesia (*Tim Laman*).

Genus *Prionodura*

A monotypic genus restricted to the wet tropics of North Queensland, the smallest of the bowerbirds, and differing from all others in the far more delicate build and longer tail structure, with a relatively small bill. This is not one of the bulky, stout species like *Amblyornis* or the Tooth-billed Bowerbird. One genetic study of mtDNA (Kusmierski *et al.* 1997) generated a phylogenetic tree that placed *Prionodura* and *Archboldia* among members of the *Amblyornis* lineage, despite those genera having dramatic differences in structure, coloration, bower design and vocalisations, and illustrating the pitfalls of relying solely on one small suite of characters to erect a phylogeny without reference to the wider picture. Regrettably, some reference books and checklists adopted the change, but it has now reverted, with *Prionodura* and *Archboldia* restored.

Etymology Genus name is derived from the Greek *prionodes*, meaning serrated, and *oura*, tail.

GOLDEN BOWERBIRD
Prionodura newtoniana   Plate 32

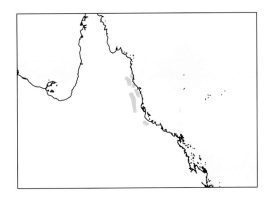

Prionodura newtoniana De Vis, 1883, *Proceedings of the Linnaean Society of New South Wales* **7**, 562. Tully River Scrubs, North Queensland, Australia.
Other English names Newton's Bowerbird, Newton's Golden Bowerbird, Queensland Gardener, Newton's Gardener-bowerbird, Meston's Bowerbird
Etymology Named for Alfred Newton, renowned English zoologist and ornithologist.
The Golden Bowerbird was named in 1883 by D Vis, the curator of the Queensland Museum, on the basis of a female-plumaged specimen, notable for its 'plentiful lack of beauty'. In 1889, however, the collector Archibald Meston discovered the striking fully plumaged male, and D Vis, not realising that it was of the same species until slightly later in that year, named it *Corymbicola mestoni*, Meston's Bowerbird. The bower also was described by Meston in that same year, and caused something of a sensation.

FIELD IDENTIFICATION The male is very distinctive, having bright golden-yellow underparts and undertail with patches of the same colour on the hindcrown and nape, appearing very vivid in the dark rainforests. The upperparts are of a curious yellowish olive-brown, unlike any other species. The female is more nondescript and resembles a whistler, olive-brown above and ashy grey below, with a pale yellow eye. **SIMILAR SPECIES** Male Australian Golden Whistler (*Pachycephala pectoralis*) inhabits the same forest and has yellow underparts, but in addition has a distinctive white chin and throat and black pectoral band and head. Female Australian Golden Whistler is similar in size but grey-brown above and greyish-buff below, the nominate race (found in the same habitat as this bowerbird) showing pale yellow undertail-coverts. Both Bower's (*Colluricincla boweri*) and Little Shrike-thrushes (*C. megarhyncha*) are far more rufous or greyish above than the female and immature Golden Bowerbird, and are rusty below.

RANGE Restricted-range endemic of north-east Australia, found only in the uplands of the wet tropics of north-east Queensland from the Thornton Range and Mt Windsor Tableland south through the tablelands area to the Seaview–Paluma Range north

of Townsville. **MOVEMENTS** This is a largely resident species; in the Paluma Range, a marked adult female was seen 140m from her original point of capture *c*.14 years later. Small-scale altitudinal movement during winter months, when some individuals in female-like plumage move to lower altitudes, *c*.350m being the lowest.

DESCRIPTION The smallest bowerbird and the lightest in weight, with ten primaries, 12 secondaries (including four tertials) and 12 rectrices. Sexually dimorphic. *Adult Male* 23–25cm; weight *c*.80g. Has the head, chin and upperparts a singular yellowish olive-brown, with an orange-yellow short erectile subcrest on hindcrown and a golden-yellow patch on the nape. The underparts are a rich golden-yellow with a glossy sheen, more greenish-olive on the flanks, the underside of the tail a bright golden-yellow with dusky-olive on the two central feathers. The rich colouring of the bird is enhanced by a pearly opalescence that glows on the edges of the feathers. *Iris* an odd dull pale yellow; *bill* dark olive-brown with pale grey bases to the mandibles; *legs and feet* brownish-grey. *Adult Female* Smaller than the male, with a noticeably shorter tail. Upperparts and tail are dull brownish-olive; the underparts are ashy grey, with paler chin and throat and slightly brownish sides of breast and flanks. *Iris* brownish-yellow, and *bill* darker than that of male. *Juvenile* Darker than female, with less yellow wash on the flight-feathers, and some browny-grey down on the head. *Iris* dark brown; *bill* pale grey with yellow gape, *mouth-lining* yellow-orange; *legs* blue-grey. Assumes adult plumage over 5–6 years.

Immature Resembles female but has a darker brown iris which becomes paler with age. **Subadult Male** Washed yellow on the chest; gradually moults into full adult plumage, older birds (presumably around four or five years old) being as adult but with some dark feathers on the tail and a darker face.

TAXONOMY AND GEOGRAPHICAL VARIATION

Monotypic within its restricted range. Some individual variation in colour of hindneck, from shining golden yellow to deep orange-yellow.

Synonym *Corymbicola mestoni* De Vis, 1889, *The Queenslander* 30 Mar, Mt Bartle Frere.

Prionodura newtoniana fairfaxi Mathews, 1915, *Austral. Avian Record* **2**: 133. This is another of Mathews's many proposed subspecies, long since synonymised.

VOCALISATIONS AND OTHER SOUNDS The different subpopulations of this species are isolated on the peaks of the Queensland Wet Tropics, and each has developed its own dialect, playback of songs from different areas eliciting no response in some cases. There is a parallel here with Chowchilla (*Orthonyx spaldingii*) in the same habitat and geographical range, which also exhibits marked local dialects and similarly shows a lack of response to playback of songs from different areas. The male Golden Bowerbird has a characteristic short rattling advertising call, often repeated and a good indicator that a bower is nearby, see xeno-canto XC334919 (by P. Gregory), XC172311 (by M. Anderson) and XC97332 (by P. Åberg). This advertising call is often followed by harsh wheezy rattles interspersed with **mimicry** of other birds such as Lewin's Honeyeater (*Meliphaga lewinii*), Bridled Honeyeater (*Caligavis frenatus*), Bower's Shrike-thrush (*Colluricincla boweri*), Chowchilla, Pied Currawong (*Strepera graculina*), Grey-headed Robin (*Heteromyias cinereifrons*), Yellow-throated Scrubwren (*Sericornis citreogularis*) and Atherton Scrubwren (*S. keri*). These calls may be given from the bower perch. A frog-like call recorded at Mt Lewis is at XC170755 (by H. Krajenbrink) and may well be an example of mimicry, and sounds of human origin can also be copied, such as crosscut saws and the chopping of wood.

HABITAT Upper montane tropical rainforest with a marked wet season, from 600m to *c*.1,500m. The habitat is often very muddy and supports many leeches (Hirudinea) when wet.

HABITS Generally seen singly, mostly in the canopy and middle levels of forest. Occasional groups of up to four individuals, often in immature plumage, may be found foraging together, sometimes accompanied by other species. The Golden Bowerbird drinks regularly from cavities in trees, and some moisture is obtained also from wet foliage in the very moist habitat. This species is not unduly shy and will at times tolerate observers quite close to the bower. It often sits still for some minutes near the bower site, and it is easily overlooked unless calling.

FOOD AND FORAGING Primarily frugivorous, with a wide range of species recorded in its diet, but especially fruits of giant pepper vine (*Piper novaehollandiae*),

Indian mulberry (*Morinda umbellata*), white supplejack (*Ripogonum album*) and *Polyscias* species. The diet is supplemented with flowers, buds, insects gleaned from foliage or caught in the air, and spiders. Chicks are fed with fruit, especially quandongs (*Elaeocarpus*), supplemented by insects, especially cicadas. The male has the curious habit of caching fruit near the bower or nest site, presumably to maximise time spent at the bower. D. Hobcroft (*in litt.*) saw one caching a large piece of reddish fruit in a tree hollow; the bird pulled it out of one hollow, ate some on the ground in front of the observers, and then stashed it in another hollow. Cache sites include tree cavities, and spaces beneath fallen logs or under leaf litter.

BREEDING BEHAVIOUR As with other members of the family, the polygynous, promiscuous male decorates his bower, in this case a double maypole structure, while the female builds and attends the nest alone. Pilfering of decorations from nearby bowers is frequent. The estimated mean home range of eight adult males at Paluma was 7ha. The immature male visits many bower sites for up to five years, before gaining his own. Some traditional bower sites have been attended for at least 40 years, although not always by the same male, and the mean nearest-neighbour distance at a site near Paluma was 150m (F&F 2004).

BOWER This species builds one of the largest bowers of any member of the family and the largest in Australia. It is of a unique double maypole design and is rather variable in size and layout, constant features being a more or less horizontal display perch in the form of a branch, vine, stick or tree root, which is kept bare in the central section. Bower decorations consist of grey-green lichen (*Usnea*) and *Melacope* flowers and creamy-white fruit, plus other creamy-white flowers (such as *Dendrobium* orchids) placed on a platform at the base of the tower abutting the display perch. Some bowers have twin pyramidal stick-tower structures nearly equal in height, some have one large pyramidal stick-tower structure and one much smaller, and occasionally a bower may be built in branches some way above ground. The bower can be 2–3m tall, large ones extending to 3m in diameter, and built around saplings, vines or trees with the sticks becoming fused together by a sort of fungus, giving rise to early reports of their being glued together. Some bowers may be in use for at least 40 years, and individual males have been known to attend a bower for 16 years, although seven years seems to be more typical. The basic bower structure remains, but the decorations are ephemeral and are replaced each season. Bowers are built on fairly flat ground or sloping hillsides, and there are sometimes small interwoven piles of sticks scattered around almost like miniature huts. These stick weavings can also be in small saplings at around head height. Immature males may build or attend **rudimentary bowers** during the non-breeding season, a sort of practice bower, and these are often located near an active bower (which they sometimes visit.) As with other species of bowerbird, Golden Bowerbird males will raid and steal materials and decorations from the bowers of rival males when the latter are absent. Bower sites are attended from about July to January, with a secondary period in March–May after moulting. Young birds sometimes take over complete bowers when the owner disappears, but can also maintain the sub-bowers nearby. On Mt Bartle Frere in the 1890s, one favoured location had 15 bowers of varying sizes within a radius of 100yds (c.91m) – perhaps including the small hut-like structures? – where the collector W. S. Day 'from time to time' shot no fewer than 30 birds at one of the bowers. This same collector shot hundreds of Golden Bowerbirds during his nine years in the area (Chisholm & Chaffer 1956). Bowers do sometimes appear in groups almost like clan arenas, within auditory and maybe visual contact of other males, but many bowers are stand-alone isolates. Kahlpahlim Rock, near Kuranda, also had a smaller cluster of up to half-a-dozen Golden Bowerbird bowers comparatively close together. Tooth-billed Bowerbird stages are often found in the same general area, as also sometimes are bowers of the northern race of Satin Bowerbird. A vivid description of a large bower is given in Chisholm & Chaffer (1956):

'*The most extraordinary bower seen was an immense structure in which the main wall reached a height of 7 feet 6 inches above ground level. This wall was built mainly in the branches of several saplings, but a column of sticks about twelve inches in diameter extended to the ground. An immense quantity of sticks went to form this pile and I estimated that the volume occupied by the main wall alone was approximately fifty cubic feet. A foot or so above the main wall a small heap of sticks was located in the branches of saplings. A living horizontal branch five feet above ground level served as the display-perch and connected the huge mass of the main wall to another substantial pile of sticks, built entirely in the branches of a couple of saplings. A few feet away from the main wall a small group of saplings supported a scattered group of sticks in the branches some five feet above the ground. What a vast amount of labour must have been expended in collecting this relatively immense pile! The feat becomes more impressive when one recalls that the bird is only about the size of the Grey Thrush (Shrike-thrush). A liberal heap of lichen was deposited on each side of the display-perch and numerous Melacope flowers were thrust into the main wall. The bower had evidently been in use for a number of years, since many of the sticks in the lower portion of the structure were much decayed and crumbled readily.*'

COURTSHIP Polygynous, the male will mate opportunistically with more than one female per season. The display starts around July–August, depending on local conditions. Males deliver the rapid buzzy advertising phrase from a horizontal perch 5–10m up near the bower, any nearby males countersinging in response. When females approach, the male bows forwards to show off his golden-yellow hindcrown and erect subcrest, with wings partly outstretched and the head sometimes bobbed up and down and shaken from side to side. The male makes a practice of hovering in a vertical position, with bill pointed directly forwards and the yellow of the tail very prominent. He may hold a decoration in his bill during this display, and the tail is repeatedly fanned to show off its yellow feathers. Hovering is always near the trunk of a tree, and sometimes near the higher wall of the bower. He vocalises during this performance, using mimicry and a variety of sounds, and he may also hide behind a tree trunk and peek out at the female, analogous to some of the grey *Chlamydera* and gardener bowerbird *Amblyornis* displays. If the female lands on his display perch, he chases her away while calling loudly; and, in contrast to other bowerbirds' behaviour, mating seems to occur away from the actual bower. **NEST/NESTING** The female alone constructs the nest, which takes up to 25 days. It is a deep, bulky, open cup placed on a loose platform of twigs, the outer part made of dry leaves and bark bound with dry tendrils and rootlets, and the internal cup lining of fine supple tendrils and occasionally feathers. A new nest may be built atop an older one from a previous year. The nest is located 1–5m above ground within a crevice in a tree trunk, sometimes in a tree fork or between tree buttresses. The average distance of 37 active nests from an active bower site was 97m (F&F 2004). **EGGS & INCUBATION** Eggs are unmarked pale cream to white, size c.35 × 25mm. Clutch size 1–3 (but usually two, with one egg slightly larger than the other), eggs laid on alternate days. Incubation solely by female, as is usual with the family, period 21–23 days. **NESTLING, BROODING AND FLEDGING PERIOD** Nestling has long dark brownish-grey down and orange-buff patches of bare skin; bill pale grey with a yellow gape. Female ceases brooding when nestling(s) 14–15 days old (irrespective of brood size), nestling period 17–20 days. The female will perform distraction display to lure predators away from nest. **NEST SUCCESS** Overall success rate of 29 nests was 28%, with average of one fledged young produced per female per season. **TIMING OF BREEDING** The breeding season is in the austral spring and summer, from September to February, with peak egg-laying in November–December, nestlings in December–January, a typical season lasting c.four months. The display season is July/August–January, ceasing with the onset of the wet-season rains. There is also some post-moult bower activity (often by immature males) March to early May. **LONGEVITY** Males have survived for in excess of 23 years, with one female at least 14 years old (F&F 2004).

HYBRIDS No hybrids known.

MOULT Peak months for moult December–March; males may start slightly later than females, and mostly complete by May–June (F&F 2004).

STATUS AND CONSERVATION Classified as Least Concern by BirdLife International. Some of the species' habitat has long since been cleared for agriculture, but a large area remains, much of it in National Parks or World Heritage Areas, and all commercial logging has now ceased. There has been some fragmentation of its habitat, but some rainforest is recolonising. A considerable amount of collecting of specimens took place after 1889, when the male was first discovered, and later of the eggs, the collector Sharp having 30 sets of eggs in 1908. A great concern to many scientists and laymen alike is that the areas predicted to suffer the greatest climatic change correspond to some of the most important biodiversity areas in Australia, such as the Wet Tropics rainforests, east-coast wet forests, and the tropical far north of Cape York. Global warming is expected to alter the structure and composition of all forests, with restricted-range montane taxa at particular risk as there is simply nowhere else suitable

for the species adapted to these particular habitats to colonise. There is also an increased frequency and intensity of cyclones predicted to occur with climate change. Populations of Golden Bowerbirds were severely affected by Cyclones Larry and Yasi, 2006, 2011. By 2018 bowerbird populations at a number of locations e.g. Paluma, Kirrama, had only partially recovered (D. Chaplin).

Golden Bowerbird, immature male, Atherton Tablelands, Queensland, Australia (*Jun Matsui*).

Golden Bowerbird, male with fruit, Atherton Tablelands, Queensland, Australia (*Jun Matsui*).

Golden Bowerbird, double maypole bower, Atherton Tablelands, Queensland, Australia (*Jun Matsui*).

Genus *Ptilonorhynchus*

This genus is thought to be closely related to *Chlamydera*, the grey bowerbirds, but its single species is an altogether stouter and shorter-tailed bird, with a heavier bill and pronounced sexual dimorphism. The striking violet-blue male, with its remarkable violet-blue eye, is a well-known bird in some parts of eastern Australia, and the bowers of this species with their predominance of blue objects are likewise locally familiar. This is one of the avenue-builders, and was the first to be discovered to possess the unusual habit of 'painting' its bower walls and using a form of tool to do so (Gannon 1930). The formal discovery of the bower is credited to John Gould's brother-in-law Charles Coxen, who collected one and donated it to the Sydney Museum. The renowned naturalist John Gould coined the now familiar term 'bower' in 1840, following his initial amazement at the structure.

Etymology The scientific name of this monotypic genus is derived from the Greek *ptilon*, a feather, and *rhunkhos*, meaning bill, referring to the base of the bill being covered by feathers.

SATIN BOWERBIRD
Ptilonorhynchus violaceus   Plate 33

Pyrrhocorax violaceus Vieillot, 1816, *Nouveau Dictionnaire d'Histoire Naturelle edition* **6**, 569. Type specimen from 'Nouvelle Hollande' = Sydney, New South Wales (Mayr 1962).

Other English names Satin Bower-Bird, Purple Satin, Satin-bird, Satin Bird, Satinbird, Satin Grackle; Lesser or Northern Bowerbird (race *minor*)

Etymology The species epithet *violaceus* is Latin and means violet-coloured or violaceous.

This species was initially thought to be a new kind of chough, from the corvid family. It is now the best-known and most familiar member of the family Ptilonorhynchidae, being found close to several large conurbations and thus easily accessible. There are a number of long-term studies such as those by Marshall (1954), Chaffer (1959) and Gilliard (1969) and the extensive and valuable research of Drs C. B and D. W. Frith. Exhaustive accounts of the species' biology may be found in HANZAB Vol. 7, Part A (Higgins *et al.* 2006), and the Friths (2004, 2008, 2016). What follows is largely a summary of their findings.

FIELD IDENTIFICATION A very distinctive species. The male is a large, stout, big-headed, deep glossy blue-black bird with a pale bill and a violet eye; females and young males are greenish above with reddish-brown wings and uppertail, and well barred dark below, with striking greyish (younger birds) or blue eyes. The wings are rather short and quite rounded. This species can become confiding at some sites, such as O'Reilly's at Lamington NP, Queensland, but males tend to be shyer than the green individuals and away from food sources can be quite elusive. **SIMILAR SPECIES** The male is virtually unmistakable. Males of the Victoria and Paradise Riflebirds have a long decurved bill, very rounded wings and a short tail, make a loud rustling sound in flight, and have a bright blue uppertail. Females and young males of the Satin Bowerbird might be mistaken for Green or Spotted Catbirds, but those have reddish eyes and are distinctively entirely green above, with no reddish-brown on wings or tail, and are heavily spotted with pale. Spotted Catbird in North Queensland has blackish ear-coverts, and the underparts of the catbirds lack the distinct barred or scalloped pattern of the green-plumaged Satin Bowerbird.

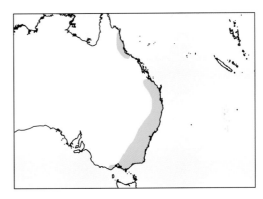

RANGE Occurs in a broad coastal band of eastern Australia, from the Bunya Mts of South East Queensland south to the Otway Ranges in southern Victoria. An isolated population, race *minor*, is found in the uplands of the Wet Tropics in northern Queensland, from Cooktown south through the tablelands to the Paluma Range near Townsville; here it overlaps in range with Spotted Catbird, Tooth-billed Bowerbird and Golden Bowerbird. **MOVEMENTS** Largely resident, with some localised winter movement to lower elevations. Some may move from woodland to more open habitats in winter. Two records of marked individuals found *c.*68km from site of initial capture.

DESCRIPTION Sexually dimorphic. *Adult Male* 24–35cm; weight *c.*230g for nominate, *c.*190g for race *minor*. Adult male can appear uniformly black, but in better light the plumage has a metallic sheen giving a deep satiny blue appearance, duller and more blackish on the flight-feathers, wing-coverts, lower abdomen and the centre of the uppertail. The outer edges and tips of the tail are glossed violet, with the undertail and underwings more bronze. Has ten primaries, 14 secondaries (including tertials) and 12 rectrices. *Iris* a striking vivid purple-violet, tending towards blue at edges; *bill* pale greenish-yellow, appears quite short and thick (forehead feathering extends over base); *legs and feet* dull yellowish-brown with faint blue tinge. Note that males do not attain mature adult plumage until seven or eight years old, by which time they have moulted completely into their characteristic blue-black adult plumage. *Adult Female* Slightly smaller and longer-tailed than male, with plumage very different, maturing at two to three years of age. Upperparts grey-green with

365

a light blue tinge, the primaries dark brown, and the upperwing-coverts and secondaries dull reddish-brown with greyish-green wash. The inner flight-feathers are white-tipped, and the feathers of the back, rump and uppertail are washed yellowish-green, with the tail rich brown above and greenish-brown below. The ear-coverts, cheeks and throat are dark with pale buff shaft streaks. The entire underparts and underwings are creamy buff with a yellowish-green wash, and heavily scaled with greenish crescents or chevrons. *Iris* deep blue; *bill* blackish; *legs* usually rather pale, flesh-coloured to grey. *Juvenile* On leaving the nest, the juvenile retains some down on the head, which is soon moulted out to reveal greenish-brown feathers, while the back becomes more greenish; the underparts are pale cream, broadly scalloped with greenish-brown. Iris is brown and ear-coverts are streaked cream-buff (Rowland 2008). *Immature* First-year to third-year birds resemble female, but with more whitish underparts. First-years have more prominent pale tipping on the upperwing-coverts, pale grey legs, and a dark brownish-grey iris. Iris becomes lilac-blue and purple by the second year, and the sides of the head become greener, especially on males. The underparts of the male become greener by the third year, while the females remain unaltered except for leg colour, which becomes more yellow. *Subadult Male* During fourth year the male is still like the female in plumage, but has a green band on the breast and the bill changes to pale yellowish-green with blue base. The male then moults progressively into mature adult plumage, acquiring first signs of the adult dress with some blue-black feathering in the fifth year, and completing the transformation usually in sixth and seventh years.

TAXONOMY AND GEOGRAPHICAL VARIATION
There are two subspecies, the nominate in the main part of the range.
1. **P. v. violaceus** (Vieillot, 1816), Sydney. Occupies southern part of the range in South East Queensland, eastern New South Wales and eastern Victoria.
2. **P. v. minor** Campbell, 1912, *Emu* **12**, 19. Type from Herberton Range. Isolated in north-east Queensland, where restricted to rainforest on the tablelands. Smaller than nominate race and about 20% lighter in weight, with dull bluish-grey cast to the mantle on female-type birds.
Synonym: *Ptilonorhynchus violaceus dulciae* Mathews, 1912, is one of Mathews's many novel creations based on very minor differences.

VOCALISATIONS AND OTHER SOUNDS This is a very vocal, noisy species with a great range of vocalisations, which vary somewhat in frequency, duration and format in different populations. A common presumed bower-advertisement call of the adult male of both races is a loud, whistled, descending *queoooow* as at xeno-canto XC146114 (by M. Anderson); this call is given from perches near the bower. There is also a raspy scratchy song: see XC2011468 and XC171965 (by M. Anderson) and loud descending whistles, as at XC171963 (also by M. Anderson). Satin Bowerbirds often utter harsh churring notes (see XC98437, by P. Åberg), as well as harsh alarm calls (XC155104, by F.

Deroussen) and harsh scolding sounds (XC329815, by A. Spencer). Other vocalisations include **mimicry** of a long list of other species, including Laughing Kookaburra (*Dacelo novaeguineae*), Lewin's Honeyeater (*Meliphaga lewinii*), Yellow-tailed Black (*Calyptorhynchus funereus*) and Glossy Black Cockatoos (*C. lathami*), Crimson Rosella (*Platycercus elegans*), Superb Lyrebird (*Menura novaehollandiae*), Golden Whistler (*Pachycephala pectoralis*), Paradise Riflebird (*Ptiloris paradiseus*), Rufous Scrub-bird (*Atrichornis rufescens*) and White-browed Scrubwren (*Sericornis frontalis*). The **courtship song** is quieter than the advertising song and includes much mimicry interspersed with odd ticking or tapping notes. Harsh loud churring notes are also given singly or in twos and threes by the male perched near the bower. The males at the start of the courtship display utter soft, high-pitched, plaintive squeaking and spluttering notes, reminiscent of calls of the Common Starling (*Sturnus vulgaris*) (F&F 2004). Males also give an odd constant mechanical whirring buzz in the first phase of the courtship display. Loud harsh calls are emitted as a threat vocalisation, and the species has one of the largest and most varied vocal repertoires in the entire family. Both sexes mimic, but the female repertoire seems less extensive than that of the male. Nesting females can mimic the calls of potential predators and may perform distraction displays to lure predators away. Mimicry of cats meowing is quite common (although there is some debate as to whether this call may be innate), and imitation of a postman whistling is recorded. Flocks are also quite noisy in the non-breeding season.

HABITAT Satin Bowerbirds prefer the tall wetter forests and woodlands, and nearby open areas, although those living around the Atherton Tableland are largely inhabitants of upland rainforest above 600m to about 1,425m, mostly at 900–1,100m. This species shows a strong preference for the edges or gaps and treefalls for feeding, nesting and bower construction, and is found also in adjacent tall sclerophyll woodlands with a sapling understorey. The forest in the north of Queensland has many lianas, and an understorey of climbing pandans, ferns, saplings, seedlings and lawyer vines (*Calamus*). Southern, nominate populations occur in a wider range of forests, including vine forest, warm-temperate rainforest, subtropical rainforest, wet or dry sclerophyll forest and drier monsoon forest, from sea level to about 1,100m. Figs (*Ficus*) of various species are often common in these habitats and are an important component of the diet at times. Since the late 1980s this species has colonised suburban Canberra, and is now one of the most conspicuous birds in some gardens. **Winter flocks** occur also in more open habitats such as parks, fruit orchards and gardens and readily exploit exotic species as food items. Some may move to lower altitudes in winter, vacating the higher levels, northern green-plumaged individuals of race *minor* being seen rarely at Kuranda, Queensland, at 350m.

HABITS Mature male Satin Bowerbirds are mostly solitary, but the green-plumaged individuals are often seen in groups or quite large flocks, with parties of

50–200 recorded; these often include some adult males as well as green-plumaged birds, although they do form single-sex feeding flocks in some districts. In the winter non-breeding season, many move to more open country and occasionally enter gardens, orchards and pastures, where they graze on grassy shoots, white-clover leaves and herbaceous plants, adult males sometimes joining the 'green' bird flocks. Of some 600 individuals ringed in a garden at Leura, in New South Wales, over a number of years, very few were adult males, suggesting some degree of territoriality. Males have been known to **roost** above their bowers, but they roost also in wooded gullies and fly to the bower each morning. A winter roost of >50 birds was in the dense foliage of garden *Pittosporum* and cypress shrubs, extending over about 100m (Holland 2000). There are several distinct types of **agonistic behaviour**. The *Warble-waggle Display* is usually given on the ground, two birds facing each other and, with quivering spread primaries, one giving a warbling sound while the other sleeks the body parallel to the ground with neck extended and tail held horizontal; once the warble has finished, the body is waggled from side to side and both birds may jump up to strike each other with their feet. If the display is performed in trees, the birds crouch on the branches and then hop between perches while waggling the body. This display is typically directed against other males, adult or immature, but may be given towards female-plumaged visitors to the bower, and is used also in winter congregations or foraging flocks. In the *Puff Threat* the male puffs up his feathers while standing. The *Open-beak Threat* involves the male partly opening the bill (without calling) while approaching the intruding bird. In the *Wing-flip Threat* the bird quickly moves the wings back and forth from the body. The *Vocal Threat* is the action of birds directing loud harsh calls at each other. Older males generally dominate the younger birds at feeding sites. Females approach the bower quietly and unobtrusively, and may be frightened off by aggressive posturing of the males.

FOOD AND FORAGING Satin Bowerbirds are largely frugivorous throughout the year, though they also eat leaves (during the winter months) and a small amount of seeds and insects. Figs are a significant component, such as strangler fig (*Ficus watkinsoniana*), sandpaper fig (*F. coronata*) and deciduous fig (*F. superba*). During the summer breeding season, the diet is supplemented with a large number of insects, while leaves are often eaten. In addition, this species will kill and eat small lizards and birds, the latter sometimes being taken from their nests. Much of the foraging is arboreal, but can occur from the ground upwards. Birds of the race *minor* have been recorded as foraging in figs, eucalypts (*Eucalyptus*) and paperbark trees (*Melaleuca*) in late May at Julatten, Queensland (*c.*7km from rainforest). Satin Bowerbirds are sometimes quite aggressive when foraging, trying to displace other birds from food sources. At Julatten, in North Queensland, for example, a subadult male fought with and dominated up to three Great Bowerbirds (Frith *et al.* 2017). They often forage with other frugivores, and may at times

displace them from food sources or themselves be displaced. They have been seen to displace Crimson Rosella from a sap scar made by Yellow-bellied Glider (*Petaurus australis*) (Chapman *et al.* 1999). They were seen to drive a Spangled Drongo (*Dicrurus bracteatus*) off *Banksia* flowers, but the bowerbirds were dominated by Little Wattlebirds (*Anthochaera chrysoptera*) at this site. Potential predators such as Australian Magpie (*Gymnorhina tibicen*), Pied Currawong (*Strepera graculina*) and Laughing Kookaburra (*Dacelo novaeguineae*) are not tolerated and are summarily driven away from nest trees. This species is very adaptable, and has taken readily to some of the great variety of plant species introduced since European settlement. It is a major **dispersal agent** for a number of problematic invasive plants, such as Camphor Laurel (*Cinnamomum camphora*), the Mediterranean Olive (*Olea europaea*) and various species of privet (*Ligustrum*). Bowers have been found in plantations of Hoop Pine (*Araucaria cunninghamii*) bisected by narrow strips of remnant vine forest in South East Queensland. They may be persecuted by horticulturalists because they are great opportunists and frequently raid fruit and vegetable crops. Large flocks have been seen feeding on grass shoots, advancing steadily together as they fed, and grass leaves and shoots are plucked or pecked off. This species seems not to use its feet to rake aside leaf litter when foraging, and most of the diet is obtained by gleaning or sallying, though the proportions of such techniques will vary with the food types available. Invertebrates form a significant component of the diet, the feeding techniques used depending on the species involved. Small beetles are struck against branches and then swallowed, whereas larger prey such as cicadas, stick insects and big beetles are dismembered before being eaten or being taken in pieces to the young. Large insects grabbed when flying are broken up, with wings and legs removed, but without the use of the feet. Leaves form a significant part of the diet in winter, when invertebrates are less readily available and fruit supplies reduced (Lenz 1999). The birds drink regularly, attending bird baths and water bowls in some areas, and they also like to bathe. Adult males often forage within 50m of their bowers in the breeding season, but in winter venture farther afield, up to 350m from the bower. At another site adult males ranged from 100m to 1,500m from their bower sites, the males having extensively overlapping foraging ranges. The differences between the two rainforest and woodland habitats reflect dietary variations, with woodland birds more folivorous than rainforest birds (Donaghey 1981).

BREEDING BEHAVIOUR The male Satin Bowerbird is the best-known and best-documented of all the bowerbirds in Australia, much of its population being close to large cities and some of the striking bowers becoming tourist attractions. The polygynous, promiscuous male seasonally builds and decorates his terrestrial avenue bower; as is usual with the family, the female alone undertakes all nest-building, incubation and nestling-care duties. Immature males build bowers and display to prospective mates, and are capable of successful copulation as five-year-old subadults, sometimes invading bowers and trying to force copulation with a female when the owner male is displaying. A series of video recordings by T. Laman of Satin Bowerbird at the bower is available at the Macaulay Library of Cornell University (ML456304 and ML456313). David Attenborough also has some remarkable footage in his documentary *The Natural World: Bowerbirds – The Art of Seduction* (2000). **BOWER** Males build a single-avenue stick bower consisting of two parallel walls that curve inwards at the apex, though rarely a bower may be a three-walled variant. It is built on a platform of woven sticks and can be completed remarkably quickly, sometimes in a day, and may comprise up to 2,000 sticks and contain from 36 to 200 or more decorations. **Orientation** is generally north–south, which presumably allows for best illumination of prized objects. These bowers are famous for being decorated with blue objects, although the species will utilise yellow and shiny objects if these are available. Objects include fruits, berries, flowers, and even ballpoint pens, toothbrushes, marbles, drinking straws and other discarded plastic items such as clothes pegs and drink-container rings (condoms and their wrappers have also been noted). As the males mature they use more blue objects than the other colours. Prior to European colonisation they would have used naturally occurring blue objects such as quandong fruits, blue berries and flowers, and the blue tail feathers of Crimson Rosella, blue feathers in general being highly prized. Yellow and brown items occur, such as flowers, leaves, snakeskins, cicada nymph cases (an important part of the diet) and brown snail shells, and small-mammal and reptile skulls are also used occasionally. The bower-owner meticulously maintains his bower throughout the year. Traditional bower sites are evenly dispersed through suitable rainforest and woodland. Theft of decorations and bower destruction are common if and when the owner is absent. A mixture of chewed vegetable matter (including tree bark and foliage), charcoal and fruits is mixed with saliva and used to paint the walls of the bower. This species was the first to be discovered using material to paint that can qualify as a kind of **tool**, as it utilises a piece of vegetable matter as brush or sponge. The bird masticates a piece of bark or fibrous matter into a kind of pellet, which it uses to apply the paint (Chaffer 1959), as with both Great and Fawn-breasted Bowerbirds, these three belonging to a very small select list of species known to use tools. Repeated painting may result in a powdery blackish coating 1–3mm thick (Chaffer 1984). Quite why they paint is another matter. Is it a signal that the bower is fresh and being maintained? Does it have an olfactory or taste component, as females often peck at or inspect it? Is it a visual signal to other bowerbirds? Immature males have even been known to paint the bowers of a rival male! Much remains to be learned – and just how widespread is the habit among the other avenue-builders and, indeed, the maypole-builders? **Bowers of the race *minor*** are smaller and have the floor of the bower lined with green moss, which may also be hung from the inner avenue-wall sticks and sometimes

extends out to the platform as well. There appears to be no particular orientation for these northern-race bowers. Chaffer (1984) made some pertinent observations: 'The bower of the Satin Bower-bird was neatly made but rather smaller than those of southern birds. The walls, which ran in an east/west direction, were about seven inches [*c.*18cm] long, three inches [7.6cm] thick and spaced four inches [*c.*10cm] apart. A fairly substantial platform was placed at each end of the bower, that on the eastern end being composed of fine sticks and that on the western end mainly of grass-stems. Decorations were not numerous; they consisted of a few snail-shells, cicada cases, yellow leaves, blue berries, blue feathers, and blue flowers of wild tobacco. In common with those found last year, this bower included in its decoration four bleached skulls of small mammals. Pieces of yellow-green moss were also included and a sprinkling of the moss was scattered through the runway, and two pieces were hung up on the sticks of the inner walls. The bower was painted, although saliva appeared to be mainly used. A. J. Marshall quotes Stresemann as suggesting that the painting habit is confined to the southern bird since no painted bower had been recorded from North Queensland. However, it will be noted that the bower mentioned above and the two seen in 1953 were all painted. With regard to the unusual decorations used by the bird, Marshall [pers. comm. to Chaffer] considers it would be of great interest if one found a constant difference in decorations selected by the northern and southern races of *Ptilonorhynchus*.' The addition of a moss lining to the bower floor by race *minor* does make it different from the unlined bower interior of the nominate race. This could reflect the perhaps more plentiful occurrence of moss in the more humid northern rainforests. Females visit the bowers and mate with the male of their choice. They may have a preference for those bowers with the most and best blue objects, which may be associated with the older, more mature males, which over time will drive the selection choice and display. This species is thought not to defend an all-purpose territory, rather just the vicinity of the bowers, with agonistic encounters between males rare away from such sites. A study during 1975–1984 at Moruya, in south-east New South Wales, found that bowers were *c.*500m apart and suggested that adult males have a territory of about 20ha. Generally, bowers are fairly evenly distributed and not clustered into any form of lek (Donaghey 1981). Bower sites have been recorded as being occupied for more than 15 successive years, and the death of a long-established male will mean fierce competition for a change of ownership and consequent adjustments at nearby bowers. **COURTSHIP** The males may attend bowers and display from May onwards, the northern race more usually in July, but most start in the austral spring in August–September and finish in late January–early February. They advertise their bowers by loud rising-and-falling whistled calls given from an exposed perch nearby. When a female arrives, the male gives soft notes and goes into a ritualised display with wings outstretched above his back, often with bill gaping and repeatedly shaking his head, and picking up and putting down a blue bower decoration; this is accompanied by various buzzing and rattling calls. The male then raises his tail high in the air and performs a series of wing-flicks, using the wing on the opposite side of the body to the bower and moving the tail up and down with each rapid wing-flick. He hops and strides around the bower with stiff legs and his body rigidly contorted, accompanied all the while by chatterings, rattles, creaks and mimicry, and then strides back and forth across the bower entrance, often with an object in his bill, and ends up where he began the display. These intense behavioural displays can be interpreted as threat by the females, and sometimes drive them away. The female may decide to move into the bower avenue for mating, crouching down and quivering her wings; copulation is usually within the bower, but can be on the platform outside. She then stands and flaps her wings, the male aggressively chasing her away before readying himself for courting more prospective females (Rowland 2008). According to recent research, female mate choice takes place in three stages: (i) visits to the bowers, before nests have been built, while the males are absent; (ii) visits to the bowers, before nests have been built, while the males are present and displaying; (iii) visits to a selection of the bowers, after nests have been built, leading to copulation with (typically) a single male. Experimental manipulation of the ornaments around the bowers has revealed that the choices of younger females (those in their first or second year of breeding) are influenced mainly by the appearance of the bowers, and hence by the first stage of this process. Older females, less affected by the threatening aspect of the male's displays, make their choice on the basis more of the male's dancing displays. It has been hypothesised that, as males mature, 'their colour discrimination develops and they are able to select more blue objects for the bower. It is not yet known whether this description would also hold true for other species of bowerbird' (Coleman *et al.* 2004). NEST/NESTING The female builds a bulky, loose, shallow saucer-shaped nest of sticks or twigs, with an inner lining of *Eucalyptus* or *Acacia* leaves which turn brown as the eggs are laid (and may serve as camouflage), the structure being up to 30cm wide with an internal cup *c.*15cm wide and 40cm deep, slightly smaller in the race *minor*; the construction work takes about two weeks. The nest is sited up to 30–35m above the ground in the fork of a tree branch, in a mistletoe clump, in vine tangles, atop a broken tree trunk or in the outer foliage of a tree. The same location may be used in successive years. Distance between nests varies with the habitat and population density, and can range from *c.*100m to less than 50m. In the 1975–1984 study at Moruya, in New South Wales, 22 nests were all roughly within an area bounded by the bowers of three adult males (Marchant *et al.* 1993). EGGS & INCUBATION Clutch one to three eggs, usually two, dark cream-buff, streaked with purple-grey and spotted with grey and dark brown, the markings concentrated at the broader end. The eggs are oval in shape and measure about 43 × 29mm, those of race *minor* slightly smaller and lacking the wavy lines seen on the eggs of nominate race. They are much larger

than typical for a bird of its size, weighing around 19g. Eggs are laid on alternate days and hatch asynchronously after 17–21 days of incubation. The female leaves the nest from just after dawn to 07:00 hours, and again in the early evening. NESTLING Newly hatched young has orange skin and orange-brown down, developing into greenish-brown feathers but retaining long brown down on the crown. Nestlings are fed largely on a high-protein diet of beetles, grasshoppers and cicadas, and a good supply of such is essential to successful fledging, with cicadas and Christmas beetles (*Anoplognathus*) especially significant. Young are able to fly three weeks after hatching, but remain dependent on the female for another two months, finally dispersing at the beginning of the southern winter (May or June). TIMING OF BREEDING The breeding season is mainly in the austral spring and summer, from August through to about February, although preliminary displays may begin in May–July. Nest-building and incubation usually take place between October and February. LONGEVITY Ringing recoveries indicate individuals living for ten years to at least 18 years, and birds of both sexes can live for 20–30 years.

HYBRIDS There are three known examples of a hybrid known as **Rawnsley's Bowerbird**, the result of intergeneric hybridisation with the Regent Bowerbird. The first was a specimen from near Brisbane in 1867, followed by one seen and photographed in south-east Queensland in 2003, and finally an individual at Kalang, New South Wales, in 2014. The specimen was described as being in adult male plumage, mainly the glossy blue-black colouring of the adult male Satin Bowerbird, but with a conspicuous and extensive yellow wing patch, yellow tipping to some tail feathers, with a paler iris colour than the Satin Bowerbird, and intermediate in size between the two putative parent species.

MOULT Annual moult of males and non-breeding females takes place from November to March, but that of breeding females occurs mainly during February–May (F&F 2004).

STATUS AND CONSERVATION Classified as Least Concern by BirdLife International, with a range of about 523,000km^2. This species is common to reasonably abundant but with much habitat lost to human exploitation, as with forest loss in the Hunter valley, and some populations are in decline. The population in the Otway Range is now cut off from the main one west of Melbourne. Numbers are killed every year because of their depredations on orchards and gardens, and many also are lost because they associate with such species as pigeons, which are illegally hunted. One minor problem was that of individuals being choked by the blue rings from plastic milk-containers, which became stuck around the bird's bill or neck and prevented feeding; the suppliers of these containers have now changed the colour from blue, which is particularly attractive to this species (see BOWERS, under BREEDING BEHAVIOUR, above), and there has been some publicity about cutting the plastic prior to disposal. Feral cats may also be a problem locally, and climate-change impacts on upland forests are an unknown variable. Until the early 1950s the Satin Bowerbird occurred in the Australian Capital Territory (ACT), and it has reappeared there in the recently developed suburban gardens and parks: initially there were only winter visits from birds in female-type plumage, but subsequently individuals came in summer and there are now adult males, with bower-building and bower attendance recorded; most bower activity is in winter (when more birds are present), with successful breeding there November–December 2000 (Frith *et al.* 2017). This is a popular and well-known species, and some of its bowers are the focus of much local interest, as has historically also been the case.

Satin Bowerbird, female, Lamington, Queensland, Australia (*Jun Matsui*).

Satin Bowerbird, bower of race *minor*, Atherton Tablelands, Queensland, Australia (*Jun Matsui*).

Satin Bowerbird, male, Lamington, Queensland, Australia (*Phil Gregory*).

Genus *Sericulus*

Sericulus Swainson, 1825, *Zool. Journ.* 1(4): 476. Type, by monotypy, *Meliphaga chrysocephalus* Lewin.

The species of this genus, known also as silky bowerbirds because of their silken appearance and feather texture, range from south-east Australia to the lowland and hill forests of western and central New Guinea. They are among the slighter and more delicately built bowerbirds, characterised by the very distinctive sexual dimorphism in the adult plumages. Males are remarkably striking with their golden or orange head and nape feathering, the remaining plumage being varying combinations of black and gold. Females of all species are plain brownish above and paler below, with variable amounts of dark scalloping on throat or breast, or barring on the underparts. It is noteworthy that the more striking the plumage is, the less elaborate is the bower, all of this colourful genus building simple avenue-type bowers with minimal decoration.

Etymology *Sericulus* derives from the Greek word *serikon*, meaning silk (or perhaps the Latin *sericus*, silky), the genus being known as the 'silky bowerbirds', a reference to the vivid silken plumage of its members.

MASKED BOWERBIRD
Sericulus aureus   Plate 34

Coracias aurea Linnaeus, 1758, *Syst. Nat.*, ed. 10, **1**: 108. Bird's Head, north-west New Guinea.
Other English names Black-faced Bowerbird, Black-faced Golden Bowerbird, New Guinea Golden Bowerbird, Northern Flame Bowerbird, Golden Regentbird
Etymology The specific name *aureus* is Latin and means golden, a reference to the colour of the male of the species.
This was the very first bowerbird to be discovered and described. A specimen was purchased by Lesson, who obtained trade skins in the Dorey Bay area of Dutch New Guinea (Papua), and described by Linnaeus, the father of modern systematics, in 1758.

FIELD IDENTIFICATION A very distinctive species, the male having rich red, gold and yellow plumage with a black face, wings and tail, the female being brownish above but quite yellow below, with scallopings on the throat and breast and not readily confused with anything else. **SIMILAR SPECIES** Golden Myna (*Mino anais*) has a black mantle and a black patch on the nape, plus yellow eye and bill, and in flight shows a white wing patch. Flame Bowerbird has a very small and marginal range overlap in south-west West Papua, but the male lacks the black face of Masked Bowerbird and the female is less mottled on the chin, throat and upper breast.

RANGE West and north New Guinea: occurs in mountains of the Vogelkop (Bird's Head), Bird's Neck (mountains of Wandammen Peninsula, and likely to occur in Fakfak Mts), north scarp of western, border

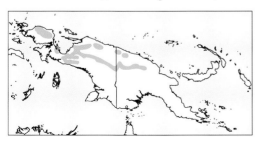

and eastern ranges (Weyland Mts east to Jimi River), Foya Mts, PNG north coastal ranges, Torricelli Mts and Prince Alexander Mts. **MOVEMENTS** Resident.

DESCRIPTION Sexually dimorphic. *Adult Male* 24–25cm; weight 175–180g. The head, nape, mantle and side of neck are highly glossy brilliant deep sunburst-orange, with contrasting black facial mask, chin and throat. Long, filamentous deep orange upper-neck feathers form a small mane that can be enlarged during intense display. Mantle largely suffused orange, and mid-back, rump and most uppertail-coverts entirely deep yellow. The upperwing is brilliant deep yellow, with some black spotting just above the carpal joint, extending to alula and primaries. The outer primaries are blackish-brown, the bases having a very small amount of paler deep yellow, becoming progressively more extensive across the primaries; secondaries are deep yellow, tipped blackish-brown, the darker tipping becoming progressively smaller towards body. The uppertail is dark blackish-brown, narrowly tipped with deep yellow. The underparts are a slightly paler deep yellow than the upperparts, and both paler and washed out on the undertail-coverts; the undertail is more brownish than the blackish uppertail, and the thighs are black. *Iris* appears strikingly large, and is pale lemon-yellow in colour (it is not yet known if the size changes during display as with that of Flame Bowerbird); *bill* is pale bluish-grey with extensive black tip; *legs and feet* greyish. *Adult Female* c.25cm; weight 165–175g. Less heavy than male, with a slightly longer tail. Plumage is olive-brown above and golden yellow below, with chin, throat and upper breast washed olive-brown, with obvious scalloping on the chest feathers. *Iris* dark brown; *bill* brown; *legs and feet* grey. *Juvenile* Undescribed. *Immature Male* Like female, and rectrices more pointed than those of subadult/adult male. *Subadult Male* Also similar to female, but with variable amounts of orange, yellow and black plumage developing, beginning with a few feathers and gradually changing to full adult.

TAXONOMY AND GEOGRAPHICAL VARIATION
Monotypic. May form a superspecies with Flame Bowerbird, Fire-maned Bowerbird and Regent Bowerbird. Formerly treated as conspecific with Flame Bowerbird, with which it has hybridised (see below).

VOCALISATIONS AND OTHER SOUNDS Poorly known. Utters occasional raspy *ksssh* notes quite similar to those of Flame Bowerbird: see xeno-canto XC26337 (by F. Lambert). Also quiet *chuk* sounds as at xeno-canto XC62367 (by A. Spencer), and odd buzzing notes given during courtship.

HABITAT Inhabits the canopy of hill forest at 850–1400m, very occasionally lower, down to *c*.500m.

HABITS Generally uncommon and shy. Usually seen singly, but sometimes in twos or threes, especially at fruiting trees, including figs (*Ficus*). Seen mostly in the canopy and middle stratum, where it sometimes perches up on bare branches for short periods; tends to fly just above the canopy. This species has been seen in mixed-species flocks, and together with Vogelkop Bowerbird, Magnificent Bird of Paradise and Beautiful Fruit-dove (*Ptilinopus pulchellus*) in the same fruiting tree.

FOOD AND FORAGING Little known. Undoubtedly mainly frugivorous, being especially fond of figs (*Ficus*), and will attend fruiting trees with mixed-species flocks. Known also to take some invertebrates, insects having been found in stomach contents. Forages mostly in the canopy and middle stratum.

BREEDING BEHAVIOUR The polygynous, promiscuous male builds and decorates his terrestrial avenue bower during the display season from August to November, which is similar to that of the Flame Bowerbird. Presumably the female alone undertakes nest-building, incubation and nestling duties, but the species remains little known. **BOWER** In Vogelkop (Tamrau Mts and Arfak Mts) bowers are active during August–November. They are of the avenue type, much as with congeners. One sited in forest atop a rounded ridge at 1,000–1,200m in the Tamrau Mts measured 18cm along central avenue and 26cm from ground to highest of avenue wall sticks; the inside of the bower was decorated with five oval blue berries, a black shelf (bracket) fungus and a small shell.

Other bower decorations noted elsewhere include purple fruits, snail shells, and yellow to bronze leaves; it was unconfirmed if decoration fruits that disappeared were stolen by rivals or eaten by visiting Arfak (Spotted) Catbirds. Masked Bowerbirds are known to 'paint' the inside of bower walls, probably using regurgitated fruit matter. One bower was sited almost directly beneath a lek of Lesser Bird of Paradise. Bowers are sometimes abandoned for no obvious reason, as with congeners the Regent and Flame Bowerbirds. **COURTSHIP** Very little known. Some postures are similar to those of Regent Bowerbird, with the head and bill being turned away from the bower. Other postures recall *Chlamydera* bowerbirds, with wings open and head turned to pull the partly fanned tail up, also wing-flicks, leaping up and making sinuous, rather serpentine head movements. Presumed immature males will visit the bowers occupied by adult males, which attend bowers for varying time intervals ranging from 2–3 minutes to 30 minutes at peak construction time. Clear sunny mornings are favoured for bower attendance, early morning and late afternoon being the peak times. **NEST, EGGS & NESTLING** Undescribed. **TIMING OF BREEDING** Insufficient information.

HYBRIDS This species has hybridised, rarely, with Flame Bowerbird in an area of marginal range overlap on the Wataikwa River system, in south-west New Guinea.

MOULT Wing moult is recorded in April, May and September (Frith & Frith 2010).

STATUS AND CONSERVATION Classified as Least Concern by BirdLife International. This is an uncommon and wary species that is seldom seen, inhabiting remote and rather inaccessible areas for the most part. Its wide range should ensure that, at least in the short term, it does not face any imminent threats beyond local forest damage for cultivation and logging.

Masked Bowerbird, male, Arfak Mts, West Papua, October 2017 (*Andrew Livermore*).

Masked Bowerbird, male at bower, Arfak Mts, West Papua (*Ross Gallardy*).

FLAME BOWERBIRD
Sericulus ardens   **Plate 34**

Xanthomelus ardens D'Albertis and Salvadori, 1879, *Ann. Mus. Civ. Genova* **14**: 113. Upper Fly River, 430m, New Guinea.

Other English names Golden Bowerbird, Golden Regent Bowerbird, Yellow-throated Golden Bird

Etymology The species name *ardens* is Latin and means burning or glowing (from *ardere*, to burn), a reference to the species' vivid glowing coloration.

FIELD IDENTIFICATION The male is one of the most striking of the family, being a sunburst-orange and yellow, with black on the wings and tail, not readily confused with any other species. The female is subtler in coloration and less obvious, but the combination of the stocky build and the distinctive yellow underparts should easily distinguish it. Sɪᴍɪʟᴀʀ sᴘᴇᴄɪᴇs None in range. This species overlaps marginally in range with Masked Bowerbird in south-west West Papua, the latter readily told by the black face (male) or by having more mottling on the throat (female).

RANGE South New Guinea from Wataikwa–Mimika Rivers and upper Noord–Endrich Rivers east to upper Fly, Strickland and Nomad Rivers, and to near Ludesa Mission (Mt Bosavi); occurs also in Tarara–Morehead area and inland from Merauke (between Kumbe and Merauke River). Mᴏᴠᴇᴍᴇɴᴛs Little known. Presumed to be resident.

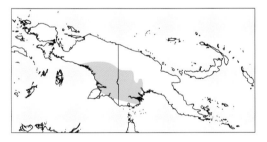

DESCRIPTION Sexually dimorphic. *Adult Male* 25–26cm; two captive males weighed 120g and 140g, respectively. Generally like Masked Bowerbird, but slightly smaller, shorter-tailed, and lacking a black face and throat, with the bill finer, paler and slightly longer. Has crown, nape and mantle rich and dark glossy sunburst flame-scarlet; the elongate deep orange upper-neck plumage forms a small mane, which can be enlarged during display. Chin and throat deep yellow. Mantle becomes increasingly suffused with orange, and mid-back to uppertail-coverts are deep yellow. The upperwing is a brilliant deep yellow, with a small area of black spotting just above the carpal joint (forming a black patch); the uppertail is black, with fine yellow edging on some or all rectrices, and very faint yellow tips (can be lacking). The underparts are deep yellow, but paler than upperparts, and the undertail-coverts more pale and washed out. *Iris* appears large and bright yellow, and is very striking, a prominent feature

in display when size of coloured area changes as the pupil dilates or shrinks; *bill* pale grey; *legs* blackish. *Adult Female* c.25cm; five females 127–168g. Similar in size to male, except for longer tail. Olive-brown above, redder about the head, and orange-yellow below, the chin to upper breast washed olive-brown; slightly darker feather edging on chest feathers (forming only faint scalloping). *Iris* dark brown. *Newly Fledged Juvenile* Has crown down silvery grey, bare face, bill base and throat bright flesh-pink; underpart plumage soft downy white with faintest yellow wash, thereafter like female but paler, bright yellow underpart feathering intruding into softer, white plumage on each side of central sternum. *Iris* dark grey-brown. *Immature Male* Similar to adult female. *Subadult Male* Resembles adult female, but varies with age from having few to almost all feathers of adult male plumage intruding into female-like plumage (Frith & Frith 2009).

TAXONOMY AND GEOGRAPHICAL VARIATION Monotypic. May form a superspecies with Masked Bowerbird, Fire-maned Bowerbird and Regent Bowerbird. Formerly this species was often treated as conspecific with Masked Bowerbird (see below).

VOCALISATIONS AND OTHER SOUNDS Poorly known. The male can be quite noisy when near a bower, often giving harsh raspy *ksssh* notes quite similar to those of Masked Bowerbird, and scolding quiet rasps, as well as a disyllabic querulous *kweyh*, see xeno-canto XC148745 (by P. Gregory). In addition, utters raucous higher-pitched *kwee*-type notes, which can be given as series, and odd rattling notes as at xeno-canto XC140444 (by B. van Balen). Also reported near bowers are a quieter hissing note and a subdued scolding, distinct from the raspy call: see XC112338 (by F. Verbelen).

HABITAT Inhabits lowland and foothill rainforest and also tall secondary forest, including mosaic-type habitat where some trees have been cleared; occurs also in savanna in the Trans-Fly, mainly beneath paperbarks (*Melaleuca*).

HABITS Generally uncommon and shy, this bowerbird is usually seen singly, sometimes in twos or threes, especially at fruiting trees, including figs (*Ficus*). Seen mostly in the canopy and mid-stratum, it sometimes perches up on bare branches for brief periods. It tends to fly quite high, often just above the canopy. Has been seen in mixed-species flocks, but is not a core member of them.

FOOD AND FORAGING Little recorded. This species is known to eat fruit and presumably takes insects. It will associate loosely with *Ptilinopus* fruit-doves at fruiting trees, and can occur in groups of two or three birds. Forages mostly in the canopy and middle storey,

BREEDING BEHAVIOUR Polygynous, promiscuous male seasonally builds and decorates an avenue bower which is very similar in structure to those of Regent Bowerbird and Masked Bowerbird. Presumably female alone performs nest-building, incubation and nestling duties. **Bᴏᴡᴇʀ** The male's bower is of avenue type, neatly constructed, with thicker, shorter sticks in outside bases of walls and finer longer ones within, these thinner sticks almost meeting in an arch over the bower and a very tight fit for the bird, which touches each side when standing within the avenue. Average dimensions of three bowers were 23cm long, 16cm wide and 19cm high externally, with an avenue 17cm long and 8cm wide, and sometimes a slight platform extending out from one end. **Compass orientations** in this small sample varied (355°, 55° and 30°); decorations seem to be generally sparse but can number up to about ten objects, which include blue, red, purple and brown fruits or nuts, purple and white flowers, snail shells and yellow-brown leaves within the avenue, and blackish glossy leaves both in and outside it; decoration theft is unrecorded as yet, but bower destruction can occur (one male destroyed a bower, of which he was himself suspected to be the owner). **Bower-painting** is also known, the 'paint' (which may be regurgitated fruit matter) being applied to inside avenue walls. New bowers may be built within short distances of old ones; they are sited within swamp forest or secondary forest in places where the undergrowth is not too dense and the floor either level or gently sloping. Some are on gentle ridges or located near treefalls where the canopy is thinned out. The birds will tidy up the bowers and remove leaf litter, they give the harsh raspy *kssh* call when approaching them, and may give more hissing calls as well. They are wary and easily scared by movements or noise: a bird at a bower along Ketu Creek (Elevala River) was frightened away by the noise of a camera shutter, and flash usage may also be problematic. **Courtship** Very little known. Courtship was recently observed at Kiunga, in north-east of range, and filmed by the BBC as part of the David Attenborough series *Life Story*. When the male sighted the female, who flew down into the bower, he collected some nearby sticks in his bill and began to dance, initially trying to make himself appear as small as possible, and then to look bigger by 'pumping' himself up (enhancing the effect by spreading his wings and enlarging the orange mane). He gives a wheezing, hissing call series throughout, this becoming deeper-toned as the performance intensifies. Curiously, the male is able to 'pulse' the size of the pupils in the very striking yellow eyes, and this was a part of the preliminary warm-up when he was perched in front of the female, the pupils diminishing to pin-prick size and then pulsing alternately to larger size and creating a very striking appearance. The feathering around the eye is not erectile like the surrounding plumage, so the eye appears in a kind of U-shaped hollow when seen from in front. The male then commences dancing, spreading the yellow and black wings like a matador's cape and slowly waving them while bobbing up and down, sometimes with just one wing extended and shaken rapidly. At times the performance slows and he stretches up with legs extended and wings slowly waved. In the video the male's dancing intensified and he grabbed a blue berry which had been picked up by the female, performing with it in his mouth and coming into the bower to butt his breast against that of the female, the two touching bills as well. If undisturbed this high-intensity display would undoubtedly lead to mating, but in the video a rival male appeared

and the female left. Birds visit the bowers for periods of around 10–15 minutes, most usually in early morning and again late in the afternoon. Bowers are sometimes used for 2–4 weeks and then abandoned for no obvious reason, something reported also for the Regent Bowerbird. **Nest/Nesting, Eggs & Nestling** Undescribed in the wild. Captive birds of this species laid a single egg, which was incubated for 21–22 days, the juvenile feeding independently 50–56 days after hatching. **Timing of Breeding** Little information. Display season in east of range is mostly May–July at Nomad River and Elevala River and August–November near Strickland River.

HYBRIDS This species has on rare occasions hybridised with Masked Bowerbird in an area of marginal range overlap on the Wataikwa River system, in south-west New Guinea.

MOULT Wing moult is recorded for specimens taken in September and December.

STATUS AND CONSERVATION Classified as Least Concern by BirdLife International. A little-known species, inhabiting remote and lightly settled regions. It is not uncommon in the Wataikwa River area in west of range and, to the east, in the upper Fly River region. Historically, people of Fly River wore dried skins of adult males as decoration, with skins traded to the highlands and coast; skins of adult males occasionally decorate vehicle rear-view mirrors in Kiunga area today (pers. obs.). Increased logging activity may be of local concern in this area, but overall the species should be secure.

Flame Bowerbird, male at bower, Elevala River, Western Province, PNG (*Brian J. Coates*).

Flame Bowerbird, male, Elevala River, Western Province, PNG (*Markus Lilje*).

FIRE-MANED BOWERBIRD
Sericulus bakeri   Plate 35

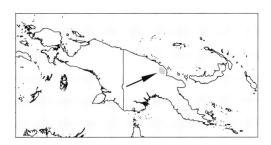

Xanthomelus bakeri Chapin, 1929, *Amer. Mus. Novit.* **367**: 1. 'Madang, Astrolabe Bay,' = Adelbert Mts at *c*.915m, above Maratambu village, north-east Sepik–Ramu.

Other English names Adelbert Regent Bowerbird, New Guinea Regent Bowerbird, Baker's Bowerbird, Beck's Bowerbird, Madang Bowerbird, Macloud Bowerbird, Madang Golden Bird

Etymology The species *bakeri* is named for G. F. Baker (Jr), one of the backers of Beck's expedition and a trustee of the AMNH.

This species was discovered in 1929 by the collector Rollo Beck, though the idea (originating from Gilliard) that he kept the locality secret as he did not want rivals to go there seems to be incorrect and is simply an artefact of the type of labelling which he did (LeCroy & Diamond 2017); hence the type was listed as originating from Madang.

FIELD IDENTIFICATION The male is a very striking bird and perhaps momentarily reminiscent of Golden Myna (*Mino anais*), appearing quite myna-like in its structure as well. It shows a broad yellow band across the flight-feathers when seen from below. The female is subtle but equally distinct, the only similarly sized sympatric species that has similar barred underparts being Magnificent Bird of Paradise, which has much darker upperparts, a blue-grey bill and a flattened head shape. **SIMILAR SPECIES** Golden Myna and female-plumaged Magnificent Bird of Paradise, as detailed above.

RANGE Endemic to a small area of PNG. This species is found locally in the Adelbert Mts of north-east New Guinea, north-north-west of Madang, in a narrow elevational band between 900m and 1,450m. There are intriguing reports from an experienced local guide of a bowerbird resembling this species at similar heights on Mt Bosavi, and this would be well worth following up. **MOVEMENTS** Presumably sedentary.

DESCRIPTION Sexually dimorphic. *Adult Male* 27cm; weight 178–183g. Has a rich deep glossy flame-scarlet crown finely spotted and smudged black (due to

semi-concealed black centres of feathers), while the vivid and striking nape and extensive highly glossy filamentous silky mane or cape (arising from the lower nape) are flame-scarlet at base to mostly deep orange near tips. The upperwing and tail are black, the deep yellow bases of the primaries and secondaries forming a large and conspicuous yellow wing patch, which shows as a band across the feathers of the underwing when seen in flight. Entire rest of plumage glossy jet-black, this extending in a narrow band over eye and lower lores to base of upper mandible. *Iris* pale yellowish and quite striking in the dark face; *bill* blue-grey with extensive black tip; *legs and feet* blackish. *Adult Female* 27cm; weight 164–184g. Similar in size to male, except for a longer tail. Generally brownish-olive above, a little paler on head, and paler below. The underparts are dirty whitish, washed light buff, the chin and throat finely barred or scalloped, otherwise broadly barred with variable dark brownish. *Iris* and *bill* dark brown; *legs and feet* brownish to lead-grey. *Juvenile* Undescribed. *Immature Male* Similar to adult female. *Subadult Male* Variable, ranging from some like adult female but with a few feathers of adult male plumage intruding to others which are almost wholly like adult male, these latter presumably the older individuals.

TAXONOMY AND GEOGRAPHICAL VARIATION Monotypic. May form a superspecies with the congeneric Flame Bowerbird, Masked Bowerbird and Regent Bowerbird.

VOCALISATIONS AND OTHER SOUNDS Poorly known. Sharp rasping notes: see xeno-canto XC267869 (by F. Lambert), bird not seen but call identified by local landowner who knows this species better than anyone. An adult male near his bower gave short churrs and mimicked the call of Magnificent Riflebird and Magnificent Bird of Paradise. Nest-building female gave quite loud *k-zzz k-zzz* or *tk-sss tk-sss* from above nest site and nearby trees, and repeated a piping note several times. One call likened to (harsh and rasping) song of Tan-capped Catbird. Other notes said to resemble the hissing and rasping vocalisations of Yellow-breasted Bowerbird.

HABITAT Hill forest, also second growth at rainforest edges, from 1,200m to 1,450m, occasionally down to 900m.

HABITS Encountered singly, in twos, or in small groups which, as is usual with this family, contain many more individuals in female-like plumage than adult males. This species sometimes perches quite prominently on high branches. Often, however, it is rather shy, and is most easily seen when attending fruiting trees.

FOOD AND FORAGING Primarily frugivorous, taking fruits including figs (*Ficus*) and berries, but will also take arthropods, including ants (Formicidae). This species forages singly or in small groups from middle storey up to canopy; it visits fruiting trees in native gardens, including one right by a landowner lodge (Keki Lodge, in the Adelbert Mts).

BREEDING BEHAVIOUR Little known. Presumably polygynous, promiscuous male seasonally building and decorating avenue bower, and female building and attending nest alone. **BOWER** Male builds a small avenue bower like that of congeners; the average external dimensions of five bowers were length 28cm, width 26cm and height 33cm, the avenue being 14cm long and 7cm wide, and mean **compass orientation** 251°. Bower decorations can number up to *c*.50 and include blue and purple fruits (*Dianella ensifolia* and *Elaeocarpus* noted) and a yellowish-brown leaf. The first bower ever described (in 1986) was sited on a gentle ridge slope *c*.60m below its crest in slightly disturbed primary hill forest, the broken canopy permitting more light to reach the bower than is typical elsewhere

in the habitat; a second bower had no low foliage canopy above it. Bower-painting is as yet unrecorded for this species. Bowers are known to be attended in September and December, though for many years this species was thought not to build a bower, and it remains little known. **COURTSHIP** Undescribed. One adult male called at his bower in September. **NEST** Little known. It appears that only one nest has been described, this built about 15m above ground in the crown of a small epiphytic fern on a semi-dead tree near a track. A female-plumaged individual brought dried leaves, vines and twiggy sticks to the nest. **EGGS & NESTLING** Undescribed. **TIMING OF BREEDING** Few data. One nest in September and female with enlarged ovary in February.

HYBRIDS None known.

MOULT No information.

STATUS AND CONSERVATION Classified as Near Threatened by BirdLife International; previously considered Vulnerable. This restricted-range species has one of the smallest areas of occurrence in New Guinea, being present only in the Adelbert and Huon Ranges EBA. It is generally uncommon but can be locally fairly common, although perhaps sparser at the lower end of its altitudinal range. The total extent of its range is estimated at about 570km², and the current global population is put at fewer than 10,000 individuals. The Adelbert Mts do not have a high human-population density, but the elevational band in which this bowerbird occurs is within the local agricultural zone, where subsistence hunting of birds and mammals occurs, although at least in some places this species is not hunted. Primary forest has been extensively modified or cleared, but mature secondary forest appears to be suitable for this bowerbird. Much of the range of this striking species is comparatively inaccessible, and therefore less vulnerable to logging, but with increasing human populations some clearance is inevitable. It is to be hoped that small-scale clearance for garden agriculture will not pose a major problem, but the seeming disappearance of Wahnes's Parotia in the same area gives some cause for concern.

Fire-maned Bowerbird, Keki, Adelbert Mts, PNG (*Steve Wood*).

Fire-maned Bowerbird, in flight, Keki, Adelbert Mts, PNG (*Steve Wood*).

REGENT BOWERBIRD
Sericulus chrysocephalus   Plate 35

Meliphaga chrysocephala Lewin, 1808, Patterson's River [= Hunter River], New South Wales, Australia.

Other English names Regentbird, Australian Regentbird, Australian Regent Bowerbird, Golden Regent, Golden Regentbird

Etymology *Chrysocephalus* means golden-headed, derived from the Greek *khrusos*, gold, and *kephal*, head. An artist called Skottowe named the bird 'The Regent' as he obtained a specimen on the day when the news of the removal of the Regency Restrictions on HRH the Prince Regent (the future King George IV of the UK and Ireland) reached him. This prince was notably fond of the colour yellow, often sporting a yellow waistcoat, hence the glowing yellow Regent Parrot (*Polytelis anthopeplis*) and profusely yellow-spotted Regent Honeyeater (*Anthochaera phrygia*) also were named in his honour. This was the second species in the family to be discovered, and the first from Australia.

FIELD IDENTIFICATION This is a distinctive, rather delicate bowerbird, being slighter with a finer bill than others of the family, with a particularly striking black and golden-yellow male plumage. The female also is distinctive in a subdued way, being barred dark below with whitish spots on the mantle, as well as a small dark cap. It is one of the best-known and most distinctive of the Australian bowerbird species, becoming remarkably tame at a few sites. **SIMILAR SPECIES** The male is distinctive and cannot be confused with any other species. The female is less obvious but still distinctive, and unlikely to cause confusion. Although the pattern of the female plumage is superficially similar to that of the Satin Bowerbird, the Regent is a far smaller and more delicate species with a quite different shape, being half the weight of a Satin.

RANGE Endemic to the coastal ranges and foothills of central eastern Australia, with one isolated population in the Connors and Clarke Ranges of the Eungella Plateau inland from Mackay, then a gap across the Fitzroy River basin inland of Rockhampton before the main distribution, which extends through south-east Queensland and east New South Wales to just north of Sydney. **MOVEMENTS** This species is basically

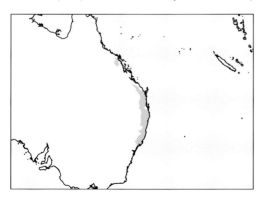

resident. Localised movement to lower altitudes may occur during winter months; individuals in female-like plumage are said to make such movements 3–4 weeks earlier than adult males. Of 22 marked individuals recaptured, the longest distance travelled was 4km. Extremely dry and hot conditions can cause some sporadic occurrence outside the core range, but this is very rare.

DESCRIPTION Sexually dimorphic. *Adult Male* 24cm; weight 76–110g. One of the most striking of the family, with a dramatic jet-black and golden-yellow plumage which takes seven years to attain. The crown, nape and upper mantle are bright golden-yellow, as are the tertials, secondaries and most primaries, contrasting with the jet-black outer primaries, mantle, rump, tail and underparts. Has a variably sized patch of orange on the forecrown, not quite touching the base of the bill. The face, chin and throat are black, black extending in a narrow line from the lores to just above the eye, where it meets a wedge of yellow cutting across to the eye. *Iris* pale yellow; *bill* orange-yellow; *legs and feet* blackish-brown to grey. *Adult Female* 25cm; weight 91–138g. Complex subtle plumage very different from that of male. The upperparts are dull olive-brown, with the mantle spotted off-white, and the head finely marked with streaks and mottling and a dull blackish central crown patch; underparts are broadly barred blackish-brown, with a dark central throat line. *Iris* dark brown; *bill* blackish; *legs and feet* grey. **Immature Male** Initially resembles female, gradually acquiring adult plumage; iris and bill coloration gained mostly during fourth and fifth years, accounting for female-type birds with yellow eyes and orange-yellow bill. *Subadult Male* By sixth year plumage is variable, from like immature male with few feathers of adult plumage intruding, to like adult male with few feathers of immature plumage remaining. Fully adult plumage is gained in the seventh year. (F&F 2009)

TAXONOMY AND GEOGRAPHICAL VARIATION Monotypic. Form *S. c. rothschildi* Mathews, 1912, Blackall Range, south Queensland, was proposed on basis of a supposedly richer orange crown of male, but studies indicate no constant geographical difference in this character (and Mathews was a renowned arch splitter).

VOCALISATIONS AND OTHER SOUNDS This is one of the least vocal of the bowerbirds except when at the bower. When displaying, the male gives a lengthy series of harsh grating notes typical of the family, and also a low and soft but continuous rattly subsong; see xeno-canto XC311771 (by A. Spencer). Mimicry of Spectacled Monarch (*Symposiachrus trivirgatus*) and White-browed Scrubwren (*Sericornis frontalis*) is reported, but mimicry is nothing like so frequent as with the Tooth-billed Bowerbird. Soft coughing noises are reportedly made by individuals feeding on figs.

HABITAT Inhabits subtropical rainforest, associated sclerophyll woodland, and more open habitats, including cultivated country and urban gardens. Occurs from sea level to 900m, altitudinal limits varying across range. Traditional bower sites are dispersed through appropriate ridgetop habitat, and tend to be within rainforest on flat or less sloping ground, with adjacent liana-thicket cover available for concealment and protection. Some very local winter dispersal to adjacent habitats such as shrubland and low woodlands with *Banksia* or tea-trees. This species forages at the edge of rainforest habitats and in tall, fairly open wet sclerophyll forests; it will visit modified habitats such as farmland with fruiting trees, orchards and suburban parks and gardens.

HABITS Regent Bowerbirds can be quite shy and elusive in the forest, but they may also become habituated to human activity and are tame and readily seen at feeding stations where fruit is available, as at O'Reilly's at Lamington NP, Queensland. Here they can often be seen with Satin Bowerbirds, and occasionally a Green Catbird makes an appearance. The species is recorded as sunbathing occasionally, and often perches prominently atop tall pines or other emergents to catch the early-morning sun, but tends to be inconspicuous otherwise. Not much is known about predator-avoidance strategies; when frightened, captive juveniles 1–2 months old adopted a frozen posture, drawing themselves rigidly upwards.

FOOD AND FORAGING Primarily frugivorous, but diet also includes buds, petals, nectar from flowers, seeds, leaves, and arthropods (mostly insects). The slender delicate bill may be an adaptation for this kind of less robust diet. Leaves may be eaten over the July–October non-breeding winter and early-spring period when food may be scarce. The diet of nestlings consists of fruits and insects, which include cicadas (Cicadidae), caterpillars, katydids (Tettigoniidae) and beetles (Coleoptera), cicadas representing a significant dietary component. This species tends to forage singly or in twos or small flocks, occasionally in larger aggregations of up to 70 in winter. Such flocks comprise mainly 'brown birds', being females or immatures, although adult males may join and are sometimes seen in flocks of up to 12 individuals. It will also on occasion form foraging flocks with Satin Bowerbirds and Pied Currawongs (*Strepera graculina*). The Regent Bowerbird forages mostly in the upper levels of trees, catching arthropods primarily by gleaning and hawking. It often forages alongside other bowerbirds and other fruit-eating bird species, in winter including such frugivores as Satin Bowerbirds, Green Catbirds, Australian Figbirds, Olive-backed Oriole (*Oriolus sagittatus*), Wompoo Fruit-doves (*Megaloprepia magnifica*) and *Ptilinopus* fruit-doves, and it will displace other species from fruiting trees. Females dominate males there, but the two sexes associate quite well at fruit feeders. Regent Bowerbirds come regularly to bird-feeding sites to take fruit, O'Reilly's at Lamington NP being a famous location, and this species indeed being a main part of the corporate logo here. The visiting birds are tame during the austral spring and will even land on the heads and shoulders of people, though the males seem to be the more confiding. They tend to be inconspicuous otherwise, but have the habit of perching prominently on tall pines or other emergents in the early morning, when they catch the sun.

BREEDING BEHAVIOUR As with its relatives, the polygynous, promiscuous male builds and decorates a small avenue bower, while the female builds and attends the nest alone. This species is non-territorial, except for the defence of the bower site by the male, which may attend the same site for at least three seasons. The mean nearest-neighbour distance for 24 traditional bower sites in Sarabah Range, South East Queensland, was 195m. This species has an unusually restricted short period of bower attendance, often around ten days, and this is one reason why the bowers are relatively seldom seen. The female may first breed in her third year, but more often in the fourth year. **BOWER** This species constructs a sparse, frail avenue of sticks (sticks touching each other often become fused by whitish fungus). The mean dimensions of 33 bowers (Sarabah Range) were length 23cm and width 20cm wide externally, with the avenue 8cm wide and 18cm high, and the average deviation of 44 avenues from N–S **compass orientation** was 38°. The male spends only *c*.3% of daylight time at the bower site; bower decorations (up to 30 or so) include green leaves, pale flowers and petals, seeds, fruits, snail shells, cicada-nymph exoskeletons, and man-made blue items. This is another of the species that 'paints' the bower, using masticated fruit mixed with saliva. Remarkably, the male sometimes paints on this concoction by using leaves, one of the few avian instances of a form of **tool usage**. Theft of bower decorations is known to occur, as is quite common with the family. Rather curiously, the bower structure itself is often short-lived (ten days or fewer), as it is regularly destroyed by rival males, or by the owner himself when a rival is known to have found it. A new bower takes *c*.3–4 hours to build and is usually beneath a different thicket, although within the same general site. **COURTSHIP** The male attracts females in the upper forest canopy by means of his bright plumage (rather than, as with many other bowerbirds, by vocalisations), and he then leads them to the bower site. Male courtship consists of three display elements, the *Initial Bower Display* at the avenue entrance to bower, the *Peripheral Bower Display* around the bower, and a *Central Bower Display* within the bower avenue. Display postures include wing-flicking, gaping and nape-presentation postures, while the male also charges at the female and makes brief vertical 'flights', all designed to show off and enhance his striking plumage coloration. Bowers are well hidden under overhanging foliage and are remarkably hard to find, as well as being attended for only a short period of 2–3 weeks during the mating season, unlike the behaviour of other Australian bowerbirds. The males spend only short amounts of time there, around 3.2% of daylight hours in a study at 33 active bowers in the Sarabah Range (Lenz 1994). The distinctive buzzing calls given by the males are one of the best ways of locating the bowers. **NEST/NESTING** The nest, a frail shallow saucer of loosely placed sticks, an egg-cup lining of finer twigs and occasionally a few leaves, is cryptically sited 2–31m above ground in dense foliage of a clump of vines, mistletoes (Loranthaceae) or other plants; mean distance of four active nests from nearest active bower was 266m (closest 20m). Females approach the nest very cautiously. **EGGS, INCUBATION**

& Nestling Clutch one to three eggs, mostly two, in captivity laid on alternate days; in captivity, replacement clutch laid in new nest after failure. Incubation period is *c*.17–21 days. Nestling period 17 days. **Nest Success** The overall success of seven nests in the Sarabah Range was 43% (Lenz 1994). **Timing of Breeding** Season September–February, peak in egg-laying November–December, fledging late February to early March, nesting cycle *c*.3–4 months. Male's display season September–January on Sarabah Range, earlier at some other localities. **Longevity** The maximum recorded longevity of a male was at least 23 years.

HYBRIDS There are three documented cases of intergeneric hybridisation with Satin Bowerbird, the hybrid being known as **Rawnsley's Bowerbird**. The first was a specimen from near Brisbane in 1867, followed by one seen and photographed in South East Queensland in 2003 (Blunt & Frith 2004) and one in New South Wales in 2014. The specimen was described as being in adult male plumage, mainly the glossy blue-black colouring of the adult male Satin Bowerbird (Blunt & Frith 2004), but with a conspicuous and extensive yellow wing patch, yellow tipping to some tail feathers, and a paler iris colour than the Satin Bowerbird, and intermediate in size between the two putative parent species.

MOULT Vellenga (1980) noted that nestlings begin to moult their down in December–January and attain first-year plumage by a post-juvenile moult in January–March. Second-year plumage follows the first annual moult, which starts in about July and is finished by the end of the year. Annual moult thereafter occurs in November–March for males and non-breeding females, but for breeding females is mainly during February–May.

STATUS AND CONSERVATION Classified as Least Concern by BirdLife International. Restricted-range species: present in Eastern Australia EBA. This bowerbird is locally common to moderately common in the larger forest blocks, but uncommon in small remnants. Its numbers have diminished in several areas as a result of the loss and degradation of its habitat. In Queensland, this species was known to occur around Rockhampton in the early 20th century but is now long gone, and numbers at Nanango, in South East Queensland, have greatly declined following habitat modification, with vine forest replaced by Hoop Pine (*Araucaria cunninghamii*) plantations. It was formerly reported as being common about Sydney but had largely gone by the 1880s, and the southern limit of the range is now the Gosford/Hawkesbury River area of New South Wales. In the north of the main range there is a break in the species' distribution from near Casino to Woolgoolga, with small and now isolated populations at Iluka and Washpool. Its absence from the Comboyne Plateau and Hunter valley (New South Wales) are probably the result of the clearing of rainforest. Adult males were once hunted for mounting as household novelties, and these striking birds were common in Victorian and Edwardian cabinets of stuffed and mounted bird specimens. They were also subject to hunting for the millinery trade in London in the early years of the 20th century. Most populations of the Regent Bowerbird appear fairly stable today, and there are reasonable populations within some national parks, such as Lamington NP.

Regent Bowerbird, female, Lamington NP, Queensland, Australia (*Jun Matsui*).

Regent Bowerbird, male, Lamington NP, Queensland, Australia (*Phil Gregory*).

Genus *Chlamydera*

There are five species in the genus *Chlamydera*, with three endemic to Australia, one found in both Far North Queensland and New Guinea, and one only in New Guinea; all have a distinctively spotted albeit dull plumage, diffuse streaking on the throat and chest, lack marked sexual dimorphism, and build variants on an avenue-structure bower with a basal platform, which is far more substantial than the small bowers of the *Sericulus* or silky bowerbirds. All the *Chlamydera* species have ten primaries, 14 secondaries (including 5 tertials) and 12 rectrices.

Etymology The name *Chlamydera* is from the Greek words *khlamus*, meaning a short cloak, and *dera*, the neck, referring to the lilac-pink erectile nuchal crest of the Spotted Bowerbird, the first to be placed in this genus.

GREAT BOWERBIRD
Chlamydera nuchalis   **Plate 36**

Ptilonorhynchus nuchalis Jardine and Selby, 1830, *Ill. Ornith.* **2**: pl.103. No locality = Cobourg Peninsula, Northern Territory.
Other English names Great Bower-bird or Bower Bird, Great-Grey Bowerbird; Queensland Bowerbird or Eastern Bower-bird (race *orientalis*); Western Great Bowerbird (for the synonymised race *oweni* of Mathews).
Etymology The species name *nuchalis* is Latin, meaning of the nape (nuchal).

FIELD IDENTIFICATION This is the largest of all the bowerbirds and is distinctive, appearing slender and elongated, with a small head, a beady dark eye and quite a long neck, and 'trousered' thighs above the long legs. Appears grey-brown and spotted above, with broad well-fingered wings and paler flight-feathers in flight, which is rather undulating. SIMILAR SPECIES Fawn-breasted Bowerbird is considerably smaller and much less grey-brown, with rufous on the underparts, and overlaps only in Cape York. Spotted Bowerbird is smaller still, and much more richly coloured, with large buffy spots above.

RANGE This species is endemic to tropical northern Australia, where it is found in North Queensland (mainly north of 21° south), in the north of Northern Territory and in northernmost Western Australia in the Kimberley Division. It is known from a number of offshore islands such as Groote Eylandt, Melville Island and the Sir Edward Pellew Group, all in Northern Territory, and small islands off the Kimberley coast. MOVEMENTS Sedentary throughout the breeding

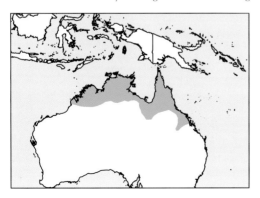

season. At other times more nomadic: has occurred as a vagrant at Trephina Gorge, near Alice Springs, and at Tennant Creek, Northern Territory, well south of the usual range, and there are also rare sporadic records from Cairns, in the humid tropics.

DESCRIPTION Sexual dimorphism greatly reduced. *Adult Male* 32–37cm; weight 180–265g (*c.*200g). Has dull greyish head, the crown more silver-grey and sides of head and neck more brownish-grey. A small erectile nuchal crest is an unusual mauve-pink in colour and surrounded by silvery-white feathers, but it is usually concealed except in display or excitement; when erect it appears somewhat like a lavender daisy on the back of the head. The upperparts are entirely brownish-grey, the feathers tipped paler grey and giving a distinctive spotted appearance, the uppertail-coverts with extensive dark barring; the primary-coverts and flight-feathers are paler brown with pale grey outer edges and tips; tail dusky brown with a broad off-white tip. The underparts are buff-grey, paler on the underwing and with cream-buff bases and centres on the underwing-coverts; the undertail is a pale dull creamy grey, with many narrow dusky-brown bars on undertail-coverts. *Iris* dark brown; *bill* black, *gape* grey-black to brownish-yellow; *legs and feet* olive-brown to olive-grey. *Adult Female* (weight 153–215g) and *Second- or perhaps Third-year Immature Male* are difficult or impossible to distinguish in the field. The plumage is similar to that of adult male, with faint barring on the abdomen, flanks and undertail-coverts. The mauve-pink nuchal crest is either lacking or greatly reduced to just a few feathers. Female is also somewhat smaller (notably in wing length) and less heavy than adult male. *Juvenile* Plumage is softer, looser and fluffier than adult's, with pale buff spots on crown, back and rump, throat and breast dark grey, becoming paler on the abdomen with narrow dusky barring on the flanks. The rectrices are much narrower and more pointed than those of adults. *Iris* brown; *bill* greyish-black. *First Immature* Retains narrow pointed juvenile rectrices and exhibits a contrast between worn secondary-coverts and the fresh inner coverts. *Subadult Male* Has less conspicuous barring on breast, abdomen and flanks, and a small nuchal crest.

TAXONOMY AND GEOGRAPHICAL VARIATION There is considerable variation across the range, western birds being larger, paler and more uniformly coloured than those to the east. Two subspecies are recognised.

1. **C. n. nuchalis** (Jardine and Selby, 1830); synonym *C. n. oweni* Mathews, 1912. Occurs from the Kimberley region of north-west Western Australia across Northern Territory to the south-west corner of Cape York Peninsula, and occurring on Melville and Groote Eylandt, where both populations show variations in colour and size. Intergrades with the following race at the south-west corner of the Gulf of Carpentaria, in north-west Queensland.

2. **C. n. orientalis** Gould, 1879; synonym *C. n. yorki* Mayr and Jennings, 1952. Ranges from much of Cape York Peninsula south to the Mackay area around 20° south, but avoiding the more humid coastal areas around Cairns and much of the eastern watershed of the Wet Tropics from Cardwell to Tully, though common at Townsville in the drier zone. This taxon has more obvious black and white variegations on the upperparts, especially the forecrown, and the females show more barring below.

VOCALISATIONS AND OTHER SOUNDS A vocal species, the Great Bowerbird utters a great variety of harsh discordant scolding calls, with rasps, cackles, hisses, whistles and churrings, and a harsh advertising song. Calls of an individual of race *orientalis* at the bower are archived on xeno-canto at XC326726 (by P. de Groot Boersma), while the song of this race is at XC158481 (by F. Deroussen), and typical scolds of two birds of nominate race at XC237185 (by N. Jackett). This species is vocal only during the display season and near the bower (F&F 2004). The peak display times near Darwin are in July onwards, when the male sings intensely for about five weeks but is silent during the construction or bower-maintenance phases. Females begin to visit the bowers and he displays, with vocalisations, some three or four times a day, often singing close to the bower at these times. At a bower near Townsville the male was present for about 51% of the day, of which some 9% was spent vocalising (Veselovsky 1979). Loud hissing notes are given during the peripheral display, and ticking notes when showing off the nape coloration. **Mimicry** seems to be used when some kind of threat is perceived, as well as when advertising the bower. Both sexes are accomplished mimics, of both avian and other sounds, with a long list of species mimicked which includes Wedge-tailed Eagle (*Aquila audax*), Black Kite (*Milvus migrans*), Whistling Kite (*Haliastur sphenurus*), Pacific Baza (*Aviceda subcristata*) and Laughing Kookaburra (*Dacelo novaeguineae*). Also a noted mimic of ambient sounds such as the barking of dogs, cat meows, squeaking gates, crumpling paper, farm machinery and human voices. Females are noted as using vocal mimicry to lure predators away from the young, using the alarm calls of such species as Blue-faced Honeyeater (*Entomyzon cyanotis*).

HABITAT Inhabits the edges of rainforest, vine thickets, riparian woodlands, open savanna, eucalypt and *Melaleuca* woodlands, and occurs also in suburban gardens and mangroves. Ranges from sea level to around 600m, in areas with annual rainfall between 500mm and 1500mm. Seldom found far from water, which seems to be significant for this species in terms of both its food sources and its propensity for bathing.

HABITS Often seen singly, sometimes in twos or small groups, but outside the breeding season loose aggregations of up to 20–30 individuals may be seen at favoured food sources, such as orchards and gardens. Not unduly shy, and can become quite confiding in some settings, such as caravan parks and gardens, and known to raid food from homes at times. Known also to take keys, nails, shiny objects, jewellery etc., which are sometimes retrieved from the bower. This bowerbird has a liking for bathing in water and will bathe under lawn sprinklers. It is known to drink from birdbaths, springs, puddles etc.

FOOD AND FORAGING This species is primarily frugivorous, as with congeners, figs (*Ficus*) being an important dietary component, as also are the fruits of the native and widely planted Carpentaria Palm (*Carpentaria acuminata*). Exotic species, such as the fruits of Neem (*Azadirachta indica*) are utilised, plus nectar from flowering trees, seeds and flowers. This bowerbird also consumes insects and spiders, and is known to take worms and small lizards, and in the past was reported as eating hens' eggs. Nestling diet includes far more animal matter (see below). Males will store surplus fruit in the forks of shrubs above the bower (F&F 2004). Great Bowerbirds will take meat scraps, and have been known to pilfer dough from outback ranch verandas, as well as breaking the cap of milk bottles to get at the cream (Chaffer 1984). When they become a nuisance around dwellings the birds may be illegally killed, but one hopes that this is now rare. They glean fruit and insects much as congeners do, and will hold fruit against a branch as they extract the seeds to eat. This species forages in fruiting trees, especially figs, but also in shrubs and on the ground. It is seen mainly singly or in twos and threes, but can form small flocks or aggregations of up to 30 birds at orchards and gardens where food is plentiful. It often forages with other opportunist frugivores such as Black-faced Cuckooshrike (*Coracina novaehollandiae*), White-bellied Cuckooshrike (*C. papuensis*), Australian Figbird (*Sphecotheres viridis*) and Red-winged Parrot (*Aprosmictus erythropterus*).

BREEDING BEHAVIOUR Polygynous, promiscuous male builds an avenue bower at which he displays, with different females visiting. Female builds nest and undertakes all nest duties alone, as is usual with the bowerbirds (but not catbirds). Males spend up to 63% of their time at the bower during the display season, and are heavily involved in maintenance and the removal of fallen leaves from in and around it. Repairing damage and painting can take up to 39% of the daily routine, with bower-building also significant. The owner may destroy some bowers, after a rival has discovered them. Density of bowers depends on the local circumstances, but at Townsville there were 21 simultaneously active bowers in an area of some 6km² (600ha/c.1,480 acres), with an average separation of about 790m; a study of some 28 bowers in Darwin found none closer than 21km to the next (Veselovsky 1979). **BOWER** The male constructs an avenue bower atop a platform of sticks, with old bowers often nearby, although with this species the same bower is often refurbished from year to year, rather than built anew.

The bower can be quite conspicuous and may be in a quite public location (e.g. school grounds, public parks, car parks, etc). Bowers are often used over a period of years if undisturbed, and there is some variation in how they are constructed, some having ill-defined and poorly decorated central depressions and larger sticks than usual (maybe made by less experienced birds?). Immature birds gather at rudimentary bowers that are just a platform of sticks or the base of a tree trunk, or a poorly made avenue, often near the bower of an experienced male; they will also attend the bowers of such males in their absence, particularly before and after the peak display season (Wieneke 2000). The males build a large two-walled avenue-type stick bower measuring about 50–100cm long, 15–25cm wide and 30–40cm tall, with the walls about 20cm thick; there is regional variation in bower measurements, Darwin bowers in one study averaging 10% shorter, 15% wider and with walls 19% thicker than Townsville ones. The bower is built on a platform of sticks up to 2m long and 1.5m wide, usually orientated north–south along the path of the sun, and sited in shady cover under a dense bush or shrub or on the fringes of thick vegetation. The **bower orientation** may be aligned to give the best angle of illumination of the display objects. The quantity of sticks used is considerable, with 4,000–5,000 at some Darwin sites, averaging 29.2mm long and 2.4mm thick in the lower wall. Bower-construction sites are cleared of leaves, twigs and stones before construction begins. Some males took three weeks to make a rough bower at the start of the season; others have completed new ones within the original was destroyed in 2.5 days (Veselovsky 1979). A bower in Kakadu NP which was burnt had the shell decorations moved on the next day to the new bower site 40–50m away (Breeden & Wright 1989). There is a saucer-like central depression, and occasionally the curving side stick walls can meet to form a tunnel, while often the tips of the sticks will meet above the centre of the bower. Occasionally a bower may be entirely out in the open without nearby vegetation cover, though still under a larger tree, from which the bird will vocalise. Bowers are often sited near water sources such as creeks or billabongs, but occasionally they can be built along the crest of beaches in mangroves or other vegetation. The Friths' Townsville study found bowers used for a mean of 4.4 years, with seven used for up to ten years, and many are active for upwards of 18–20 years, maybe longer. The same bowers may be refurbished, or new ones built within the bower site. Some sites are known to have 8–11 old bower structures adjacent to the active bower, and the birds appear very site-faithful, sometimes remaining even if the site has been cleared of vegetation. The floor of the avenue is often covered with **display objects**, such as small white stones and bleached bones. The two platform areas outside the bower are also utilised and can have extensive collections of objects, often dozens to hundreds of bleached white snail shells or, in coastal regions, seashells and coral. One end usually contains more decorations than the other, and up to 1,000 objects are known from some large well-established bowers. White predominates, but green objects are also used and seed

pods, berries, leaves, flowers and fruit are utilised, as also are man-made items such as green glass, cloth or metal. Grey, red and silvery items are also used at times, but most bright colours are avoided. Used in the Friths' Townsville study of five bowers were 120 white objects, 94 green, nine red and four lilac. Curiously, males may use lilac-coloured objects in their nape-presentation display. One well-known bower in the grounds of Mt Molloy primary school, in North Queensland, had a green plastic hand grenade in its trove! Bowers close to human habitation often contain plastic, pens, glass, metal and gun cartridges. In such locations unsympathetic council-flowerbed maintenance may disrupt them, and human vandalism does occur occasionally. In Northern Territory, bowers seem to have a predominance of white objects, followed by green, with a few odd items of other colours, which seem to set off or contrast and complement the main colour scheme; for example, several bowers in 2017 had just single small red plastic items in the main platform of white and green. In contrast, **vandalism by rival birds** is frequent, neighbouring males pilfering objects and damaging bower walls when the owner is away. A video of this behaviour by N. Doerr is at the IBC site. The Darwin study by Veselovsky found that the total weight of bower objects ranged from 6.2kg to 12.1kg, individual objects weighing from 0.2g to 40g and the numbers of decorations on bower sites ranging from 5,000 to 12,000 items. **Bower-painting**, well known but seeming not to occur in all areas, involves the male wiping its bill up and down the inner-avenue twigs with a sticky substance held in its bill and the mandibles glistening with saliva (Warham 1962a). The 'paint' consists of masticated vegetable matter, sometimes with kangaroo or wallaby dung, mixed with saliva and applied with the bill. This forms a discoloured tea-brown stain (similar staining occurs also with Fawn-breasted, Spotted and Satin Bowerbirds). This species provides one of the few avian instances of a kind of tool usage, being one of a very select few known to use material to paint that can qualify as a kind of tool. It utilises a piece of vegetable matter as a sort of brush, as do both Fawn-breasted and Satin Bowerbirds. In one three-month study at Darwin birds were not seen to paint their bowers, but bower-painting is common around Townsville. A study by Kelley & Endler (2012) revealed that the males use **visual illusions** when constructing their bowers. They do so by arranging the objects covering the floor of the court in a particular way, so that they increase in size as the distance from the bower increases. This positive size–distance gradient creates a forced perspective, resulting in false perceptions of the geometry of the bower, which is visible to the female when she is standing in the avenue. From her point of view, all of the objects in the court appear to be of the same size. When the decorative objects in the bower were moved to reverse the gradient (i.e. with the large objects placed closest to the bower and the smaller ones farther away), within three days the birds had rearranged the objects to restore the original pattern and thus the positive gradient. The researchers were unsure about the function of the illusion, but speculated that the female, by making the court appear smaller, may perceive the male to be bigger, hence it may be important for the female's selection of a suitable mate. This study, which involved the observation over a two-month period of bowers built by 20 males in Queensland, used motion-sensitive digital video recorders. By so doing, the researchers collected over 1,600 hours of footage, containing 129 courtship displays and 23 matings supplemented by still photographs for later analysis. These data revealed that the geometry of the bower was directly related to the mating success of its builder. The most successful males were the ones that had arranged the objects to form the most regular patterns on the floor of the court and which enhanced the strength of the forced perspective illusion. This was achieved by males repeatedly going in and out of the avenue, to rearrange objects to obtain the desired effect. Whether this results in an improvement in the illusory elements of the design is still under investigation. The illusion created by a regular pattern may make both the court seem smaller and the male's displayed objects stand out more (as a regular background pattern is less distracting). During his courtship display, the male waves objects towards the female, causing their apparent size to increase. The researchers suggest that this 'Ebbinghaus illusion' could enhance the apparent size increase, making the display objects even more conspicuous to the female. Further investigation and analysis of the video footage will seek to confirm and quantify this effect. A recent study by N. Doerr (2018) has shown that the more submissive non-alpha males may choose a specific tree trunk, similar in diameter to the width of the normal bower, and then clear the ground and decorate around it as they would with a normal bower. This is less liable to be raided and destroyed than a bower would be. It is also possible to choose such a site much closer to traditional bowers than would usually be tolerated. This has the great advantage of allowing passing females the chance to visit and perhaps permit opportunistic mating, and also allows practice of display performances, another step in the evolutionary arms race for passing on genes. **COURTSHIP** Courtship displays occupied about 14% of the time, while perching silently and preening, bill-wiping or looking about took 17% (Higgins *et al.* 2006). A remarkable video by Natalie Doerr of a male displaying and then copulating at his bower is on the IBC site http://www.hbw.com/ibc/video/great-bowerbird-chlamydera-nuchalis/male-displaying-and-attempting-mate-visiting-female, as is another showing a failed mating attempt which results in a conflict at the bower. The male advertises at his bower by giving a harsh discordant song (see XC158481 by F. Deroussen), and when a female is attracted two main displays are involved, which can be performed in any order. In the **Central Display**, the male raises himself up to his full height, with the mouth open and the tongue flicking in and out, and often calling. He stretches forwards with the nuchal crest partly open, and may offer green fruit to the female, and pick up other ornaments; he may high-step forwards during this action. He also pecks at the end wall of the bower and presents his nape by lowering and twisting his head to show the mauve-pink crest; during the nape presentation he may droop

his wings and repeatedly pick up and throw down bower decorations, usually from the platform end that has the most such, generally the northern end. During the **Peripheral Display** the male struts around the perimeter of his bower with his wings drooped, the tail outstretched and partially raised, and his nuchal crest opened. He begins to hop or bound as the intensity increases, while the female stands just outside the avenue partially or fully concealed. He vocalises during this display with clicks and perhaps some mimicry, the female responding with hissing. The male tries to lure her into the avenue with a display which resembles a young bird begging for food. Copulation may result, either inside or just outside the avenue. **NEST/NESTING** The nest, built by the female alone, is a loosely constructed shallow cup composed of slender sticks and twigs and lined with finer twigs and a few leaves, measuring *c.*200mm across and 125mm deep, with the egg cup *c.*60mm. It is located within foliage 2–10m above ground in the fork of a tree, vine tangle or clump of mistletoe. **EGGS & INCUBATION** The glossy eggs have a pale grey-green, olive-green or creamy-white ground colour and a maze of dark vermiculations and spots, mostly concentrated on the middle of the egg. Clutch usually a single egg, sometimes two, rarely a third, size around 41 × 29mm. The female alone incubates, the incubation period being around 21 days. **NESTLING, PARENTAL CARE & FLEDGING PERIOD** On hatching, the chick has dark purplish skin sparsely covered in grey down, and has bluish-grey legs. The iris is brown and the mouth-lining yellow. The nestling is fed a diet rich in animal matter, such as grasshoppers, crickets, cockroaches, beetles, caterpillars, mantids and spiders (also flies, skinks and moths); larger arthropod prey such as grasshoppers are stripped of head, wings and legs before being fed to the young. Fruit is also presented, often figs. Feeding seems not to involve regurgitation, at least during the Townsville study, where all food was carried in the bill. The female swallows the faeces and faecal sacs of the young. The young leaves the nest at around 20 days Females may continue to feed the young after fledging, but for how long is uncertain. **TIMING OF BREEDING** Breeds mainly during the wet season from September to February, with peak egg-laying October–November, although breeding can be opportunistic dependent on local conditions and can be at any time of the year. The main rainy-season peak coincides with the arrival of the monsoon trough and probably an increase in the amount of insect prey available at that time. **PREDATORS** Predation by raptors is recorded, with Red Goshawk (*Erythrotriorchis radiatus*) and Collared Sparrowhawk (*Accipiter cirrhocephalus*) known to have taken juveniles; and Oenpelli Rock Python (*Morelia oenpelliensis*) is thought to be a predator. Green Tree Ants (*Oecophylla smaragdina*) have been seen to attack young bowerbirds in trees shared with the latter (F&F 2004).

HYBRIDS Hybridisation with Spotted Bowerbird is known from south of Charters Towers along the Cape River, where the two species are sympatric (Frith & Frith 1995b). The hybrids are basically intermediate between the two parent species. The Great Bowerbird comes into contact sporadically with the Fawn-breasted

Bowerbird on Cape York around Iron Range, and hybridisation may be possible, although not yet recorded.

MOULT Moult appears to occur in most months, with a peak in October–November (F&F 2004).

STATUS AND CONSERVATION Classified as Least Concern by BirdLife International. This species is widespread and quite common in much of the drier areas of tropical northern Australia. It has become much commoner at Townsville, in North Queensland, where it was scarce in the early 1960s, but it avoids the humid tropical coast around Cairns and is uncommon in the wetter tablelands around Atherton. In areas where crops are grown commercially, this species can cause some damage by feeding on green vegetables and soft fruits and in such places it is often persecuted by local farmers.

Great Bowerbird, race *orientalis*, Mareeba, Queensland, Australia (*Jun Matsui*).

Great Bowerbird, male, race *orientalis* at bower, Mareeba, Queensland, Australia, November 2013 (*Jun Matsui*).

Great Bowerbird, nominate avenue bower, Rum Jungle, Northern Territory, Australia, October 2017 (*Phil Gregory*).

Great Bowerbird, avenue bower, Mareeba, Queensland, Australia, November 2015 (*Jun Matsui*).

WESTERN BOWERBIRD
Chlamydera guttata   **Plate 37**

Chlamydera guttata Gould, 1862, *Proceedings of the Zoological Society of London* 1862, 162. The type specimen is from north-west Australia and was collected by the explorer Francis Gregory in 1861, in the East Murchison region.
Other English names Spotted Bowerbird, Yellow-spotted, Large-spotted or Pale-spotted Bowerbird, Mimic-bird, Guttated Bower-bird or Bower Bird
Etymology The epithet *guttata* is Latin for speckled, dappled or spotted.

FIELD IDENTIFICATION A typical stocky, medium-sized bowerbird with a rounded head, richly coloured ochre-brown and heavily spotted buff both above and below, with a pale buff tail tip. SIMILAR SPECIES This is the only bowerbird species within its range, and is smaller than the allopatric Great Bowerbird, and far more richly coloured with heavily buff-spotted plumage. It is not known to overlap with the likewise allopatric and very similar but slightly larger and paler Spotted Bowerbird, whose range is farther east.

RANGE Western Bowerbird occurs in a broad band in two seemingly disjunct regions of mid-western and Central Australia, from the Everard Ranges of northern South Australia to about 300km north-east of Alice Springs, and extending west to the Pilbara in the north and Leonora in the south of Western Australia, with a gap apparently in the centre of this range (being absent from much of Simpson Desert). It is allopatric with Spotted Bowerbird, with both species reportedly absent from a band of *c.*100km between 137° east and 138° east, at the inland extremities of their ranges. It was previously believed that the Western Bowerbird's range was coincident with that of the Rock Fig (*Ficus platypoda*), which is undoubtedly an important food source across much of the range, but it has subsequently been found to occur beyond the range of that species and into the south-east interior of Western Australia. MOVEMENTS Some purely local movements occur, these related to rainfall and the abundance of Rock Figs. Wandering outside this range is almost unknown, but there is a single anomalous record from Corny Pt, in the south-west part of the Yorke Peninsula of South Australia, in June 1985.

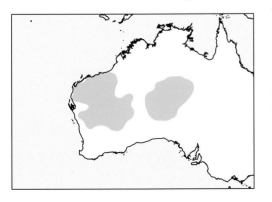

DESCRIPTION Sexes more or less alike. Length 24–28cm; weight *c.*130g. *Adult Male* The head, neck and upperparts are mainly dark blackish-brown, heavily marked with buff to rufous spots; the crown has a silvery sheen, and a large lilac-pink nuchal crest is usually visible only during display. The creamy-yellowish underside is heavily marked with dark-edged buff spots on the chin and throat, becoming more reddish on the thighs, flanks and undertail. The underwing is buff-cream, with black barring on the coverts. *Iris* brown; *bill* black, slightly downcurved and hooked at tip; *legs and feet* greenish-horn, the long legs being feathered at the top. *Adult Female* Very like male, but nuchal crest usually smaller (or invisible), tail on average slightly longer, and throat more spotted; older individuals may be much as male. *Juvenile* Resembles immature, but spotting of upperparts paler and much less distinct, and nuchal crest absent. *Immature* Resembles adult female but paler and greyer, and lacks the silver gloss on the head, while the lilac-pink nuchal crest is absent or much reduced and the bill is brownish. *Subadult Male* Resembles female.

TAXONOMY AND GEOGRAPHICAL VARIATION This species, formerly treated as conspecific with Spotted Bowerbird, is now regarded as an example of east–west species pairs, as are many other Australian taxa, e.g. wedgebills (*Psophodes*) and quail-thrushes (*Cinclosoma*). Two subspecies are recognised.
1. *C. g. guttata* Gould, 1862. The nominate race occurs throughout most of the range, northern birds being slightly darker than the southern birds.
2. *C. g. carteri* Mathews, 1920. A rare restricted-range taxon found only on North West Cape of mid-western Western Australia, this is a smaller bird with a smaller and narrower nuchal crest.

VOCALISATIONS AND OTHER SOUNDS The vocalisations of this species resemble those of the Spotted Bowerbird and include frequent harsh scolds and churrings, as with a male at a bower archived on xeno-canto at XC327267 (by P. de G. Boersma). It also gives metallic clinks, harsh grinding sounds, hisses, clicks and cackles, and a cat-like meowing. Song seems to be given throughout the year at North West Cape, Western Australia, with a peak in September–November and after rainy periods. As with other bowerbirds, the song advertises the bower and ownership thereof. The species is an accomplished **mimic**, frequently imitating the calls of other birds, especially when there is a predator nearby. Birds mimicked include such typical arid-country vocal species as Whistling Kite (*Haliastur sphenurus*), Blue-winged Kookaburra (*Dacelo leachii*), Brown Falcon (*Falco berigora*), Australian Ringneck (*Barnardius zonarius*), Pied Butcherbird (*Cracticus nigrogularis*), White-browed Babbler (*Pomatostomus superciliosus*), Yellow-throated Miner (*Manorina flavigula*), Singing Honeyeater (*Gavicalis virescens*), Spiny-cheeked Honeyeater (*Acanthagenys rufogularis*), Black-faced Cuckooshrike (*Coracina novaehollandiae*), Little Crow (*Corvus bennetti*) and Zebra Finch (*Taeniopygia guttata*). Further, the Western Bowerbird has been heard to mimic the sound of rattling metal, a horse whinnying, barking dogs, a person coughing, and the bleating of a goat.

HABITAT Occurs in riverine woodland, near gum-fringed creeks, in wooded savanna and dense vegetation such as *Casuarina* and *Acacia* in rocky areas. Found also in dense vegetation near water sources such as bores and dams, as well as in gardens, parks, camping sites and picnic areas. It is mainly a lowland species, ranging from sea level to 500+m.

HABITS Seen singly, in twos or in small groups at food sources, wary but can become fairly confiding at picnic sites. Rarely found far from water, where it drinks regularly. May make local movements in search of food, but largely resident.

FOOD AND FORAGING This species is primarily frugivorous, feeding on the fruits of Rock Fig, Sandal-wood (*Santalum spicatum*), Snake Gourd (*Trichosanthes cucumerina*) and mistletoes (Loranthaceae). It will also take nectar and eat flowers, as well as sundry insects such as moths, beetles, ants and grasshoppers, and also eats spiders. It is reported that during times of drought it can subsist on dry seeds. Small flocks visit homesteads and orchards to feed on garden fruits and vegetables, when they may be something of a nuisance. Some 80 birds were shot in a garden near Wiluna, in Western Australia, over *c.*90 days in the 1970s (Forshaw & Cooper 1977). Drinks regularly from waterholes and springs, and during times of drought from tins of water in campsites and buckets in gardens. Often drinks during the heat of the day, scooping or briefly sipping and then tipping the head back to let water run down the throat.

BREEDING BEHAVIOUR Polygynous, promiscuous male builds an avenue bower at which he displays; nest-building and care of young are entirely by the female, as is usual with the bowerbirds. Non-territorial, except for defence of bower site by male. **BOWER** The bower is of the avenue type, with two open U-shaped parallel walls of twigs built upon a platform of sticks, and partly concealed under shady vegetation such as bushes and trees, although some are more obvious than others. The platform is about 15–20cm high and is not orientated in any specific direction. The walls measure 25–35cm, and are 20–25cm high, 10–12cm thick and about 15cm apart. A series of some nine videos by P. de Groot Boersma of the bird at the bower can be seen at the IBC site, as can one by M. Gilfedder of Western Bowerbirds at Olive Pink Botanic Garden, in Alice Springs (accessible at hbw.com/ibc/1339245). This species favours **white or green objects**, such as berries, fruits, seedpods, leaves, bones, snail shells and pebbles. **Man-made objects** such as gun cartridges, glass and miscellaneous metal items may also be acquired. The number of objects varies with the locality and availability of decorations, but can total hundreds of different items. One at Exmouth Gulf had the following: 136 mulga pods, 18 green *Datura leichardtii* berries, 11 clusters of *Jasminium* fruits, four *Acacia* pods and an unidentified pod, 91 brass cartridge cases, 2 brass shotgun cartridges, 20 calcite crystals, several bleached bones, and numerous whitish pebbles (Serventy 1955). **Bower-painting**

has been recorded, a male using the pulp of the red fruits of an *Eremophila* species to paint extensively the walls of his bower (F&F 2004). The total weight of decorations on a bower near Alice Springs was 7.4kg, the items including 1,472 bone fragments, 174 snail shells and the rest pebbles, glass and metal fragments (Veselovsky 1979). A bower may be used for 2–3 years before a new one is built, either on the same site or nearby, and recycling materials from the old bower. This is rather different from the situation with the Spotted Bowerbird, the bowers of which seem to be used for far longer periods, but this may reflect lack of knowledge of the present species. The bower is often well concealed beneath overhanging Rock Figs and usually sited quite near water sources. Spacing will depend on the habitat, but eight bowers at North West Cape were 1–2km apart. **Bower damage** by rival males is recorded, but theft of objects has not yet been noted though it must surely occur. Groups of 3–7 visiting birds, presumably immature males, are not permitted on to the court (Bradley 1987). **COURTSHIP** The Western Bowerbird's courtship behaviour is not well known, but it appears to be similar to that of the Spotted Bowerbird. Males advertise the bower with harsh vocalisations, and display to a female by ritualised dances. These include jumping upwards while flicking and opening the wings, and tail-fanning. The feathers of the lilac-pink nuchal crest are erected and the nape presented to the female, with the tail fanned and held sideways. The male also holds a bower decoration in his bill and shakes the head vigorously, giving harsh calls and whistles and mimicry as he does so. **NEST/NESTING** The nest is a loose platform of dry sticks which supports a shallow cup lined with fine twigs, grasses and *Casuarina* needles, the nest about 150mm in diameter and the egg cup 110mm wide and 50mm deep. It is placed 2–6m above ground in foliage in the fork of a tree or bush branch or in a clump of mistletoe. Old nests are quite often found nearby. **EGGS & INCUBATION** The oval-shaped eggs are slightly glossy, pale grey-green to buff, and heavily marked with dark scribbles, most concentrated on the wider end, size 38 × 26mm; they closely resemble the eggs of Spotted Bowerbird. The typical clutch is of two eggs, but may at times be just a single egg. The incubation period is unknown, but likely to be similar to that of Spotted Bowerbird. **NESTLING & FLEDGLING DEPENDENCY** The nestling is covered in dusky or blackish down and resembles that of Spotted Bowerbird. Fledging period is unknown, but likely to be similar to that of Spotted Bowerbird. Fledglings are reported to be dependent on the parent for *c.*2 months. **TIMING OF BREEDING** The breeding season is July–December, with the main activity August–October, though bowers may be attended throughout the year. Activity increases with rainfall, which causes a flush of plant growth and insect numbers.

HYBRIDS None known, this species and its congeners being allopatric.

MOULT Poorly known, but many specimens show evidence of wing moult in October (F&F 2004).

STATUS AND CONSERVATION Classified as Least Concern by BirdLife International. This is a widespread and quite common, albeit patchily distributed species over much of its extensive and remote range. The race *C. g. carteri* is classified as Near Threatened (Garnett & Crowley 2000) owing to its small population size (total *c*.2,000 individuals) and restricted range (500km²). Major threats may include inappropriate fire regimes resulting in the burning of the species' habitat, and predation by cats and foxes.

Western Bowerbird, decorating bower with green leaf, Olive Pink Botanic Gardens Alice Springs, Northern Territory, Australia (*Don Hadden*).

Below: **Western Bowerbird**, carrying fruit at bower, Olive Pink Botanic Gardens, Alice Springs, Northern Territory, Australia (*Don Hadden*).

SPOTTED BOWERBIRD
Chlamydera maculata   Plate 37

Calodera maculata Gould, 1837, *Synops. Birds Aust.* **1**: pl. 6 and text. New Holland = Liverpool Plains, New South Wales.

Other English names Large-frilled Bowerbird, Cabbage-bird or Mimic-bird.

Etymology The specific name *maculata* refers to the rufous-spotted upperparts, being the feminine form of the Latin *maculatus*, meaning spotted.

FIELD IDENTIFICATION A typical medium-sized, stocky bowerbird with a rounded head, blackish-brown upperparts with heavy pale buff spotting, and a pale buff tail tip. Its flight is undulating on broad upswept wings. **SIMILAR SPECIES** The darker and more richly coloured Western Bowerbird is allopatric. Great Bowerbird is considerably larger and much paler, with far paler, less obvious spotting, and overlaps in range only in Central Queensland.

RANGE Widely distributed in the arid and semi-arid zones of eastern Australia from about Rockhampton, Queensland, west to the Georgina River and Eyre Creek catchment, extending south through central New South Wales. **MOVEMENTS** Largely sedentary and resident, with no large-scale movements, though some vagrancy due to drought-related conditions, and local movements associated with ripening fruit. Occasional vagrant to coastal parts of Queensland and New South

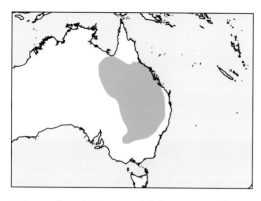

Wales, as from Brisbane and Sydney; reported from Roma (Queensland) only in winter.

DESCRIPTION Sexes almost alike; plumage may vary considerably according to state of moult and degree of feather wear. *Adult Male* 28–31cm; weight *c.*140g. Has a pale rufous head that is streaked with grey-brown, and a filamentous lilac-pink nuchal crest which is usually seen only in display (or as a small colourful patch). The upperparts are blackish-brown and marked extensively with amber spots, while the paler underparts are cream with greyish scalloping and barring, the lower belly and undertail having a slightly yellow shade. *Iris* dark brown to grey-brown; *bill* black, with pink gape and *mouth-lining*; legs and feet olive-brown. *Adult Female* Almost identical to male, but nuchal crest is smaller or absent. *Juvenile* Like immature, but with nape feathers heavily streaked pale buff. *Immature* Resembles adult female but paler above and below, with grey barring and less pronounced spotting. Gape yellow. *Subadult Male* Differs from adult in having a smaller nuchal crest and paler brown streaking on the crown.

TAXONOMY AND GEOGRAPHICAL VARIATION Monotypic. Formerly treated as conspecific with Western Bowerbird, but now regarded as an example of an east–west species pair, as are various other Australian avian taxa, e.g. wedgebills (*Psophodes*) and quail-thrushes (*Cinclosoma*). Synonyms: *Chlamydodera occipitalis* Gould, 1875; *Chlamydera maculata clelandi* Mathews, 1912; *Chlamydera maculata sedani* Mathews, 1913.

VOCALISATIONS AND OTHER SOUNDS This species is renowned for its **mimicry**, and besides the usual harsh churrings, clicks and hisses (see XC287091, by M. Anderson) it seems able to imitate almost any other sound. A typical raspy song type is archived on xeno-canto at XC287093 (by M. Anderson). Cat-like meowing may be mimicry, and barking dogs, animals walking through scrub, wood-chopping, diesel engines, whip-cracks and twanging wire fences are all noted. Bird species imitated are the typical (and more vocal) arid-zone assemblage, including Whistling Kite (*Haliastur sphenurus*) (see XC287086, by M. Anderson), Wedge-tailed Eagle (*Aquila audax*), Blue-winged Kookaburra (*Dacelo leachii*), the whistling flight of Crested Pigeon (*Ocyphaps lophotes*), Apostlebird (*Struthidea cinerea*), Noisy Miner

(*Manorina melanocephala*), Grey Butcherbird (*Cracticus torquatus*) and Australian Magpie (*Gymnorhina tibicen*). Mimicry is often used during responses to predators or people near the bowers or nests.

HABITAT This species shows a strong preference for riverine vegetation within open, dry sclerophyll–eucalypt savanna woodland; also areas of the much-reduced brigalow (*Acacia harpophylla*) habitat in north-east Queensland. It is found from sea level to *c.*500m. Bower sites seem often to be within fruit-bearing or thorny habitats such as Wild Lemon (*Canthium oleifolium*) and currant bush (*Apophyllum*) in New South Wales (Rowland 2008). Habitat with a dense understorey of bushes, shrubs and grasses is often favoured.

HABITS Seen singly, in twos or in small groups at food sources. Wary, but can become fairly confiding at picnic sites, camp grounds and homesteads. Groups of up to 16 immature Spotted Bowerbirds may visit a bower, but will be driven away by the owner if he is present. When not attending bowers or nests, they can form flocks of up to 30 individuals, occasionally up to *c.*50. Males often call and display on the ground, as well as on branches. Associates with other fruit-eating species such as honeyeaters, and bower-owning males have been seen to eat the fruits, flowers and leaves which were used to conceal their bowers, which may explain their preference for certain species.

FOOD AND FORAGING This species is primarily frugivorous, as with other bowerbirds, supplementing the diet with seeds, nectar, leaves, insects, spiders and small lizards. Favourite fruits include White Cedar (*Melia azedarock*), Native Orange (*Caparis mitchelli*), Peppercorn Tree (*Schinus molle*), Mangrove Boobialla (*Myoporum acuminatum*), Currant Bush (*Carissa ovata*), Conker Berry (*Carissa lanceolata*) and Turkey Bush (*Eremophila deserti*). During times of drought seeds, particularly those of Kurrajong (*Brachychiton populneus*), become a much more significant part of the diet, as also do flowers and nectar (Rowland 2008). This bowerbird is known also to take sugar put out for honeyeaters during times of drought (F&F 1995b), and it is known to drink, visiting homesteads in search of water during such times; indeed, 20–30 individuals were reported as drinking at a garden in New South Wales. It forages from the ground to the canopy.

BREEDING BEHAVIOUR As with its congeners, the Spotted Bowerbird is a polygynous species with promiscuous males, while nest-building and care of the young are done entirely by the female, as is usual with the bowerbirds. It is not a territorial species, apart from the male's defence of his bower site. Distance between bowers depends on the habitat, but is generally no more than 2.5km and often within 1km. **BOWER** A series of videos of the bird at the bower by P. de Groot Boersma can be seen at the IBC site. Bowers are generally built under large, thorny bushes that provide shelter and fruit. Some bower sites may be used for upwards of 20 years, and are rebuilt each year by a succession of males. This seems rather different from what is known of the Western Bowerbird, whose bowers seem to be used for

2–3 years, but this may reflect lack of knowledge of the latter species. Bowers are built mainly during late winter–early spring, July–August, and are maintained throughout the display season until December/January. This species favours grass stems for bower construction, but sticks may be used at the base of the outer walls. Males may **paint the walls** of bowers using masticated grass and saliva, which leaves a dark band, much as is the case with congeners. They will raid nearby bowers to steal items or even destroy the bower; Borgia & Mueller (1992) recorded one male as destroying 85% of a bower before being chased off by the owner, then reappearing some 218 minutes later to steal two objects from the now mostly rebuilt structure. Theft of decorations does, however, seem to be much less frequent than with Satin Bowerbird. An avenue measuring 2.1m long is recorded in North (1901–14), this made from stems of blue grass (*Andropogon sericeus*). An odd bower platform was built atop the horizontal branch of a gum some 4.6m above ground (Favaloro 1940). The avenue walls measure 40–70cm in length, and are 20–30cm high, 10–25cm thick and 15–25cm apart. **Bower orientation** appears random, with no strong preference evident. This species decorates the avenue bower and platform with white, grey, pale green, amber and mauve objects, but apparently rejects those that are red, yellow and blue. Items used include bleached bones, eggshells, snail shells, pebbles, insect casings, seedpods, flowers and berries, green *Solanum* berries being highly prized. Man-made items include glass, wire, keys, plastic, bottle caps, can rings, coins and cartridge cases. The Spotted Bowerbird is well known for **stealing shiny trinkets** such as jewellery from campsites and homesteads: one individual stole the ignition keys from a vehicle, leaving the farmer stranded, and another was reported to have stolen a bushman's glass eye from a glass of water in which it was stored overnight, the eye being recovered 'staring vacantly up from among the bones and shells on the display ground!' (Chisholm 1929). A male near the homestead at Bowra Station, in south Queensland, had dozens of silvery-grey metal-roofing iron bolts as objects in the bower, from which the farmer would collect them as required for maintenance (pers. obs.). Circular objects are often hung from the avenue walls, and they are carefully arranged, the male spending much time in inspecting and correcting any deficiencies. The theft of decorations appears to be uncommon, but bower destruction and theft of items are recorded. Borgia & Mueller (1992) suggested that Spotted Bowerbirds prefer decorations that match their lilac-pink nuchal crest, and they believed that bower quality affects mate-choice decisions by females (as with *Amblyornis* bowerbirds). **Courtship** Male Spotted Bowerbirds tend, watch and sing over their bowers during almost any time of the day, though with early-morning and late-afternoon peaks. Females are attracted to the bowers by the calling of the males, which then perform quite vocal elaborate courtship displays that consist of central and peripheral display elements, much as with the Great Bowerbird. *Central Displays* are performed in the immediate vicinity of the bower and usually begin with the male holding one of his prized objects in his bill and constantly dropping it and picking it up again or gathering a new one. His movements are jerky and appear aggressive and involve an upright posture, the male standing to one side of the main avenue entrance and the female keeping one wall of the bower between her and the male. The male raises and lowers his wings while jumping to and fro and fanning the tail, while the nuchal crest is erected and displayed. He may also pick up a decoration and crouch, with the bill slightly open, while jerking the head up and down. The *Peripheral Display* consists of the males walking around their bower in wide circles with raised head, open bill, cocked tail and drooped wings. Courtship displays can last for minutes or sometimes more than an hour, and copulation in or nearby the bower follows successful courtship. Relative quality of bower and reproductive success are thought to be related to the age of the males, bowers having many bones and glass objects also having greater success (Borgia 1995). In a study at Taunton NP, in eastern Queensland, it was found that males attended the bower for some 54% of daylight hours, with peaks in the early morning and to a lesser extent in the late afternoon (F&F 2004). **Nest** The nest, about 200mm in diameter, has a bulky dead-stick foundation, with a shallow saucer-shaped cup of fine twigs and tendrils, often lined with dry grass or leaves, the egg cup 40–50mm deep. It is placed 2–12m above ground in dense foliage in the crown or fork of a tree or bush, or in a clump of mistletoe. **Eggs & Incubation** The eggs are slightly glossy and oval-shaped, pale grey-green to buff and heavily marked with dark scribbles, concentrated on the wider end, size 39×27mm. They closely resemble the eggs of Western Bowerbird. The typical clutch is of two eggs, but may be one or three. The incubation period seems not to have been recorded. **Nestling & Fledging Period** The chick has dark maroon-grey skin and is covered with long grey down, especially dense on the head and back; iris brown, bill greyish-black with a yellowish gape. The difference in the size of nestlings suggests asynchronous hatching and a laying interval of 2 days, much as with other bowerbirds. The faeces of the nestlings are eaten by the female, as is usual in the family. The fledging period is about 20 days. **Timing of Breeding** The breeding season is July–January, with the main activity August–October, but may begin as early as May and end as late as April, dependent on local conditions. Bowers may be attended throughout the year, and activity increases with rainfall, which causes a flush of plant growth and an increase in insect numbers. Eggs are recorded from July to March. **Predators** Predators such as Brown Falcon (*Falco berigora*) and *Accipiter* hawks are known to take adults occasionally, and Pied Butcherbird (*Cracticus torquatus*) has been seen to prey on young. Goannas (*Varanus*) are also great predators and are believed occasionally to take young birds of the present species.

HYBRIDS The Spotted Bowerbird has hybridised with the Great Bowerbird along the Cape River 100km south of Charters Towers, where the two species' habitats are adjacent (F&F 1995b). The hybrids are basically intermediate between the two parent species.

MOULT January–February appears to be the peak period for wing moult (F&F 2004).

STATUS AND CONSERVATION Classified as Least Concern by BirdLife International. This is a widespread and quite common, albeit patchily distributed species across the extensive range in eastern Australia, though with some local declines owing to habitat loss. It is now extinct in South Australia and also in its small former range in north-west Victoria, with no records for those areas in the New Atlas of Australian Birds (Barrett *et al.* 2003). Causes of this decline may be illegal shooting and poisoning by farmers, who consider Spotted Bowerbirds a pest in soft-fruit orchards, and predation by introduced species such as domestic cats (*Felis catus*) and Red Foxes (*Vulpes vulpes*). Loss of habitat caused by land clearance, modification and

fragmentation is also a problem locally. In the past, the species' habit of feeding on fruit crops resulted in indiscriminate shooting by orchard-owners. For example, Favaloro (1940) states that as many as 30 or more birds were destroyed each week at one homestead near Mildura. Other processes likely to have contributed to the decline of the Spotted Bowerbird and its habitat include persistent grazing by domestic stock, kangaroos and rabbits, inappropriate fire regimes, and the clearing of understorey vegetation by removing timber (Flora and Fauna Guarantee Action Statement State of Victoria Dept of Sustainability and Environment 2003).

Spotted Bowerbird, sexes similar, Lake Moondarra near Mt Isa, Queensland, Australia, September 2003 (*Don Hadden*).

Spotted Bowerbird, sexes similar, Lake Moondarra near Mt Isa, Queensland, Australia, September 2003 (*Don Hadden*).

YELLOW-BREASTED BOWERBIRD
Chlamydera lauterbachi   Plate 38

Chlamydodera [sic] *lauterbachi* Reichenow, 1897, *Orn. Monatsber.* **5**: 24. Yagei, upper Ramu, north-east New Guinea.
Other English names Lauterbach's Bowerbird/Bowerbird/Bower Bird, Yellow-bellied Bowerbird
Etymology The specific name *lauterbachi* commemorates the German botanist, explorer and collector Carl Lauterbach, who discovered the species in 1896.

FIELD IDENTIFICATION A quite slender bowerbird of the montane grassland and forest ecotone, including wooded gardens with casuarinas and the edges of coffee plantations where shrubs and trees survive to give cover. This is a shy species, slightly smaller than the Fawn-breasted Bowerbird and much richer yellow beneath. It can at times be quite conspicuous and fairly noisy, perching up high in trees, and the yellowish underparts are clearly visible even in flight.
SIMILAR SPECIES Fawn-breasted Bowerbird, at its upper

altitudinal limit sympatric with Yellow-breasted in a few areas, is a slightly larger and less gracile species with rufous on the underparts, entirely lacking the rich yellow coloration.

RANGE Occurs across central New Guinea from east of Geelvink Bay (Siriwo River) east to upper Ramu River and to Okapa–Aiyura area of Eastern Highlands, and to the north scarp of the Huon Peninsula; also perhaps the eastern Sepik and Baiyer Rivers. This species is very patchily and locally distributed in the western ranges and in central New Guinea: localities include Digul River, Puwani River, the north scarp of the Bewani Mts, the middle Sepik–Wahgi valley, Sepik–Wahgi Divide, Minj, Awande, Asaro valley and Mt Rondon. The Yellow-breasted Bowerbird is sympatric with Fawn-breasted at Aiome (in Ramu valley) and in the Baiyer valley, and probably also in the foothills of the north Finisterre Range (Gilliard & LeCroy 1968) and at Marienburg, on the lower Sepik. There are single records for Bewani Mts, in north coastal range, and in lower Jimi valley. **MOVEMENTS** Resident, with some short-

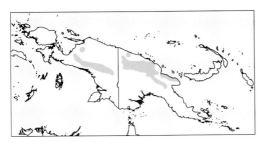

chilp chilp chilp. A nasal rasping *chewp* reminiscent of the call of White-bellied Cuckooshrike (*Coracina papuensis*) is sometimes uttered. Grating noises and sharp clicks are given, and **mimicry** of ambient sounds is heard, but, oddly, avian mimicry is so far unrecorded in the wild and the species gives the impression of being considerably less vocal than the Fawn-breasted Bowerbird. Like that species, it is most vocal in the early mornings and late afternoons.

HABITAT Inhabits lowland and mid-montane secondary growth from sea level to 1800m. This species occupies remnant forest patches and forest edges, overgrown gardens and adjacent bushy grasslands, pit-pit grass (*Miscanthus floridulus*), canegrass-swamps, coffee plantations, and *Casuarina* stands. Bower sites are located amid dense vegetation under large bushy trees just within the edge of forest or forest patches, and nearby kunai (*Imperata*) or pit-pit grassland with dispersed shrubs and trees (F&F 2004). The abundance of this species in the anthropogenic mid-mountain grasslands may be correlated with the deforestation and subsequent modification of the area by its ancient inhabitants (Gilliard 1969).

HABITS A shy species, this bowerbird seems less confiding and less conspicuous than the Fawn-breasted Bowerbird. It can be quite vocal and may be heard giving harsh tearing calls from cover. Usually seen singly or in twos; sometimes in small groups at fruiting trees.

FOOD AND FORAGING Little known. Diet consists primarily of fruits taken from a variety of species, but also insects, including caterpillars and beetles (*Coleoptera*); one young was fed insects. Typically occurs singly, in twos or in small parties. Forages at fruiting trees and shrubs.

BREEDING BEHAVIOUR Polygynous, promiscuous male builds and decorates elaborate avenue bower; female builds and attends nest alone. Non-territorial except for defence of bower sites by male, as is usual in the family. The mean distance of four active nests from the nearest active bower was 204m (Baiyer valley) (F&F 2004). Gilliard suggested that bowers may be locally concentrated and form exploded leks, but this remains to be confirmed. **BOWER** The bower of the nominate race is as yet undescribed, but limited data on its display (derived from two birds displaying in the bower of a Fawn-breasted Bowerbird!) suggest that it is similar to that of *uniformis*. Bowers of the race *uniformis* are sited *c.*0·5–1km apart, in shady sites beneath large bushy trees or bushes, or inside the edge of forest adjacent to grassland. They are usually built on level ground, sometimes atop a small rise, or in marshy areas (where the substantial stick bower base may serve as an island). One bower site was used for more than ten years. The avenue bower of this species is unique in having **four walls** built upon a substantial stick-and-cane base. There is the usual *Chlamydera*-type avenue-bower structure in the centre, flanked at right angles by two stick walls about 50cm away from the ends of the central

distance local movement likely. A *Chlamydera* species at Tabubil, in Western Province, was a wanderer, most likely of this species, which is the more proximate of the two Papuan species.

DESCRIPTION Adults exhibit relatively limited sexual dimorphism. ***Adult Male*** 26.5–29cm; weight *c.*160g. Resembles adult male Fawn-breasted Bowerbird, but forehead and forecrown not streaked and head, back and rump have a yellowish wash, with the cap and sides of head a reddish-copper (nominate subspecies) or greenish (race *uniformis*). Upper breast is more heavily streaked brown, and the abdomen yellow with sides barred brown (fainter in the nominate); tail is tipped whitish. *Iris* dark brown; *bill* black, *mouth-lining* orange-pink; *legs and feet* variable, yellowish-brown, yellowish-olive, dull olive or greyish. ***Adult Female*** Similar to male but duller, with less yellow-olive on crown, slightly paler upperparts, more buff-yellow below; also lighter in weight and slightly shorter-winged. ***Juvenile*** Newly fledged juvenile is like adult, but more streaked above, and has chin to breast, flanks and abdomen soft downy off-white with grey mottling, and pale yellow wash on central abdomen. ***Immature Male*** Like adult female. ***Subadult Male*** Has few to nearly all reddish feathers intruding crown (nominate race).

TAXONOMY AND GEOGRAPHICAL VARIATION
Two subspecies are recognised. They appear to meet in the Baiyer valley.
1. ***C. l. lauterbachi*** Reichenow, 1897, is endemic to PNG. It occurs in a fairly small area, from the Ramu valley east to the north scarp of the Huon Peninsula, also perhaps the eastern Sepik and Baiyer Rivers.
2. ***C. l. uniformis*** Rothschild, 1931, *Novit. Zool.* **36**: 250. Siriwo River, 72km from mouth, west Mamberamo basin, north-west lowlands. More widespread, being found in central and western New Guinea.

The males of the two races are distinguished mainly by the crown colour, either a bright copper (nominate) or yellowish olive-green (*uniformis*). The two come into contact in the Baiyer valley (see Coates 1990, photos 406 & 410); this is not surprising, as the Baiyer drains into the eastern Sepik, and it may be a zone of overlap, as it is for the Lesser and Raggiana Birds of Paradise. (Beehler & Pratt 2016)

VOCALISATIONS AND OTHER SOUNDS Short rasps and chucks of typical *Chlamydera* bowerbird type, as at xeno-canto XC38127 (by F. Lambert). Gilliard (1969) notes an alarm call like the sound made by rapping on a cardboard box, and this species also gives rattling sounds and hisses and a sharp lisping rattle, bower, forming an H-shape with a long cross piece when viewed from above. The main walls are angled outwards (rather than forming a vertical or arched-over avenue), while each end of the avenue platforms is built up into additional outcurved walls that form cross-passages at right angles to the central avenue. A bird in the central chamber is almost invisible when viewed from the exterior. The bower may consist of up to 3,000 sticks, lined with 1,000 or more strands of brownish grass; mean dimensions of bower 64cm long, 67cm wide and 43cm high, of main avenue 23cm long and 10cm wide, mean size of exposed stick platform at each end of avenue 84cm long and 17cm deep. The bowers are tightly meshed together and may be picked up without disintegrating. The total mass with the decorations ranges from 2.9kg to 7.48kg (Gilliard 1969). No favoured **orientation** of the central avenue is apparent, at least in the Baiyer and Wahgi valleys, orientation being highly variable. Immature birds may make much less complex bowers, some resembling the ordinary avenue shape. Bower **decorations** (up to 1,000 or more) are often mostly grey to blue-grey river-washed pebbles in total weighing more than 5kg, and large spherical blue quandong fruits (*Elaeocarpus*), which are placed against or pushed among the sticks of the inner end walls and in the central avenue. The stones and pebbles are placed in the end passages, where up to 600 may be located, and in the centre of the main passageway, where Gilliard found from two to 99 stones, which may be inserted into the inner surfaces of the end walls to form a kind of miniature stone wall, a unique feature of the bowers of this species. Smaller red berries are sometimes placed on the bower floor, and charcoal can be used if stones are hard to find. In the middle Sepik, bower decorations are red, green, blue, black and grey tree fruits and hard berries, plus much charcoal in lieu of stones, which are very scarce here. A bower in the Wahgi valley, PNG, in July 2018 had five blue quandong fruits, one blue berry, plus three small grey stones, two larger stones, one pebble and a blue thread fragment as decoration at each end, with six small grey stones placed in the bower. **Bower-painting** is well known for this species, and nicely described by Bell (1967) from a hide by a bower: 'Even when standing still, looking suspiciously at the hide, it continued to masticate something held in its bill. Then it was seen to run its bill up and down the fine fibres of the bower lining, masticating between movements.' It did this for 30 seconds and then flew off, and, when the bower was examined, the inner lining was found to have been 'painted' along a 50mm-wide band the full length of the avenue and clearly visible at a range of 3m. The 'painting' material was a whitish substance almost like a weak flour-and-water paste, presumably masticated plant matter mixed with saliva. Painting bouts are quite frequent, can last for 10–15 minutes and may result in extensive areas of the bower becoming stained. When the owner is away foraging or gathering bower materials, bower destruction and pilfering occur as with congeners: rivals enter the central avenue and vigorously tear out the painted sticks by

their bases. The male Yellow-breasted Bowerbird is exceedingly wary at the bower, and photographers need to have long-established hides and be able to sit very still and silent to observe the behaviour of this species. **Courtship** The courtship of this species seems relatively simplistic compared with that of some other bowerbirds. When females arrive in the bower they stay in the central avenue, keeping still but sometimes pecking at the inner 'painted' wall. The male, positioned by the central-avenue wall, hops on the spot while giving soft grating and low-pitched calls, and then approaches the avenue entrance while uttering plaintive notes and churrs. He stays to one side of the entrance, mostly out of view but peeking with his head and neck forward so that they are visible to the female. Like Fawn-breasted Bowerbird, he presents his unadorned nape to the female, frequently picking up a blue or red fruit or sometimes a stone, and often jerking his head up and down in a somewhat reptilian fashion. Occasionally he may present a decoration to the female while flicking his tongue in and out and making soft calls. If not holding a decoration, the bill is gaped widely to show the brightly coloured orange mouth-lining, still tongue-flicking and often with sharp jerky head movements. Soft plaintive notes and churring calls continue. The later stage of this nape presentation involves him in twisting back towards his tail, which is now fanned and pulled around towards his bill, and occasionally jerking his body to flick his wing on the side facing the bill and tail (F&F 2004). Great Bowerbird has a very similar display posture. The male Yellow-breasted may also adopt an upright posture and sharply flick one or both wings while giving a loud hissing, with the bill gaping towards the female. Gilliard (1969) several times noted a red fruit being held in the bill and presented towards the female. The male may interrupt his performance to step out of the female's view, continuing to vocalise and then coming back into sight. He may run and enter the bower to mount and mate with the female, presumably as she solicits him. Courtship may last for a few minutes or up to 36 minutes, and is broken off to chase intruders away, the female remaining in the bower. After mating, the male may give a sharp *chit* call when perched near the bower, this contrasting with the soft, low calls given during display. Some aspects of the display are reminiscent of the behaviour of congeners, particularly the nape-presentation display, while the berry presentation and hide-and-seek by the bower can recall both Regent Bowerbird and MacGregor's Bowerbird. **Nest/Nesting** The nest, a neat, compact shallow cup of fine twigs, vine tendrils, dried grasses and bark placed upon a sparse, somewhat haphazard foundation of dry sticks, is built 1–4m above ground and located in a tree, bush, sapling, bamboo, sugar cane or dense cane-grass (*Saccharum spontaneum*), sometimes in or close to native gardens. It is not known if the same nest site is used in subsequent years. **Eggs, Incubation & Nestling Period** Clutch consists of a single egg, pale sea-green or pearly grey in colour, scribbled with variable brown or black vermiculations concentrated at the larger end. There

is no information on incubation and nestling periods.
TIMING OF BREEDING Breeds in all months except May;
eggs found in April and June–January. Display season
appears to be April–January.

HYBRIDS None reported, but a male Yellow-breasted
Bowerbird was seen courting a female-plumaged bird
of the same species at a bower of the Fawn-breasted
Bowerbird at Aiome, in the Ramu valley (Gilliard
& LeCroy 1968), where the two species overlap.
Hybridisation is undoubtedly possible but would be
very hard to detect.

MOULT There are no data on seasonality of moult.

STATUS AND CONSERVATION Classified as Least
Concern by BirdLife International. This species can
be locally fairly common and conspicuous, yet scarce
elsewhere. It is common in the mid-mountain valleys
of the Wahgi and Baiyer Rivers, but uncommon on
the Sepik River; and common also in savanna and
woodland of the drier Ramu valley. Its distribution is
oddly patchy in montane grasslands in Weyland Mts,
Snow Mts and Star Mts and in Central and Eastern
Highlands, including Mt Hagen vicinity.

Yellow-breasted Bowerbird, race *uniformis*, Rondon Ridge,
PNG, July 2018 (*Doug Gochfeld*).

Yellow-breasted Bowerbird, typical four-walled bower, race *uniformis*, Kama, Enga Province, PNG, July 2018 (*Phil Gregory*).

FAWN-BREASTED BOWERBIRD
Chlamydera cerviniventris   Plate 39

Chlamydera cerviniventris Gould, 1850, *Jardine's Contrib. Orn.*: 106 (also *Proc. Zool. Soc. London*, 1851 [1850]: 201). Cape York, North Queensland, Australia.
Other English names Fawn-breasted Bower Bird/Bower-bird, Buff-breasted Bowerbird, Meyer's Bowerbird
Etymology The specific name is derived from the Latin *cervinus*, meaning relating to a stag or deer, and refers to the brownish colouring of the underparts, *ventris* meaning of the abdomen or belly.

The species was discovered in 1849 by John MacGillivray, the naturalist on the survey ship HMS *Rattlesnake*, who shot a male at his bower on Cape York. He had been told, probably by a local Aborigine, of a bower in a patch of scrub half a mile (*c.*800m) from the beach. He went ashore before dawn with a local helper and a large plank on which to carry the quite substantial structure back to the ship, after having collected the bird on the bower. The specimen and its bower were sent to Gould in London, who subsequently wrote the formal description.

FIELD IDENTIFICATION This is a distinctive mid-sized (like a large thrush), grey-brown spotted and streaked bird with rufous on the underparts and a stout, blunt black bill. **SIMILAR SPECIES** Female Australasian Figbird (*Sphecotheres vieilloti*) is heavily streaked below and lacks any rufous. Friarbirds (*Philemon*) are uniform grey-brown above, have a more slender and longer bill often with a casque, and are uniformly pale below.

RANGE Largely confined to coastal areas of eastern New Guinea west to Jayapura in the north and in the south to the Trans-Fly (Dolak Island east to Wasur NP, extending north to the Bian Lakes); in PNG ranges from Bensbach east to the Aramia Range and north to Lake Daviumbu, and is locally common around Port Moresby and at Alotau, in Milne Bay Province. There are also some isolated and very poorly known populations in the Vogelkop, from Ransiki and the Kebar valley (Hoogerwerf 1964), and a few interior populations in mountains of the Huon and the Southeast Peninsulas. Some of the upland populations may involve colonisation of areas now modified by

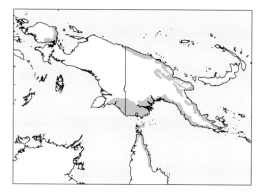

agriculture. A *Chlamydera* species at Tabubil, in Western Province, was a wanderer, most likely of Yellow-breasted Bowerbird, which is the more proximate of the two Papuan species. An old (1891) specimen record purportedly from Tagula Island, in the Louisiades, is presumably in error, and the present species is not known from the Torres Strait islands. In Queensland, it is local around the coastal lowlands of northern Cape York, from the tip of the peninsula south to about Silver Plains (east of Coen).

DESCRIPTION Sexes alike, but female slightly smaller than male. *Adult* 28–30cm; weight *c.*175g. Forehead, lores, sides of head and throat are grey-brown, densely streaked pale buffish-white; crown to hindneck and upperparts grey-brown to fawn-brown, whitish to pale buffy tips of back feathers giving a spotted appearance. The upper breast has an indistinct and incomplete grey-brown pectoral band, which is heavily streaked pale buffy; remainder of underparts rufous-buff, with indistinct brown mottling on flanks. *Iris* dark brown; *bill* black, *mouth-lining* a deep orange-yellow; *legs and feet* greyish-brown. *Juvenile* Resembles first-year immature, but has a featherless nape with dark pinkish-brown skin showing, more streaking on the upperparts, and pale grey-buff underparts with some faint dusky barring. *Immature* Resembles adult, but with more streaking on the forehead and crown, large buffy spots on mantle, back and wing-coverts, wing and tail feathers tipped pale buff and more brightly coloured underparts. Full adult plumage is attained in the second year.

TAXONOMY AND GEOGRAPHICAL VARIATION Monotypic. Proposed taxon *Alphachlamydera cerviniventris nova* Mathews, 1915, is a synonym.

VOCALISATIONS AND OTHER SOUNDS This is quite a vocal species, especially near bowers, emitting distinctive long drawn-out harsh rasps, wheezes and slurred whistles. It is also **a noted mimic** and quite ventriloquial, while it is hard to credit that the volume of noise near the bower is being produced by just a single bird. Mimics many human-derived noises with amazing accuracy, including those of hammering, sawing, electric motors, truck engines and the sound of voices, and also various bird sounds and grasshopper- or cicada-like noises. Species mimicked include Sulphur-crested Cockatoo (*Cacatua galerita*), Orange-footed Scrubfowl (*Megapodius reinwardt*), Eclectus Parrot (*Eclectus roratus*), Blue-winged Kookaburra (*Dacelo leachii*), Black-backed Butcherbird (*Cracticus mentalis*), Willie Wagtail (*Rhipidura leucophrys*) and Noisy Friarbird (*Philemon corniculatus*). Sounds made at the bower include loud churring noises, throaty rolling notes followed by a short harsh scream, sharp rattles and hisses, sputtering rasps interspersed with whistles, explosive notes including a *kaa kaaa kaa* sequence and a cat-like whine (Gilliard 1969; Coates 1990; pers. obs.). Recordings of four birds calling loudly to one another is at xeno-canto XC152535 (by P. Gregory), and a bird whistling and scolding at a bower is at XC38125 (by F. Lambert). There is a loud rasping **Threat Call**, which is often given constantly when people are close to the bower, the bird remaining hidden nearby and, it seems, almost swearing at the intruders!

HABITAT This species inhabits patches of scrub and trees in savanna, grassland and forest edge, also mangroves near grassland and wooded suburban gardens and parks in Port Moresby; it visits teak plantations. Occurs from sea level to 1,400m, but locally up to a maximum of 1,800m at Wau, PNG (Coates & Peckover 2001). It is mainly a lowland species, preferring the ecotone between grassland and woodland thickets, or natural savanna with clumps of trees and brush. Although generally absent from true forest, it is found in riverine thickets and mangroves. In Queensland, Australia, it occupies mangroves and *Melaleuca* bordering them, as well as open and closed forest (where it meets the Great Bowerbird); occurs from sea level to 100m. Traditional bower sites are beneath low bushes in open savanna or woodland, just inside the edge of gallery forest or mangroves, or within tall secondary growth.

HABITS Can be quite conspicuous and fairly noisy, perching high in trees, but also shy and wary. In dry conditions it visits water to drink and bathe. Can be quite common but patchy and scarce or absent from large areas. Seen singly, in twos or in small loose groups of up to a dozen birds. Breeding males maintain large avenue bowers built on stick platforms concealed in dense shrubbery. Quite vocal and can be heard from some distance away. Known to perform **distraction displays** to lure intruders away from bowers: gives harsh alarm calls and hops slowly and feebly on the ground with wings fluttering and drooping and giving calls of potential predators, especially Blue-winged Kookaburra (*Dacelo leachii*) and Black-backed Butcherbird (*Cracticus mentalis*) (Bell 1969).

FOOD AND FORAGING Primarily frugivorous, feeding on a variety of fruits from trees and shrubs, including small red and fiery hot chillies, also the Kangaroo Vine/Native Grape (*Cissus antarcticus*). Insects, including Green Tree Ants (*Oecophylla smaragdina*), are also taken opportunistically. Nestlings are fed with fruits and insects, including caterpillars and beetles (Coleoptera) (Coates 1990).

BREEDING BEHAVIOUR The polygynous, promiscuous male builds and decorates his avenue bower while the female alone builds and attends the nest, as is usual with the family. Non-territorial, except for defence of the bower site by the male; the average distance between bowers is about 500m, but may vary with local conditions. When defending a nestling, the female may use a *Distraction Display*. Bell (1969) reported that a female flew to a nearby tree and gave a harsh Threat Call, and then hopped to the ground, where it moved slowly and feebly away with lowered fluttering wings while mimicking the calls of various nest predators. **BOWER** Subadult males may build rudimentary practice bowers quite close to established bowers. Up to six birds, presumably some young males as well as females, have been seen attending a bower. Younger males may attend and build or decorate in the bowers of older males, or build rudimentary

practice bowers between established bowers (F&F 2004). The male usually constructs a new bower every season at the same site or close to the previous one, or he may refurbish the old bower, as do some Great Bowerbirds. Bell (1982) reported a population average of 1.3 birds and one active bower in a 10ha area of coastal hill savanna near Port Moresby over a 12-month period. The species is locally quite common in this region. The bower avenue is more elaborate and much larger than those of the *Sericulus* avenue-builders, and is rather variable in size, if not structure. The **two-walled avenue** is built atop a deep long and untidy interwoven base of sticks, which extends beyond the avenue entrance in the form of elevated platforms. One is usually larger and thicker than the other, while size and dimensions vary, but typically bowers are 100–175cm long, 50–80cm wide and 20–40cm tall, with the bower avenue around 50–100cm in length and 20–30cm tall with 8–10cm between the avenue walls (Rowland 2008). One unusually small example was just 50cm long (Peckover 1970). The bower of this species is rather different from that of Great Bowerbird, the whole structure being built upon a raised platform of sticks several inches above ground. The side walls are not so high as those of Great Bowerbird, and the sticks do not arch over the central runway, which is open along the entire length. Decorations are also less elaborate, and usually feature green berries or fruits, and never white shells or stones. Like its congeners, the Fawn-breasted Bowerbird is quite noisy at the bower. Avenue **compass orientation** is rather variable, but around Port Moresby most are within 30° of an east–west orientation, the main decorated platform being at the western end and the southern avenue wall slightly larger than the northern one, although occasionally this can be reversed (Peckover 1970). Bower **decorations** in New Guinea typically include green fruits, usually in small bunches, seed pods, unopened flower buds and fresh green leaves, and are generally placed on the larger platform or on the avenue floor, or hung or placed atop it. The number of objects can exceed 100, but is sometimes many fewer. Birds at Iron Range use a variety of plants that have small green fruits, and as these decay and become discoloured they are removed and replaced, the old fruits being placed in a kind of midden nearby. **Bower-raiding** by other bowerbirds is common, and bower destruction occurs, too. Should a male die, it is likely that other males will step in almost immediately, destroy the bower and build their own version on the site, as surplus males without bowers seem to be quite readily available. Competition for sites is fierce and bowers are readily pilfered or even destroyed if the male is away for too long. This species has the unusual habit of the male '**painting**' the interior of the bower, a habit known also for congeneric species, *Sericulus* bowerbirds and the Satin Bowerbird. Green vegetable matter is chewed up and mixed with saliva and then wiped on to the avenue walls with the bill, newly painted sticks being green but drying to dull olive-grey (Rowland 2008). The male paints by gently and repeatedly pushing his bill between the sticks of the bower wall, and some bowers on Cape York at Iron

Range can have the floor twigs appearing red in colour, but not so the sticks on the wall (F&F 2004). Painting sessions can last for over 30 minutes and, as with Yellow-breasted Bowerbird, the habit is quite frequent. As with the Great, Spotted and Satin Bowerbirds, this species provides one of the few avian instances of tool usage, as it employs a piece of vegetable matter as a sort of brush or sponge. Bowers at Iron Range, in Queensland, are active at least from June onwards, but activity ceases with the start of the heavy rains in December–January. In New Guinea, bowers seem to be active for some 8–9 months, but usage again is outside the main wet periods; renewed building can start in February, but such bowers seem often to be temporary and are frequently abandoned, with activity at Port Moresby very limited in April but picking up from May onwards. Bowers are often destroyed at the end of the display season, but it is not certain if this is by the builder or by rival males. In 1849, John MacGillivray collected a bower (which was shipped back to England), but a replacement was completed by the bird 17 days later (MacGillivray 1918). **COURTSHIP** Males advertise their sites with loud repetitive vocalisations given from an exposed branch high in a dead tree. In PNG the following display behaviours have been noted. Males will wing-preen, tail-preen, flash their flanks, make presentation displays, hop sideways, make hunched runs, hop with the head held high, show the tongue and pretend to bill-wipe (Coates 1990). There is a **Peripheral Display** in which the male crouches low and moves quickly and jerkily about the bower, and constantly zigzags in and out of ground cover at the site but maintaining the bower as the focal point, sometimes moving along the avenue itself. Once attracted, the female may perch above the bower or come on to the east end of the platform. The actual display is not well known, but the male starts to run in and out of the vegetation around the bower and will 'freeze' at times. Once the female enters the avenue he hops up and down on the same spot, giving soft churring calls and perhaps some mimicry. A **Nape Presentation Display** is given whereby he peers at the female while holding a green decoration, usually a berry or bunch of berries, in his bill, which he opens to show the orange-yellow interior. Curiously, he also twists his head and presents his nape, even though this species lacks the colourful nuchal patch present in some congeners such as the Great Bowerbird. He may also open and close his wings abruptly, and bend his fanned tail upwards (Rowland 2008). He may make churring noises while executing this display, and include mimicry. He jerks the head and neck and may swing the tail towards his turned head and bill, the tail sometimes partly fanned and sometimes raised almost over the head, or lowered to the ground. He can also alternate between crouching low on the platform and standing stiffly erect, with an exaggerated jumping accompanied by a sputtering vocalisation (F&F 2004). **Mating** occurs at the bower, often with a Nape Presentation Display or the mouth and tongue being shown to the female, and he may present a fruit or twig to the female. Mating can occur more than once per visit, the female fluttering her wings after copulation

(Coates 1990). **NEST** The nest is a fairly large (up to 40cm or more) open cup made from sticks and vine tendrils, sometimes with strips of bark, lined with finer twiglets, a few curly vine tendrils and sometimes dry grass stems; the cup is about 10cm wide and 6cm deep. It is built up to 14m above ground and concealed in a fork in a large tree, bush or shrub, or sometimes a mangrove. In one study, the mean distance of six active nests from the nearest active bower was 143m; the nest site may be reused habitually over some years. **EGGS** The clutch consists of one egg, sometimes two in Australia, the egg slightly glossy creamy-coloured, tinged with greenish-buff, and marked with long brown and black vermiculations, these concentrated more at the wider end; mean size 40.8 × 27.9mm in New Guinea, while Australian clutches have slightly larger dimensions. No information on the incubation period. **NESTLING** As mentioned above, all parental care is by the female. The nestling is downy, coloured grey-brown on crown and body. It is fed with fruits and insects, including caterpillars and beetles (Coleoptera) (Coates 1990). One nestling was in the nest in Port Moresby for 21 days before leaving it (F&F 2004). **TIMING OF BREEDING** Breeds throughout the year in New Guinea, with some regional variation in peak months. Bower construction often coincides with the start of the dry season (April) around Port Moresby. Australian birds breed in the austral spring, September–December, with egg-laying mostly in November, but this may depend on the annual rainfall pattern. The display season lasts around 8–9 months, but is mainly from June to December.

HYBRIDS None reported, but a male Yellow-breasted Bowerbird (of nominate race) was displaying to a female-plumaged individual at a bower of a Fawn-breasted Bowerbird at Aiome, in the Ramu valley. Hybridisation is possible but would be very hard to detect. Similarly, the species comes into contact with Great Bowerbird on Cape York, and despite habitat differences, the Great inhabiting open forest, mixed pairings may be possible.

MOULT Little studied, but individuals in wing moult from October to March.

STATUS AND CONSERVATION Classified as Least Concern by BirdLife International. This is a fairly common species over much of its fairly extensive range in New Guinea, although the Vogelkop outlying population is curiously isolated and very poorly known. In Australia, however, it is classified as Near Threatened as the species is very localised in some areas of Cape York, in extreme North Queensland. The Australian Bird Atlas project received only 99 records of Fawn-breasted Bowerbird between September 1998 and July 2012, raising the question of why these bowerbirds are more frequently encountered in New Guinea than in Australia. The paucity of birdwatchers on Cape York may be part of the answer, but this bowerbird does seem far less obvious here than in parts of New Guinea, where it can be a suburban species, as in some areas of Port Moresby. The Fawn-breasted Bowerbird feeds mostly on fruits, so there was some concern that the regular, intensive dry-season burning on the Cape might be depleting some of this food resource. Another theory was that competition from Great Bowerbirds (absent in New Guinea) might be driving them out (despite habitat differences between the two species), but neither theory is as yet substantiated. M. Tarburton (at a site near Port Moresby) notes: 'During a lapse in cat and dog trapping with our eco-traps, on two occasions I saw a cat hiding near and directing its attention towards a bower at PAU. As cats are much more widespread in Australia I would posit that cats might be part of the problem.' Preliminary results of surveys to follow up on distribution and population trends indicate that the Fawn-breasted Bowerbird appears to be most numerous within heathlands that have an overstorey of tall, emergent trees. The Action Plan for Australian Birds 2010 upgraded the conservation status of the Fawn-breasted Bowerbird from Least Concern to Near Threatened, this based on a total estimated population of fewer than 10,000 individuals and which may be declining (BirdLife Australia 2016).

Fawn-breasted Bowerbird, Iron Range National Park, Queensland, Australia, September 2018 (*Jun Matsui*).

Fawn-breasted Bowerbird, avenue bower, Iron Range National Park, Queensland, Australia, September 2018 (*Jun Matsui*).

BIBLIOGRAPHY

Aggerbeck, M. Fjeldså, J. L. Christidis, L. Fabre, P. H. & Jonsson, K. A. (2014) Resolving deep lineage divergences in core corvoid passerine birds supports a proto-Papuan island origin. *Mol. Phyl. & Evol.* 70: 272–285.

Aruah, A. & Yaga, A. (1992) Acquisition of adult male plumage in some birds of paradise at Baiyer River sanctuary. *Muruk* 5(2): 49–52.

Attenborough, D. (1996) *Attenborough in Paradise*. BBC, Bristol.

Attenborough, D. & Fuller, E. (2012) *Drawn from Paradise: The Discovery, Art and Natural History of the Birds of Paradise*. Collins, London.

Bailey, S. F. (1996) The perfect love "nest". *Pacific Discovery* 49: 41–43.

Barker, K. F. Barrowclough, G. F. & Groth, J. G. (2001) A phylogenetic analysis for passerine birds: taxonomic and biogeographic implications of an analysis of nuclear DNA sequence data. *Proc. Royal Soc. London (B)* 269: 295–308.

Barker, F. K. Cibois, A. Schikler, P. Feinstein, J. & Cracraft, J. (2004) Phylogeny and diversification of the largest avian radiation. *Proc. Natl. Acad. Sci. USA* 101: 11040–11045.

Barnard, H. G. (1911) Field notes from Cape York. *Emu* 11: 17–32.

Barraclough, T. G. Harvey, P. H. & Nee, S. (1995) Sexual selection and taxonomic diversity in passerine birds. *Proc. Royal Soc. London (B Biol. Sci.)* 259: 211–215.

Barrett, G. *et al.*, (2003) *The New Atlas of Australian Birds*. Birds Australia (RAOU), Melbourne.

Baylis, T. (2015) *Audiowings* Vol. 1(2): 12–14.

Beccari, O. (1878) The gardener bird and a new orchid. *The Gardener's Chronicle* 16: 332.

Beehler, B. (1981) Paradise birds. *Wildlife* (London) 23(9): 15–17.

Beehler, B. M. (1983a) The behavioral ecology of four birds of paradise. PhD dissertation, Princeton University; 209pp. From *Diss. Abstr. Int. B Sci. Eng.* 44(4): 998. October 1983.

Beehler, B. M. (1983b) Frugivory and polygamy in birds of paradise. *Auk* 100: 1–12.

Beehler, B. M. (1983c) Lek behavior of the Lesser Bird of paradise. *Auk* 100: 992–995.

Beehler, B. M. (1985) Adaptive significance of monogamy in the Trumpet Manucode *Manucodia keraudrenii* (Aves, Paradisaeidae). *Orn. Monogr.* 37: 83–99.

Beehler, B. M. (1987a) Ecology and behaviour of the Buff-tailed Sicklebill, (Paradisaeidae, *Epimachus albertisi*). *Auk* 104: 48–55.

Beehler, B. M. (1987b) Birds of paradise and mating system theory – predictions and observations. *Emu* 87: 78–89.

Beehler, B. M. (1988) Lek behavior of the Raggiana Bird of Paradise. *Natl. Geogr. Res.* 4: 343–358.

Beehler, B. M. (1989) Patterns of frugivory and the evolution of birds of paradise. In *Acta XIX Congressus Internationalis Ornithologica* pp. 816–818. University of Ottawa Press, Ottawa.

Beehler, B. M. (1991) *A Naturalist in New Guinea*. University of Texas Press, Austin.

Beehler, B. M. & Beehler, C. H. (1986) Observations on the ecology and behaviour of the Pale-billed Sicklebill. *Wilson Bull.* 98: 505–515.

Beehler, B. M. & Dumbacher, J. P. (1996) More examples of fruiting trees visited predominantly by birds of paradise. *Emu* 96: 81–88.

Beehler, B. M. & Finch, B. (1985) *Species Checklist of the Birds of Papua New Guinea*. RAOU, Melbourne.

Beehler, B. M. & Pratt, T. K. (2016) *Birds of New Guinea: Distribution, Taxonomy and Systematics*. Princeton University Press, Princeton.

Beehler, B. M. & Prawiradilaga, D. M. (2010) New taxa and records of birds from the north coastal ranges of New Guinea. *Bull. Brit. Orn. Club* 130: 277–285.

Beehler, B. & Pruett-Jones, S. (1983) Display dispersion and diet of birds of paradise: a comparison of nine species. *Behav. Ecol. Sociobiol.* 13: 229–238.

Beehler, B. M. & Swaby, R. J. (1991) Phylogeny and biogeography of the *Ptiloris* riflebirds (Aves, Paradisaeidae). *Condor* 93: 738–745.

Beehler, B. M. Diamond, J. Kemp, N. Scholes, E. Milensky, C. & Laman, T. 2012. Avifauna of the Foja Mountains of western New Guinea. *Bull. Brit. Orn. Club* 132: 84–101.

Beehler, B. Pratt, T. K. & Zimmerman, D. (1986) *Birds of New Guinea*. Princeton University Press, Princeton.

Bell, H. L. (1967) Bower painting by Lauterbach's Bowerbird. *Emu* 66: 353–356.

Bell, H. L. (1969) Field notes on the birds of Ok Tedi River drainage, New Guinea. *Emu* 69: 193–211.

Bell, H. L. (1970) Additions to the avifauna of Goodenough Island, Papua. *Emu* 70: 179–182.

Bell, H. L. (1982a) A bird community of lowland rainforest in New Guinea. 1. Composition and density of the avifauna. *Emu* 82: 24–41.

Bell, H. L. (1982b) A bird community of lowland rainforest in New Guinea. 2. Seasonality. *Emu* 82: 65–74.

Bell, H. L. (1982c) A bird community of lowland rainforest in New Guinea. 3. Vertical distribution of the avifauna. *Emu* 82: 143–161.

Bell, H. L. (1983) A bird community of lowland rainforest in New Guinea. 5. Mixed-species feeding flocks. *Emu* 82(Suppl.): 256–275.

Bell, H. L. (1984) A bird community of lowland rainforest in New Guinea. 6. Foraging ecology and community structure of the avifauna. *Emu* 84: 142–158.

Berggy, J. (1978) Bird observations in the Madang Province. *PNG Bird Soc. Newsletter* 148: 9–20.

Bergman, S. (1956) On the display and breeding of the King Bird of Paradise, *Cicinnurus regius rex* (Scop) in captivity. *Nova Guinea* 7: 197–205.

Bergman, S. (1957a) On the display and breeding of the King Bird of Paradise, *Cicinnurus regius rex* in captivity. *Avicult. Mag.* 63: 115–124.

Bergman, S. (1957b) On the display of the six-plumed bird of paradise *Parotia sefilata* (Pennant). *Nova Guinea*, new series 8: 81–86.

Bergman, S. (1958) On the display of the Six-plumed Bird of Paradise, *Parotia sefilata* (Pennant). *Avicult. Mag.* 64: 3–8.

Bergman, S. (1961) *My father is a Cannibal*. Hale, London.

Bergman, S. (1968) *Mina paradisfåglar*. Bonnier, Stockholm.

Bernstein, H. A. (1864) Ueber einen neuen Paradiesvogel und einige andere neue Vögel. *J. Orn.* 12: 401–410.

Beruldsen, G. (2003) *Australian Birds their Nests and Eggs*. Beruldsen, Benmore Hills.

BirdLife International (2000) *Threatened Birds of the World*. Lynx Edicions and BirdLife International Barcelona and Cambridge.

BirdLife International (2017) *IUCN Red List for birds*. Downloadable from http://www.birdlife.org

Bishop, K. D. (1984) Notes on Wallace's Standardwing *Semioptera wallacii*. *Bull. Brit. Orn. Club* 104: 118–120.

Bishop, K. D. (1987) Interesting bird observations in Papua New Guinea. *Muruk* 2: 52–57.

Bishop, K. D. (1992) The Standardwing Bird of Paradise *Semioptera wallacii*. *Emu* 92: 72–78.

Bishop, K. D. (2005a) *A Review of the Avifauna of the TransFly Ecoregion: the Status, Distribution, Habitats and Conservation of the Region's Birds*. TransFly Ecoregion Action Program, WWF Project no. 950739.02.

Bishop, K. D. (2005b) *A Review of the Avifauna of New Guinea's Southern Lowland Forests and Freshwater Swamp Forests: the Status, Distribution, Habitats and Conservation of the Region's Birds*. High Conservation Value Forests, WWF Project no. PG0033.0.

Bishop, K. D. & Frith, C. B. (1979) A small collection of eggs of birds-of-paradise at Baiyer River Sanctuary, Papua New Guinea. *Emu* 79: 140–141.

Blakers, M. *et al.*, (1984) *The Atlas of Australian Birds*. RAOU and Melbourne University Press, Melbourne.

Blunt, D. & Frith, C. B. (2004) *Rawnsley's Bowerbird (Satin × Regent)*. Gondwana Guides.

Bock, W. J. (1963) Relationships between the birds of paradise and the bowerbirds. *Condor* 65: 91–125.

Bock, W. J. (1994) History and nomenclature of avian family-group names. *Bull. Amer. Mus. Nat. Hist.* 222: 1–281.

Boehm, E. M. (1967) Successful breedings at the Eduard Marshall Boehm aviaries in 1966. *Avicult. Mag.* 73: 116–120.

Borgia, G. (1985) Bower quality, number of decorations and mating success of male satin bowerbirds (*Ptilonorhynchus violaceus*): an experimental analysis. *Anim. Behav.* 33: 266–271.

Borgia, G. (1986) Sexual selection in bowerbirds. *Sci. Amer.* 254: 70–79.

Borgia, G. (1995) Threat reduction as a cause of differences in bower architecture, bower decoration and male display in two closely related bowerbirds *Chlamydera nuchalis* and *C. maculata*. *Emu* 95: 1–12.

Borgia, G. & Mueller, U. (1992) Bower destruction, decoration stealing and female choice in the Spotted Bowerbird *Chlamydera maculata*. *Emu* 92: 11–18.

Borgia, G. Pruett-Jones, S. & Pruett-Jones, M. (1985) The evolution of bower-building and the assessment of male quality. *Z. Tierpsychol.* 67: 225–236.

Bradley, J. M. (1987) Vocal behaviour and annual cycle of the Western Bowerbird *Chlamydera guttata*. *Austral. Bird Watcher* 12: 83–90.

Breeden, S. & Breeden, K. (1970) *A Natural History of Australia: 1. Tropical Queensland*. Collins, Sydney.

Breeden, S. & Wright, B. (1989) *Kakadu: Looking After the Country – the Gagudju Way*. Simon & Schuster, Sydney.

Buller, W. L. (1887) *A History of the Birds of New Zealand*, vol. 1. The Author, London.

Campbell, A. J. (1901) *Nests and Eggs of Australian Birds Including the Geographical Distribution of the Species and Popular Observations Thereon*. Parts 1–2. Published privately, Melbourne, Victoria.

Campbell, B. & Lack, E. (1985) *A Dictionary of Birds*. T. & A. D. Poyser, London.

Campbell, R. (1977) Magnificent Bird of Paradise *Diphyllodes magnificus*. *New Guinea Bird Soc. Newsletter* 138: 6.

Chaffer, N. (1945) The Spotted and Satin bower-birds, a comparison. *Emu* 44: 161–181

Chaffer, N. (1958a) Additional observations on the Golden Bower-bird. *Emu* 49: 19–25.

Chaffer, N. (1958b) "Mimicry" of the "Stage-maker". *Emu* 58: 53–55.

Chaffer, N. (1959) Bower building and display of the Satin Bower-bird. *Austral. Zool.* 12: 295–305.

Chaffer, N. (1984) *In Quest of Bowerbirds*. Rigby, Adelaide.

Chisholm, A. H. (1929) *Birds and Green Places*. Dent, London.

Chisholm, A. H. (1937) The problem of vocal mimicry. *Ibis* 14: 703–721.

Chisholm, A. H. (1951) More about vocal mimicry. *Emu* 51: 75–76.

Chisholm, A. H. (1957) Concerning the Golden Bower-bird. *Emu* 57: 52.

Chisholm, A. H. (1959) The history of anting. *Emu* 59: 101–130.

Chisholm, A. H. & Chaffer, N. (1956) Observations on the Golden Bower-bird. *Emu* 56: 1–38.

Christidis, L. & Boles, W. E. (1994) *The Taxonomy and Species of Birds of Australia and its Territories*. Monograph 2. RAOU, Melbourne.

Christidis, L. & Boles, W. E. (2008) *Systematics and Taxonomy of Australian Birds*. CSIRO Publishing, Melbourne.

Christidis, L. & Norman, J. A. (2010) Evolution of the Australasian songbird fauna. *Emu* 110: 21–31.

Christidis, L. & Schodde, R. (1991) Relationships of Australo-Papuan songbirds – protein evidence. *Ibis* 133(3): 277–285.

Christidis, L. & Schodde, R. (1992) Relationships among the birds-of-paradise (Paradisaeidae) and bowerbirds

(Ptilonorhynchidae), protein evidence. *Austral. J. Zool.* 40: 343–353.

Christidis, L. & Schodde, R. (1993) Sexual selection for novel partners: a mechanism for accelerated morphological evolution in the birds-of-paradise (Paradisaeidae). *Bull. Brit. Orn. Club* 113: 169–172.

Christidis, L. Leeton, P. R. & Westerman, M. (1996) Were bowerbirds part of the New Zealand fauna? *Proc. Natl. Acad. Sci. USA* 93: 3898–3901.

Clapp, G. E. (1986) Birds of Mt Scratchley summit and environs: 3520 m asl in south-eastern New Guinea. *Muruk* 1: 75–84.

Clements, J. F. (2007) *The Clements Checklist of the Birds of the World.* Christopher Helm, London.

Clench, M. H. (1992) Pterylography of birds-of-paradise and the systematic position of MacGregor's Bird-of-paradise (*Macgregoria pulchra*). *Auk* 109: 923–928.

Coates, B. J. (1973a) Birds observed on Mt Albert Edward, Papua. *New Guinea Bird Soc. Newsletter* 84: 3–7.

Coates, B. J. (1973b) Magnificent rifle birds in display. *New Guinea Bird Soc. Newsletter* 87: 3.

Coates, B. J. (1985) *The Birds of Papua New Guinea,* vol. 1 *Non-passerines.* Dove Publications, Alderley, Queensland.

Coates, B. J. (1990) *The Birds of Papua New Guinea:* vol. 2. Dove Publications, Alderley, Queensland.

Coates, B. J. & Bishop, K. D. (1997) *A Guide to the Birds of Wallacea.* Dove Publications, Alderley, Queensland.

Coates, B. J. & Lindgren, E. (1978) *Ok Tedi birds.* Report of a preliminary survey of the avifauna of the Ok Tedi area, Western Province, PNG. Unpublished report prepared for the Ok Tedi Environmental Task Force, Ok Tedi Development Co. and Office of Environment and Conservation, PNG.

Coates, B. J. & Peckover, W. S. (2001) *Birds of New Guinea and the Bismarck Archipelago. A Photographic Guide.* Dove Publications, Alderley, Queensland.

Coleman, S. W, Patricelli, G. L. & Borgia, G. (2004) Variable female preferences drive complex male displays. *Nature* 428: 742–745.

Coleman, S. W. Patricelli, G. L. *et al.,* (2007) Female preferences drive the evolution of mimetic accuracy in male sexual displays. *Biol. Lett.* 3: 463–466.

Collar, N. J. *et al.,* (1994) *Birds to Watch 2: The World List of Threatened Birds.* BirdLife International, Cambridge, UK.

Conway, C. J. & Martin, T. E. (2000) Effects of ambient temperature on avian incubation behaviour. *Behav. Ecol.* 11: 178–188.

Cooper, P. (1995) Observations of parent-rearing behaviour in the Lesser Bird of paradise. *Avicult. Mag.* 101: 194–199.

Cracraft, J. (1981) Toward a phylogenetic classification of the recent birds of the world. *Auk* 98: 681–714.

Cracraft, J. (1981) The use of functional and adaptive criteria in phylogenetic systematics. *Amer. Zool.* 21: 21–36.

Cracraft, J. (1992) The species of the birds-of-paradise (Paradisaeidae), applying the phylogenetic species concept to a complex pattern of diversification. *Cladistics* 8: 1–43.

Cracraft, J. & Feinstein, J. (2000) What is not a bird of paradise? Molecular and morphological evidence places *Macgregoria* in the Meliphagidae and the Cnemophilinae near the base of the corvoid tree. *Proc. Royal Soc. London (Ser. B Biol. Sci.)* 267: 233–241.

Cracraft, J. & Helm-Bychowski, K. (1993a) Recovering phylogenetic signal from DNA sequences: relationships with the corvine assemblage (Class Aves) as inferred from complete sequences of the mtDNA cytochrome b gene. *Mol. Biol. Evol.* 10: 1196–1214.

Crandall, L. S. (1931) *Paradise Quest, a Naturalist's Experience in New Guinea.* Scribner, New York.

Crandall, L. S. (1932) Notes on certain birds of paradise. *Zoologica* 11: 77–87.

Crandall, L. S. (1937a) Further notes on certain birds of paradise. *Zoologica* 22: 193–195.

Crandall, L. S. (1937b) Position of wires in the display of the Twelve-wired Bird of Paradise. *Zoologica* 22: 307–310.

Crandall, L. S. (1940) Notes on the display forms of Wahnes' Six-plumed Bird of Paradise. *Zoologica* 25: 257–259.

Crandall, L. S. (1941) Description of an egg of the Long-tailed bird of Paradise. *Zoologica* 26: 41–48.

Crandall, L. S. (1946) Further notes on display forms of the Long-tailed Bird of Paradise, *Epimachus meyeri meyeri* Finsch. *Zoologica* 31: 9–10.

Crandall, L. S. & Lester, C. W. (1937) Display of the Magnificent Rifle Bird. *Zoologica* 22: 311–314.

D'Albertis, L. M. (1880) *New Guinea, What I Did and What I Saw.* 2 volumes. Sampson Low, London.

David, N. & Gosselin, M. (2002) Gender agreement of avian specific names. *Bull. Brit. Orn. Club* 122: 14–49.

David, N. *et al.,* (2009) Justified corrections to avian names under Article 32.5.1.1. of the International Code of Zoological Nomenclature. *Zootaxa* 2217: 56–66.

Davis, W. E. Jr, & Beehler, B. M. (1994) Nesting behavior of a Raggiana bird of paradise. *Wilson Bull.* 106: 522–530.

Day, L. B. Westcott, D. A. & Olster, D. H. (2005) Evolution of bower complexity and cerebellum size in bowerbirds. *Brain Behav. Evol.* 66: 62–72.

Dharmakumarsinhji, K. S. (1943) Notes on the breeding of the Empress of Germany's Bird of Paradise in captivity. *Zoologica* 28: 139–144.

Diamond, J. M. (1969) Preliminary results of an ornithological exploration of the North Coastal Range, New Guinea. *Amer. Mus. Novit.* 2362: 1–57.

Diamond, J. M. (1972) *Avifauna of the Eastern Highlands of New Guinea.* Publications of the Nuttall Ornithological Club 12. Cambridge, Massachusetts.

Diamond, J. M. (1981) *Epimachus bruijnii,* the Lowland Sickle-billed Bird-of Paradise. *Emu* 81: 82–86.

Diamond, J. M. (1982a) Rediscovery of the yellow-fronted gardener bowerbird. *Science* 216: 431–434.

Diamond, J. M. (1982b) Rediscovery of the bowerbird *Amblyornis flavifrons. PNG Bird Soc. Newsletter* 187–8: 38–39.

Diamond, J. M. (1982c) Evolution of bowerbirds, animal origins of the aesthetic sense. *Nature (London)* 297: 99–102.

Diamond, J. M. (1985) New distributional records and taxa from the outlying mountain ranges of New Guinea. *Emu* 85: 65–91.

Diamond, J. M. (1986) Biology of birds of paradise and bowerbirds. *Ann Rev. Ecol. Syst.* 17: 17–37.

Diamond, J. M. (1987) Flocks of brown and black New Guinea birds, a bicoloured mixed-species foraging association. *Emu* 87: 201–211.

Diamond, J. M. (1987a) Bower building and decoration by the bowerbird *Amblyornis inornatus*. *Ethology* 74: 177–204.

Diamond, J. M. (1988) Experimental study of bower decoration by the bowerbird *Amblyornis inornatus*, using colored poker chips. *Amer. Natur.* 131: 631–653.

Diamond, J. M. & Bishop, K. D. (1994) New records and observations from the Aru Islands, New Guinea region. *Emu* 94: 41–45.

Diamond, J. & Bishop, K. D. (2015) Avifaunas of the Kumawa and Fakfak Mountains, Indonesian New Guinea. *Bull. Brit. Orn. Club* 135: 292–336.

Dickinson, E. C. (ed.) (2003) *The Howard and Moore Complete Checklist of the Birds of the World*. Christopher Helm, London.

Dickinson, E. C. & Christidis, L. (eds) (2014) *The Howard and Moore Complete Checklist of the Birds of the World*. Vol. 2 *Passerines*. Aves Press, Eastbourne, UK.

Dickinson, E. C. & Remsen, J. V. (eds) (2013) *The Howard and Moore Complete Checklist of the Birds of the World*. Vol. 1 *Non-passerines*. Aves Press, Eastbourne, UK.

Doerr, N. (2018) Male Great Bowerbirds perform courtship display using a novel structure that rivals cannot destroy. *Emu* https://doi.org/10.1080/015841 97.2018.1442228

Donaghey, R. H. (1981) *The Ecology and Evolution of Bowerbird Mating Systems*. PhD thesis, Department of Zoology, Monash University, Melbourne, Victoria.

Donaghey, R. H. (1996) Bowerbirds. Pp. 138–187 in: Strahan, R. (ed.) *Finches, Bowerbirds and Other Passerines of Australia*. Angus & Robertson, Sydney & London.

Donaghey, R. H. Frith, C. B. & Lill, A. (1985) Bowerbird. Pp. 60–62 in: Campbell & Lack (1985).

Doucet, S. M. & Montgomerie, R. (2003a) Bower location and orientation in Satin Bowerbird: optimizing the conspicuousness of male displays. *Emu* 103: 105–109.

Doucet, S. M. & Montgomerie, R. (2003b) Multiple sexual ornaments in satin bowerbirds: UV plumage and bowers signal different aspects of male quality. *Behav. Ecol.* 14: 503–509.

Draffan, R. D. W. (1978) Group display of the Emperor of Germany Bird-of-Paradise *Paradisaea guilielmi* in the wild. *Emu* 78: 157–159.

Draffan, R. (1979) A defence of his display ground by the Magnificent Bird of paradise, *Diphyllodes magnificus*. *PNG Bird Soc. Newsletter* 156: 2–3.

Draffan, R. D. W. Garnett, S. T. & Malone, G. J. (1983) Birds of the Torres Strait: an annotated list and biogeographical analysis. *Emu* 83: 207–234.

Dwyer, P. Minnegal, M. & Thomson, J. (1985) Odds and ends, bower birds as taphonomic agents. *Austr. Archaeology* 21: 1–10.

Eastwood, C. (1989) Recent observations July–September 1988. *Muruk* 4: 25–37.

Eastwood, C. & Gregory, P. 1995. Interesting sightings during 1993 & 1994. *Muruk* 7: 128–142.

Elliot, D. G. (1873) *A Monograph of the Paradiseidæ, or Birds of Paradise*. Published privately, London.

Endler, J. Endler, L. & Doerr, N. R. (2010) Great bowerbirds create theaters with forced perspective when seen by their audience. *Current Biology* 20(18): 1679–1684.

Ensley, P. K. & Osborn, K, (1993) Hemosiderosis in a wild caught bird of paradise and other species in Papua New Guinea. *Annu. Proc. Am. Assoc. Zoo Vet.* p. 27.

Ericson, P. G. P. Christidis, L. Cooper, A. Irestedt, M. Jackson, J. Johansson, U. S. & Norman, J. A. (2002) A Gondwanan origin of passerine birds supported by DNA sequences of the endemic New Zealand wrens. *Proc. Royal Soc. London (B)* 269: 235–241.

Ericson, P. G. P. *et al.*, (2014) Dating the diversification of the major lineages of Passeriformes (Aves). *BMC Evol. Biol.* 14:8 doi:10.1186/1471-2148-14-8.

Erritzøe, J. Kampp, K. Winker, K. & Frith, C. B. (2007) *The Ornithologist's Dictionary*. Lynx Edicions, Barcelona.

Everett, M. (1979) *The Birds of Paradise and Bowerbirds*. New Burlington Books, London.

Everitt, C. (1965) Breeding the Magnificent Bird of paradise. *Avicult. Mag.* 71: 146–148.

Everitt, C. (1973) *Birds of the Edward Marshall Boehm Aviaries*. Boehm, Trenton.

Favaloro, N. (1940) The Spotted Bower-bird in Victoria. *Emu* 39: 273–277.

Finch, B. W. (1983) Birds of the Vanapa–Veimauri–Kanosia–Cape Suckling regions. *PNG Bird Soc. Newsletter* 199–200: 17–40.

Finsch, O. (1874) Zusätze und Berichtigungen zur Revision der Vögel Neuseelands. *J. Orn.* 22: 167–224.

Forbes, W. A. (1882) On the convoluted trachea of two species of manucode with remarks on similar structures in other birds. *Proc. Zool. Soc. London* 1882: 353.

Ford, J. (1977) Taxonomic status of the Spotted Catbird on Cape York Peninsula. *Sunbird* 8: 61–64.

Forshaw, J. M. & Cooper, W. T. (1977) *The Birds of Paradise and Bower Birds*. Collins, Sydney.

Fortune-Hopkins, H. C. (1988) Some feeding records for birds of paradise. *Muruk* 3(1): 12–13.

Friedmann, H. (1934) The display of Wallace's Standard-wing Bird of Paradise in captivity. *Scientific Monthly* (39): 52–55.

Frith, C. B. (1968) Some displays of Queen Carola's Parotia. *Avicult. Mag.* 74: 85–90.

Frith, C. B. (1970) Sympatry of *Amblyornis subalaris* and *A. macgregoriae* in New Guinea. *Emu* 70: 196–197.

Frith, C. B. (1970a) The nest and nestling of the Short-tailed Paradigalla *Paradigalla brevicauda* (Paradisaeidae). *Bull. Brit. Orn. Club* 90: 122–124.

Frith, C. B. (1971) Some undescribed nests and eggs of New Guinea birds. *Bull. Brit. Orn. Club* 91: 46–49.

Frith, C. B. (1974) Observations on Wilson's Bird of Paradise. *Avicult. Mag.* 80: 207–212.

Frith, C. B. (1976) Displays of the Red Bird of Paradise *Paradisaea rubra* and their significance, with a discussion on displays and systematics of other Paradisaeidae. *Emu* 76: 69–78.

Frith, C. B. (1979) Ornithological literature of the Papuan Subregion 1915 to 1976: an annotated bibliography. *Bull. Amer. Mus. Nat. Hist.* 164: 379–465.

Frith, C. B. (1980) Copulation by the Green Catbird. *Emu* 80: 39.

Frith, C. B. (1981) Displays of Count Raggi's Bird-of-Paradise *Paradisaea raggiana* and congeneric species. *Emu* 81: 193–201.

Frith, C. B. (1987) An undescribed plumage of Loria's Bird of Paradise *Loria loriae*. *Bull. Brit. Orn. Club* 107: 177–180.

Frith, C. B. (1987) Fawn-breasted Bowerbird *Chlamydera cerviniventris* on the Lai River, Jimi Valley, Western Highlands, Papua New Guinea. *Muruk* 2: 63.

Frith, C. B. (1989) A construction worker in the rainforest. *Birds International* 1: 29–39.

Frith, C. B. (1992) Standardwing Bird of Paradise *Semioptera wallacii* displays and relationships, with comparative observations on displays of other Paradisaeidae. *Emu* 92: 79–86.

Frith, C. B. (1994) Egg-laying at long intervals in bowerbirds (Ptilonorhynchidae). *Emu* 94: 60–61.

Frith, C. B. (1994a) The status and distribution of the Trumpet Manucode *Manucodia keraudrenii* (Paradisaeidae) in Australia. *Austral. Bird Watcher* 15: 218–224.

Frith, C. B. (1994b) Adaptive significance of tracheal elongation in manucodes (Paradisaeidae). *Condor* 96: 552–555.

Frith, C. B. (1995a) Cicadas (Insecta: Cicadidae) as prey of Regent Bowerbirds *Sericulus chrysocephalus* and other bowerbird species (Ptilonorhynchidae). *Sunbird* 25(2): 44–48.

Frith, C. B. (1995b) Range extension of Splendid Astrapia *Astrapia splendidissima*, a sighting of an *A. mayeri* × *A. stephaniae* hybrid, or an unidentified *Astrapia* sp. (Paradisaeidae)? *Muruk* 7: 49–52.

Frith, C. B. (1996) Further notes on little-known plumages of the Crested and Loria's Birds of Paradise *Cnemophilus macgregorii* and *C. loriae*. *Bull. Brit. Orn. Club* 116: 247–251.

Frith, C. B. (1997) Huia (*Heteralocha acutirostris*, Callaeidae) -like sexual bill dimorphism in some birds of paradise (Paradisaeidae) and its significance. *Notornis* 44: 177–184.

Frith, C. B. (2001) *Evolutionary Studies of Bowerbirds and Birds of Paradise: Affinities and Divergence*. 2 vols. PhD thesis, Australia School of Environmental Studies, Griffith University, Brisbane, Queensland.

Frith, C. B. (2003) A nest of the Greater Bird of Paradise *Paradisaea apoda*. *Bull. Brit. Orn. Club* 123: 271–273.

Frith, C. B. (2006) A history and reassessment of the unique but missing specimen of Rawnsley's Bowerbird *Ptilonorhynchus rawnsleyi*, Diggles 1867 (Aves: Ptilonorhynchidae). *Historical Biol.* 18: 53–64.

Frith, C.B. (2016) Bowerbird display site nomenclature: The court case for the Tooth-billed Bowerbird. *North Queensland Natur.* 46: 107–114.

Frith, C. B. & Beehler, B. M. (1997) Courtship and mating behaviour of the Twelve-wired Bird of Paradise *Seleucidis melanoleuca*. *Emu* 97: 133–140.

Frith, C. B. & Beehler, B. M. (1998) *The Birds of Paradise*. Oxford University Press, Oxford.

Frith, C. B. & Coles, D. (1976) Additional notes on displays of Queen Carola's Bird of Paradise. *Avicult. Mag.* 82: 52–53.

Frith, C. B. & Cooper, W. T. (1996) Courtship display and mating of Victoria's Riflebird *Ptiloris victoriae* (Paradisaeidae) with notes on the courtship of congeneric species. *Emu* 96: 102–113.

Frith, C. B. & Frith, D. W. (1979) Leaf-eating by birds-of-paradise and bower birds. *Sunbird* 10(1): 21–23.

Frith, C. B. & Frith, D. W. (1981) Displays of Lawes's Parotia *Parotia lawesii* (Paradisaeidae), with reference to those of congeneric species and their evolution. *Emu* 81: 227–238.

Frith, C. B. & Frith, D. W. (1984) Foraging ecology of birds in an upland tropical rainforest in north Queensland. *Aust. Wildl. Res.* 11: 325–347.

Frith, C. B. & Frith, D. W. (1985a) Parental care and investment in the Tooth-billed Bowerbird *Scenopoeetes dentirostris* (Ptilonorhynchidae). *Australian Bird Watcher* 11: 103–113.

Frith, C. B. & Frith, D. W. (1985) Seasonality of insect abundance in an Australian upland tropical rainforest. *Austr. J. Ecol.* 10: 31–42.

Frith, C. B. & Frith, D. W. (1986) *Australian Tropical Rainforest Life*. Tropical Australia Graphics, Paluma, Australia.

Frith, C. B. & Frith, D. W. (1988) Discovery of nests and eggs of Archbold's Bowerbird *Archboldia papuensis* (Ptilonorhynchidae). *Australian Bird Watcher* 12: 251–257.

Frith, C. B. & Frith, D. W. (1989) Miscellaneous notes on the bowerbirds *Chlamydera cerviniventris* and *Chlamydera lauterbachi* (Ptilonorhynchidae) in Papua New Guinea. *Australian Bird Watcher* 13: 6–19.

Frith, C. B. & Frith, D. W. (1990a) Archbold's Bowerbird *Archboldia papuensis* (Ptilonorhynchidae) uses plumes from King of Saxony Bird of Paradise *Pteridophora alberti* (Paradisaeidae) as bower decoration. *Emu* 90: 136–137.

Frith, C. B. & Frith, D. W. (1990b) Nesting biology and relationships of the Lesser Melampitta *Melampitta lugubris*. *Emu* 90: 65–73.

Frith, C. B. & Frith, D. W. (1990c) Discovery of the King of Saxony Bird of Paradise *Pteridophora alberti* nest, egg and nestling with notes on parental care. *Bull. Brit. Orn. Club* 110: 160–164.

Frith, C. B. & Frith, D. W. (1990d) The nesting biology of the Spotted Bowerbird *Chlamydera maculata* (Ptilonorhynchidae). *Australian Bird Watcher* 13: 218–225.

Frith, C. B. & Frith, D. W. (1990e) Notes on the nesting

biology of the Great Bowerbird *Chlamydera nuchalis* (Ptilonorhynchidae). *Australian Bird Watcher* 13: 137–148.

Frith, C. B. & Frith, D. (1991) *Australia's Cape York Peninsula*. Frith & Frith, Malanda.

Frith, C. B. & Frith, D. W. (1992a) Annotated list of birds in western Tari Gap, Southern Highlands, Papua New Guinea, with some nidification notes. *Australian Bird Watcher* 14: 262–276.

Frith, C. B. & Frith, D. (1992b) *Australia's Wet Tropics Rainforest Life*. Frith & Frith, Malanda.

Frith, C. B. & Frith, D. W. (1992c) The nesting biology of the Short-tailed Paradigalla *Paradigalla brevicauda* (Paradisaeidae). *Ibis* 134: 77–82.

Frith, C. B. & Frith, D. W. (1993a) Results of a preliminary highland bird-banding study at Tari Gap, southern highlands, Papua New Guinea. *Corella* 17: 5–21.

Frith, C. B. & Frith, D. W. (1993b) The nesting biology of the Ribbon-tailed Astrapia *Astrapia mayeri* (Paradisaeidae). *Emu* 93: 12–22.

Frith, C. B. & Frith, D. W. (1993c) Nidification of the Crested Bird of Paradise *Cnemophilus macgregorii* and a review of its biology and systematics. *Emu* 93: 23–33.

Frith, C. B. & Frith, D. W. (1994) Discovery of nests and an egg of Loria's Bird of Paradise *Cnemophilus loriae*. *Bull. Brit. Orn. Club* 114: 182–192.

Frith, C. B. & Frith, D. W. (1995a) Notes on the nesting biology of Victoria's Riflebird *Ptiloris victoriae* (Paradisaeidae). *Emu* 95: 162–174.

Frith, C. B. & Frith, D. W. (1995b) Hybridization between the Great and Spotted Bowerbird *Chlamydera nuchalis* and *C. maculata*, the first authenticated hybrid bowerbird (Ptilonorhynchidae). *Mem. Queensland Mus.* 38: 471–476.

Frith, C.B. & Frith, D.W. (1995c) Court site constancy, dispersion, male survival and court ownership in the male Tooth-billed Bowerbird, Scenopoeetes dentirostris (Ptilonorhynchidae). *Emu* 95: 89-98.

Frith, C. B. & Frith, D. W. (1996a) Description of the unique *Parotia lawesii* × *Paradisaea rudolphi* hybrid bird of paradise (Paradisaeidae). *Rec. Austr. Mus.* 48: 111–116.

Frith, C. B. & Frith, D. W. (1996b) The unique type specimen of the bird of paradise *Lophorina superba pseudoparotia* Stresemann 1934 (Paradisaeidae): a hybrid of *Lophorina superba* × *Parotia carolae*. *J. Orn.* 137: 515–521.

Frith, C. B. & Frith, D. W. (1997a) The taxonomic status of the bird of paradise *Paradigalla carunculata intermedia* (Paradisaeidae) with notes on the other *Paradigalla* taxa. *Bull. Brit. Orn. Club* 117: 38–48.

Frith, C. B. & Frith, D. W. (1997b) *Chlamydera guttata carteri* – an overlooked subspecies of Western Bowerbird (Ptilonorhynchidae) from North West Cape, Western Australia. *Rec. West. Austr. Mus.* 18: 219–226.

Frith, C. B. & Frith, D. W. (1997c) A distinctive new subspecies of Macgregor's Bowerbird (Ptilonorhynchidae) of New Guinea. *Bull. Brit. Orn. Club* 117: 199–205.

Frith, C. B. & Frith, D. W. (1997d) Biometrics of birds of paradise (Aves: Paradisaeidae) with observations on interspecific and intraspecific variation and sexual dimorphism. *Mem. Queensland Mus.* 42: 159–212.

Frith, C. B. & Frith, D. W. (1997e) Courtship display and mating of the King of Saxony Bird of Paradise *Pteridophora alberti* (Paradisaeidae) in New Guinea and its taxonomic significance. *Emu* 97: 185–193.

Frith, C. B. & Frith, D. W. (1998a) Additional notes on the nesting biology of Victoria's Riflebird *Ptiloris victoriae* (Paradisaeidae). *Emu* 98: 138–142.

Frith, C. B. & Frith, D. W. (1998b) Hybridization between Macgregor's Bowerbird *Amblyornis macgregoriae* and the Streaked Bowerbird *A. subalaris* (Ptilonorhynchidae) of New Guinea. *Bull. Brit. Orn. Club* 118: 7–14.

Frith, C. B. & Frith, D. W. (1998c) Nesting biology of the Golden Bowerbird *Prionodura newtoniana* endemic to Australian upland tropical rainforest. *Emu* 98: 245–268.

Frith, C. B. & Frith, D. W. (1998d) Aberrant plumages in some birds of paradise. *Mem. Queensland Mus.* 42: 439–443.

Frith, C. B. & Frith, D. W. (1999a) Folivory and bill morphology in the Tooth-billed Bowerbird, *Scenopoeetes dentirostris* (Passeriformes: Ptilonorhynchidae): food for thought. *Mem. Queensland Mus.* 43: 589–596.

Frith, C. B. & Frith, D. W. (1999b) Subspeciation in the Australian-endemic Great Bowerbird *Chlamydera nuchalis* (Ptilonorhynchidae): a review and revision. *Bull. Brit. Orn. Club* 119: 177–289.

Frith, C. B. & Frith, D. W. (2000a) Bower system and structures of the Golden Bowerbird, *Prionodura newtoniana* (Ptilonorhynchidae). *Mem. Queensland Mus.* 45: 293–316.

Frith, C. B. & Frith, D. W. (2000b) Attendance levels and behaviour at bowers by male Golden Bowerbirds, *Prionodura newtoniana* (Ptilonorhynchidae). *Mem. Queensland Mus.* 45: 317–341.

Frith, C. B. & Frith, D. W. (2000c) Home range and associated sociobiology and ecology of male Golden Bowerbirds *Prionodura newtoniana* (Ptilonorhynchidae). *Mem. Queensland Mus.* 45: 343–357.

Frith, C. B. & Frith, D. W. (2000d) Fidelity to bowers, adult plumage acquisition, longevity and survival in male Golden Bowerbirds *Prionodura newtoniana* (Ptilonorhynchidae). *Emu* 100: 249–263.

Frith, C. B. & Frith, D. W. (2001a) Morphology, moult and survival of three sympatric bowerbirds in Australian Wet Tropics upland rainforest. *Corella* 25: 41–60.

Frith, C. B. & Frith, D. W. (2001b) Display behaviour of the adult male Golden Bowerbird *Prionodura newtoniana* at the bower. *Australian Bird Watcher* 19: 3–13.

Frith, C. B. & Frith, D. W. (2001c) Biometrics of the bowerbirds (Aves: Ptilonorhynchidae): with observations on species limits, sexual dimorphism, intraspecific variation and vernacular nomenclature. *Mem. Queensland Mus.* 46: 521–542.

Frith, C. B. & Frith, D. W. (2004) *The Bowerbirds – Ptilonorhynchidae*. Oxford University Press, Oxford.

Frith, C. B. & Frith, D. W. (2005) A long-term bird banding study in upland tropical rainforest, Paluma Range, north-eastern Queensland with notes on breeding. *Corella* 29(2): 25–48.

Frith, C. B. & Frith, D. W. (2010) *Birds of Paradise: Nature, Art, History*. Frith & Frith, Malanda.

Frith, C. B. & Frith, D. W. (2009) Family Ptilonorhynchidae (Bowerbirds). Pp. 350–403 in: del Hoyo, J. Elliott, A. & Christie, D. A. (eds) (2009) *Handbook of the Birds of the World* Vol. 14. Bush-shrikes to Old World Sparrows. Lynx Edicions, Barcelona.

Frith, C. B. & Frith, D. W. (2009) Family Paradisaeidae (Birds of Paradise). Pp. 404–493 in: del Hoyo, J. Elliott, A. & Christie, D. A. (eds) (2009) *Handbook of the Birds of the World* Vol. 14. Bush-shrikes to Old World Sparrows. Lynx Edicions, Barcelona.

Frith, C. & Frith, D. (2014) Mr. Carter's bowerbird. *Australian Birdlife* 3(4): 42–45.

Frith, C. B. & Harrison, C. J. O. (1989) An undescribed plumage of the Crested Bird of Paradise *Cnemophilus macgregorii*. *Bull. Brit. Orn. Club* 109: 137–139.

Frith, C. B. & McGuire, M. (1996) Visual evidence of vocal avian mimicry by male Tooth-billed Bowerbirds *Scenopoeetes dentirostris* (Ptilonorhynchidae). *Emu* 96: 12–16.

Frith, C. B. & Murphy, T. (2012) A pale 'cream' Satin Bowerbird *Ptilonorhynchus violaceus* (Family Ptilonorhynchidae). First documented evidence of any plumage aberration in bowerbirds. *Australian Field Orn.* 29: 40–44.

Frith, C. B. & Nevill, S. (1998) Sunning by an aggregation of Regent Bowerbirds *Sericulus chrysocephalus* (Ptilonorhynchidae). *Australian Bird Watcher* 17(8): 398–401.

Frith, C. B. & Poulsen, M. K. (1999) Distribution and status of the Paradise Crow *Lycocorax pyrrhopterus* and Standardwing Bird of Paradise *Semioptera wallacii*, with notes on their biology and nidification. *Emu* 99: 229–238.

Frith, C. B. Borgia, G. & Frith, D. W. (1996) Courts and courtship behaviour of Archbold's Bowerbird *Archboldia papuensis* in Papua New Guinea. *Ibis* 136: 153–160.

Frith, C. Frith, D. W. & Bonan, A. (2017) Birds of Paradise (Paradisaeidae). In: del Hoyo, J. Elliott, A. Sargatal, J. Christie, D. A. & de Juana, E. (eds) *Handbook of the Birds of the World Alive*. Lynx Edicions, Barcelona.

Frith, C. Frith, D. & Bonan, A. (2017) Bowerbirds (Ptilonorhynchidae). In: del Hoyo, J. Elliott, A. Sargatal, J. Christie, D. A. & de Juana, E. (eds) *Handbook of the Birds of the World Alive*. Lynx Edicions, Barcelona.

Frith, C. B. Frith, D. W. & McCullough, M. (1995) Great and Spotted Bowerbirds *Chlamydera nuchalis* and *C. maculata* sympatric and interacting at each others bowers. *Australian Bird Watcher* 16: 49–57.

Frith, C. B. Frith, D. W. & Moore, G. J. (1994) Home range and extra-court activity in the male Tooth-billed Bowerbird, *Scenopoeetes dentirostris* (Ptilonorhynchidae). *Mem. Queensland Mus.* 37: 147–154.

Frith, C. B. Frith, D. W. & Weineke, J. (1994) An exceptionally elaborate bower structure of the Great Bowerbird *Chlamydera nuchalis* (Ptilonorhynchidae). *Australian Bird Watcher* 15: 314–319.

Frith, C. B. Frith, D. W. & Weineke, J. (1996) Dispersion, size and orientation of bowers of the Great Bowerbird *Chlamydera nuchalis* (Ptilonorhynchidae) in Townsville City, tropical Queensland. *Corella* 20: 45–55.

Frith, C. B. Gibbs, D. & Turner, K. (1995) The taxonomic status of populations of Archbold's Bowerbird *Archboldia papuensis* in New Guinea. *Bull. Brit. Orn. Club* 115: 109–114.

Frith, D. W. & Frith, C. B. (1988) Courtship display and mating of the Superb Bird of paradise *Lophorina superba*. *Emu* 88(3): 183–188.

Frith, D. W. & Frith, C. B. (1990) Seasonality of litter invertebrate populations in an Australian upland tropical rainforest. *Biotropica* 22: 181–191.

Frith, D. W. & Frith, C. B. (1991) Say it with bowers. *Wildl. Conserv.* 94: 74–83.

Frith, D. W. & Frith, C. B. (1995) *Cape York Peninsula*. Reed, Sydney.

Fuller, E. (1979) Hybridization among the Paradisaeidae. *Bull. Brit. Orn. Club* 99: 145–152.

Fuller, E. (1995) *The Lost Birds of Paradise*. Swan Hill Press, Shrewsbury, UK.

Gannon, G. R. (1930) Observations on the Satin Bower Bird with regard to the material used by it in painting its bower. *Emu* 30: 39–41.

Garnett, S. T. & Crowley, G. M. (2000) *The Action Plan for Australian Birds 2000*. Environment Australia, Canberra.

Garnett, S. T. Szabo, J. K. & Dutson, G. (2010) *The Action Plan for Australian Birds 2010*. CSIRO Publishing, Collingwood.

Gibbs, D. (1993) *Irian Jaya, Indonesia, 21 January–12 March 1991: a site guide for birdwatchers, with brief notes from 1992*. Published by the author.

Gibbs, D. (1994) Undescribed taxa and new records from the Fakfak Mountains, Irian Jaya. *Bull. Brit. Orn. Club* 114: 1–38.

Gill, F. & Donsker, D. (eds) (2017) *IOC World Bird List* (v 7.2).

Gill, F. & Wright, M. (2006) *Birds of the World Recommended English Names*. Christopher Helm, London.

Gilliard, E. T. (1950) Notes on birds of southeastern Papua. *Amer. Mus. Novit.* 1453: 1–40.

Gilliard, E. T. (1953) New Guinea's rare birds and stone age men. *Natl. Geogr. Mag.* 103: 421–488.

Gilliard, E. T. (1955) The land of the head-hunters. *Natl. Geogr. Mag.* 108: 437–486.

Gilliard, E. T. (1956a) The systematics of the New Guinea Manucode *Manucodia ater*. *Amer. Mus. Novit.* 1770: 1–13.

Gilliard, E. T. (1956b) Bower ornamentation versus plumage characters in bower-birds. *Auk* 73: 450–451.

Gilliard, E. T. (1963) The evolution of bowerbirds. *Sci. Amer.* 209: 38–46.

Gilliard, E. T. (1969) *Birds of Paradise and Bowerbirds*.

Weidenfeld and Nicholson. London.

Gilliard, E. T. & LeCroy, M. (1961) Birds of the Victor Emanuel and Hindenburg Mountains, New Guinea – Results of the American Museum of Natural History Expedition to New Guinea in 1954. *Bull. Amer. Mus. Nat. Hist.* 123: 1–86.

Gilliard, E. T. & LeCroy, M. (1967) Annotated list of the birds of the Adelbert mountains, New Guinea. Results of the 1959 Gilliard expedition. *Bull. Amer. Mus. Nat. Hist.* 138: 51–82.

Gilliard, E. T. & LeCroy, M. (1968) Birds of the Schrader Mountain Region, New Guinea. Results of the American Museum of Natural History Expedition to New Guinea in 1964. *Amer. Mus. Novit.* 2343: 1–41.

Gilliard, E. T. & LeCroy, M. (1970) Notes on birds from the Tamrau Mountains, New Guinea. *Amer. Mus. Novit.* 2420: 1–28.

Gooddie, C. (2010) *The Jewel Hunter.* WILDGuides, Old Basing.

Goodfellow, W. (1908). Account of expedition to British New Guinea. *Bull. Brit. Orn. Club* 23: 35–39.

Goodfellow. W. (1926) Remarks on his recent journey in Papua New Guinea and on the birds of paradise met with. *Bull. Brit. Orn. Club* 46: 58–59.

Goodfellow, W. (1927) Wallace's Bird of Paradise (*Semioptera wallacei*). *Avicult. Mag.* (5)4: 57–65.

Goodwin, A. P. (1890) Notes on the Paradise-birds of British New Guinea. *Ibis* 32(2): 150–156.

Gould, J. & Sharpe, R. B. (1875–1888) *The Birds of New Guinea.* Vol. 1. Sotheran, London.

Greenway, J. C. (1942) A new manucode bird of paradise. *Proc. New England Zool. Club* 14: 15–106.

Greenway, J. C. Jr (1966) Birds collected on Batanta, off Western New Guinea, by E. Thomas Gilliard in 1964. *Amer. Mus. Novit.* 22: 1–27.

Gregory, P. A. & Johnston, G. R. (1993) Birds of the cold tropics: Dokfuma, Star Mountains, New Guinea. *Bull. Brit. Orn. Club* 113: 139–144.

Gregory, P. (1995) Further studies of the birds of the Ok Tedi area, Western Province, Papua New Guinea. *Muruk* 7: 1–38.

Gregory, P. (1996) *The birds of the Ok Tedi area.* Ok Tedi Mining, Port Moresby.

Gregory, P. (2011) An overview of recent taxonomic changes to the avifauna of New Guinea. *Muruk* 10: 2–40.

Gregory, P. (2015) *A Checklist of the Birds of New Guinea and its offshore islands v.5.1.* Sicklebill Publications, Kuranda.

Gregory, P. (2017) *A Checklist of the Birds of Australia and its Island Territories v.6.0.* Sicklebill Publications, Kuranda.

Gregory, P. (2017) *Birds of New Guinea Including Bismarck Archipelago and Bougainville.* Lynx Edicions, Barcelona.

Grzimek, B. & Schultze-Westrum, T. (1984) Birds of paradise. Pp. 487–499 in: Grzimek *et al.*, (1984) *Animal Life Encyclopaedia: Birds* 3. Vol. 9. Van Nostrand Reinhold Inc.

Gyldenstope, N. (1955) Birds collected by Dr Sten Bergman during his expedition to Dutch New Guinea, 1951. *Ark. f. Zool.* 8: 1–181.

Hadden, D. (1975) Birds seen in the Tari area from 1 August to 14 August 1975. *New Guinea Bird Soc. Newsletter* 113: 8–9.

Hallstrom, E. (1959) Some breeding results in the Hallstrom collection. *Avicult. Mag.* 65: 77–80.

Hamilton, W. D. & Zuk, M. (1982) Heritable true fitness and bright birds: a role for parasites? *Science* 218: 384–387.

Hammer, S. Jensen, S. Balzer, J. & Sandow, D. (2003) DNA sexing in birds of paradise and bowerbirds. *Intern. Zoo News* 50: 156–159.

Harrington G. (2017) If it smells like a bower and functions like a bower then... response to Frith. *North Queensland Natur.* 47: 6–7.

Harrison, C. J. O. (1985) Plumage, abnormal. In Campbell, B. & Lack, E. (eds) *A Dictionary of Birds.* Poyser, Calton.

Harrison, C. J. O. & Frith, C. B. (1970) Nests and eggs of some New Guinea birds. *Emu* 71: 85–86.

Harrison, J. M. (1964) Plumage. Pp. 639–643 in: Thomson, A. L. (ed.) (1964) *A New Dictionary of Birds.* Thomas Nelson and Sons, London.

Hartert, E. J. O. (1910) On the eggs of the Paradisaeidae. *Novit. Zool.* 17: 484–491.

Heads, M. (2001a) Birds of paradise, biogeography and ecology in New Guinea: a review. *J. Biogeogr.* 28: 893–925.

Heads, M. (2001b) Birds of paradise and bowerbirds: regional levels of biodiversity and terrane tectonics in New Guinea. *J. Zool.* 255: 331–339.

Heads, M. (2002) Birds of paradise, vicariance biogeography and terrane tectonics in New Guinea. *J. Biogeog.* 29: 261–283.

Healey. C. J. (1976) Sympatry in *Parotia lawesii* and *P. carolae. Emu* 76: 85.

Healey, C. J. (1978a) Effects of human activity on *Paradisaea minor* in the Jimi Valley, New Guinea. *Emu* 78: 149–155.

Healey, C. J. (1978b) Communal display of Princess Stephanie's Astrapia *Astrapia stephaniae* (Paradisaeidae). *Emu* 78: 197–200.

Healey, C. J. (1980) Display of Queen Carola's Parotia *Parotia carolae* (Paradisaeidae). *PNG Bird Soc. Newsletter* 163–164: 6–9.

Healey, C. J. (1986) Men and birds in the Jimi Valley. The impact of man on birds of paradise in the Papua New Guinea Highlands. *Muruk* 1: 34–71.

Heinrich G. (1956) Biologische Aufzeichnungen über Vögel von Halmahera und Batjan. *J. Orn.* 97: 31–40.

Helm-Bychowski, K. M. & Cracraft, J. (1993) Recovering phylogenetic signal from DNA sequences: relationships within the Corvine assemblage (Class Aves) as inferred from complete sequences of the mitochondrial DNA cytochrome b gene. *Mol. Biol. Evol.* 10: 1196–1214.

Hicks, R. (1987) *Checklist of the Birds of Papua New Guinea.* PNGBS, Port Moresby.

Hicks, R. K. (1988a) Feeding observations at a fruiting *Pipturus. Muruk* 3: 15.

Hicks, R. K. (1988b) Feeding observations of female Crested Bird of Paradise. *Muruk* 3: 15.

Hicks, R. (1998) *Checklist of the Birds of New Guinea*. Privately published.

Hicks, J. H. & Hicks, R. K. (1988) Display of Loria's Bird of Paradise. *Muruk* 3: 52.

Hicks, R. K. & Hicks, J. H. (1988) Feeding observations of Short-tailed Paradigalla. *Muruk* 3: 14.

Hides, J. G. (1936) *Papuan Wonderland*. Blackie, London.

Higgins, P. J. Peter, J. M. & Cowling, S. J. (eds) (2006) *Handbook of Australian, New Zealand and Antarctic Birds*. Vol. 7, Part A. Boatbill to larks. Oxford University Press, Melbourne, Victoria.

Hoogerwerf, A. (1964) On birds new for New Guinea or with a larger range than previously known, *Bull. Brit. Orn. Club* 84: 153–161.

Hoogerwerf, A. (1971) On a collection of birds from the Vogelkop, near Manokwari, north-western New Guinea. *Emu* 71: 73–83.

Hopkins, H. C. F. (1988) Some feeding records for Birds of Paradise. *Muruk* 3: 12–13.

Howe, R. W. (1986) Bird distributions in forest islands in north-eastern New South Wales. Pp. 119–129 in: Ford, H. A. & Paton, D. C. (eds) *The Dynamic Partnership: Birds and Plants in Southern Australia*. The Flora and Fauna of South Australia Handbooks Committee, Adelaide.

Hoyle, M. A. (1975) Observations on birds of paradise/ bowerbirds. *New Guinea Bird Soc. Newsletter* 110: 6.

del Hoyo, J. & Collar, N. J. (2016) *HBW and BirdLife International Illustrated Checklist of the Birds of the World*. Vol. 2: Passerines. Lynx Edicions, Barcelona.

del Hoyo, J. & Collar, N. J. (2016) In del Hoyo, J. Elliott, A. Sargatal, J. Christie, D. A. & de Juana, E. (eds) *Handbook of the Birds of the World Alive*. Lynx Edicions, Barcelona.

Hundgen, K. H. & Bruning, D. F. (1988) Propagation techniques for birds of paradise at the New York Zoological Park. *AAZPA (Amer. Assoc. Zool. Parks Aquariums) Ann. Conf. Proc.*: 14–20.

Hundgen, K. Hutchins, M. Sheppard, C. Bruning, D. & Worth, W. (1991) Management and breeding of the Red Bird of Paradise *Paradisaea rubra* at the New York Zoological Park. *Intern. Zoo Yearbook* 30: 192–199, illustr.

Hundgen, K. Sheppard, C. Bruning, D. Hutchins, M. Worth, W. & Laska, M. (1990) Management and breeding the Lesser Bird of paradise *Paradisaea minor* at the New York Zoological Park. *AAZPA (Amer. Assoc. Zool. Parks Aquariums) Ann. Conf. Proc.*

Hutton, F. W. & Drummond, J. (1904) *The Animals of New Zealand : An Account of the Colony's Air-breathing Vertebrates*. Whitcombe & Tombs, Christchurch.

Iredale, T. (1948) A check list of the birds of paradise and bowerbirds. *Austr. Zool.* 2: 161–189.

Iredale, T. (1950) *The Birds of Paradise and Bowerbirds*. Georgian House, Melbourne, Victoria.

Iredale, T. (1956) *Birds of New Guinea*; 2 Vols. Georgian House, Melbourne.

Irestedt, M. Ohlson, J. I. Zuccon, D. Källersjö, M. & Ericson, P. G. P. (2006) Nuclear DNA from old collections of avian study skins reveals the evolutionary history of the Old World suboscines (Aves, Passeriformes). *Zool. Scripta* 35: 567–580.

Irestedt, M. & Ohlson, J. I. (2008) The division of the major songbird radiation into Passerida and 'core Corvoidea' (Aves: Passeriformes) – the species tree vs. gene trees. *Zool. Scripta* 37: 305–313.

Irestedt, M. Jønsson, K. A. Fjeldså, J. L. & Ericson, P. G. P. (2009) An unexpectedly long history of sexual selection in birds-of-paradise. *BMC Evol. Biol.* 9: 235. doi:10.1186/1471-2148-9-235.

Irestedt, M. Batalha-Filho, H. Roselaar, C. S. Christidis, L. & Ericson, P. G. P. (2015) "Contrasting phylogeographic signatures in two Australo-Papuan bowerbird species complexes (Aves: *Ailuroedus*)". *Zool. Scripta*. doi:10.1111/zsc.12163

Irestedt, M. Batalha-Filho, H. Ericson, P. G. P. Christidis, L. & Schodde, R. (2017) Phylogeny, biogeography and taxonomic consequences in a bird-of-paradise species complex, *Lophorina–Ptiloris* (Aves: Paradisaeidae). *Zool. J. Linn. Soc.* 181: 439–470.

Isenberg, A. H. (1961) Nesting of the Red bird of paradise. *Avicult. Mag.* 67: 43–44.

Isenberg, A. H. (1962) Further notes on the breeding of the Red bird of paradise. *Avicult. Mag.* 68: 48.

Jackson, S. W. (1907) *Egg collecting and bird life of Australia. Catalogue and data of the Jacksonian Oological Collection*. Sydney. F. W. White, Printer, Melbourne.

Jobling, J. A. (1991) *A Dictionary of Scientific Bird Names*. Oxford University Press, Oxford.

Jobling, J. A. (2010) *Helm Dictionary of Scientific Bird Names*. Christopher Helm, London.

Jobling, J. A. (2016) *Key to Scientific Names in Ornithology*. In: del Hoyo, J. Elliott, A. Sargatal, J. Christie, D. A. & de Juana, E. (eds) (2016) *Handbook of the Birds of the World Alive*. Lynx Edicions, Barcelona.

Johnsgard, P. A. (1994) *Arena Birds – Sexual Selection and Behaviour*. Smithsonian Institution Press, Washington, D.C.

Johnston, G. R. & Richards, S. J. (1994) Notes on birds observed in the Western Province during July 1993. *Muruk* 6: 9.

Johnstone, R. E. & Storr, G. M. (2004) *Handbook of Western Australian Birds*. Vol. 2. Passerines (Blue-winged Pitta to Goldfinch). Western Australian Museum, Perth, Western Australia.

Jønsson, K. A. Fabre, P. H. Ricklefs, R. E . & Fjeldså, J. (2011) Major global radiation of corvoid birds originated in the proto-Papuan archipelago. *Proc. Natl. Acad. Sci. USA* 108: 2328–2333.

Kelley, L. A. & Endler, J. A. (2012) Illusions promote mating success in Great Bowerbirds. *Science* 335 (Issue 6066): 335–338.

Kelley, L. A. & Endler, J. A. (2017) How do great bowerbirds construct perspective illusions? *Royal Soc. Open Sci.* 4.

Kelley, L. A. & Healy, S. D. (2011) The mimetic repertoire of the spotted bowerbird *Ptilonorhynchus maculatus*. *Naturwissenschaften* DOI 10.1007/s00114-011-0794-z

Kende, P. (1982) Why we do not kill the bowerbird. Pp.

201–203 in: Morauta, L. Pernetta, J. & Heaney, W. (eds) (1982) *Traditional Conservation in Papua New Guinea: Implications for Today. An Overview and Summary of Proceedings of a Conference Organized by the Office of Environment and Conservation and the Institute of Applied Social and Economic Research in Port Moresby, 27–31 October 1980.* Monograph 16. Institute of Applied Social and Economic Research, Boroko, Papua New Guinea. xii, 392 pp.

King, B. (1979) New distributional records and field notes for some New Guinea birds. *Emu* 79: 146–148.

Kleinschmidt, O. (1897a) Beschreibung eines neuen Paradiesvogels. *Orn. Monatsb.* 5: 46.

Kleinschmidt, O. (1897b) Parotia berlepschi. *J. Orn.* 45: 174–178.

Kuah, L. (1992) Display and attempted breeding of the Lesser Bird of paradise. *Avicult. Mag.* 98(4): 178.

Kusmierski, R. Borgia, G. Crozier, R. H. & Chan, B. H. Y. (1993) Molecular information on bowerbird phylogeny and the evolution of exaggerated male characteristics. *J. Evol. Biol.* 6: 737–752.

Kusmierski, R. Borgia, G. Uy, A. & Crozier, R. H. (1997) Labile evolution of display traits in bowerbirds indicates reduced effects of phylogenetic constraint. *Proc. Royal Soc. London (Ser. B Biol. Sci.)* 264: 307–313.

Kwapena, N. (1985) *The Ecology and Conservation of Six Species of Birds of Paradise in Papua New Guinea.* Biological Resources Management, Port Moresby.

Laman, T. & Scholes, E. (2012) *Birds of Paradise: Revealing the World's Most Astonishing Birds.* National Geographic/Cornell, Washington D.C.

Lambert, F. R. (1994) Notes on the avifauna of Bacan, Kasiruta and Obi, north Moluccas. *Kukila* 7: 1–9.

Lambert, F. & Young, D. (1989) Some recent bird observations from Halmahera. *Kukila* 4: 30–33.

Lambley, P. (1990) Observations on the feeding habits of the Huon Astrapia *Astrapia rothschildi. Muruk* 4: 75.

Laska, M. S. Hutchins, M. Sheppard, C. Worth, W. Hundgen, K. & Bruning, D. (1992) Reproduction by captive unplumed male Lesser Bird of paradise *Paradisaea minor*: evidence for an alternative mating strategy? *Emu* 92: 108–111.

Laska, M. S. Hutchins, M, Sheppard, C. Dale, G. Burger, J. Worth, W. & Hundgen, K. (1994) Social interactions and display behavior in captive Lesser birds of paradise *Paradisaea minor. Bird Behaviour* 1.

LeCroy, M. (1981) The genus *Paradisaea* – display and evolution. *Amer. Mus. Novit.* 2714: 1–52.

LeCroy, M. & Diamond, J. (2017) Rollo Beck's collections of birds in northeast New Guinea. *Amer. Mus. Novit.* 3873: 1–36.

LeCroy, M. & Peckover, W. (2000) Birds observed on Goodenough and Wagifa Islands, Milne Bay Province. *Muruk* 8: 41–44.

LeCroy, M. Kulupi, A. & Peckover, W. S. (1980) Goldie's bird of paradise: display, natural history, and traditional relationships of people to the bird. *Wilson Bull.* 92: 289–301.

LeCroy, M. Peckover, W. S. Kulupi, A. & Manseima, J. (1984) Bird observations on Normanby and

Fergusson, D'Entrecasteaux islands–Papua New Guinea. *Wildlife in Papua New Guinea* 831: 1–7.

Lenz, N. (1999) Evolutionary ecology of the Regent Bowerbird *Sericulus chrysocephalus. Ökol. Vögel* 22(Suppl.): 1–200.

Lever, C. (1987) *Naturalized Birds of the World.* Longman Scientific and Technical.

Lever C. (2005) *Naturalized Birds of the World,* second edition. T. & A. D. Poyser, London

Linsley, M. D. (1995) Some bird records from Obi, Maluku. *Kukila* 7: 142–151.

Loke Wan Tho (1957) *A Company of Birds.* Michael Joseph, London.

Macdonald, J. D. (1987) *The Illustrated Dictionary of Australian Birds.* Reed, Sydney.

MacGillivray, W. (1914) Notes on some north Queensland birds *Emu* 13: 132–86.

MacGillivray, W. (1918) Ornithologist in North Queensland Part 3. *Emu* 18: 180–212.

Mack, A. L. (1992) The nest, egg and incubating behaviour of a Blue Bird of Paradise *Paradisaea rudolphi. Emu* 92: 132–86.

Mack, A. & Wright, D. (1996) Notes on occurrence and feeding of birds at Crater Mountain Biological Research Station, PNG. *Emu* 96: 89–101.

Mack, A. & Wright, D. (2000) Notes on the Crested *Cnemophilus macgregorii* and Yellow-breasted *Loboparadisaea sericea* Birds of Paradise. *Bull. Brit. Orn. Club* 120: 186–189.

Mackay, M. (1981) Display behavior by female birds of paradise in captivity. *PNG Bird Soc. Newsletter* 185–186: 5.

Mackay, R. D. (1966) Men and birds of Nomad River. *PNG Bird Soc. Newsletter* 12: 1.

Mackay, R. (1981) Display behaviour by female birds of paradise in captivity. *PNG Bird Soc. Newsletter* 185–186: 5.

Mackay, R. D. (1989) The bower of the Fire-maned Bowerbird *Sericulus bakeri. Australian Bird Watcher* 13: 62–64.

Mackay, R. (1990) Variation in the display of Magnificent Riflebird *Ptiloris magnificus. Muruk* 4: 65–66.

Mackinnon, J. L. *et al.,* (1995) Halmahera '94, a University of Bristol Expedition to Indonesia (Final Report). Unpublished.

Madden, J. R. (2001) Sex, bowers and brains. *Proc. Royal Soc. London (Ser. B Biol. Sci.)* 268: 833–838.

Majnep, I. S. & Bulmer, R. (1977) *Birds of My Kalam Country.* Auckland University Press & Oxford University Press, Auckland.

Marchant, S. & Higgins, P. J. (1993) *Handbook of Australian, New Zealand and Antarctic Birds,* Vol. 2. Oxford University Press, Melbourne.

Marsden, S. J. Symes, C. T. & Mack, A. L. (2006). The response of New Guinean avifauna to conversion of forest to small-scale agriculture. *Ibis* 148: 629–640.

Marshall, A. J. (1954) *Bower-birds, their Displays and Breeding Cycles – a Preliminary Statement.* Oxford University Press, Oxford.

Mathews, G. M. (1923) Additions and corrections to my

lists of the birds of Australia. *Austral Avian Record* 5: 42.

Mauersberger, G. (1976) On the display of Wilson's Bird-of-paradise. *Emu* 76: 90.

Maxmen, A. (2010) Bowerbirds trick mates with optical illusions. *Nature* doi:10.1038/news.2010.458

Mayr, E. (1930) *Loboparadisaea sericea aurora* subsp. nova. *Ornithologische Monatsberichte* 38: 147–148.

Mayr, E. (1931) Die Vögel des Saruwaged und Herzoggebirges (NO-Neuguinea). *Mitt. Zool. Mus. Berlin* 17: 639–723.

Mayr, E. (1941) *List of New Guinea Birds. A Systematic and Faunal List of the Birds of New Guinea and Adjacent Islands.* American Museum of Natural History, New York.

Mayr, E. (1945) *Birds of Paradise.* Man and Nature Publications Science Guide 127. American Museum of Natural History, New York.

Mayr, E. (1962) Family Paradisaeidae, birds of paradise. Pp. 181–204 in: Mayr, E. & Greenway, J. C. (eds) (1962) *Check-list of Birds of the World. A Continuation of the Work of James L. Peters.* Vol. 15. Museum of Comparative Zoology, Cambridge, Massachusetts.

Mayr, E. (1962) Family Ptilonorhynchidae, bowerbirds. Pp. 172–181 in: Mayr & Greenway (1962).

Mayr, E (1964) *Systematics and the Origin of Species.* Dover, New York.

Mayr, E. Amadon, D. *et al.*, (1951) A classification of Recent birds. *Amer. Mus. Novit.* no. 1496.

Mayr, E. & Gilliard, E. T. (1954) Birds of central New Guinea. Results of the American Museum of Natural History Expeditions to New Guinea in 1950 and 1952. *Bull. Amer. Mus. Nat. Hist.* 103: 311–374, pls 14–34.

Mayr, E. & Greenway, J. C. (eds) (1962) *Check-list of the Birds of the World*, Vol. 15. Cambridge, Massachusetts.

Mayr, E. & Rand, A. L. (1937) Results of the Archbold Expeditions. 14. Birds of the 1933–1934 Papuan Expedition. *Bull. Amer. Mus. Nat. Hist.* 73: 1–248.

McCarthy, E. M. (2006) *Handbook of Avian Hybrids of the World.* Oxford University Press, Oxford.

McCoy, D. (2018) Unpublished PhD research on the colour black in paradisaeids, in *Australian birdlife* 7(2): 28.

McGill, A. R. (1951) Proceedings of the annual congress of the R. A. O. U. Sydney, 1950. *Emu* 50: 240–250.

Meek, A. S. (1913) *A Naturalist in Cannibal Land.* Unwin, London.

Mees, G. F. (1964) Notes on two small collections of birds from New Guinea. *Zool. Verhand.* 66: 1–37.

Mees, G. F. (1965) The avifauna of Misool. *Nova Guinea, Zoology* 31: 39–203.

Mees, G. F. (1982) Birds from the lowlands of southern New Guinea (Merauke and Koembe). *Zool. Verhand.* 191: 3–188.

Melville, D. (1979) Ornithological notes on a visit to Irian Jaya. *PNG Bird Society Newsletter* 161: 3–22.

Meyer, A. B. (1894) [*Parotia carolae*, sp. n.]. *Bull. Brit. Orn. Club* 4: 6.

Mittermeier, J. C. *et al.*, (2015) Obi, North Moluccas, Indonesia: a visitor's guide to finding the endemics. *BirdingASIA* 23: 66–76.

Moorhouse, R. J. (1996) The extraordinary bill dimorphism of the Huia (*Heteraclocha acutirostris*): sexual selection or intersexual competition? *Notornis* 43: 19–34.

Morcombe, M. (2004) *Field Guide to Australian Birds.* Steve Parish Publishing, Archerfield.

Morrison-Scott, T. (1936) Display of *Lophorina superba minor. Proc. Zool. Soc. London* 1936: 809.

Muller, K. A. (1974) Rearing the Count Raggi's Bird of Paradise *Paradisaea raggiana* at Taronga Zoo, Sydney. *Intern. Zoo Yearbook* 14: 102–105.

Muzinic, J. Bogdan, J. F. & Beehler, B. (2009) Julije Klovic: the first colour drawing of Greater Bird of Paradise *Paradisaea apoda* in Europe and its model. *J. Orn.* 150: 645–649.

Newton, A. (1895) *A Dictionary of Birds.* A. & C. Black, London.

North, A. J. (1892) Notes on the nidification of *Manucodia comrii*, Sclater (Comrie's Manucode). *Records of the Australian Museum* 2: 32.

North, A. J. (1901–14) *Nests and Eggs of Birds Found Breeding in Australia and Tasmania.* 2nd edition. Vol. 1(1). Australian Museum Special Catalogue 1. Trustees of the Australian Museum, Sydney.

Novotny, V. (2009) "*Notebooks from New Guinea*". Oxford University Press, Oxford.

Nunn, G. B. & Cracraft, J. (1996) Phylogenetic relationships among the major lineages of the birds-of-paradise (Paradisaeidae) using mitochondrial DNA gene sequences. *Mol. Phyl. & Evol.* 5: 445–459.

Ogilvie-Grant, W. R. (1915a) Report on the birds collected by the BOU Expedition and the Wollaston Expedition in Dutch New Guinea. *Ibis,* Jubilee Supplement 2: 1–329.

Ogilvie-Grant, W. R. (1915b) Note on male *Paradisaea apoda novaeguineae* adult plumage acquisition. *Bull. Brit. Orn. Club* 36: 41.

Olson, S. L. Parkes, K. C. Clench, M. H. & Borecky, S. R. (1983) The affinities of the New Zealand passerine genus *Turnagra. Notornis* 30: 319–336.

Opit, G. (1975a) Display of Magnificent Rifle Bird. *New Guinea Bird Soc. Newsletter* 113: 15.

Opit G. (1975b) Observations (along Kokoda Trail). *New Guinea Bird Soc. Newsletter* 115: 4–5.

Parker, S. A. (1963) Nesting of the Paradise Crow *Lycocorax pyrrhopterus* (Bonaparte) and the Spangled Drongo *Dicrurus hottentotus* (Linn) in the Moluccas. *Bull. Brit. Orn. Club* 83: 126–127.

Patricelli, G. L. Uy, J. A. C. Walsh, G. & Borgia, G. (2002) Male displays adjusted to female's response: macho courtship by the satin bowerbird is tempered to avoid frightening the female. *Nature* 415: 279–280.

Patricelli, G. L. *et al.*, (2006) Male satin bowerbirds, *Ptilonorhynchus violaceus*, adjust their display intensity in response to female startling: an experiment with robotic females. *Anim.Behav.* 71: 49–59.

Peckover, W. S. (1970) The Fawn-breasted Bowerbird (*Chlamydera cerviniventris*). *Pro. 1969 Papua New Guinea Sci. Soc.* 21: 23–35.

Peckover, W. S. (1985) Seed dispersal of *Amorphophallus paeoniifolius* by birds of paradise in Papua New Guinea. *Aroideana* 8: 70–71.

Peckover, W. S. (1990) *Papua New Guinea Birds of Paradise.* Brown, Carina.

Peckover, W. S. (1995) Moult in birds of paradise. *Muruk* 7: 115–116.

Peckover, W. S. & LeCroy, M. K. (undated) National animals: birds of paradise. Department of Lands, Surveys & Environment, Division of Wildlife, Papua New Guinea.

Pizzey, G. & Doyle, R. (1980) *A Field Guide to the Birds of Australia.* Collins, Sydney.

Pizzey, G. & Knight, F. (1997) *The Field Guide to the Birds of Australia.* Angus and Robertson, Sydney.

Pizzey, G. & Knight, F. (2012) *The Field Guide to the Birds of Australia.* 9th edition. Angus and Robertson, Sydney.

Poulsen, B. O. & Frolander, A. 1994. *Birding Irian Jaya, Indonesian New Guinea.* Periplus Editions & BirdLife International, Hong Kong.

Pratt, T. K. (1982) Additions to the avifauna of the Adelbert Range, Papua New Guinea. *Emu* 82: 117–125.

Pratt, T. K. (1984) Examples of tropical frugivores defending fruit-bearing plants. *Condor* 86: 123–129.

Pratt, T. K. & Beehler, B. M. (2015) *Birds of New Guinea*: second edition. Princeton University Press, Princeton.

Price, D. & Nielsen, L. (1991) Bird list for Karawari Lodge and area, East Sepik Province. *Muruk* 5: 23–24.

Pruett-Jones, M. A. & Pruett-Jones, S. G. (1982) Spacing and Distribution of bowers in MacGregor's Bowerbird (*Amblyornis macgregoriae*). *Behav. Ecol. Sociobiol.* 11: 25–32.

Pruett-Jones, M. A. & Pruett-Jones, S. G. (1985) Food caching in the tropical frugivore Macgregor's Bowerbird (*Amblyornis macgregoriae*). *Auk* 102: 334–341.

Pruett-Jones, S. G. & Pruett-Jones, M. A. (1986) Altitudinal distribution and seasonal activity patterns of birds of paradise. *Natl. Geogr. Res.* 2: 87–105.

Pruett-Jones, S. G. & Pruett-Jones, M. A. (1988a) A promiscuous mating system in the Blue Bird of Paradise *Paradisaea rudolphi. Ibis* 130: 373–377.

Pruett-Jones, S. G. & Pruett-Jones, M. A. (1988b) The use of court objects by Lawes' Parotia. *Condor* 90: 538–545.

Pruett-Jones, S. G. & Pruett-Jones, M. A. (1994) Sexual competition and courtship disruptions: why do male bowerbirds destroy each other's bowers? *Anim. Behav.* 47: 607–620.

Pruett-Jones, S. G. Pruett-Jones, M. A. & Jones, H. I. (1990) Parasites and sexual selection in birds of paradise. *Amer. Zool.* 30: 287–298.

Pycraft, W. P. (1907) Contributions to the osteology of birds. Part ix. Tyranni; Hirundines; Muscicapae, Lanii, and Gymnorhines. *Proc. Zool. Soc. London* 1907: 352–379.

Quammen, D. (1996) *The Song of the Dodo. Island Biogeography in an Age of Extinction.* Pimlico, London.

Ramsay, J. (1919) Notes on birds observed in the upper Clarence River District, N. S. W. Sept–Dec 1918. *Emu* 19: 2–9.

Rand, A. L. (1938) Results of the Archbold Expeditions No. 22. On the breeding habits of some birds of paradise in the wild. *Amer. Mus. Novit.* 993: 1–8.

Rand, A. L. (1940a) Results of the Archbold Expeditions No. 26. Breeding habits of the birds of paradise, *Macgregoria* and *Diphyllodes. Amer. Mus. Novit.* 1073: 1–14.

Rand, A. L. (1940b) Courtship of the Magnificent Bird of Paradise. *Nat. Hist. Mag.* 45: 55.

Rand, A. L. (1942a) Results of the Archbold Expeditions No. 42. Birds of the 1936–37 New Guinea Expedition. *Bull. Amer. Mus. Nat. Hist.* 79: 289–366.

Rand, A. L. (1942b) Results of the Archbold Expeditions No. 43. Birds of the 1938–39 New Guinea Expedition. *Bull. Amer. Mus. Nat. Hist.* 797: 425–516.

Rand, A. L. & Gilliard, E. T. (1967) *Handbook of New Guinea Birds.* Weidenfeld & Nicolson, London.

Rand, A. L. & Frith, C. B. (1985) Bird-of-paradise. Pp. 55–56 in: Campbell & Lack (1985).

Rimlinger, D. (1984) The Empress of Germany's Bird of Paradise *Paradisaea raggiana augustaevictoriae. Zoonooz* 57(2): 10–14.

Rimlinger, D. Azua, J. & Lewins, E. (1997) Observations on the breeding and hand-rearing of the Empress of Germany's Bird of Paradise. *Zool. Garten N.F.* 67(1/2): 71–80.

Ripley, S. D. (1950) Strange courtship of birds of paradise. *Natl. Geogr. Mag.* 100: 247–278.

Ripley, S. D. (1957) The display of the Sickle-billed Bird of Paradise. *Condor* 59: 207.

Ripley, S. D. (1959) Birds from Djailolo, Halmahera. *Postilla* 41: 1–8.

Ripley, S. D. (1964) A systematic and ecological study of birds of New Guinea. *Bull. Peabody Mus. Nat. Hist.* 19: 1–87.Rothschild, W. (1898) Paradiseidae. Issue 2 in: Schulze, F. E. (ed.) (1898) *Das Tierreich.* R. Friedländer und Sohn, Berlin.

Rothschild, W. (1906) *Two New Birds of Paradise, by Professor F. Foerster and the Hon. Walter Rothschild, Ph.D.* Special pamphlet of the Zoological Museum, Tring, 3 pp.

Rothschild, W. (1908) [Exhibition of an adult male example of *Parotia berlepschi*]. *Bull. Brit. Orn. Club* 23: 42–43.

Rothschild, Lord W. (1930) Exhibition of eggs of the Paradise-Crow (*Lycocorax pyrrhopterus*) and *Phonygammus keraudrenii. Bull. Brit. Orn. Club* 51: 9.

Rowland, P. (2008) *Bowerbirds.* Australian Natural History Series. CSIRO Publishing, Collingwood, Victoria.

Safford, R. J. & Smart, L. M. (1996) The continuing presence of MacGregor's Bird of Paradise *Macgregoria pulchra* on Mount Albert Edward, Papua New Guinea. *Bull. Brit. Orn. Club* 116: 186–188.

Saito, M. (1997) Breeding Raggi's Bird of Paradise. *Animals in Zoos,* no. 5. Tokyo Zoological Park Society.

Salvadori, T. (1880–82) *Ornitologia della Papuasia e della Molucche.* Stamperia Reale Della Ditta G. B. Paravia e Comp. Torino.

Schmid, C. K. (1993) Birds of Nokopo. *Muruk* 6(2): 1–61.

Schodde, R. (1975) *Interim List of Australian Songbirds: Passerines.* RAOU, Melbourne.

Schodde, R. (1976) Evolution in the birds-of-paradise and bowerbirds, a resynthesis. Pp. 137–149 in: Frith, H. J. & Calaby, J. H. (eds) (1976) *Proc. 16th Intern. Orn. Congr. 1974, Canberra.* Australian Academy of Science, Canberra.

Schodde, R. & Mason, I. J. (1999) *The Directory of Australian Birds: Passerines.* CSIRO, Collingwood.

Schodde, R. & McKean, J. L. (1972) Distribution and taxonomic status of *Parotia lawesii helenae* De Vis. *Emu* 72: 113–114.

Schodde, R. & McKean, J. (1973) Distribution, taxonomy and evolution of the gardener bowerbirds *Amblyornis* spp. in eastern New Guinea, with descriptions of two new subspecies. *Emu* 73: 51–60.

Schodde, R. & McKean, J. (1973) The species of the genus *Parotia* (Paradisaeidae) and their relationships. *Emu* 73: 145–156.

Schodde, R. & Tidemann, S. (1988) *Reader's Digest Complete Book of Australian Birds.* Reader's Digest, Sydney.

Scholes, E. (2006) Courtship ethology of Carola's Parotia (*Parotia carolae*). *Auk* 123(4): 967–990.

Scholes, E. (2008) Structure and composition of the courtship phenotype in the bird of paradise *Parotia lawesii* (Aves: Paradisaeidae). *Zoology* 111: 260–278.

Scholes, E. (2017) Dance moves support evidence for new bird of paradise species. *Cornell Lab of Ornithology Press Release* June 29 2017.

Scholes, E. Beehler, B. M. & Laman, T. G. (2017) Taxonomic status of *Parotia berlepschi* Kleinschmidt, 1897 based on analysis of external appearance, voice and behavior (Aves: Paradisaeidae). *Zootaxa* 4329: 6.

Schram, B. (2000) Notes on the behaviour of the Curl-crested Manucode, *Manucodia comrii. Muruk* 8: 75.

Searle, K. C. (1980) Breeding Count Raggi's Bird of Paradise *Paradisaea raggiana salvadorii* at Hong Kong. *Intern. Zoo Yearbook* 20: 210–214.

Sekhran, N. & Miller, S. E. (1996) *Papua New Guinea Country Study on Biological Diversity.* Waigani, Papua New Guinea: Papua New Guinea Department of Environment and Conservation.

Serventy, V. (1955) Notes on the Spotted Bower-bird (*Chlamydera maculata*). *West Australian Natur.* 5: 5–8.

Seth-Smith, D. (1923a) The Birds of Paradise and Bower Birds. *Avicult. Mag.* (4)1: 41–60.

Seth-Smith, D. (1923b) On the display of the Magnificent Bird of Paradise *Diphyllodes magnifica hunsteini. Proc. Zool. Soc. London* 1923: 609–613.

Severin, T. (1997) *The Spice Islands Voyage.* Abacus, Indonesia.

Sharpe, R. B. (1891–98) *Monograph of the Paradisaeidae, or Birds of Paradise, and Ptilonorhynchidae or Bower-birds.* Parts 1–8. H. Sotheran & Co. London.

Sharpe, R. B. (1898) *Monograph of the Paradiseidae or Birds of Paradise and Ptilonorhynchidae or Bowerbirds.* Taylor & Francis, London.

Sibley, C. G. & Ahlquist, J. E. (1985) The phylogeny and classification of the Australo-Papuan passerine birds. *Emu* 85: 1–14.

Sibley, C. G. & Ahlquist, J. E. (1987) The Lesser Melampitta is a bird of paradise. *Emu* 87: 66–68.

Sibley, C. G. & Ahlquist, J. E. (1990) *Phylogeny and Classification of Birds: a Study in Molecular Evolution.* Yale University Press, New Haven & London.

Sibley, C. G. & Monroe, B. L. (1993) *A Supplement to Distribution and Taxonomy of Birds of the World.* Yale University Press, New Haven & London.

Simpson, C. C. (1942) Across the Owen Stanley Range. *Victorian Natur.* 59: 98–104.

Sims, R. W. (1956) Birds collected by Mr F. Shaw-Mayer in the central highlands of New Guinea 1950–51. *Bull. Brit. Mus. (Nat. Hist.)* 3: 389–438.

Simson, C. C. (1907) On the habits of the birds-of-paradise and bower-birds of British New Guinea. *Ibis* 1: 380–387.

Slater, P. *et al.,* (2009) *The Slater Field Guide to Australian Birds.* New Holland.

Smyth, H. (1970) Hand-rearing and observing birds of paradise. *Avicult. Mag.* 76: 67–70.

Solomon, S. Minnegal, M. & Dwyer, P. (1986) Bower birds, bones and archaeology. *J. Archaeol. Sci.* 13: 307–318.

Stattersfield, A. J. & Capper, D. R. (eds) (2000) *Threatened Birds of the World.* Lynx Edicions & BirdLife International, Barcelona & Cambridge.

Stein, G. H. W. (1936) Beiträge zur Biologie papuanischer Vögel. *J. Orn.* 84: 21–57.

Steinheimer, F. D. (2005) The type specimens of Paradisaeidae, Cnemophilidae and Ptilonorhynchidae (Aves) in the Museum für Naturkunde of the Humboldt-University of Berlin. *Zootaxa* 1072: 1–25.

Stonor, C. R. (1936) The evolution and mutual relationships of some members of the Paradiseidae. *Proc. Zool. Soc. London* 1936: 1177–1185.

Stonor, C. R. (1937) On the systematic position of the Ptilonorhynchidae. *Proc. Zool. Soc. London* B 107: 425–490.

Stonor, C. R. (1938) Some features of the variation of the birds of paradise. *Proc. Zool. Soc. London* B 108: 417–481.

Stott, K. W. Jr (1981) New Guinea's blushing gargoyles. *Zoonooz* 54(1): 5–9.

Stresemann, E. (1924) Neue Beiträge zur Ornithologie Deutsch-Neuguineas. *J. Orn.* 72: 424–428.

Stresemann, E. (1930) Welche Paradiesvogelarten der Literatur sind hybriden Ursprungs? *Novit. Zool.* 36: 6–15.

Stresemann, E. (1934) Aves. In Kükenthal, W. & Krumbach, T. (eds) *Handbuch der Zoologie.* Vol. 7, Part 2. Walter de Gruyter, Berlin.

Stresemann, E. (1934a) Vier neue Unterarten von Paradiesvögeln. *Orn. Monatsb.* 42: 144–147.

Sujatnika, Jepson, P. Soehartono, T. R. Crosby, M. J. & Mardiastuti, A. (1995) *Conserving Indonesian Biodiversity: the Endemic Bird Area Approach.* BirdLife International Indonesia Programme, Bogor.

Swadling, P. Wagner, R. & Laba, B. (1996) *Plumes from Paradise: trade cycles in outer Southeast Asia and their impact on New Guinea and nearby islands until 1920.* Coorparoo DC, Queensland, Australia: PNG National Museum in association with Robert Brown & Associates.

Thair, S. & Thair, M. (1977) Report on display of Magnificent Bird of paradise. *PNG Bird Soc. Newsletter* 128: 13.

Thibault, M. *et al.*, (2013) New and interesting records for the Obi archipelago (north Maluku, Indonesia), including field observations and first description of the vocalization of Moluccan Woodcock *Scolopax rochussenii. Bull. Brit. Orn. Club* 133: 83–115.

Timmis, W. H. (1968) Breeding of the Superb Bird of paradise at the Chester Zoo. *Avicult. Mag.* 74: 170–172.

Todd, W. & Berry, R. J. (1980) Breeding the Red Bird of Paradise *Paradisaea rubra* at the Houston Zoo. *Intern. Zoo Yearbook* 20: 206–210.

Tolhurst, L. (1989) Extension of the known range of Splendid Astrapia *Astrapia splendidissima. Muruk* 4: 20.

Uy, J. A. C. (2002) Say it with bowers. *Natural History* 111: 78–82.

Uy, J. A. C. & Borgia, G. (2000) Sexual selection drives rapid divergence in bowerbird display traits. *Evolution* 54: 273–278.

Uy, J. A. C. Patricelli, G. I. & Borgia, G. (2000) Dynamic mate-searching tactic allows female satin bowerbirds (*Ptilonorhynchus violaceus*) to reduce searching. *Proc. Royal Soc. London (B)* 267: 251–256.

Uy, J. A. C. Patricelli, G. I. & Borgia, G. (2001) Loss of preferred mates forces female satin bowerbirds (*Ptilonorhynchus violaceus*) to reduce searching. *Proc. Royal Soc. London (B)* 268: 633–638.

Uy, J. A. C, Patricelli, G. L. & Borgia, G. (2001) Complex mate searching in the satin bowerbird (*Ptilonorhynchus violaceus*). *Amer. Natur.* 158: 530–542.

Valentijn, F. (1724–26) *Oud en Nieuw Oost-Indiën.* 111, Dordrecht.

Van den Bergh, M. O. L. Kusters, K. & Dietz, A. J. T. (2013) Destructive attraction: factors that influence hunting pressure on the Blue Bird-of-paradise *Paradisaea rudolphi. Bird Conserv. Int.* 23: 221–231.

van Grouw, H. (2012) Plumage aberrations in Australian birds: a comment on Guay *et al.,* (2012) and Frith & Murphy (2012). *Australian Field Orn.* 29: 210–214.

Vellenga, R. E. (1980) Moults of the Satin Bowerbird. *Emu* 80: 49–54.

Veselovsky, Z. (1979) A field study of Australian Bower Birds. *Acta scientarum naturalium Academiae scientarum bohemosloveae.*

Vetter, J. (2009) *Impacts of Deforestation on the Conservation Status of Endemic Birds in the North Maluku Endemic Bird Area from 1990–2003.* Nicholas School of the Environment, Duke University.

Wahlberg, N. (1992) Observations of birds feeding in a fruiting fig *Ficus* sp. in Varirata National Park. *Muruk* 5(3): 109–110.

Wallace, A. R. (1869) *The Malay Archipelago.* Macmillan, London.

Warham, J. (1962a) Field notes on Australian bower-birds and cat-birds. *Emu* 62: 1–30.

Warham, J. (1962b) Bird islands within the Barrier Reef and Torres Strait. *Emu* 52: 111.

Watson, J. D. Wheeler, W. R. & Whitbourn, E. (1962) With the RAOU in Papua New Guinea, October 1960. *Emu* 62: 31–50, 67–98.

Whitney, B. M. (1987) The Pale-billed Sicklebill *Epimachus bruijnii* in Papua New Guinea. *Emu* 87: 244–246.

Wieneke, J. (2000) *Where to find birds in North-east Queensland.* The author, Townsville.

Winkler, D. W. Billerman, S. M. & Lovette, I. J. (2015) *Bird Families of the World: An Invitation to the Spectacular Diversity of Birds.* Lynx Edicions, Barcelona.

Wojcieszek, J. M. *et al.*, (2006) Theft of bower decorations among male Satin Bowerbirds (*Ptilonorhynchus violaceus*): why are some decorations more popular than others? *Emu* 106: 175–180.

Worth, W. Hutchins, M. Sheppard, C. Bruning, D. Gonzalez, J. & McNamara, T. (1991) Hand-rearing, growth, and development of the Red Bird of Paradise (*Paradisaea rubra*) at the New York Zoological Park. *Zoo Biol.* 10(1): 17–23.

Yealland, J. J. (1969) Breeding of Princess Stephanie's Bird of Paradise at London Zoo. *Avicult. Mag.* 75: 50–51.

Zuccon, D. & Ericson, P. G. P. (2012) Molecular and morphological evidences place the extinct New Zealand endemic *Turnagra capensis* in the Oriolidae. *Mol. Phyl. & Evol.* 62: 414–426.

Web Resources

Australian Bird Image Database (aviceda.org/abid)

Bird Checklists for 610 Melanesian Islands (birdsofmelanesia.net)

Birdforum – One of the major discussion sites in birding (birdforum.net)

BirdLife International – One of the major bird conservation sites (birdlife.org)

Cloudbirders – Site for trip reports from around the world, many from New Guinea (Cloudbirders.com)

The Cornell Lab's Birds-of-Paradise Project (birdsofparadiseproject.org) is a research and education initiative to document, interpret, and protect the birds-of-paradise, their native environments, and the other biodiversity of the New Guinea region, one of the largest remaining tropical wildernesses on the planet.

Kukila – the Journal of Indonesian Ornithology (Kukila. org/index.php/KKL)

Geonames – Site for geographical locality names (geonames.org)

Internet Bird Collection (IBC) – Open-access, massive collection of video, photo and sound clips for most species (ibc.lynxeds.com/)

IOC World Bird List – Open access, updated every 3 months (worldbirdnames.org)

IUCN Red List of Threatened Species (iucnredlist.org)

Muruk 1985–2011 – Journal for ornithology in New Guinea. Archived collection (sicklebillsafaris.com/ index.php/muruk)

Oriental Bird Images – Database of the Oriental Bird Club (orientalbirdimages.org)

Surfbirds – Repository for trip reports, lists etc (Surfbirds. com)

Taxonomy in flux (Tif) – An amazing website, keeping up with the latest taxonomic changes (http://jboyd.net/ birds.html)

xeno-canto – Open-access community site with a huge range of sound recordings, many from this region (http://www.xeno-canto.org/collection/area/ australia)

INDEX

Numbers in **bold** refer to plate numbers, page numbers in *italic* refer to the caption text in the plate section. Other page numbers refer to the first occurrence of species and subspecies names in the relevant species account. Some subspecies have detailed sub-entries later in the species account.